SolidWorks 2023 Black Book (Colored)

By
Gaurav Verma
Matt Weber
(CADCAMCAE Works)

Edited by
Kristen

ISBN # 978-1-77459-094-2

NOTICE TO THE READER

DEDICATION

To teachers, who make it possible to disseminate knowledge
to enlighten the young and curious minds
of our future generations

To students, who are the future of the world

THANKS

To my friends and colleagues

To my family for their love and support

Training and Consultant Services

At CADCAMCAE WORKS, we provide effective and affordable one to one online training on various software packages in Computer Aided Design(CAD), Computer Aided Manufacturing(CAM), Computer Aided Engineering (CAE), Computer programming languages(C/C++, Java, .NET, Android, Javascript, HTML and so on). The training is delivered through remote access to your system and voice chat via Internet at any time, any place, and at any pace to individuals, groups, students of colleges/universities, and CAD/CAM/CAE training centers. The main features of this program are:

Training as per your need

Highly experienced Engineers and Technician conduct the classes on the software applications used in the industries. The methodology adopted to teach the software is totally practical based, so that the learner can adapt to the design and development industries in almost no time. The efforts are to make the training process cost effective and time saving while you have the comfort of your time and place, thereby relieving you from the hassles of traveling to training centers or rearranging your time table.

Software Packages on which we provide basic and advanced training are:

CAD/CAM/CAE: CATIA, Creo Parametric, Creo Direct, SolidWorks, Autodesk Inventor, Solid Edge, UG NX, AutoCAD, AutoCAD LT, EdgeCAM, MasterCAM, SolidCAM, DelCAM, BOBCAM, UG NX Manufacturing, UG Mold Wizard, UG Progressive Die, UG Die Design, SolidWorks Mold, Creo Manufacturing, Creo Expert Machinist, NX Nastran, Hypermesh, SolidWorks Simulation, Autodesk Simulation Mechanical, Creo Simulate, Gambit, ANSYS and many others.

Computer Programming Languages: C++, VB.NET, HTML, Android, Javascript and so on.

Game Designing: Unity.

Civil Engineering: AutoCAD MEP, Revit Structure, Revit Architecture, AutoCAD Map 3D and so on.

We also provide consultant services for Design and development on the above mentioned software packages

For more information you can mail us at:
cadcamcaeworks@gmail.com

Table of Contents

Training and Consultant Services iv
Preface xvi
About Authors xviii

Chapter 1 : Starting with SolidWorks

Installing SolidWorks 2023 1-2
Starting SolidWorks 2023 1-2
Starting A New Document 1-3
Part Mode CommandManagers 1-6
Assembly Mode CommandManagers 1-10
Drawing Mode CommandManagers 1-11
Opening a Document 1-11
Closing a Document 1-12
Saving the Documents 1-13
Printing Document 1-15
Applying Page Setup 1-16
Setting Header/Footer for Print Pages 1-17
Print Margins 1-17
Print Range and other Parameters 1-18
Publishing to eDrawings 1-18
Packing files for Sharing 1-20
Sending Files in Email 1-21
Reloading Files 1-21
Replacing Component in Assembly 1-22
Checking Summary of Document 1-24
Customizing Menus 1-24
Basic Settings of SolidWorks 1-25
Mouse Button Function 1-27
Loading Add-Ins 1-27
Search Tools 1-28
Login to SolidWorks 1-30
Workflow in SolidWorks 1-30

Chapter 2 : Sketching Basic to Advanced

Basics for Sketching 2-2
Sketching Plane 2-3
Relation between sketch, plane, and 3D model 2-4
Starting Sketch 2-4
Heads-up View Toolbar 2-5
Sketch Creation tools 2-7
Creating Line Tools 2-7
Creating Rectangle 2-10
Creating Slot 2-13
Creating Circle 2-15
Creating Arc 2-16
Creating Polygon 2-17

Creating Spline **New** 2-18
Creating Ellipse 2-22
Creating Sketch Fillet 2-26
Creating Sketch Chamfer 2-27
Creating Text 2-27
Creating Point 2-29
Sketch Editing tools **2-29**
Trimming Entities 2-30
Extending Entities 2-32
Offsetting Entities 2-33
Mirroring Entities 2-34
Creating Linear Sketch Pattern 2-35
Creating Circular Sketch Pattern 2-36
Moving Entities 2-38
Copying Entities 2-39
Rotating Entities 2-39
Scaling Entities 2-40
Stretching Entities 2-41
Segment Tool 2-42
Splitting Entities 2-44
Creating Jog Lines 2-44
Dynamic Mirroring 2-45
Replacing Sketch Entities 2-46
Placing Picture in Sketch 2-47
Area and Hatch Fill in Sketch 2-48
Relations **2-49**
Dimensional Constraints (Dimensions) 2-49
Geometric Constraints 2-52
Fully Defined Sketch 2-57

Chapter 3 : Advanced Dimensioning and Practice

Dimensioning and its Relations **3-2**
Dimension Style **3-2**
Style Rollout 3-2
Tolerance/Precision Rollout 3-4
Tolerance modifier Rollout 3-8
Dual Dimension Rollout 3-10
Witness/Leader Display Rollout 3-11
Leader/Dimension Line Style Rollout 3-12
Extension Line Style Rollout 3-12
Custom Text Position Rollout 3-12
Override Units Rollout 3-13
Text Fonts Rollout 3-13
Options Rollout 3-13
Horizontal/Vertical Dimensioning Between Points **3-13**
Sketch Ink **3-13**
Starting Sketch 3-14
Selecting the sketch color 3-14
Using Draw tool for Sketching 3-15
Erasing Sketch Ink 3-16

Power Modify	3-17
Practical 1	3-18
Practical 2	3-22
Practical 3	3-29

Chapter 4 : 3D Sketch and Solid Modeling

3D Sketching	**4-2**
Convert Entities	4-3
Silhouette Entities	4-4
Extruded Boss/Base Tool	**4-5**
Revolved Boss/Base Tool	**4-10**
Swept Boss/Base Tool	**4-12**
Using Guide Curves	4-14
Applying Twist in Sweep Feature	4-15
Circular Profile Sweep	4-16
Reference Geometry	**4-17**
Plane	4-17
Plane Parallel to Screen	4-19
Axis	4-20
Coordinate System	4-22
Point	4-23
Center of Mass	4-24
Mate Reference	4-25
Bounding Box	4-26
Lofted Boss/Base Tool	**4-27**
Boundary Boss/Base Tool	**4-30**
Removing Material from Solid Objects	**4-33**
Extruded Cut	4-33
Revolved Cut	4-34
Swept Cut	4-34
Hole Wizard	4-36
Advanced Hole	4-39
Thread	4-41
Stud Wizard	4-44

Chapter 5 : Solid Editing and Practical

Introduction	**5-2**
Fillet	**5-2**
Constant Size Fillet	5-2
Variable Radius Fillet	5-4
Face fillet	5-5
Full round fillet	5-6
FilletXpert	5-6
Chamfer	**5-7**
Linear Pattern	**5-9**
Circular Pattern	**5-10**
Mirror	**5-12**
Curve Driven Pattern	**5-13**
Sketch Driven Pattern	**5-14**

Table Driven Pattern **5-15**
Fill Pattern **5-17**
Variable Pattern **5-19**
Rib **5-21**
Draft **5-22**
Using Draft PropertyManager 5-23
Using DraftXpert PropertyManager 5-24
Shell **5-25**
Wrap **5-26**
Intersect **5-27**
Equations **5-28**
Adding Global Variables 5-30
Suppressing/Un-suppressing Features 5-30
Setting Dimension Equations 5-31
Design Table **5-32**
Applying Material to Part **5-34**
Mass Properties of Part **5-35**
Practical 1 5-36
Practical 2 5-40
Practical 3 5-45

Chapter 6 : Advanced Solid Modeling

Introduction **6-2**
Creating Dome **6-2**
Free Forming Solid **6-3**
Deforming Solid **6-4**
Deforming Model using Point Option 6-4
Deforming Model using Curve to Curve Option 6-6
Deforming Model using Surface push Option 6-7
Creating Indent on Solid Model **6-7**
Applying Flex Transformation **6-8**
Bending 6-9
Combining Bodies **6-10**
Splitting a Solid Body **6-11**
Moving and Copying Objects **6-12**
Deleting Bodies **6-13**
Converting to Mesh Bodies **6-14**
Applying 3D Texture **6-15**
Importing Model in Part Environment **6-17**
Fastening Features **6-17**
Creating Mounting Boss 6-17
Creating Snap Hook 6-18
Creating Snap Hook Groove 6-19
Creating Vent 6-21
Creating Lip and Groove 6-22
Block Designing **6-24**
Creating Block Using Sketch 6-24
Saving the Block 6-26
Editing a Block 6-26

Exploding Block 6-28
Macros **6-28**
Creating Macros 6-28
Recording Macros in SolidWorks 6-29
Running a Macro 6-29
Editing Macro 6-30
Mesh Modeling **6-31**
Decimating Mesh Body 6-31
Segmenting Mesh Body 6-32
Comparing Bodies 6-34

Chapter 7 : Assembly and Motion Study

Assembly **7-2**
Inserting Base Component **7-3**
Inserting Components in Assembly **7-3**
Assembly Constraints (Mates) **7-5**
Coincident 7-6
Parallel 7-7
Perpendicular 7-8
Tangent 7-8
Concentric 7-9
Lock 7-9
Distance 7-10
Angle 7-10
Advanced Mates **7-11**
Profile Center 7-11
Symmetric 7-12
Width 7-13
Path Mate 7-13
Linear/Linear Coupler 7-15
Advanced Distance 7-16
Advanced Angle 7-17
Cam 7-18
Slot 7-18
Hinge 7-19
Gear 7-20
Rack Pinion 7-21
Screw 7-21
Universal Joint 7-22
Exploded View **7-22**
Explode Lines 7-23
Bill of Materials **7-24**
Mate Controller **7-25**
Motion Study **7-27**
Playing Motion Study 7-29
Bottom Up Approach and Top Down Approach **7-29**
Creating Parts in Assembly (Top Down Approach) 7-30

Chapter 8 : Advanced Assembly Practical and Practice

Introduction **8-2**
Assembly Editing Tools **8-2**
Creating Hole Series 8-2
Creating Weld Bead in Assembly 8-4
Creating Belt/Chain in Assembly 8-5
Smart Fasteners **8-6**
Toolbox **8-8**
Creating Grooves 8-10
Creating Cams 8-11
Beam Calculator and Bearing Calculator 8-14
Magnetic Mates **8-15**
Asset Publisher 8-15
Creating Configurations 8-16
Creating Assembly with Magnetic Mates 8-18
Practical 1 8-20
Practical 2 8-24
Practical 3 8-26

Chapter 9 : Surfacing and Practice

Surfacing **9-2**
Surfacing tools similar to Solid creation tools **9-2**
Extruded Surface 9-2
Revolved Surface 9-3
Swept Surface 9-4
Lofted Surface 9-4
Boundary Surface 9-5
Filled Surface 9-6
Freeform 9-7
Special Surfacing Tools **9-8**
Planar Surface 9-8
Offset Surface 9-9
Ruled Surface 9-10
Surface Flatten tool 9-12
Surface Editing Tools **9-13**
Delete Face 9-14
Replace Face 9-14
Delete Hole 9-15
Extend Surface 9-16
Trim Surface 9-16
Untrim Surface 9-17
Knit Surface 9-17
Thicken 9-18
Thickened Cut 9-19
Cut with Surface 9-19
Practical 1 9-20
Practical 2 9-24

Chapter 10 : Drawing and Practice

Introduction **10-2**
Drawing Sheet Selection **10-2**
Adding Views to Sheet **10-5**
Model View 10-5
Projected View 10-9
Auxiliary View 10-9
Section View 10-10
Removed Section 10-14
Detail View 10-15
Relative View 10-16
Standard 3 View 10-17
Broken-out Section 10-18
Break 10-19
Crop View 10-19
Creating Alternate Position View 10-20
Creating Empty View 10-22
Creating Predefined View 10-22
Replacing Model 10-23
Adding Annotations to View **10-23**
Smart Dimension 10-23
Angular Running Dimension 10-23
Chamfer Dimension 10-24
Model Items 10-25
Note 10-26
Flag Notes 10-27
Surface Finish 10-30
Datum Feature 10-32
Datum Target 10-33
Geometric Tolerance 10-34
Weld Symbol and Hole Callout 10-38
Generating Exploded View of Assembly **10-39**
Generating Bill of Material 10-39
Generating Balloons for Bill of Material New 10-40
Editing Title Block **10-40**
Practical 10-42

Chapter 11 : AnalysisXpress

Introduction **11-2**
SimulationXpress Analysis Wizard **11-2**
Fixture Setting 11-4
Load Setting 11-6
Material Setting 11-7
Changing Mesh Density 11-8
Running Simulation 11-9
Results 11-10
Optimizing 11-12
FloXpress Analysis **11-15**
Preparing Model 11-16

Starting Flow analysis 11-17
DFMXpress Analysis 11-20
Costing 11-23
Sustainability 11-25
DriveWorksXpress 11-27

Chapter 12 : Mold Tools

Starting the Mold Tools 12-2
Analyzing the Model 12-4
Draft Analysis 12-4
Undercut Analysis 12-5
Parting Line Analysis 12-7
Preparing Model for Mold 12-8
Splitting Faces using Split Line tool 12-8
Applying draft using Draft tool 12-11
Increasing/Decreasing thickness of walls using the Move Face tool 12-12
Scaling the model to allow shrinkage in part 12-13
Inserting Mold Folder 12-14
Parting Line 12-14
Shut-off surfaces 12-16
Parting surfaces 12-18
Tooling Split 12-19
Core 12-20

Chapter 13 : Sheetmetal and Practice

Sheet Metal Introduction 13-2
Creating Base Flange/Tab 13-2
Setting Parameters for Base Flange/Tab with Open sketch 13-3
Creating Lofted-Bend 13-4
Sheet Metal Design Terms 13-6
Bend Allowance 13-7
K-Factor 13-7
Creating Edge Flange 13-8
Flange Parameters Rollout 13-9
Miter Flange 13-16
Hem 13-19
JOG 13-20
Sketched Bend 13-21
Cross-Break 13-21
Swept Flange 13-22
Close Corner 13-23
Welded Corner 13-23
Break-Corner/Corner-Trim 13-24
Corner Relief 13-25
Forming Tool 13-26
Sheet Metal Gusset 13-26
Tab and Slot 13-27
Extrude Cut 13-29

Vent 13-29
Normal Cut 13-31
Unfold Tool and Fold Tool 13-32
Unfold Tool 13-32
Fold Tool 13-33
Convert to Sheet Metal 13-34
Rip Tool 13-35
Insert Bends Tool 13-36
Flatten 13-37
Flat Pattern Properties 13-37
Inserting Flat Pattern in drawing 13-38

Chapter 14 : Weldments and Markup Tools

Introduction 14-2
Welding Symbols and Representation in Drawing 14-2
Butt/Groove Weld Symbols 14-2
Fillet and Edge Weld Symbols 14-3
Miscellaneous Weld Symbols 14-4
Weldment tool 14-8
Structural Member 14-8
End Cap tool 14-9
Weld Bead 14-11
Inserting Welding data in drawing 14-14
Inserting the Cut list 14-16
Markup Tools 14-17
Setting Color and Thickness for Markup 14-17
Drawing Markup 14-18
Erasing Markup 14-18

Chapter 15 : 3D Printing and Model Based Definition (MBD)

3D Printing 15-2
Part Preparation for 3D Printing 15-2
3D Printing Processes 15-2
Part Preparation for 3D Printing 15-7
Print3D 15-8
Model Based Definition (MBD) 15-10
Auto Dimension Scheme 15-11
Location Dimension 15-13
Size Dimension 15-14
Applying Angle Dimension 15-14
Basic Location Dimension/Basic Size Dimension 15-15
Datum 15-15
Geometric Tolerance 15-16
Pattern Feature 15-16
Show Tolerance Status 15-18
Other DimXpert Tools 15-18
Inserting Tables 15-18
Section View 15-19
Model Break View 15-21

Capture 3D View 15-22
Dynamic Annotation Views 15-23
Publish to 3D PDF 15-24
3D PDF Template Editor 15-26
3D PMI Compare 15-28
SolidWorks Inspection 15-29
Starting New Inspection Project 15-29
Adding/Editing Inspection Balloons 15-33
Removing Inspection Balloons 15-33
Selecting Inspection Balloons 15-33
Sequencing Inspection Balloons 15-33
Updating Inspection Project 15-34
Launching Template Editor 15-34
Defining Inspection Methods 15-36
Editing Operations for Inspection Dimensions 15-36
Exporting Inspection Sheet 15-37

Chapter 16 : Introduction to SolidWorks CAM

Introduction 16-2
Defining Machine 16-2
Setting Tool Crib 16-3
Adding Tools 16-3
Editing Tool 16-5
Removing Tool from the Crib 16-5
Updating tool in Database 16-5
Saving Tool Crib 16-6
Creating and Saving New Tool in Library 16-7
Setting Postprocessor Parameters 16-7
Posting Options 16-8
Setting Coordinate System 16-9
Defining or Editing Stock 16-10
Setting Milling Operation Parameters 16-12
Extracting Machinable Features 16-13
Generating Operation Plan 16-14
Generating Tool Paths 16-14
Simulating Tool Paths 16-15
Visualizing Toolpaths 16-16
Saving Toolpaths 16-17
Post Processing Toolpaths 16-17
Manually Creating Mill Operations 16-18

Chapter 17 : Rendering

Introduction 17-2
Editing Appearance 17-2
Copying and Pasting Appearances 17-4
Editing Scene 17-4
Editing Decal 17-4
Selecting Display States Target 17-5
Integrated Preview 17-5

Adding a Camera 17-6
Activating Perspective View 17-8
Rendering Options 17-8
Preview Window 17-9
Render Region 17-10
Final Render 17-10

Index I-1

Preface

SolidWorks 2023 is a parametric, feature-based solid modeling tool that not only unites the three-dimensional (3D) parametric features with two-dimensional (2D) tools, but also addresses every design-through-manufacturing process. The continuous enhancements in the software has made it a complete PLM software. The software is capable of performing analysis with an ease. Its compatibility with CAM software is remarkable. Based mainly on the user feedback, this solid modeling tool is remarkably user-friendly and it allows you to be productive from day one.

The **SolidWorks 2023 Black Book** is the 10th edition of our series on SolidWorks. With lots of additions and thorough review, we present a book to help professionals as well as learners in creating some of the most complex solid models. The book follows a step by step methodology. In this book, we have tried to give real-world examples with real challenges in designing. We have tried to reduce the gap between university use of SolidWorks and industrial use of SolidWorks. In this edition of book, we have included many new features of SolidWorks as per the latest enhancements of software. The book covers almost all the information required by a learner to master the SolidWorks. The book starts with sketching and ends at advanced topics like Mold Design, Sheetmetal, Weldment, SolidWorks CAM, Rendering, and MBD. Some of the salient features of this book are :

In-Depth explanation of concepts

Every new topic of this book starts with the explanation of the basic concepts. In this way, the user becomes capable of relating the things with real world.

Topics Covered

Every chapter starts with a list of topics being covered in that chapter. In this way, the user can easy find the topic of his/her interest easily.

Instruction through illustration

The instructions to perform any action are provided by maximum number of illustrations so that the user can perform the actions discussed in the book easily and effectively. There are about 1350 illustrations that make the learning process effective.

Tutorial point of view

At the end of concept's explanation, the tutorial make the understanding of users firm and long lasting. Almost each chapter of the book has tutorials that are real world projects. Moreover most of the tools in this book are discussed in the form of tutorials.

Project

Projects and exercises are provided to students for practicing.

For Faculty

If you are a faculty member, then you can ask for video tutorials on any of the topic, exercise, tutorial, or concept. As faculty, you can register on our website to get electronic desk copies of our latest books, self-assessment, and solution of practical. Faculty resources are available in the **Faculty Member** page of our website (**www.cadcamcaeworks.com**) once you login. Note that faculty registration approval is manual and it may take two days for approval before you can access the faculty website.

New Addition

If anything is added in this edition but is not available in the previous editions, then it is displayed with symbol **New** in table of content.

Formatting Conventions Used in the Text

All the key terms like name of button, tool, drop-down etc. are kept bold.

Free Resources

Link to the resources used in this book are provided to the users via email. To get the resources, mail us at ***cadcamcaeworks@gmail.com*** with your contact information. With your contact record with us, you will be provided latest updates and informations regarding various technologies. The format to write us mail for resources is as follows:

Subject of E-mail as ***Application for resources of .... book***.
Also, given your information like
Name:
Course pursuing/Profession:
Contact Address:
E-mail ID:

Note: We respect your privacy and value it. If you do not want to give your personal informations then you can ask for resources without giving your information.

About Authors

The author of this book, Matt Weber, has written more than 15 books on CAD/CAM/CAE available in market. He has coauthored SolidWorks Simulation, SolidWorks Electrical, SolidWorks Flow Simulation, and SolidWorks CAM Black Books. The author has hands on experience on almost all the CAD/CAM/CAE packages. If you have any query/doubt in any CAD/CAM/CAE package, then you can contact the author by writing at cadcamcaeworks@gmail.com

The author of this book, Gaurav Verma, has written and assisted in more than 16 titles in CAD/CAM/CAE which are already available in market. He has authored Autodesk Fusion 360 Black Book, AutoCAD Electrical Black Book, Autodesk Revit Black Books, and so on. He has provided consultant services to many industries in US, Greece, Canada, and UK. He has assisted in preparing many Government aided skill development programs. He has been speaker for Autodesk University, Russia 2014. He has assisted in preparing AutoCAD Electrical course for Autodesk Design Academy. He has worked on Sheetmetal, Forging, Machining, and Casting designs in Design and Development departments of various manufacturing firms.

For Any query or suggestion

If you have any query or suggestion, please let us know by mailing us on ***cadcamcaeworks@gmail.com***. Your valuable constructive suggestions will be incorporated in our books and your name will be addressed in special thanks area of our books on your confirmation.

Chapter 1

Starting with SolidWorks

Topics Covered

The major topics covered in this chapter are:

- ***Installing SolidWorks 2023.***
- ***Starting SolidWorks 2023.***
- ***Starting a new document.***
- ***Terminology used in SolidWorks.***
- ***Opening a document.***
- ***Closing documents.***
- ***Basic Settings for SolidWorks***
- ***Workflow in Industries using the SolidWorks***

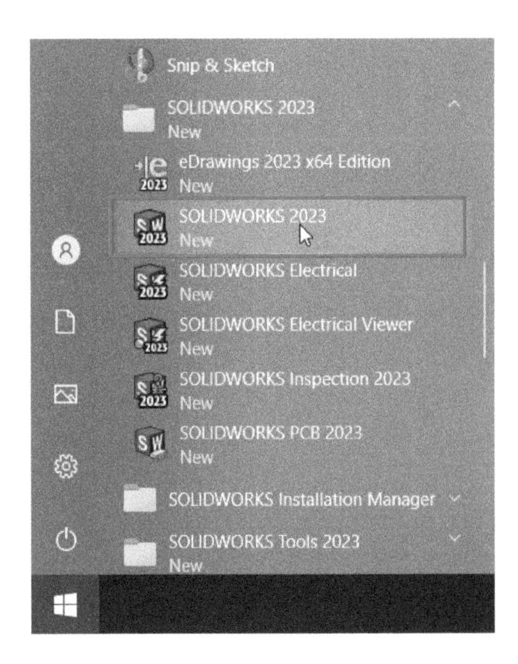

INSTALLING SOLIDWORKS 2023

You can get SolidWorks software installation files in two ways, downloading from website or getting DVD of software from your reseller.

- If you are installing SolidWorks using the CD/DVD provided by Dassault Systemes then go to the folder containing **setup.exe** file and right-click on **setup.exe** in the folder. A shortcut menu is displayed on the screen; refer to Figure-1.

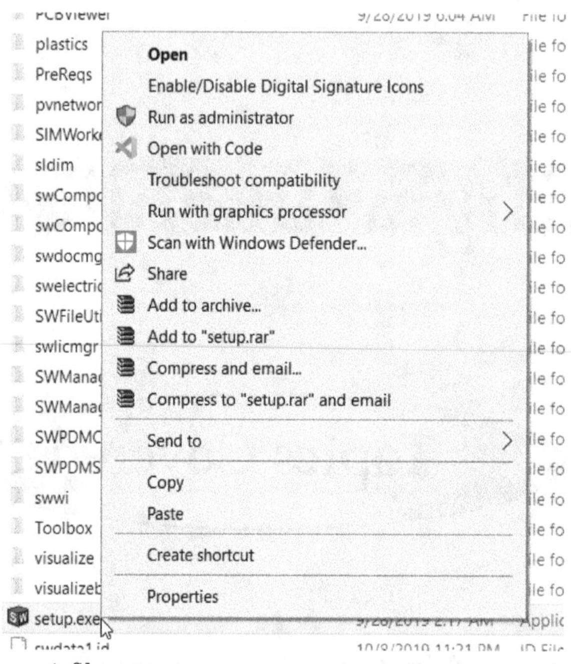

Figure-1. Shorcut_menu

- Select the **Run as Administrator** option from the menu displayed; refer to Figure-1.
- Select the **Yes** button from the dialog box displayed. The **SolidWorks 2023 Installation Manager** will be displayed. Follow the instructions given in the dialog box. Note that you must have the **Serial Number** with you to install the application. To get more about installation, double-click on the **Read Me** documentation file in the Setup folder.
- If you have downloaded the software from Internet, then you are required to browse in the **SolidWorks Download** folder in the **Documents** folder of Computer. Open the folder of latest version of software and then run **setup.exe**. Rest of the procedure is same.

STARTING SOLIDWORKS 2023

- To start SolidWorks in Windows 10 from **Start** menu, click on the **Start** button in the Taskbar at the bottom left corner and click on the **SolidWorks 2023** folder. In this folder, select the SolidWorks 2023 icon; refer to Figure-2.

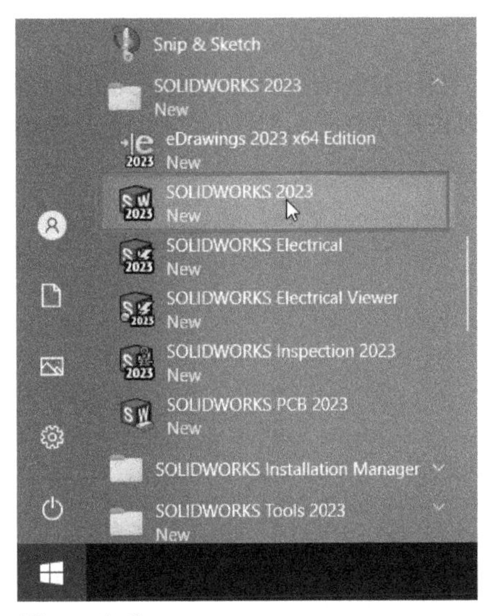

Figure-2. Start menu

- While installing the software, if you have selected the check box to create a desktop icon then you can double-click on that icon to run the software.
- If you have not selected the check box to create the desktop icon but want to create the icon on desktop, then drag and drop the **SolidWorks 2023** icon from the Start menu on the desktop.

After you perform the above steps, the SolidWorks 2023 application window will be displayed; refer to Figure-3.

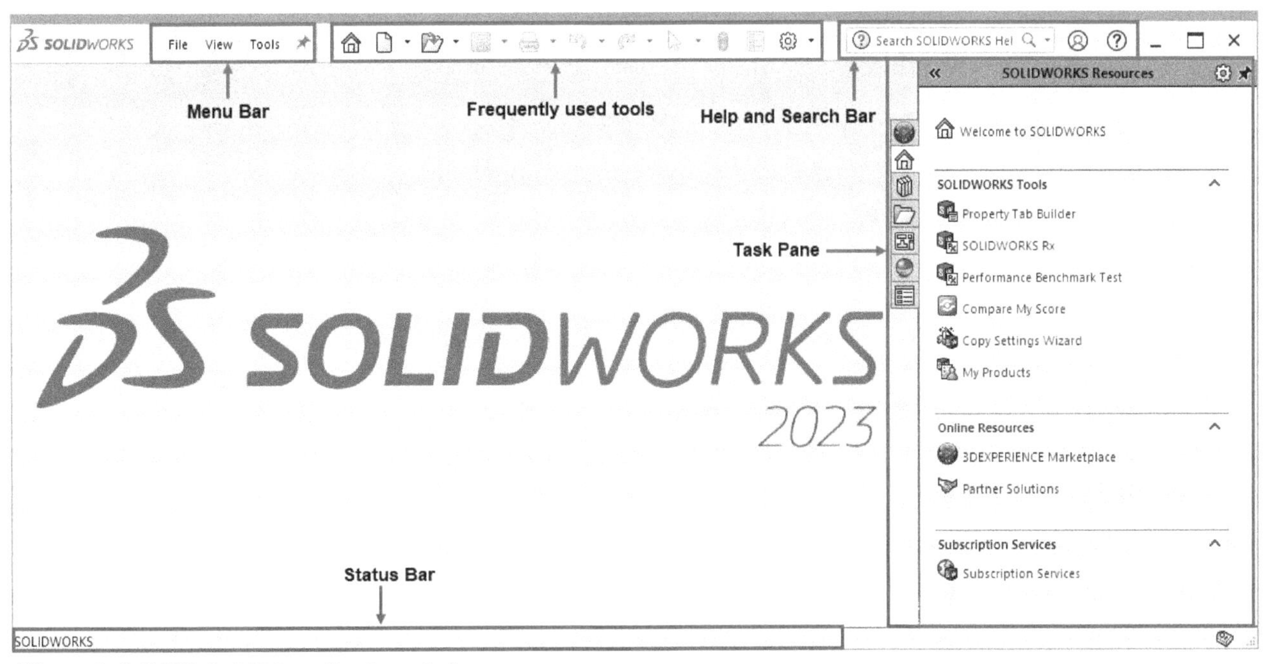

Figure-3. SolidWorks 2023 application window

STARTING A NEW DOCUMENT

There are various methods to start a new document in SolidWorks. The procedure to start a new document in SolidWorks is given next.

1. Click on the **New** button in the **Menu Bar** .

Or

2. Move the cursor on the left arrow near the **SolidWorks icon**; refer to Figure-4 and then click on the **File** menu button. The **File** menu will be displayed, click on the **New** button; refer to Figure-4.

Figure-4. File_menu

Or

3. Press **CTRL** and **N** together from the Keyboard.

- After performing any of the above steps, the **New SOLIDWORKS Document** dialog box will be displayed as shown in Figure-5.

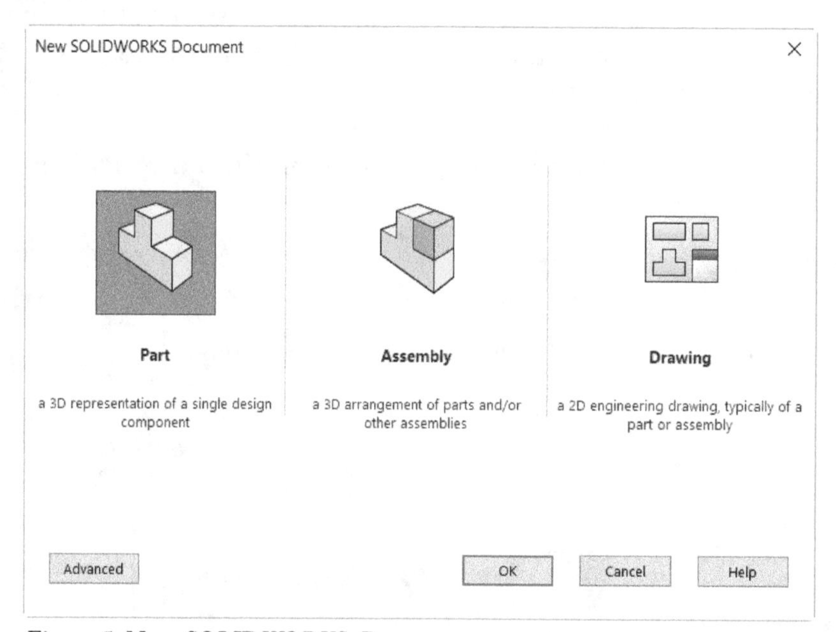

Figure-5. New_SOLIDWORKS_Document

- If you are creating first document after installing SolidWorks then **Units and Dimension Standard** dialog box will be displayed; refer to Figure-6.

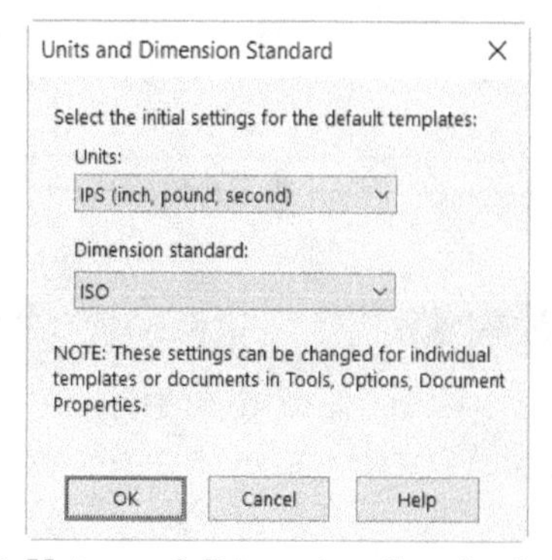

Figure-6. Units_and_Dimension_Standard_dialog_box

- Select the unit system and dimension standards that you want to use while creating documents in SolidWorks. These options will be set as default for later documents. You can change these parameters later using **System Options**. Click on the **OK** button from the dialog box.

There are three buttons available in this dialog box; **Part**, **Assembly**, and **Drawing**.

The **Part** button is used to create Solid, Surface, Sheetmetal, Mesh and other types of models.

The **Assembly** button is used to create Assemblies.

The **Drawing** button is used to create drawings from the part models or assemblies.

You will learn more about solids, surfaces, assemblies, and drawings later in the book.

Note that the building blocks of CAD are solid models. In SolidWorks, solid models are created by using the tools available in the **Part** mode. You can start with the **Part** mode by selecting the **Part** button in the **New SOLIDWORKS Document** dialog box.

- Double-click on the **Part** button to start the part modeling environment of SolidWorks. On doing so, the application interface will be displayed as shown in Figure-7.

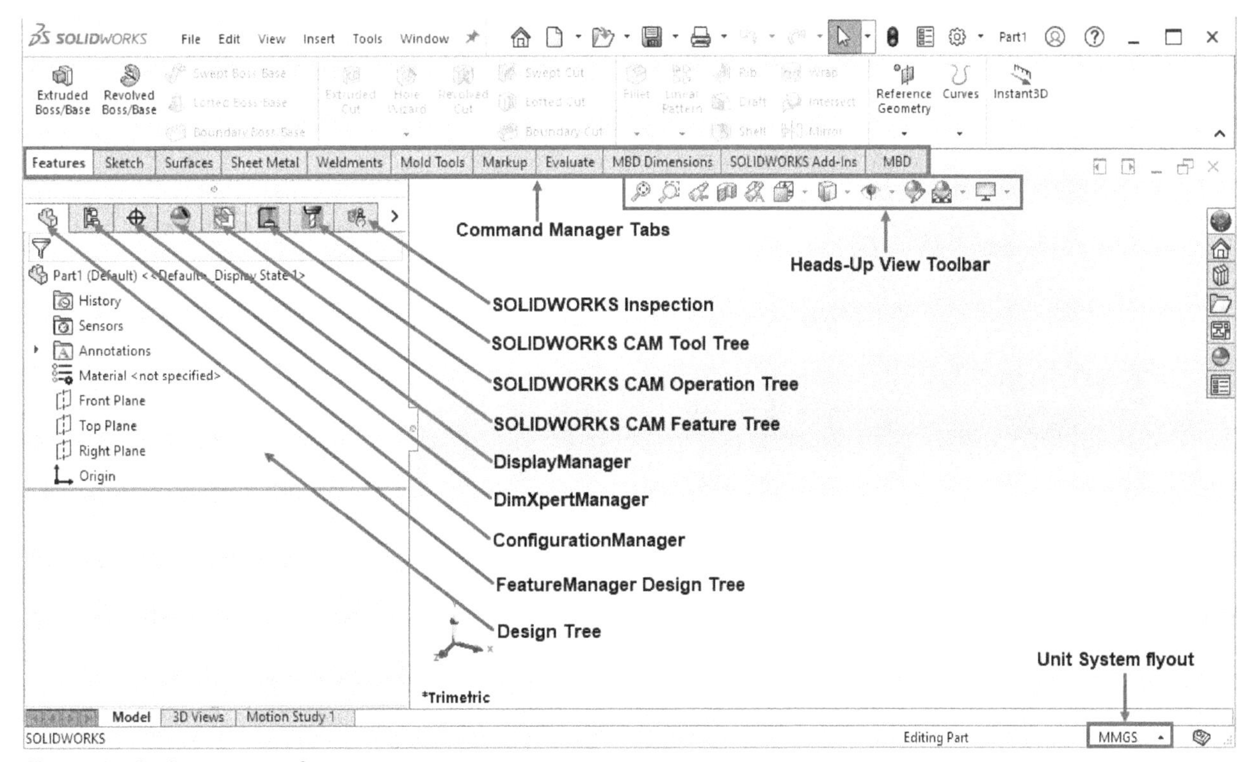

Figure-7. Application interface

The tools available in the **Part**, **Assembly**, and **Drawing** mode are compiled in the form of **CommandManagers**. To display or hide a **CommandManager** from **Ribbon**, right-click on a **CommandManager** in the **Ribbon**. A shortcut menu will be displayed. Hover the cursor over **Tabs** option, the list of **CommandManagers** will be displayed as shown in Figure-8. Select the **CommandManager** that you want to add to or remove from the **Ribbon**. Note that a tick mark is displayed before name of **CommandManager** in the list if the **CommandManager** is there in the **Ribbon**. Various **CommandManagers** available in SolidWorks will be discussed next.

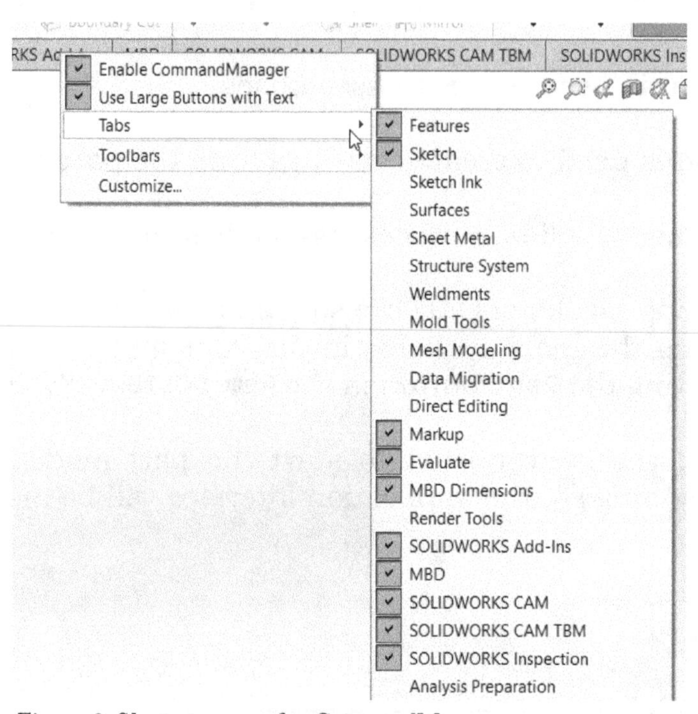

Figure-8. Shortcut_menu_for_CommandManager

Part Mode CommandManagers

A number of **CommandManagers** can be invoked in the **Part** mode. These **CommandManagers** with their functioning are described next.

Sketch CommandManager

The tools available in this **CommandManager** are used to draw sketches for creating solid/surface models. This **CommandManager** is also used to add relations and smart dimensions to the sketched entities. The **Sketch CommandManager** is shown in Figure-9.

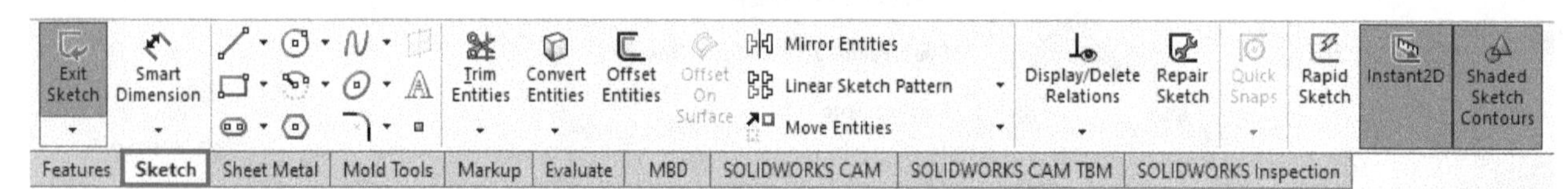

Figure-9. Sketch_CommandManager

Features CommandManager

This **CommandManager** provides all modeling tools that are used for feature-based solid modeling. The **Features CommandManager** is shown in Figure-10.

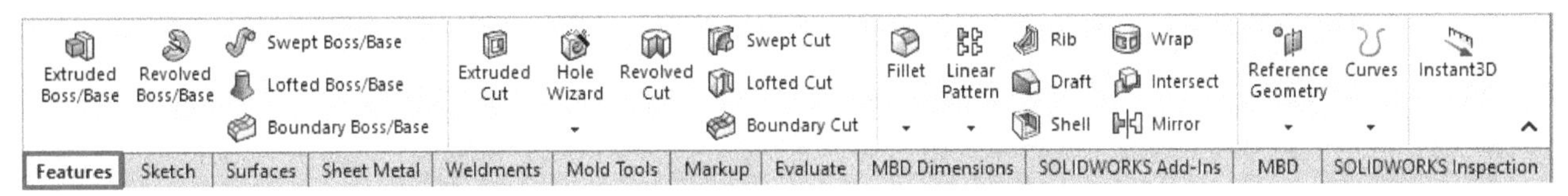

Figure-10. Features CommandManager

MBD Dimensions CommandManager

This **CommandManager** is used to add dimensions and tolerances to the features of a part. The **MBD Dimensions CommandManager** is shown in Figure-11.

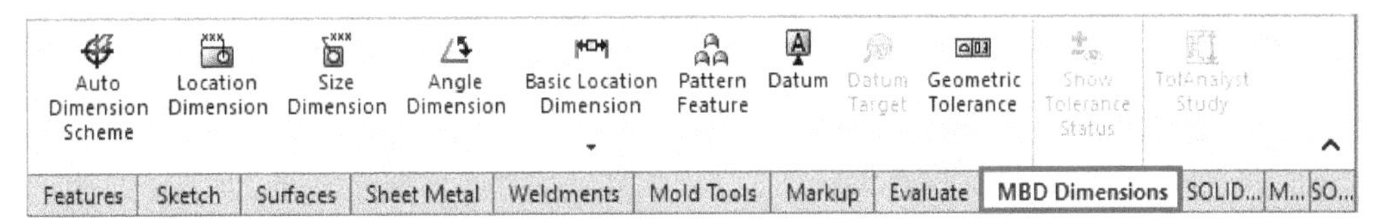

Figure-11. MBD Dimensions CommandManager

Sheet Metal CommandManager

The tools in this **CommandManager** are used to create the sheet metal parts. The **Sheet Metal CommandManager** is shown in Figure-12. If this **CommandManager** is not added in the **Ribbon**, then right-click on any of the **CommandManager** tab and select the **Sheet Metal** option from the menu; refer to Figure-13.

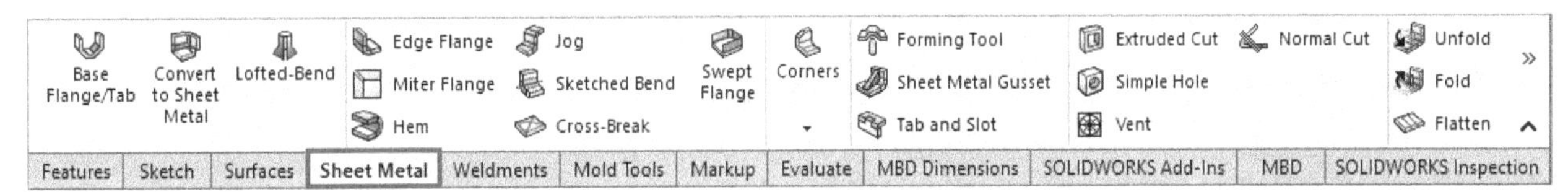

Figure-12. Sheet Metal CommandManager

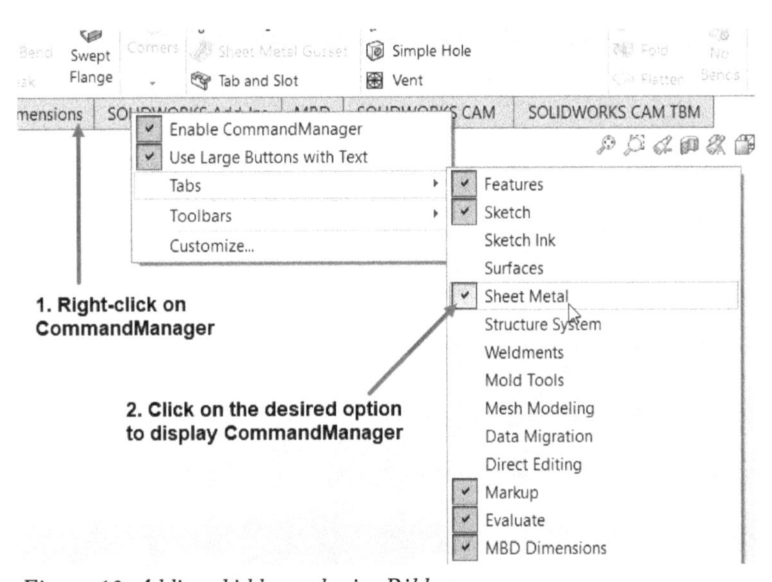

Figure-13. Adding_hidden_tabs_in_Ribbon

Mold Tools CommandManager

The tools in this **CommandManager** are used to design a mold and split core & cavity steel. The **Mold Tools CommandManager** is shown in Figure-14.

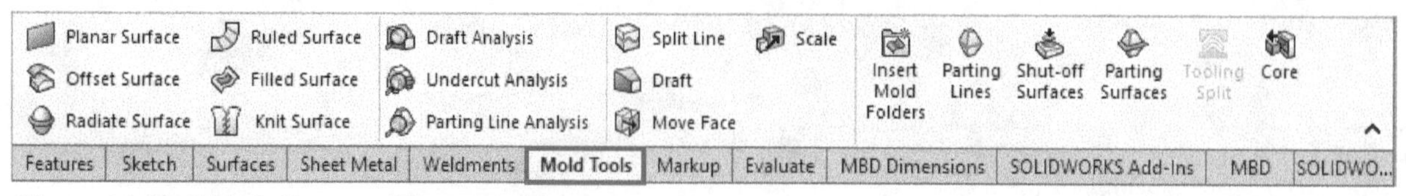

Figure-14. Mold Tools CommandManager

Evaluate CommandManager

This **CommandManager** is used to measure entities, perform analysis, and so on. The **Evaluate CommandManager** is shown in Figure-15.

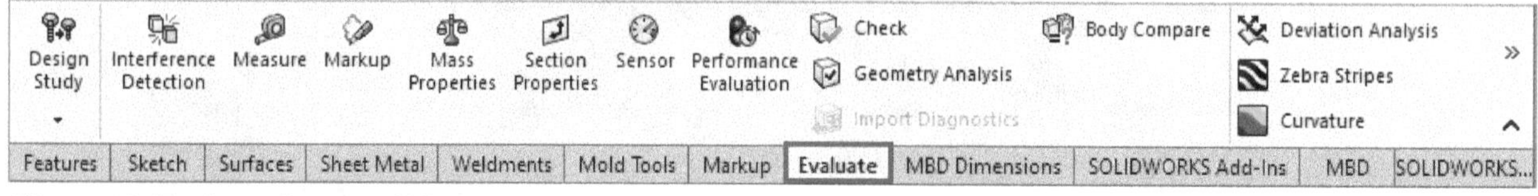

Figure-15. Evaluate CommandManager

Surfaces CommandManager

This **CommandManager** is used to create complicated surface features. The **Surfaces CommandManager** is shown in Figure-16.

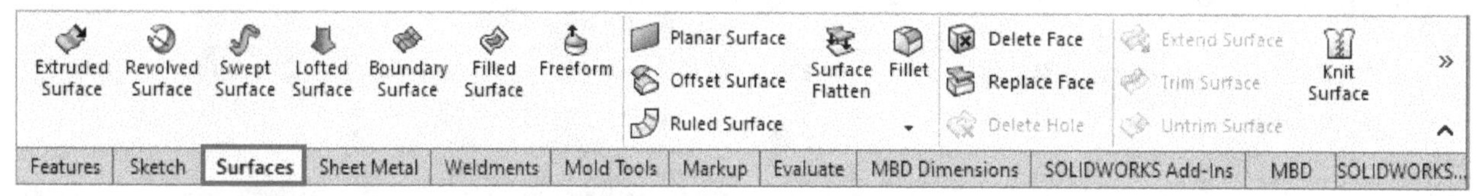

Figure-16. Surfaces CommandManager

Direct Editing CommandManager

This **CommandManager** consists of tools (Figure-17) that are used for editing a feature.

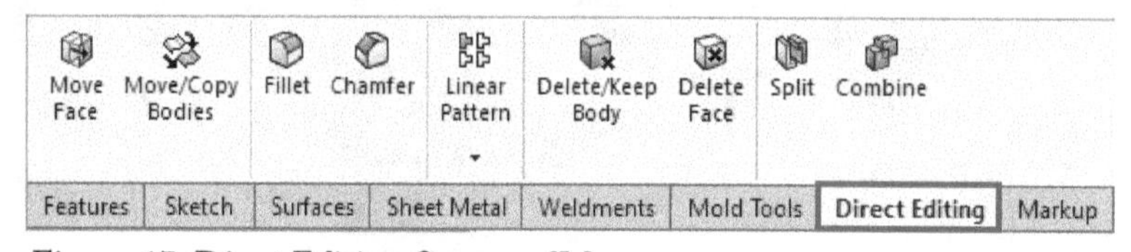

Figure-17. Direct Editing CommandManager

Data Migration CommandManager

This **CommandManager** consist of tools (Figure-18) that are used to work with the models created in other packages or in different environments.

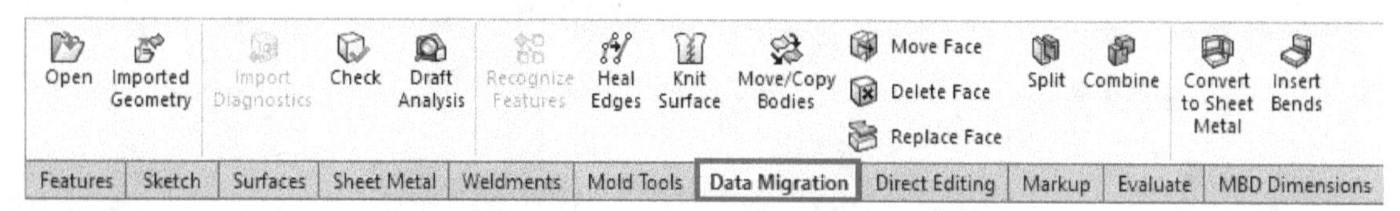

Figure-18. Data Migration CommandManager

Weldments CommandManager

This **CommandManager** is used to create welding joints in the model and assembly. The **Weldments CommandManager** is shown in Figure-19.

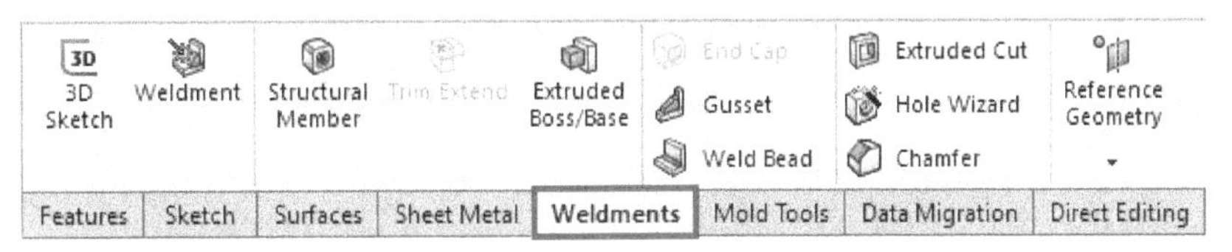

Figure-19. Weldments CommandManager

SOLIDWORKS MBD CommandManager

This **CommandManager** is used to apply Model Based Dimension which means the dimensions are directly applied to model while skipping the steps of generating drawings. Use of SolidWorks MBD in manufacturing industry requires electronic gadgets at shop floor for production and quality checks. These gadgets should be capable of displaying CAD eDrawings. The **CommandManager** is shown in Figure-20.

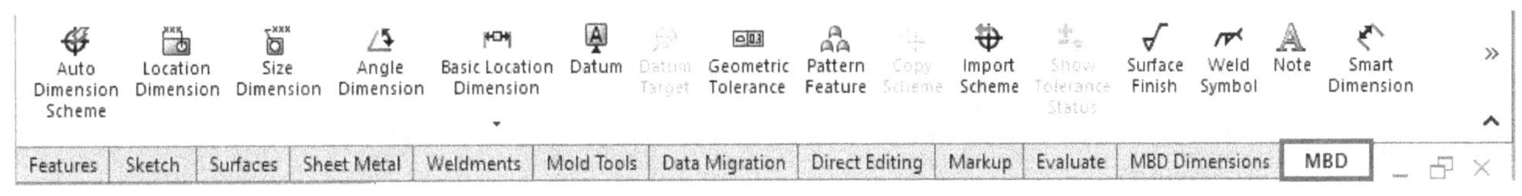

Figure-20. SOLIDWORKS MBD CommandManager

Render Tools CommandManager

The tools in **Render Tools CommandManager** are used to render the image of current model using appearance parameters specified. The **CommandManager** is shown in Figure-21.

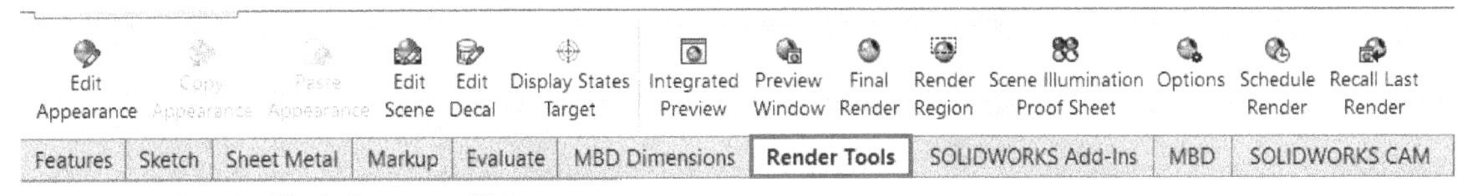

Figure-21. Render Tools CommandManager

SOLIDWORKS CAM CommandManager

The tools in **SOLIDWORKS CAM CommandManager** are used to generate NC programs for CNC machines. The **CommandManager** is shown in Figure-22.

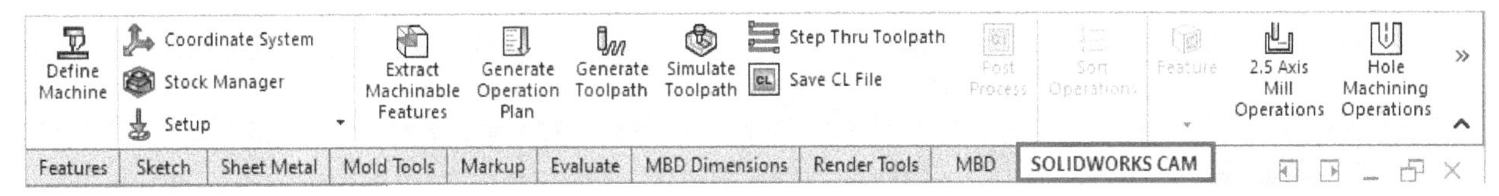

Figure-22. SOLIDWORKS CAM CommandManager

SOLIDWORKS CAM TBM CommandManager

The tools in **SOLIDWORKS CAM TBM CommandManager** are used to automatically generate toolpaths and operation settings based on specified tolerances and dimension in the MBD; refer to Figure-23.

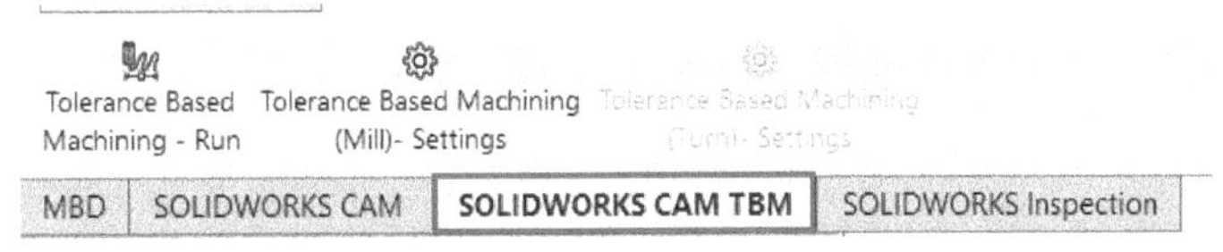

Figure-23. SOLIDWORKS CAM TBM CommandManager

SOLIDWORKS Inspection CommandManager

The tools in **SOLIDWORKS Inspection CommandManager** are used to create inspection drawings and reports for various standards and production part approval process(PPAP); refer to Figure-24. You can also use SolidWorks Inspection for First Article Inspection (FAI) and in process inspection applications.

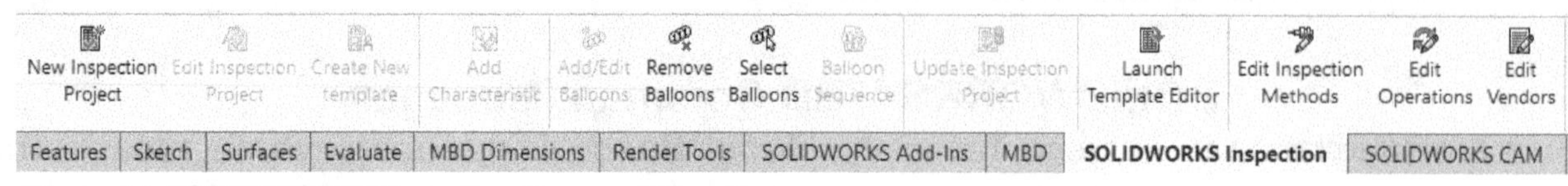

Figure-24. SOLIDWORKS Inspection CommandManager

Sketch Ink CommandManager

The tools in **Sketch Ink CommandManager** are create sketch free-hand using either stylus or fingertips on touch screen devices; refer to Figure-25.

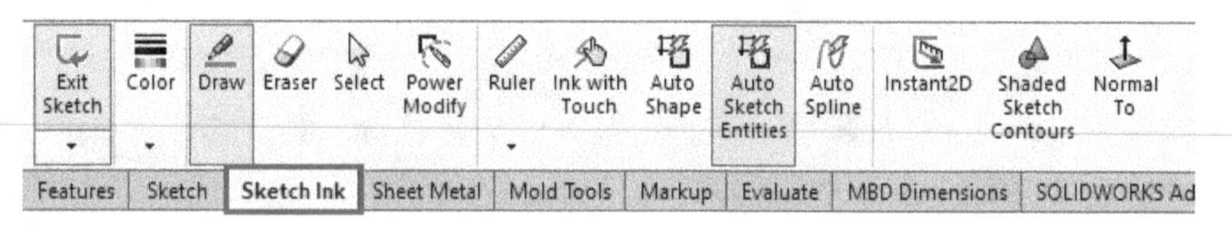

Figure-25. Sketch Ink CommandManager

Markup CommandManager

The tools in **Markup CommandManager** are used to create free hand markings in the part, assembly, or drawing; refer to Figure-26.

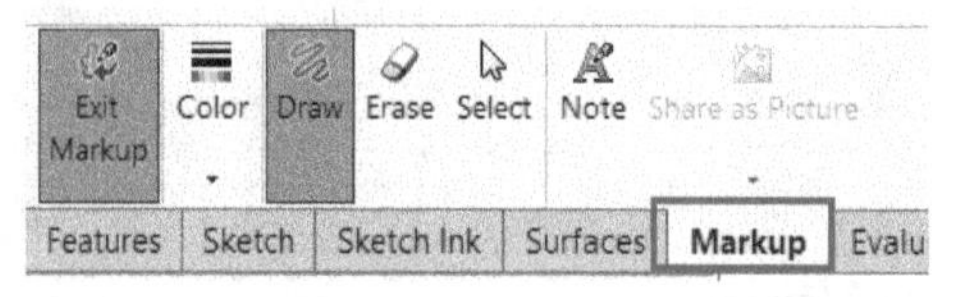

Figure-26. Markup CommandManager

Assembly Mode CommandManagers

The tools in **CommandManagers** of the **Assembly** mode are used to assemble the components. The **CommandManagers** in the **Assembly** mode are discussed next.

Assembly CommandManager

This **CommandManager** is used to insert a component and apply various types of mates to the assembly. The **Assembly CommandManager** is shown in Figure-27.

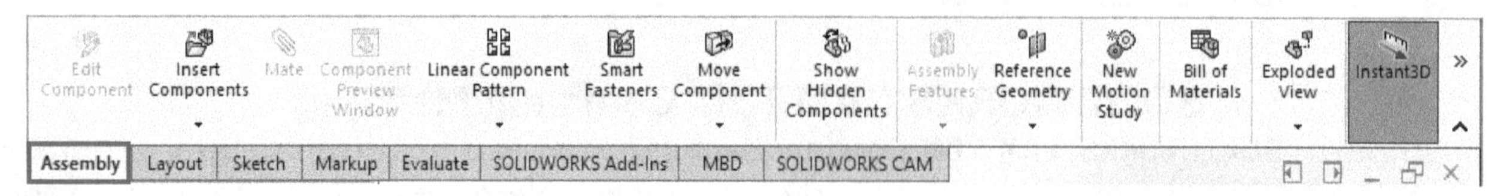

Figure-27. Assembly CommandManager

Layout CommandManager

The tools in this **CommandManager** (Figure-28) are used to create and edit blocks.

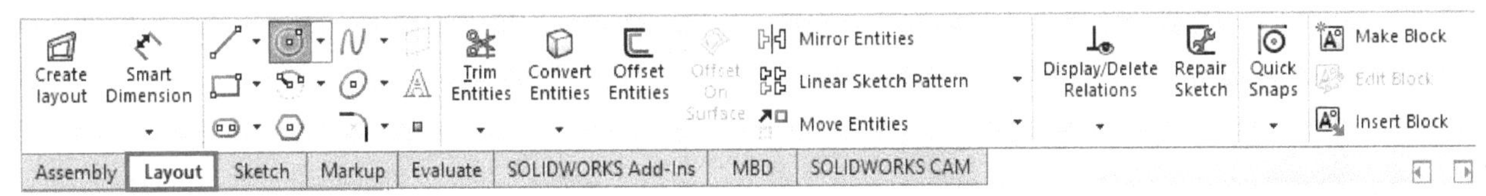

Figure-28. Layout CommandManager

Drawing Mode CommandManagers

You can invoke a number of **CommandManagers** in the **Drawing** mode. The **CommandManagers** that are extensively used during the designing process in this mode are discussed next.

Drawing CommandManager

This **CommandManager** is used to generate the drawing views of an existing model or an assembly. The **Drawing CommandManager** is shown in Figure-29.

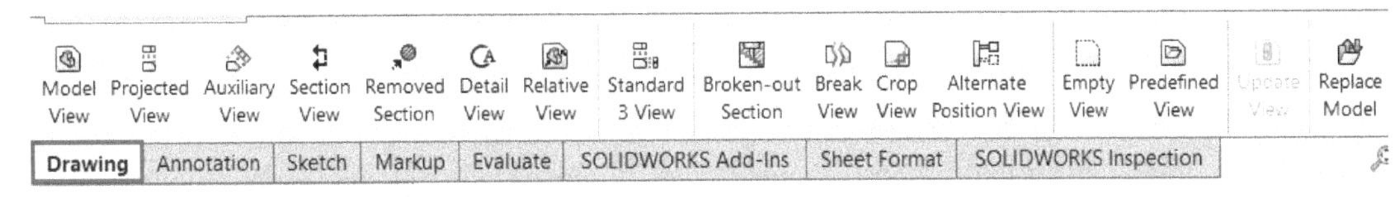

Figure-29. Drawing CommandManager

Annotation CommandManager

The **Annotation CommandManager** is used to generate the model items and to add notes, balloons, geometric tolerance, surface finish symbols, and so on to the drawing views. The **Annotation CommandManager** is shown in Figure-30.

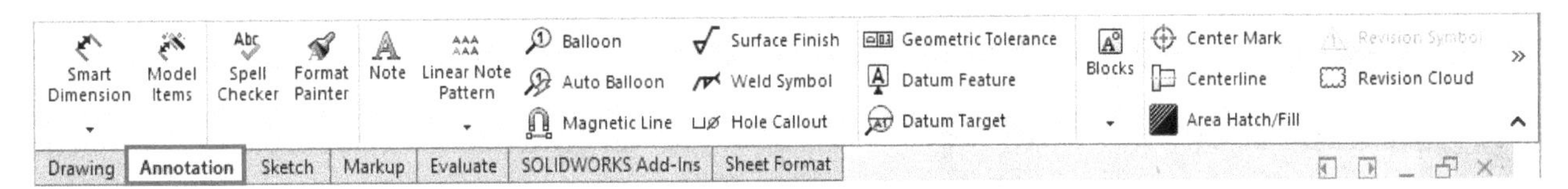

Figure-30. Annotation CommandManager

The commands available in these **CommandManagers** will be discussed one by one later in this book.

OPENING A DOCUMENT

Like creating new documents, there are many ways to open documents. Some of them are discussed next.

- Click on the **Open** button in the **Menu Bar** or move the cursor on the **SolidWorks icon** and then click on the **File** > **Open** button from the menu or press **CTRL** and **O** together from the Keyboard.
- After performing any of the above step, the **Open** dialog box will be displayed; refer to Figure-31.

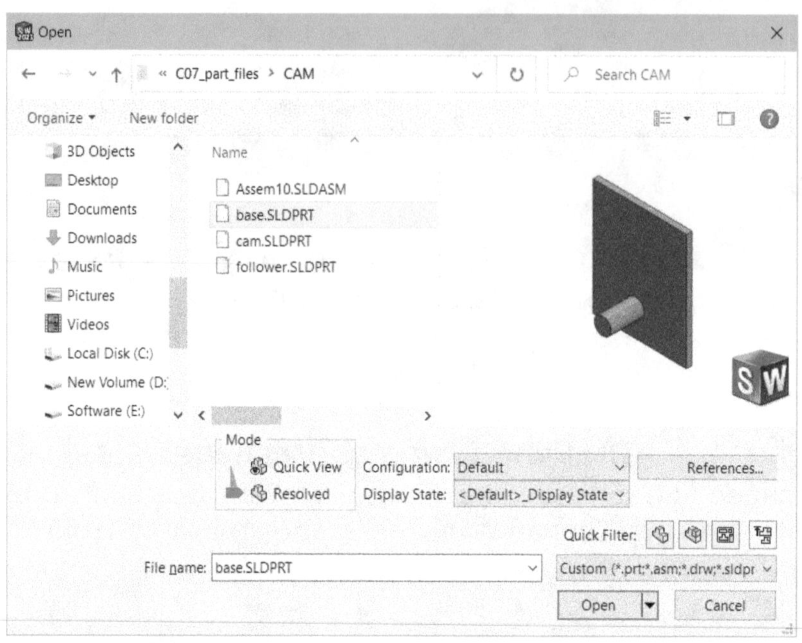

Figure-31. Open_dialog_box

- Select the file type of your file from the **File Type** fly-out Custom (*.prt;*.asm;*.drw;*.sldpr in the bottom right corner of the dialog box.
- Browse to the folder in which you have saved the file and then double-click on it to open.

Most of the time people close the application to close the current file and then restart the application to start working again. In SolidWorks, you can close the current file while the application remains open. The steps to do so are given next.

Note that there is no separate tool in SolidWorks to import models created in other software. You need to use **Open** tool to import the non native files.

CLOSING A DOCUMENT

To close a document, there are two ways displayed in Figure-32.

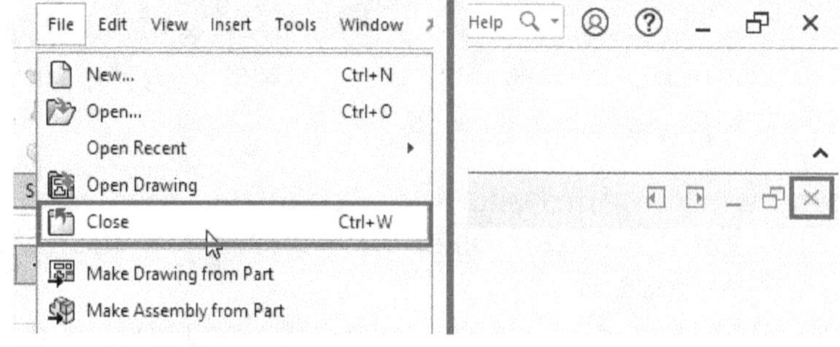

Figure-32. Closing_options

- Open the **File** menu and then click on the **Close** option from it or click on the **Close** button at the top-right of the current viewport (Viewport is the area in which the model is displayed.) or press **CTRL+W**.
- If you have done some editing in the document then a dialog box will be displayed prompting you to save the document; refer to Figure-33.

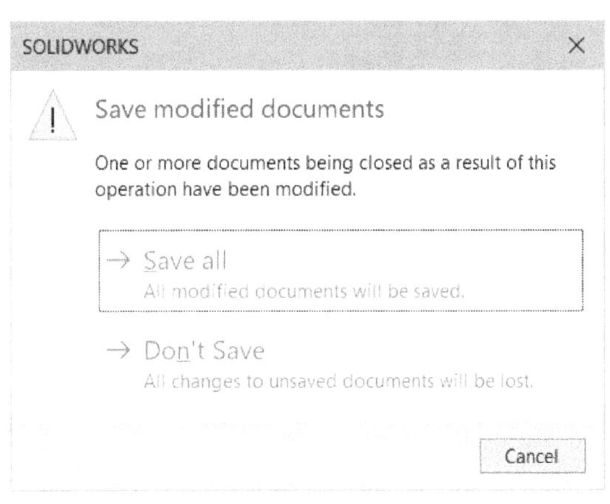

Figure-33. Save_prompt

- Click on the **Save all** button to save the changes or click on the **Don't Save** button to reject the changes.

SAVING THE DOCUMENTS

There are three tools available in **File** menu to save document in SolidWorks; **Save**, **Save As**, and **Save All**. If you are working on a model and click on the **Save** tool for the first time after starting the new document then the **Save** tool will work as **Save As** tool. The **Save As** tool allows you to create a new copy of the document in different formats. The **Save All** tool is used to save all the modified documents currently open in SolidWorks. The procedure to use the **Save As** tool is discussed next.

- Click on the **Save As** tool from the **File** menu. The **Save As** dialog box will be displayed; refer to Figure-34.

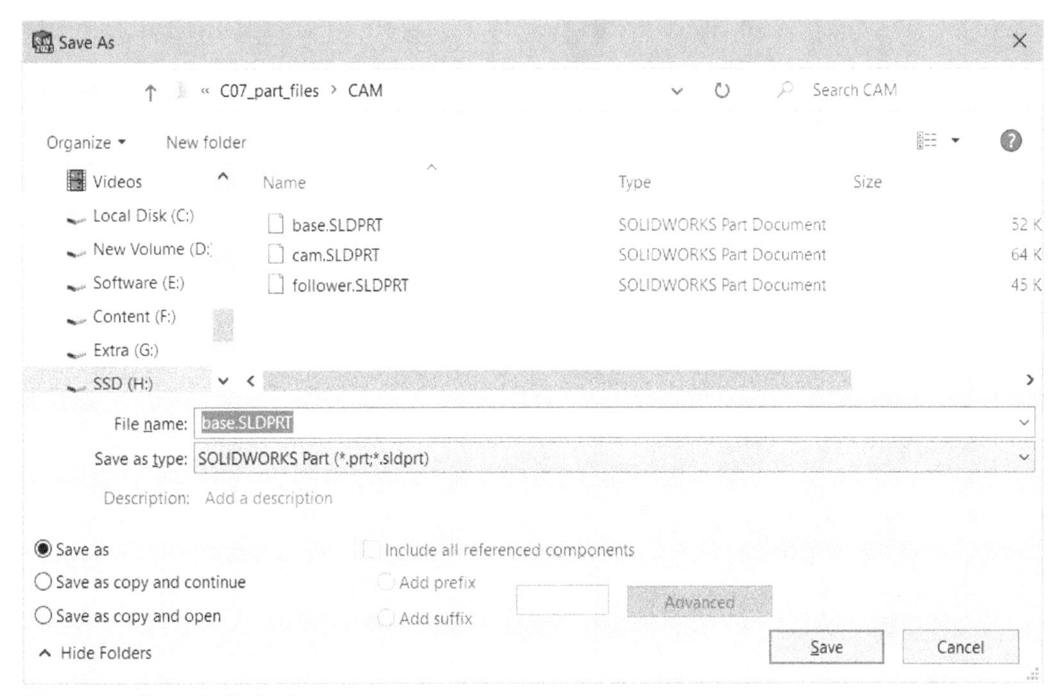

Figure-34. Save As dialog box

- Specify desired name for the file in the **File name** edit box.
- Click in the **Save as type** drop-down and select desired format for file; refer to Figure-35.

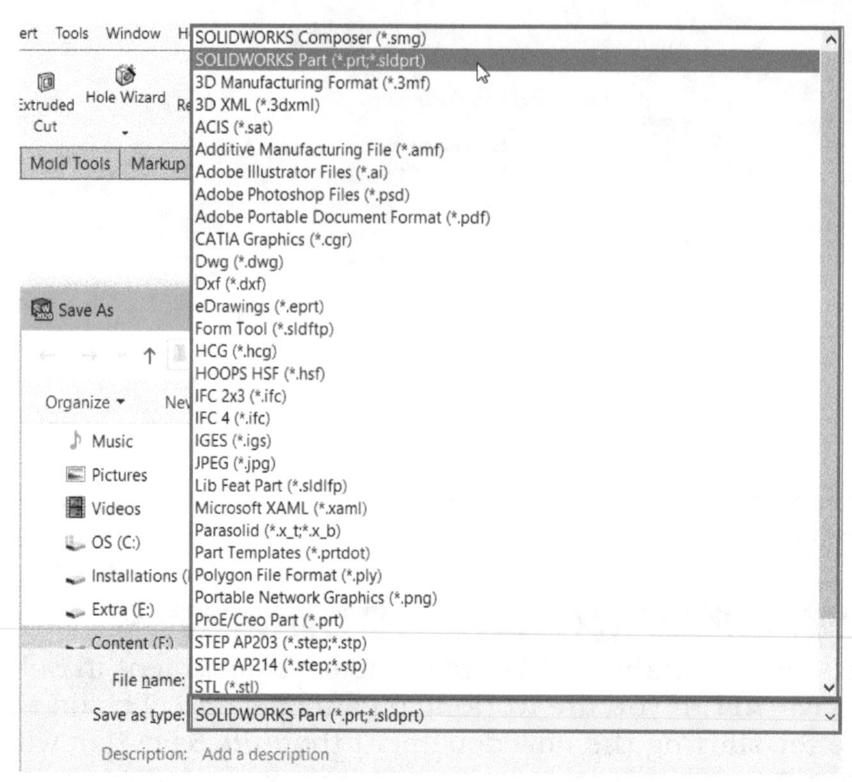

Figure-35. Save as type drop-down

- Select the **Save as** radio button to save the current open file with desired settings. Select the **Save as copy and continue** radio button to save a new copy of currently open file and do not open the new copy while leaving the original file unchanged. Select the **Save as copy and open** radio button to save a new copy of currently open file and open the new copy while leaving the original file unchanged.
- Click in the **Description** field and specify desired description for the file.
- If you are saving an assembly file then **Include all reference components** check box will be active. Select this check box to add a prefix/suffix to the assembly components names and define other parameters for saving assembly components files. The related options will become active. Select desired radio button; **Add prefix** or **Add suffix** and then specify desired text in the next edit box. Click on the **Advanced** button. The **Save As with References** dialog box will be displayed; refer to Figure-36. Select the **Nested view** or **Flat view** radio button to check assembly components in nested view or flat view.
- Select the **Include broken references**, **Include Toolbox parts**, and/or **Include virtual components** check boxes to include respective parts while saving them.
- Click in the **Specify folder for selected items** edit box and specify the location where you want to save assembly components.
- Select the **Save all as copy (opened documents remain unaffected)** check box if you want to save new copy of all the assembly files in specified folder location.
- Double-click in desired field of table in the dialog box to change it.
- Click on the **Save All** button from the dialog box to save assembly files.

Note that once you have saved SolidWorks assembly or part file then next time clicking on **Save** tool or pressing **CTRL+S** will not display the **Save As** dialog box. It will directly save the files.

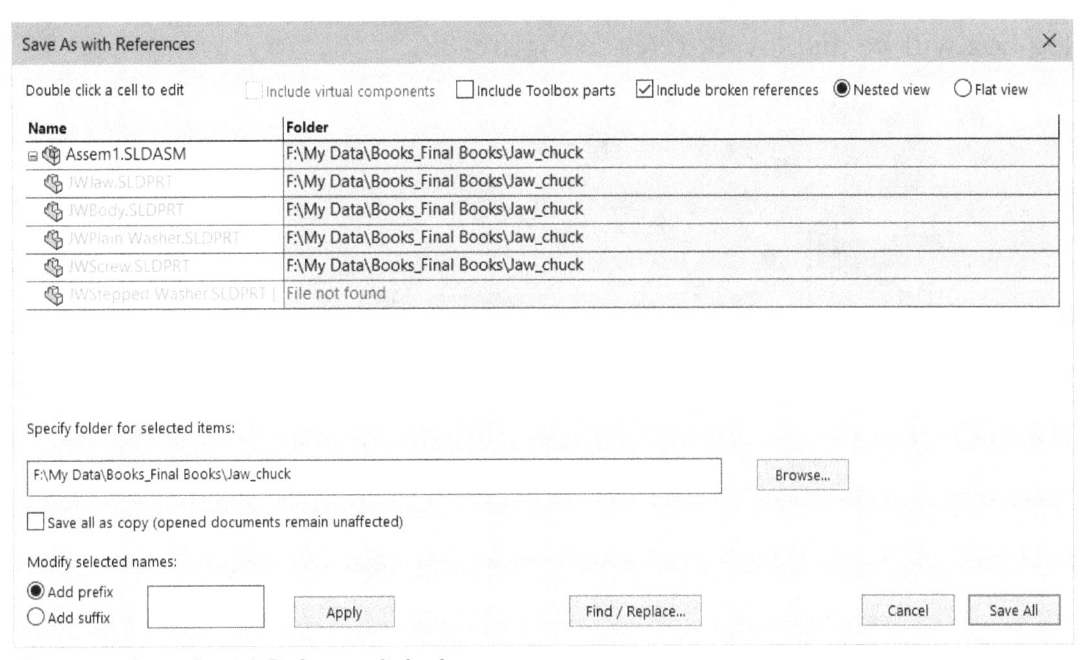

Figure-36. Save As with References dialog box

PRINTING DOCUMENT

The **Print** tool in **File** menu is used to print documents on paper or in different formats. The procedure to use this tool is given next.

- Click on the **Print** tool from the **File** menu or press **CTRL+P**. The **Print** dialog box will be displayed; refer to Figure-37.

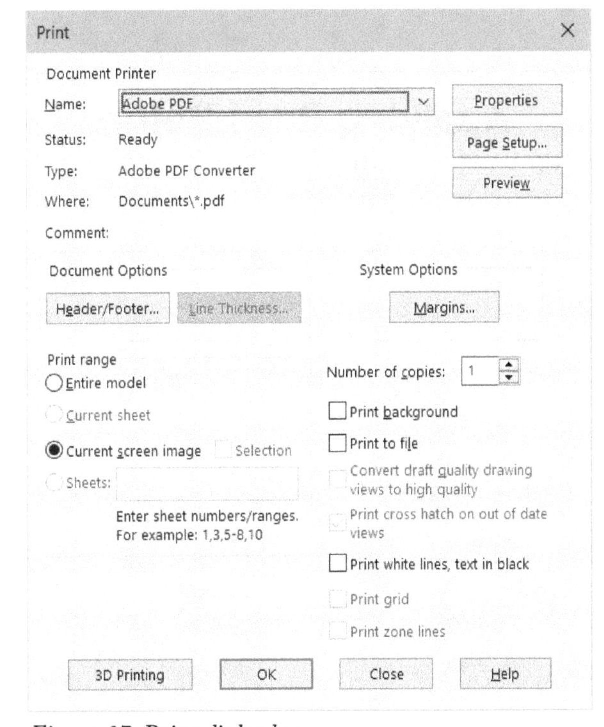

Figure-37. Print dialog box

- Select desired printer from the **Name** drop-down in the **Document Printer** area of the dialog box.

- Click on the **Properties** button from the dialog box. The respected Properties dialog box will be displayed; refer to Figure-38.

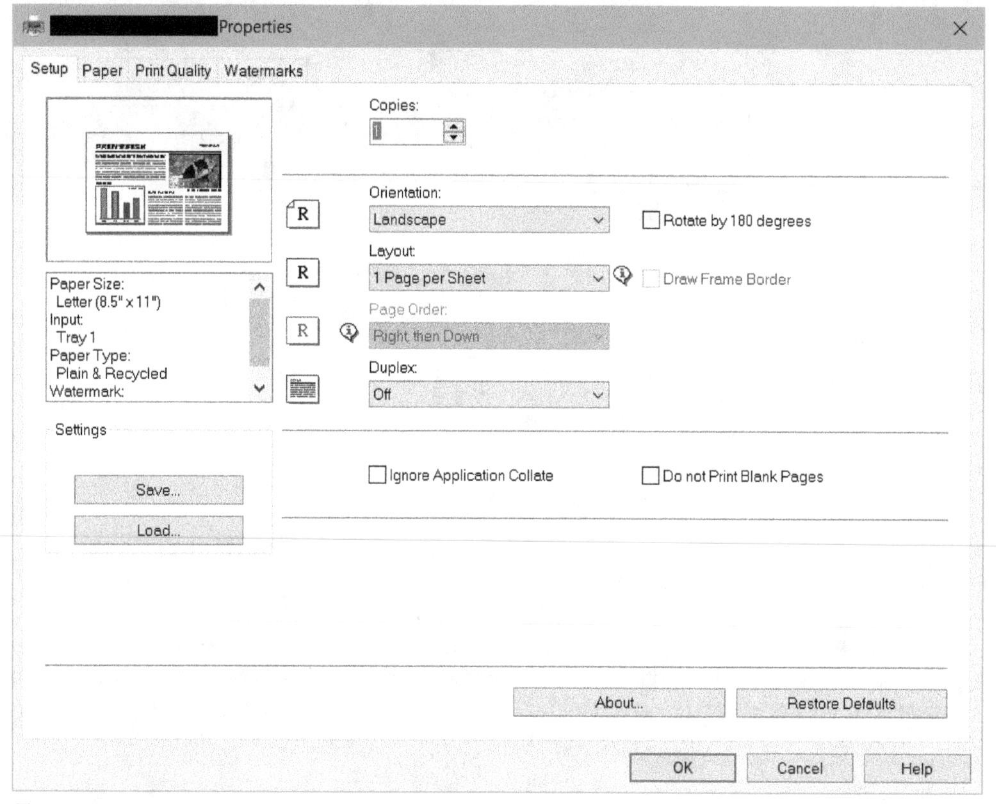

Figure-38. Printer Properties dialog box

- Set the other parameters as desired for printing like number of copies, paper size, orientation, resolution, and so on. Click on the **OK** button from the dialog box to apply properties.

Applying Page Setup

- Click on the **Page Setup** button from the **Print** dialog box. The **Page Setup** dialog box will be displayed; refer to Figure-39.

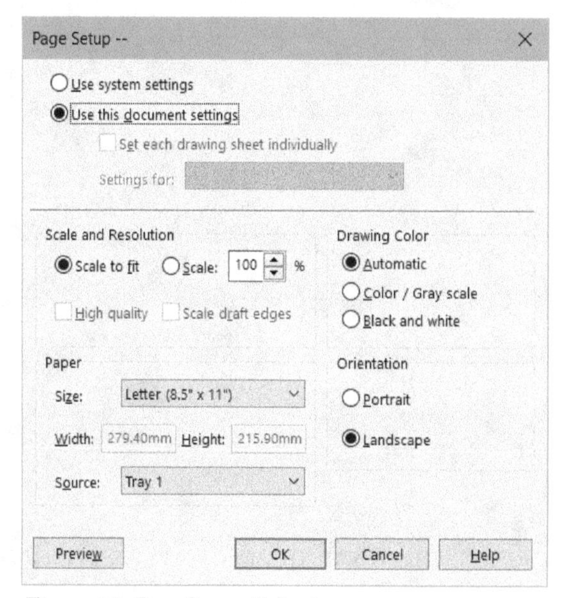

Figure-39. Page Setup dialog box

- Select the **Use system settings** radio button or the **Use this document settings** radio button to use respective page settings for printing.
- Set the other parameters as desired like print page source, paper size, scale, resolution, drawing color, and so on.
- Click on the **OK** button from the dialog box. The **Print** dialog box will be displayed again.

Setting Header/Footer for Print Pages

- Click on the **Header/Footer** button from the **Document Options** area of the dialog box. The **Header/Footer** dialog box will be displayed; refer to Figure-40.

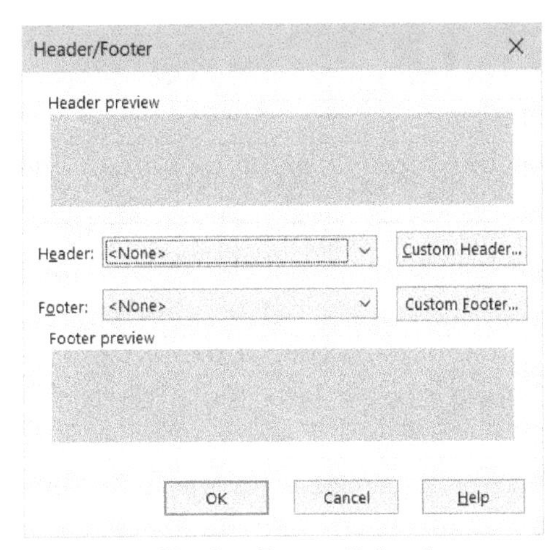

Figure-40. Header_Footer_dialog_box

- Select desired options from the **Header** and **Footer** drop-downs. If you want to create a custom header/footer then click on the **Custom Header** button. The **Custom Header** dialog box will be displayed; refer to Figure-41.

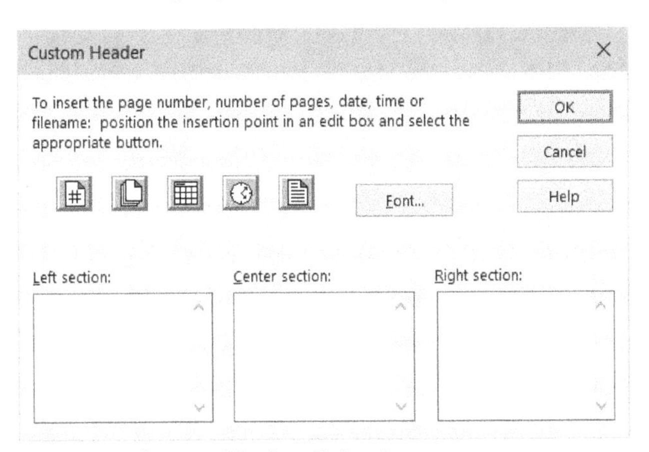

Figure-41. Custom_Header_dialog_box

- Click in desired section and specify the custom text to be displayed in header or footer. You can also use the buttons like **Page Numbers**, **Date**, and so on to specify custom data. After setting desired parameters, click on the **OK** button. The **Print** dialog box will be displayed again.

Print Margins

- Click on the **Margins** button from the **System Options** area of the dialog box. The **Margins** dialog box will be displayed; refer to Figure-42.

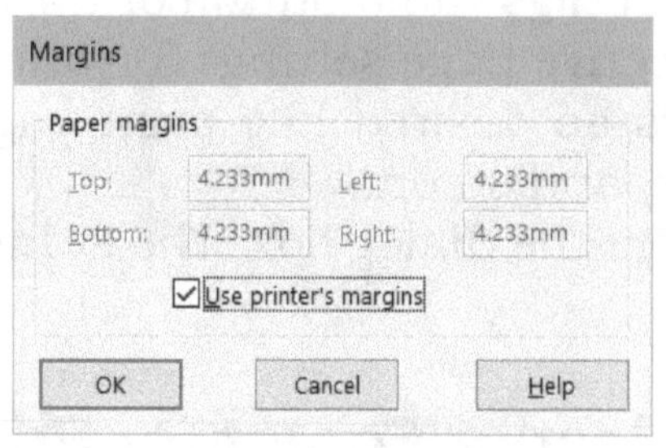

Figure-42. Margins dialog box

- Clear the **Use printer's margins** check box if you want to specify custom margins to keep object within paper boundary and specify desired parameters in the **Paper margins** area.
- Click on the **OK** button from the dialog box to apply parameters.

Print Range and other Parameters

- Select desired radio button from the **Print range** area and define respective parameters. If your document has multiple sheets for printing generally in case of printing drawings then **Current sheet** and **Sheets** radio button will be displayed.
- Specify desired number for prints to be produced in the **Number of copies** edit box.
- Select the **Print background** check box from the dialog box if you want to print the background also.
- Select the **Print to file** check box if you want to save a printer file.
- Click on the **3D Printing** button from the dialog box if you want to use a 3D printer to create the model. You will learn more about 3D printing later in Chapter 15.
- Similarly, set the other parameters and click on the **OK** button from the **Print** dialog box to start printing the document.

Publishing to eDrawings

The **Publish to eDrawings** tool is used to publish model for eDrawings Viewer. eDrawings is a software developed by Dassault Systemes Corporation for sharing models with marketing and senior management. The procedure to publish model is given next.

- Click on the **Publish to eDrawings** tool from the **File** menu. The **Save Configurations to eDrawings file** dialog box will be displayed; refer to Figure-43.

Figure-43. Save Configurations to eDrawings file dialog box

- Select the check boxes for configurations to be included in the eDrawing file.
- Click on the **Options** button from the dialog box. The **System Options** dialog box will be displayed; refer to Figure-44.

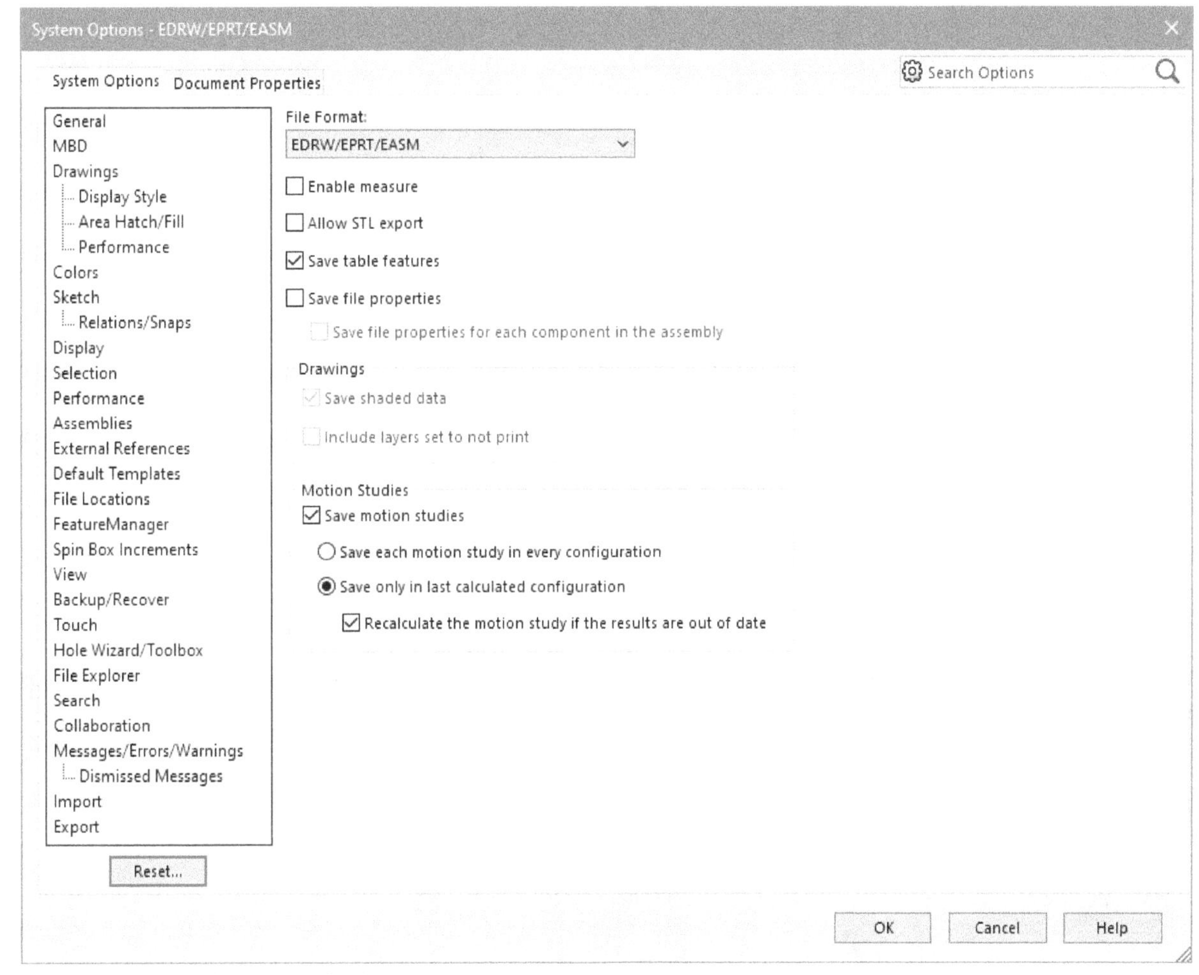

Figure-44. System Options dialog box

- Select the **Enable measure** check box if you want to allow measurement of geometry in eDrawing.
- Select the **Allow STL export** check box to allow saving of eDrawing as STL from the eDrawing Viewer software.
- Similarly, set the other parameters as desired in the dialog box and click on the **OK** button.

Specifying Password

- If you want to set password for opening eDrawing file then click on the **Password** button from the **Save Configurations to eDrawings file** dialog box. The **Password** dialog box will be displayed; refer to Figure-45.

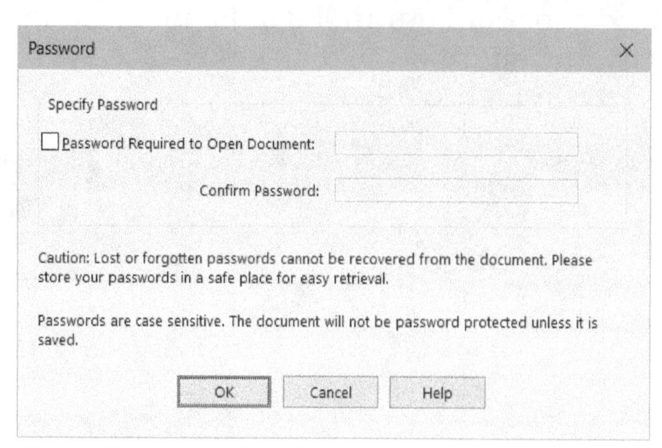

Figure-45. Password dialog box

- Select the **Password Required to Open Document** check box. The edit box to define password will be activated.
- Specify desired password and input the same in **Confirm Password** edit box. Click on the **OK** button from the dialog box.

After specifying desired parameters, click on the **OK** button from the dialog box. The model will open in eDrawing Viewer; refer to Figure-46. The options of eDrawings Viewer are out of scope of this book.

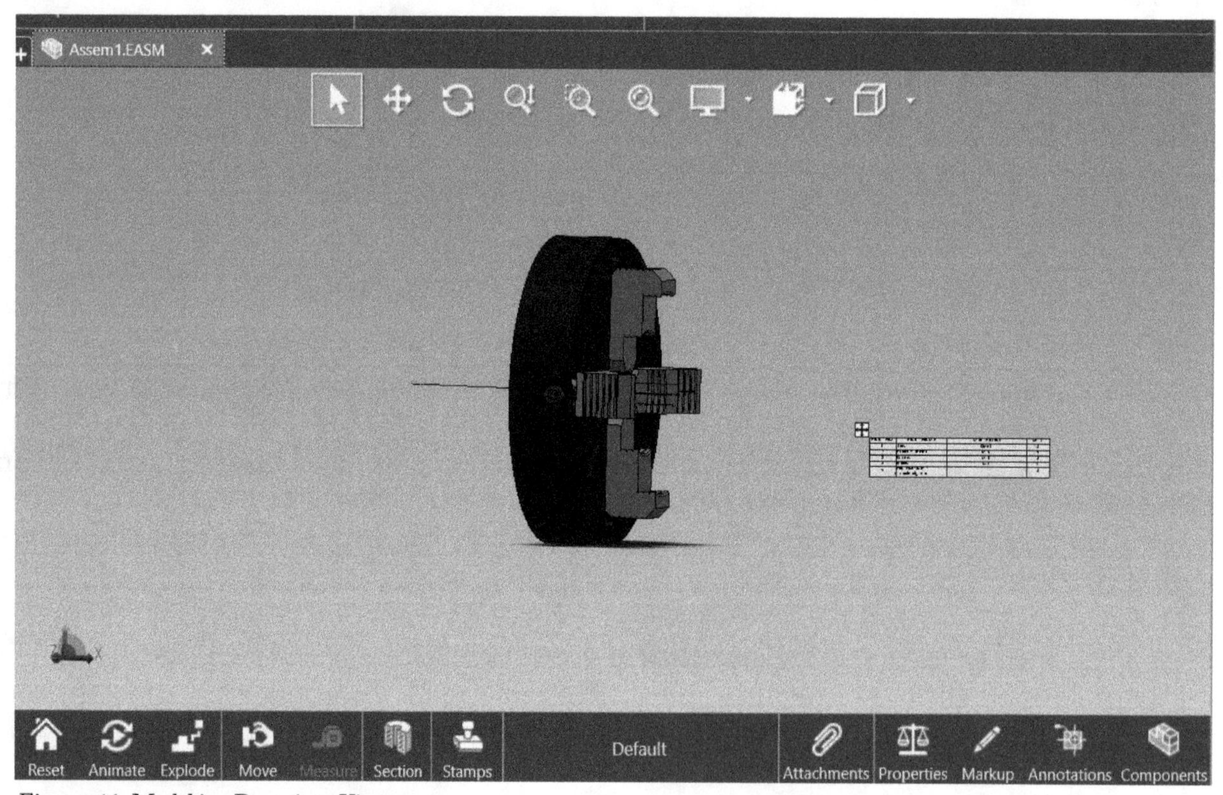

Figure-46. Model in eDrawings Viewer

PACKING FILES FOR SHARING

The **Pack and Go** tool in **File** menu is used to package the files for sharing with colleagues or client. The procedure to use this tool is given next.

- Click on the **Pack and Go** tool from the **File** menu. The **Pack and Go** dialog box will be displayed; refer to Figure-47.

Figure-47. Pack and Go dialog box

- Select check boxes for documents to be included in the package.
- Select the **Save to folder** radio button if you want to create a new folder for documents or select the **Save to Zip file** radio button. Specify desired location for file or folder in adjacent edit box.
- Select the **Add prefix** and **Add suffix** check boxes if you want to add prefixes and suffixes. After selecting the check boxes, specify desired text in adjacent edit boxes.
- Select the **Email after packaging** check box if you want to email your client after packaging files.
- Click on the **Save** button to save the file.

SENDING FILES IN EMAIL

The **Send To** tool in **File** menu is used to send current model file as attached in an E-mail. Make sure you have an E-mail program set default for sending e-mail.

RELOADING FILES

The **Reload** tool in **File** menu is used to reload the files that have been modified outside the SolidWorks software or the files that have not been updated automatically. The procedure to use this tool is given next.

- Click on the **Reload** tool from the **File** menu. The **Reload** dialog box will be displayed; refer to Figure-48.

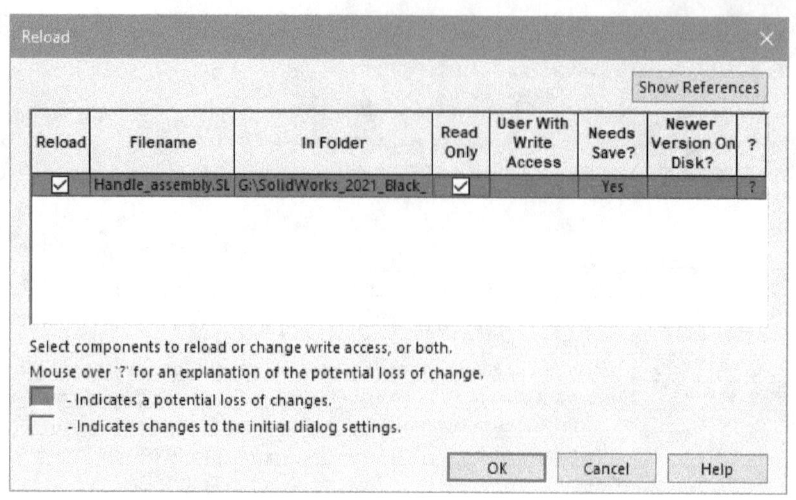

Figure-48. Reload_dialog_box

- Click on the **OK** button from the dialog box. An information box will be displayed; refer to Figure-49.

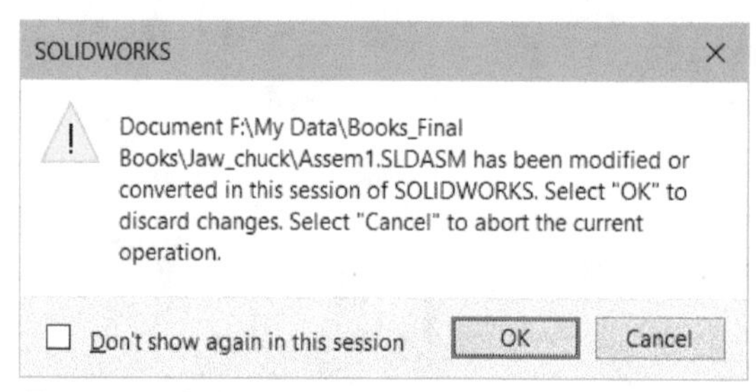

Figure-49. Information box for reloading files

- Click on the **OK** button from the dialog box.

REPLACING COMPONENT IN ASSEMBLY

The **Replace** tool in **File** menu is used to replace selected component in the assembly. The procedure to use this tool is given next.

- Click on the **Replace** tool from the **File** menu. The **Replace PropertyManager** will be displayed; refer to Figure-50.
- Select the component to be replaced from **FeatureManager Design Tree**.
- If there are multiple instances of the same component and you want to replace all the instances then select the **All instances** radio button from the **Instances to replace** area. If you want to replace all instances in same parent assembly then select the **All in same parent assembly** radio button.
- Click on the **Browse** button from the **With this one** area of **PropertyManager**. The **Open** dialog box will be displayed.
- Select desired component by which original component will be replaced and click on the **Open** button.
- Expand the **Thumbnail Preview** rollout to check the preview of model by which part will be replaced.
- Select desired radio button from the **Options** rollout. If you want to select configuration of new component with name matching to original one then select the **Match name** radio button. Select the **Manually select** radio button if you want to manually select configuration of component.

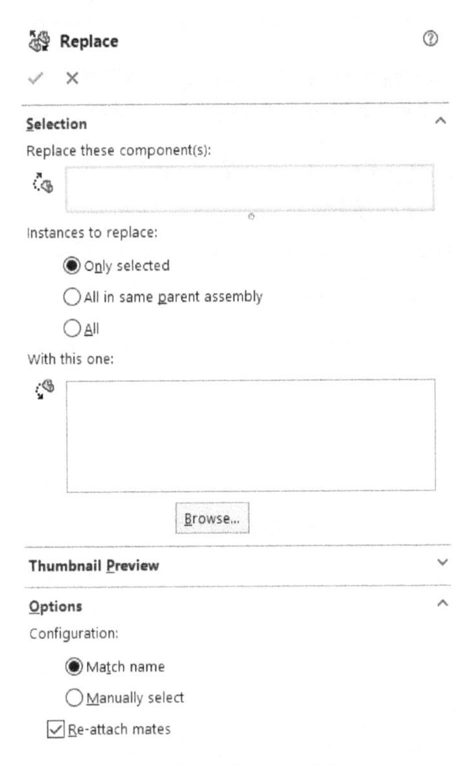

Figure-50. Replace_PropertyManager

- Select the **Re-attach mates** check box if you want to reapply the mates. If you have selected this check box then you will be asked to select references for different mates.
- After setting desired parameters, click on the **OK** button from the **PropertyManager**. The **Mated Entities PropertyManager** will be displayed; refer to Figure-51.

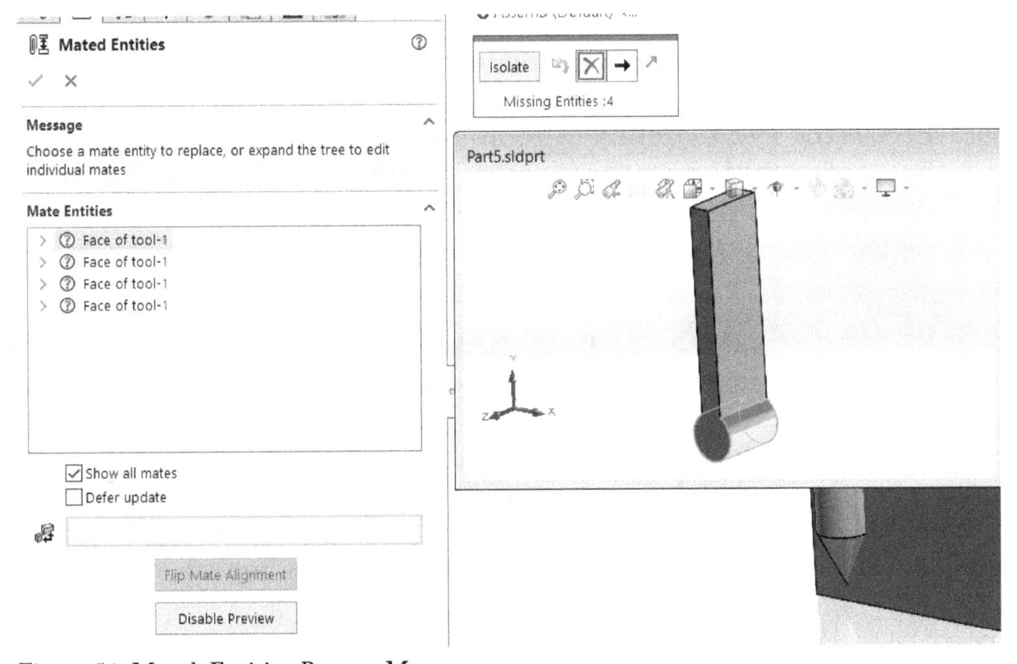

Figure-51. Mated_Entities_PropertyManager

- Select the replacement face/geometry from graphics area for reference shown in preview window. Similarly, set the other references as desired and click on the **OK** button from **PropertyManager**. The mates will be applied. You will learn more about mates later in the book.

You can use the **Find References** tool from the **File** menu in the same way.

CHECKING SUMMARY OF DOCUMENT

The **Properties** tool in **File** menu is used to check and manage summary information of file. The procedure to use this tool is given next.

- Click on the **Properties** tool from the **File** menu. The **Summary Information** dialog box will be displayed; refer to Figure-52.

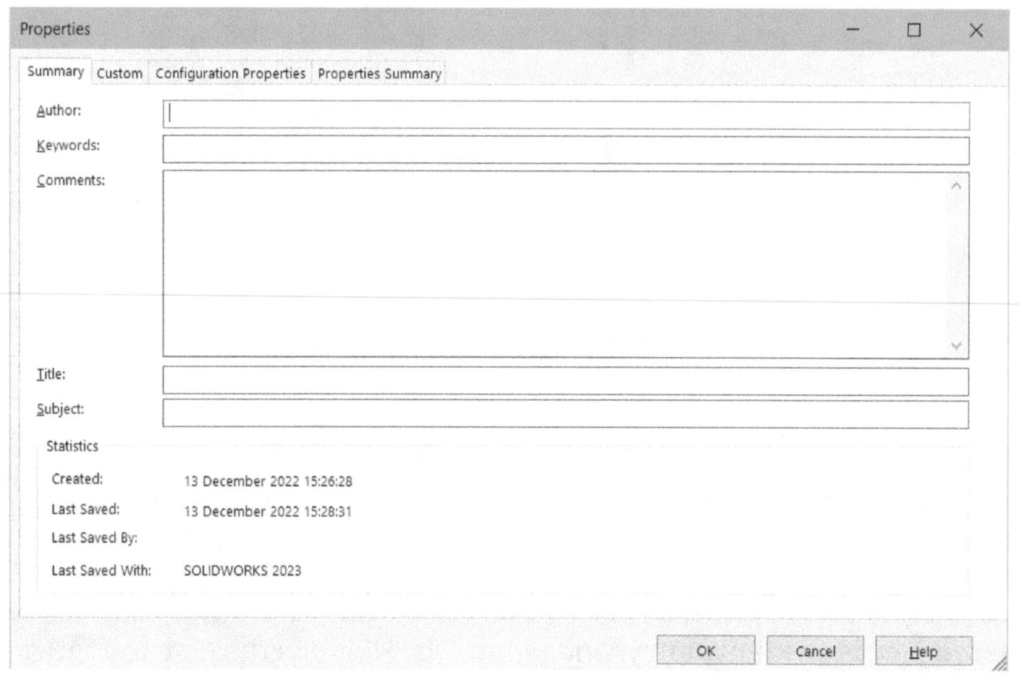

Figure-52. Summary Information dialog box

- Specify the parameters like name of design author, title of design, comments, and so on for the model.
- Click on the **Custom** tab to define custom summary data.
- Click on the **Configuration Properties** tab from the dialog box to define various bill of materials related parameters like material name, cost of material, and so on.
- Click on the **OK** button after setting desired parameters.

CUSTOMIZING MENUS

You can customize the menu by selecting the **Customize Menu** option from the **File** menu. The options to customize menu will be displayed; refer to Figure-53. Clear the check boxes for options to be removed from the menu. After setting desired options for menu, click in the empty drawing area. The menu will be updated accordingly.

To exit the software, click on the **Exit** tool from the **File** menu.

Till this point, you have learned the basic file handling operations and you have some idea about the interface of SolidWorks. Now, you will learn about some basic settings that are required for easy working with SolidWorks.

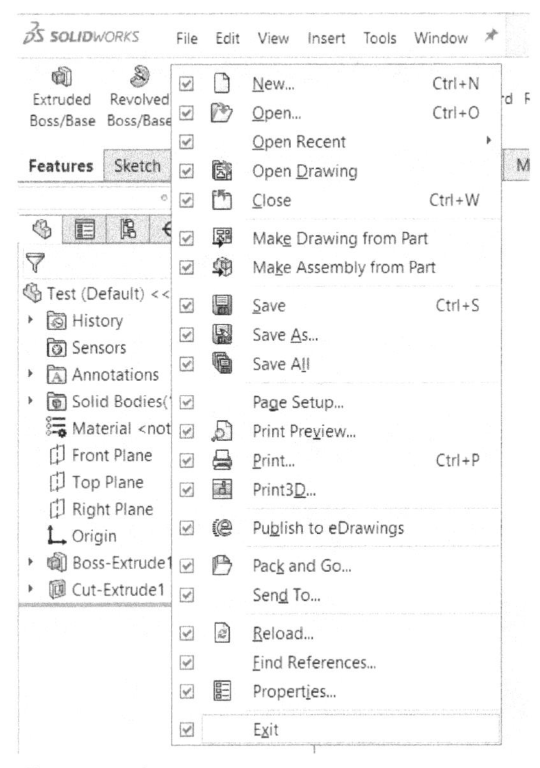

Figure-53. Customizing menu

BASIC SETTINGS OF SOLIDWORKS

All the settings of SolidWorks are compiled in the **Options** dialog box. The steps to change the settings for SolidWorks are given next.

- Click on the **Options** option in the **Tools** menu or click on the **Options** button from the **Menu Bar**.
- On performing the above step, the **System Options** dialog box will be displayed as shown in Figure-54.

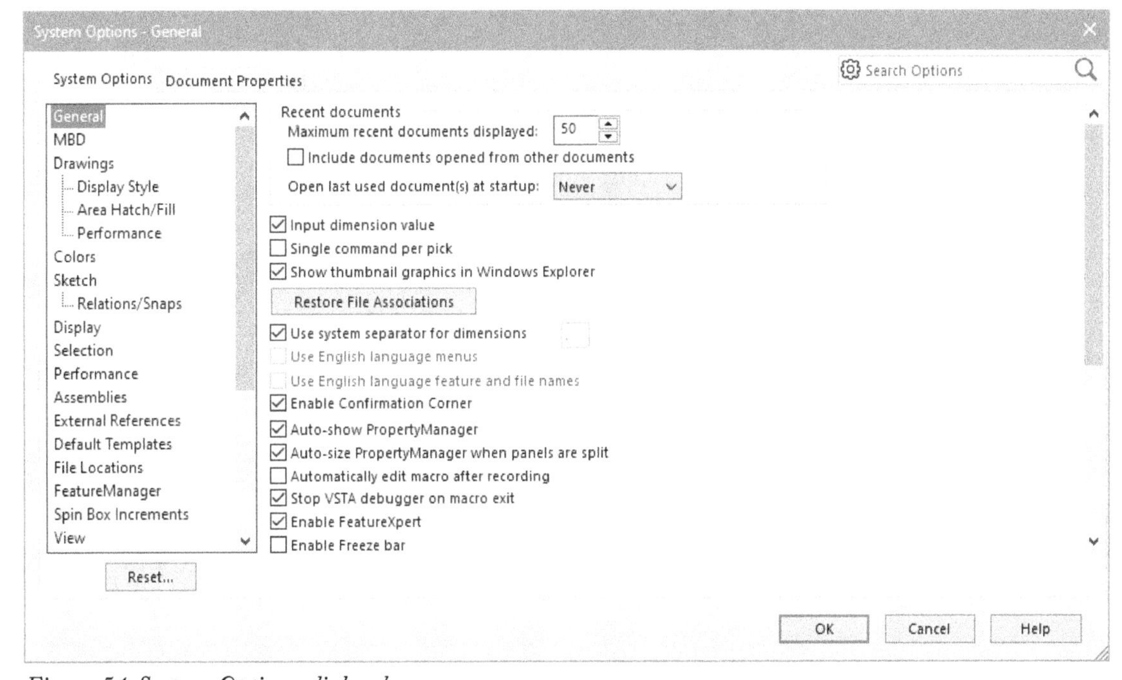

Figure-54. System_Options_dialog_box

Note that if you have a document opened then the **Document Properties** tab is also added with the **System Options** tab. To get the detail about each and every option, you need to refer to SolidWorks Help Documentation. In this section, we will discuss about some of the important options that are generally required.

- Click on the **Sketch** option in the left of the dialog box, select the **Enable on screen numeric input on entity creation** check box from the right to enter dimensions while creating the sketch. Also, select the **Create dimension only when value is entered** check box to create dimensions only when you have manually entered the dimension value.
- If you want to use only fully defined sketches for creating features in SolidWorks then select the **Use Fully defined sketches** check box. Fully defined sketches are those sketch which have all their entities dimensioned or constrained.
- Select the **Auto-rotate view normal to sketch plane on sketch creation and sketch edit** check box to automatically make the sketching plane parallel to screen.
- Click on **Relations/Snaps** in the left of the dialog box and then select the **Enable snapping** check box to enable auto snapping to the key points.
- Click on the **Document Properties** tab if you have any document opened in the viewport. Click on the **Units** option from the left. The **Options** dialog box will be displayed as shown in Figure-55.

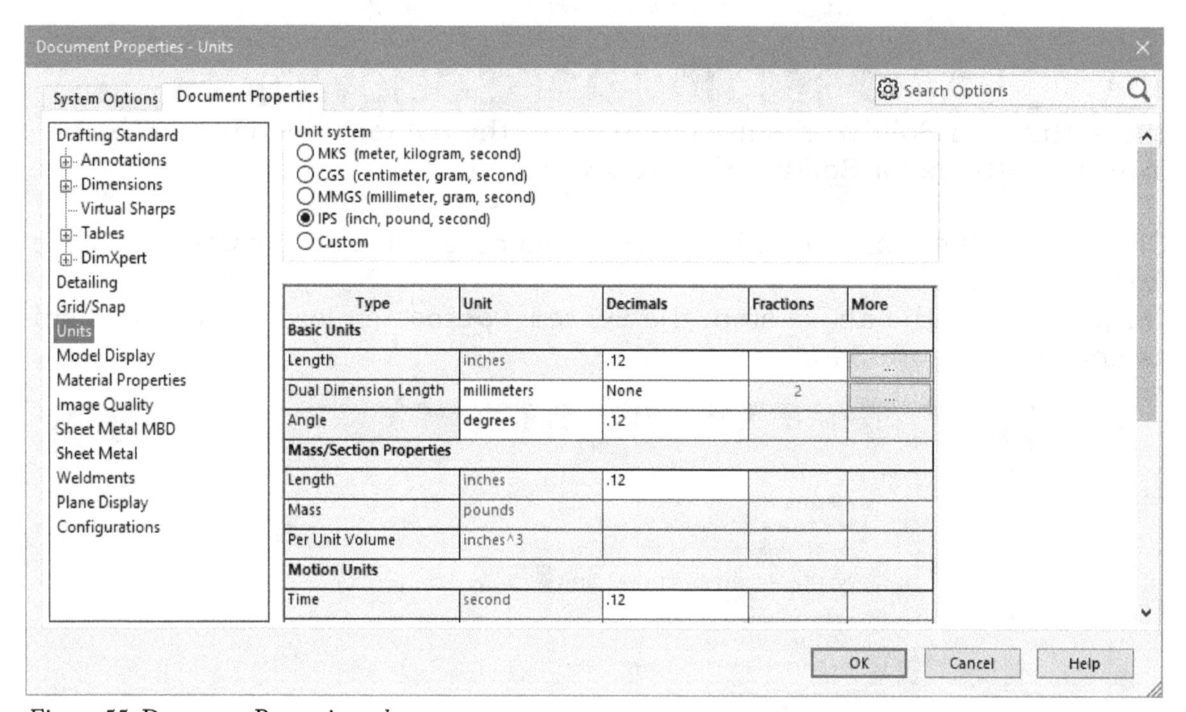

Figure-55. Document_Properties_tab

- Select desired radio button from the right to set the unit system for the current document.
- Click on the **Drafting Standard** option from the left area of the dialog box and select desired dimensioning standard from the **Overall drafting standard** drop-down; refer to Figure-56.

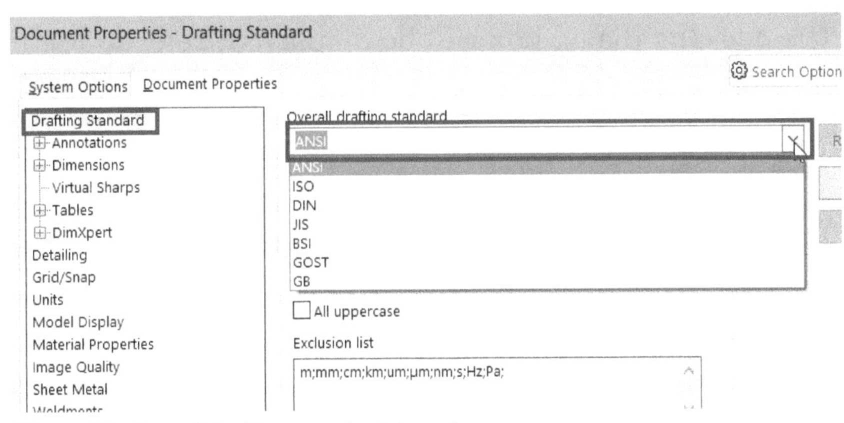

Figure-56. Overall drafting standard drop-down

- Similarly, you can set the other options in the dialog box. Note that we will be revisiting this dialog box many times in this book.
- Click on the **OK** button from the bottom of the dialog box to save the settings.

You can also change the units of document by selecting desired option from the list displayed on clicking on the **Unit system** flyout at the bottom of the viewport; refer to Figure-57.

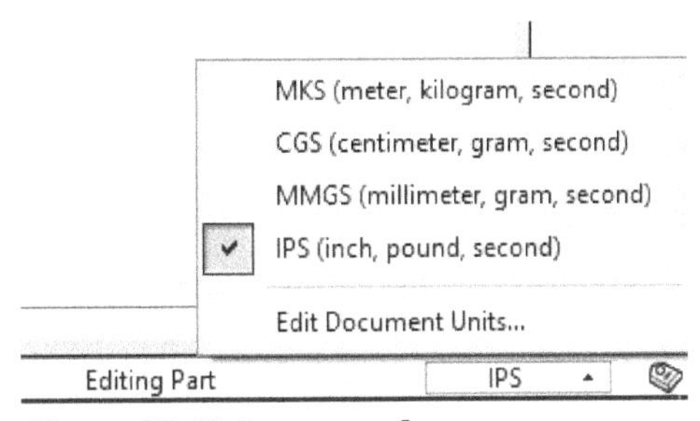

Figure-57. Unit_system_flyout

The other options of the **Options** dialog box are discussed in appendix pages.

MOUSE BUTTON FUNCTION

Rotate View (parts and assemblies only)

- To rotate the model view: Drag with the middle mouse button.
- To rotate about a vertex, edge, or face: Middle-click a vertex, edge, or face; then middle-drag the pointer.

Pan

Hold down **CTRL** key and drag with the middle mouse button. (In an active drawing, you do not need to hold down **CTRL** key.)

Zoom In/Out

Hold down **Shift** and drag the middle mouse button up/down or scroll the mouse wheel downward/upward to zoom in/out.

LOADING ADD-INS

Add-Ins are used to allow external functions in SolidWorks. Like, you can use SolidWorks Electrical, SolidWorks Simulation, SolidWorks PCB, MasterCAM, etc. The procedure to load Add-Ins is given next.

- Click on the **Add-Ins** option from the **Options** drop-down in the **Menu Bar**; refer to Figure-58. The **Add-Ins** dialog box will be displayed; refer to Figure-59.

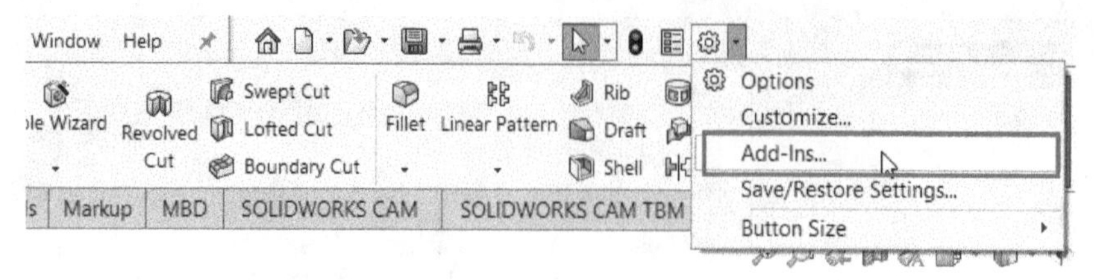

Figure-58. Add-Ins option

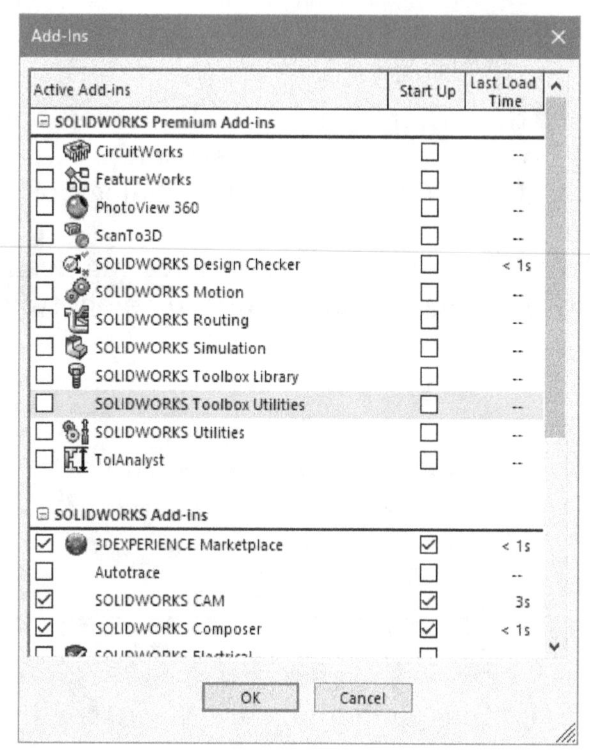

Figure-59. Add-Ins box

- Select the left side check box for Add-In that you want to be loaded now. If you want to load an Add-In at startup of SolidWorks then select the right-side check box for it.
- Click on the **OK** button from dialog box to apply the settings.

SEARCH TOOLS

The **Search box** at the top-right corner of the application window is a multipurpose tool. You can use this search box to search for help content, commands, cad models, training files, etc. The method to use **Search box** for searching commands is given next. You can apply the same method to search other things.

- Click on the **Commands** option from the **Search** cascading menu of **Help** menu; refer to Figure-60.

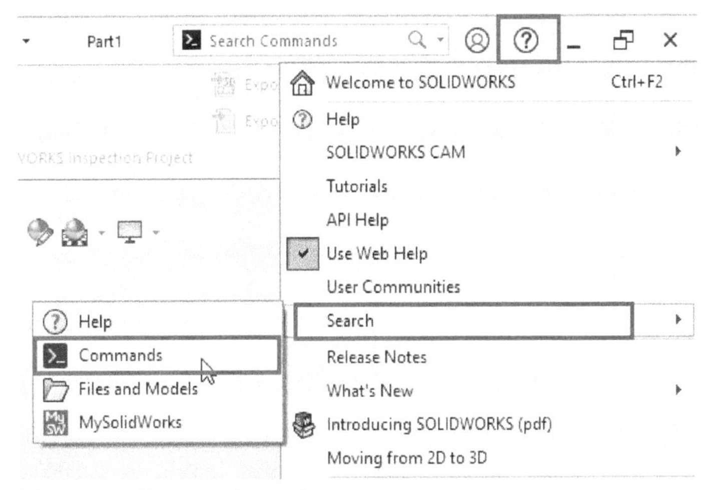

Figure-60. Commands search option

- Click in the **Search box** and type few characters of command that you are searching. A list of tools with typed characters will be displayed; refer to Figure-61.

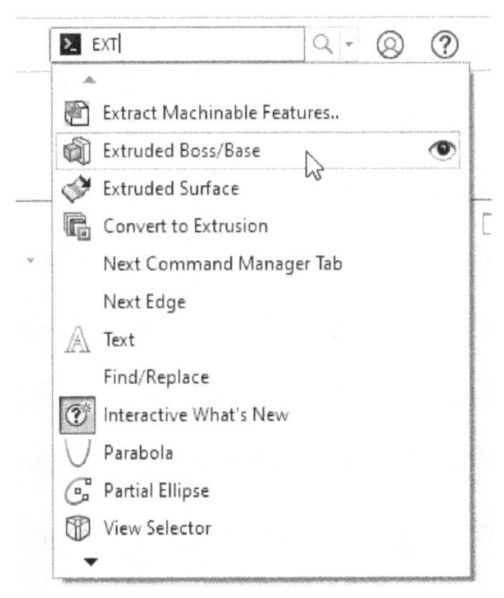

Figure-61. List of tools in search box

- Click on the eye icon displayed next to tool in the list. The cursor will move to the tool location and an arrow will be displayed pointing to the tool; refer to Figure-62. If you click on the tool in the list then it will be activated directly.

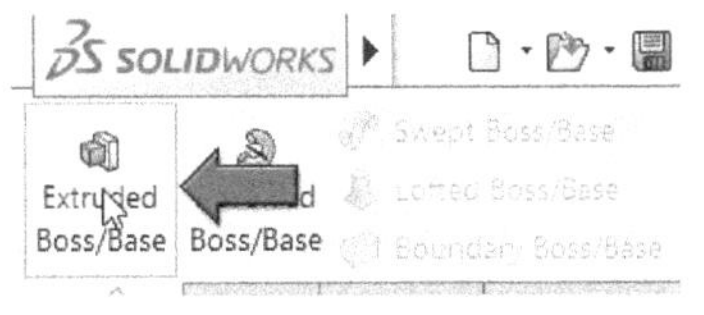

Figure-62. Location of tool displayed

Note that you can pin the menu bar to display all the menus by clicking on the **Pin** button; refer to Figure-63.

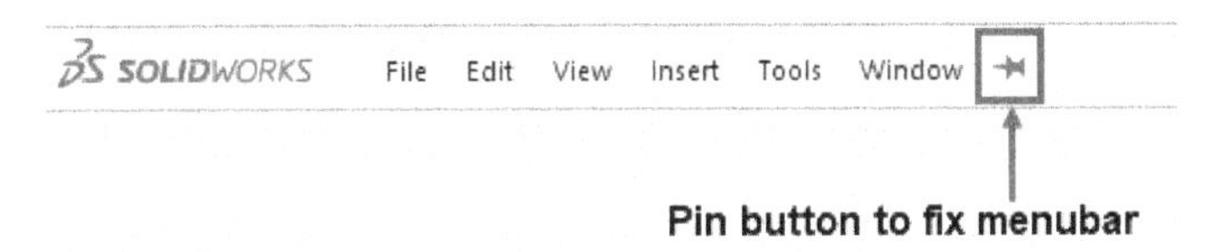

Figure-63. Pinned menubar

Login to SolidWorks

The **Login to SOLIDWORKS** button in the **Search Bar** is used to login to your SolidWorks account using the software; refer to Figure-64. Using this tool to login allows you to automatically log you into SOLIDWORKS websites, such as: MySolidWorks, SOLIDWORKS Forum, Customer Portal, and Get Support.

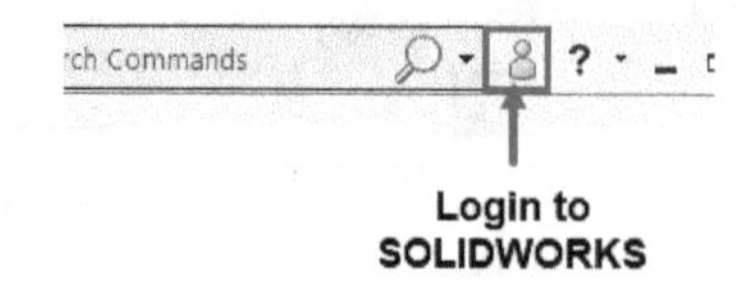

Figure-64. Login to Solidworks button

WORKFLOW IN SOLIDWORKS

The first step in SolidWorks is to create a sketch. After creating sketch of desired feature, we create solid or surface model from that sketch. After doing desired operations on the solid/surface model, we go for assembly or analyses. After, we are satisfied with the assembly/ analyses, we create the engineering drawings from the model to allow manufacturing of the model into a real world object.

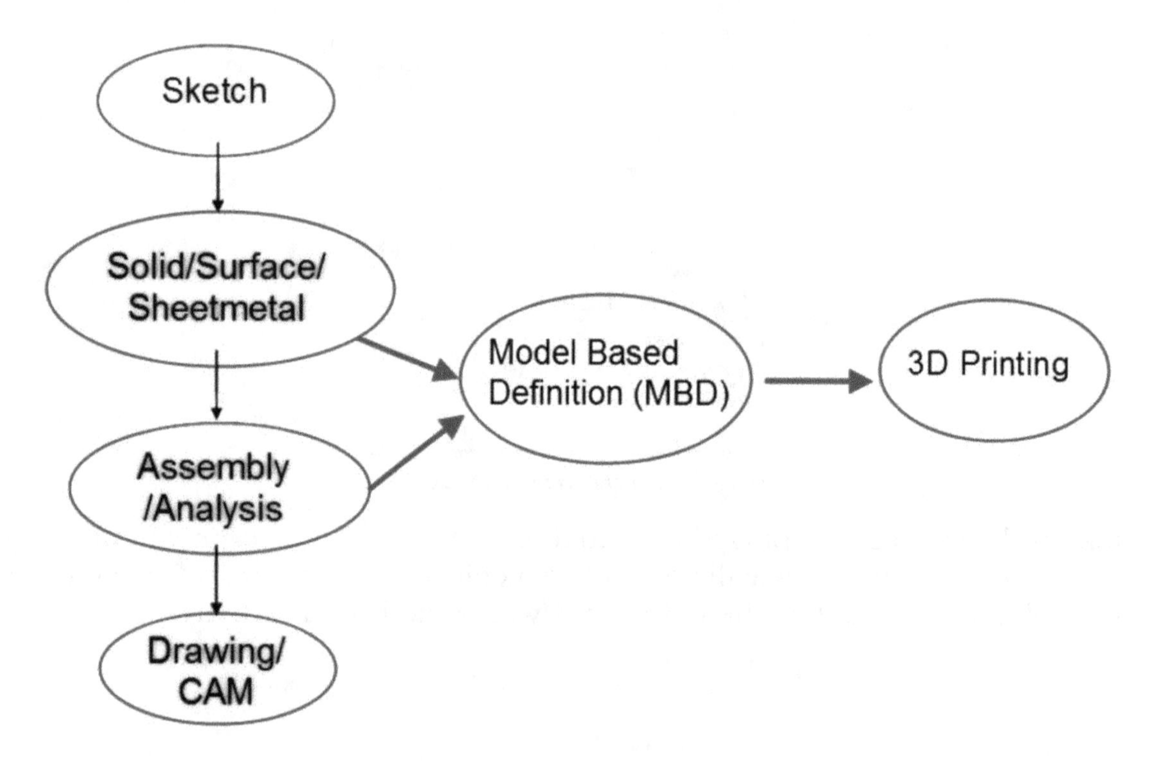

SELF-ASSESSMENT

Q1. If you have downloaded the SolidWorks Setup files from Internet then the files will be available in Downloads folder of Windows by default. (T/F)

Q2. We cannot create the desktop icon of SolidWorks if we have not opted for it while installing. (T/F)

Q3. Status Bar in SolidWorks window also displays the tips related to current tool. (T/F)

Q4. Which of the following operation results in display of **New SolidWorks Document** dialog box?
(a) Press CTRL+N
(b) Click on **New** button from Menu Bar
(c) Click on **New Document** link button from Task Pane.
(d) All of the above

Q5. Which of the following is not an option in the **New SolidWorks Document** dialog box?
(a) Part
(b) Assembly
(c) Drawing
(d) Sketch

Q6. The .... **CommandManager** provides all modeling tools that are used for feature-based solid modeling.

Q7. The .... **CommandManager** is used to add dimensions and tolerances to the features of a part.

Q8. The .... **CommandManager** is used to insert a component and apply various types of mates to the assembly.

Q9. Write down the steps to close the current document in SolidWorks.

Q10. How can we change the unit system of current SolidWorks document?

Q11. What is the purpose of Add-Ins in SolidWorks?

Q12. The .... check box forces SolidWorks to use only fully defined sketches for creating features.

Answer to Self-Assessment:
1. T, **2.** F, **3.** T, **4.** d, **5.** d, **6.** Features, **7.** MBD, **8.** Assembly, **12.** Use fully defined sketches

Chapter 2

Sketching Basic to Advanced

Topics Covered

The major topics covered in this chapter are:

- ***Basics for Sketching***
- ***Entity Creation tools***
- ***Entity Editing tools***
- ***Dimensioning and Constraining***
- ***3D Sketching***
- ***Printing and exporting sketch***

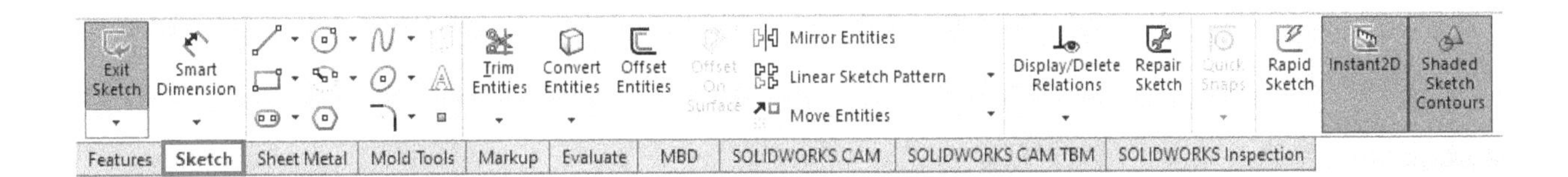

BASICS FOR SKETCHING

In Engineering, sketching do not mean sketches of birds or animals. It means sketches that are based on real dimensions of real-world objects. In this chapter, we will work with geometric entities like; line, circle, arc, ellipse, and so on which you have used in your school and engineering classes. But this time, we will be using the software in place of traditional geometric tools. Note that sketch is the base for most of 3D Models so you should be proficient in sketching.

To start with sketching, we must have a good understanding of planes in SolidWorks. Figure-1 shows the names of planes and their respective faces in Isometric View.

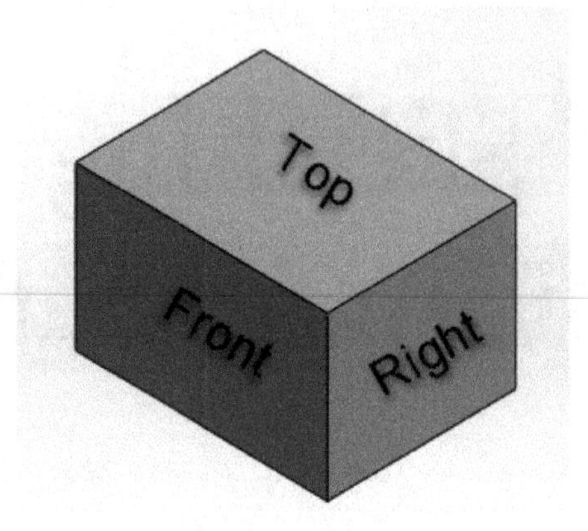

Figure-1. Planes

In SolidWorks, the planes are displayed in the same orientation as shown in the above figure. To check the planes of SolidWorks, click on the planes (Front Plane, Top Plane, Right Plane) in the **FeatureManager Design Tree**; refer to Figure-2.

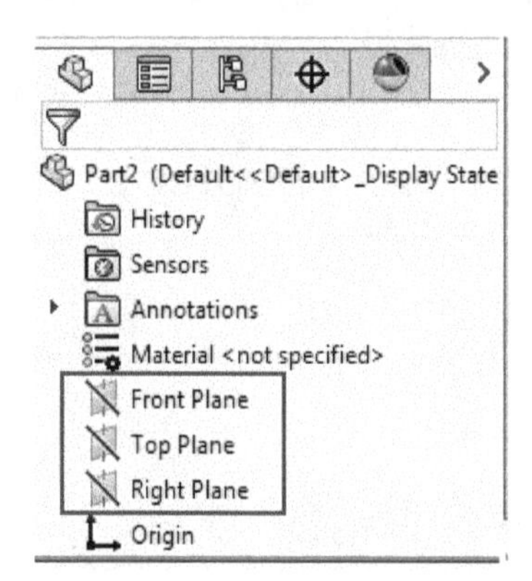

Figure-2. Planes_in_PropertyManager

- To show these planes, select them one by one while holding the **CTRL** key and right-click. A shortcut menu will be displayed; refer to Figure-3.

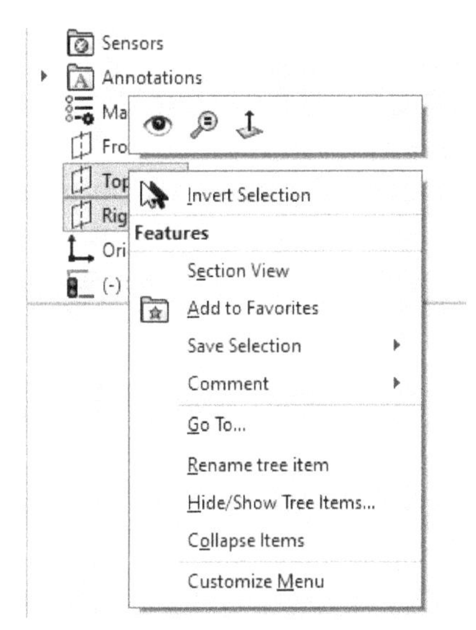

Figure-3. Shortcut menu on right clicking on planes

- Select the **Show/Hide** button from the shortcut menu; the planes will be displayed. To hide planes, click on the **Show/Hide** button again.

SKETCHING PLANE

In a CAD software, everything is referenced to other entity like a line created must be referenced to any other geometry so that you can clearly define the position of line in infinite space. But, what if there is no geometry in the sketch to reference from. In these cases, we have tools to create reference geometries like reference planes, axes, points, curves, etc. Out of these reference geometries, the sketching plane acts as foundation for other geometries. By default, there are three plane perpendicular to each other in SolidWorks; refer to Figure-4.

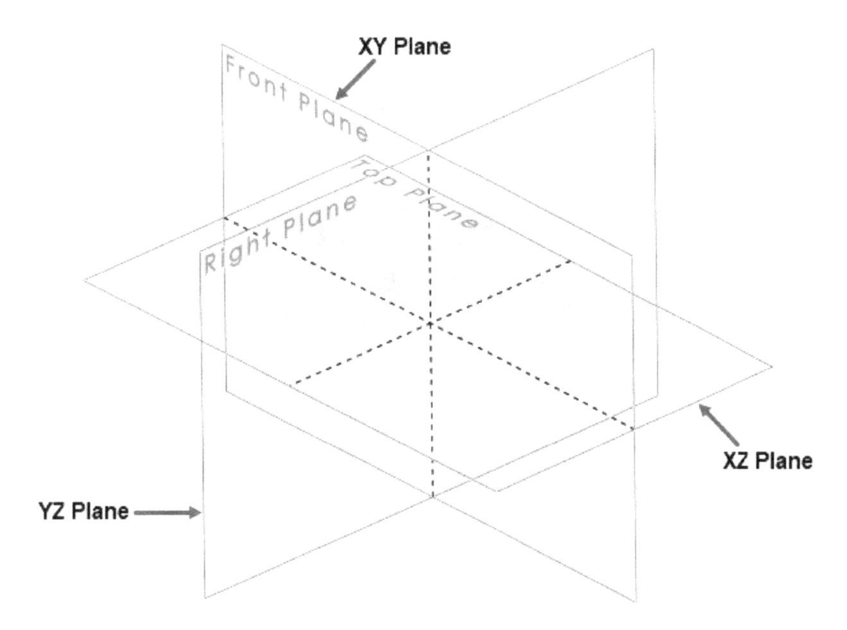

Figure-4. Sketching planes in SolidWorks

Relation between sketch, plane, and 3D model

Sketch has a direct relationship with planes and the outcome which is generally a 3D model; refer to Figure-5. In this figure, rectangle is created on the XY Plane which is also called **Front** plane. A circle is created on the YZ Plane which is also called **Right** plane. A polygon is created on the XZ plane which is also called **Top** plane. In a 3D model, the geometry seen from the Front view should be drawn on the **Front** plane. Similarly, geometry seen from the Right view should be drawn at **Right** plane and geometry seen from the Top view should be drawn at the **Top** plane.

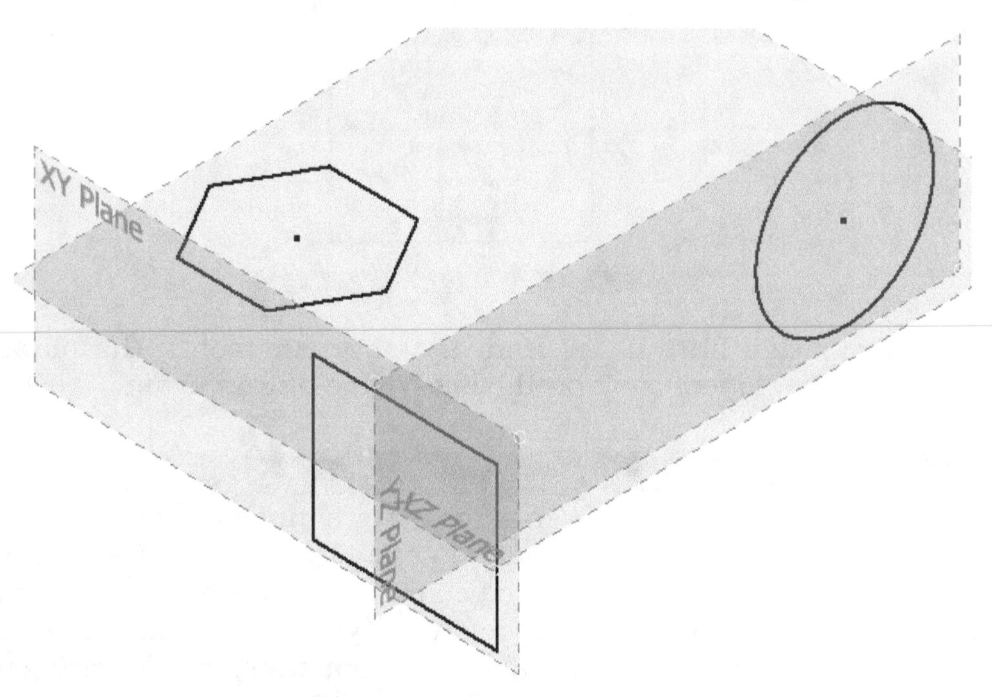

Figure-5. Sketches created on different planes

We will learn more about planes at the beginning of 3D Modeling. We will now start with Sketching tools.

Start SolidWorks, create a new Part document, and then click on the **Sketch** tab. The tools for sketching are displayed in the **Sketch CommandManager**.

STARTING SKETCH

- Click on the **Sketch** button at the left in the **Sketch CommandManager**. Three main planes are displayed with their names in the viewport; refer to Figure-6.

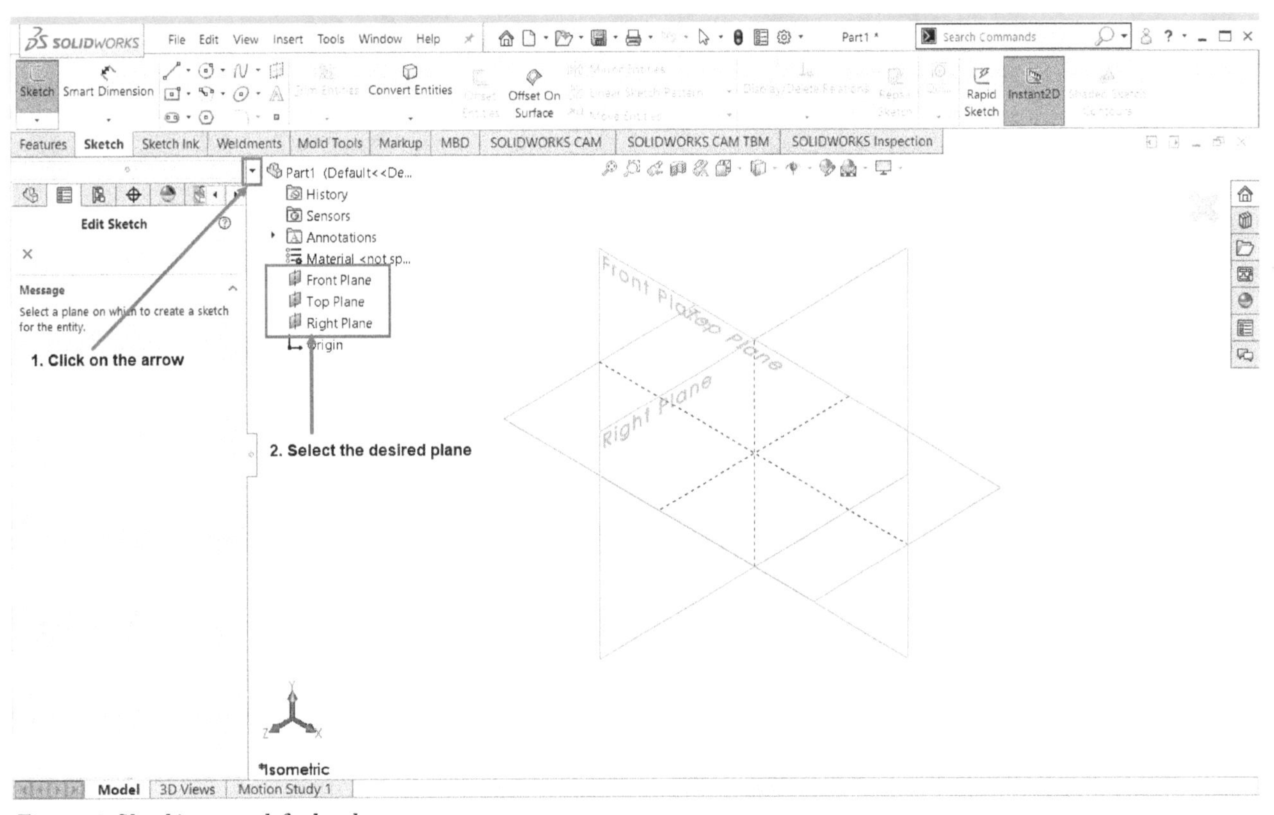

Figure-6. Sketching_on_default_planes

- Click on desired plane from the viewport or click on the arrow displayed in the top-left corner of the viewport and select the name of the plane from the list of features; refer to Figure-6.
- On clicking on a plane, the selected plane will become parallel to the screen. Now, we are ready to draw sketch on the plane. The **Sketch CommandManager** is divided into sections depending on the functions of the tools. Figure-7 shows the **Sketch CommandManager** and the division of tools.

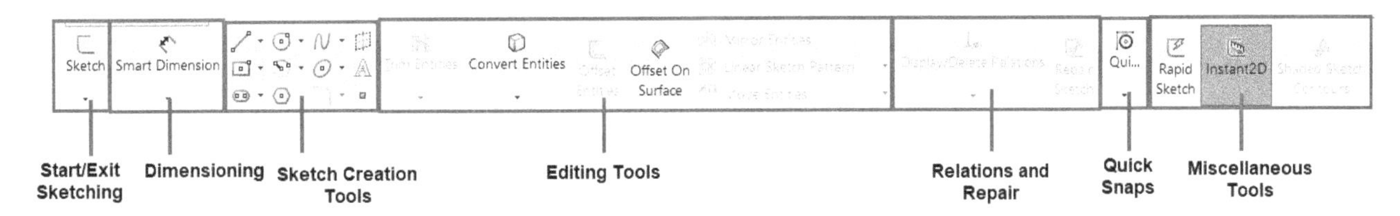

Figure-7. Sketch_CommandManager

We will start with the sketch creation tools and then one by one we will discuss other tools. But, before that it is important to understand the tools in **Heads-Up View Toolbar**.

HEADS-UP VIEW TOOLBAR

The **Heads-up View Toolbar** contains tools to change the view and orient the modeling area; refer to Figure-8.

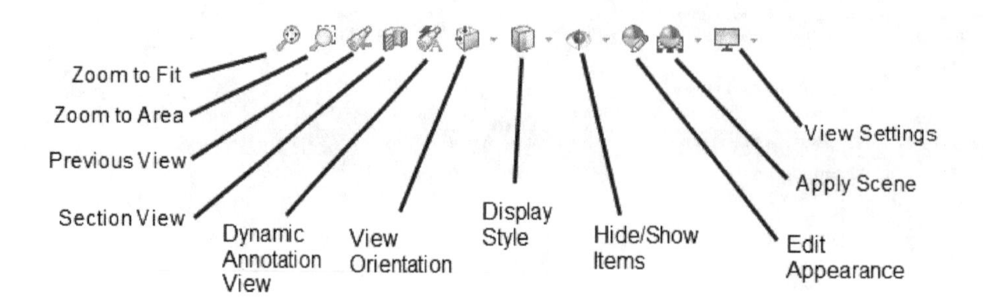

Figure-8. Heads-up_View_toolbar

The functions of the tools displayed in above figure are discussed next.

Zoom to Fit : This tool is used to display all the objects created in the viewport. To use this tool, click on the tool once. The objects will automatically fit in the current viewport.

Zoom to Area : This tool is used to display a specific area in the viewport zoomed to the full extent. To use this tool, click on it. The cursor will change to a zoom box selection cursor and you are asked to create a boundary box surrounding the entities you want to zoom in. Click to specify the starting point of the zoom box and then drag to the point till where you want to complete the zoom box. The area in the box will zoom automatically. Figure-9 shows a zoom box drawn to zoom.

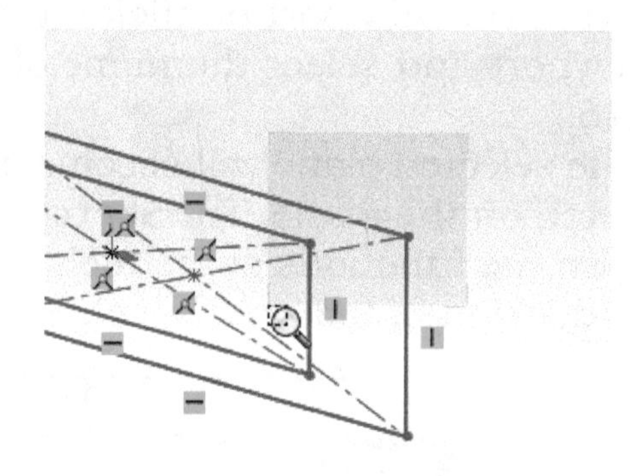

Figure-9. Zoom_box

Previous View : This tool is used to zoom to the previous level. To use this tool, click on it once. The viewport will be displayed at the previous level.

Section View : This tool is used to display section of a solid model. (Section is created when you cut a solid from a plane. It is mainly used to see the inside of the model.)

Dynamic Annotation View : This tool is used to display or hide the annotations applied to the model in Part/Assembly environment. You will learn more about this tool in Solid Modeling.

View Orientation : This tool is used to change the view orientation of the model. When you click on this button, a toolbox will be displayed as shown in Figure-10. The buttons perform the action mentioned in the figure.

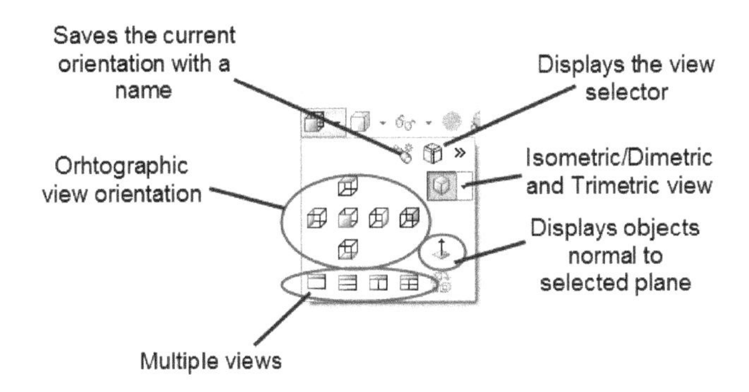

Figure-10. View_orientation_toolbox

Display Style : These tools are used to display the model in different styles, like shaded, hidden, no hidden, and so on.

Hide/Show Items : When you click on this button, a tool box will be displayed with various toggle buttons. These buttons allow to display or hide key feature like, geometric relations, center lines, annotations, grids, and so on. To enable or disable the view of a key feature, select the respective button.

Next three buttons will be discussed later in the book.

SKETCH CREATION TOOLS

The standard tools to draw sketch entities are available in this section of **Ribbon**. These tools are discussed next.

Creating Line Tools

There are three tools in the **Line** drop-down to create different type of lines; **Line**, **Centerline**, and **Midpoint Line**; refer to Figure-11.

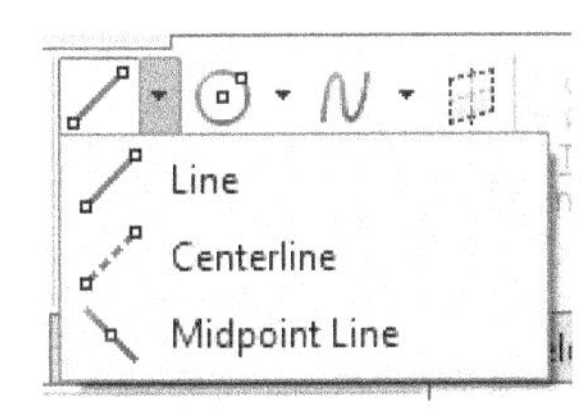

Figure-11. Line_drop-down

Line

We use this tool to create every type of lines required in creating sketch. The procedure to create line is explained in the next steps:

- Click on the **Line** tool from the drop-down. The **Insert Line PropertyManager** will be displayed; refer to Figure-12.

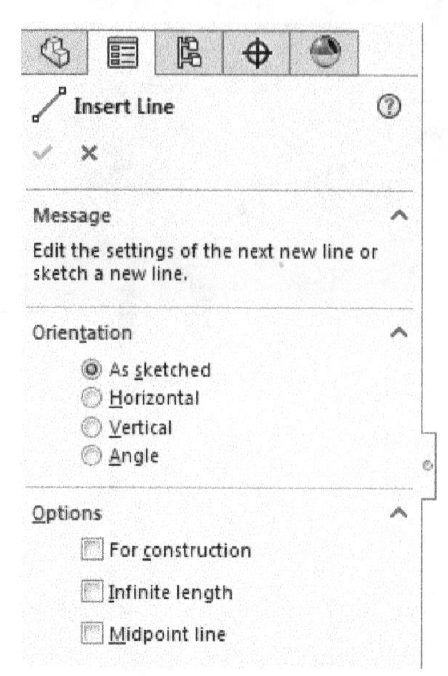

Figure-12. Insert_Line_PropertyManager

- The radio buttons in the **Orientation** rollout are used to set the orientation of the line before drawing it. There are four radio buttons available in this rollout; **As sketched**, **Horizontal**, **Vertical**, and **Angle**. The **As sketched** radio button is selected by default. So, you do not need to define the orientation of line, you can just start creating line. Click in the drawing area to create the line.
- Select the **Horizontal** radio button, if you want to create horizontal line. On selecting this radio button, the **Parameters** rollout will be displayed at the bottom of the rollout. Specify the length of the line in the edit box displayed in the **Parameters** rollout and click to specify start point of line to create the line with specified length.
- Select the **Vertical** radio button, if you want to create the vertical line. This works in the same way as the horizontal option works.
- Select the **Angle** radio button, if you want to create the lines at the specified angle. On selecting this radio button, two edit boxes will become available. Specify the length of the line and angle of the line in the respective edit boxes; refer to Figure-13.

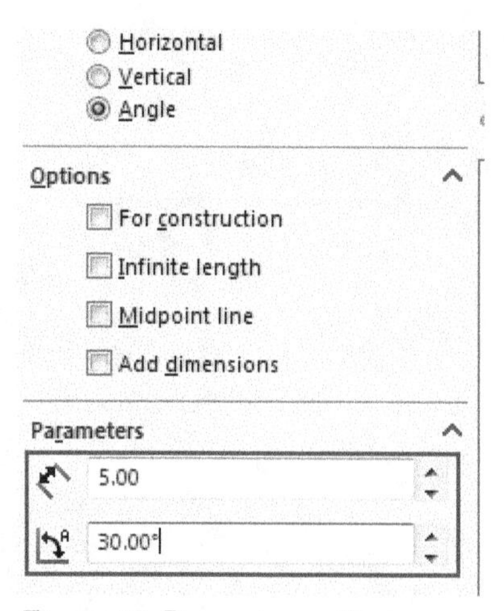

Figure-13. Parameters_rollout

- The check boxes in the **Options** rollout are used to modify the properties of line while creating it. On selecting the **For construction** check box, you will create a line in construction mode. On selecting the **Infinite length** check box, you can create a line of infinite length. The **Midpoint line** check box is used to create a line with the help of mid point and end point. The **Add dimensions** check box is used to add dimensions while creating the line. Note that the **Add dimensions** check box is displayed only when you have selected the **Enable on screen numeric input on entity creation** check box from the **Options** dialog box as discussed in previous chapter.
- Most of the time, we use the **As sketched** radio button to create lines in sketch. Select the **As sketched** radio button.
- When you move the cursor in the viewport. By default the cursor snaps to the key points like horizontally/vertically aligned to coordinate system, coincident to the coordinate system, and so on. Click to specify the start point on the screen. Figure-14 shows the creation of line.

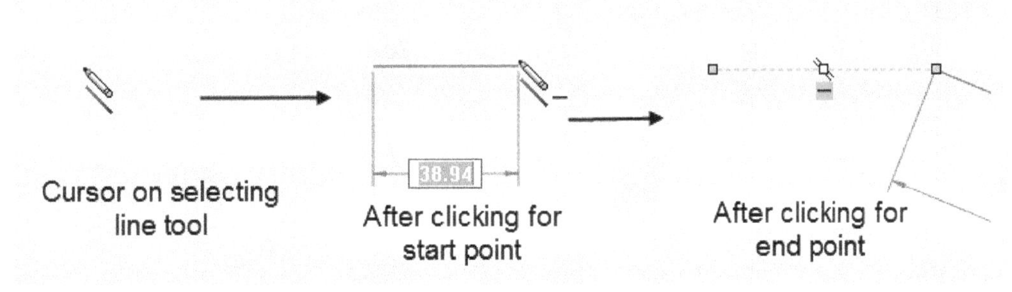

Figure-14. Line_creation

- Press **ESC** button from the keyboard to exit the tool.

Now, stop reading the book and first practice on the tool by using all the options one by one. From now onwards, you should practice on the tool in every possible manner after it has been discussed because we are not in theory business!!

Centerline

The **Centerline** tool is used to create center line in the viewport. The procedure to create a center line is the same as line creation. Click on this tool, specify the start point of the center line and then specify the end point of the center line. This centerline is used to create revolve feature using sketch which will be discussed later.

Midpoint Line

The **Midpoint Line** tool was added in SolidWorks 2015. Generally, we draw a line in SolidWorks by using the start point and end point but we can also draw a line by using the mid point and end point of the line. The method to use this tool is given next.

- Click on the **Midpoint Line** tool from the **Line** drop-down. The **Insert Line PropertyManager** will be displayed as shown in Figure-15.

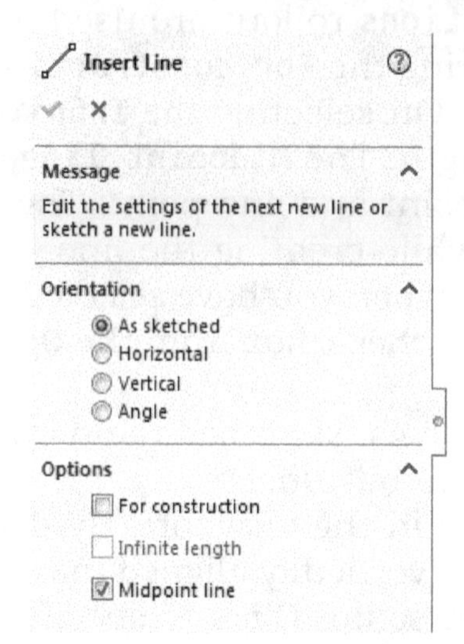

Figure-15. Insert_Line_PropertyManager

- Click to specify the mid point of the line. The end point of line will get attached to the cursor.
- Move the cursor in desired direction and click to specify the end point of line. You can also enter the value in the edit box displayed with the cursor while specifying end point; refer to Figure-16.

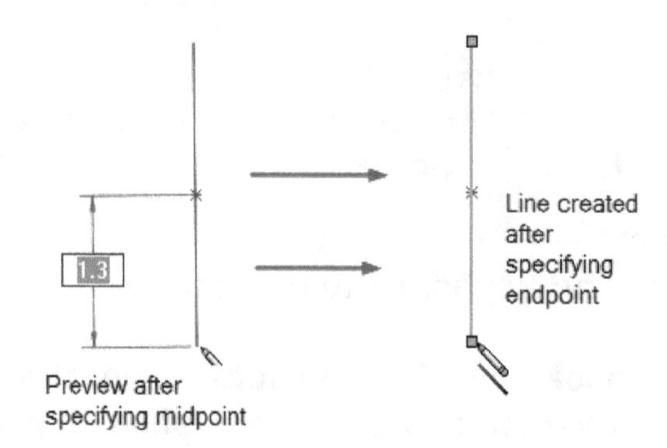

Figure-16. Midpoint_line_creation

Creating Rectangle

There are five tools in the **Rectangles** drop-down; **Corner Rectangle**, **Center Rectangle**, **3 Point Corner Rectangle**, **3 Point Center Rectangle**, and **Parallelogram**; refer to Figure-17.

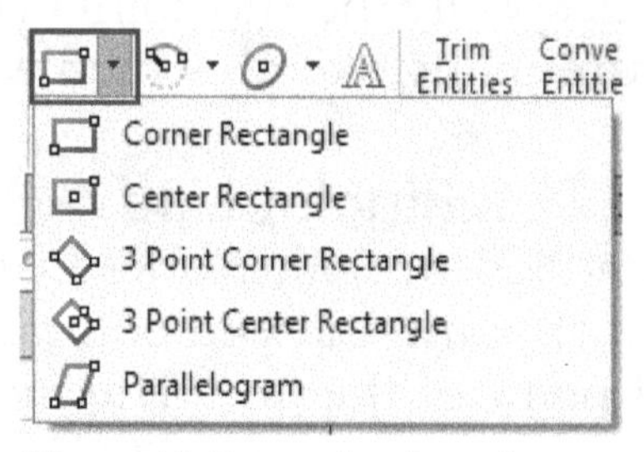

Figure-17. Rectangles_drop-down

The procedures to use these tools are discussed next.

Corner Rectangle

- Click on the **Corner Rectangle** tool from the drop-down. The **Rectangle PropertyManager** will be displayed as shown in Figure-18.

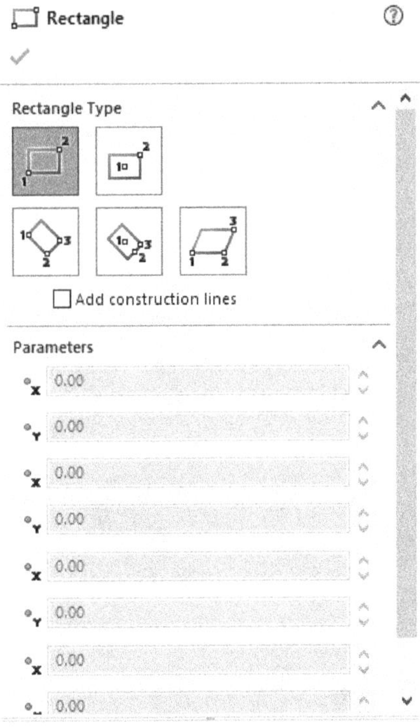

Figure-18. Rectangle PropertyManager

- Click in the viewport to specify the first point. You will be asked to specify the second point.
- Move the cursor away and click at desired location in the viewport to specify the second point. The rectangle will be created.
- You can add the construction lines for rectangle being created by selecting the **Add construction lines** check box. There are two ways by which you can add the construction lines in a rectangle; refer to Figure-19.

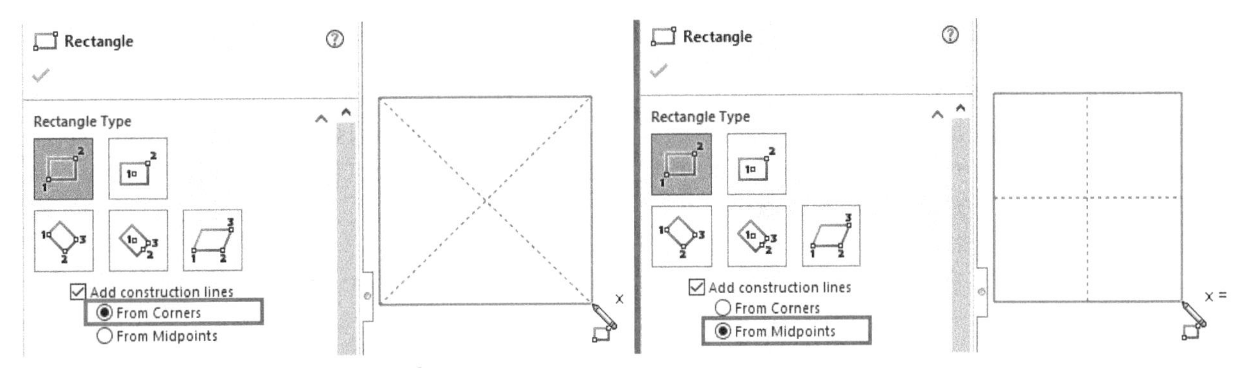

Figure-19. Rectangle with construction lines

- You can switch to the other types of rectangle by using the five buttons available in the top section of the **PropertyManager**.
- Note that SolidWorks 2017 onwards, the closed loop sketches are displayed as shaded like in Figure-20.

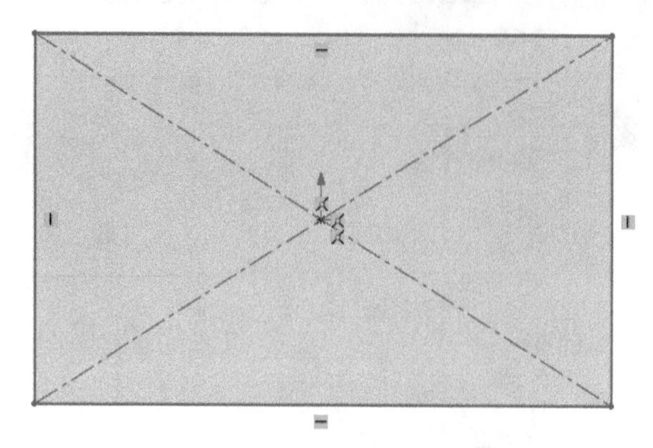

Figure-20. Closed_loop_sketch_displayed_shaded

- If you do not want the closed loops to be displayed shaded, then click on the **Shaded Sketch Contours** toggle button from the **Sketch** tab in the **Ribbon**; refer to Figure-21.

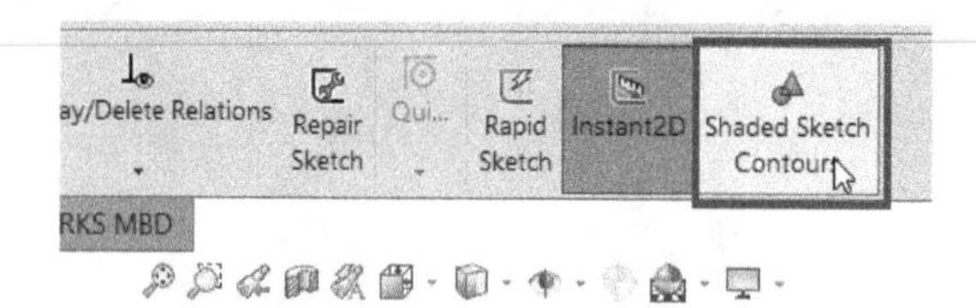

Figure-21. Shaded_Sketch_Contour_button

Center Rectangle

- Click on the **Center Rectangle** tool from the drop-down. The **Rectangle PropertyManager** will be displayed similar to the one displayed earlier. In this **PropertyManager**, the **Center Rectangle** button is selected by default.
- Click to specify the center point of the rectangle.
- Specify the corner point of the rectangle. You can specify any of the corner point by moving the cursor in desired direction.

3 Point Corner Rectangle

- Click on the **3 Point Corner Rectangle** tool from the drop-down. The **Rectangle PropertyManager** will be displayed similar to the one displayed earlier. In this **PropertyManager**, the **3 Point Corner Rectangle** button is selected by default.
- Click to specify the starting point of the rectangle.
- Click to specify the end point of the base line.
- Click to specify the end point for the height. Refer to Figure-22 for procedure.

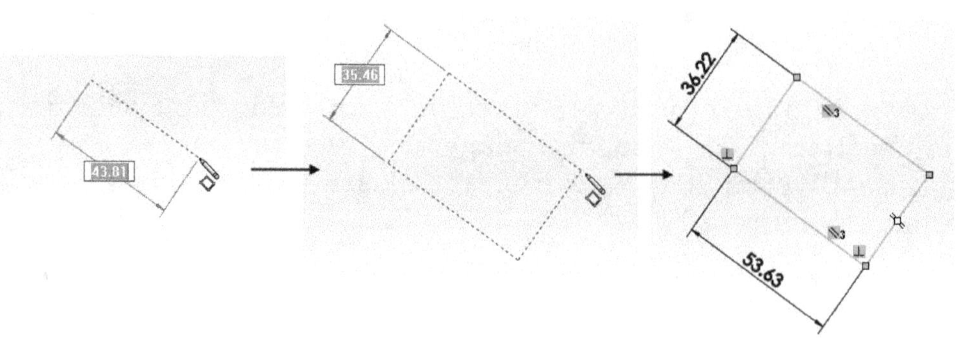

Figure-22. 3Point_Rectanlge_Creation

3 Point Center Rectangle

- Click on the **3 Point Center Rectangle** tool from the drop-down. The **Rectangle PropertyManager** will be displayed similar to the one displayed earlier. In this **PropertyManager**, the **3 Point Center Rectangle** button is selected by default.
- Click to specify the center point of the rectangle.
- Click to specify the half length of the base line
- Click to specify the half length of vertical line. Refer to Figure-23 for the procedure.

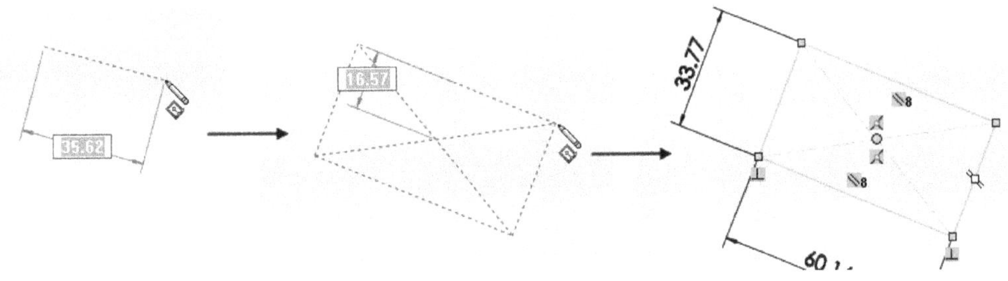

Figure-23. 3Point_Center_Rectangle_creation

Parallelogram

- Click on the **Parallelogram** tool from the drop-down. The **Rectangle PropertyManager** will be displayed similar to the one displayed earlier. In this **PropertyManager**, the **Parallelogram** button is selected by default.
- Click to specify the start point of the base line.
- Click to specify the end point of the base line.
- Click to specify the end point of the line defining height of the parallelogram. While defining this line, you can move the cursor in left/right and vertical direction.

Creating Slot

There are four tools in the **Slot** drop-down; **Straight Slot**, **Centerpoint Straight Slot**, **3 Point Arc Slot**, and **Centerpoint Arc Slot**; refer to Figure-24.

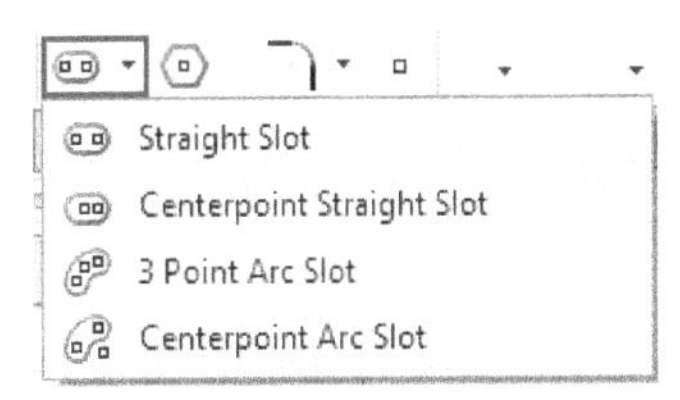

Figure-24. Slots_drop-down

The tools in this drop-down are discussed next.

Straight Slot

- Click on the **Straight Slot** tool from the drop-down. The **Slot PropertyManager** will be displayed as shown in Figure-25.

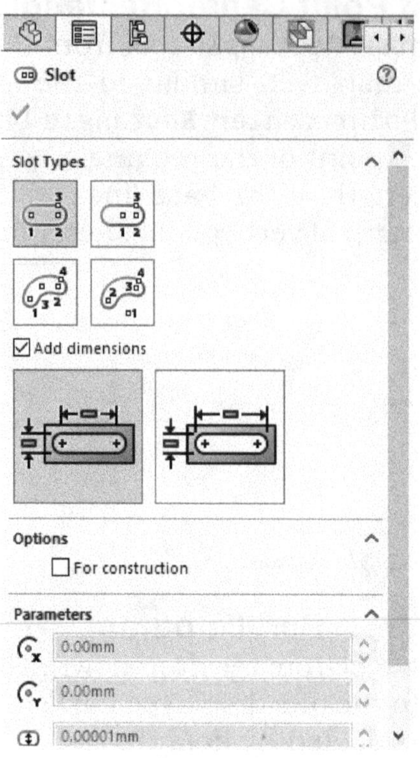

Figure-25. Slot PropertyManager

- Click on the **Center to Center** button or **Overall Length** button to set the length dimension style for slot. By default, the **Center to Center** button is selected in **PropertyManager**.
- Click in the viewport to specify the center of the first semi-circle of slot.
- Click to specify the center to center distance of the two end semi-circles.
- Click to specify the width of the slot. Figure-26 shows the procedure of creating straight slot.

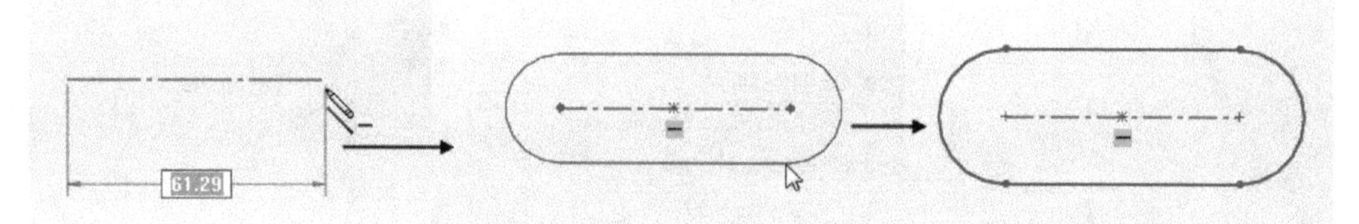

Figure-26. Straight_Slot_creation

Centerpoint Straight Slot

- Click on the **Centerpoint Straight Slot** tool from the drop-down. The **Slot PropertyManager** will be displayed as discussed earlier. In this **PropertyManager**, the **Centerpoint Straight Slot** button is selected by default in the top section.
- Click to specify the center of the slot.
- Click to specify the half length of the center to center distance of the slot.
- Click to specify the length of the slot. Figure-27 shows the procedure of creating the centerpoint straight slot.

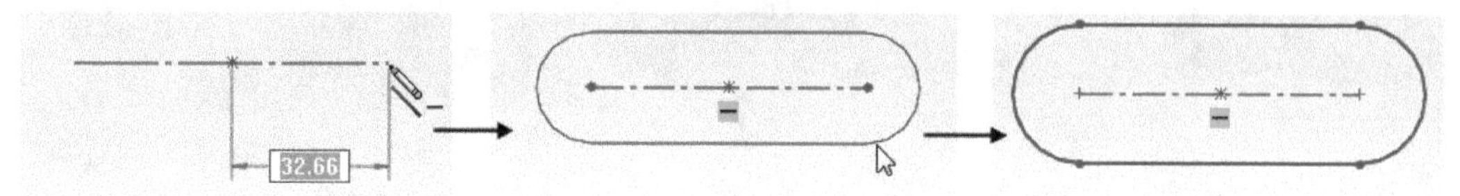

Figure-27. Centerpoint_Straight_Slot_creation

3 Point Arc Slot

- Click on the **3 Point Arc Slot** tool from the drop-down. The **Slot PropertyManager** will be displayed as earlier. In this **PropertyManager** the **3 Point Arc Slot** button is selected by default in the top section.
- Click to specify the start point of the center arc.
- Click to specify the end point of the center arc.
- Click to specify the radius of the slot.
- Click to specify the width of the slot.

Centerpoint Arc Slot

- Click on the **Centerpoint Arc Slot** tool from the drop-down. The **Slot PropertyManager** will be displayed as earlier. In this **PropertyManager** the **Centerpoint Arc Slot** button is selected by default in the top section.
- Click to specify the center of the slot center arc.
- Click to specify the start point of the center arc.
- Click to specify the end point of the slot.
- Click to specify the width of the slot.

Creating Circle

There are two tools in this drop-down; **Circle** and **Perimeter Circle**; refer to Figure-28.

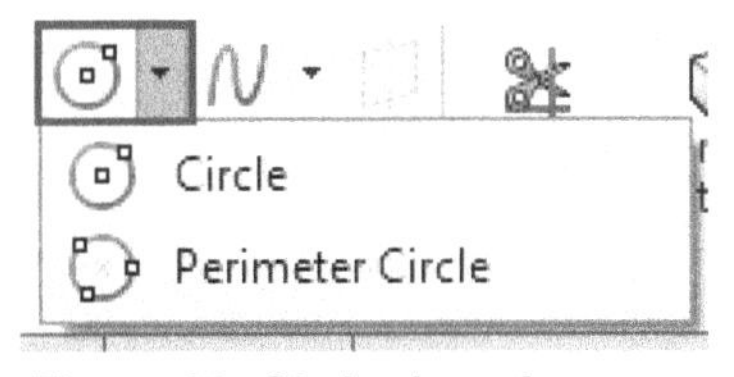

Figure-28. Circle_drop-down

The tools in this drop-down are explained next.

Circle

- Click on the **Circle** tool from the drop-down. The **Circle PropertyManager** will be displayed; refer to Figure-29.

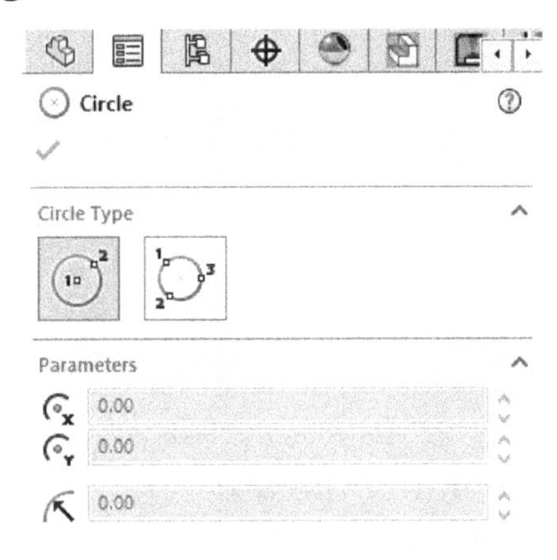

Figure-29. Circle PropertyManager

- Click to specify the center point of the circle.
- Click to specify the diameter of the circle. The circle will be created.

Perimeter Circle

- Click on the **Perimeter Circle** tool from the drop-down. The **Circle PropertyManager** will be displayed as earlier. In this **PropertyManager**, the **Perimeter Circle** button is selected by default.
- Click one by one at three locations to specify three perimeter points through which the circle should pass.
- Click on the **OK** button from the **PropertyManager** to create circle.

Creating Arc

There are three tools in this drop-down; **Centerpoint Arc**, **Tangent Arc**, and **3 Point Arc**; refer to Figure-30.

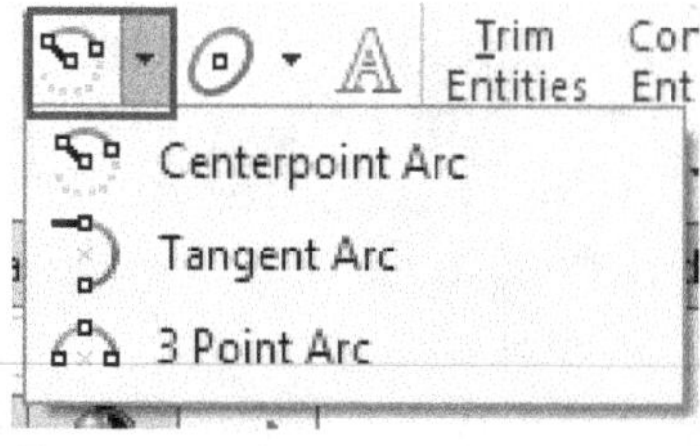

Figure-30. Arc_drop-down

The tools in this drop-down are explained next.

Centerpoint Arc

- Click on the **Centerpoint Arc** tool from the drop-down. The **Arc PropertyManager** will be displayed as shown in Figure-31.

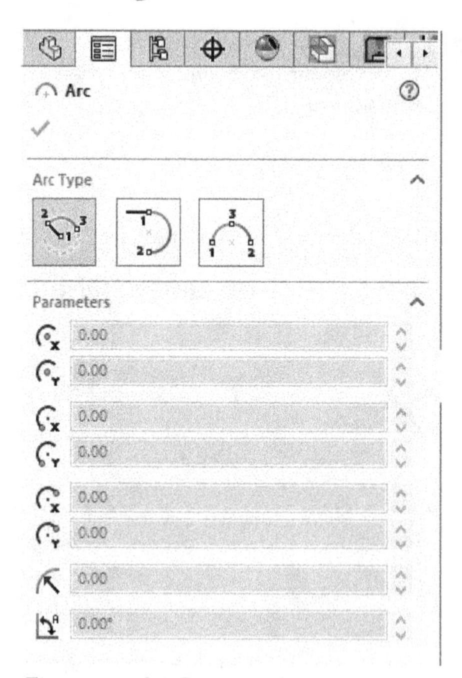

Figure-31. Arc_PropertyManager

- Click to specify the center point of the arc.
- Click to specify the start point of the arc.
- Click to specify the end point of the arc. Refer to Figure-32 for the creation of the arc.

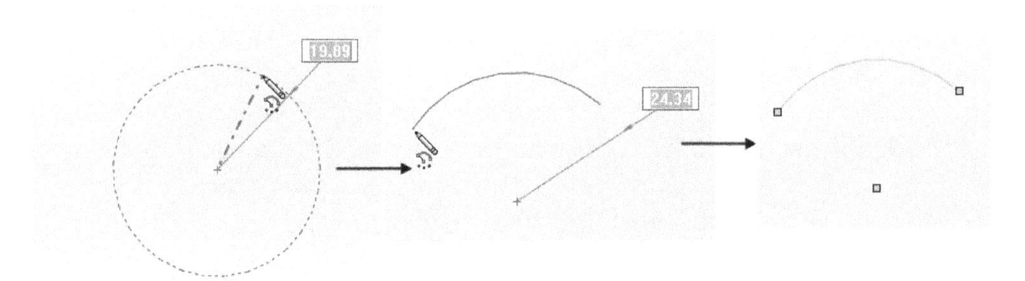

Figure-32. Arc_creation

Tangent Arc

- Click on the **Tangent Arc** tool from the drop-down. The **Arc PropertyManager** will be displayed as earlier. In this **PropertyManager**, the **Tangent Arc** button is selected by default.
- Select end point of the entity to which the arc should be tangent.
- Click to specify the end point of the arc. Note that the tangent constraint is automatically applied at the arc.

3 Point Arc

- Click on the **3 Point Arc** tool from the drop-down. The **Arc PropertyManager** will be displayed as earlier. In this **PropertyManager**, the **3 Point Arc** button is selected by default.
- Click to specify the start point of the arc.
- Click to specify the end point of the arc.
- Click to specify a point on the arc to set the radius of the arc.
- Click on the **OK** button from the **PropertyManager**.

Creating Polygon

The **Polygon** tool is used to create polygons of desired number of sides. The procedure to create polygon is explained next.

- Click on the **Polygon** tool from the **Ribbon**. The **Polygon PropertyManager** will be displayed as shown in Figure-33.

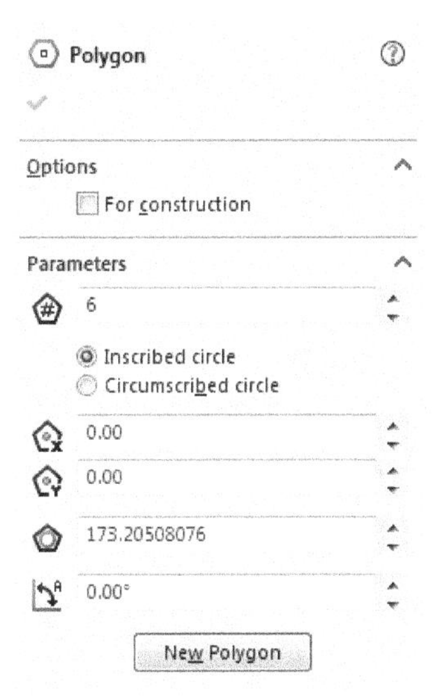

Figure-33. Polygon PropertyManager

- Specify the number of sides of the polygon by using the spinner in the **Parameters** rollout.
- Select desired radio button from the rollout. If you want to create the polygon inside the circle, then select the **Inscribed circle** radio button. If you select the **Circumscribed circle** radio button, the polygon will be drawn outside the circle. Note that polygon corner points lie on the circumscribed circle if you select the **Circumscribed circle** radio button. On the other hand, sides of polygon are tangent to the circle if you have selected the **Inscribed circle** radio button.
- Click to specify the center point of the circle.
- Click to specify the corner point of the polygon. Figure-34 shows the procedure of creating polygons.

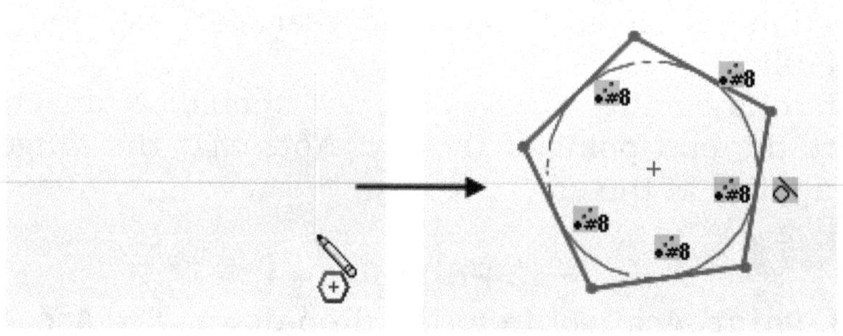

Figure-34. Polygon_creation

- Click on the **OK** button from the **PropertyManager**.

Creating Spline

The tools in **Spline** drop-down are used to create splines with different methods. Figure-35 shows the tools of this drop-down. These tools are discussed next.

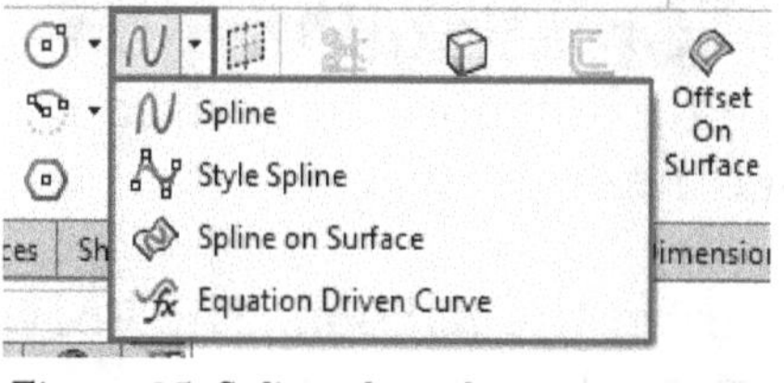

Figure-35. Spline_drop-down

Spline

- Click on the **Spline** tool from the drop-down. You are asked to specify the points through which the spline should pass.
- One by one click in the viewport to specify the points of the spline.
- To end specifying points, press the **ESC** key from the keyboard.
- Now, select the spline created to modify its parameters. The **Spline PropertyManager** will be displayed as shown in Figure-36.
- Now, you can use the options in the **PropertyManager** to modify the spline like, select the **Show curvature** check box in **Options** rollout to display curvature of spline.

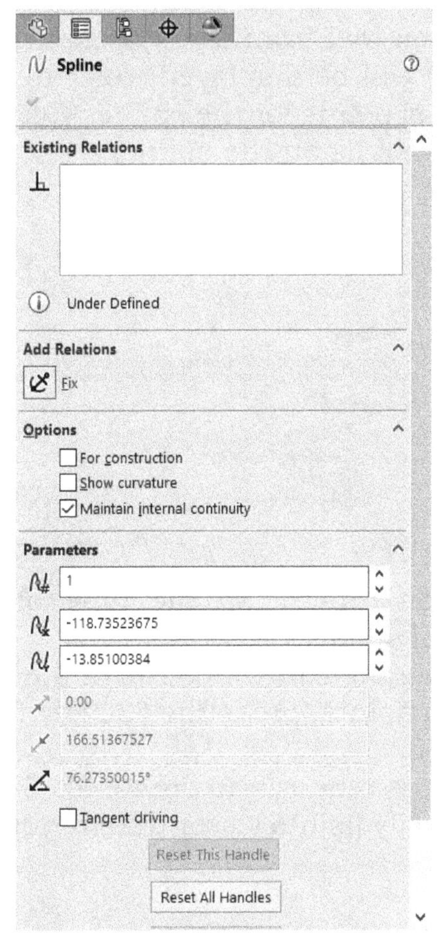

Figure-36. Spline_PropertyManager

- After performing the modifications, click on the **OK** button from the **PropertyManager**.
- If you want to add more control points on the spline, then right-click on the spline. A shortcut menu will be displayed; refer to Figure-37.

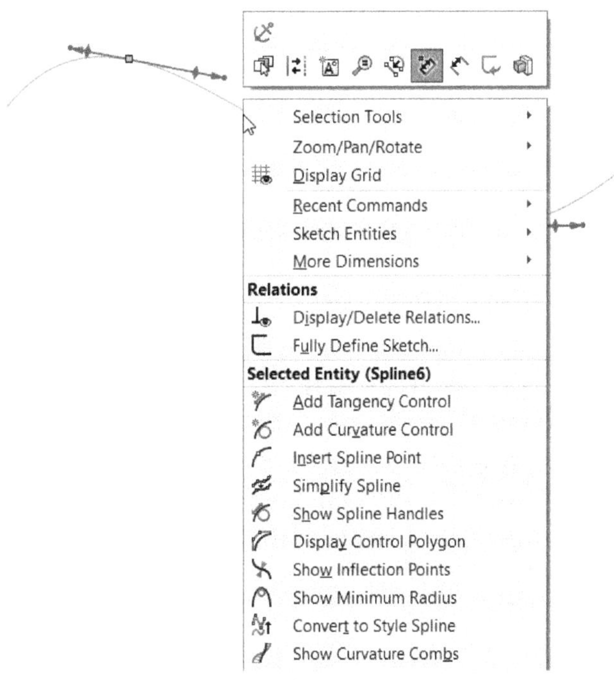

Figure-37. Shortcut_menu_for_spline

Style Spline

- Click on the **Style Spline** tool from the **Spline** drop-down. The **Insert Style Spline PropertyManager** will be displayed; refer to Figure-38. Also, you will be asked to specify the control points for the spline. (Control points control the shape of the spline but they do not lie on the spline).

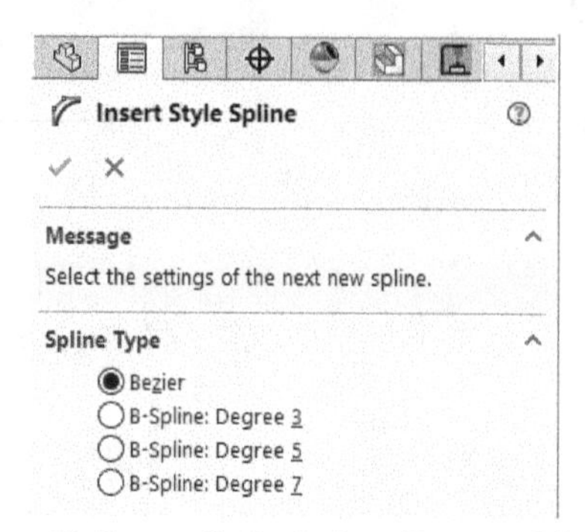

Figure-38. Insert_Style_Spline_PropertyManager

- Select desired radio button from the **Spline Type** rollout to set the spline style or say set the smoothness of spline.
- One by one click to specify the control points of the spline.
- To end specifying control points, press the **ESC** key from the keyboard.

Note that the minimum points required to create 3°, 5°, and 7° B-Splines are 4, 6, and 8 respectively. If you specify points lesser than required then you will be creating a curve rather than spline!!

Figure-39 shows the splines created.

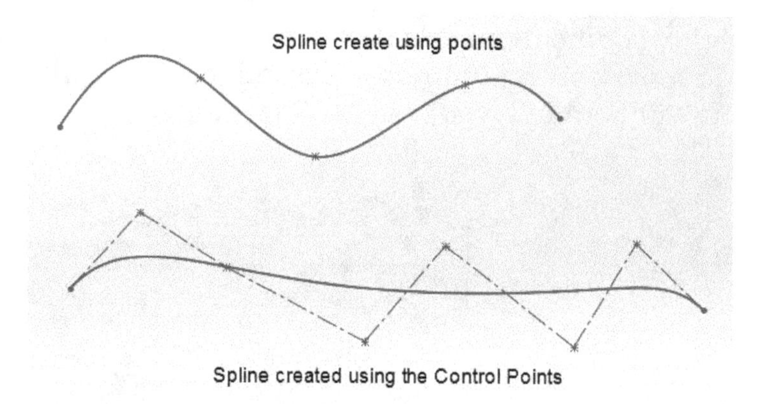

Figure-39. Spline_creation

Spline on Surface

The **Spline on Surface** tool is used to create splines on surfaces. This tool is active only after you have created a solid or surface featured in the model. After selecting this tool from **Spline** drop-down in **Sketch CommandManager** of **Ribbon**, click at desired points on a surface to create the spline. The procedure to use this tool is same as discussed for **Spline** tool; refer to Figure-40.

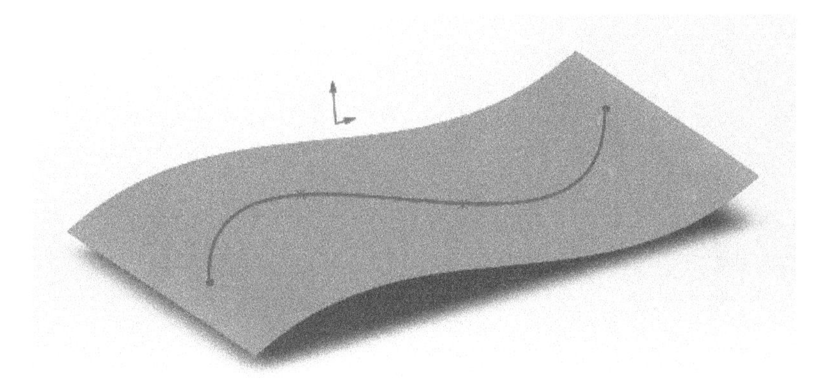

Figure-40. Spline_on_surface_created

Equation Driven Curve

The **Equation Driven Curve** tool is very helpful in creating curves using mathematical equations. Follow the procedure given next to create the curve using the equation.

- Click on the **Equation Driven Curve** tool from the **Spline** drop-down. The **Equation Driven Curve PropertyManager** will display as shown in Figure-41.

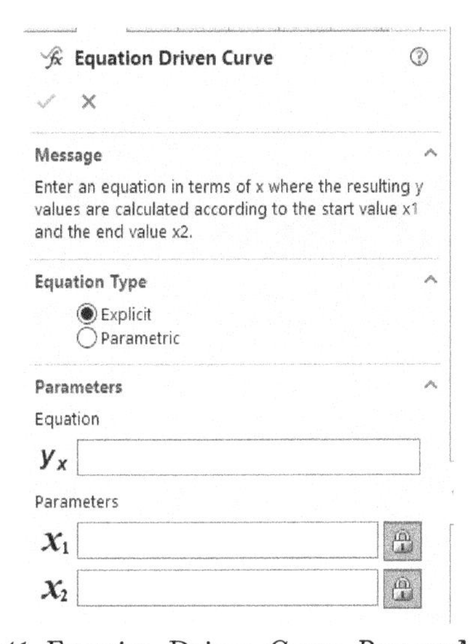

Figure-41. Equation_Driven_Curve_PropertyManager

- Click on desired radio button to create the **Explicit** or **Parametric** equation. **Explicit** means **Y** is function of **x** and **Parametric** means **X** and **Y** are functions of **t**.
- Specify the parameters for y_x. For example, in the Y_x edit box specify **$x^2/10$**.
- Click in the **x1** edit box and specify the starting value. For example, specify the value as **0**.
- Click in the **x2** edit box and specify the ending value. For example, specify the value as **20**.
- Figure-42 shows the output spline created by specifying the above parameters.
- If you have selected the **Parametric** radio button from the **PropertyManager** then specify the values of formulae for X_t and Y_t. Next, specify the limits in t_1 and t_2 edit boxes as discussed earlier.

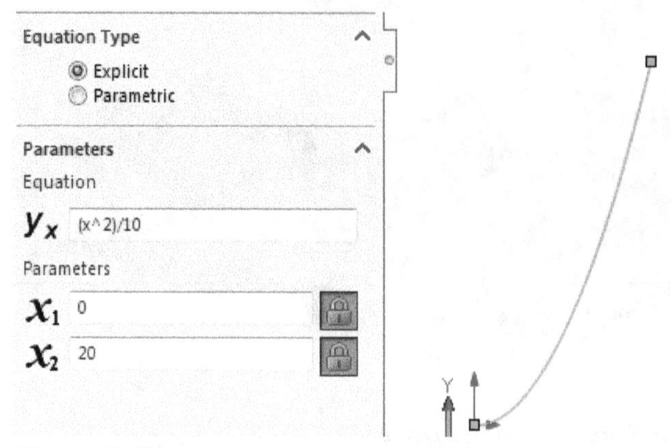

Figure-42. Equation_type

Creating Ellipse

The tools in the **Ellipse** drop-down are used to create geometric profiles like; ellipse, partial ellipse, parabola, and conic; refer to Figure-43. The tools in the **Ellipse** drop-down are discussed next.

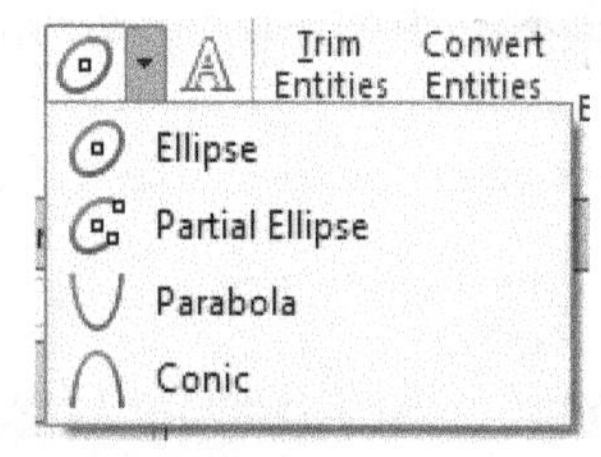

Figure-43. Ellipse_drop-down

Ellipse

The **Ellipse** tool is used to create ellipses in the sketch. The procedure to create ellipse is discussed next.

- Click on the **Ellipse** tool. You are asked to specify the center of the ellipse.
- Click to specify the center. You are asked to specify the radius along the major axis.
- Click to specify the radius or specify it by entering value in the **PropertyManager**.
- Click to specify the radius along the minor axis. Figure-44 shows the process of creating ellipse. Once you click to specify the radius of minor axis, the **Ellipse PropertyManager** will be displayed; refer to Figure-45. Note that you can also specify the values in the edit boxes in **PropertyManager**.

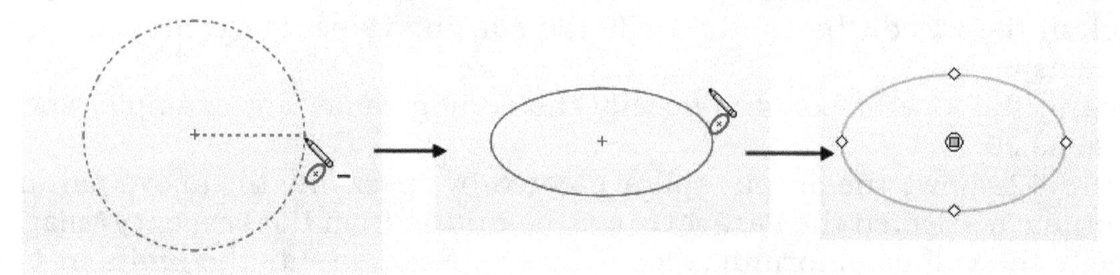

Figure-44. Ellipse_creation

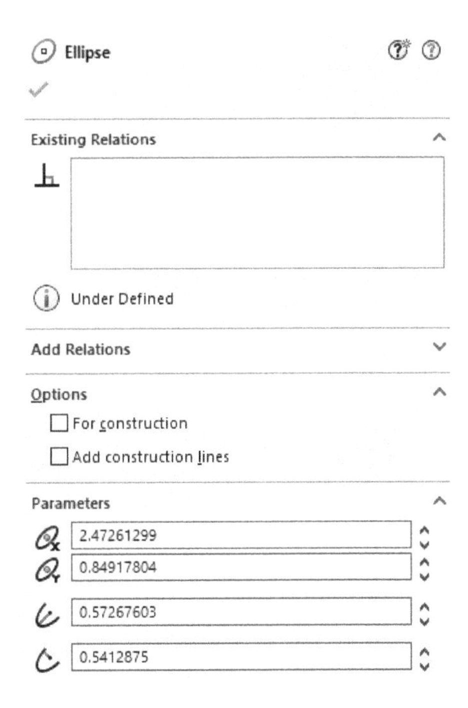

Figure-45. Ellipse_PropertyManager

Partial Ellipse

The **Partial Ellipse** tool is used to create partial ellipses in the sketch. The procedure to create partial ellipse is given next.

- Click on the **Partial Ellipse** tool from **Ellipse** drop-down. You are asked to specify the center point of the ellipse.
- Click to specify the center point. You are asked to specify the radius along major axis.
- Click to specify the radius. You are asked to specify the radius along minor axis.
- Click to specify the radius along minor axis. The point where you specify the radius of minor axis will become the starting point of the partial ellipse.
- Click to specify the end point of the partial ellipse. Figure-46 shows the process of creating partial ellipse.

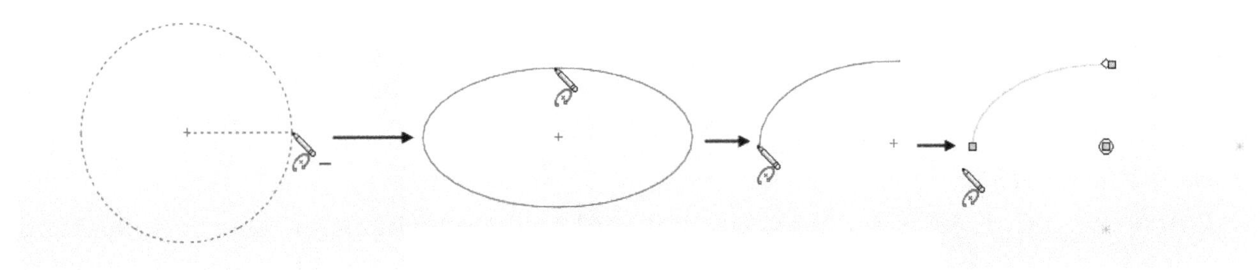

Figure-46. Partial_ellipse_creation

- After specifying the end point of the partial ellipse, the **Ellipse PropertyManager** displays as shown in Figure-47.

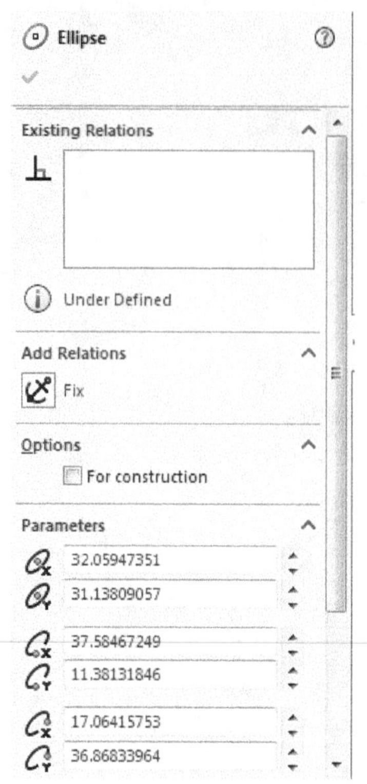

Figure-47. Partial_Ellipse_PropertyManager

Parabola

The **Parabola** tool is used to create parabola in the sketch. The procedure to create parabola is explained next.

- Click on the **Parabola** tool from the drop-down. You will be prompted to specify the focal center of the ellipse.
- Click to specify the focal center. You are asked to specify the distance of focal center from the directrix.
- Click to specify the distance. You are asked to specify the starting point of the parabola segment.
- Click on the dashed curve to specify the start point and then click to specify the end point of the parabola. Figure-48 shows the process of creating parabola.

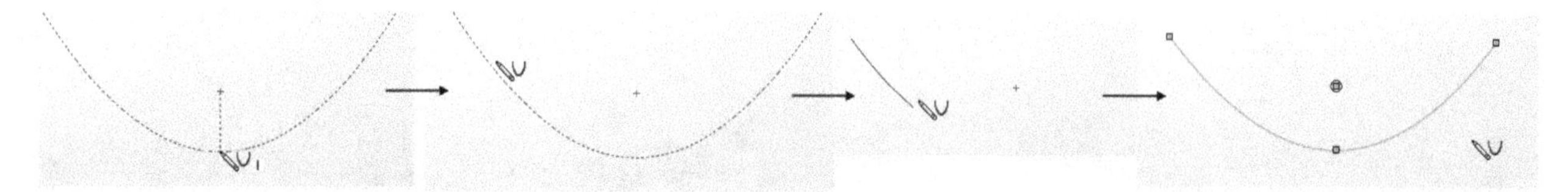

Figure-48. Parabola_creation

- Once the parabola is created, the **Parabola PropertyManager** will be displayed; refer to Figure-49. You can change the parameters as per your requirement by using the options of **PropertyManager**.

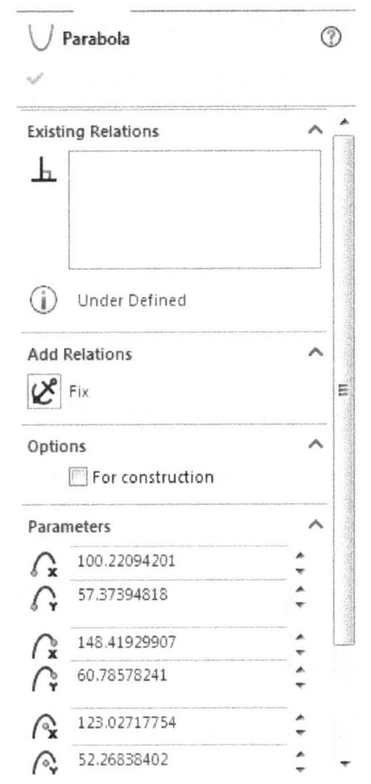

Figure-49. Parabola_PropertyManager

Conic

The **Conic** tool is used to create conic curves in the sketch. The procedure to create a conic curve is explained next.

- Click on the **Conic** tool from the drop-down. You will be prompted to specify the start point of the conic.
- Click to specify the start point of the conic curve. You will be asked to specify the end point of the conic curve.
- Click to specify the end point. You will be asked to specify the top vertex of the curve.
- Click to specify the vertex. You will be asked to specify the **Rho** value for the curve. Specify desired value of **Rho**. Figure-50 shows the process of conic curve creation. Once the curve is created, the **Conic PropertyManager** is displayed as shown in Figure-51.

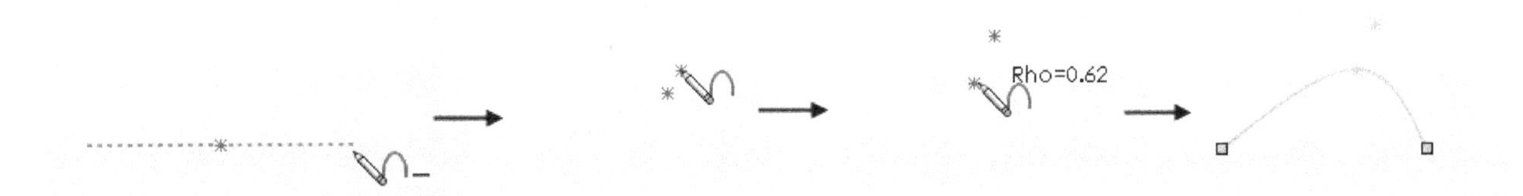

Figure-50. Conic_creation

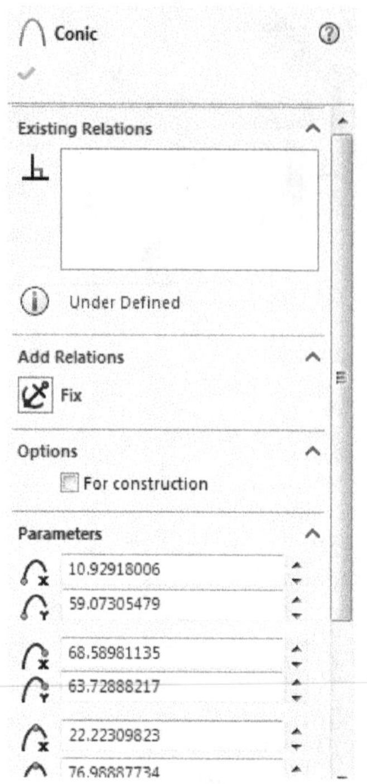

Figure-51. Conic_PropertyManager

Creating Sketch Fillet

The **Sketch Fillet** tool is used to create fillets at the corners created by intersection of two entities. Fillet is sometimes also referred to as round. Generally it is not advised to use this tool first if there are some major changes going to occur later in the model. You should create the complete sketch first and then apply fillets to corners. The procedure to create fillet is given next.

- Click on the **Sketch Fillet** tool from the **Sketch Fillet** drop-down. The **Sketch Fillet PropertyManager** will be displayed; refer to Figure-52.

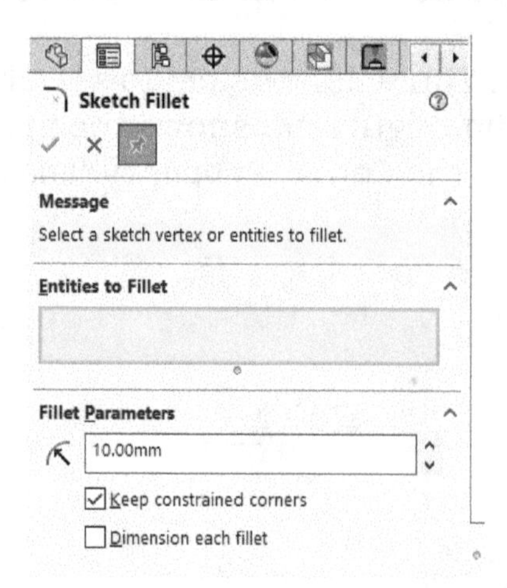

Figure-52. Sketch_Fillet_PropertyManager

- Enter the radius value in the **Fillet Radius** spinner of **Fillet Parameters** rollout. Select the first entity. You will be asked to select the second entity.
- Select the second entity. The preview of fillet is displayed.

- Click on the **OK** button from the **PropertyManager** or select the next two entities between which you want to create the fillet. Figure-53 shows the process of fillet creation.

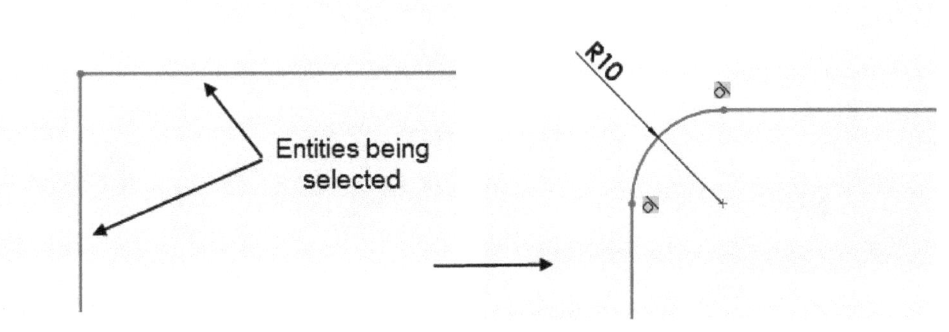

Figure-53. Fillet_Creation

Creating Sketch Chamfer

The **Sketch Chamfer** tool is used to create chamfer at the corners created by intersection of two entities. The procedure to create chamfers is explained next.

- Click on the **Sketch Chamfer** tool from the **Sketch Fillet** drop-down. The **Sketch Chamfer PropertyManager** will be displayed; refer to Figure-54.

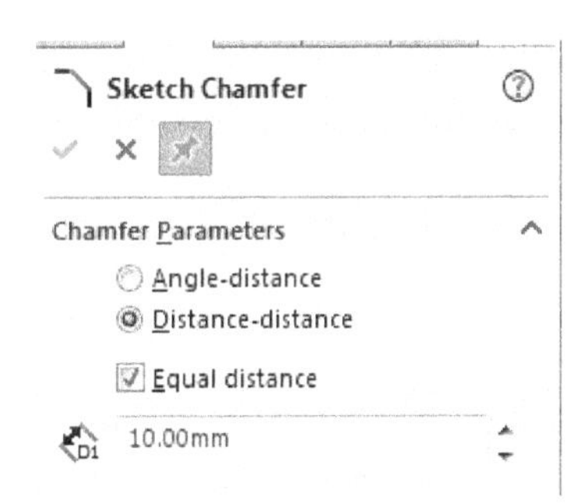

Figure-54. Sketch_Chamfer_PropertyManager

- By default, **Distance-distance** radio button is selected in the **PropertyManager**. Specify the chamfer length for first side. If the **Equal distance** check box is selected then it will be applied for the both sides. If you clear the check box then you can specify the chamfer length for both sides by using the respective edit boxes.
- You can also create chamfer by specifying angle and distance. To do so, select the **Angle-distance** radio button. Specify the parameters in the **Chamfer Parameters** rollout.
- Select the first line and then select the second line. The chamfer will be created between both the lines.

Creating Text

The **Text** tool is used to create text which can be used for embossing/engraving on a solid face. The **Text** tool is also used to provide notes and other information for the model. The procedure to create text is explained next.

- Click on the **Text** tool. The **Sketch Text PropertyManager** will be displayed as shown in Figure-55.

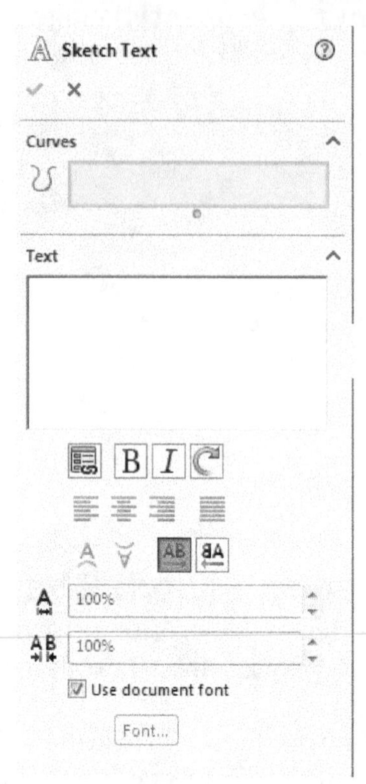

Figure-55. Sketch_Text_PropertyManager

- Click in the **Text** box and enter desired text you want to use in the sketch. The text will be created at the default coordinate system.
- Select a curve along which you want to create the text. The text will be placed along the curve. Figure-56 shows the procedure of creating text.

Figure-56. Text_along_curve

Note that if the curve length is smaller than the text specified then the text that can be spaced over the curve will only be displayed. If you want to dissolve the text for converting it into separate sketch entities then select the text, right-click on it, and select the **Dissolve Sketch Text** tool from the shortcut menu displayed; refer to Figure-57.

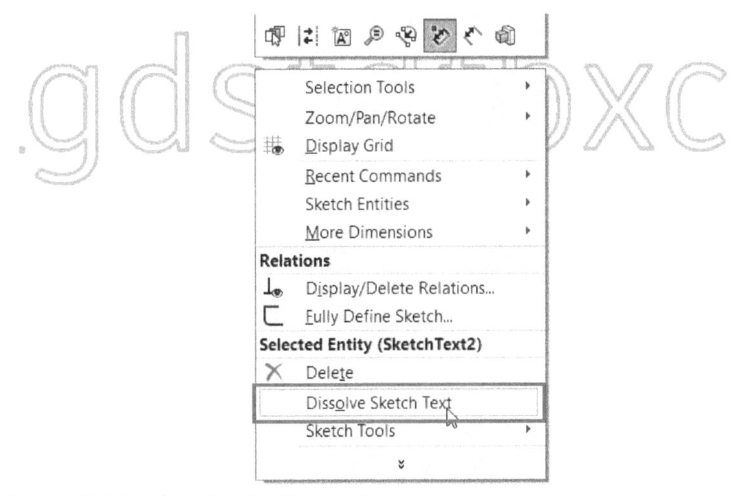

Figure-57. Dissolve_Sketch_Text_tool

Creating Point

The **Point** tool is used to create sketch point in the viewport. The point is a very important entity and finds its major usage when you start creating surfaces. The points give the flexibility to parametrically change the surface design. The procedure to create points is discussed next.

- Click on the **Point** tool from the **Ribbon**. You will be asked to click in the viewport to specify the location of the point.
- Click in the viewport or at desired location on an entity in viewport. The **Point PropertyManager** will be displayed as shown in Figure-58.

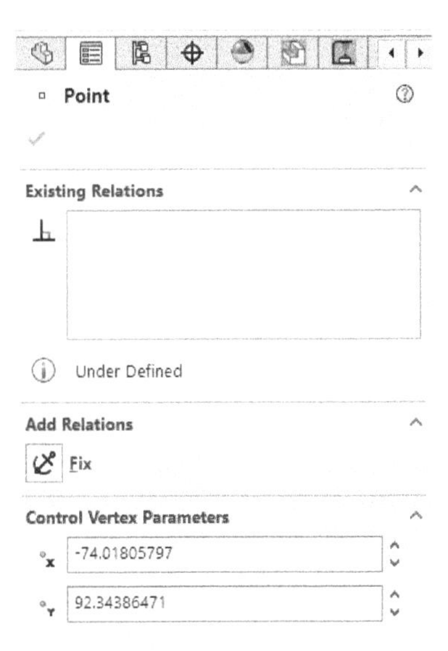

Figure-58. Point_PropertyManager

- Click in the spinners of the **PropertyManager** to specify the parameters of the points.

SKETCH EDITING TOOLS

The standard tools to edit sketch entities are categorized in this section. The tools in this section are discussed next.

Trimming Entities

The **Trim Entities** tool is available in the **Trim Entities** drop-down of the **Ribbon**. This tool is used to remove unwanted part of a sketch entity. While removing the segments, the tool considers intersection point as the reference for trimming. The procedure to use this tool is explained next.

- Click on the **Trim Entities** tool. The **Trim Entities PropertyManager** will display as shown in Figure-59.

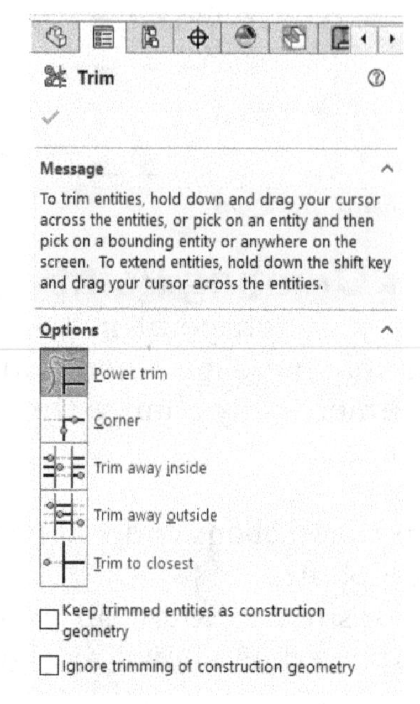

Figure-59. Trim_PropertyManager

By default, the **Power trim** button is selected. This button is having the functionality of all the other buttons displayed in the **PropertyManager**. So, we will be explaining the procedure of using this button first.

Power Trim

- Make sure the **Power Trim** button is selected in the **PropertyManager**.
- Click on the entity you want to trim. Note that the selected side of the entity will be removed.
- Click on the entity that you want to use as reference for trimming.
- The entity will be trimmed. Figure-60 shows the procedure of trimming.

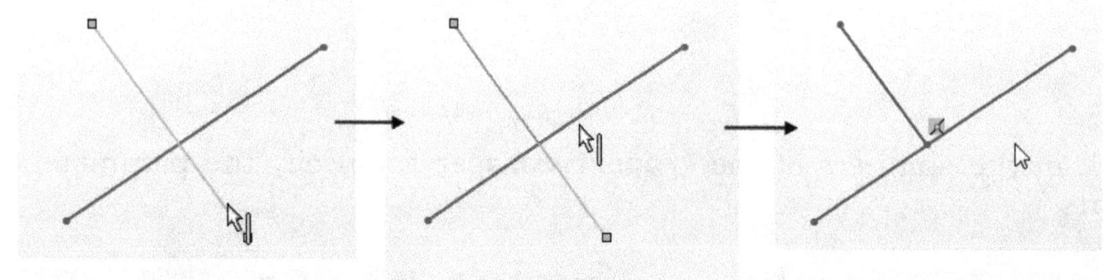

Figure-60. Trimming_procedure

Or

- Select the **Trim Entity** tool, make sure the **Power Trim** button is selected.

- Click and hold the mouse button, and drag the cursor over the portion of entities you want to be removed; refer to Figure-61.

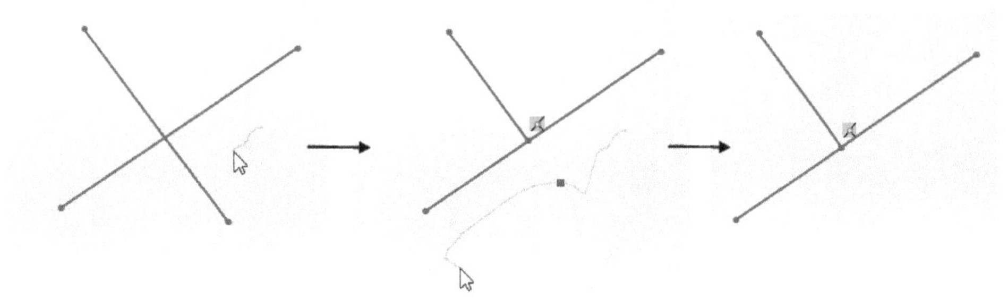

Figure-61. Trimming_by dragging

Corner

The **Corner** tool is used to trim two intersecting entities while forming corner at the intersection point. The procedure is given next.

- Click on the **Corner** button from the **Options** rollout of **Trim PropertyManager**. You will be asked to select the entities.
- Select the two intersecting sketch entities. The corner trim will be created; refer to Figure-62.

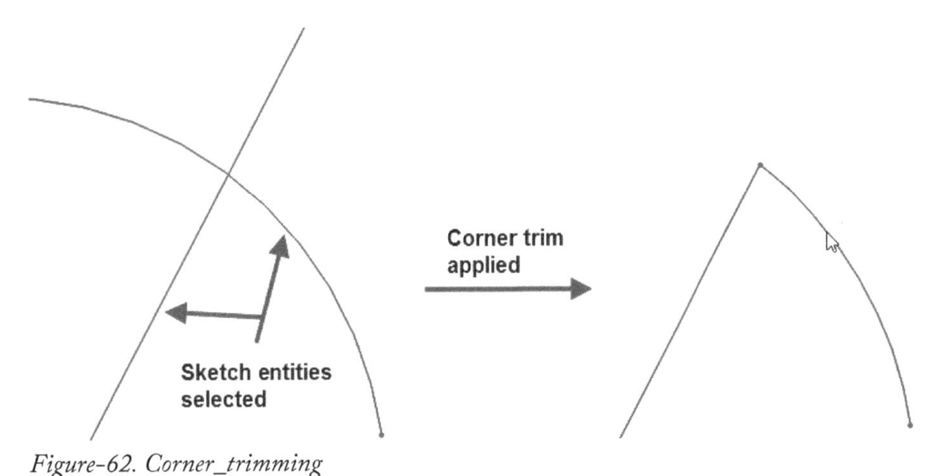

Figure-62. Corner_trimming

Trim away inside

The **Trim away inside** button is used to trim all the entities falling inside the selected boundaries. The procedure to use this option is given next.

- Click on the **Trim away inside** button from the **Trim PropertyManager**. You will be asked to select two bounding entities.
- Select a closed loop or select two bounding entities which are open loop like lines, arcs, etc. You will be asked to select or box-select the entities that are to be trimmed. The entities that are inside the selected boundaries will be trimmed; refer to Figure-63.

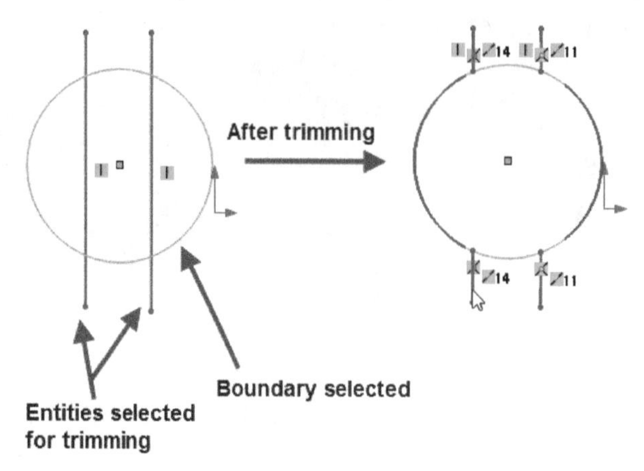

Figure-63. Using_Trim_away_inside_button

- Select the **Keep trimmed entities as construction geometry** check box to convert trimmed portion of entities as construction geometry.
- Select the **Ignore trimming of construction geometry** check box to ignore the trimming of construction geometry. Click on the **OK** button from the **PropertyManager** to apply changes and exit the tool.

In the same way, you can use the **Trim away outside** and **Trim to closet** button in the **Trim PropertyManager**.

Extending Entities

The **Extend Entities** tool does the reverse of **Trim Entities** tool. This tool is available in the **Trim Entities** drop-down. This tool extends the sketch entities up to the nearest intersecting entity. The procedure to use this tool is discussed next.

- Click on the **Extend Entities** tool from the **Trim Entities** drop-down.
- Hover the cursor on the entity that you want to extend. Preview of the extension will be displayed.
- Click on the entity if the preview is as per your requirement. If you want the entity to be extended in reverse direction then click on the other portion of the entity. Figure-64 shows the process of extending entities.

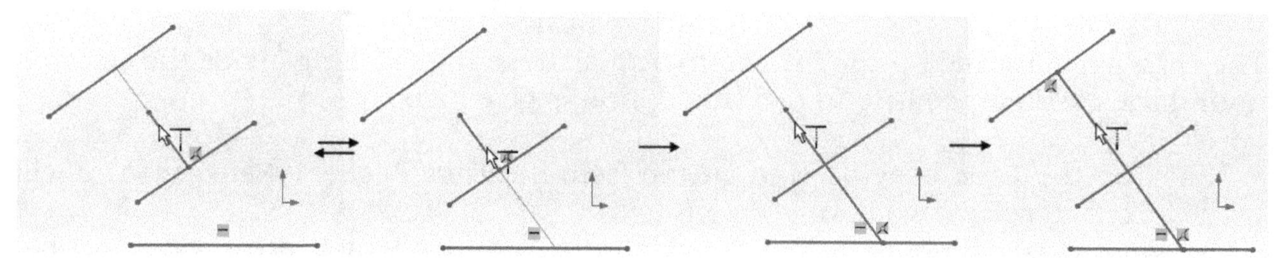

Figure-64. Extending_process

The **Convert Entities**, **Intersection Curves**, and **Silhouette Entities** tools will be explained in the Modeling section of the book, later.

Offsetting Entities

The **Offset Entities** tool is used to create copy of the selected entities at a specified distance from them. If you are the user of AutoCAD then you know that this tool is the most common tool being used while creating layouts. The procedure to use this tool is given next.

- Click on the **Offset Entities** tool from the **Ribbon**. The **Offset Entities PropertyManager** will be displayed; refer to Figure-65.

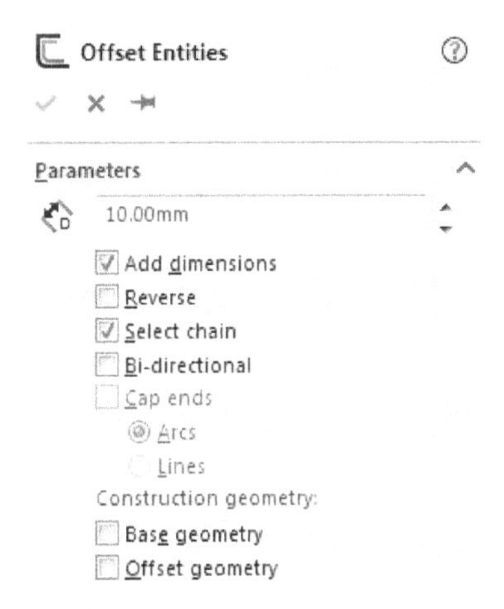

Figure-65. Offset_Entities_PropertyManager

There are various options in this **PropertyManager** that are linked to the output of the tool. So, we will discuss them one by one.

Offset Distance

This spinner is used to set the distance value for offset. You can also enter desired value in the edit box.

Add dimensions

Select this check box if you want to create the dimensions while creating offset.

Reverse

Select this check box if you want to reverse the direction of offset being displayed.

Select chain

Select this check box if you want to select the chain of entities connecting to the selected entity.

Bi-directional

Select this check box if you want to create the offset entities on both sides of the selected entity.

Cap ends

Select this check box if you want to close the ends of offset entities. This check box is active when the **Bi-directional** check box is selected and entity is selected. After selecting the **Cap ends** check box, you can close the offset entities by using the arcs or lines. For using arcs or lines, select the respective radio button.

Construction Geometry

There are two check boxes in this section, **Base geometry** and **Offset geometry**, to make the base and offset geometry as construction entity. Select desired check box/ boxes from the section.

- After selecting desired options, select the entity from the viewport. Preview will be displayed in yellow color.
- Click on **OK** button to create the offset. Figure-66 shows the process of creating offset entities.

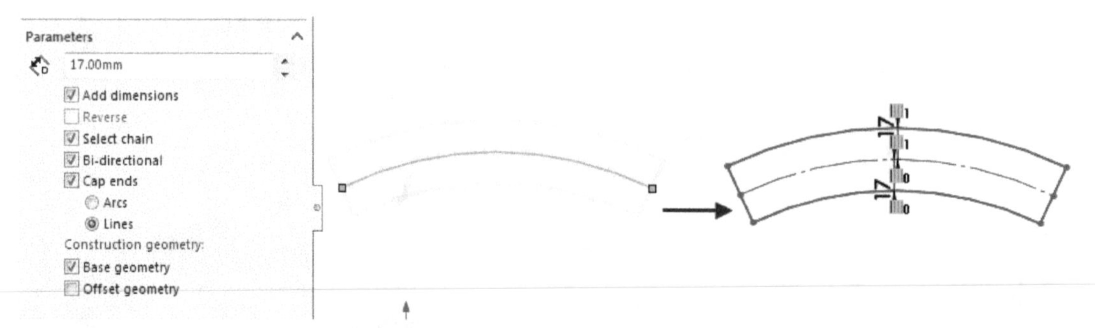

Figure-66. Offset_entities_creation

Mirroring Entities

The **Mirror Entities** tool is used to create mirror copy of the selected entities with respect to a reference called mirror line. The procedure to create mirror entities is given next.

- Click on the **Mirror Entities** tool from the **Ribbon**. The **Mirror PropertyManager** will be displayed; refer to Figure-67.

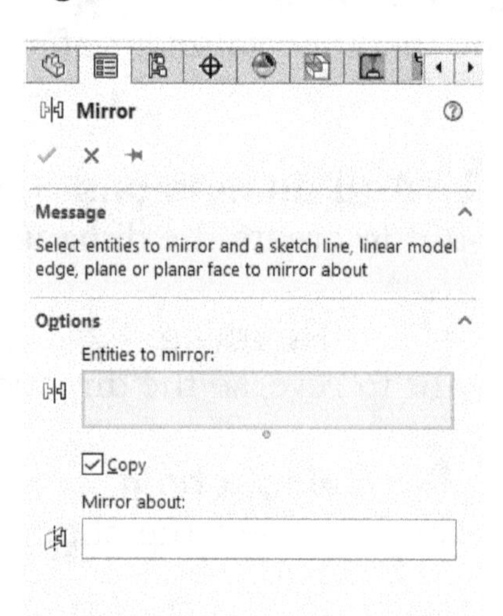

Figure-67. Mirror_PropertyManager

- Select the entity/entities you want to create mirror copy of.
- Deselect the **Copy** check box from the **PropertyManager** if you want to delete the original entities after creating mirror copy.
- Click in the **Mirror about** box and select the reference line that you want to use as mirror line. The mirror line can be an edge of a solid, sketch line, or a centerline. Figure-68 shows the process of creating mirror entities.

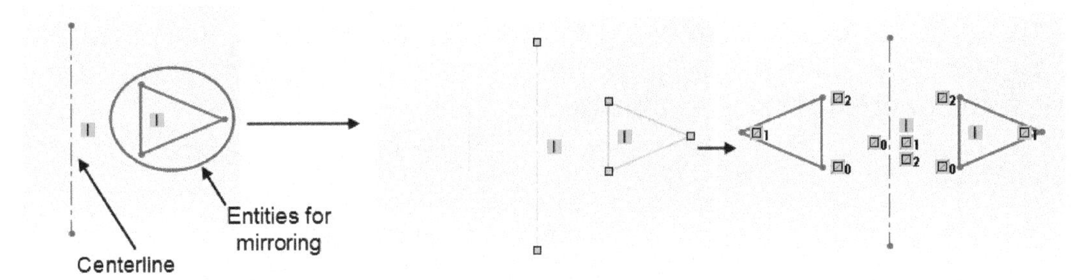

Figure-68. Mirror_entities_creation

Sometimes, we need to create multiple copies of the sketch entities. Like in sketch of a keyboard or piano. For such cases, SolidWorks provides two tools, **Linear Sketch Pattern** and **Circular Sketch Pattern**. These tools are discussed next.

Creating Linear Sketch Pattern

The **Linear Sketch Pattern** tool is used to create multiple copies of an entity in linear directions. You can create pattern in two linear directions at a time. The procedure to create linear sketch pattern is given next.

- Click on the **Linear Sketch Pattern** tool from the **Linear Sketch Pattern** drop-down of the **Ribbon**. The **Linear Pattern PropertyManager** will be displayed as shown in Figure-69.

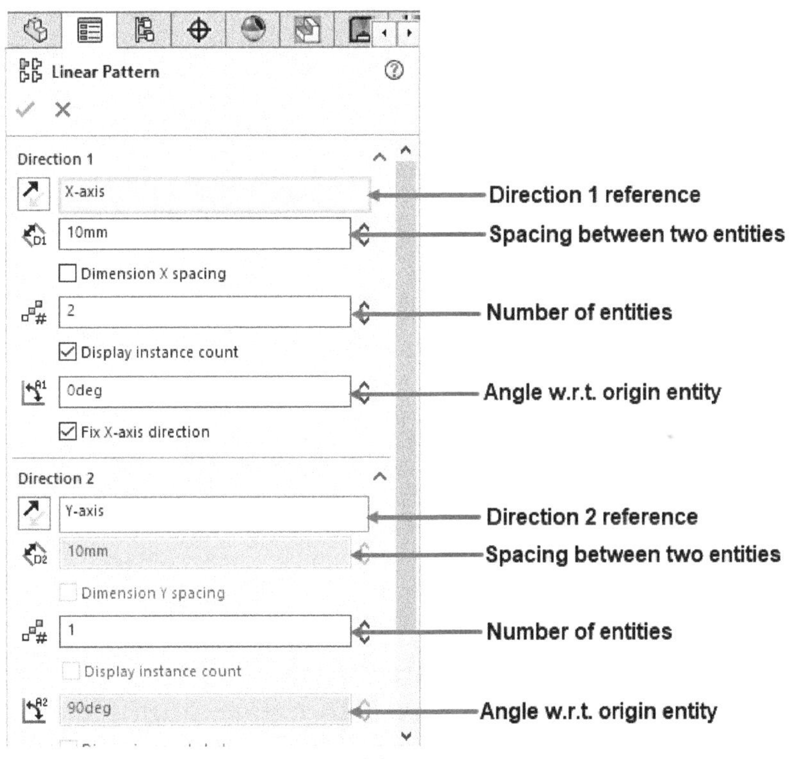

Figure-69. Linear Pattern PropertyManager

- Select the entity that you want to pattern. Specify the parameters as per your requirement.
- If you want to create pattern along an axis or line then click in the **Direction 1** reference box or **Direction 2** reference box as per your need. Now, select desired axis or line to specify the direction reference.

- If you want to skip any of the entity created in pattern then expand the **Instances to skip** rollout at the bottom of the **PropertyManager** and click in the **Instances to skip** box. Now, click on the pink dot for the entities from the preview that you do not want to create; refer to Figure-70.

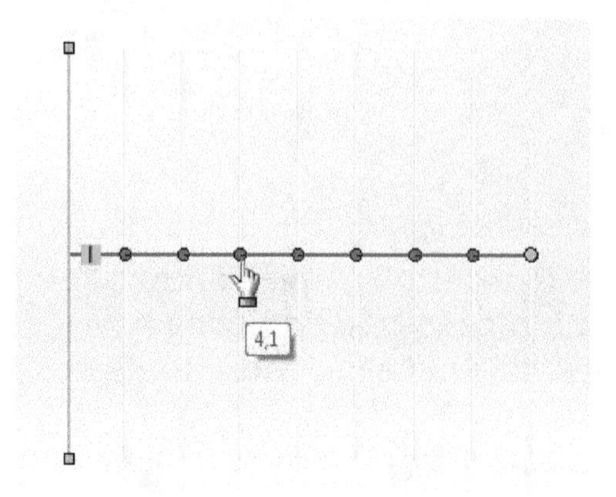

Figure-70. Instances_to_be_removed

- Increase the number of entities in the **Direction 2** rollout to activate the options in the rollout. Figure-71 shows the pattern created for a circle and its respective options in the **PropertyManager**.

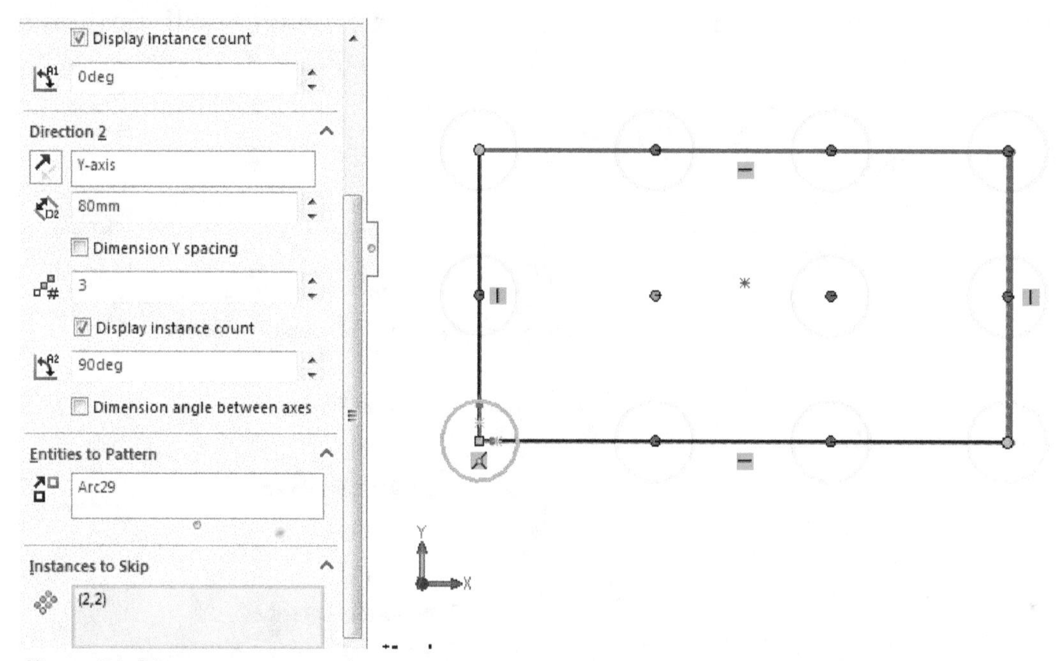

Figure-71. Linear_pattern_creation

Creating Circular Sketch Pattern

The **Circular Sketch Pattern** tool is used to create multiple copies of an entity in circular fashion. You can create pattern in two linear directions at a time. The procedure to create circular sketch pattern is given next.

- Click on the **Circular Sketch Pattern** tool from the **Linear Sketch Pattern** drop-down. The **Circular Pattern PropertyManager** will be displayed; refer to Figure-72.

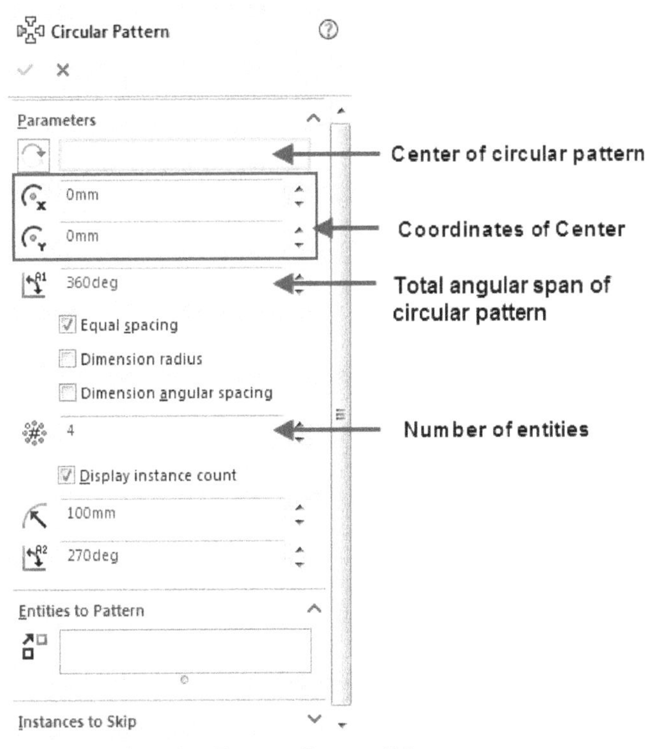

Figure-72. Circular_Pattern_PropertyManager

- Select the entity that you want to pattern and click in the **Center of circular** pattern box to specify the center of the pattern.
- Select the point to be the origin, specify the number of entities, and click on the **OK** button to create the pattern.
- You can skip the entities as you did for Linear pattern. Figure-73 shows a circular pattern created with its parameters specified.

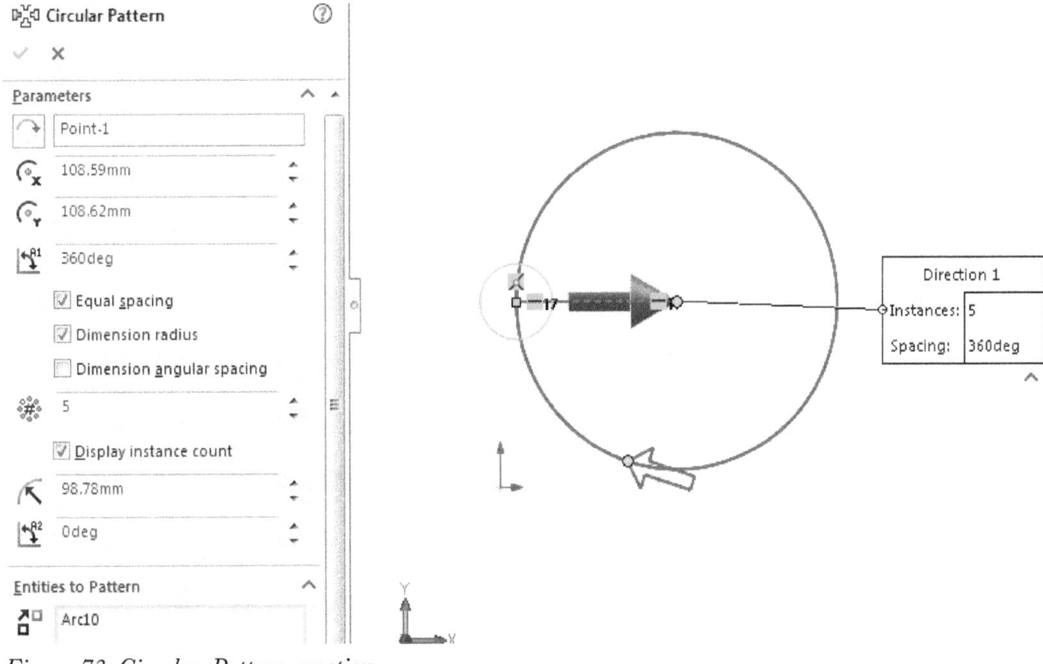

Figure-73. Circular_Pattern_creation

Moving Entities

The **Move Entities** tool is used to move the entities from one position to another position. You can move the entities either by coordinate values or by clicking. The procedure to move entities is discussed next.

- Select the **Move Entities** tool from the **Ribbon**. The **Move Entities PropertyManager** will display as shown in Figure-74.

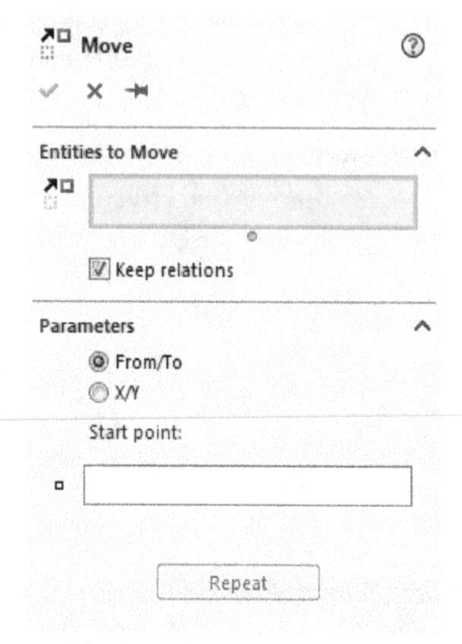

Figure-74. Move_Entities_PropertyManager

- Select the entity that you want to move.
- Make sure **From/To** radio button is selected. Now click on the point you want to use as base point.
- Click at the destination point where you want to place the entity. The entity will move to specified place.

Or

- Click on the **X/Y** radio button and enter the distance in X and Y direction to place the sketch entity.

Or

- Select the entity, you want to move and drag it from the key point/curve. Note that circular entities like circle, arc, ellipse are dragged from center for moving them. For the other entities, select the curve from the location which is not a key point. Figure-75 shows a spline and line moved together by dragging.

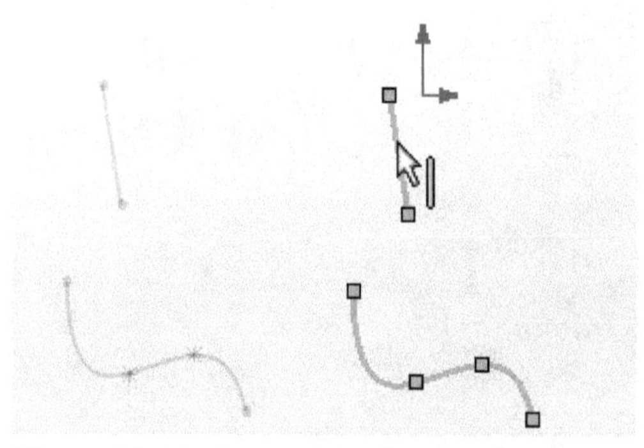

Figure-75. Moving_entities_by_dragging

Copying Entities

The **Copy Entities** tool is available in the **Move Entities** drop-down. This tool is used to copy the entities by specifying position. This tool works in the same way as the **Move Entities** tool does. The only difference is that it does not move the entity but it creates the entities. Figure-76 shows the entities copied by this tool.

Figure-76. Copying_entities_using_Copy_Entities_tool

Rotating Entities

The **Rotate Entities** tool is available in the **Move Entities** drop-down. This tool is used to rotate the entities by specified angle. The procedure to use this tool is given next.

- Click on the **Rotate Entities** tool from the drop-down. The **Rotate PropertyManager** will be displayed as shown in Figure-77.

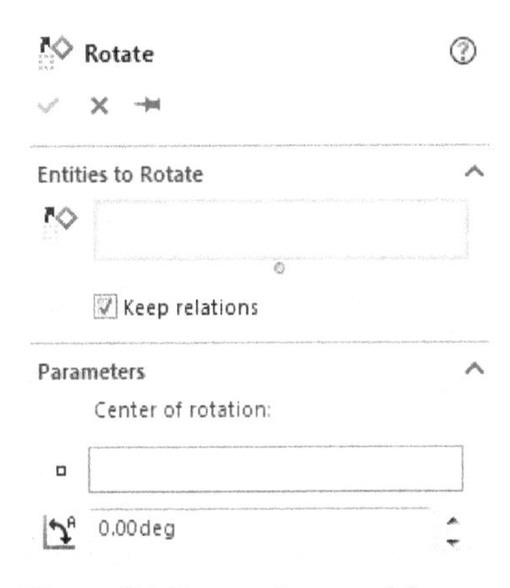

Figure-77. Rotate_PropertyManager

- Select the entity that you want to rotate.
- Click in the **Center of rotation** box and click on the point that you want to make center of rotation.
- Specify angle in the **Angle** spinner. The object will rotate by specified value; refer to Figure-78.

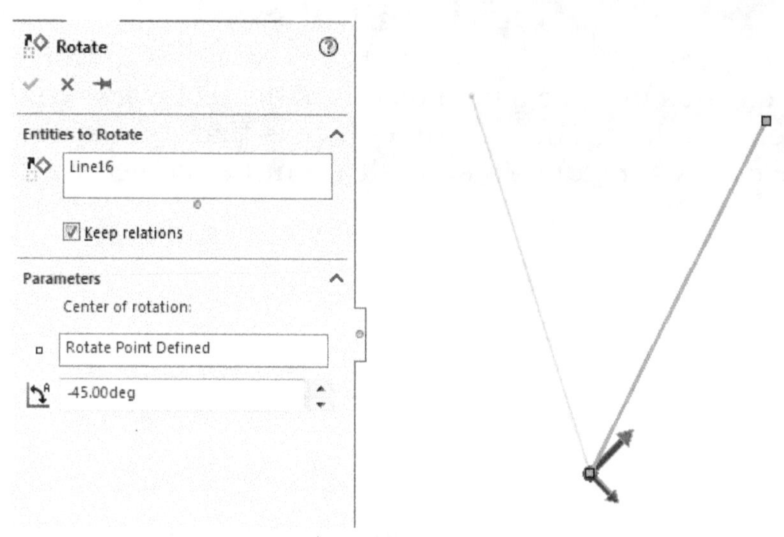

Figure-78. Rotating_an_entity

Scaling Entities

The **Scale Entities** tool is available in the **Move Entities** drop-down. This tool is used to increase or decrease the size of an entity by specified scale value. The procedure to use this tool is given next.

- Click on the **Scale Entities** tool from the drop-down. The **Scale PropertyManager** will be displayed as shown in Figure-79.

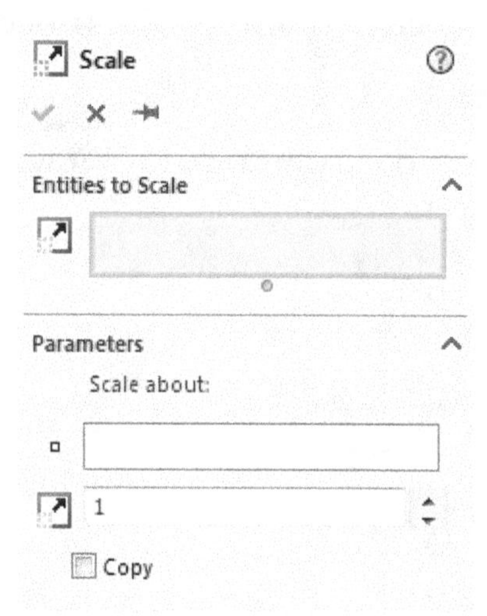

Figure-79. Scale_PropertyManager

- Select the entities that you want to scale up or scale down.
- Click in the **Scale about** box and select the base point about which you want to scale the entities.
- Enter the scale value in the **Scale Factor** spinner. You can create a copy of entities scaled to specified value by selecting the **Copy** check box. In that case, the selected entities will remain unchanged. Refer to Figure-80.

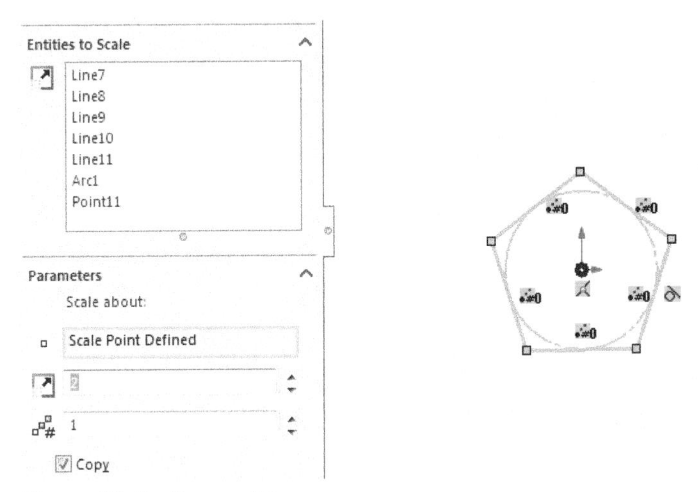

Figure-80. Scaling_entities

Stretching Entities

The **Stretch Entities** tool is available in the **Move Entities** drop-down. This tool is used to stretch any sketched entity. The procedure to use this tool is given next.

- Click on the **Stretch Entities** tool from the drop-down. The **Stretch PropertyManager** will be displayed as shown in Figure-81.

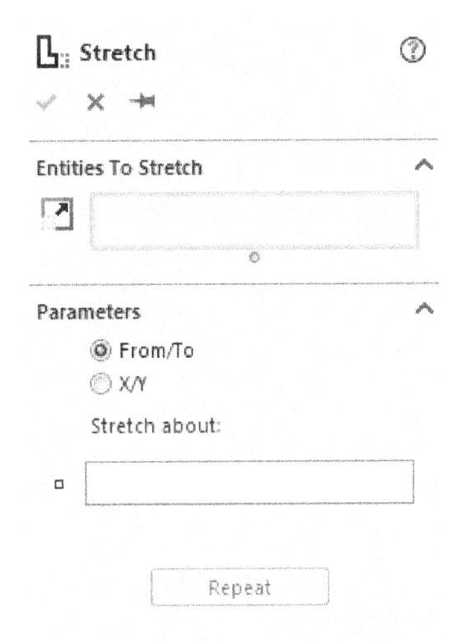

Figure-81. Stretch_PropertyManager

- By using the cross-rectangle selection, select the portion of entities that you want to stretch.
- Click in the **Stretch about** box and then select the base point on the sketch.
- Move the cursor to desired position and click to stretch the entities. Figure-82 shows the procedure of stretching the entities.

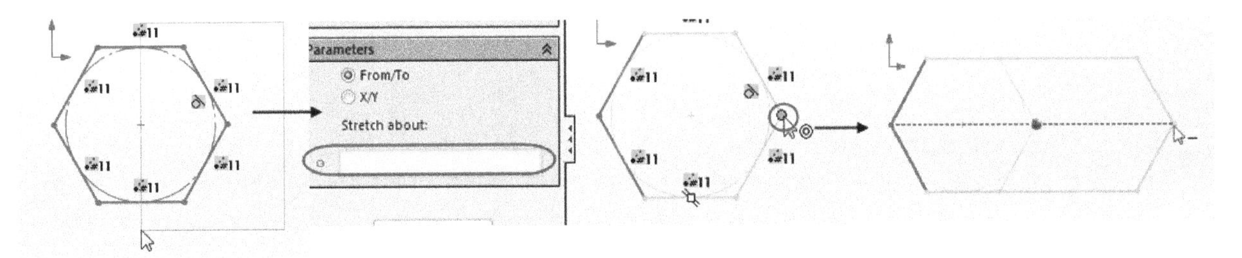

Figure-82. Stretching_entities

Segment Tool

The **Segment** tool is used to create segments of circles, arcs, spline, parabolas, etc. By default, the **Segment** tool is not available in the **CommandManager** and you need to add the tool manually. The procedure to add the tool to **CommandManager** is given next.

Adding Segment Tool in CommandManager

- Right-click on any tool in the **CommandManager** and select the **Customize** option from the shortcut menu displayed; refer to Figure-83. The **Customize** dialog box will be displayed; refer to Figure-84.

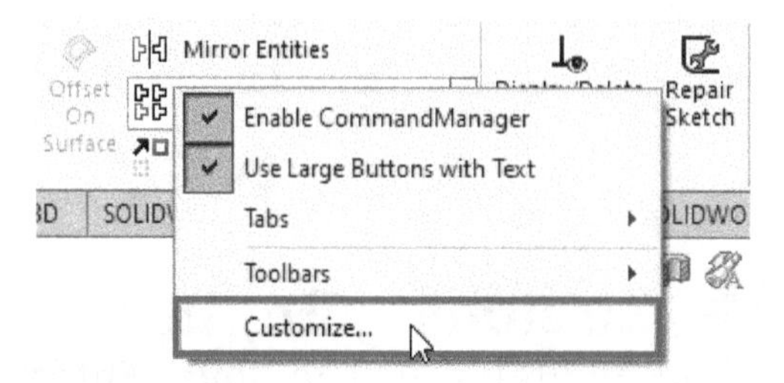

Figure-83. Customize option in shortcut menu

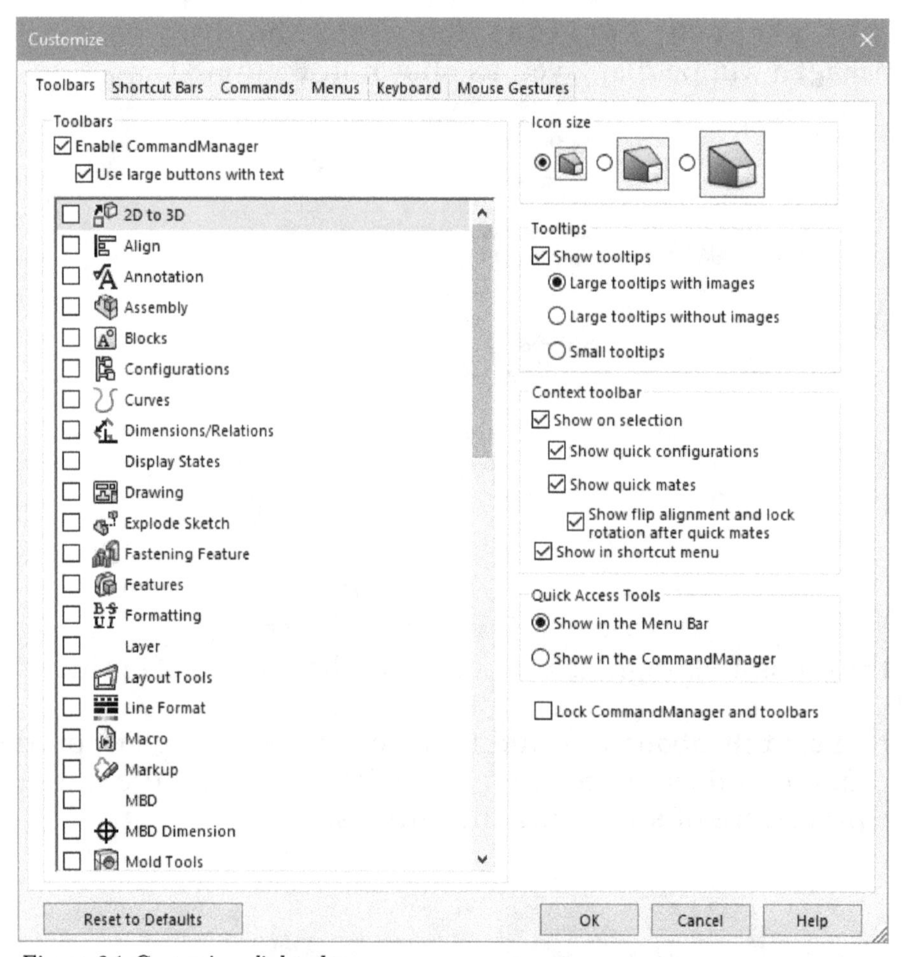

Figure-84. Customize_dialog_box

- Click on the **Commands** tab in the dialog box and select the **Sketch** option from the **Categories** area. The buttons of the **Sketch** categories will be displayed; refer to Figure-85.

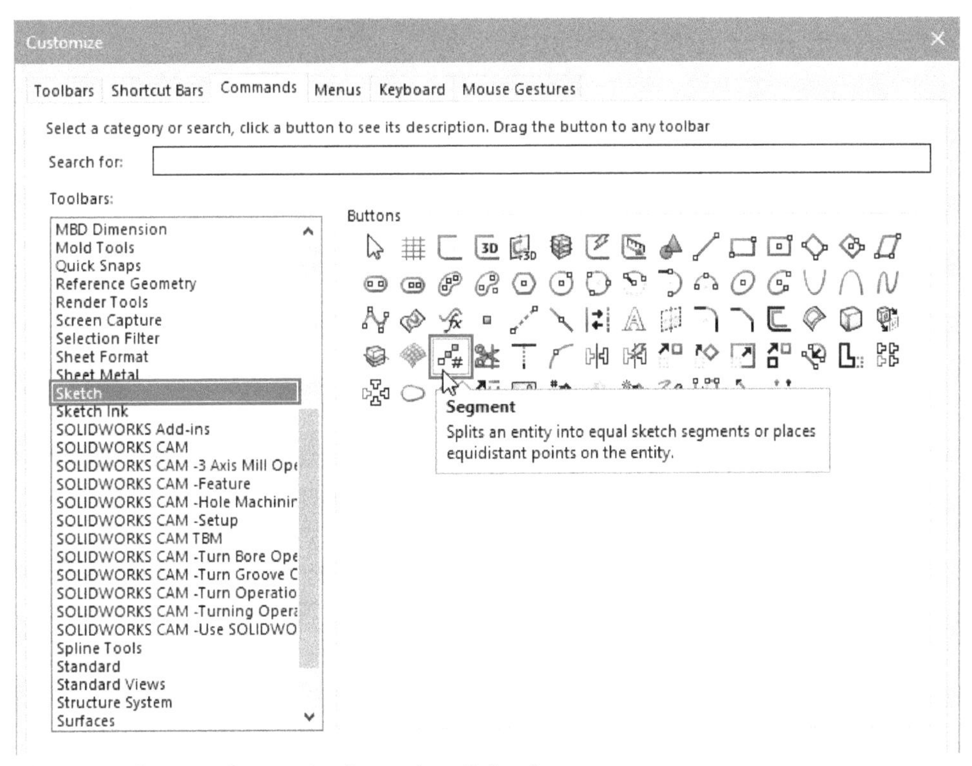

Figure-85. Segment_button_in_Customize_dialog_box

- Select the **Segment** button from the dialog box and drag-drop it in the **Sketch CommandManager**; refer to Figure-86.

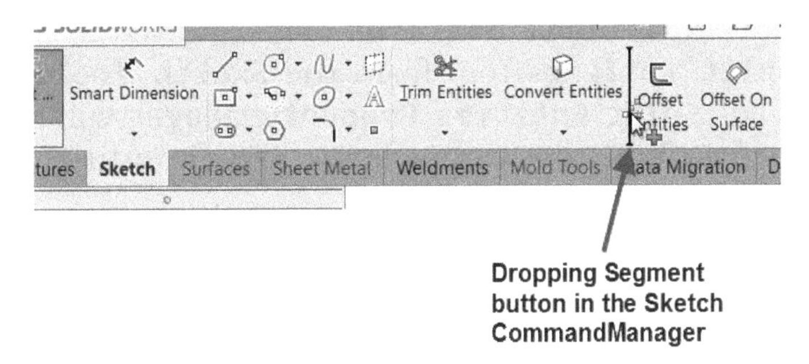

Figure-86. Dropping_Segment_button_in_CommandManager

- Click on the **OK** button from the dialog box.

Creating Segments of Sketch Entity

- Click on the newly added **Segment** tool from the **Sketch CommandManager** or select the **Segment** tool from the **Sketch Tools** cascading menu of the **Tools** menu. The **Segment PropertyManager** will be displayed; refer to Figure-87.

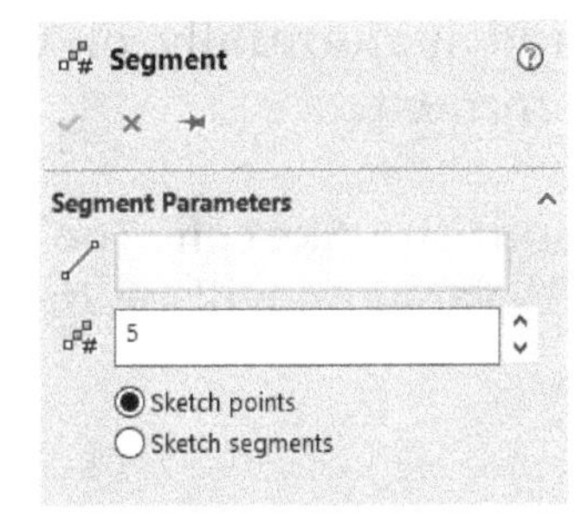

Figure-87. Segment_PropertyManager

- Select the entity that you want to break into equal segments. Preview of segment will be displayed; refer to Figure-88.

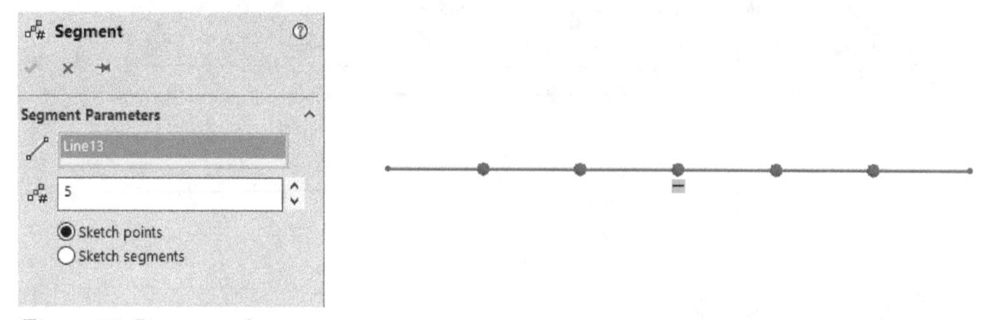

Figure-88. Preview_of_segment

- By default, the **Sketch points** radio button is selected in the **PropertyManager** and hence the points are created on the line at equal distance. Select the **Sketch segments** radio button to create segments of entity in place of only points.
- Set the number of segments in the spinner displayed in **PropertyManager**.
- Click on the **OK** button from the **PropertyManager** to create the segments/points.

Splitting Entities

The **Split Entities** tool is used to split the selected sketch entities at specified points. By default, this tool is not available in the **CommandManager** so you need to add this tool in the same way as you have added the **Segment** tool in the **Sketch CommandManager**. The procedure to use this tool is given next.

- Click on the newly added **Split Entities** tool from the **Sketch CommandManager** or click on the **Split Entities** tool from the **Sketch Tools** cascading menu of the **Tools** menu. The **Split Entities PropertyManager** will be displayed; refer to Figure-89. Also, you will be asked to select a sketch entity to be split.

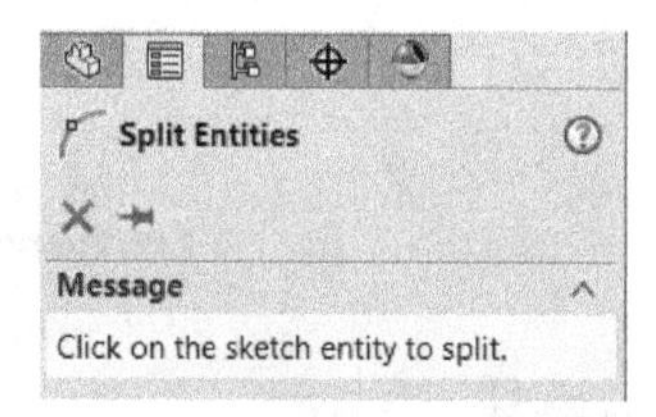

Figure-89. Split_Entities_PropertyManager

- Click at desired location on the sketch entity to split it at that location.
- Specify as many split points as you want and then press **ESC** from keyboard.

Creating Jog Lines

The **Jog Line** tool is used to create jogged lines. This tool is not available by default in the **Ribbon** so you need to add it manually or you can use it from menu. The procedure to use this tool is given next.

- Click on the **Jog Line** tool from the **Sketch Tools** cascading menu of the **Tools** menu. The **Jog Line PropertyManager** will be displayed; refer to Figure-90.

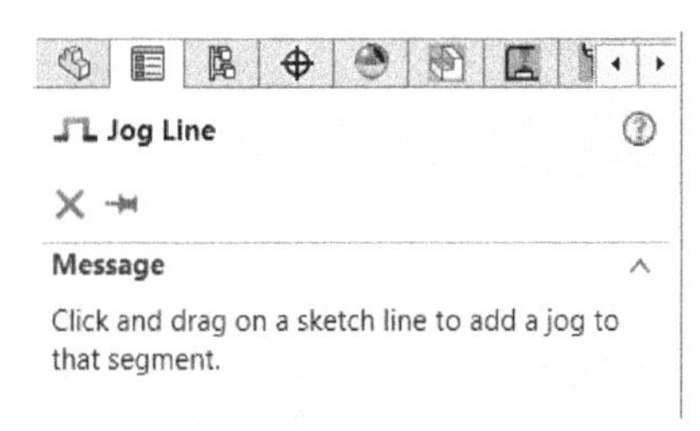

Figure-90. Jog Line PropertyManager

- Click at desired location on line from where you want to start the jog line. Move the cursor at the distance upto which you want to create jog line. Preview of jog line will be displayed; refer to Figure-91.

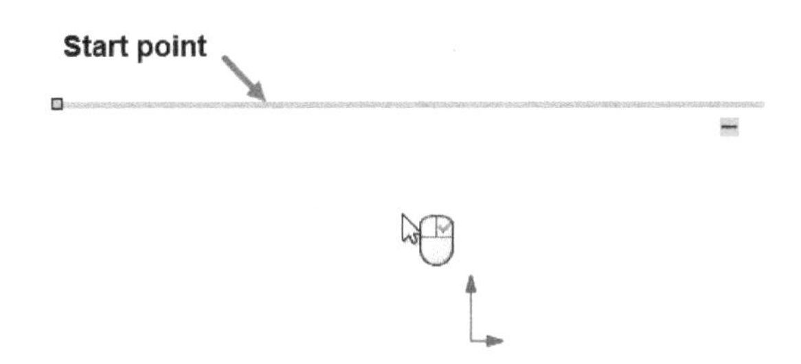

Figure-91. Preview_of_jog_line

- Click at desired location to create the jog line. You can create multiple jogs in the same line. After creating jog line, click on the **Close** button from the **PropertyManager**.

Dynamic Mirroring

The **Dynamic Mirror Entities** tool is used to dynamically create mirror copy of the entities while they are being created in the sketch. The procedure to use this tool is given next.

- Make sure a sketch line or linear model edge is there in the sketch. Click on the **Dynamic Mirror Entities** tool from the **Sketch Tools** cascading menu of the **Tools** menu. The **Mirror PropertyManager** will be displayed; refer to Figure-92.

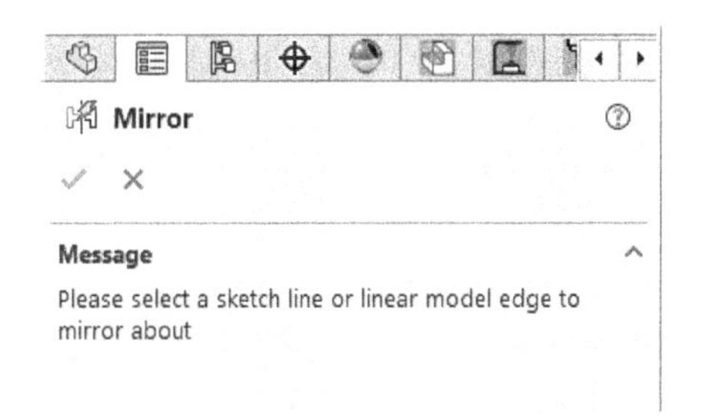

Figure-92. Mirror_PropertyManager_for_dynamic_mirror

- Select the sketch line to be used for dynamic mirroring. The line will be modified as shown in Figure-93.

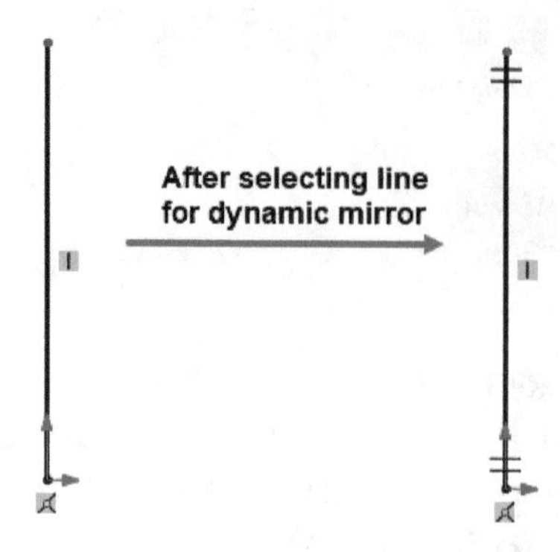

Figure-93. Line_selected_for_dynamic_mirroring

- Now, create sketch entities on one side of the dynamic mirror line. You will notice that a mirror copy of the same sketch entities is automatically created on the other side of line; refer to Figure-94.

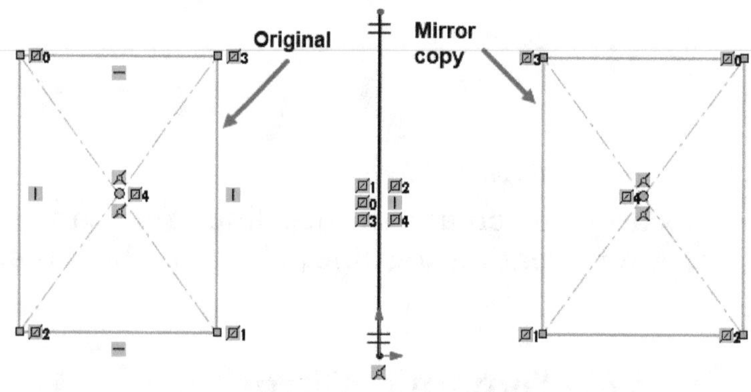

Figure-94. Dynamic_mirror_copy

- Click on the **Dynamic Mirror Entities** tool again to exit the tool.

Replacing Sketch Entities

The **Replace Entity** tool is used to replace selected sketch entity by another entity. The procedure to use this tool is given next.

- Click on the **Replace Entity** tool from the **Sketch Tools** cascading menu of the **Tools** menu. The **Replace Entity PropertyManager** will be displayed; refer to Figure-95.

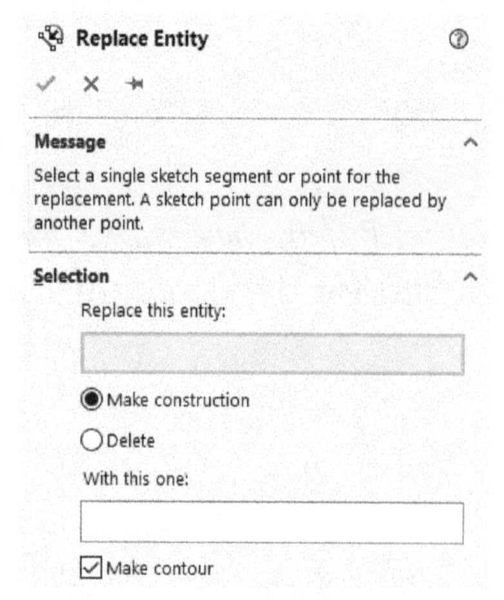

Figure-95. Replace_Entity_PropertyManager

- Select desired radio button from the **PropertyManager**. Select the **Make construction** radio button if you want to convert the original entity as construction. Select the **Delete** radio button if you want to delete the original entity after replacement.
- Select the entity to be replaced. You will be asked to select the entity by which it should be replaced.
- Select desired entity and click on the **OK** button from the **PropertyManager**.

Placing Picture in Sketch

The **Sketch Picture** tool is used to insert a picture/image on the sketching plane to use as reference for sketch. The procedure to use this tool is given next.

- Click on the **Sketch Picture** tool from the **Sketch Tools** cascading menu in the **Tools** menu. The **Open** dialog box will be displayed; refer to Figure-96.

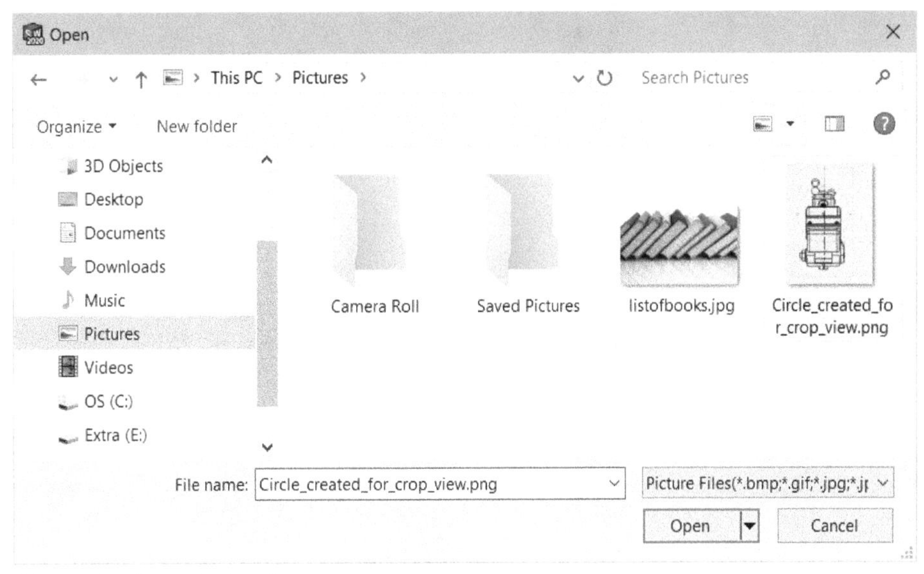

Figure-96. Open_dialog_box

- Select desired image file and click on the **Open** button. Preview of image will be displayed with **Sketch Picture PropertyManager**; refer to Figure-97.

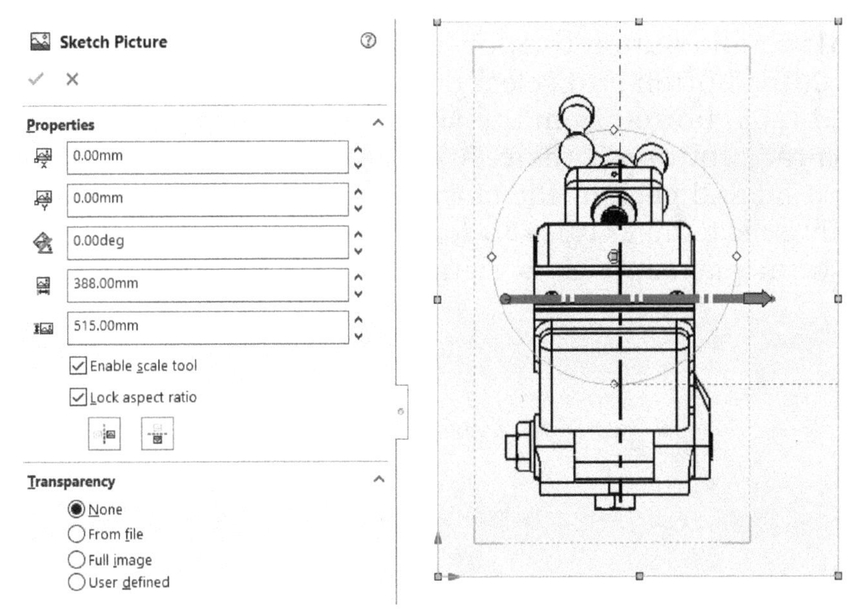

Figure-97. Sketch Picture PropertyManager

- Set the parameters as desired in the **PropertyManager** like origin location, angle, size of image, and other parameters.
- Click on the **OK** button from the **PropertyManager**.

Area and Hatch Fill in Sketch

The **Area Hatch/Fill** tool is used to create hatching and area fills in closed regions of sketch. The procedure to use this tool is given next.

- Click on the **Area Hatch/Fill** tool from the **Sketch Tools** cascading menu of **Tools** menu. The **Area Hatch/Fill PropertyManager** will be displayed; refer to Figure-98.

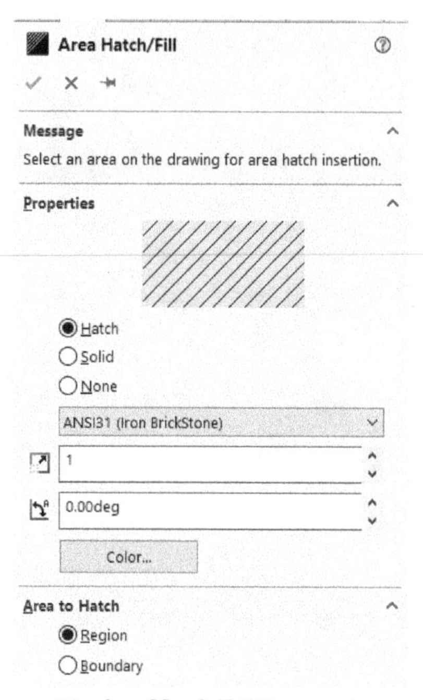

Figure-98. Area Hatch Fill PropertyManager

- Select desired radio button to define the type of fill to be made. Select the **Hatch** radio button to apply desired hatching pattern. Select desired option from the **Hatch Pattern** drop-down. Set desired values of **Hatch Pattern Scale** and **Hatch Pattern Angle** edit boxes.
- Select the **Solid** radio button to apply solid fill inside selected region/boundary. Click on the **Color** button and select desired color for fill.
- Select desired radio button from the **Area to Hatch** rollout of **PropertyManager**. Select **Region** radio button if there is a closed loop in which you want to apply hatching/area fill and click in the closed loop. Select the **Boundary** radio button to individually select the entities to form closed boundary. Preview of hatch or solid fill will be displayed; refer to Figure-99.

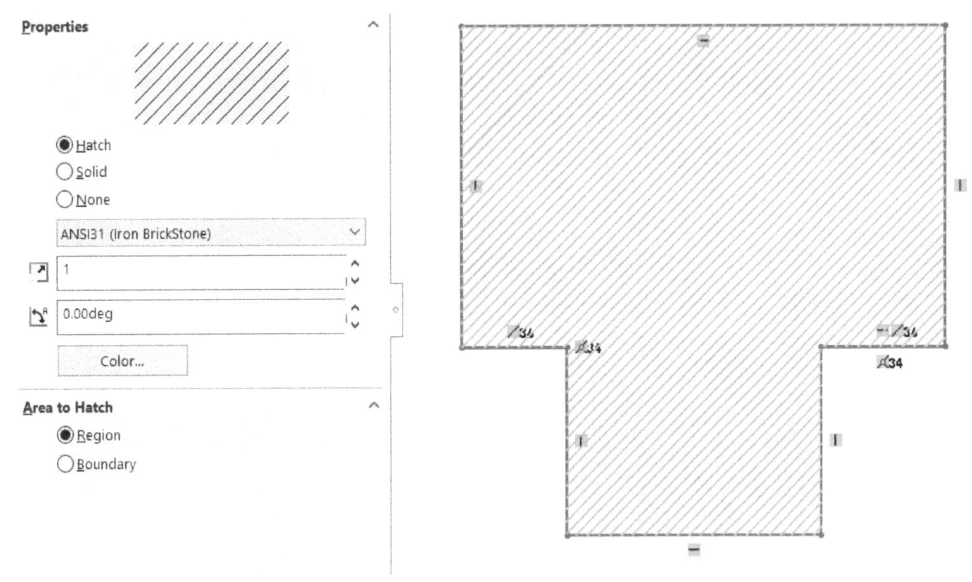

Figure-99. Preview_of_hatch

- Click on the **OK** button from the **PropertyManager** to create hatch.

RELATIONS

Relations are used to constrain the sketch entities dimensionally and/or geometrically. In SolidWorks, you can also constrain any sketch entity by using dimensions or geometrical constraints. Both type of constraints are explained next.

Dimensional Constraints (Dimensions)

Dimensions are used to limit size of the sketch entities. For example; specifying length of the line, specifying diameter of the circle, and so on. The tools to dimension sketch entities are given in the **Smart Dimension** drop-down. The tools that are commonly used for dimensioning are discussed next.

Smart Dimension

The **Smart Dimension** tool is used to dimension the entities automatically. This tool can create different type of dimensions like, horizontal, vertical, or inclined dimensions. The procedure to create dimensions are explained next.

- Click on the **Smart Dimension** tool from the **Smart Dimension** drop-down in the **Ribbon**. You are asked to select entities to dimension.
- Select the entity you want to dimension and then click at desired distance to place dimension.

Selection pattern for dimensioning various entities after selecting the **Smart Dimension** tool is given next.

Dimensioning a line

Click on the line to be dimensioned and then click at desired position to place the dimension.

Dimensioning inclined line

Click one by one at the end points of the line. Move the cursor perpendicularly above the line to create inclined dimensions. If you want to create horizontal dimension of a line then move the cursor vertically downward or upward beyond the limit of the line. Similarly, move the cursor towards left or right of the line to create vertical dimension.

Dimensioning arcs/circle

Click on the arc/circle to specify its radius/diameter.

Dimensioning elliptical arcs/ellipse

Click on the end points of the elliptical arcs/ellipses to dimension it.

Figure-100 shows various entities dimensioned.

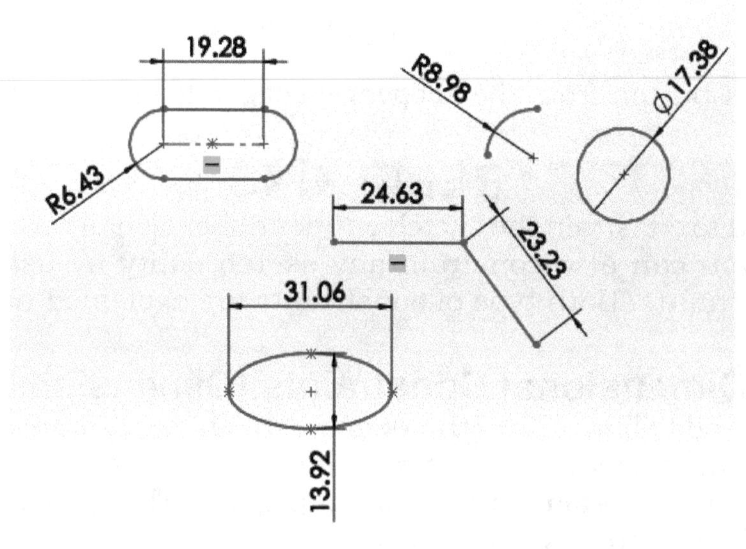

Figure-100. Dimensioning

Ordinate Dimension

You can also use Ordinate dimensioning if you are dimensioning for CNC coordinates. The procedure to create ordinate dimensions is given next.

- Click on the **Ordinate Dimension** tool from the **Smart Dimension** drop-down.
- Select the first reference that you want to make zero reference.
- Click to place the zero reference.
- Select the next line for which you want to specify the dimension.
- After you have specified the dimensions in one direction. Press the **ESC** key and then restart the tool.
- Now, select the next zero reference and repeat the procedure. Figure-101 shows a sketch dimensioned by ordinates.

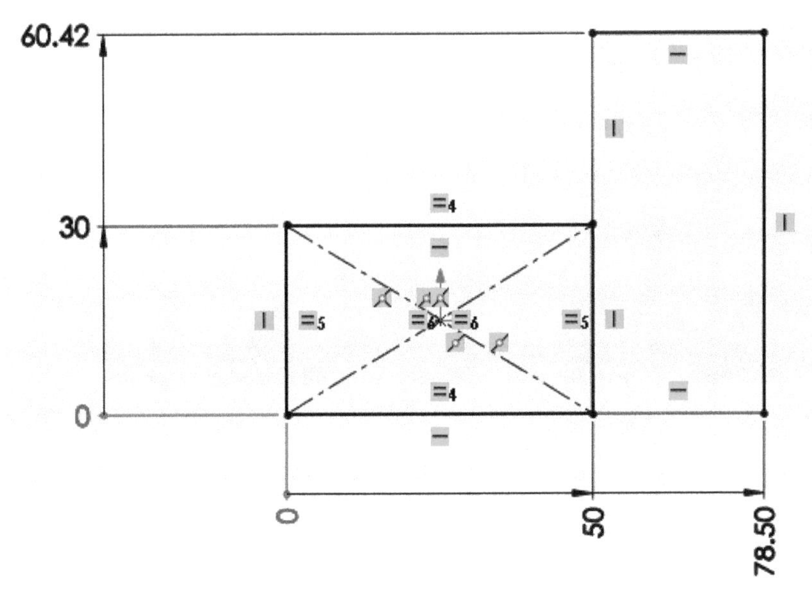

Figure-101. Ordinate_dimensioned_sketch

Note that you can use the **Horizontal Ordinate Dimension** and **Vertical Ordinate Dimension** tools in the **Dimension** drop-down in the same way.

Path Length Dimension

The **Path Length Dimension** tool is used to dimension the total length of selected entities. The procedure to use this tool is given next.

- Click on the **Path Length Dimension** tool from the **Smart Dimension** drop-down in the **Sketch CommandManager** of **Ribbon**. The **Path Length PropertyManager** will be displayed and you will be asked to select the sketch entities.
- Select the curve(s) whose path length is to be dimensioned and then click on the **OK** button from the **PropertyManager**. The Path Length dimension will be attached to the curve(s); refer to Figure-102.

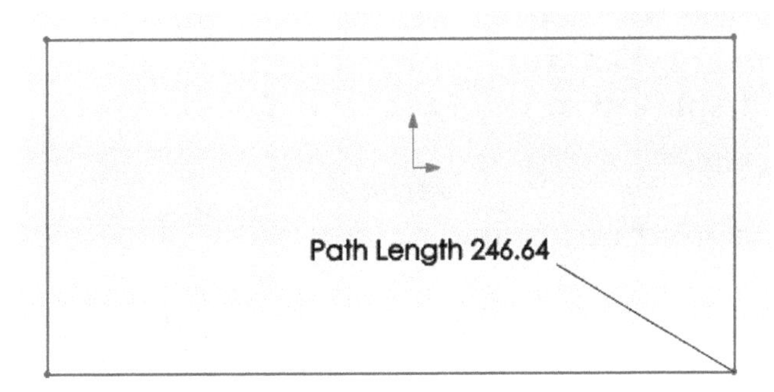

Figure-102. Path_length_dimension_created

Note that the information about path length, angle, and distance are also displayed in the status bar on selecting the entities; refer to Figure-103.

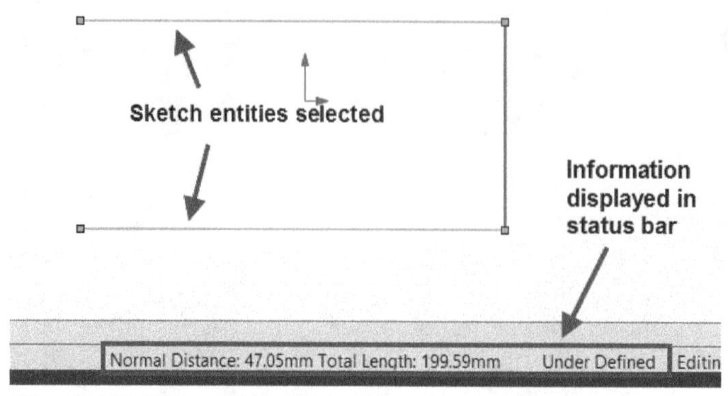

Figure-103. Information_displayed_in_status_bar

Geometric Constraints

These constraints are used to constrain the shape/position of sketch entities with respect to other entities.

- To apply the geometric constraints, click on the **Add Relation** button in the **Display/Delete Relations** drop-down.

The list of constraints that can be applied in SolidWorks are discussed next.

Horizontal

The **Horizontal** constraint makes one or more selected lines or center lines to become horizontal. You can also select an external entity such as an edge, plane, axis, or sketch curve on an external sketch that will act as a line to apply this constraint. You can also make two or more points to become horizontal using the **Horizontal** constraint. A point can be a sketch point, a center point, an endpoint, a control point of a spline, or an external entity such as origin, vertex, axis, or point in an external sketch. To apply this constraint, invoke the **Add Relations PropertyManager**. Select the entity or entities to apply the **Horizontal** constraint. Choose the **Horizontal** button from the **Add Relations** rollout in the **Add Relations PropertyManager**. You will notice that the name of the horizontal constraint will be displayed in the **Existing Constraints** rollout.

Vertical

The **Vertical** constraint makes one or more selected lines or centerlines to become vertical. You can force two or more points to become vertical using the **Vertical** constraint. To apply this constraint, invoke the **Add Relations PropertyManager** and select the entity or entities to apply the **Vertical** constraint. Choose the **Vertical** button from the **Add Relations** rollout. You will notice that the name of the vertical constraint is displayed in the **Existing Constraints** rollout.

Collinear

The **Collinear** constraint makes the selected lines to lie on the same infinite line; refer to Figure-104. To use this constraint, select the lines to apply the **Collinear** constraint. Choose the **Collinear** button from the **Add Relations** rollout.

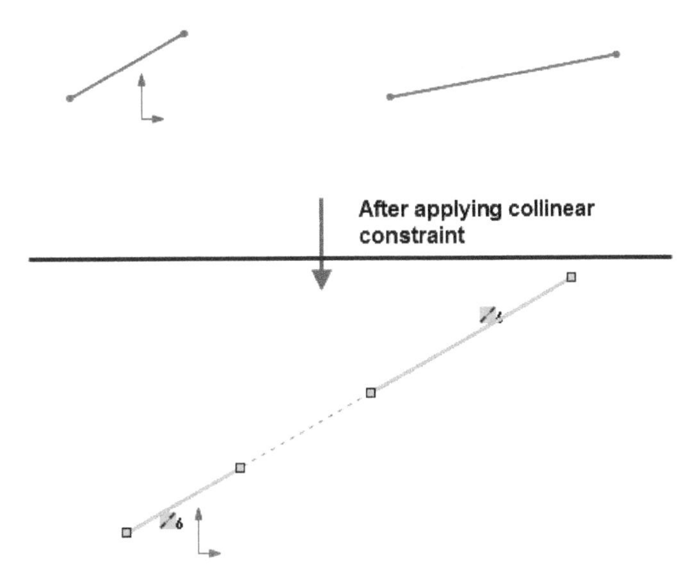

Figure-104. Applying_collinear_constraint

Coradial

The **Coradial** constraint makes the selected arcs or circles to share the same radius and the same center points; refer to Figure-105. You can also select an external entity that projects as an arc or a circle in the sketch to apply this constraint. To use this constraint, invoke the **Add Relations PropertyManager**. Select two arcs or circles, or an arc and a circle to apply the **Coradial** constraint. Choose the **Coradial** button from the **Add Relations** rollout.

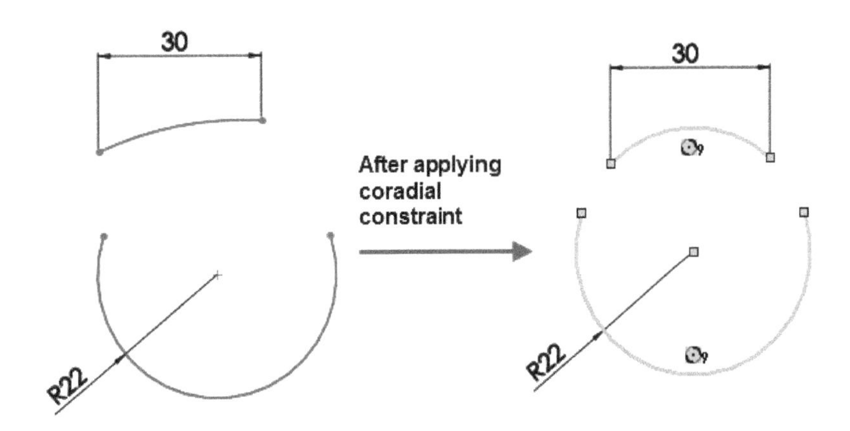

Figure-105. Applying_coradial_constraint

Perpendicular

The **Perpendicular** constraint makes the selected lines/curves to become perpendicular to each other. To apply this constraint, invoke the **Add Relations PropertyManager**. Select two lines and choose the **Perpendicular** button from the **Add Relations** rollout. Figure-106 shows two lines before and after applying the **Perpendicular** constraint.

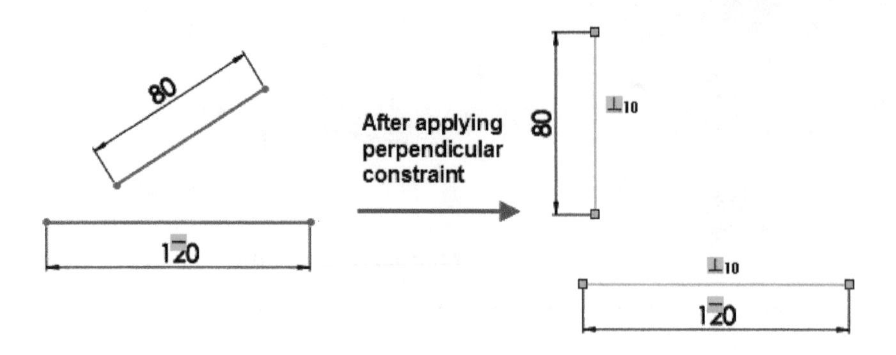

Figure-106. Applying_perpendicular_constraint

Parallel

The **Parallel** constraint makes the selected lines to become parallel to each other. To apply this constraint, invoke the **Add Relations PropertyManager**. Select two lines and choose the **Parallel** button from the **Add Relations** rollout.

ParallelYZ

The **ParallelYZ** constraint makes a line in the three-dimensional (3D) sketch to become parallel to the YZ plane with respect to the selected plane. To apply this constraint, invoke the **Add Relations PropertyManager**. Select a line in the 3D sketch and then select a plane. Next, choose the **ParallelYZ** button from the **Add Relations** rollout.

ParallelZX

The **ParallelZX** constraint makes a line in the 3D sketch to become parallel to the ZX plane with respect to the selected plane. To apply this constraint, invoke the **Add Relations PropertyManager**. Select a line in the 3D sketch and then select a plane. Next, choose the **ParallelZX** button from the **Add Relations** rollout.

Along X

The **AlongX** constraint makes a line in the 3D sketch to become parallel to the X-axis. To apply this constraint, invoke the **Add Relations PropertyManager**. Select a line in the 3D sketch and then choose the **Along X** button from the **Add Relations** rollout; the selected line will be oriented along the X axis.

Along Y

The **Along Y** constraint makes a line in the 3D sketch to become parallel to the Y-axis. To apply this constraint, invoke the **Add Relations PropertyManager**. Select a line in the 3D sketch and then choose the **Along Y** button from the **Add Relations** rollout; the selected line will be oriented along the Y axis.

Along Z

The **AlongZ** constraint makes a line in the 3D sketch to become parallel to the Z-axis. To apply this constraint, invoke the **Add Relations PropertyManager**. Select a line in the 3D sketch and then choose the **Along Z** button from the **Add Relations** rollout; the selected line will be oriented along the Z axis.

Normal

The **Normal** constraint makes a line in the 3D sketch to become normal to the selected plane; refer to Figure-107. To apply this constraint, invoke the **Add Relations PropertyManager**. Select a line in the 3D sketch and then select a plane. Next, choose the **Normal** button from the **Add Relations** rollout; the selected line will be oriented normal to the plane.

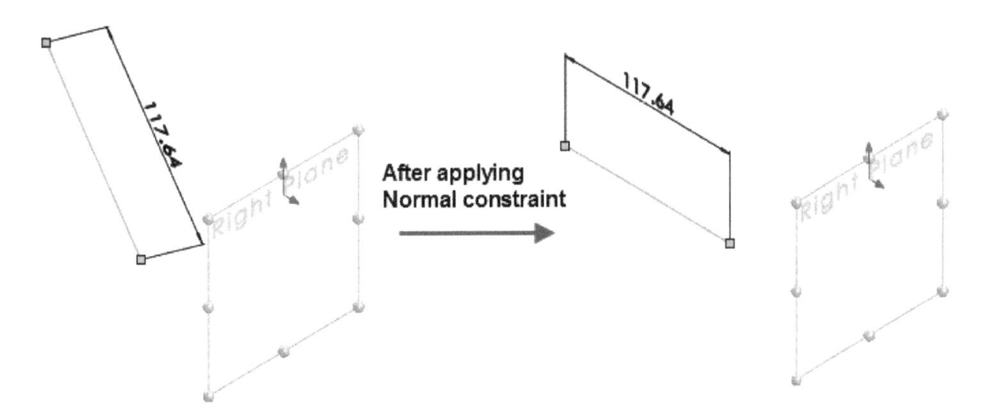

Figure-107. Applying_Normal_constraint

On Plane

The **On Plane** constraint makes a line in the 3D sketch to become parallel and places on the selected plane. To apply this constraint, invoke the **Add Relations PropertyManager**. Select a line in the 3D sketch and then select a plane. Next, choose the **On Plane** button from the **Add Relations** rollout; the selected line will be oriented parallel to the selected plane and places on it.

Tangent

The **Tangent** constraint makes a selected arc, circle, spline, or ellipse to become tangent to other arc, circle, spline, ellipse, line, or edge. To apply this constraint, invoke the **Add Relations PropertyManager**. Select two entities and then choose the **Tangent** button from the **Add Relations** rollout.

Concentric

The **Concentric** constraint makes a selected arc or circle to share the same center point with other arc, circle, point, vertex, or circular edge. To apply this constraint, invoke the **Add Relations PropertyManager**. Select the required entity to apply the **Concentric** constraint and then choose the **Concentric** button from the **Add Relations** rollout.

Equal

The **Equal** constraint makes the selected lines to have equal length and the selected arcs, circles, or arc and circle to have equal radii. To apply this constraint, invoke the **Add Relations PropertyManager**. Select the required entity to apply the **Equal** constraint and choose the **Equal** button.

Intersection

The **Intersection** constraint makes a selected point to move at the intersection of two selected lines. To apply this constraint, invoke the **Add Relations PropertyManager**. Select the required entity to apply the **Intersection** constraint. Choose the **Intersection** button from the **Add Relations** rollout.

Coincident

The **Coincident** constraint makes a selected point to be coincident with a selected line, arc, circle, or ellipse. To apply this constraint, invoke the **Add Relations PropertyManager**. Select the required entity to apply the **Coincident** constraint. Choose the **Coincident** button from the **Add Relations** rollout.

Midpoint

The **Midpoint** constraint makes a selected point to move to the midpoint of a selected line. To apply this constraint, invoke the **Add Relations PropertyManager**. Select the point and the line to which the midpoint constraint has to be applied. Choose the **Midpoint** button from the **Add Relations** rollout.

Symmetric

The **Symmetric** constraint makes two selected lines, arcs, points, and ellipses to remain equidistant from a centerline. This constraint also makes the entities to have the same orientation. To apply this constraint, invoke the **Add Relations PropertyManager**. Select the required entity to apply the **Symmetric** constraint and select a center line. Choose the **Symmetric** button from the **Add Relations** rollout.

Fix

The **Fix** constraint makes the selected entity to be fixed at the specified position. If you apply this constraint to a line or an arc, its location will be fixed but you can change its size by dragging the endpoints. To apply this constraint, invoke the **Add Relations PropertyManager**. Select the required entity and choose the **Fix** button.

Merge

The **Merge** constraint makes two sketch points or endpoints to merge in a single point. To apply this constraint, invoke the **Add Relations PropertyManager**. Select the required entities to apply the **Merge** constraint and choose the **Merge** button from the **Add Relations** rollout.

Pierce

The **Pierce** constraint makes a sketch point or an endpoint of an entity to be coincident with an entity of another sketch. To apply this constraint, invoke the **Add Relations PropertyManager**. Select the required entities to apply the **Pierce** constraint and choose the **Pierce** button from the **Add Relations** rollout.

Torsion Continuity

The **Torsion Continuity** constraint makes two curves (splines and arcs) in continuity. To apply this constraint, select the two connected splines or arcs and click on the torsion continuity button; refer to Figure-108.

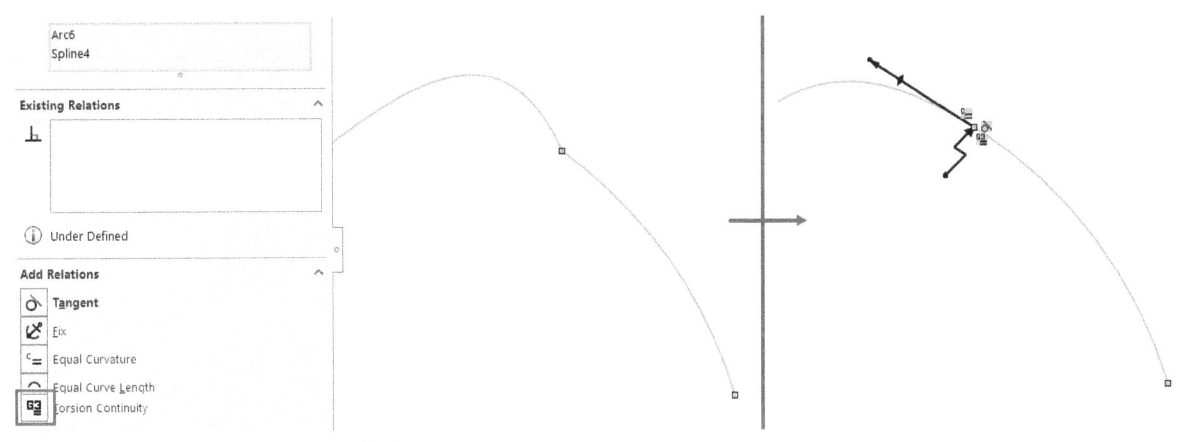

Figure-108. Torsion continuity applied

Fully Defined Sketch

Fully defined sketch is the one whose specifications can not be changed unintentionally. In some complex sketches, when you change dimension of one entity, the dimension of other entity gets changed automatically.

If you have the sketch fully defined then the dimension of the entities will not change unintentionally. In technical terms, a fully defined sketch is the one in which entities have zero degree of freedom. In SolidWorks, the sketch that is fully defined will be displayed in bold black color; refer to Figure-109.

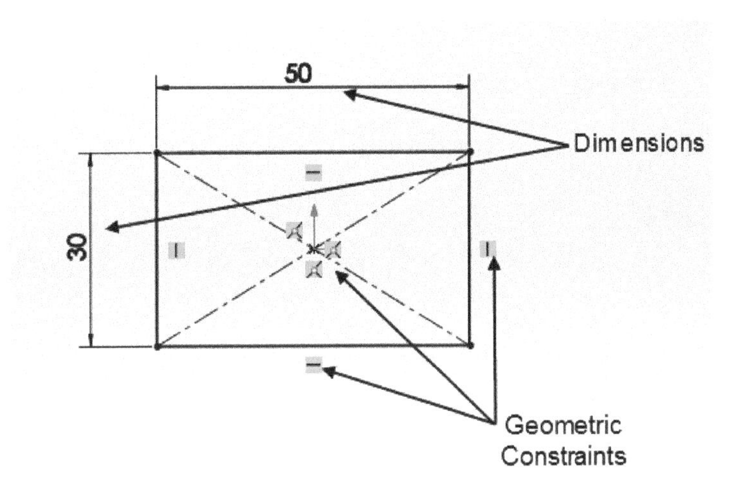

Figure-109. A_fully_defined_sketch

You can fully define a sketch either by manually applying the dimensions and geometric constraints or you can do it by using the **Fully Define Sketch** button. This button is available in the **Display/Delete Relations** drop-down. The procedure to use this tool is discussed next.

- Click on the **Fully Define Sketch** button in the **Display/Delete Relations** drop-down of the **Ribbon**. The **Fully Define Sketch PropertyManager** will be displayed as shown in Figure-110. Note that all the rollouts in the **PropertyManager** are expanded.

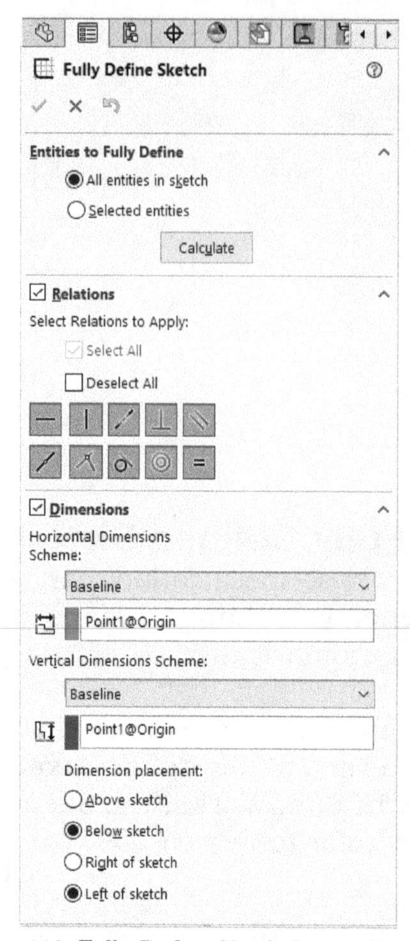

Figure-110. Fully_Define_Sketch_PropertyManager

- Select the type of dimensions and constraints that you want to use while making a sketch fully defined.
- Click on the **Calculate** button and then click on **OK** button to fully define the sketch. The dimensions and constraints will be automatically applied. Now, change the values as per your requirement; refer to Figure-111.

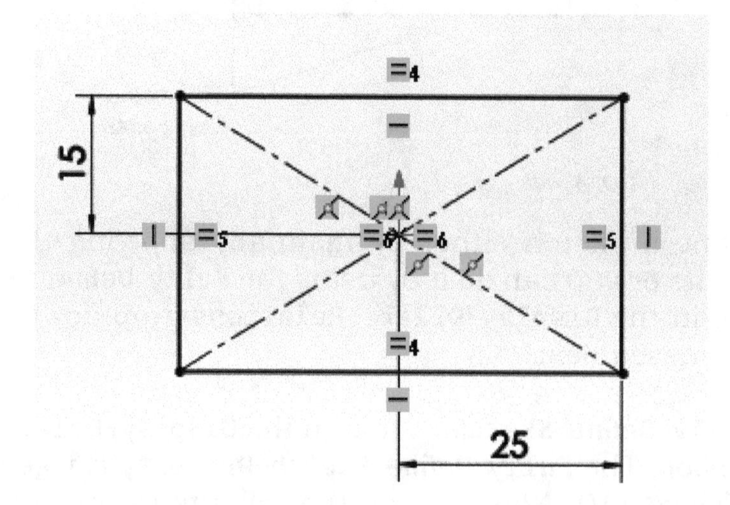

Figure-111. Fully_defined_sketch

If you add dimensions or constraints that are more than required then such sketches are called over defined sketches. The dimensions, constraints, and sketches that are over defined are displayed in yellow color; refer to Figure-112. In such cases, you need to delete the conflicting dimensions or make them driving. You can also use the **SketchXpert** to do the modifications. The procedure is given next.

- Click on the **Over Defined** message in the **Information Bar** at the bottom of the viewport. Refer to Figure-112.

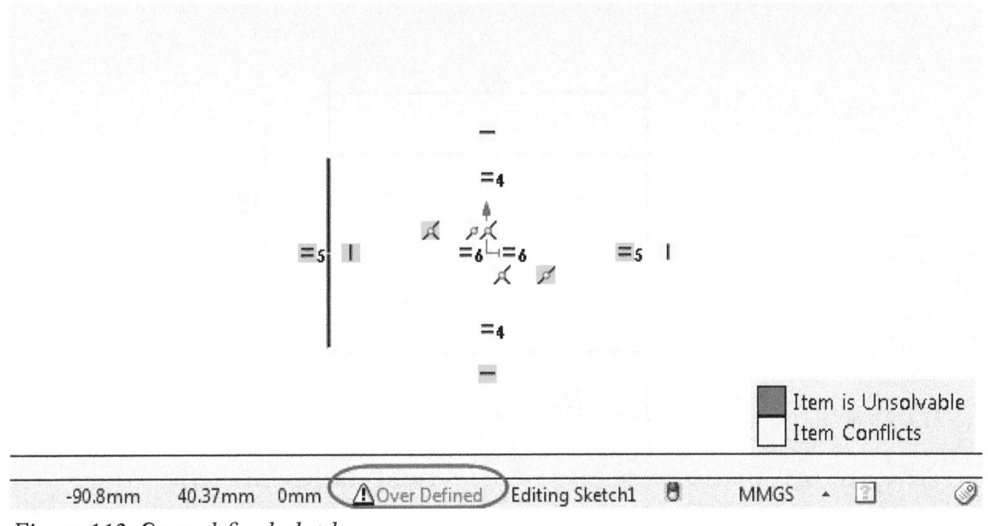

Figure-112. Over_defined_sketch

- The **SketchXpert PropertyManager** will be displayed as shown in Figure-113.

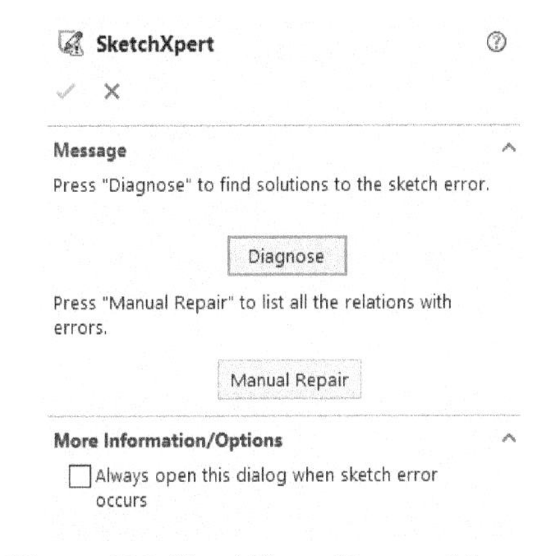

Figure-113. SketchXpert_PropertyManager

- Click on the **Diagnose** button to find automatic solutions. The modified **SketchXpert PropertyManager** will display and interfering dimension/constraint will be displayed crossed; refer to Figure-114.

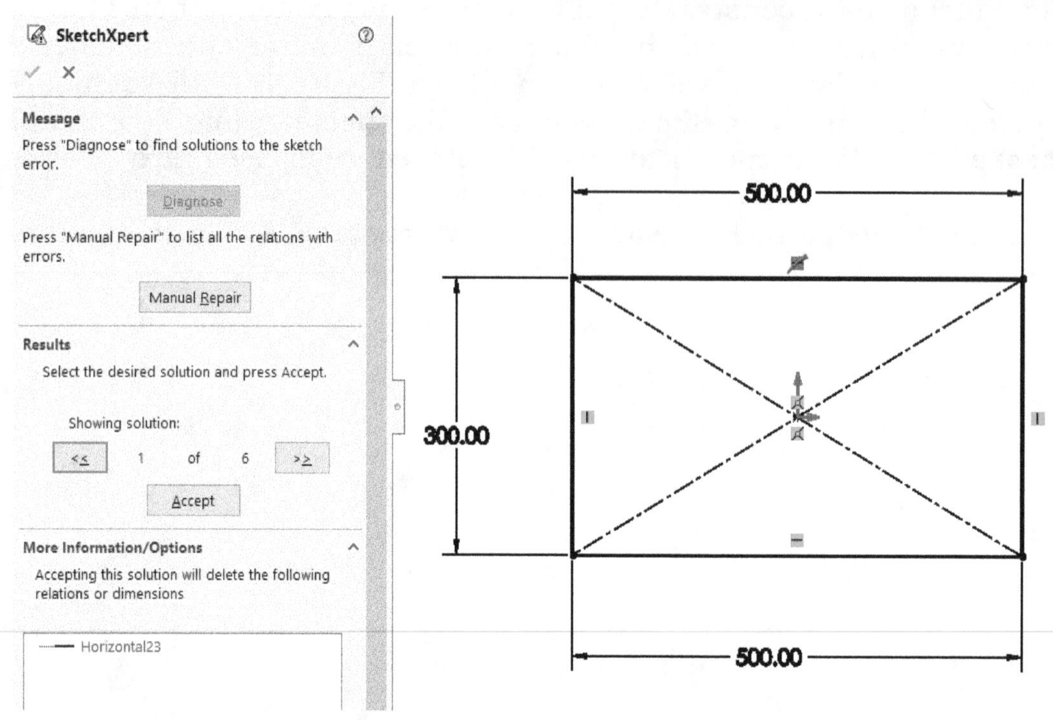

Figure-114. Diagnosed_sketch

- Click on the arrow button to check the next solution and once you find the appropriate solution, click on the **Accept** button from the **Results** rollout.
- Click on **OK** button from the **PropertyManager** to accept the changes.

Till this point, we have completed the 2D sketching of SolidWorks. In next chapter, we will practice on the tools discussed till now using some examples.

SELF ASSESSMENT

Q1. Discuss about Planes and their placement in 3D space in the classroom. Use the example of a closed room to define different planes.

Q2. Which of the following is not the default plane available in SolidWorks?

a. Front Plane
b. Right Plane
c. Top Plane
d. Left Plane

Q3. Which of the following is selected by default in the **Insert Line PropertyManager** while creating a line?

a. As sketched
b. Horizontal
c. Vertical
d. Angle

Q4. How many tools are available in SolidWorks to create a rectangle?

a. 3
b. 6
c. 5
d. 4

Q5. Which of the following tool is not available in SolidWorks to create an arc?

a. Tangent Arc
b. 3 Point Arc
c. Centerpoint Arc
d. 2 Point and Angle Arc

Q6. If you select the **Circumscribed circle** radio button while creating polygon then, the polygon will be drawn outside the circle. (T/F)

Q7. If you want to create the polygon inside the circle then select the **Circumscribed circle** radio button from the **PropertyManager**. (T/F)

Q8. The **Equation Driven Curve** tool is available in the...........drop-down.

Q9. Discuss the use of **Trim away inside** button in the **Trim PropertyManager**.

Q10. Discuss the difference between Linear pattern and circular pattern.

Q11. Discuss the use of **Smart Dimension** tool with example.

Q12. What is the difference between dimensional constraining and geometrical constraining?

FOR STUDENT NOTES

Answer to Self-Assessment:

2. d, **3.** a, **4.** c, **5.** d, **6.** F, **7.** T, **8.** Spline

Chapter 3

Advanced Dimensioning and Practice

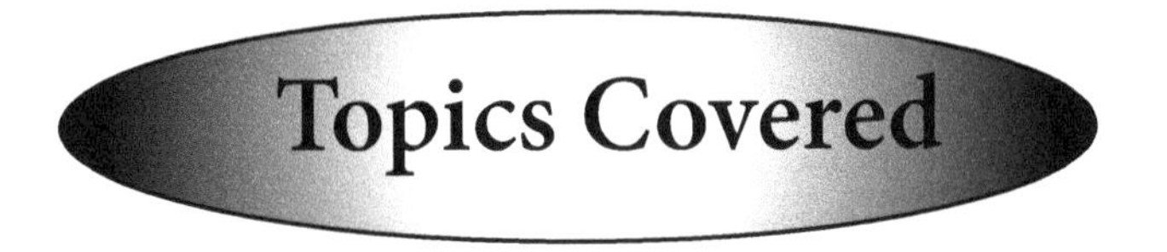

The major topics covered in this chapter are:

- ***Dimensioning and its relations with drawing***
- ***Dimension Style***
- ***Practical 1***
- ***Practical 2***
- ***Practice Drawings***

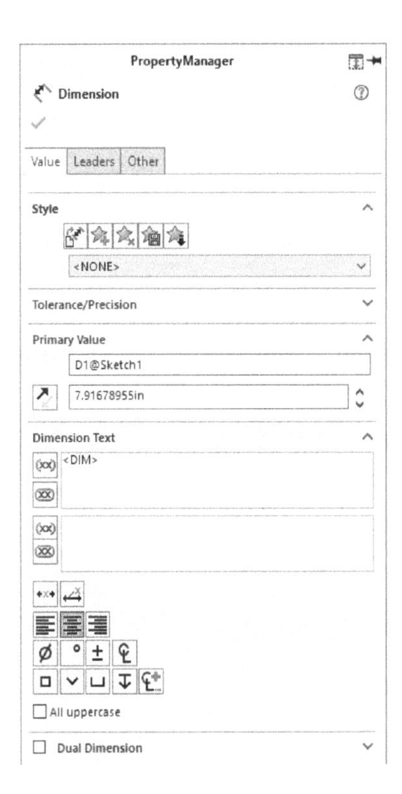

DIMENSIONING AND ITS RELATIONS

When we create a dimension in a sketch it is not confined to that sketch only. You will learn later that it also affects the parameters in 3D model and drafting environment. At that time, the style and dimensions that we apply in sketch will be reflected in the drawing. So, it is very important to understand dimension styles here. In the previous chapters, we have worked on basic dimensions. In this chapter, we will discuss the dimensions and styles in detail.

DIMENSION STYLE

Select a dimension that you have applied in the sketch. The **Dimension PropertyManager** will be displayed; refer to Figure-1.

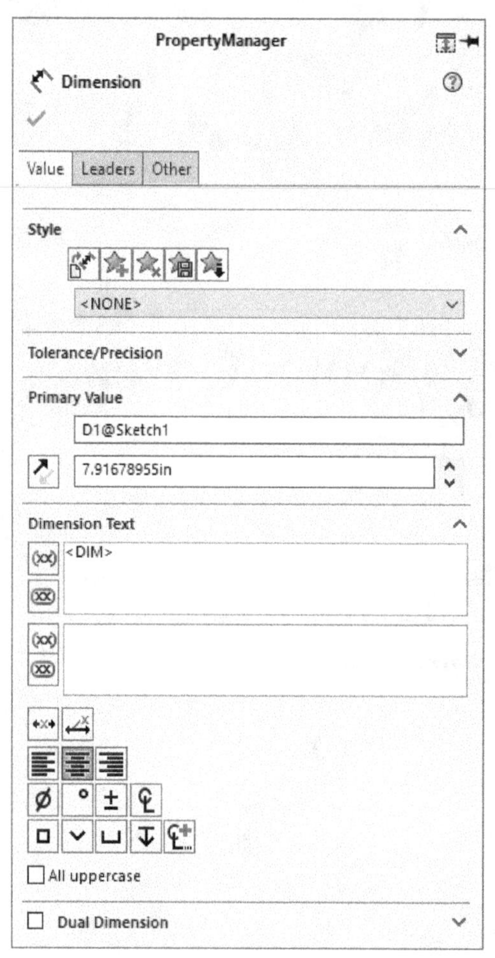

Figure-1. Dimension_PropertyManager

The options in the **PropertyManager** are discussed next.

Style Rollout

The **Style** rollout is used to create, save, delete, and retrieve the dimension style in the current document; refer to Figure-2. You can also use the dimension styles saved in other documents using this rollout. The options in this rollout are discussed next.

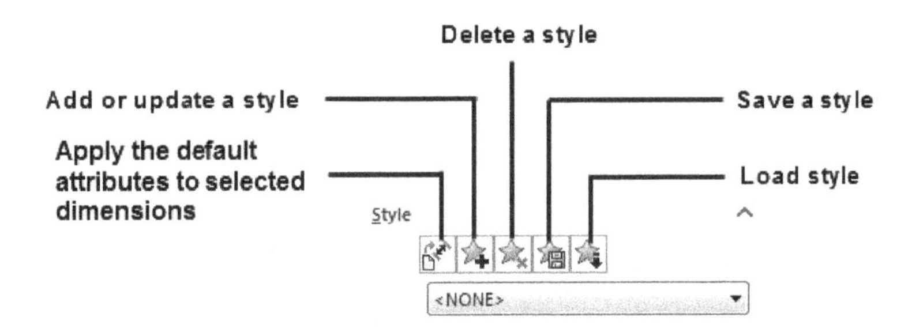

Figure-2. The_Style_rollout

Apply the default attributes to selected dimensions

The **Apply the default attributes to selected dimensions** button is used to apply the default attributes to the selected dimension or dimensions. The attributes can be dimension text, tolerance, precision, arrow style, and so on.

Add or Update a Style

The **Add or Update a Style** button is used to add a dimension style to the current document for a selected dimension. After invoking the **Dimension PropertyManager**, set the attributes using various options provided in this **PropertyManager**. Next, choose the **Add or Update a Style** button. The **Add or Update a Style** dialog box will be displayed, as shown in Figure-3. Enter the name of the dimension style in the edit box and press **ENTER**; the dimension style will be added to the current document.

You can apply a new dimension style to the selected dimension by selecting a dimension style from the **Set a current Style** drop-down list in the **Style** rollout. You can also update a dimension style. To do so, select the dimension and set the options of the dimension style according to your need. Next, choose the **Add or Update a Style** button to invoke the **Add or Update a Style** dialog box. Select the dimension style to update from the drop-down list in the dialog box; the two radio buttons in this dialog box will be enabled. Select the **Update all annotations linked to this Style** radio button and choose the **OK** button to update all the dimensions linked to the selected **Style**. If you select the **Break all links to this Style** radio button and choose the **OK** button, then the link between the other dimensions having the same style and the selected **Style** will be broken.

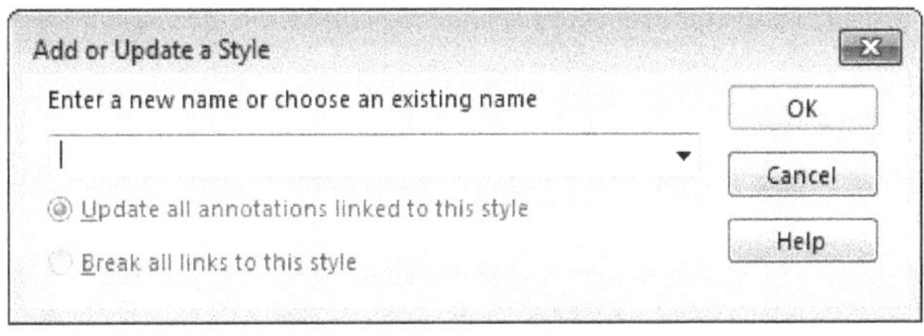

Figure-3. Add or update a style dialog box

Delete a Style

The **Delete a Style** button is used to delete a dimension style. Select a dimension style from the **Set a current Style** drop-down list and then choose the **Delete a Style** button.

Save a Style

The **Save a Style** button is used to save a dimension style so that it can be retrieved in some other document. Select a dimension style from the **Set a current Style** drop-down list and choose the **Save a Style** button. The **Save As** dialog box will be displayed. Browse to the folder in which you want to save the style and enter its name in the **File name** edit box. Choose the **Save** button from the **Save As** dialog box. The style file will be saved with the extension **.sldstl.**

Load Style

The **Load Style** button is used to open a saved style in the current document. The properties of that favorite will be applied to the selected dimension. To load a style, choose the **Load Style** button to invoke the **Open** dialog box. Browse to the folder in which the style is saved. Now, select the file with the extension **.sldstl** and choose the **Open** button; the **Add or Update a Style** dialog box will be displayed. Choose the **OK** button from this dialog box.

Tolerance/Precision Rollout

The **Tolerance/Precision** rollout shown in Figure-4 is used to specify tolerance and precision in dimensions. The options in this rollout are discussed next.

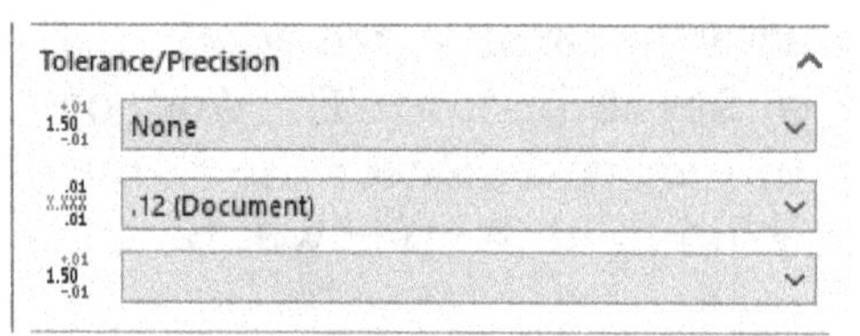

Figure-4. Tolerance_Precision_rollout

Tolerance Type

The **Tolerance Type** drop-down is used to apply tolerance to a dimension. By default, the **None** option is selected. Therefore, no tolerance is applied to the dimensions. The other tolerance types available in this drop-down list are discussed next.

Basic

The basic dimension is the dimension taken as reference for other features. To display the basic dimension, select the dimension that you want to display as the basic dimension and then select the **Basic** option from the **Tolerance Type** drop-down list. On doing so, the dimension is enclosed in a rectangle; refer to Figure-5.

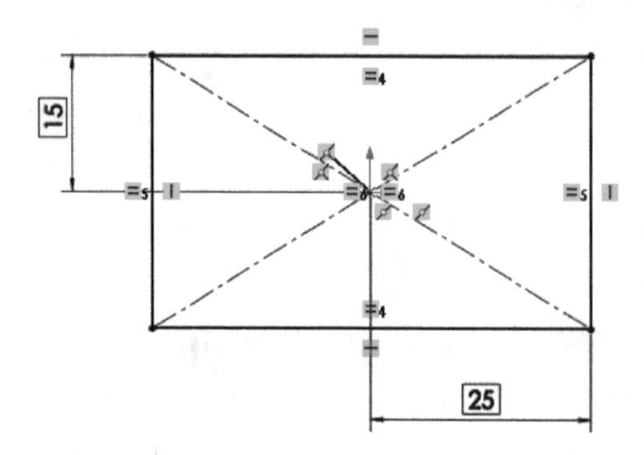

Figure-5. Basic_dimension

Bilateral

The **Bilateral** option is used to display the minimum and maximum limit of tolerance for a dimension. To apply the bilateral tolerance, select the dimension and then select the **Bilateral** option from the **Tolerance Type** drop-down; the **Maximum Variation** and **Minimum Variation** edit boxes will be enabled, where you can enter the maximum and minimum variations for a dimension. Also, the **Show parentheses** check box will be displayed. If you select this check box, the bilateral tolerance will be displayed with parentheses. The dimension with a bilateral tolerance is shown in Figure-6.

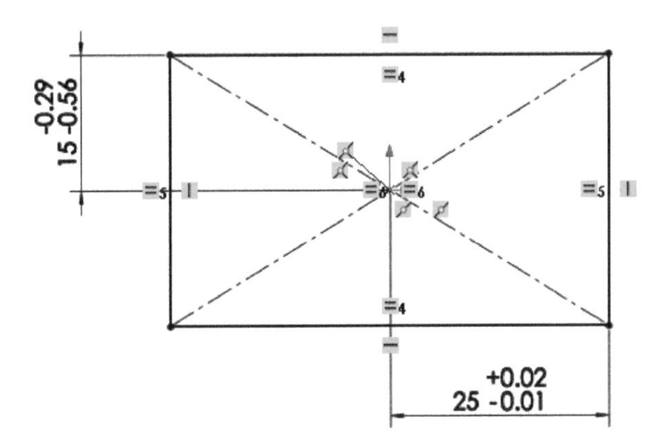

Figure-6. Bilateral_tolerance

Limit

The maximum and minimum permissible dimensional values of an entity are displayed on selecting the **Limit** option. To apply this tolerance type, select the dimension to be displayed as the limit dimension and select the **Limit** option; the **Maximum Variation** and **Minimum Variation** edit boxes will be enabled. Enter the values of the maximum and minimum variations. The dimension along with the limit tolerance is shown in Figure-7.

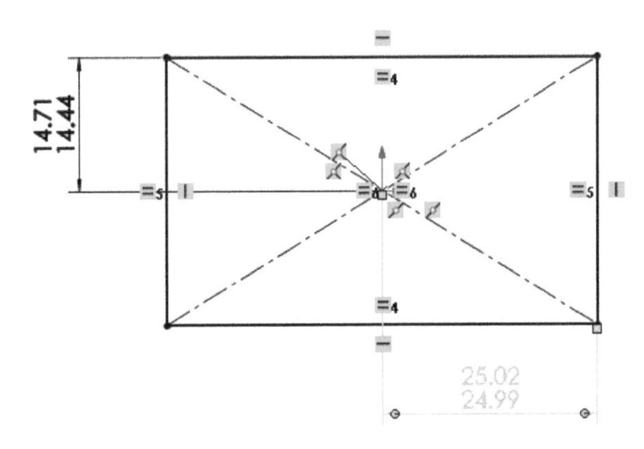

Figure-7. Limit_tolerance

Symmetric

The symmetric tolerance when the variation is equal in both positive and negative direction. To use this tolerance, first select the dimension and then select the **Symmetric** option; the **Maximum Variation** edit box will be displayed. Enter the value of the tolerance in this edit box. Also, you can select the **Show parentheses** check box to show the tolerance in parentheses. The dimension along with the symmetric tolerance is shown in Figure-8.

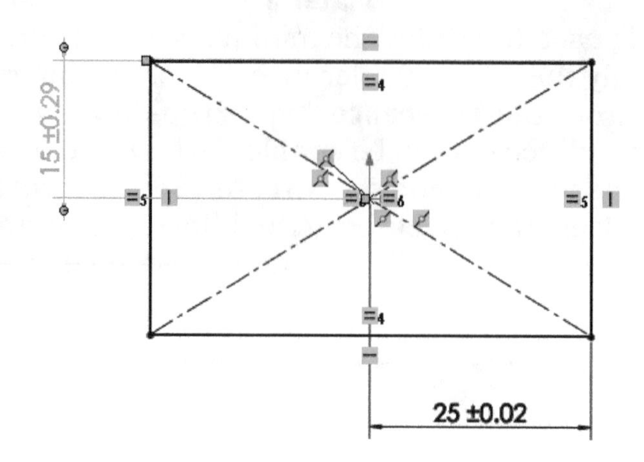

Figure-8. Symmetric_tolerance

MIN

In this dimensional tolerance, the **MIN.** symbol is added to dimension as suffix. This implies that the dimensional value is the minimum value that is allowed in the design. To display this dimensional tolerance, select a dimension and then the **MIN** option from the **Tolerance Type** drop-down list. The dimension along with the minimum tolerance is shown in Figure-9.

MAX

In this dimensional tolerance, the **MAX.** symbol is added to dimension as suffix. This implies that the dimensional value is the maximum value that is allowed in the design. To display this dimensional tolerance, select a dimension and then the **MAX** option from the **Tolerance Type** drop-down list. The dimension along with the maximum tolerance is shown in Figure-9.

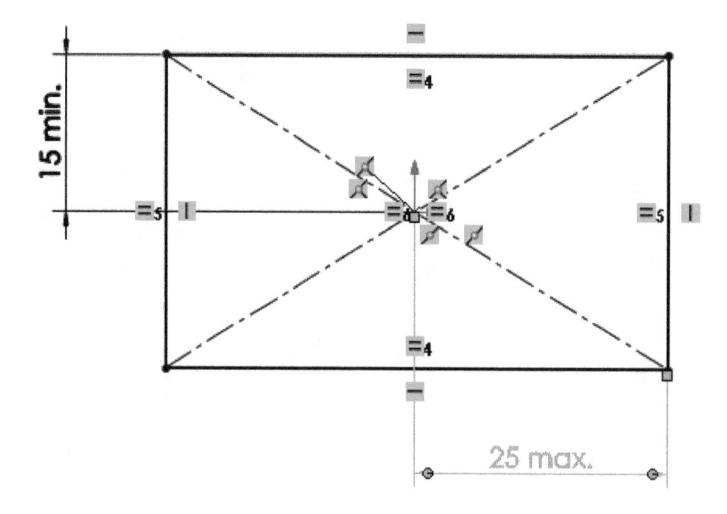

Figure-9. Min_max_tolerance

Fit

The **Fit** option is used to apply fit according to the Hole Fit and Shaft Fit systems. The **Tolerance/Precision** rollout with the **Fit** option selected in the **Tolerance Type** drop-down list is shown in Figure-10. Select the type of fit from the **Classification** drop-down list. The **Classification** drop-down list is used to define the **User Defined** fit, **Clearance** fit, **Transitional** fit, or **Press** fit. To apply a fit using the Hole Fit system or the Shaft Fit system, select the dimension and then the **Fit** option from the **Tolerance Type** drop-down list. The **Classification**, **Hole Fit**, and **Shaft Fit** drop-down lists will be displayed below the **Tolerance Type** drop-down list. Select the required fit from the **Classification** drop-down list and select the fit standard from the **Hole Fit** drop-down list or the **Shaft Fit** drop-down list. If you select the **Clearance**, **Transitional**, or **Press** option from the **Classification** drop-down list and the fit standard from the **Hole Fit** drop-down list, then only the standards matching the selected hole fit will be displayed in the **Shaft Fit** drop-down list and vice versa. However, if you select the **User Defined** option from the **Classification** drop-down list, you can select any standard from the **Hole Fit** and **Shaft Fit** drop-down lists. The **Stacked with line display** button is chosen to display the stacked tolerance with a line. You can also display the tolerance as stacked without a line using the **Stacked without line display** button. If you choose the **Linear display** button, the tolerance will be displayed in the linear form. The dimension along with the hole fit and shaft fit is shown in Figure-11.

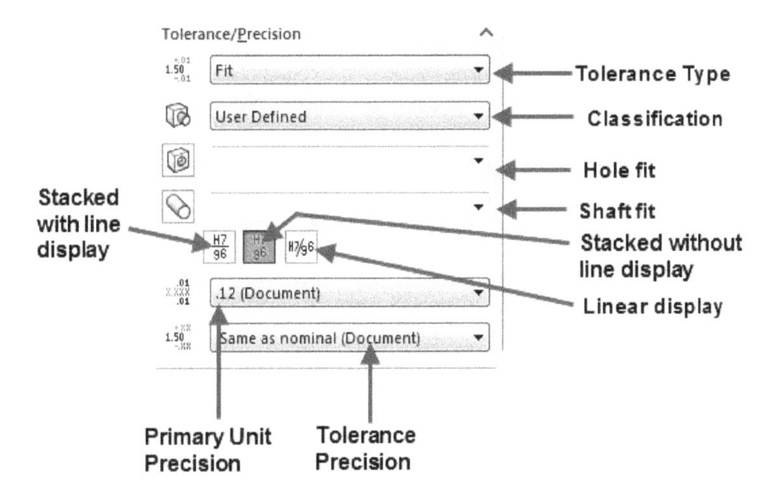

Figure-10. The_Tolerance_Precision_rollout_with_the_Fit_option_selected_in_the_Tolerance_Type_drop-down_list

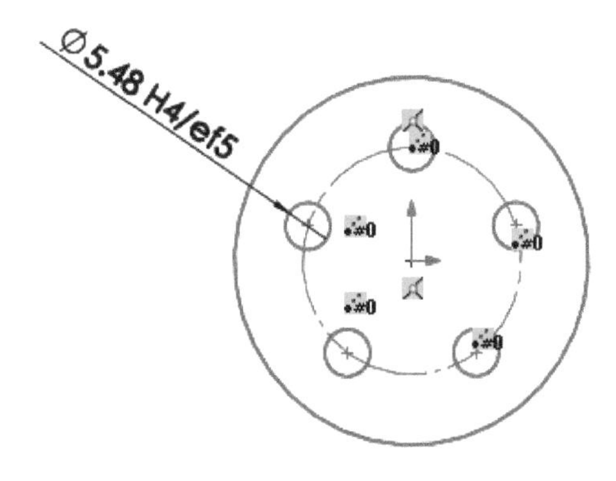

Figure-11. Hole_fit_and_shaft_fit

Fit with tolerance

The **Fit with tolerance** option in the **Tolerance Type** drop-down list is used to display tolerance along with the hole fit and shaft fit in a dimension. To apply fit with tolerance, select a dimension, and then select the **Fit with tolerance** option from the **Tolerance Type** drop-down list. Select the type of fit from the **Classification** drop-down list. Next, select the fit standard from the **Hole Fit** or the **Shaft Fit** drop-down list. Tolerance will be displayed with the fit standard only if you select a fit system from the **Hole Fit** or **Shaft Fit** drop-down list. Tolerance will be displayed along with the fit standard in the drawing area. In SolidWorks, tolerance is calculated automatically, depending on the type and standard of the fit selected. The **Show parentheses** check box can be selected to show tolerance in parentheses. The dimension along with the fit and tolerance is shown in Figure-12.

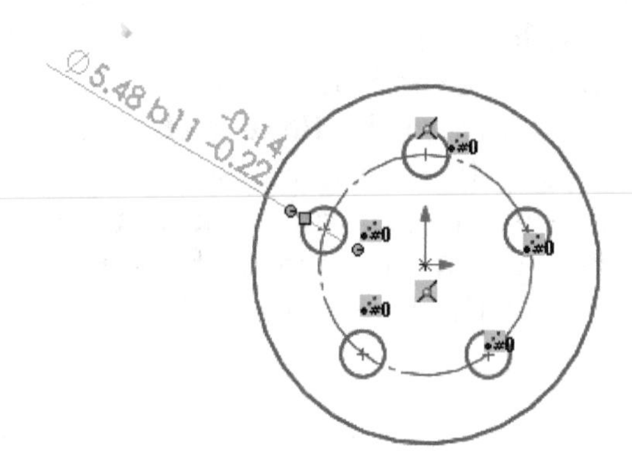

Figure-12. Dimensioning along with the fit and tolerance

Fit (tolerance only)

The **Fit (tolerance only)** option in the **Tolerance Type** drop-down list is used to display the tolerance in a dimension based on the hole fit or shaft fit.

None

The **None** option in the **Tolerance Type** drop-down list is used to display the dimensional value without any tolerance.

Unit Precision

The **Unit Precision** drop-down is used to specify the precision of the number of places after the decimal for dimensions. By default, the selected precision is two places after the decimal.

Tolerance Precision

The **Tolerance Precision** drop-down is used to specify the precision of the number of places after the decimal for tolerance. By default, the selected precision is two places after the decimal. This drop-down list will not be available, if the **None** option is selected in the **Tolerance Type** drop-down list.

Tolerance modifier Rollout

The options in **Tolerance modifier** rollout are used to add symbols and other information with the dimension; refer to Figure-13.

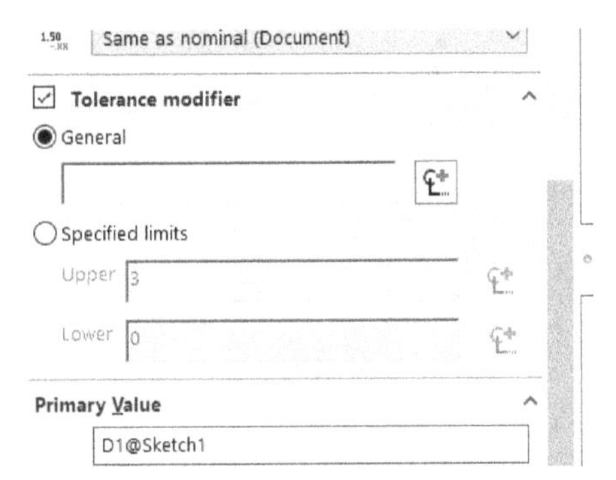

Figure-13. Tolerance modifier rollout

- Select the **General** radio button if you want to specify a common tolerance modifier for both upper and lower limits. Specify desired text in the edit box next to the radio button. If you want to add symbols then click on the **More Symbols** button.
- If you want to specify different tolerance modifier for upper and lower limits then select the **Specified limits** radio button. Specify desired values in respective edit boxes.

Dimension Text Rollout

The **Dimension Text** rollout, as shown in Figure-14, is used to add text and symbols to dimension. The text box in this rollout is used to add text to dimension. The **<DIM>** text displayed in the text box symbolizes the dimensional value. You can add text before or after the dimension value. There are two text boxes in SolidWorks to write text above and below the dimension line. There are four buttons at the left of these text boxes. Choose the **Add Parentheses** button to enclose the dimension text in parentheses. Choose the **Inspection Dimension** button to enclose the dimension text in an obround shape and this dimension will be checked during inspection. Choose the **Center Dimension** button to place the dimension at the center of the dimension line. If you need to place the text at a distance from the dimension line, choose the **Offset Text** button and drag the dimension to the required location. Note that you can also choose all four buttons for a specific dimension.

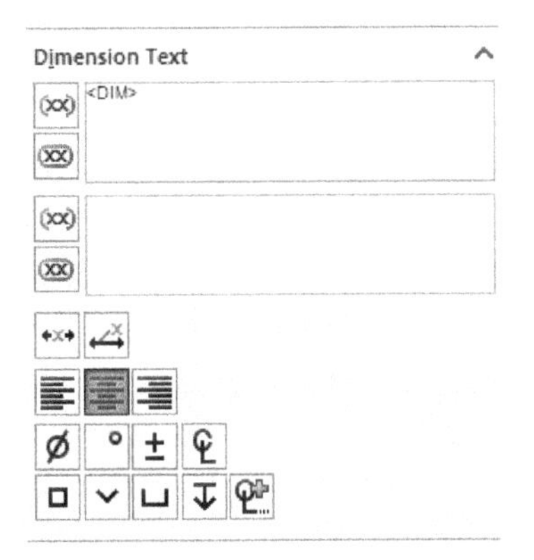

Figure-14. Dimension Text rollout

This rollout also provides buttons to modify text justification and add symbols such as diameter, degree, plus/minus, centerline, and so on to the dimension text. You can add more symbols by choosing the **More Symbols** button from the **Dimension Text** rollout and click on the **More Symbols** button from the flyout displayed. On choosing this button, the **Symbol Library** dialog box will be displayed, as shown in Figure-15.

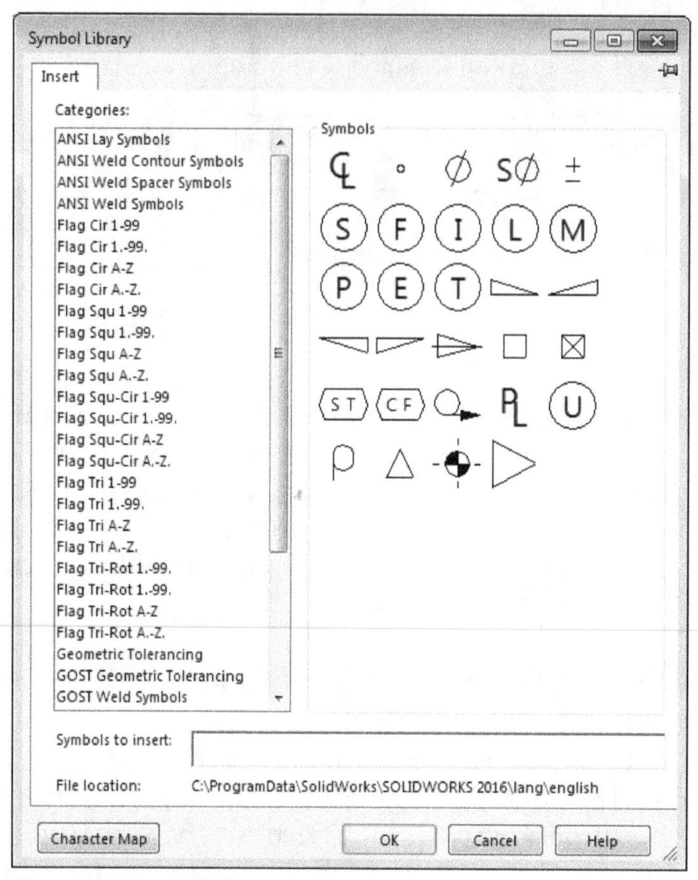

Figure-15. Symbol Library dialog box

Select desired symbol and click on the **OK** button.
Select the **All uppercase** check box to make the dimension text in uppercase.

Dual Dimension Rollout

You need to select the check box in the **Dual Dimension** rollout to enable the options in this rollout, refer to Figure-16. The options in this rollout are used to display the alternative dimension value. Note that the alternative dimension value is displayed in square brackets, as shown in Figure-17. The options in this rollout are similar to those discussed in the earlier sections. Note that the alternative unit is set in the **Dual Dimension Length** cell in the **Document Property - Units** dialog box. To invoke this dialog box, choose **Tools > Options** from the SolidWorks menus; the **System Options** dialog box will be displayed. Choose the **Document Properties** tab; the name of this dialog box will be changed to the **Document Property - Drafting Standard** dialog box. In this dialog box, select the **Units** option from the area on the left to display the options for setting units. Note that on selecting the **Unit** option, the name of this dialog box will be changed to the **Document Property - Units** dialog box.

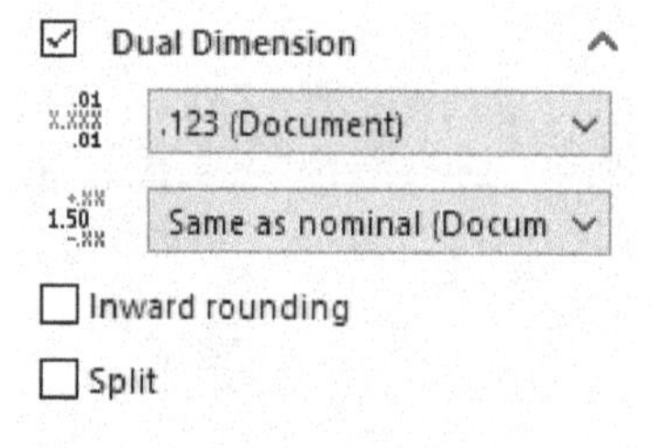

Figure-16. Dual Dimension rollout

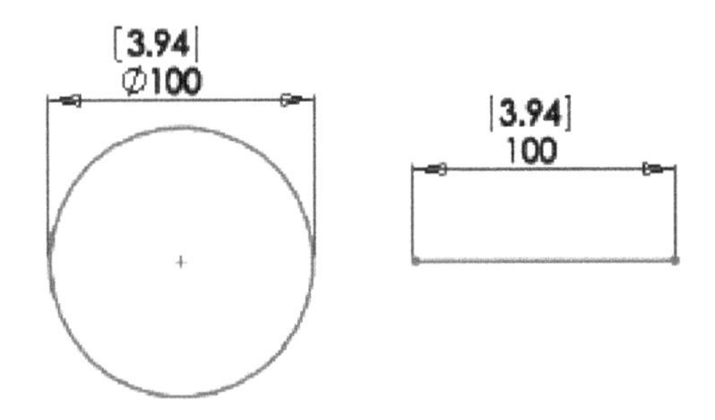

Figure-17. Entities_with_dual_dimension

Sometimes, you may need to change the type of arrowheads or place the dimension at a distance from the entity because of space constraint. In SolidWorks, these actions can be performed by choosing the **Leaders** tab in the **Dimension PropertyManager**. The rollouts in this tab are discussed next.

Witness/Leader Display Rollout

The **Witness/Leader Display** rollout is used to specify the arrowhead style in dimensions, refer to Figure-18. The options in this rollout are discussed next.

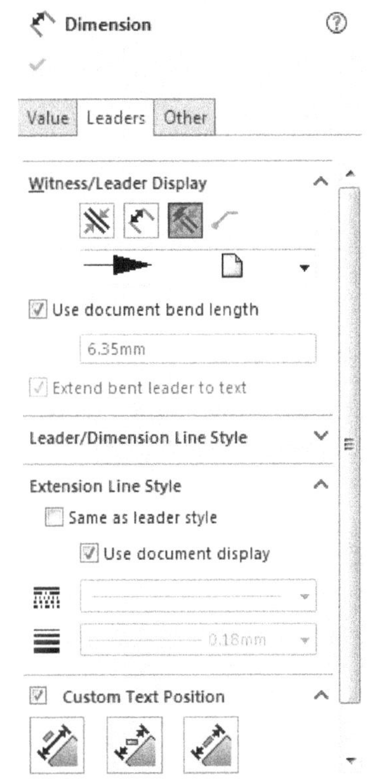

Figure-18. The_rollouts_in_the_Leaders_tab

Outside

The **Outside** button is used to display the arrows outside the extension line. To do so, select a dimension from the drawing area and choose the **Outside** button from the **Witness/Leader Display** rollout.

Inside

The **Inside** button is used to display the arrows inside the extension line. To do so, select a dimension from the drawing area and choose the **Inside** button. You can also click on the control point displayed on the arrowhead to reverse its direction. Note that you can also click on the control point displayed on the arrowhead to reverse the direction of the arrowhead.

Smart

The **Smart** button is chosen by default and the arrows are displayed inside or outside the extension line, depending on the space available between the extension lines.

Directed Leader

This button is chosen to change the leader style of a dimension created on a surface by using the **DimXpert**.

Style

The **Style** drop-down list is used to select the style of the arrowhead. The unfilled triangular arrow is selected by default. You can select any arrowhead style for a particular dimension or dimension style. To change the arrowhead style, select a dimension from the drawing area and then the arrowhead style from the **Style** drop-down list.

Use document bend length

To change the length of a leader line after the bend, clear this check box and specify the length in the edit box below the check box. By default, the value specified in the **Document Properties - Detailing - Dimensions** dialog box will be displayed in the edit box given below this check box.

Leader/Dimension Line Style Rollout

To enable the options in the **Leader/Dimension Line Style** rollout, you need to clear the **Use document display** check box in it. After clearing this check box, the **Leader Style** and **Leader Thickness** drop-down lists will be enabled. The **Leader Style** drop-down list is used to specify the leader style and the **Leader Thickness** drop-down list is used to specify the thickness of the leader. You can also select the **Custom size** option from the **Leader Thickness** drop-down list and specify the thickness of the leader in the spinner available below this drop-down list as per your requirement.

Extension Line Style Rollout

To enable the options in the **Extension Line Style** rollout, you need to clear the **Use document display** check box and **Same as leader style** check box in it. The options in this rollout are same as discussed for **Leader/Dimension Line Style** rollout. These options modify the display of extension lines in dimensions.

Custom Text Position Rollout

The options in this rollout are used to specify the position of the text on a dimension line. Select the check box in the **Custom Text Position** rollout to enable the options in this rollout. The options in this rollout are discussed next.

Solid Leader, Aligned Text

If you choose this button, the leader line will be placed parallel to the dimension line along with the text.

Broken Leader, Horizontal Text

On choosing this button, leader line will be placed parallel to the horizontal axis along with the text.

Broken Leader, Aligned Text

Choose this button to place the dimension line to the center of the text. In this case, the text and the leader line will be placed parallel to the dimension line.

In SolidWorks, you can change the units and font of the dimension text by choosing the **Other** tab. The rollouts in this tab are discussed next.

Override Units Rollout

If you need to change the existing units of the dimension, select the check box in this rollout to expand it and select the units from the **Length Units** drop-down list.

Text Fonts Rollout

The font style set in the **Document Property - Detailing - Annotations Font** dialog box will be the default font style. To change the font style, clear the **Use document font** check box and change the font style by choosing the **Font** button in the **Text Fonts** rollout.

Options Rollout

If you select the **Read only** check box in this rollout, the dimensional value cannot be changed. If you select the **Driven** check box, the value will be the driven value.

HORIZONTAL/VERTICAL DIMENSIONING BETWEEN POINTS

As mentioned earlier, you can add a horizontal or vertical dimension between two points. To add any of these dimensions, choose the required button from the **Dimensions/Relations** toolbar or the **Smart Dimension** flyout in the **Sketch CommandManager**. Select the first point and then the second point. Next, specify a point to place the dimension; the **Modify** dialog box will be displayed. Enter a new dimension value in this dialog box and press **ENTER**.

SKETCH INK

The tools in **Sketch Ink CommandManager** are used to create sketch by using the touch screen. By default, touch is set to right hand but if you are left handed then select **Left Hand** radio button from the **Touch** page in **System Options** dialog box; refer to Figure-19.

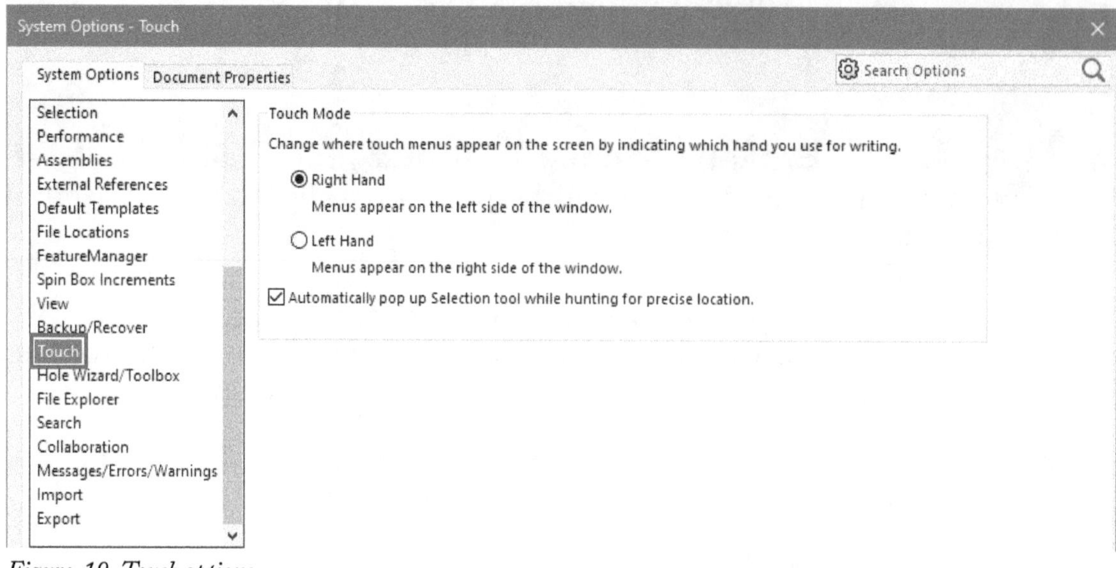

Figure-19. Touch options

Starting Sketch

Click on the **Sketch** tool from the **Sketch Ink CommandManager** and select the plane on which you want to create sketch using touch input.

Now, we will deal with real-world problems related to sketches. The tools of **CommandManager** will become active; refer to Figure-20.

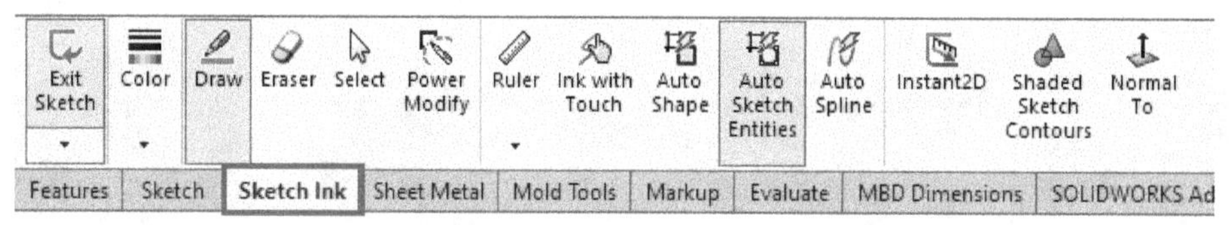

Figure-20. Sketch Ink CommandManager

Selecting the sketch color

The **Color** tool in **Sketch Ink CommandManager** is used to set the ink color and thickness. The procedure to use this tool is discussed next.

- Click on the down arrow below **Color** button in the **CommandManager**. The options to modify color and thickness of pen will be displayed; refer to Figure-21.

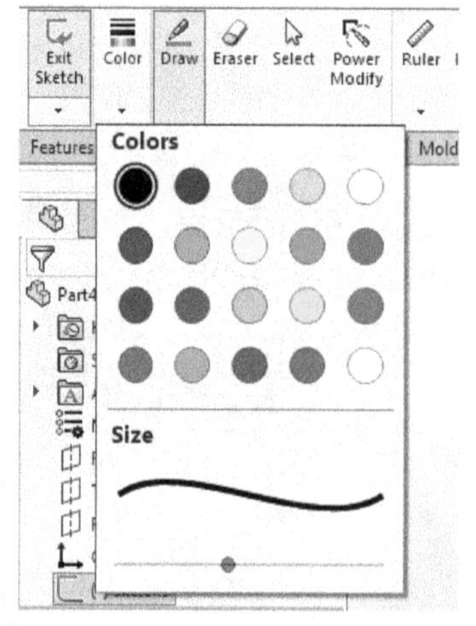

Figure-21. Pen color and size options

- Select desired color and move the slider of **Size** to define the thickness of pen for sketching.

Using Draw tool for Sketching

The **Draw** tool in **Sketch Ink CommandManager** is used to switch to sketching mode for touch input devices. The procedure to use this tool is given next.

- Click on the **Draw** button if not active by default. Using the touch pen or Stylus, create desired sketch on the screen; refer to Figure-22.

Figure-22. Sketch_created_using_touch_pen

- Click on the **Auto Shape** button to activate creation of regular shapes automatically based on sketching by touch devices. So, if you have drawn entity similar to rectangle then a regular rectangle shape will be displayed in place of hand drawn lines; refer to Figure-23.

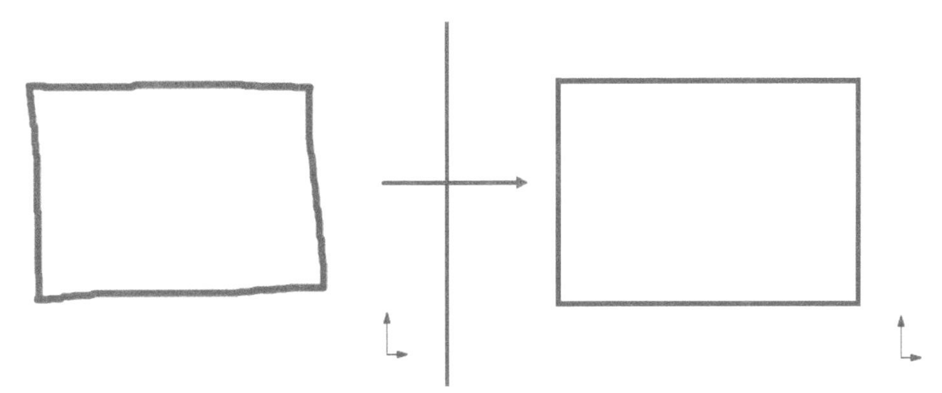

Figure-23. Auto shape generated by hand drawing

- Click on the **Auto Sketch Entities** button from the **CommandManager** to create SolidWorks sketch entities based on drawn sketch. Like if you have drawn shape similar to rectangle then SolidWorks rectangle entity will be generated; refer to Figure-24.

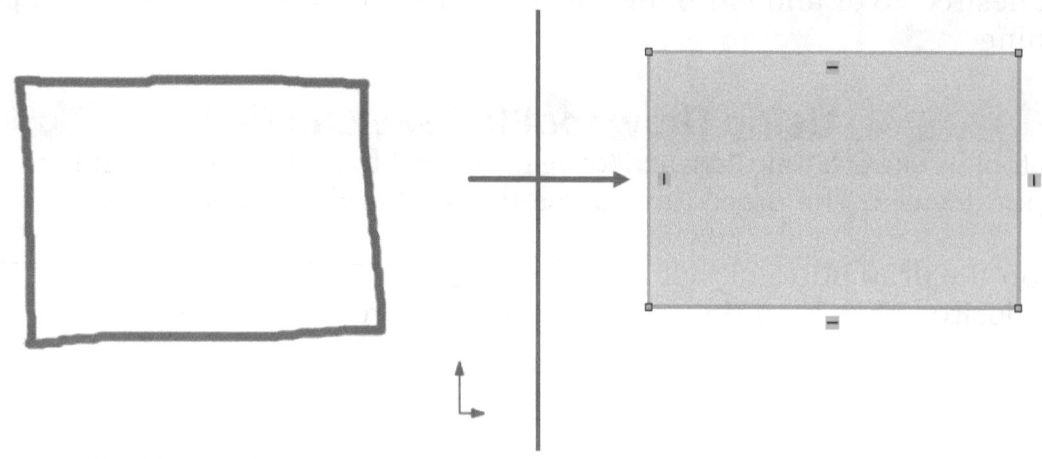

Figure-24. Hand drawn to sketch

- Select the **Auto Spline** button to automatically convert hand drawn irregular curves to splines.
- Select the **Ink with Touch** button to create sketch using finger tips in place of touch pens or stylus.
- You can use the **Ruler** tool from **CommandManager** to create hand drawn entities close to straight; refer to Figure-25. You can drag the ruler at desired location by touch device or by left click drag of mouse. If you want to rotate the ruler then twist the ruler using two fingertips or by moving cursor on ruler and then scrolling up/down.

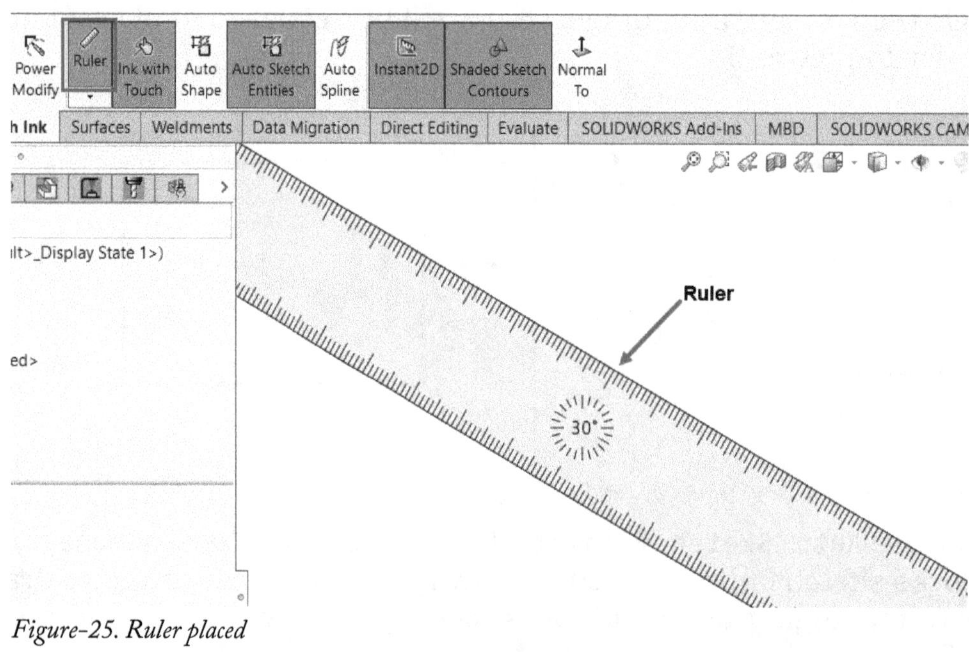

Figure-25. Ruler placed

- Similarly, you can use **Protractor** tool from the **Ruler** drop-down in **CommandManager** to check angles of created entities.

Erasing Sketch Ink

- Click on the **Eraser** tool from the **Sketch Ink CommandManager** to active erase mode.
- Tap/click on the sketch entities to be erased.

Power Modify

The **Power Modify** button is used to activate editing mode for sketch. Various touch gestures for editing are shown in Figure-26.

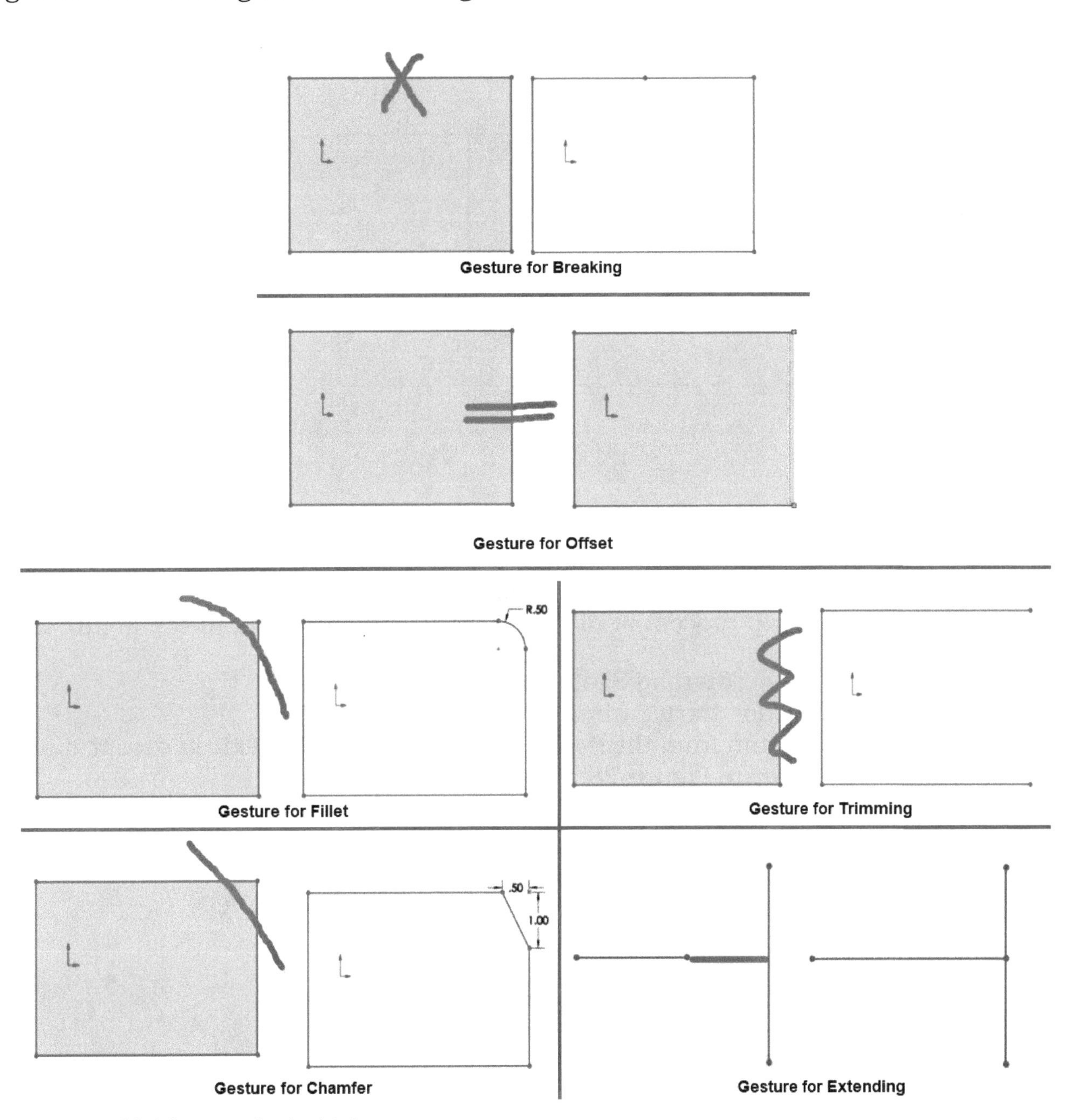

Figure-26. Modify gestures for sketch ink

Practical 1

Create the sketch as shown in Figure-27. Also, dimension the sketch as per the figure.

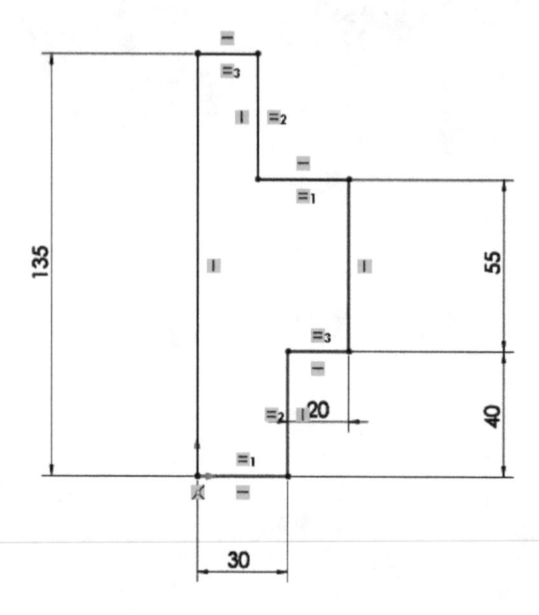

Figure-27. Practical_1

Steps to be performed:

Below is the step by step procedure of creating the sketch shown in the Figure-27.

Starting Sketching Environment

- Start SolidWorks if not started already.
- Click on the **New** button from the **Menu Bar**. The **New SOLIDWORKS Document** dialog box will display; refer to Figure-28.

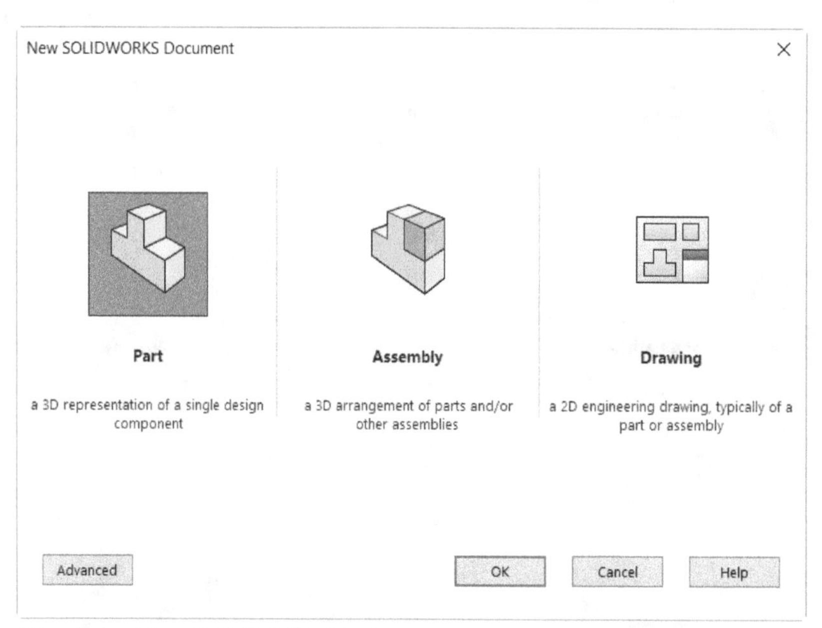

Figure-28. New_SOLIDWORKS_Document_dialog_box

- Double-click on the **Part** button. The Part environment of SolidWorks will be displayed as shown in Figure-29.

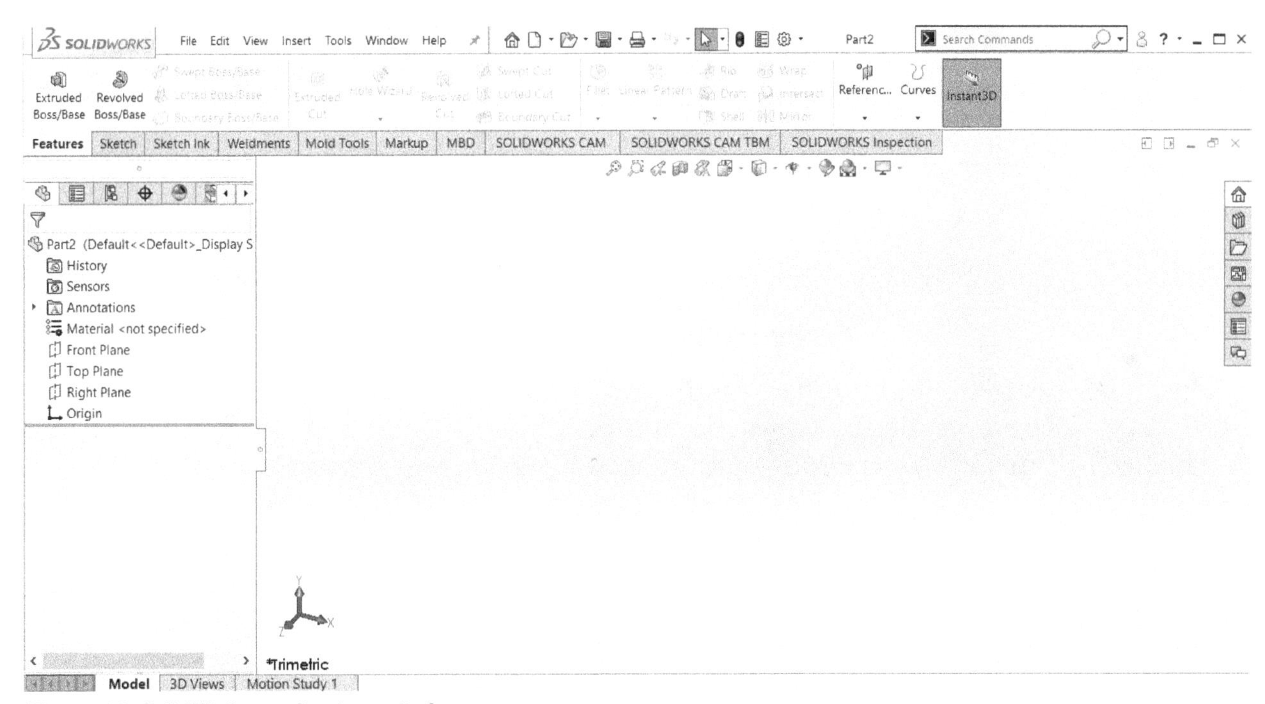

Figure-29. SolidWorks_application_window

- Click on the **Sketch** tab of the **Ribbon**; refer to Figure-29. The tools related to sketch will display in the **Ribbon**.

Starting Sketching

- Click on the **Sketch** button. Three default planes will be displayed.
- Select the **Front** plane from the viewport, refer to Figure-30. The viewport will become parallel to the view screen.

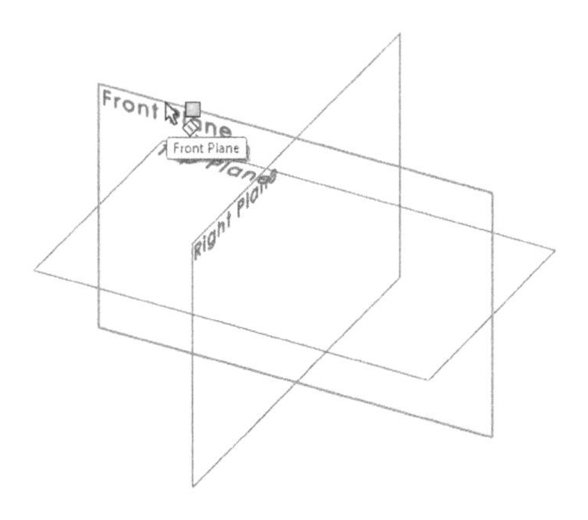

Figure-30. Selecting_front_plane

Creating Lines

- Click on the **Line** button from the **Ribbon**. The **Line** tool will become active and you are asked to select the start point.
- Click on the origin and move the cursor towards right; refer to Figure-31.

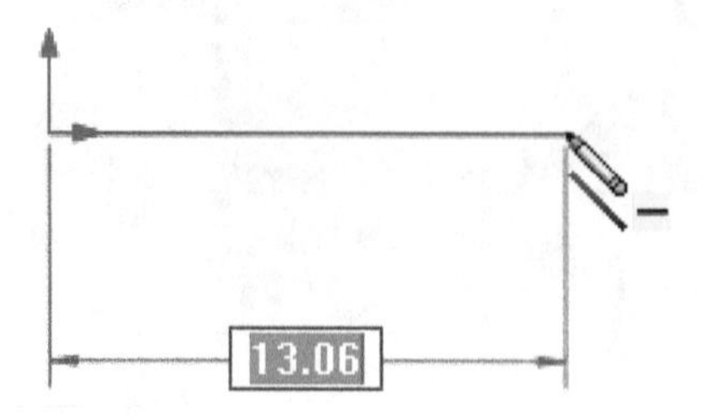

Figure-31. Starting_creation_of_line

- Enter **30** in the dimension box displayed below the line.
- Move the cursor vertically upwards and enter the value **40** in the dimension box.
- Move the cursor towards right and enter **20** in the dimension box. Refer to Figure-32.

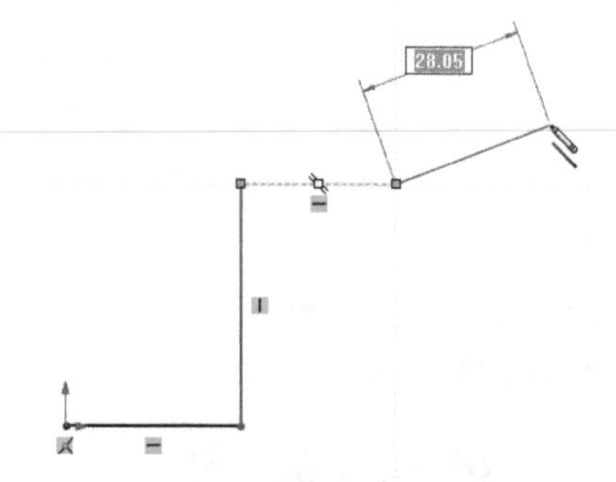

Figure-32. Sketch_after_specifying_20_value

- Move the cursor upward and specify the value as **55** in the dimension box.
- Move the cursor towards left and specify the value as **30** in the dimension box.
- Move the cursor upward and specify the value as **40** in the dimension box.
- Move the cursor towards left and specify **20** in the dimension box.
- Move the cursor downward and click on the origin to close the sketch.

The sketch after performing the above steps is displayed as shown in Figure-33.

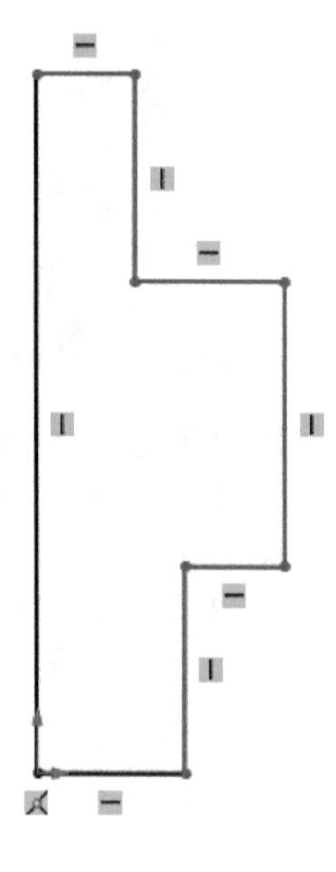

Figure-33. Completed_Sketch

Dimensioning the Sketch

- Click on the **Smart Dimension** button from the **Ribbon**. You are asked to select entities.
- Click on the bottom line joining with the origin and move the cursor downwards; refer to Figure-34.

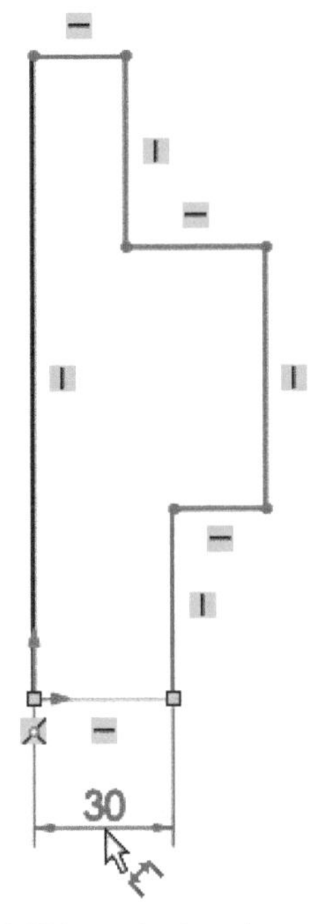

Figure-34. Dimensioning_bottom_line

- Click at the appropriate distance to place the dimension. The **Modify** input box will display.
- Enter the dimension if you want to change. In this case, we will press **ENTER** to apply the default value.
- Click on the vertical line joining the end point of the bottom line selected earlier; refer to Figure-35.

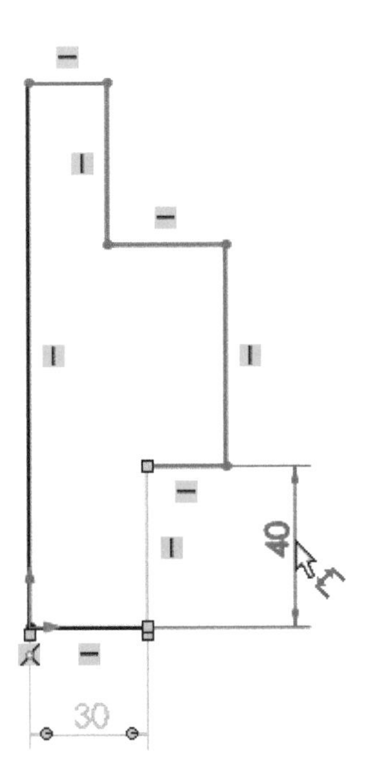

Figure-35. Vertical_dimension_to_selected

- Place the dimension at proper distance from the line and press **ENTER** at the **Modify** input box.
- Similarly, dimension other entities of the sketch.

The sketch after applying all the dimensions will be displayed as shown in Figure-36.

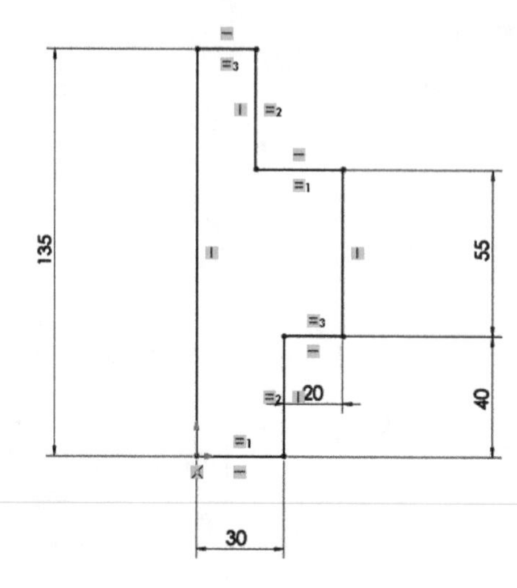

Figure-36. Final_sketch_after_applying_dimensions

Practical 2

In this practical, we will create the sketch as shown in Figure-37.

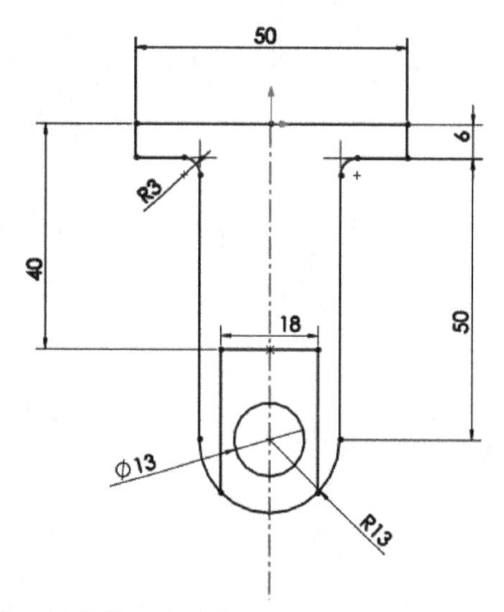

Figure-37. Practical_2

Steps to be performed:

Below is the step by step procedure of creating the sketch shown in the Figure-37.

Starting Sketching Environment

- Start SolidWorks if not started already.
- Click on the **New** button from the Menu Bar. The **New SolidWorks Document** dialog box will display.
- Double-click on the **Part** button. The Part environment of SolidWorks will display.
- Click on the **Sketch** tab of the **Ribbon**. The tools related to sketch will display in the **Ribbon**.

Starting Sketching

- Click on the **Sketch** button from the **Ribbon**. Three default planes will be displayed.
- Select the **Front** plane from the viewport. The viewport will become parallel to the view screen.

Creating Lines

- Click on the **Line** button from the **Ribbon**. You are asked to specify the start point of the line.
- Click on the coordinate system, move the cursor towards right, and enter **25** in the dimension box.
- Move the cursor down perpendicular to the previous line and enter **6** in the dimension box.
- Move the cursor to left and enter **12** in the dimension box.
- Move the cursor downwards and enter **50** in the dimension box. Till this point, our sketch should display like Figure-38.

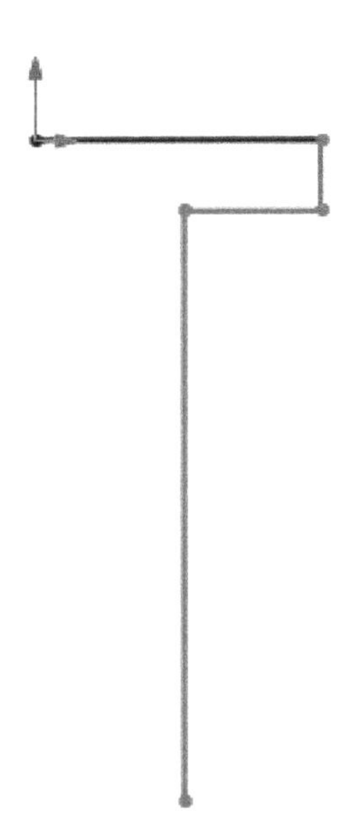

Figure-38. Sketch_after_creating_lines

Creating Arcs

- Click on the down arrow of **Arcs** drop-down and select **Tangent Arc** button from the list. You are asked to click on an end point.
- Click on the end point of vertical line recently created and move the cursor downwards on right until you get the preview as shown in Figure-39.
- When you get the preview like the Figure-39, click to create the arc.

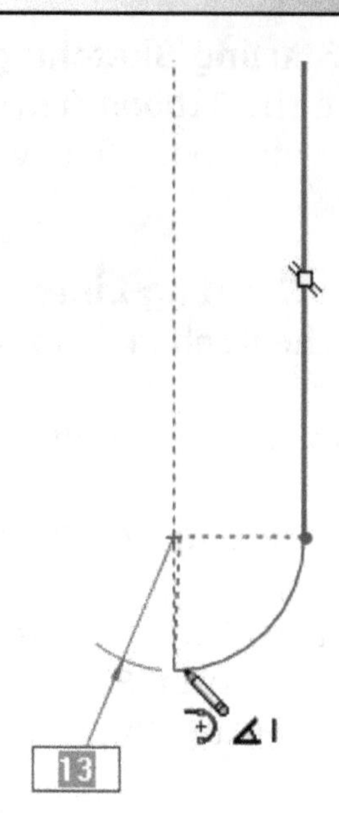

Figure-39. Arc_creation

Creating Fillet

- Click on the **Sketch Fillet** tool from the **Ribbon** . The **Sketch Fillet PropertyManager** will display as shown in Figure-40.

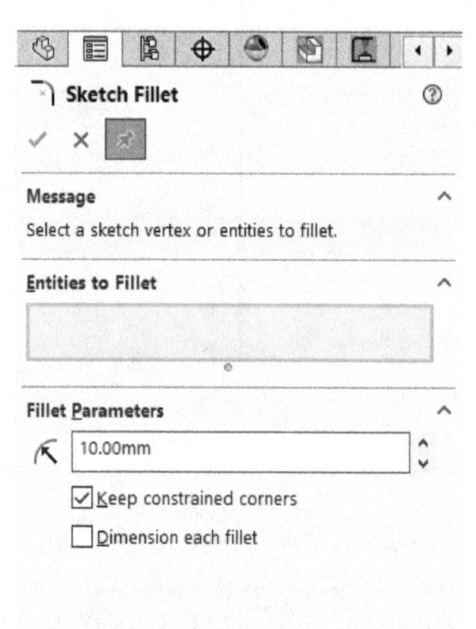

Figure-40. Sketch_Fillet_PropertyManager

- Click in the **Radius** spinner edit box in the **Parameters** rollout of the **PropertyManager** and enter the value as **3**.
- Select the lines as shown in Figure-41 for applying fillet. The fillet will be created between the two lines.
- Click on the **OK** button from the **PropertyManager**.

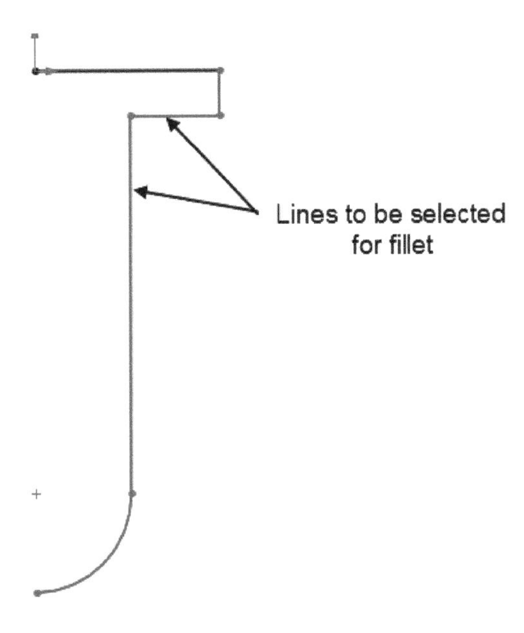

Figure-41. Lines_selection_for_fillet

Creating Mirror Copy

- Click on the **Centerline** tool from the **Line** drop-down. The **Line PropertyManager** will display as shown in Figure-42.

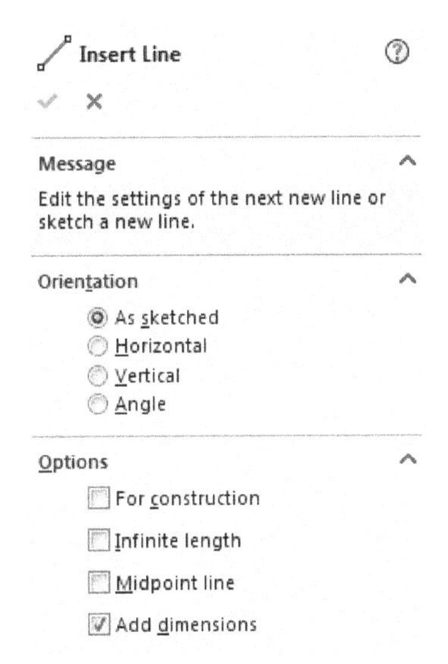

Figure-42. Line_PropertyManager

- Click on the coordinate system (start point of the sketch line) and then the end point of the arc. A center line will be created.
- Press **ESC** from **Keyboard** and exit the tool.
- Select the **Mirror Entities** tool from the **Ribbon** and select all the entities we have sketched except center line; refer to Figure-43.

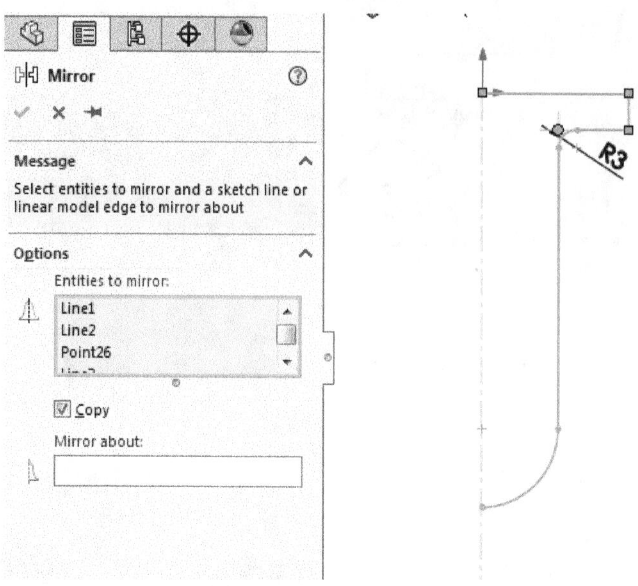

Figure-43. Entities_selected_for_creating_mirror

- Click in the **Mirror about** box and select the center line from the sketch. Preview of mirror will be displayed; refer to Figure-44.

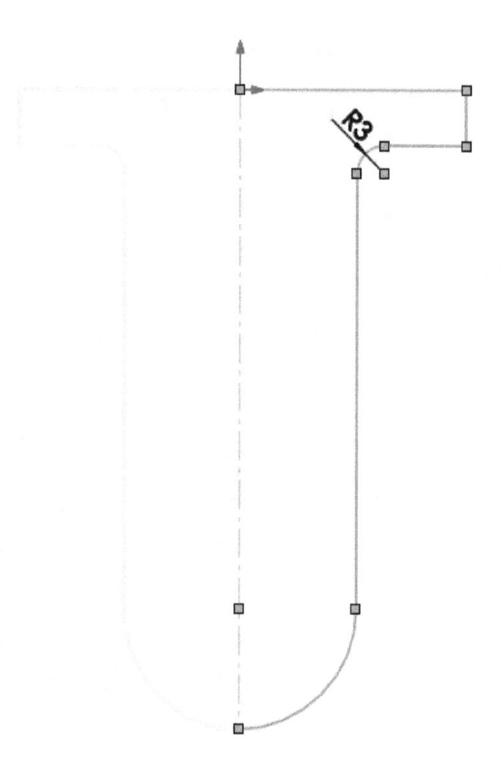

Figure-44. Mirror_preview

- Click on the **OK** button from the **PropertyManager** to create the mirror.

Creating Circle and lines to complete

- Click on the **Circle** tool from the **Circle** drop-down in the **Ribbon**. The **Circle PropertyManager** will be displayed; refer to Figure-45.

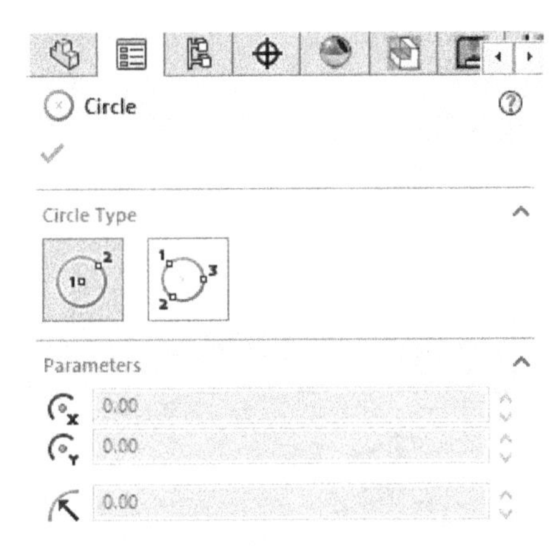

Figure-45. Circle PropertyManager

- Click on the **Diameter dimensions** check box from the **PropertyManager**.
- Click at the center point of the bottom arc and drag the cursor; refer to Figure-46.

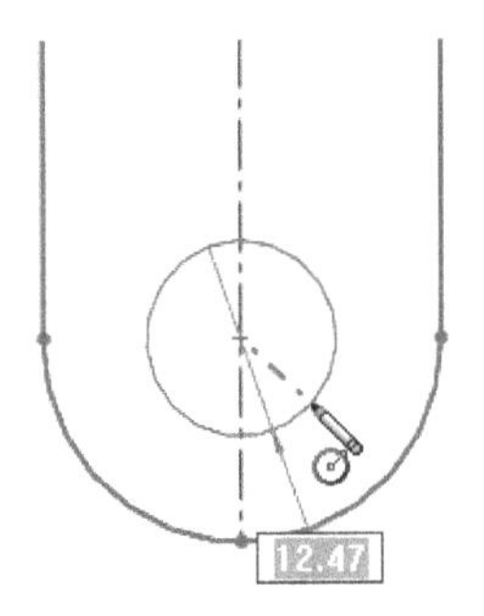

Figure-46. Circle_creation

- Enter the dimension as **13** in the input box.
- Click on the **OK** button from the **PropertyManager** to close it.
- Click on the **Line** tool from the **Ribbon** and click at any point on the center line to specify start point of the line; refer to Figure-47.

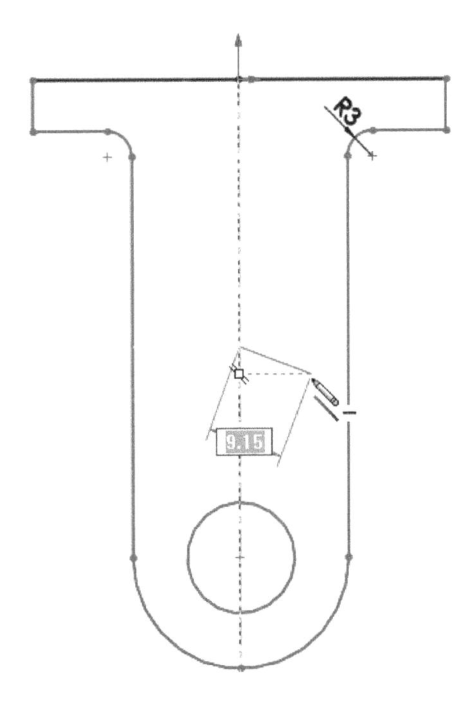

Figure-47. Point_selected_on_centerline

- Move the cursor horizontally towards right and specify the value as 9.
- Move the cursor vertically downwards and click when the cursor is on arc.
- Press **ESC** to exit the tool.
- Mirror both the lines as we did earlier. The sketch should display as shown in Figure-48.

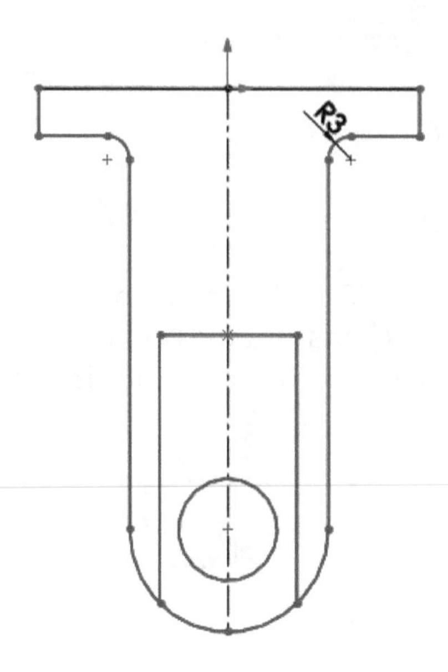

Figure-48. Sketch_after_all_sketching_operations

Dimensioning Sketch

- Click on the **Smart Dimension** tool from the **Ribbon** and select the arc. Dimension will get attached to cursor.
- Place the dimension at proper spacing. Press **ENTER** at the **Modify** input box.
- Click on the circle and place the dimension at proper place. Press **ENTER** at the **Modify** input box.
- Click on the two lines as shown in Figure-49.

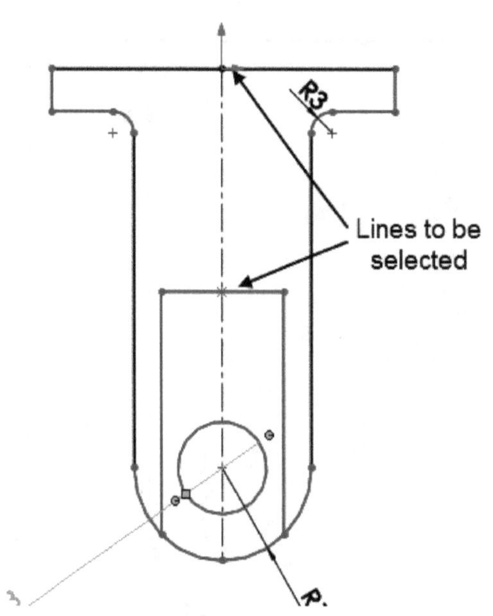

Figure-49. Lines to be selected for dimensioning

- Click to place the dimension at its proper place. In the **Modify** input box, enter the value as **40**.

In the same way, dimension all the entities in the sketch until it is fully defined. The final sketch after dimensioning will be displayed as shown in Figure-50.

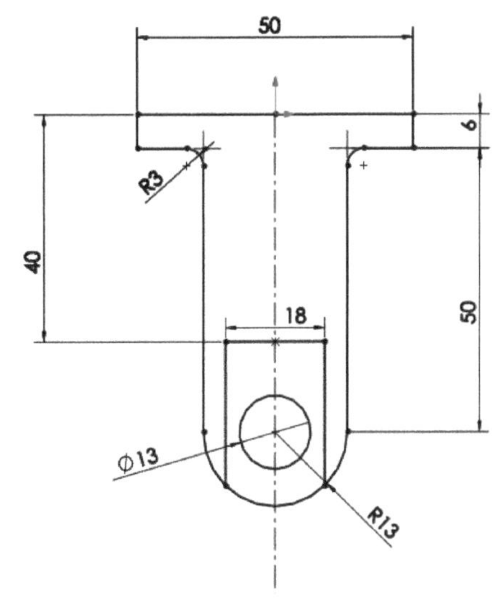

Figure-50. Final_sketch

Practical 3

Create the sketch as shown in Figure-51. Also, dimension the sketch as per the figure.

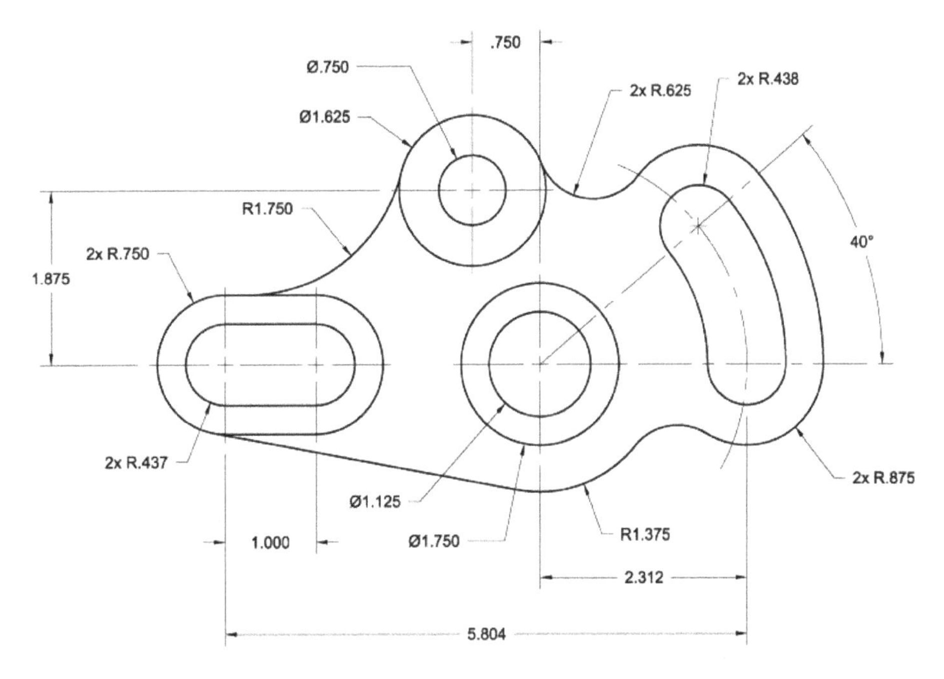

Figure-51. Practical_3

Steps to be performed:

Below is the step by step procedure of creating the sketch shown in the Figure-51.

Starting Sketching Environment

- Start SolidWorks if not started already.
- Click on the **New** button from the Menu Bar. The **New SolidWorks Document** dialog box will display.
- Double-click on the **Part** button. The **Part** environment of SolidWorks will display.
- Click on the **Sketch** tab of the **Ribbon**. The tools related to sketch will display in the **Ribbon**.

Starting Sketching

- Click on the **Sketch** button from the **Ribbon**. Three default planes will be displayed.
- Select the **Top** plane from the viewport. The viewport will become parallel to the view screen.

Creating Circles

- Click on the **Circle** button from the **Circle** drop-down in the **Ribbon**. You are asked to specify the center point of the circle.
- Select the **Diameter dimensions** check box and **Add dimensions** from the **PropertyManager**.
- Click at the center of coordinate system to specify center of the circle.
- Enter the diameter as **1.125** in the edit box displayed.
- Again, click on the center of the coordinate system and specify the diameter value as **1.75** in the edit box.
- Click at the top left of the circles created to specify the center point of other circle; refer to Figure-52.

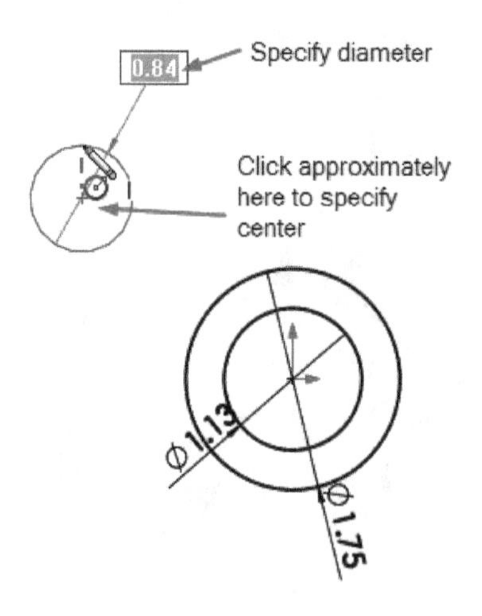

Figure-52. Center_position_for_circle

- Specify the diameter value as **0.75**.
- Click at the center of newly created circle and specify the diameter value as **1.625**.
- Clear the **Diameter dimensions** check box from the **PropertyManager** and click on the center of coordinate system to specify center of a circle.
- Specify the radius of circle as **1.375** in the edit box.

Creating Slots

- Click on the **Centerpoint Arc Slot** button from the **Slot** drop-down. You are asked to specify the center point for construction circle of arc slot.

- Select the **Add dimensions** check box from the **Slot PropertyManager**.
- Click at the center of the coordinate system. Specify the radius of construction circle as **2.312**. You are asked to specify the starting point of the slot arc.
- One by one click at the two positions displayed in Figure-53. Move the cursor and you are asked to specify the width of the slot.

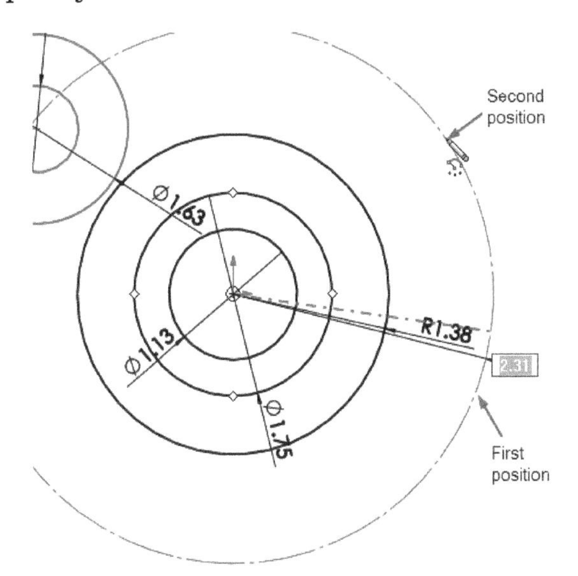

Figure-53. Positions_selected_for_slot_arc

- Click in the screen when the reading of width is approximately **0.9**; refer to Figure-54.

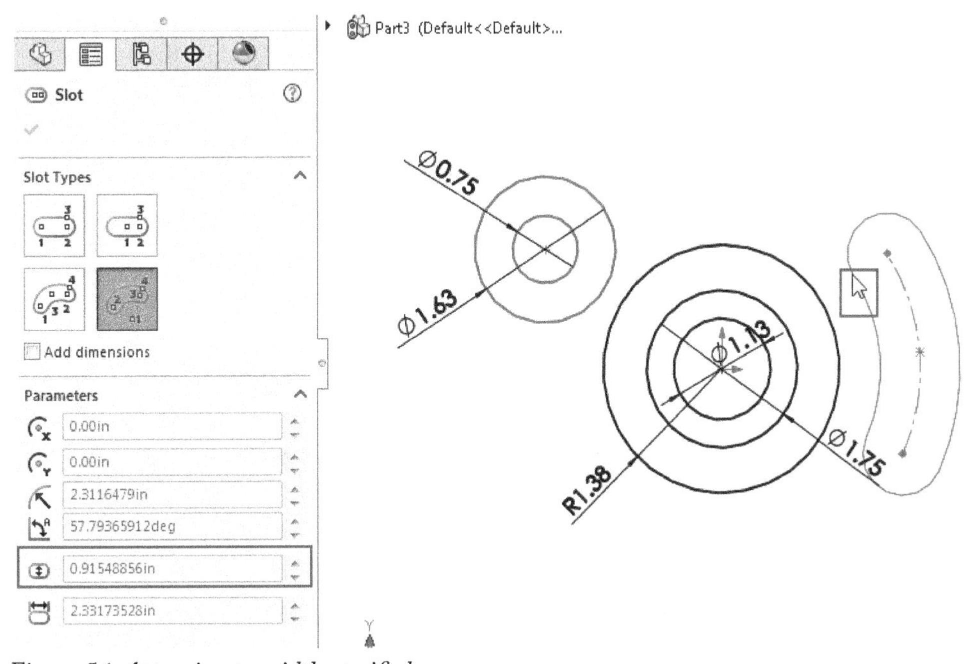

Figure-54. Approximate_width_specified

- Press the **ESC** key from the Keyboard and change the dimensions as shown in Figure-55.

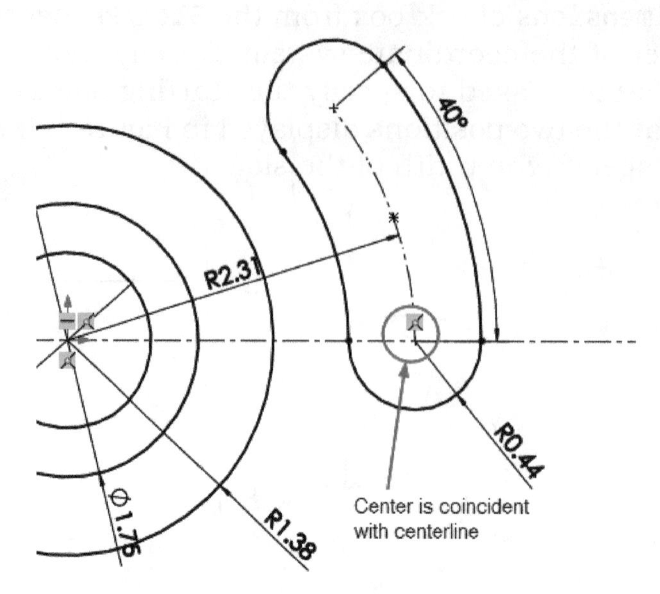

Figure-55. Dimensioned_arc_slot

- Similarly, create other arc slot at the same location but with the corner radius of **0.875**; refer to Figure-56.

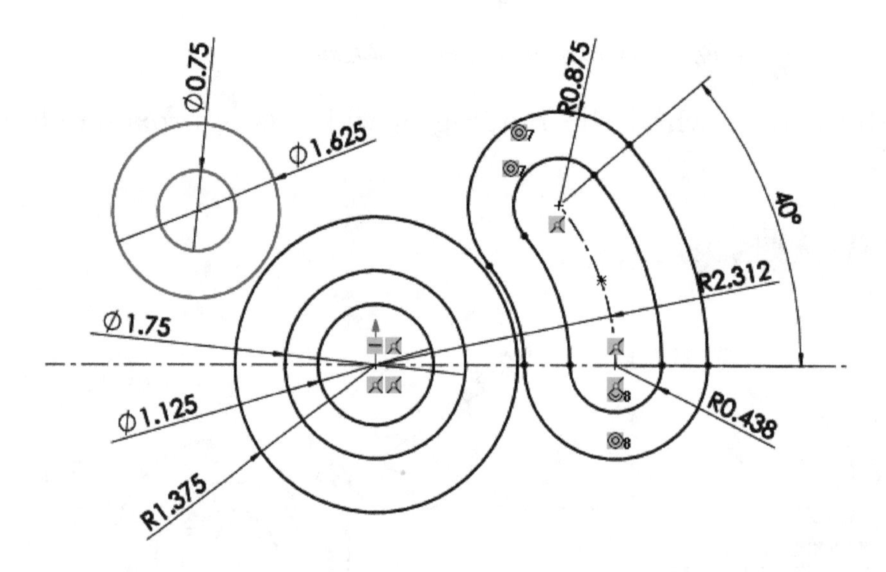

Figure-56. Sketch_after_creating_second_slot

- Click on the **Straight Slot** button from the **Slot** drop-down and create the slots as shown in Figure-57.

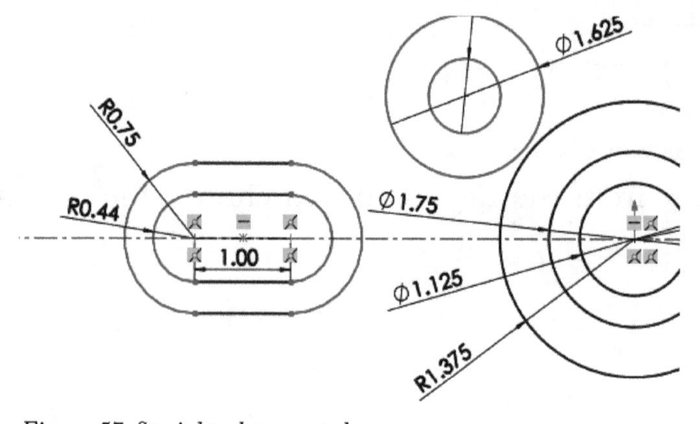

Figure-57. Straight_slots_created

Dimensioning not-dimensioned entities

- Click on the **Smart Dimension** tool and click on the center of circle with diameter **1.625**.
- Next, click on the center of coordinate system. Move the cursor vertically upward and place the dimension; refer to Figure-58.

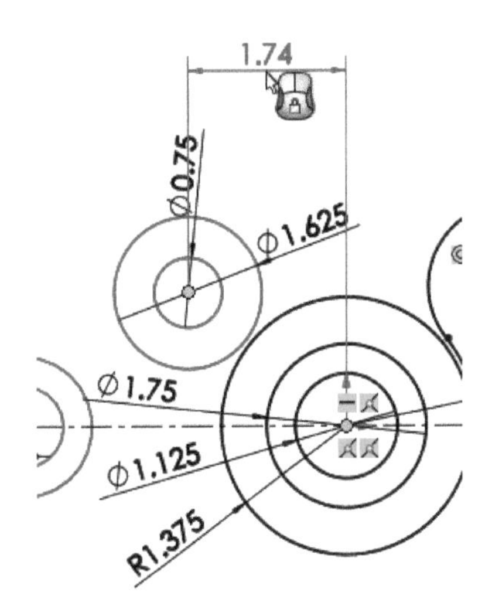

Figure-58. Placing_dimension

- Enter the distance value as **0.75**.
- Similarly, apply other dimensions; refer to Figure-59.

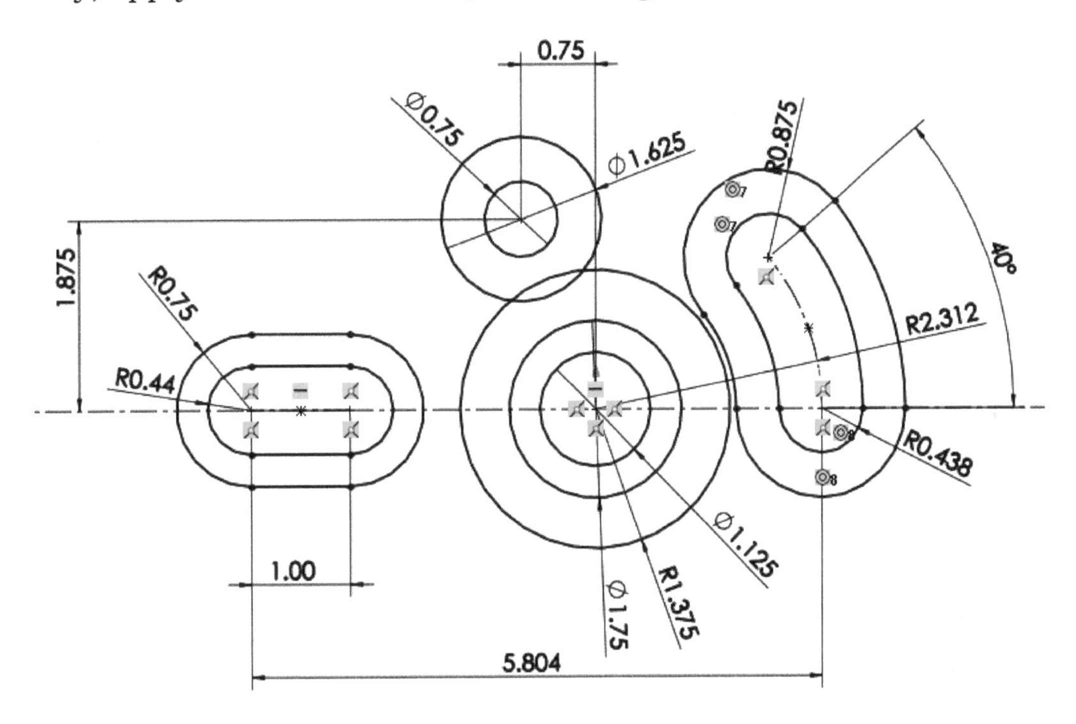

Figure-59. Sketch_after_applying_dimensions

Creating Arcs

- Click on the **3 Point Arc** tool from the **Arc** drop-down in the **Ribbon**; refer to Figure-60. The **Arc PropertyManager** will be displayed; refer to Figure-61.

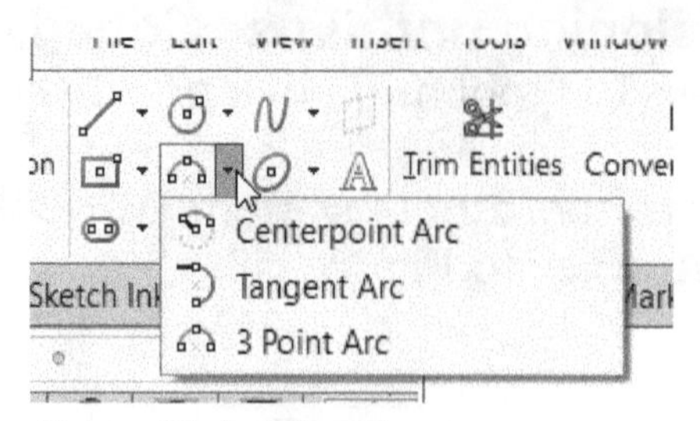

Figure-60. Arc_drop-down

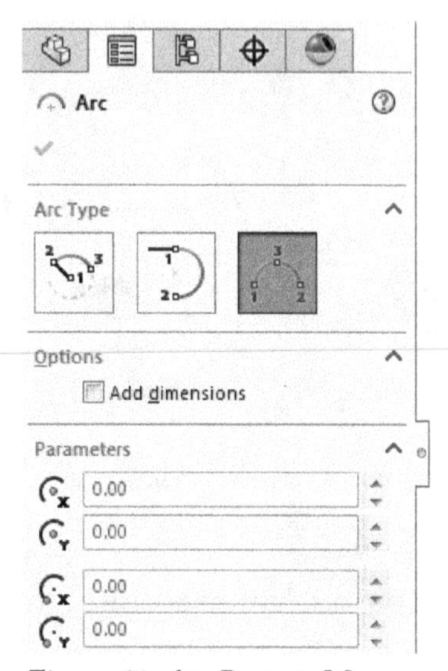

Figure-61. Arc_PropertyManager

- Click one by one at the locations displayed in the Figure-62. Click on the **OK** button from the **PropertyManager** and change the value of radius to **1.75**.

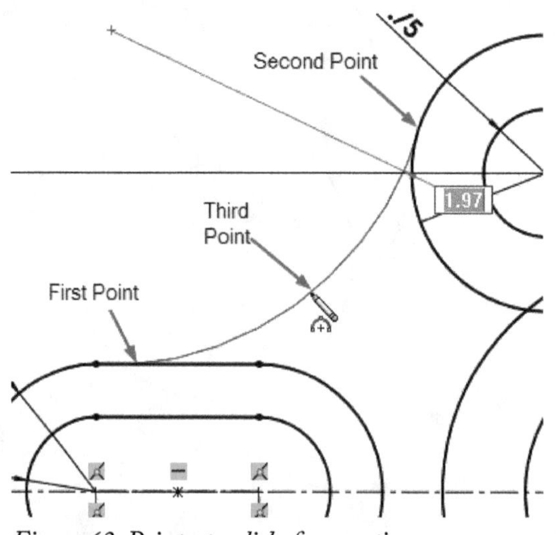

Figure-62. Points_to_click_for_creating_arc

- Select the arc and connected circle by holding the **CTRL** key, and select the **Tangent** button from the **Properties PropertyManager**; refer to Figure-63.

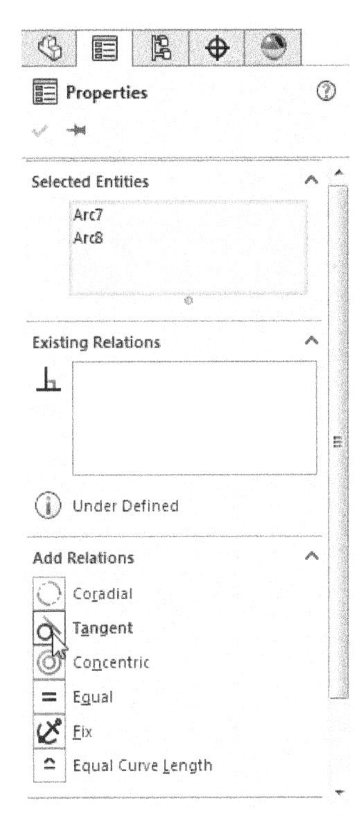

Figure-63. Tangent_button_to_be_selected

- Similarly, make the arc and slot tangent at the connecting point; refer to Figure-64.

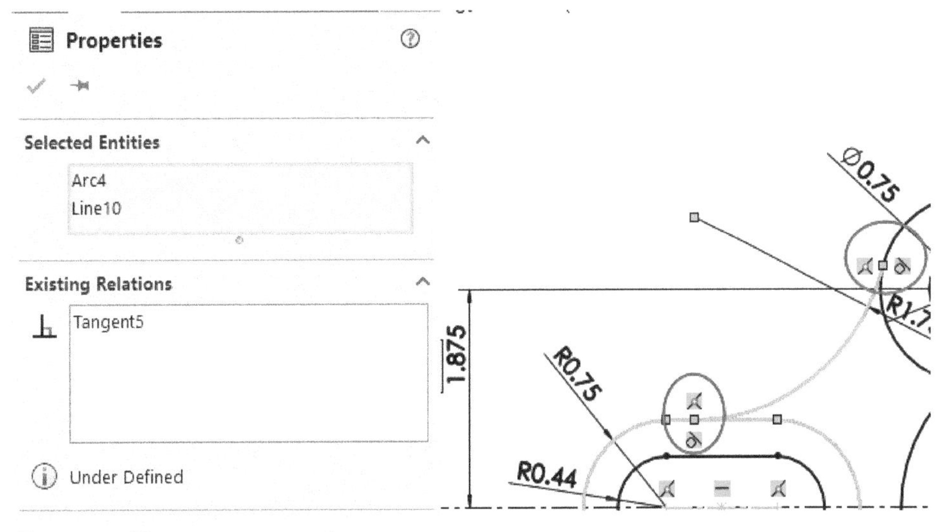

Figure-64. Tangent_arc_created

- Click on the **Three Point Arc** tool again and similarly create the other arcs; refer to Figure-65.

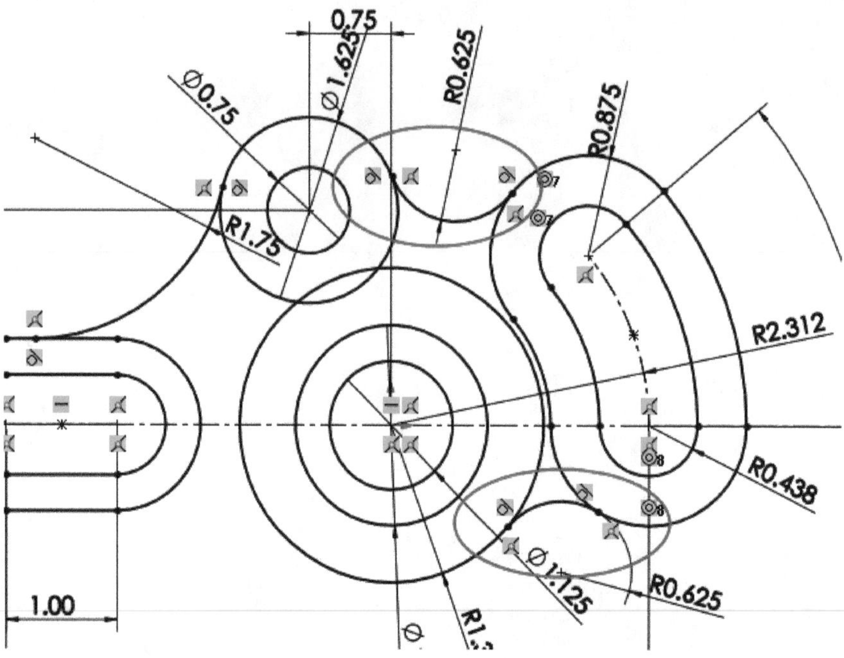

Figure-65. Arcs_to_be_created

Creating Line and Trimming

- Click on the **Line** tool from the **Line** drop-down. The **Insert Line PropertyManager** will be displayed.
- One by one click on the points shown in Figure-66.

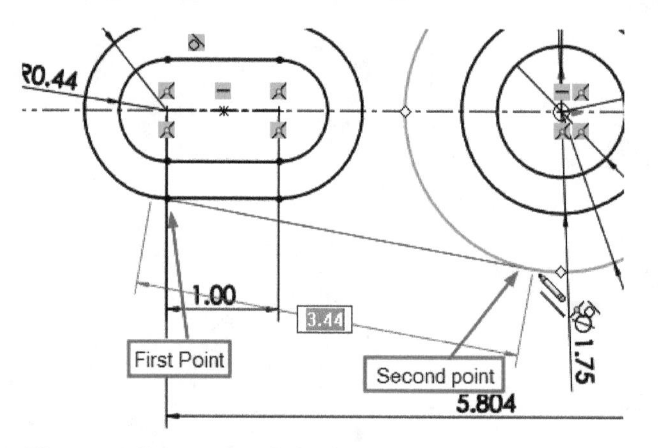

Figure-66. Points_selected_for_line

- Make the end points of the line tangent to slot and circle.
- Click on the **Trim Entities** tool from the **Trim Entities** drop-down and remove the extra sketched entities; refer to Figure-67.

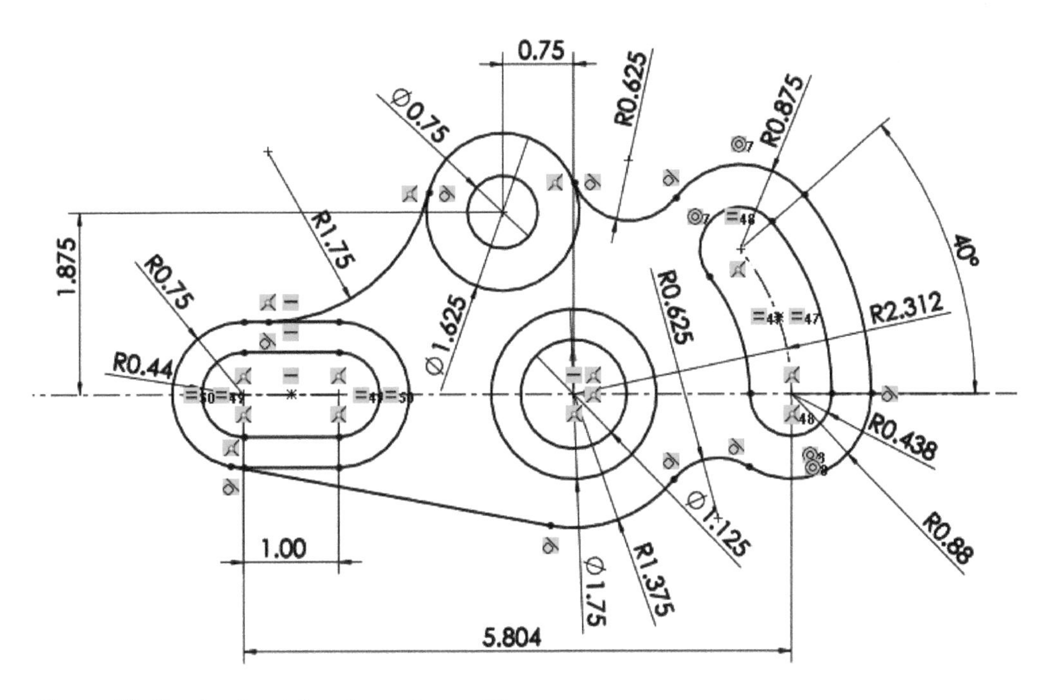

Figure-67. Sketch_after_trimming_extra_entities

Following are some sketches for practicing.

PRACTICE 1

In this practice session, we will create a sketch for the drawing given in Figure-68.

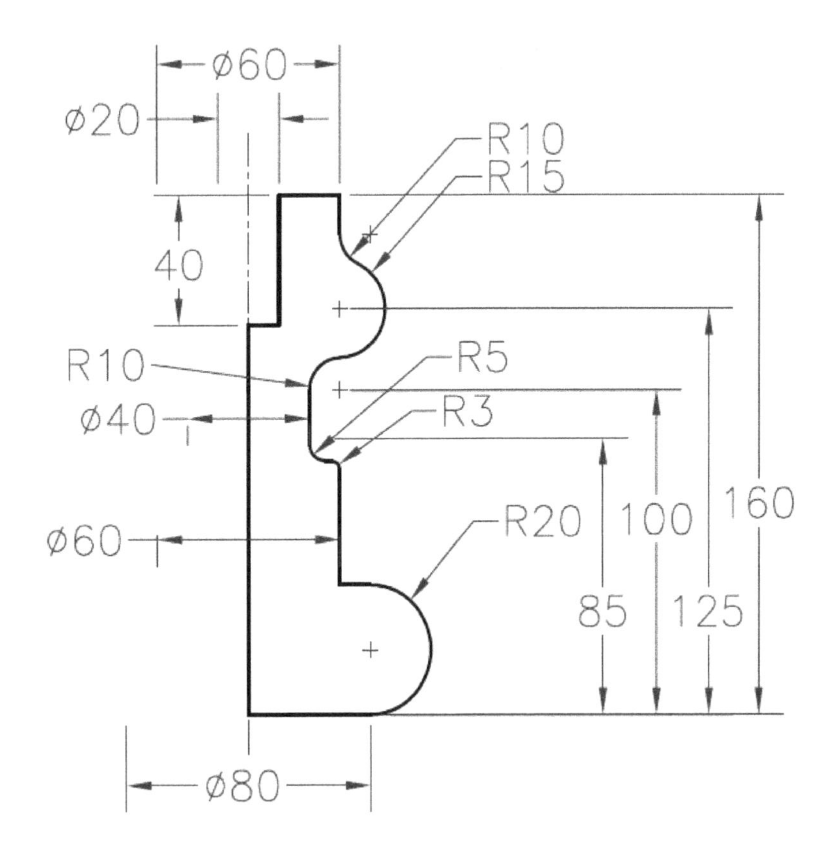

Figure-68. Practice_1

PRACTICE 2

In this practice session, we will create a sketch for the drawing given in Figure-69.

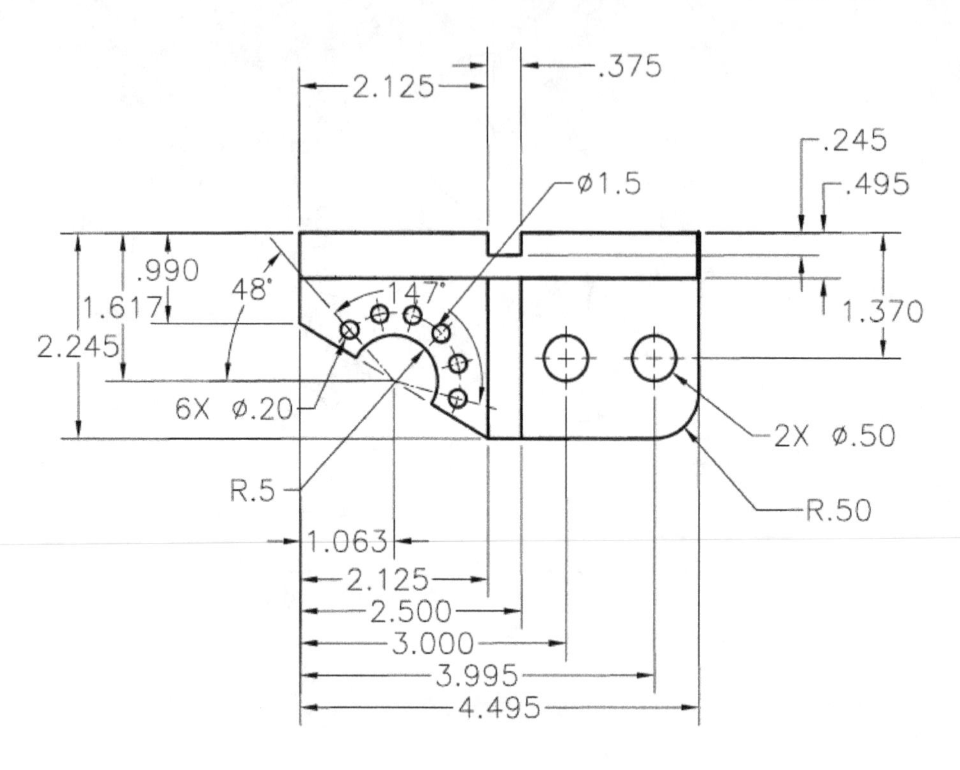

Figure-69. Practice_2

PRACTICE 3

In this practice session, we will create a sketch for the drawing given in Figure-70.

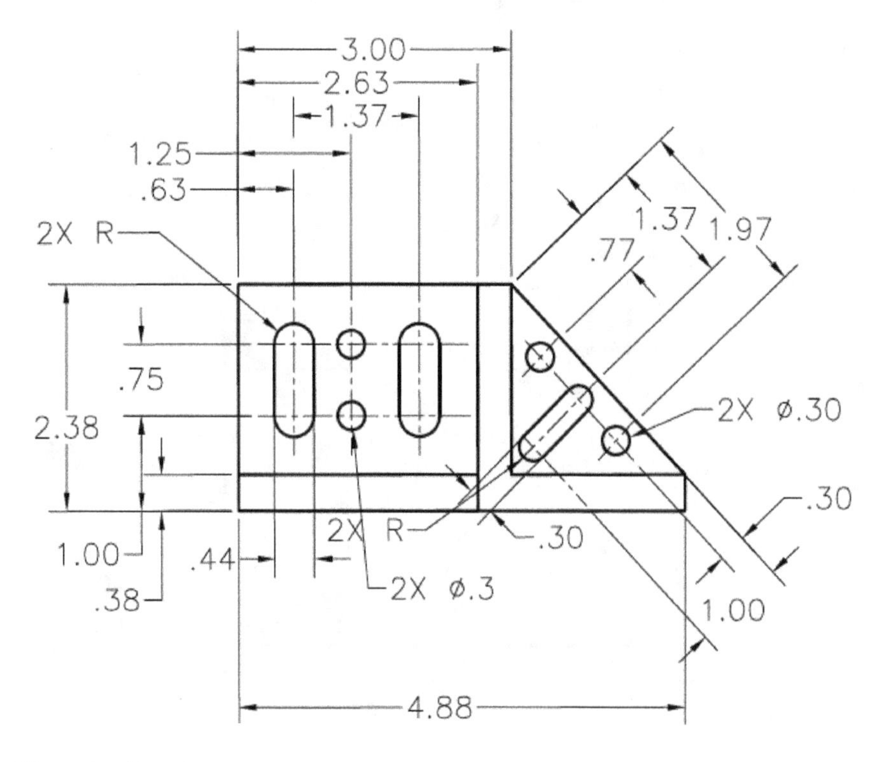

Figure-70. Practice_3

PRACTICE 4

In this practice session, we will create a sketch for the drawing given in Figure-71.

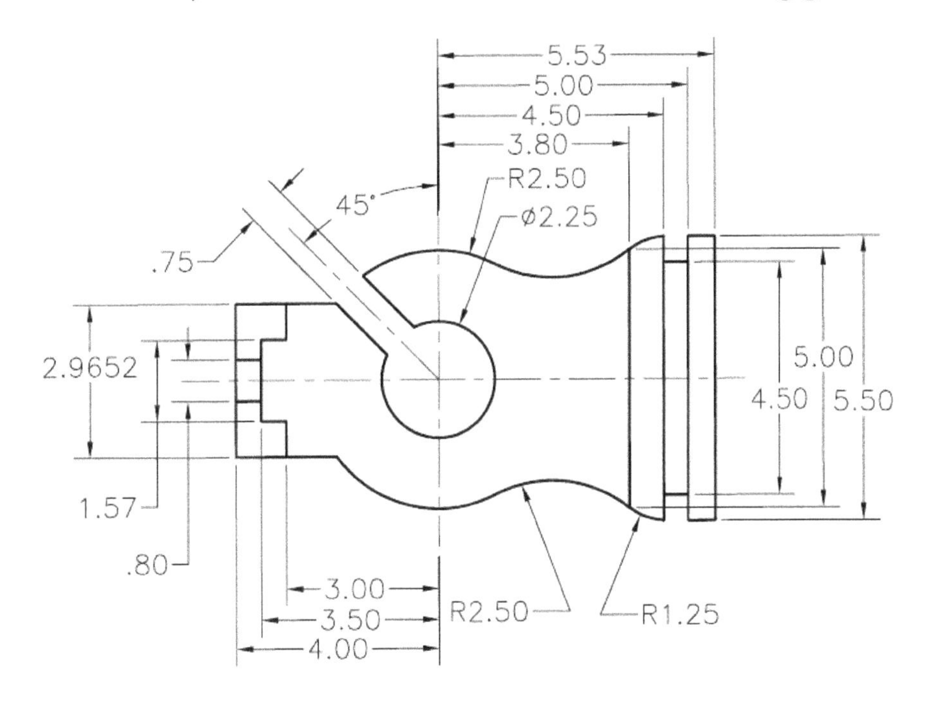

Figure-71. Practice_4

PRACTICE 5

In this practice session, we will create a sketch for the drawing given in Figure-72.

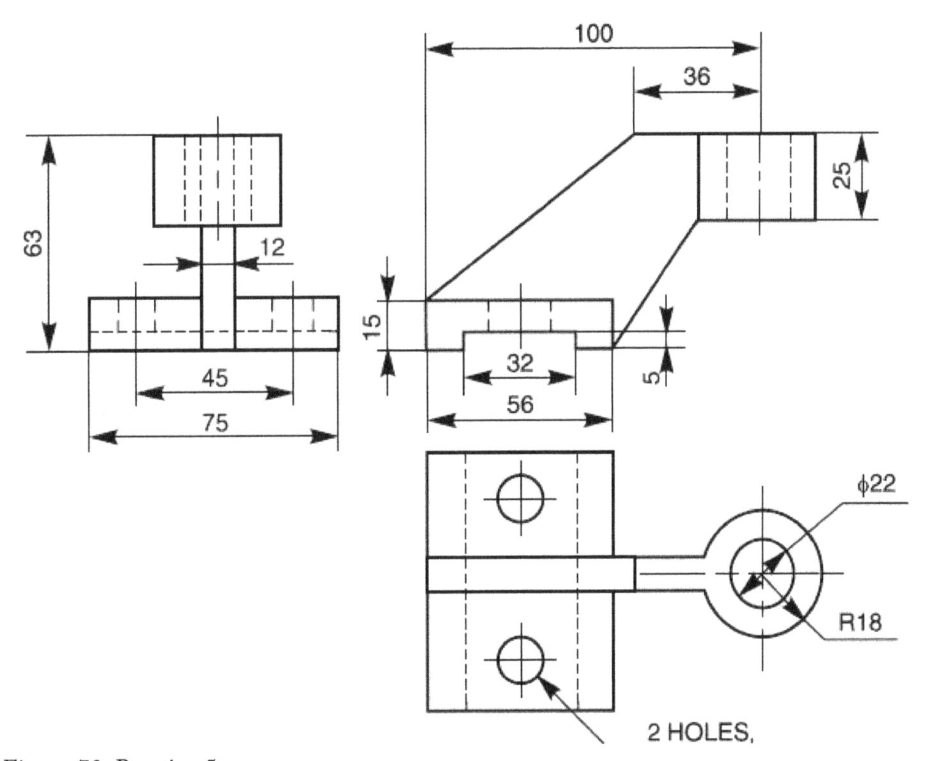

Figure-72. Practice_5

PRACTICE 6

In this practice session, we will create a sketch for the drawing given in Figure-73.

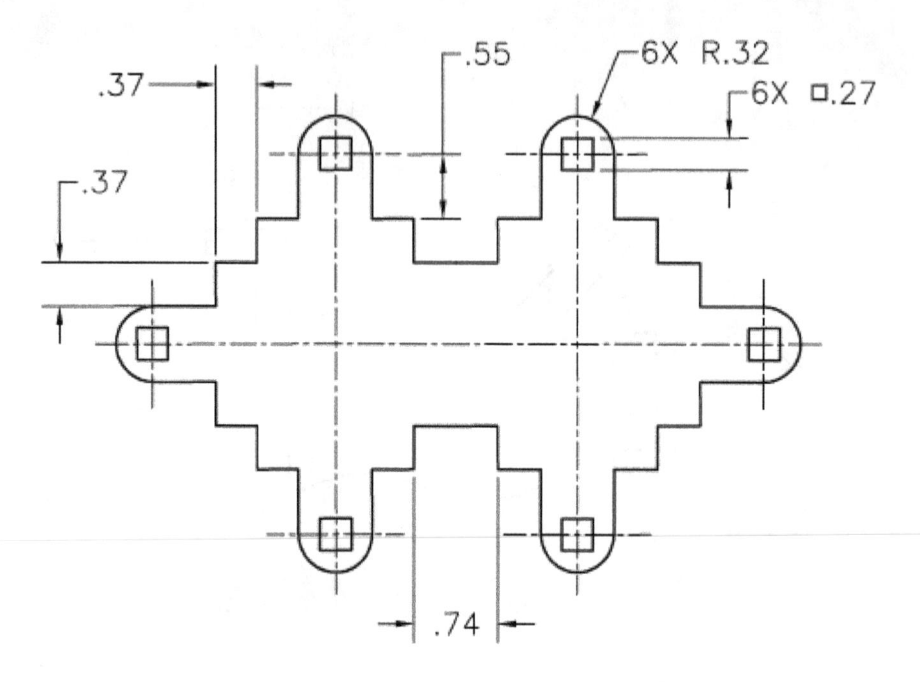

Figure-73. Practice_6

To get more exercises, mail us at cadcamcaeworks@gmail.com

SELF ASSESSMENT

Q1. The .... **PropertyManager** is used to create, save, delete, and retrieve the dimension style in the Part environment.

Q2. On selecting a dimension, the .... **PropertyManager** is displayed.

Q3. The .... button is used to delete a dimension style.

Q4. The style file will be saved with the extension ....

a. .sldstl
b. .sldprt
c. .slddrw
d. .prt

Q5. Which of the following is a type of tolerance?

a. Limit
b. MIN
c. None
d. Both a and b

Q6. The inspection dimensions are enclosed in .... shaped box.

Q7. You can display dual dimensions in sketches of SolidWorks. (T/F)

Q8. The arrows are displayed inside or outside the extension line, depending on the space available between the extension lines by using the **Smart** button in **Leader Display** rollout. (T/F)

FOR STUDENT NOTES

Answer to Self-Assessment:

1. Dimension, 2. Dimension, 3. Delete a Style, 4. a, 5. d, 6. obround, 7. T, 8. T

Chapter 4

3D Sketch and Solid Modeling

The major topics covered in this chapter are:

- ***3D Sketching and Plane Selection***
- ***Convert Entities tool***
- ***Extruded Boss/Base tool***
- ***Revolved Boss/Base tool***
- ***Swept Boss/Base tool***
- ***Creating Extra references for modeling***
- ***Lofted Boss/Base tool***
- ***Boundary Boss/Base tool***
- ***Hole Wizard***
- ***Thread Tool***
- ***Removing material using the above tools***

Extruded Boss/Base Revolved Boss/Base Swept Boss/Base Lofted Boss/Base Boundary Boss/Base Intersect Reference Geometry Curves Instant3D

Features Sketch Sketch Ink Sheet Metal Mold Tools Markup Evaluate MBD SOLIDWORKS CAM SOLIDWORKS CAM TBM SOLIDWORKS Inspection

3D SKETCHING

In previous chapter, you have worked on 2D sketches and have created them on Front plane. Now, we will come out of the 2D sketching and will explore the 3D world.

A 3D sketch is the sketch which is not confined to one plane only. The 3D sketch can be in all the planes available in the viewport. To start with 3D sketching, we are required to open the **Sketch** tab in **Ribbon**. The steps to create a 3D sketch are given next.

- Click on the down arrow below the **Sketch** button. A list of tools will display.
- Click on the **3D Sketch** button from the list; refer to Figure-1.

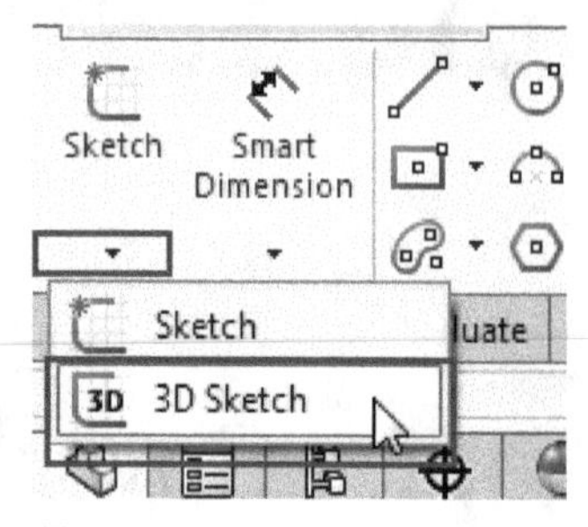

Figure-1. 3D_Sketch_button

- Click on any of the sketching tool to create entities like line, circle, rectangle, and so on. (In our case, the **Line** tool is selected.)
- The cursor will start annotating the current sketching plane. Figure-2 shows the cursor which denotes that the current sketching plane is XY.

Figure-2. Current Sketching plane

- Click to specify the start point of the line in the XY plane.
- The point in 3D space where you click will become the current sketching plane.
- Enter the parameters for the tool; refer to Figure-3.

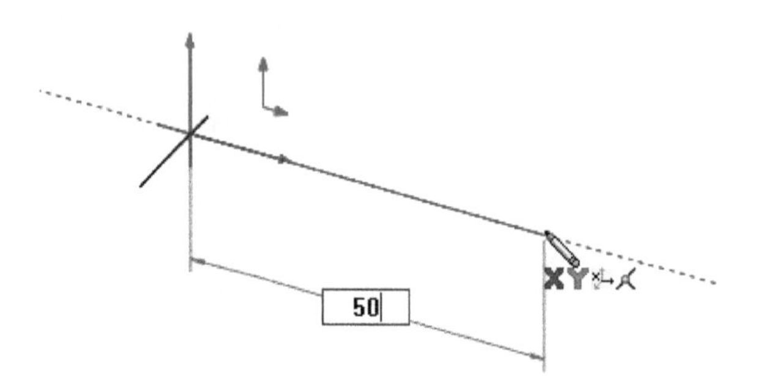

Figure-3. Entering parameters of line

- Press **TAB** from the **Keyboard** to toggle between three standard planes; refer to Figure-4.

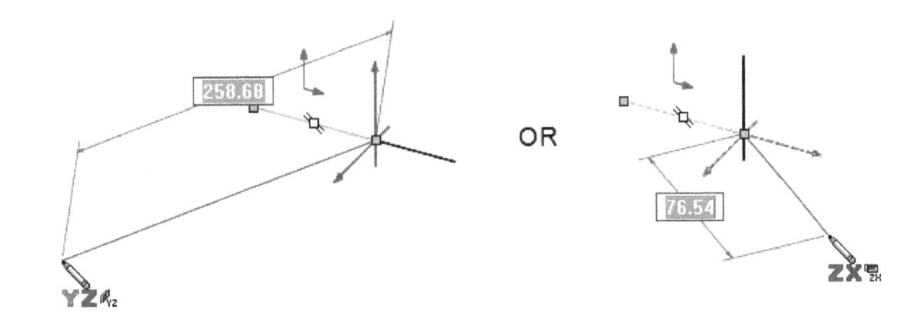

Figure-4. Toggling_between_planes

- Click or enter the parameters in desired planes to create the sketch.
- You can also join two points in different planes by using the sketching tools; refer to Figure-5.

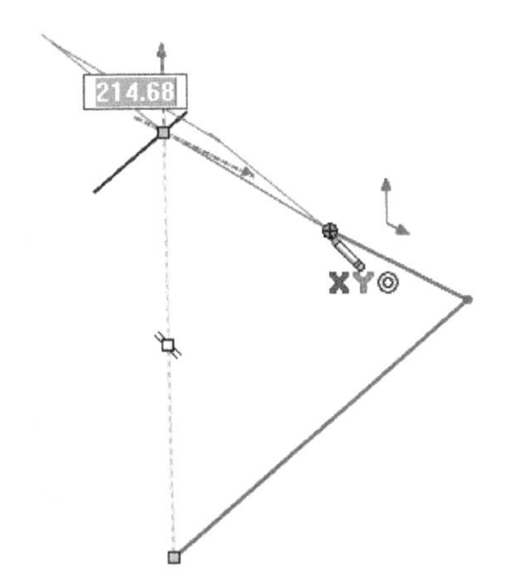

Figure-5. Connecting_entities_in_3D_space

When we will be creating surfaces, we will use the 3D sketches again.

Convert Entities

The **Convert Entities** tool is used when you need the projection of any face, edge, or sketch entity in another sketch. In this way, you can create the sketch entities from the projection of other features. The tool is available in the **Convert Entities** drop-down in the **Sketch** tab of **Ribbon**. The procedure to use this tool is given next.

- Click on the **Convert Entities** tool from the **Convert Entities** drop-down in the **Ribbon**. The **Convert Entities PropertyManager** will be displayed; refer to Figure-6.

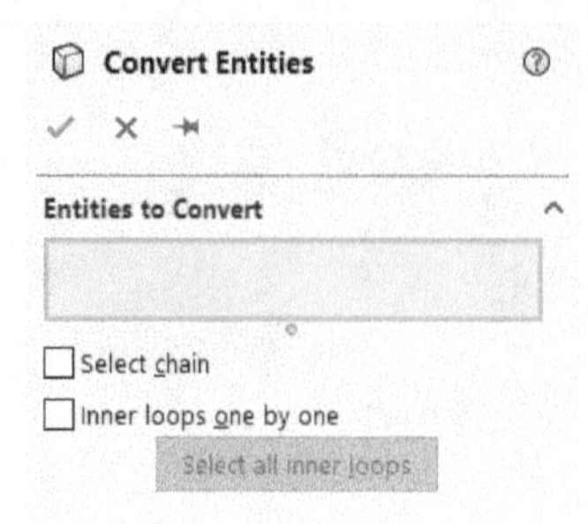

Figure-6. Convert_Entities_PropertyManager

- Select the face, edge, or curve that you want to use in your current sketch; refer to Figure-7.

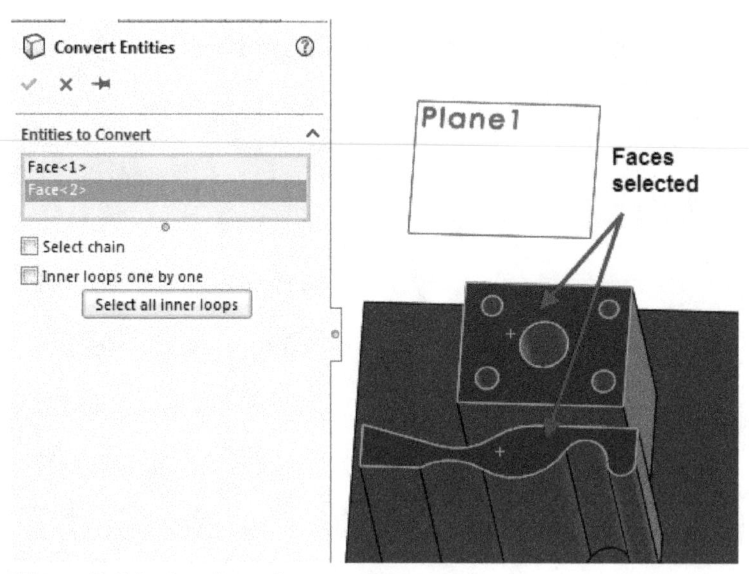

Figure-7. Selecting_faces_for_converting_entities

- Click on the **OK** button from the **PropertyManager**. The sketch entities will be generated; refer to Figure-8.

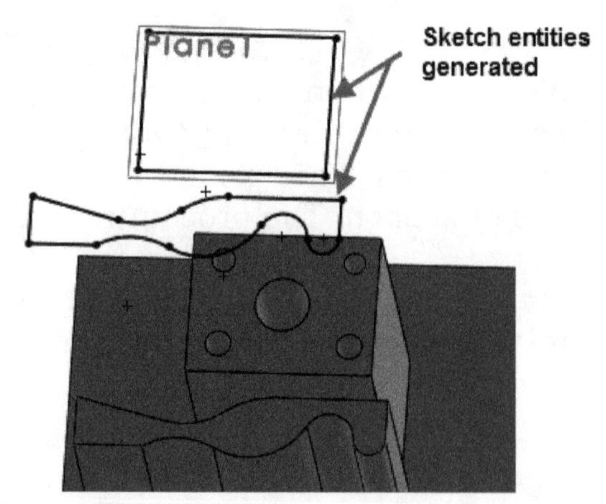

Figure-8. Entities_converted

Silhouette Entities

Silhouette Entities tool is used to project the external loops and internal loops of selected body on the sketching plane. The procedure to use this tool is given next.

- After starting sketch and selecting sketching plane, click on the **Silhouette Entities** tool from the **Ribbon**. The **Silhouette Entities PropertyManager** will be displayed; refer to Figure-9.

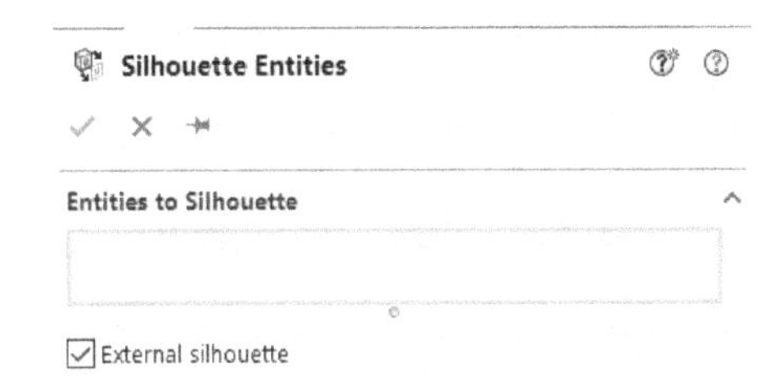

Figure-9. Silhouette Entities PropertyManager

- Select the body/component(s) to be projected on the sketching plane.
- Select the **External silhouette** check box if you want to project external loops. If you want to project internal loops then clear this check box.
- Click on the **OK** button from the **PropertyManager**. The entities will be projected; refer to Figure-10.

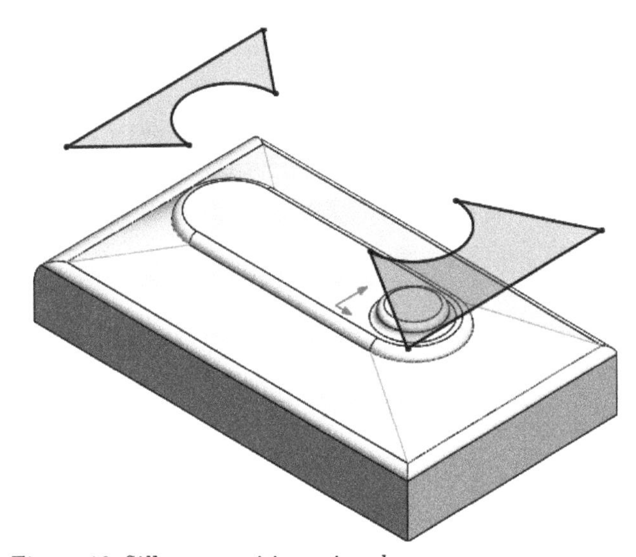

Figure-10. Silhouette entities projected

For any kind of analysis, simulation, assembly, or CAM; we need solid models. In SolidWorks, we are provided various tools to convert sketch into solid. Name of some of such tools are **Extruded Boss/Base** tool, **Revolved Boss/Base** tool, **Lofted Boss/Base** tool, and so on. These tools are discussed next.

EXTRUDED BOSS/BASE TOOL

Extruded Boss/Base tool is used to create a solid volume by adding height to the selected sketch. In other words, this tool adds material (by using the boundaries of sketch) in the direction perpendicular to the plane of sketch. In the term Boss/Base; the Base denotes the first feature and Boss denotes the feature created on any other feature. The steps to create extruded feature is given next.

- Click on the **Features** tab of the **Ribbon**. The tools related to solid modeling will display; refer to Figure-11.

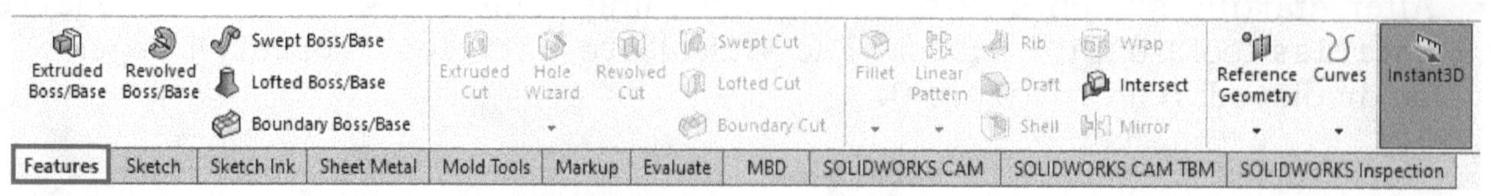

Figure-11. Features Command Manager in Ribbon

- Click on the **Extruded Boss/Base** tool from the **Ribbon**. The **Extrude PropertyManager** will display; refer to Figure-12.

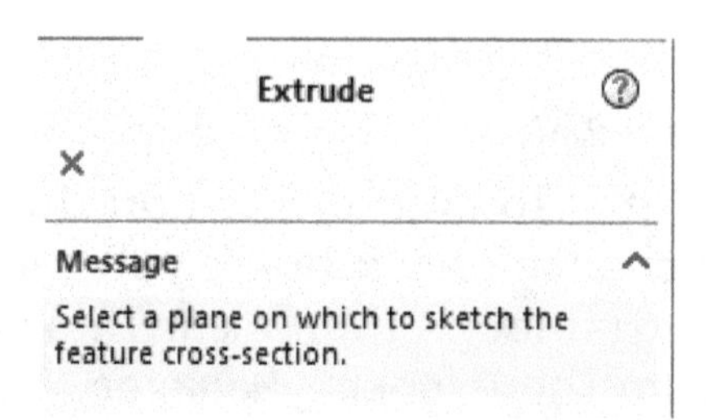

Figure-12. Extrude_PropertyManager

- Select a plane from the planes displayed in the viewport. The sketching environment will display with the sketch tools activated.
- Create a closed sketch and then click on the **Exit Sketch** button from the viewport as shown in Figure-13. You can also select the **Exit Sketch** button from the **Ribbon**. The **Boss-Extrude PropertyManager** will display as shown in Figure-14.

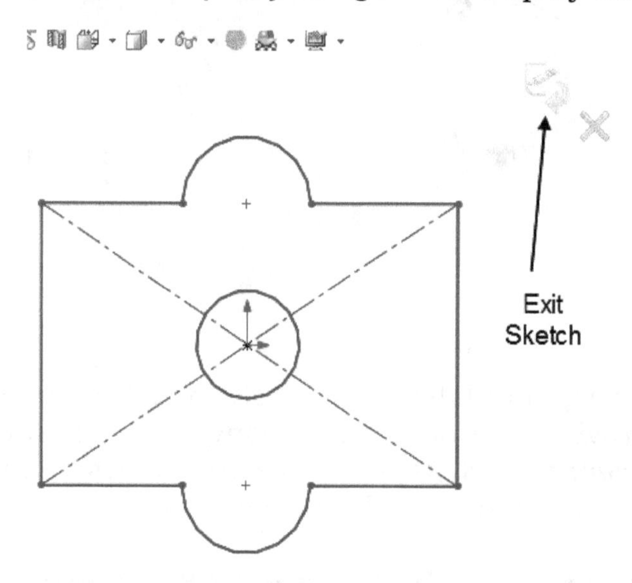

Figure-13. Sketch_environment_of_extrude

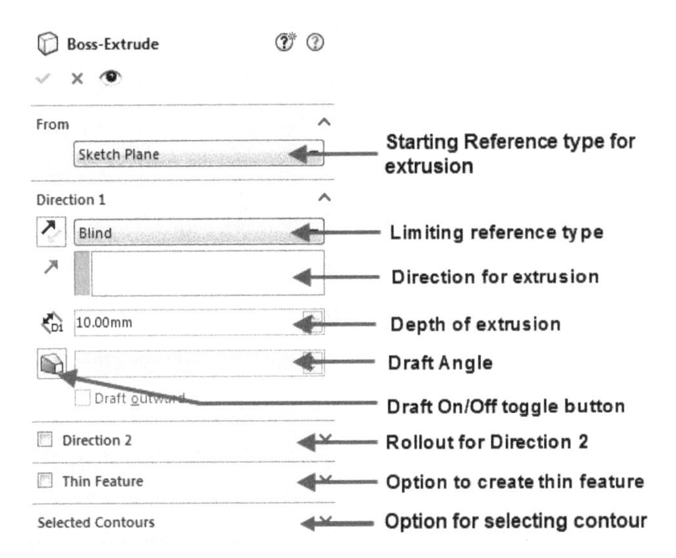

Figure-14. Boss-Extrude_PropertyManager

- Click on the **Starting reference** drop-down and select desired option.

There are four options in this drop-down; **Sketch Plane**, **Surface/Face/Plane**, **Vertex**, and **Offset**. The **Sketch Plane** is selected by default. Select this option if you want the extrusion to start from sketching plane. Select the **Surface/Face/Plane** option to start the extrusion from the selected surface/face/plane. Select the **Vertex** option to start extrusion from selected vertex. Select the **Offset** option if you want to start at specified distance from the sketching plane; refer to Figure-15.

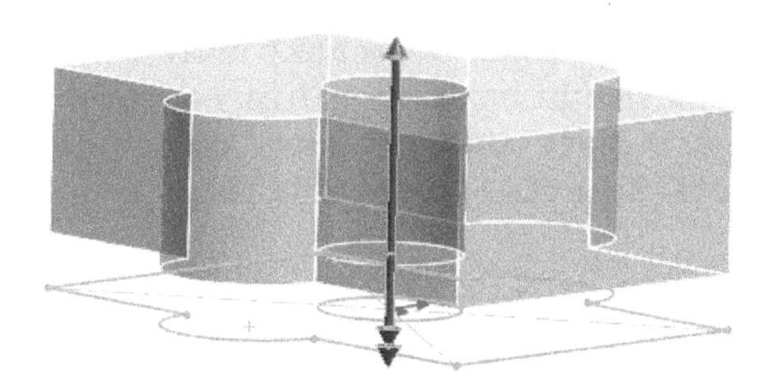

Figure-15. Preview_of_offset_option

- Click in the **Limiting reference type** drop-down and select the reference for end of extrusion.

There are six options in the drop-down; **Blind**, **Up To Vertex**, **Up To Surface**, **Offset From Surface**, **Up To Body**, and **Mid Plane**. If you have selected **Blind** or **Mid Plane** option then you need to specify the distance value in the **Height of extrusion** spinner. If you have selected any of the other option then select the respective reference from the viewport. Figure-16 shows preview of extrusion by using the **Mid Plane** option. Note that if **Mid Plane** option is selected then the **Direction 2** rollout will not display.

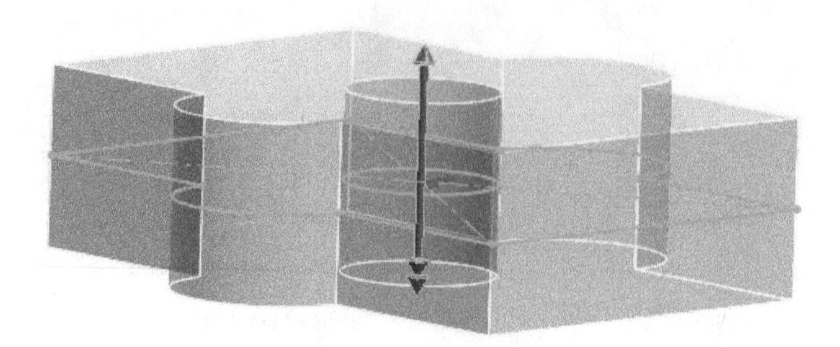

Figure-16. Mid-Plane_extrusion

- Click in the **Direction of extrusion** selection box and select the reference if you do not want to extrude perpendicular to the sketching plane and want to extrude along selected axis/plane.
- Click in the edit box for extrusion height and enter desired extrusion height or you can set the value by using spinner.
- Click on the **Draft On/Off** button to apply draft angle on the vertical faces of the model. On selecting this button, **1**° draft will be applied by default taking the sketching plane as reference. Select the **Draft outward** check box to apply draft angle outwards on the vertical faces of extrusion. Specify the draft angle in the **Draft Angle** spinner.

The parameters you specified above can also be applied to the opposite direction. To apply these parameters, select the **Direction 2** check box. The parameters for the opposite direction will display.

- Select the **Thin Feature** check box to create the thin walled extrusion. Enter the thickness in the **Thickness** edit box of the **Thin Feature** rollout. Figure-17 shows a thin featured extrusion. **Note that if open sketch is selected for extrude then this option gets selected automatically**.

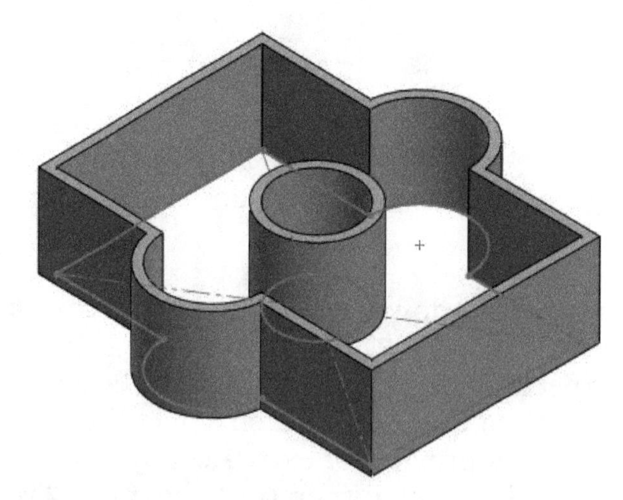

Figure-17. Thin_Feature_extrusion

- If you want to close the start and end face of the extrusion then select the **Cap ends** check box; refer to Figure-18.

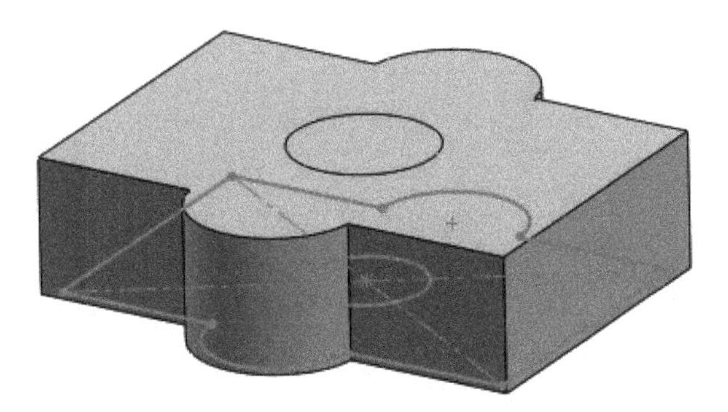

Figure-18. Extrusion_with_cap_ends

- Expand the **Selected Contours** rollout and click in the selection box inside the rollout. Now, you can select different closed regions to create extrusion. Select the desired closed regions.
- Once you have finished creating the feature, click on the **Detailed Preview** button from the **PropertyManager** to verify the feature; refer to Figure-19.

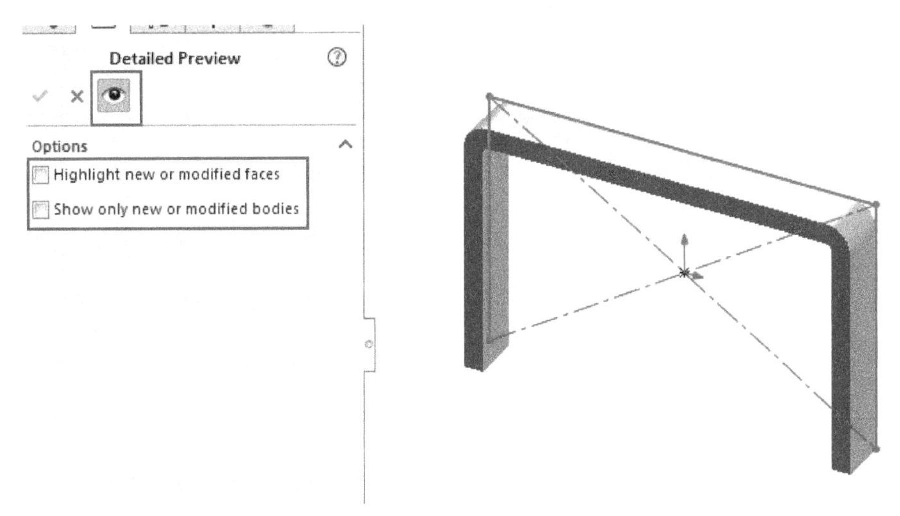

Figure-19. Detailed_Preview_button

- Select the **Highlight new or modified faces** check box if you want to highlight the new/modified faces only. Similarly, you can select the **Show only new or modified bodies** check box if you want to display only new or modified objects.

SolidWorks 2017 onwards, you can also activate the **Extrude Boss/Base** tool by **ALT** key. To do so, press the **ALT** key from keyboard and click on the shaded sketch section in the Sketching environment. The **Extrude** button will be displayed; refer to Figure-20. Click on the **Extrude** button to display **Extrude PropertyManager**. Rest of the procedure is same as discussed earlier.

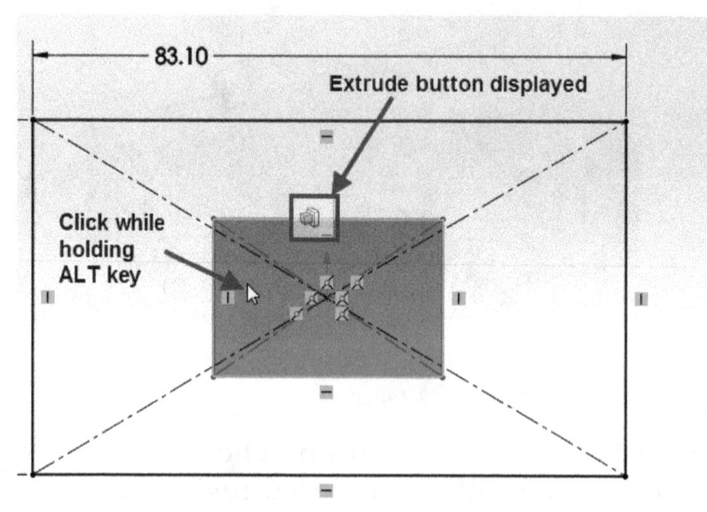

Figure-20. Alternate_method_for_extrude

REVOLVED BOSS/BASE TOOL

Revolved Boss/Base tool is used to create a solid volume by revolving a sketch about selected axis. In other words, if you revolve a sketch about an axis then the volume that is swept by revolved sketch boundary is called revolved boss/base feature. The steps to create revolved boss/base feature are given next.

- Click on the **Revolved Boss/Base** tool. If you have not selected any existing sketch, then the **Revolve PropertyManager** displays as shown in Figure-21.

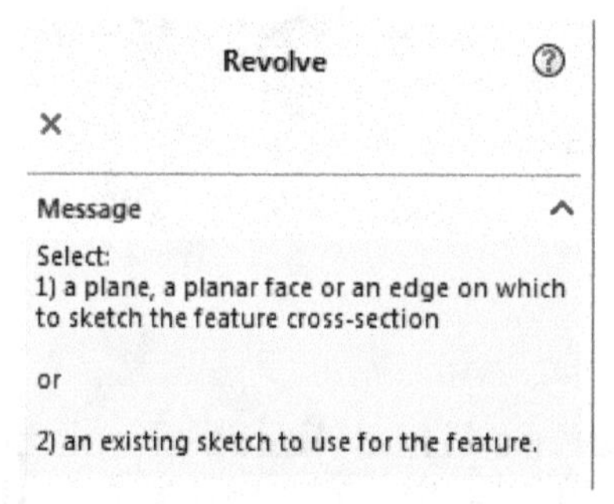

Figure-21. Revolve_PropertyManager

- Select a plane if you want to create a new sketch or select an already created sketch. In our case, we are selecting an already created sketch.
- Select the region of the sketch that you want to revolve if you have multiple loops in sketch; refer to Figure-22. The updated **Revolve ProperyManager** will display as shown in Figure-23.

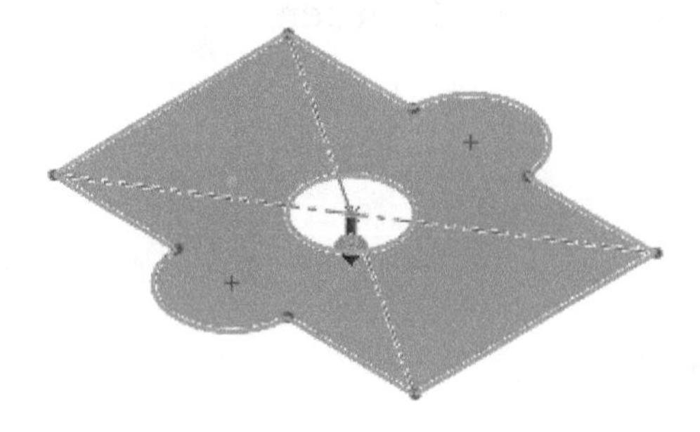

Figure-22. Region_selected_for_revolve

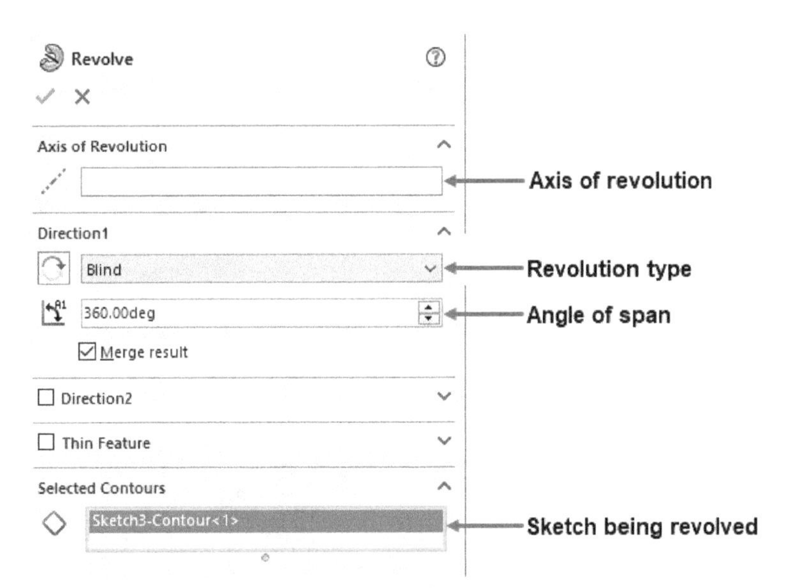

Figure-23. Updated Revolve Property Manager

- Click in the **Axis of Revolution** selection box to select the axis. Select the edge, line, or center line about which you want to revolve the sketch. Preview of the revolve feature will be displayed; refer to Figure-24.

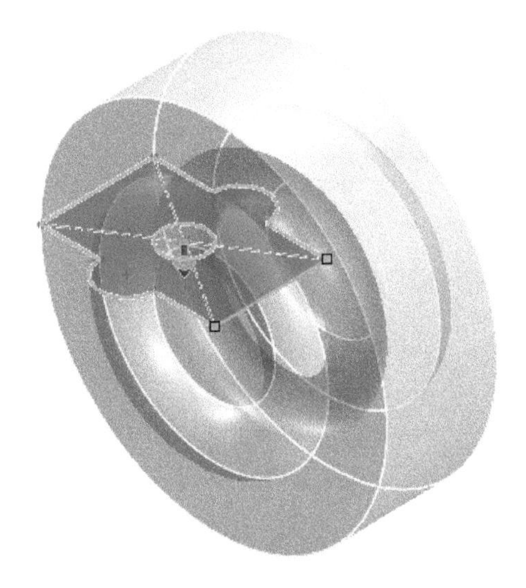

Figure-24. Preview_of_revolve_feature

- Click on the **Revolve Type** drop-down and specify the revolution limiting reference. The options in this drop-down are same as discussed for **Extruded Boss/Base** tool.
- If you have selected **Blind** option in the **Revolve Type** drop-down then specify the degrees of revolution by using the **Angle** spinner.
- Click on the **Direction 2** check box to revolve in the direction opposite the earlier on. The options in the **Direction 2** rollout are same as discussed earlier.
- You can also create thin feature by selecting the **Thin Feature** check box. **Note that if you select an open sketch then this option is automatically selected**.
- Click in the **Selected Contours** box to add more sketches for revolution and select the sketches you want to revolve.

Figure-25 shows a sketch, axis of revolution, and resulting revolve feature preview.

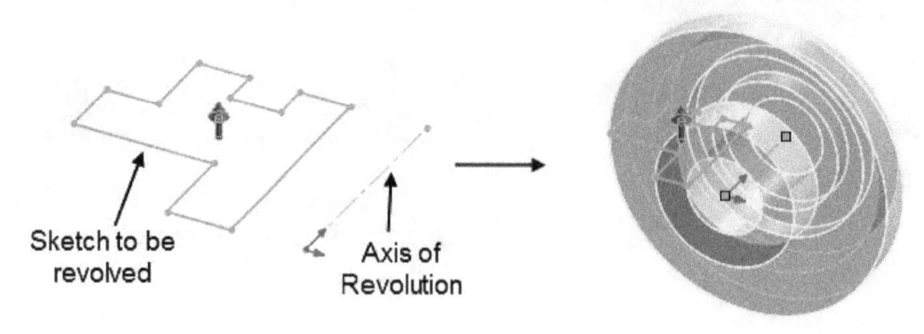

Figure-25. Revolved_feature

SWEPT BOSS/BASE TOOL

The **Swept Boss/Base** tool is used to create a solid volume by moving a sketch along the selected path. In other words, if you move a sketch along a path then the volume that is covered by moving sketch boundary is called swept boss/base feature. Note that to use this tool, you must have a sketch section and a path then only the tool will be active. The steps to create swept boss/base feature are given next.

- Click on the **Swept Boss/Base** tool. The **Sweep PropertyManager** will display as shown in Figure-26.

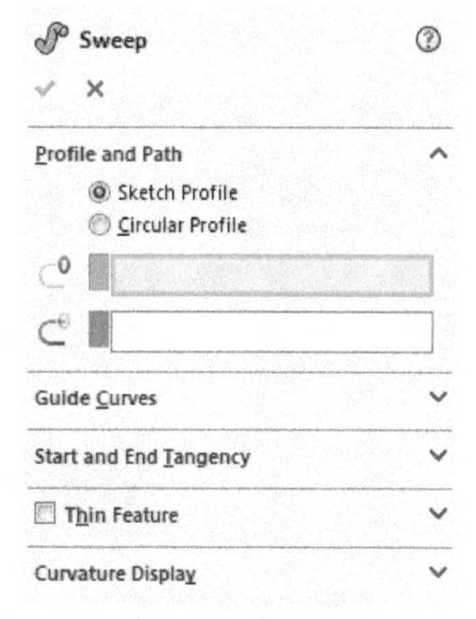

Figure-26. Sweep_PropertyManager

- By default, the **Sketch Profile** radio button is selected and you are asked to select a sketch section (profile).
- Select the close section from viewport that you want to sweep.
- On selecting the section, the **Path** selection box becomes selected automatically and you are asked to select a path.
- Select the curve that you want to use as path. If the path curve is a chain of multiple sketch entities then a toolbar will be displayed for selection; refer to Figure-27. There are five buttons in the toolbar for curve selection; **Select Closed Loop**, **Select Open Loop**, **Select Group**, **Select Region**, and **Standard Selection**. Select desired button and click on the **OK** button from the toolbar. Preview of the sweep feature will be displayed.

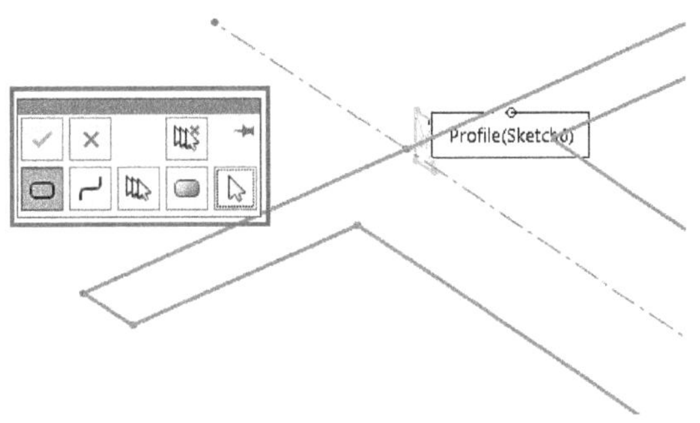

Figure-27. Toolbar for curve selection

- If your path is extended on both sides of section then three buttons will be displayed below the **Path** selection box; refer to Figure-28. Select the **Direction 1**, **Bidirectional**, and **Direction 2** button as per the requirement. Preview of sweep with these buttons is shown in Figure-29.

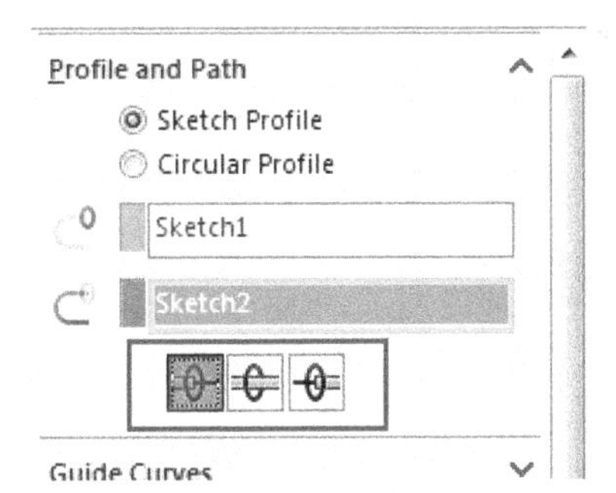

Figure-28. Buttons_for_bidirectional_sweep

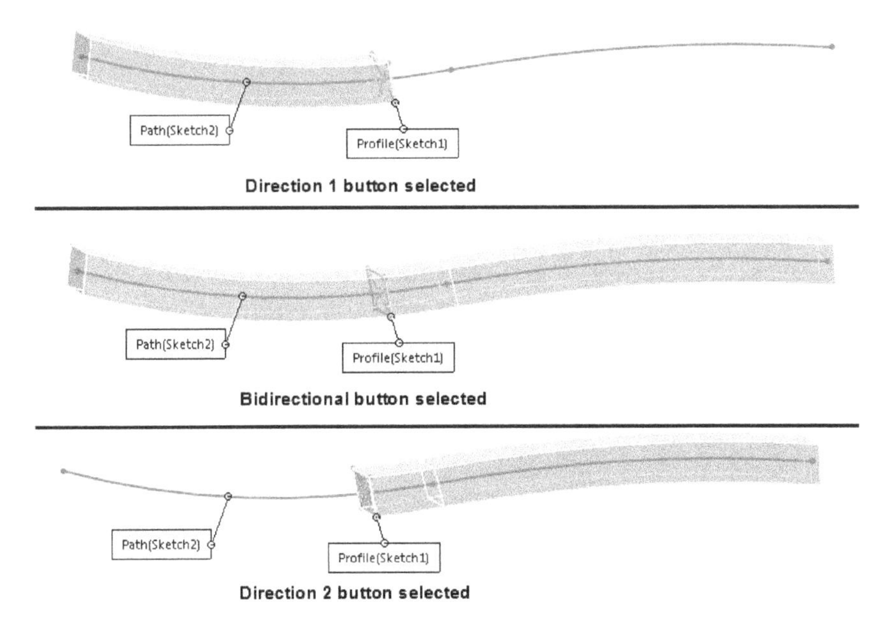

Figure-29. Sweep with directional options

- If your path is not perpendicular to the section plane then click on the **Options** rollout. The options in the rollout will display as shown in Figure-30.

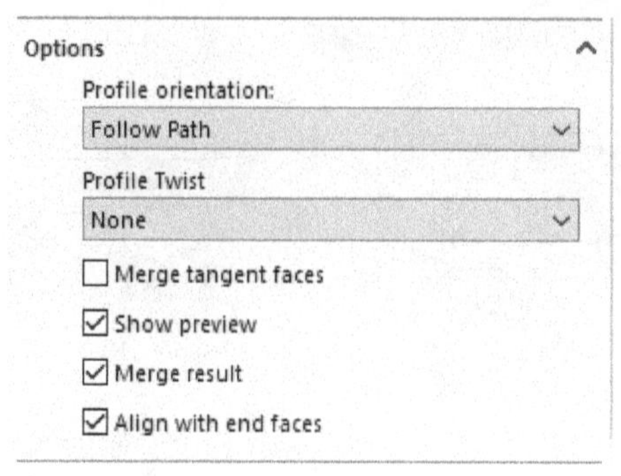

Figure-30. Options rollout

- Click on the **Profile orientation** drop-down and select **Keep Normal Constant** option to create the sweep feature. Figure-31 shows a sweep feature created by this method.

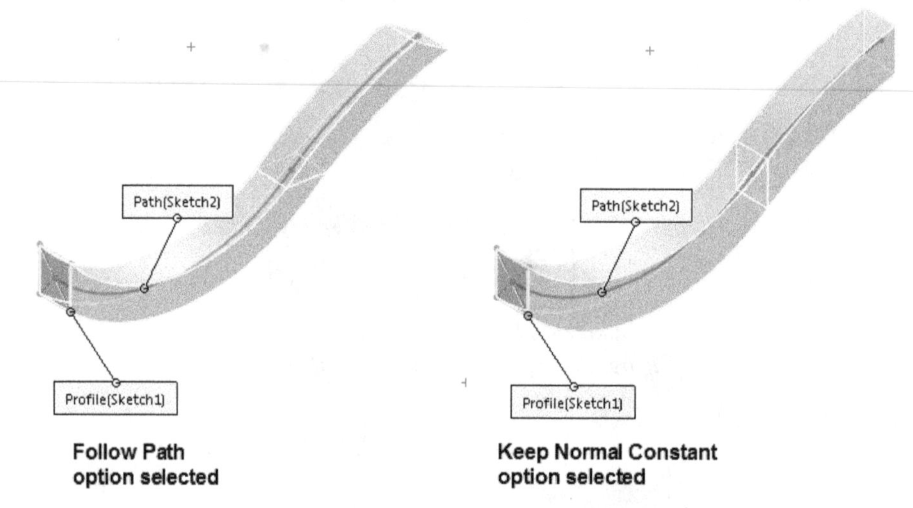

Figure-31. Sweep with Follow Path and Keep Normal Constant options

Using Guide Curves

The **Follow Path** option is selected by default in the **Profile Orientation** drop-down and as a result the swept feature is created by making the sketch section exactly follow the curve's curvature. If you want to influence the shape of swept feature with the help of a guide curve then expand the **Guide Curves** rollout and select the guide curve/curves; refer to Figure-32. Select the **Follow path and First guide curve** option from the **Profile Twist** drop-down if you want to sweep feature to follow the path and first guide curve for its shape; refer to Figure-33. Similarly, you can select the **Follow First and Second Guide Curves** option from the drop-down to make the sweep feature follow both the guide curves for its shape.

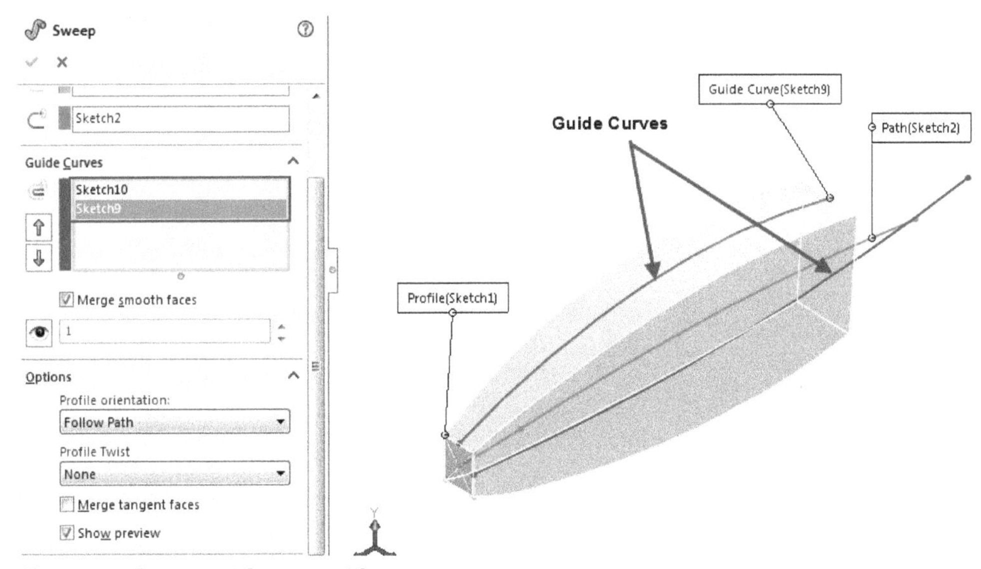

Figure-32. Sweep_with_two_guide_curves

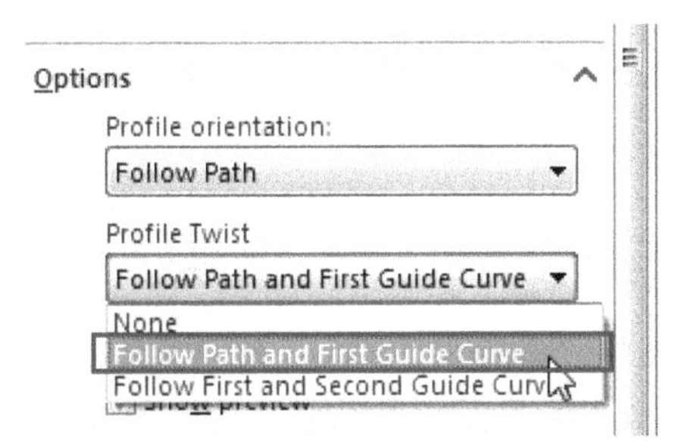

Figure-33. Profile_Twist_drop-down

Note that you need to have individual sketches for section, path, and guide curves although they can be on same plane.

Applying Twist in Sweep Feature

You can twist the section while sweeping along the path to create drill bit type of shape of conduits. To do so, select the **Specify Twist Value** option from the **Profile Twist** drop-down in the **Options** rollout after selecting the section and path. The **Options** rollout will display as shown in Figure-34.

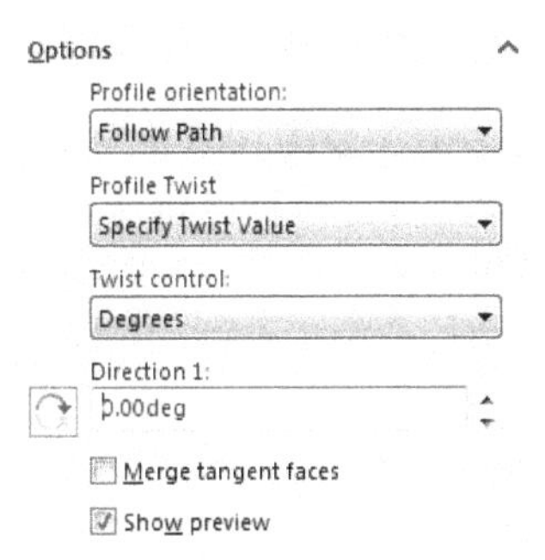

Figure-34. Options_rollout_with_twist_along_path

Click on the **Twist Control** drop-down and select desired unit to twist the section. In this case, we have selected **Revolutions** option. Specify desired number of revolutions in the **Direction 1** spinner. Click on the flip button to reverse the twisting. Figure-35 shows the preview of twisting along the path while sweeping.

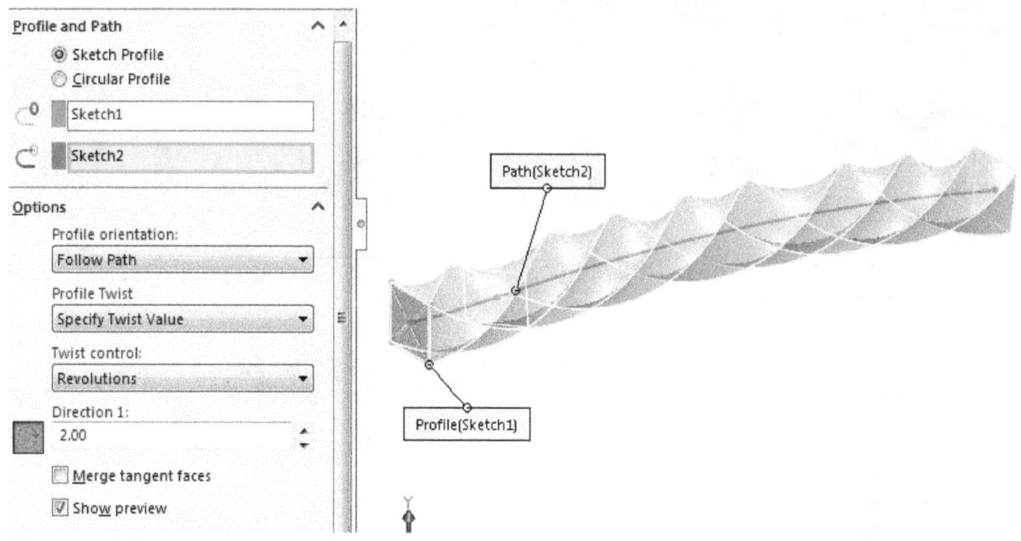

Figure-35. Preview_of_twist_along_path

You can also make a spring by using this option; refer to Figure-36.

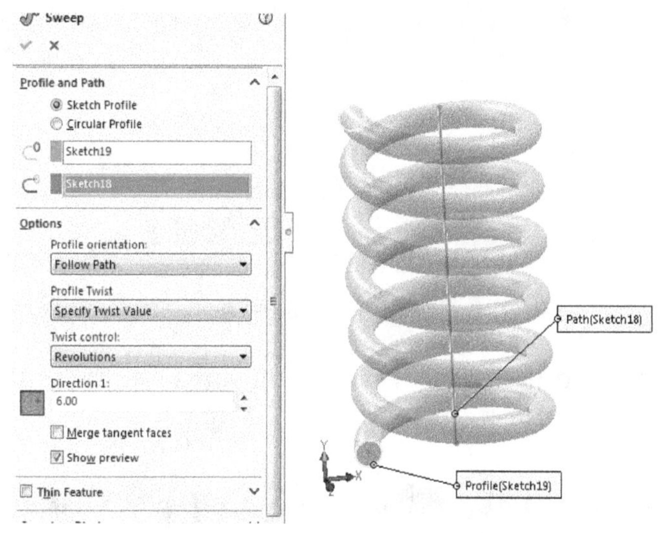

Figure-36. Spring_created_by_twist_slong_path

Circular Profile Sweep

This option is available from SolidWorks 2016 onwards version of software. Using this option, you can create round bar/rod. The procedure to use this option is given next.

- Select the **Circular Profile** radio button from the **Profile and Path** rollout of **PropertyManager**. You are asked to select a path.
- Select the path along which the circular sweep should be created. Preview of the circular sweep will be displayed; refer to Figure-37.

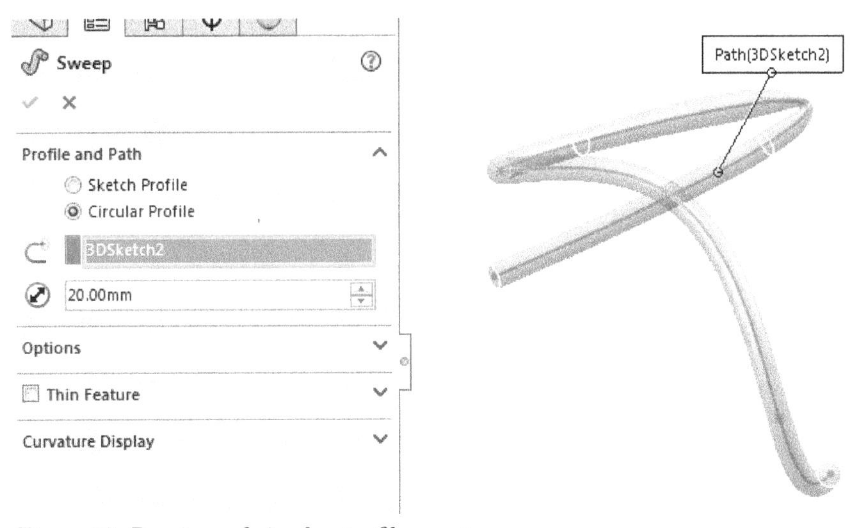

Figure-37. Preview_of_circular_profile_sweep

- Specify desired diameter in the **Diameter** spinner and click on the **OK** button from the **PropertyManager**.

Before we move to **Lofted Boss/Base** tool and other feature creation tools. We need to understand the procedure to create reference planes, axis, coordinate system, and other references as they are important for advanced modeling.

REFERENCE GEOMETRY

There are various types of reference geometries that can be created in SolidWorks. All the tools to create these reference geometries are available in the **Reference Geometry** drop-down; refer to Figure-38. The tools available in the drop-down are:

- Plane
- Axis
- Coordinate System
- Point
- Center of Mass
- Bounding Box
- Mate Reference

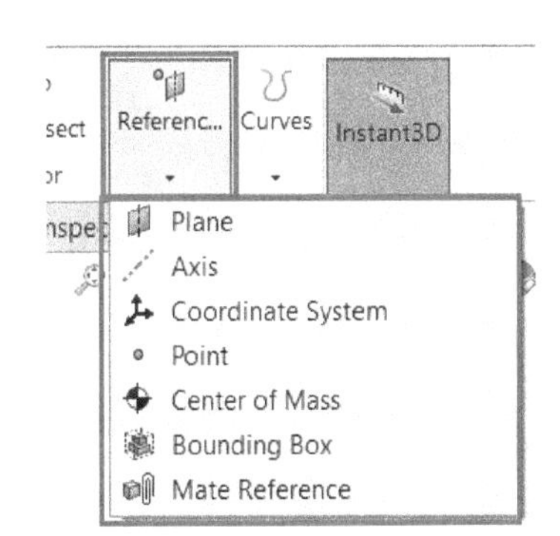

Figure-38. Reference drop-down

These tools are discussed next.

Plane

The **Plane** tool is used to create reference planes. By default, there are three planes available in SolidWorks: **Front**, **Top**, and **Right**. To create more planes, follow the steps given next.

- Click on the **Plane** tool from the **Reference Geometry** drop-down. The **Plane PropertyManager** will display as shown in Figure-39.

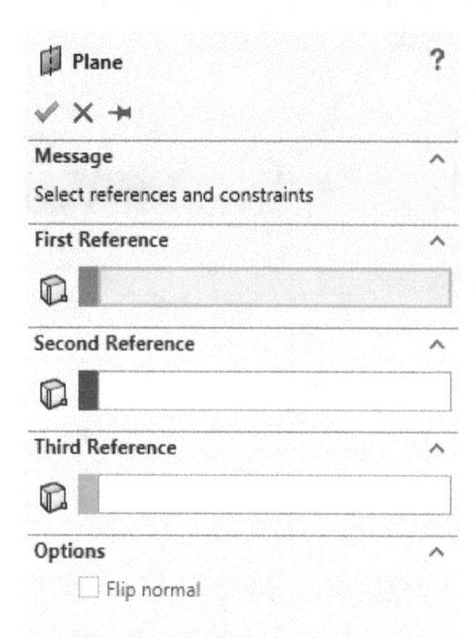

Figure-39. Plane_PropertyManager

- You can select maximum three references to create a plane. You can select plane/face, edge/axis/curve, or vertex/point. The ways in which you can create planes by using these references are discussed next.

Creating plane at a distance from plane/face

- To create a plane at a distance from a plane/face, select the plane/face. The updated **Plane PropertyManager** will display as shown in Figure-40.

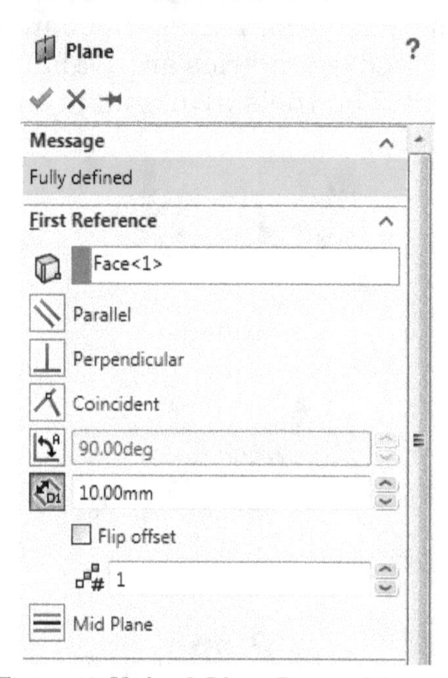

Figure-40. Updated_Plane_PropertyManager

- Specify desired distance in the spinner.
- Click on the **OK** button to create the plane.

Creating plane at an angle to plane/face

- Activate the **Plane PropertyManager** and select a plane/face to which you want to specify the angle.
- Click on the **At Angle** button to specify the angle.

- Click in the **Second Reference** box and select the edge or axis to which you want to make the plane coincident or select the two planar point through which you want the plane to pass. Figure-41 shows the plane create by both the methods discussed.

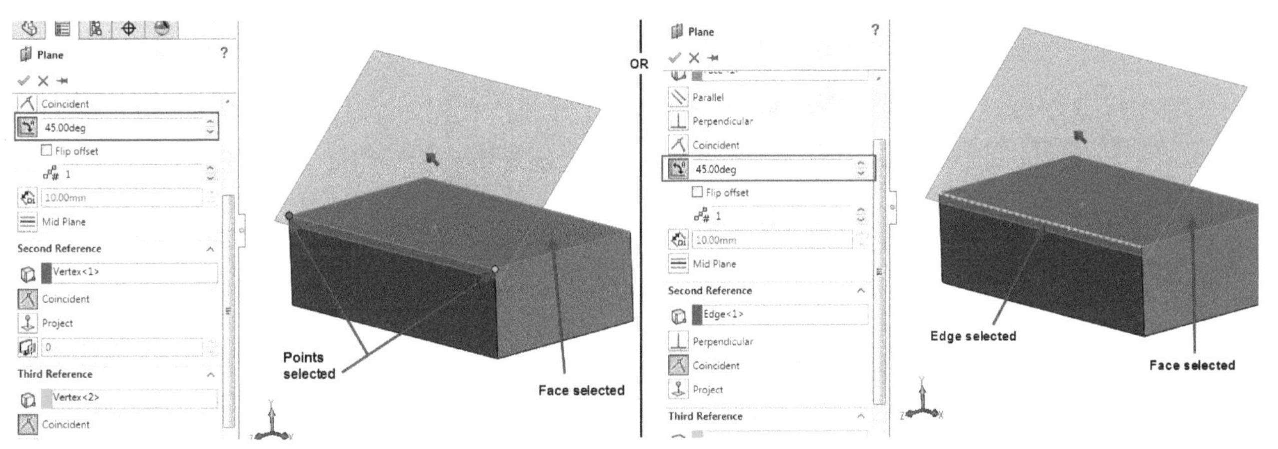

Figure-41. Plane_creation_at_angle

Creating plane passing through points

- Activate the **Plane PropertyManager** and one by one click three points through which you want the plane to pass through. Figure-42 shows the plane passing through three points.

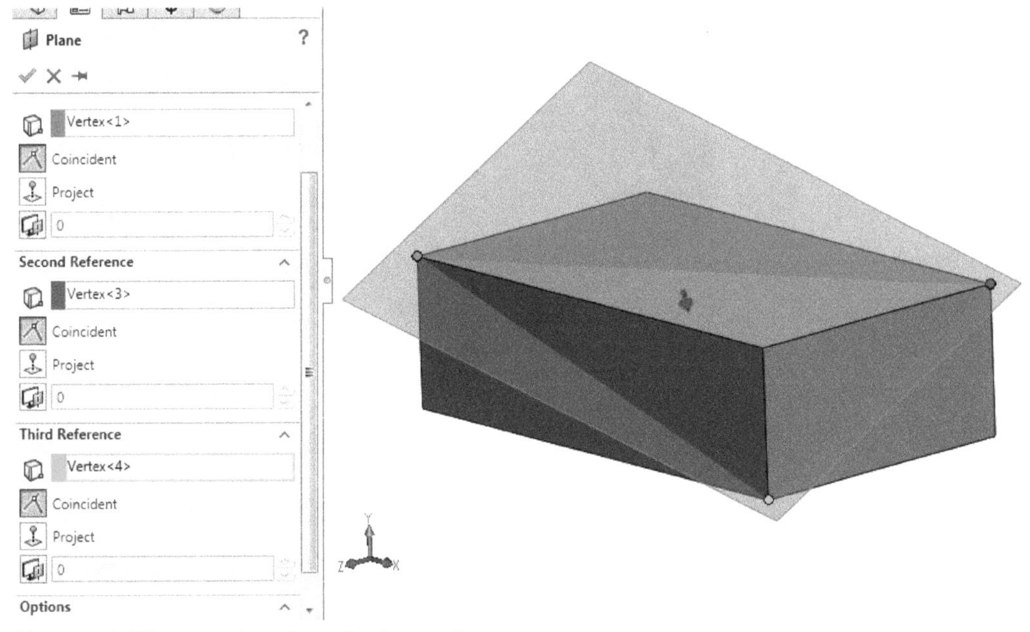

Figure-42. Plane_passing_through_three_points

Plane Parallel to Screen

This option is a new feature from SolidWorks 2016 onwards. The procedure to create plane parallel to screen is given next.

- Right-click on any face, edge, or vertex of the model. A shortcut menu will be displayed; refer to Figure-43.

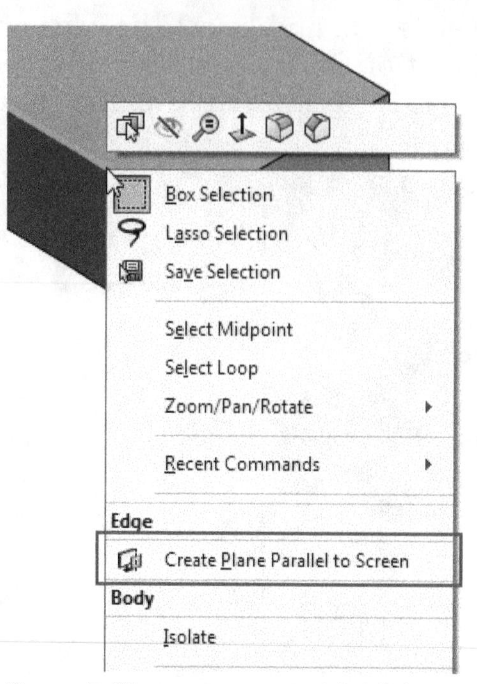

Figure-43. Shortcut_menu_on_right_clicking_on_edge

- Click on the **Create Plane Parallel to Screen** option from the shortcut menu. A plane parallel to screen will be created; refer to Figure-44.

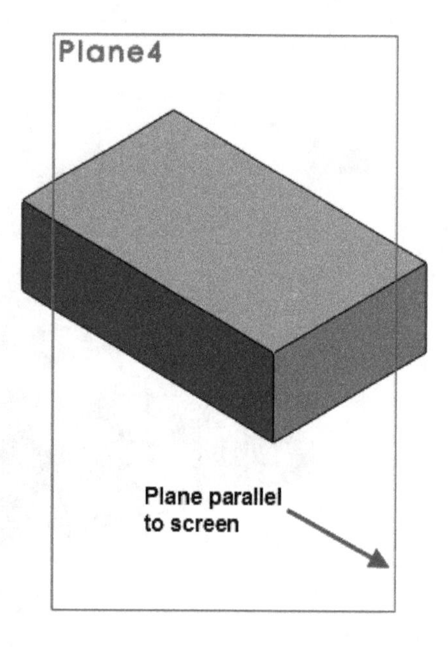

Figure-44. Plane_parallel_to_screen

Axis

The **Axis** tool is used to create reference axes. An axis is useful in creating revolve features or to create planes at angle. To procedure to create axis by using the **Axis** tool is given next.

- Click on the **Axis** tool from the **Reference Geometry** drop-down. The **Axis PropertyManager** will display as shown in Figure-45.

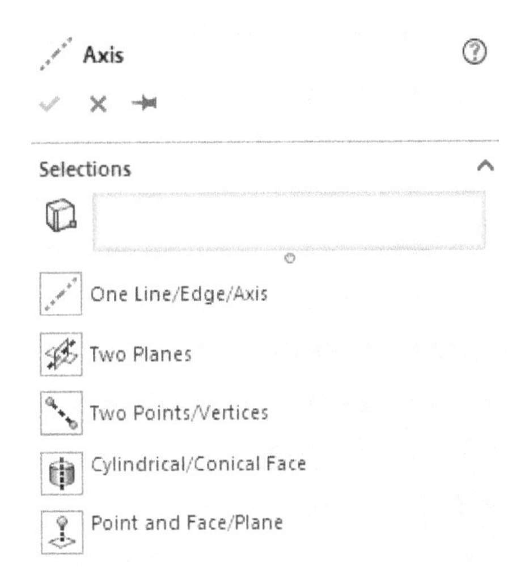

Figure-45. Axis_PropertyManager

- Select desired button from the **PropertyManager**. The buttons in this **PropertyManager** are explained next.

One Line/Edge/Axis

Select this button if you want to create an axis coincident to the selected line/edge/axis. After selecting this button, click on the line/edge/axis. The axis will be created coincident to the selected line/edge/axis; refer to Figure-46.

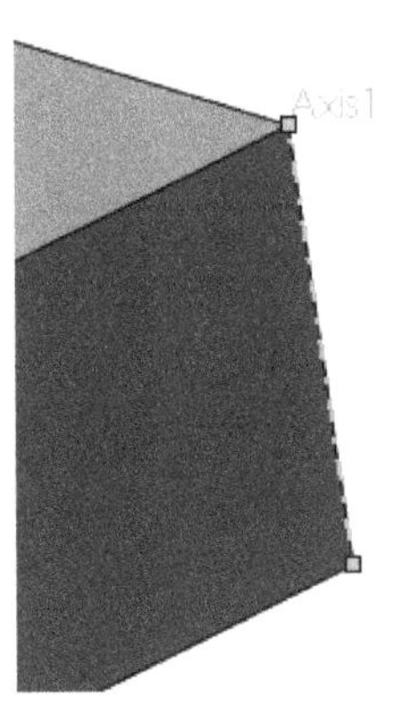

Figure-46. axis_created_on_edge

Two Planes

Select the **Two Planes** button if you want to create an axis at the intersection of the two selected planes/faces. After selecting this button, click on the two intersecting. The axis will be created at the intersection; refer to Figure-47.

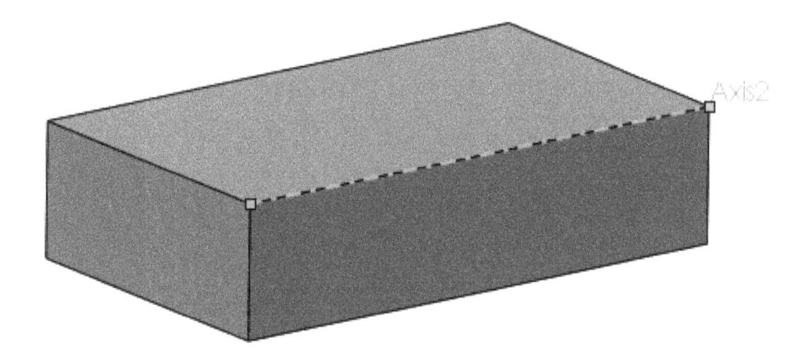

Figure-47. Axis_at_intersection_of_planes

Two Points/Vertices

Select the **Two Points/Vertices** button if you want to create axis passing through the selected two points/vertices; refer to Figure-48.

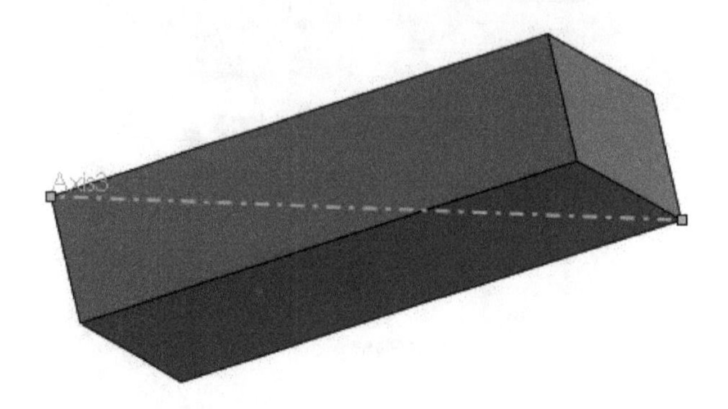

Figure-48. Axis_passing_through_two_points_or_vertices

Cylindrical/Conical Face

Select the **Cylindrical/Conical Face** button and select a cylindrical/conical face. An axis passing through center of cylindrical/conical face will be created; refer to Figure-49.

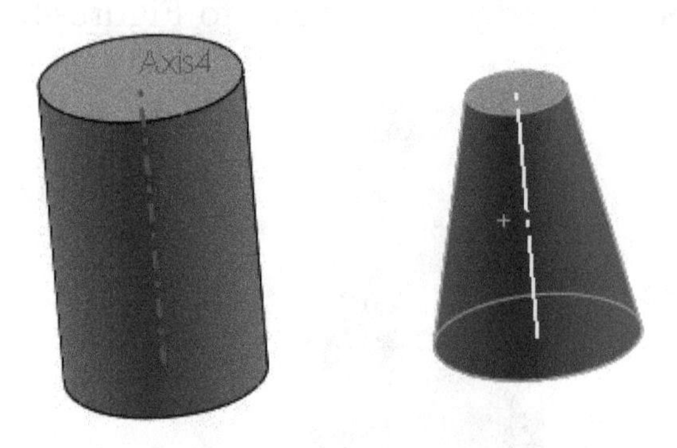

Figure-49. Axis_through_cylinder_and_conical

Point and Face/Plane

Select the **Point and Face/Plane** button if you want to create an axis passing through the selected point and perpendicular to the selected face/plane.

Coordinate System

The **Coordinate System** tool is used to create reference coordinate system. The steps to create coordinate system are explained next.

- Click on the **Coordinate System** tool from the **Reference** drop-down. The **Coordinate System PropertyManager** will display as shown in Figure-50.

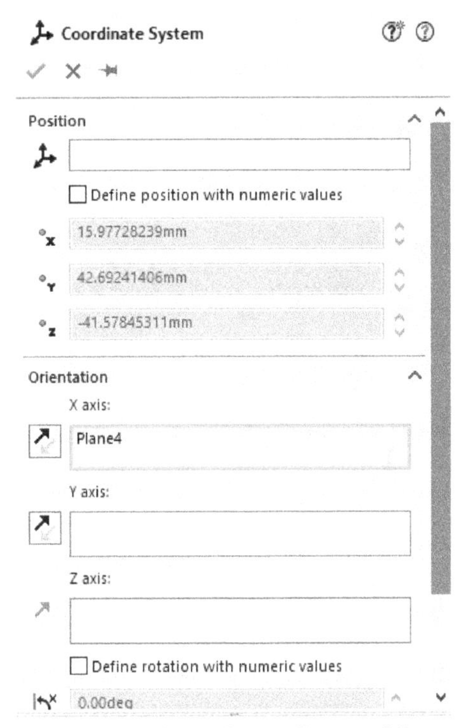

Figure-50. Coordinate System PropertyManager

- Click on the point where you want to place the coordinate system.
- Click in the box for which you want to specify direction reference and select the reference like plane, axis, and so on. Figure-51 shows a coordinate system created on the face.

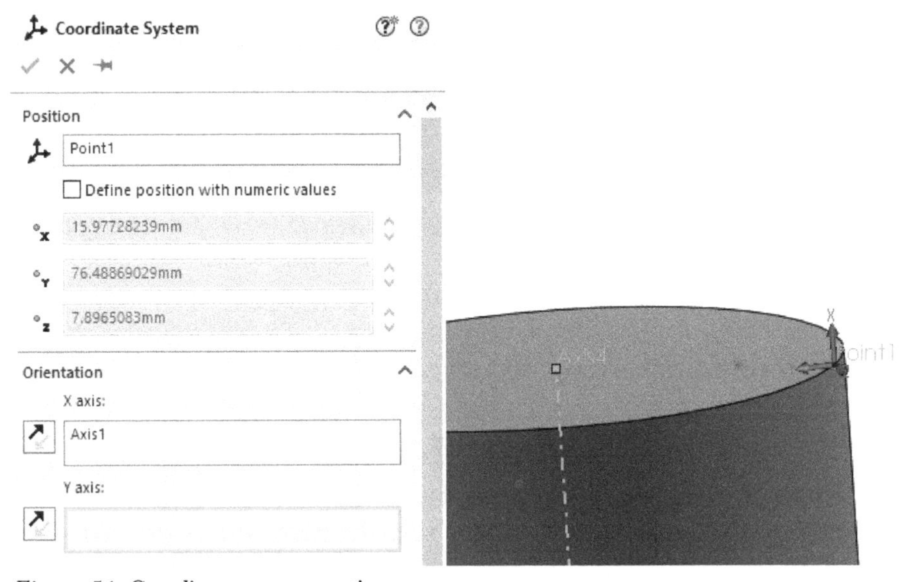

Figure-51. Coordinate system creation

Point

The **Point** tool is used to create reference points on the model. The steps to create points are given next.

- Click on the **Point** tool from the **Reference** drop-down. The **Point PropertyManager** will display as shown in Figure-52.

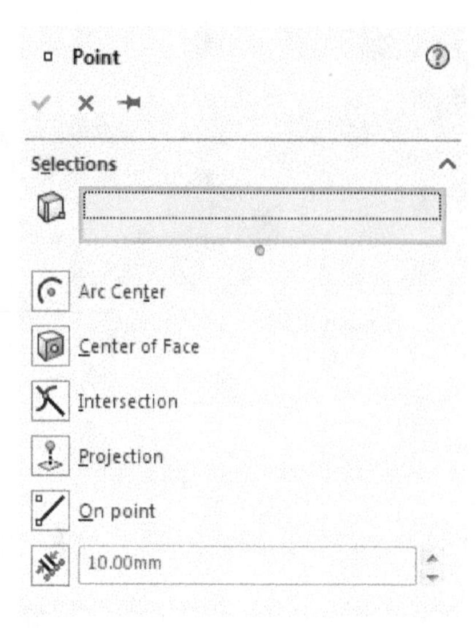

Figure-52. Point PropertyManager

- Select desired button to specify the type of point you want to create. In this case, we have selected **Center of Face** button.
- Select the reference (face of the model in this case). Preview of the point will display; refer to Figure-53.

Figure-53. Preview_of_point

- Click on the **OK** button to create the point.

You can create array of points along a curve by selecting button.

Center of Mass

The **Center of Mass** tool is used to display the center of mass of the model. The coordinates of center of mass are generally required in some calculations related to inertia of the objects. Identification of center of mass is also helpful in checking the stability of object in constraint free environment. To display the center of mass, click on the **Center of Mass** tool from the **Reference** drop-down and the center of mass will display in the viewport; refer to Figure-54.

Figure-54. Center_of_mass_of_cylinder

Mate Reference

Mate References specify one or more entities of a component to use for automatic mating. When you drag a component with a mate reference into an assembly, the SolidWorks software tries to find other combinations of the same mate reference name and mate type. The procedure to use this tool is discussed next.

- Click on the **Mate Reference** tool from the **Reference Geometry** drop-down. The **Mate Reference PropertyManager** will be displayed; refer to Figure-55.

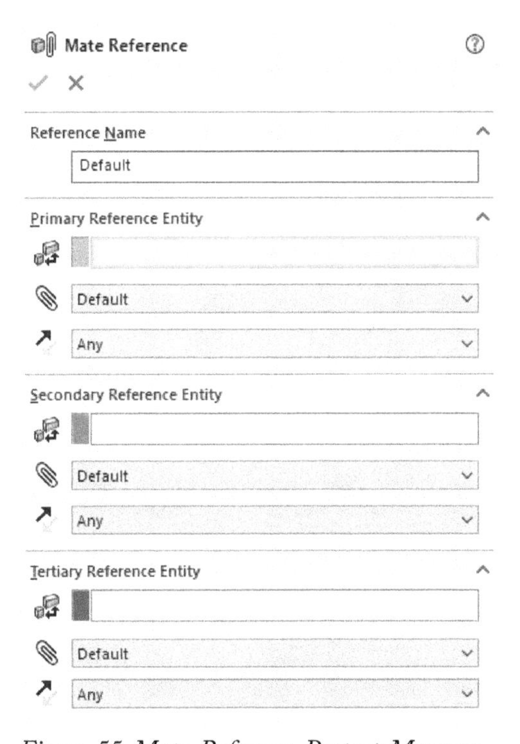

Figure-55. Mate_Reference_PropertyManager

- Specify desired name in the **Mate Reference Name** edit box from **Reference Name** rollout.
- Click in the **Primary Reference Entity** selection box from **Primary Reference Entity** rollout and select desired face, edge, vertex, or plane to be used for potential mates when dragging a component into an assembly.

- Select desired type and alignment of mate reference from **Mate Reference Type** and **Mate Reference Alignment** drop-down.
- Specify the parameters in the **Secondary Reference Entity** and **Tertiary Reference Entity** rollouts as discussed for the **Primary Reference Entity** rollout.
- After specifying desired parameters, click on the **OK** button from the **PropertyManager**. The mate reference will be added in the **Mate References** folder in the **FeatureManager Design Tree**; refer to Figure-56.

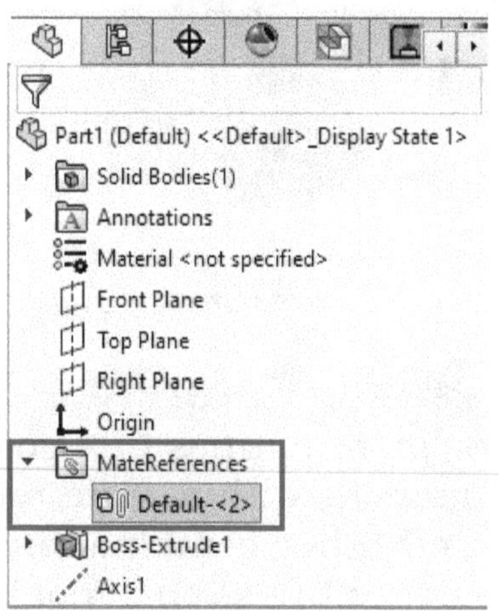

Figure-56. Mate_reference_created

Bounding Box

The **Bounding Box** tool is used to create an envelope for parts created in the drawing area. The procedure to use this tool is given next.

- Click on the **Bounding Box** tool from the **Reference Geometry** drop-down in the **Features CommandManager** of **Ribbon**. The **Bounding Box PropertyManager** will be displayed; refer to Figure-57.

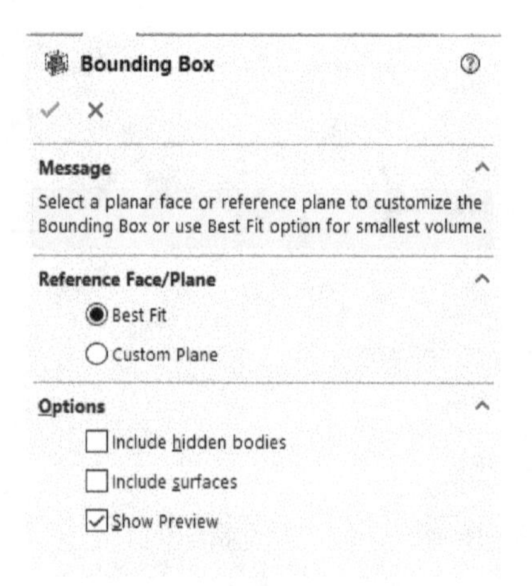

Figure-57. Bounding Box PropertyManager

- Select the **Best Fit** radio button if you want to create the bounding box automatically based on geometry of the part. Select the **Custom Plane** radio button if you want to create bounding box based on selected plane/face. The selection box for plane will be displayed. Click in the selection box and select desired face/ plane to be used as boundary reference for bounding box.
- Select the **Include hidden bodies** check box to include hidden bodies of part while creating the bounding box.

- Select the **Include surfaces** check box to include surfaces while creating the bounding box.
- Select the **Show Preview** check box to display preview while create the box.
- After setting desired parameters, click on the **OK** button from the **PropertyManager**.

LOFTED BOSS/BASE TOOL

The **Lofted Boss/Base** tool is used to create a solid volume joining two or more sketches created on different planes; refer to Figure-58. The procedure to create lofted features is given next.

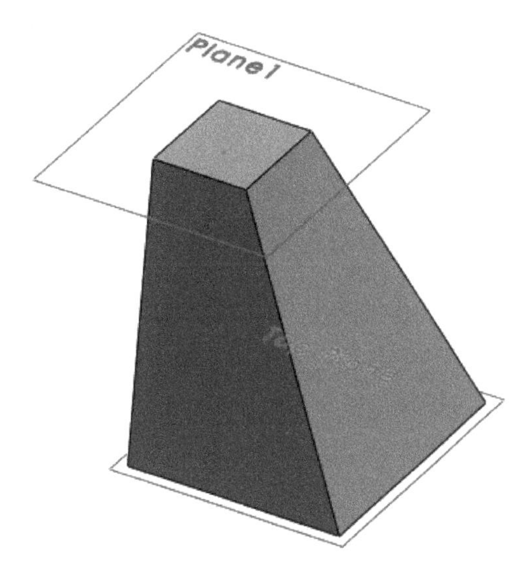

Figure-58. Lofted feature example

- Click on the **Lofted Boss/Base** tool from the **Ribbon**. The **Loft PropertyManager** will display as shown in Figure-59.

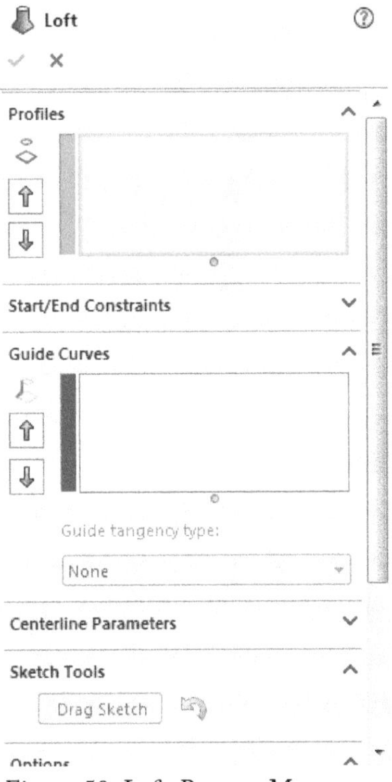

Figure-59. Loft_PropertyManager

- By default, **Profiles** selection box is active and you are asked to select sketches for lofted feature.
- Click one by one on the sketches created at different planes. Note that you need to select the sketches in the order by which they can be joined to each other successively. The preview of the lofted feature will display as shown in Figure-60.

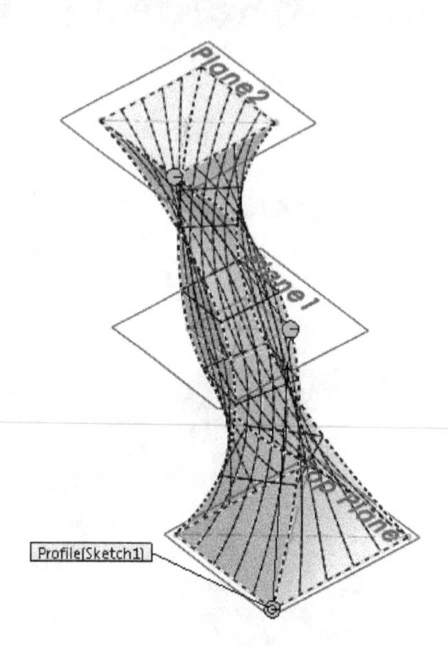

Figure-60. Preview_of_the_lofted_feature

- Drag the green handle to align edges of the lofted feature. After aligning the edges, the above figure will be displayed as shown in Figure-61.

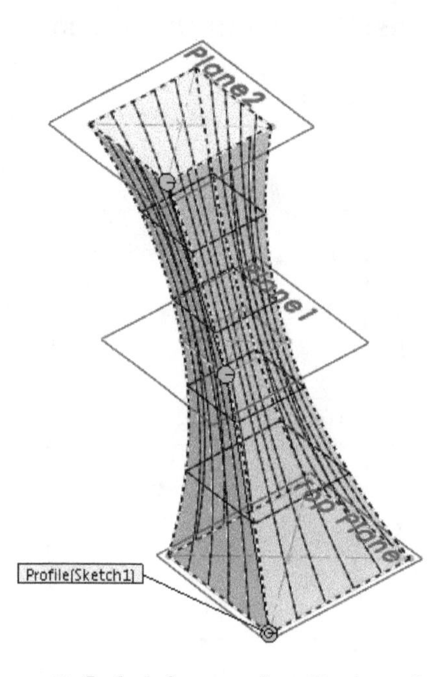

Figure-61. Lofted_feature_after_aligning_edges

- If you want to change the starting or end conditions of the loft feature then expand the **Start/End Constraints** rollout and select desired option from the **Start Constraint** and **End Constraint** drop-down. Figure-62 shows the preview of model after changing the constraints.

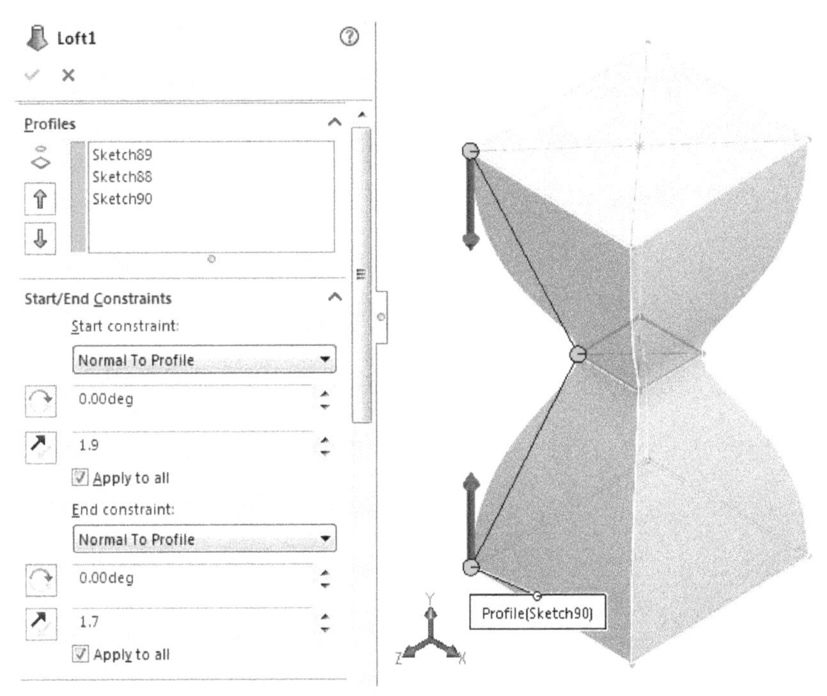

Figure-62. Preview_after_changing_constraints

- If you have guide curve/curves, then click in the **Guide Curves** selection box and select them to refine the shape of lofted feature; refer to Figure-63.

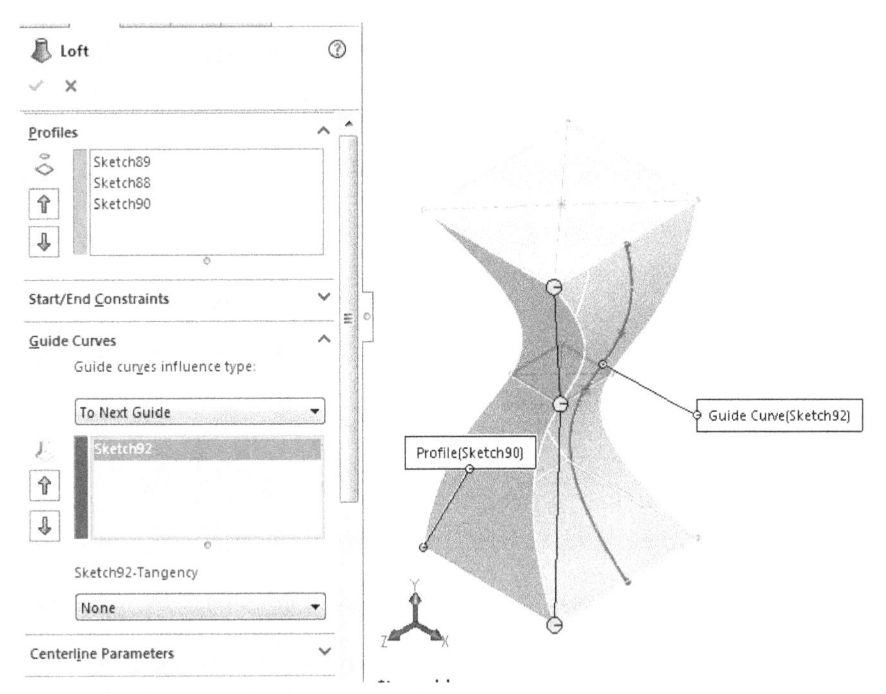

Figure-63. Preview_of_loft_after_selecting_guide_curves

- If you have a curve passing through center by which you want to control the shape of feature then expand the **Centerline Parameters** rollout and click in the selection box of this rollout. You are asked to select a center line or center curve
- Select desired curve, the lofted feature will be modified accordingly.
- Next, click in the **Options** rollout to change the basic options of the lofted feature.
- Select the **Merge tangent faces** check box to merge the lofted feature with the adjoining tangent features.

- The **Close loft** check box is selected if you want to create a closed lofted feature. Note that to use this option, you need to clear the selection of guide curves by right-clicking in the **Guide Curves** box to use this feature. Figure-64 shows a closed lofted feature.

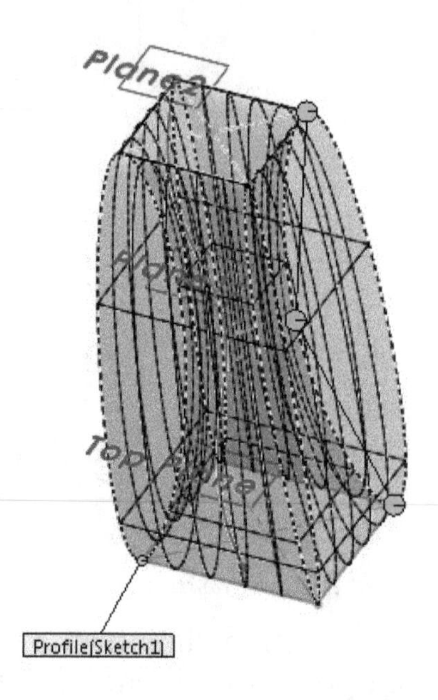

Figure-64. Preview_of_closed_lofted_feature

- Clear the **Merge Result** check box if you want to create it as an individual object and don't want to join it with other solids in the model.
- Select the **Show preview** check box to display preview of loft feature.
- Select the **Micro tolerance** check box if there are small edges in the model.
- Expand the **Curvature Display** rollout to check curvature of loft feature with different type of plots. Select the **Mesh preview** check box to check meshing preview. Select the **Zebra stripes** check box to check zebra strips preview of loft feature. Select the **Curvature combs** check box to check curvature lines on the loft feature.
- After setting desired parameters, click on the **OK** button from the **PropertyManager**.

Note that the **Merge Result** check box will be available for each feature creation tool like **Extrude** and **Revolve**, after you have created the first solid/surface feature in the viewport. As discussed earlier, you can create a thin feature by using the options in the **Thin Feature** rollout.

BOUNDARY BOSS/BASE TOOL

The **Boundary Boss/Base** tool is used to create a solid volume by joining curves in different directions. On selecting this tool, the **Boundary PropertyManager** is displayed as shown in Figure-65. The tool works in the same way as the **Lofted Boss/Base** tool do. But using this tool, you can simultaneously analyze the surface of the feature being created and make it smoother or rougher as per requirement. Figure-66 shows the preview of the boundary boss/base feature with the mesh preview, Zebra stripes, and Curvature combs.

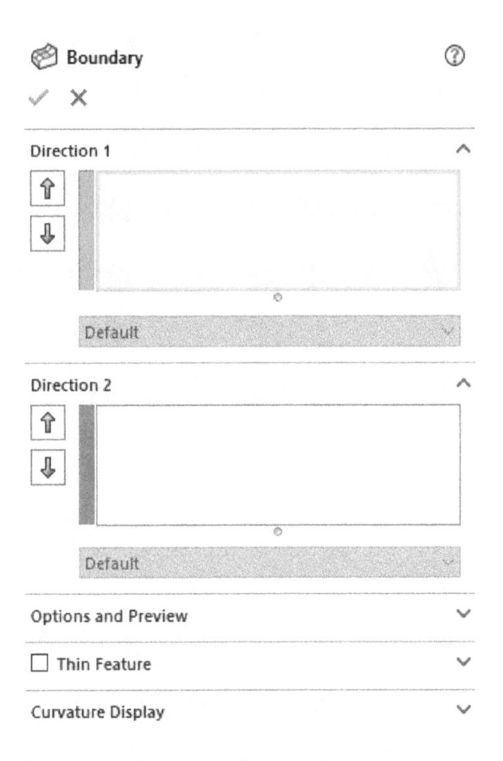

Figure-65. Boundary PropertyManager

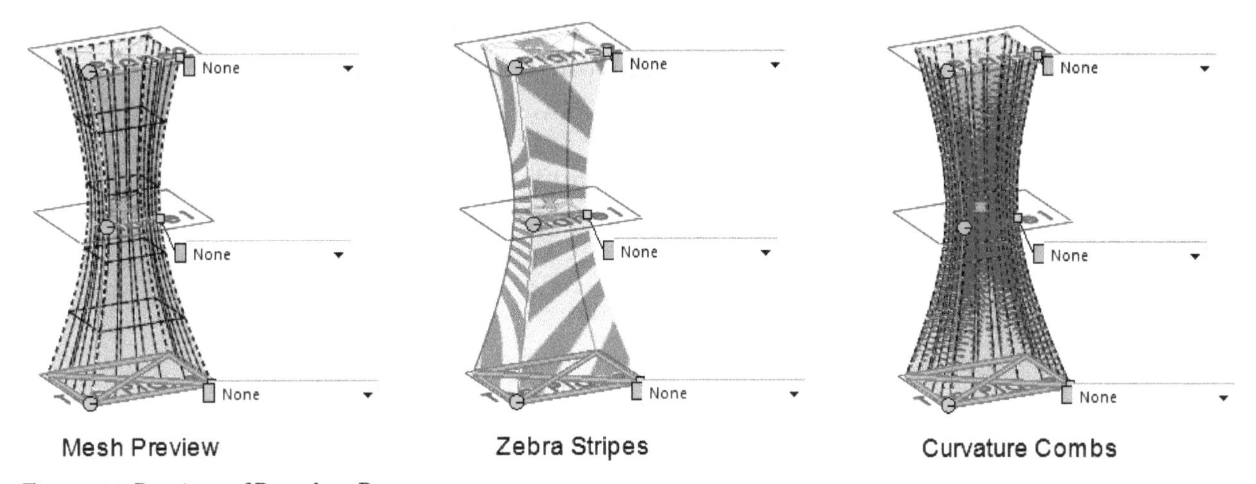

Figure-66. Previews of Boundary Boss

The procedure to use the **Boundary Boss/Base** tool is given next.

- Create at least two sketch profiles for boundary base/boss feature. Note that the profiles should be intersecting each other at one point; refer to Figure-67.

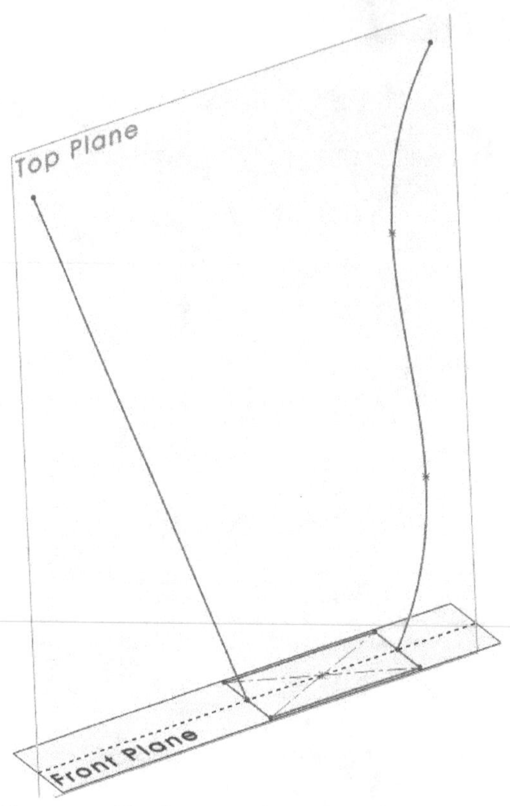

Figure-67. Sketches_created_for_Boundary_Base_tool

- Click on the **Boundary Boss/Base** tool from the **Features CommandManager**. The **Boundary PropertyManager** will be displayed; refer to Figure-65.
- Select the base sketch. The sketch will be selected in **Direction 1** selection box.
- Click in **Direction 2** selection box in **PropertyManager** and then select the guide sketches. Preview of the boundary feature will be displayed; refer to Figure-68.

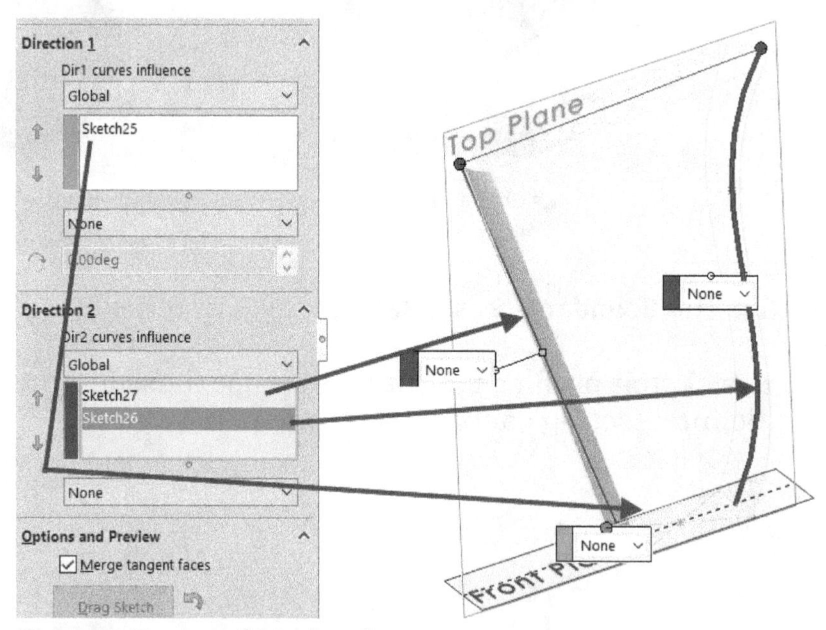

Figure-68. Preview_of_boundary_feature

- Set the other parameters in **PropertyManager** as required and then click on the **OK** button to create the feature.

REMOVING MATERIAL FROM SOLID OBJECTS

Till this point, you have created base/boss features by using various feature creation tools. But in Engineering, machining an object means removing material from it. So, it is equally important to know the ways you can remove material from objects in SolidWorks. All the tools to remove material are available in the next column to the one discussed earlier; refer to Figure-69.

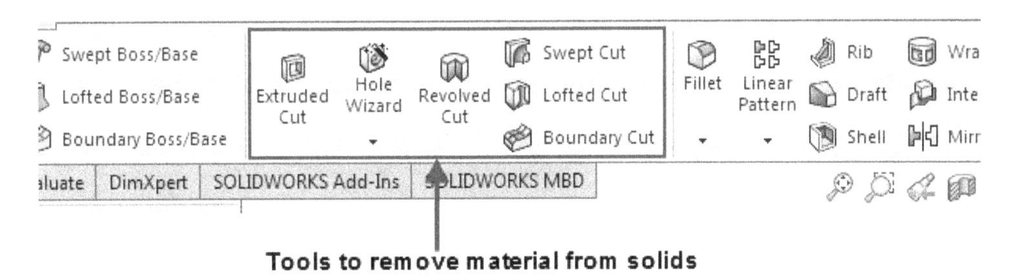

Figure-69. Tools_to_remove_material

The tools except the **Hole Wizard** work in the same way as their respective Boss/ Base tool. All these tools are explained one by one as follows:

Extruded Cut

The **Extrude Cut** tool is used to remove material by extruding the sketch. The steps to use this tool are given next.

- Click on the **Extruded Cut** tool. The **Extrude PropertyManager** will display.
- Select the face from which you want to start removing the material. The sketch environment will activate.
- Click the sketch using which you want to remove the material.
- Click on the **Exit Sketch** button from the **Ribbon**. The preview of cut feature will display; refer to Figure-70.

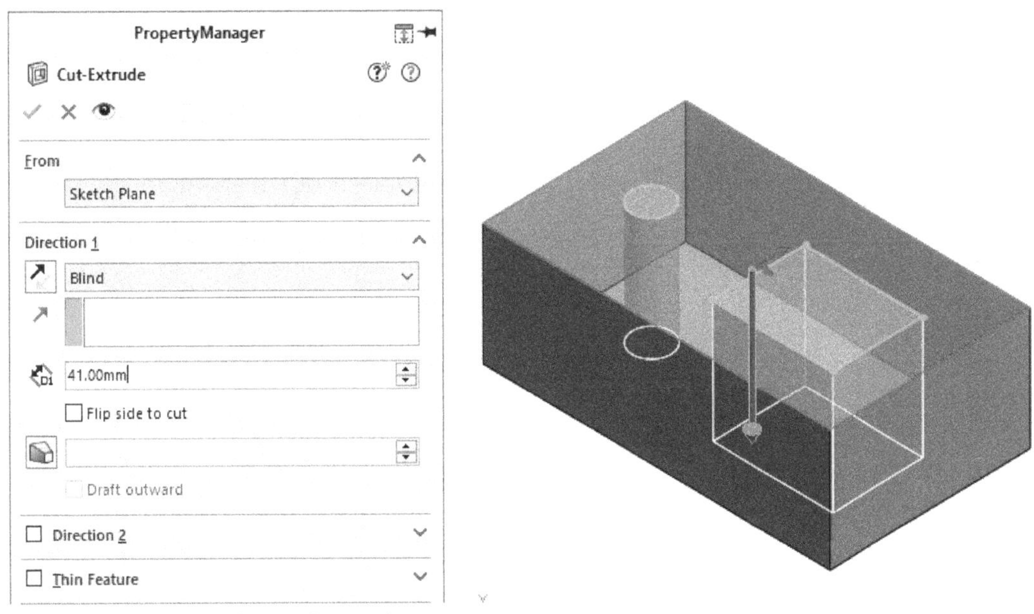

Figure-70. Preview_of_extrude_cut_feature

- Specify the depth of material removal and other parameters in the **PropertyManager** as discussed for **Extrude Boss/Base** tool and then click on the **OK** button.

Revolved Cut

The **Revolve Cut** tool is used to remove material by revolving the sketch. The steps to use this tool are given next.

- Click on the **Revolved Cut** tool. The **Revolve PropertyManager** will display.
- Select the face from which you want to start removing the material. The sketch environment will be activated.
- Create the sketch of cut feature and a center line then click on the **Exit Sketch** button. Preview of revolved cut feature will display; refer to Figure-71.

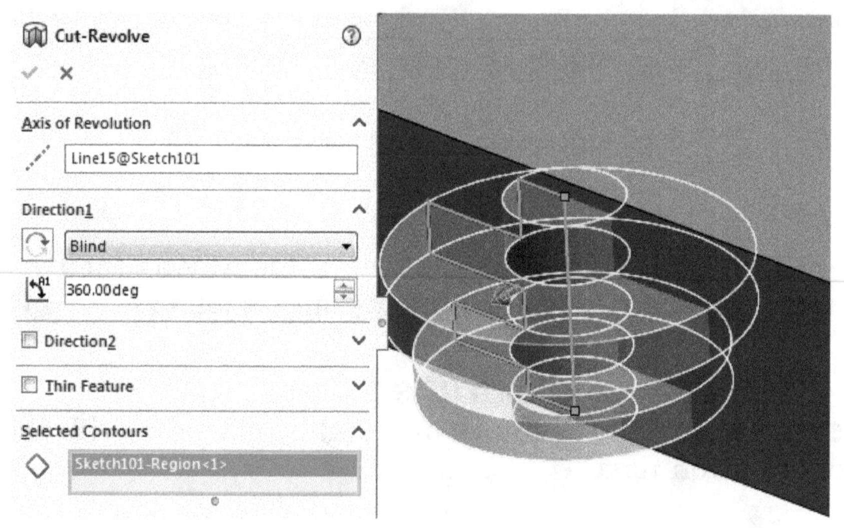

Figure-71. Preview of revolve cut feature

- Specify the parameters as discussed for **Revolved Boss/Base** tool and click on **OK** button to remove material.

Swept Cut

The **Swept Cut** tool is used to remove material by sweeping a section along the specified path. The steps to use this tool are given next.

- Click on the **Swept Cut** tool. The **Cut-Sweep PropertyManager** will display; refer to Figure-72. By default, the **Sketch Profile** radio button is selected in the **Cut-Sweep PropertyManager** and you will be asked to select a sketch section.

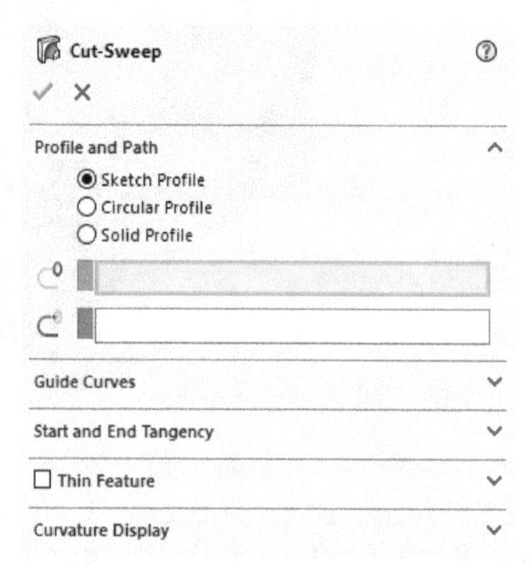

Figure-72. Cut Sweep PropertyManager

- Select a closed sketch section and then an open sketch for path.
- Click on the **OK** button from the **PropertyManager** to create the swept cut; refer to Figure-73.

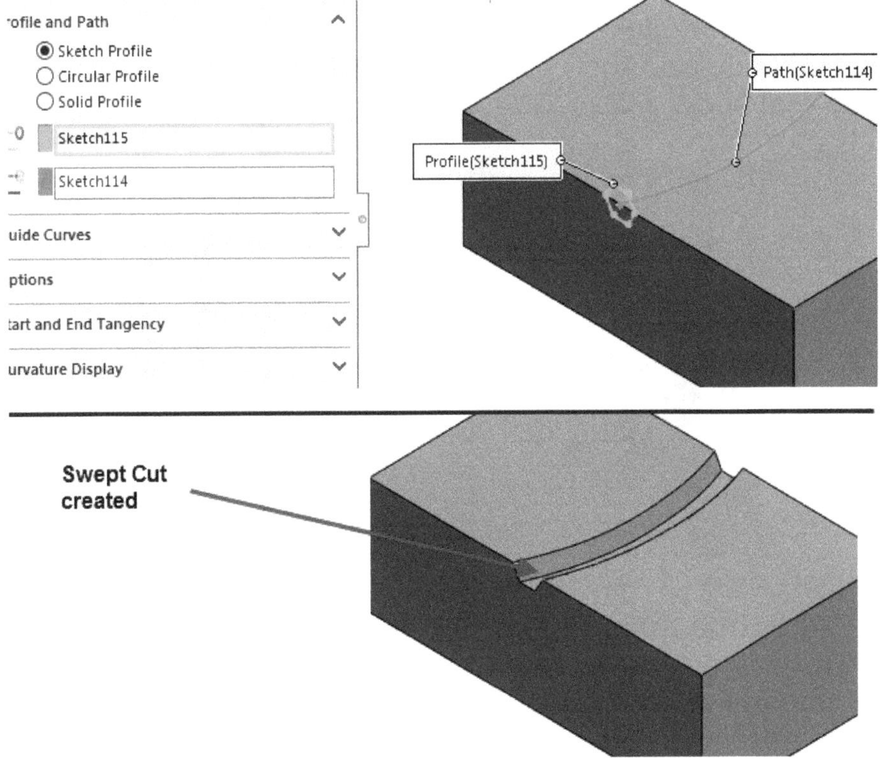

Figure-73. Swept cut with sketch profile

Swept Cut with Circular Profile

- Select the **Circular Profile** radio button from the **Cut-Sweep PropertyManager**. You are asked to select a path for the swept cut.
- Select an open sketch for the path. Preview of the cut will be displayed; refer to Figure-74.

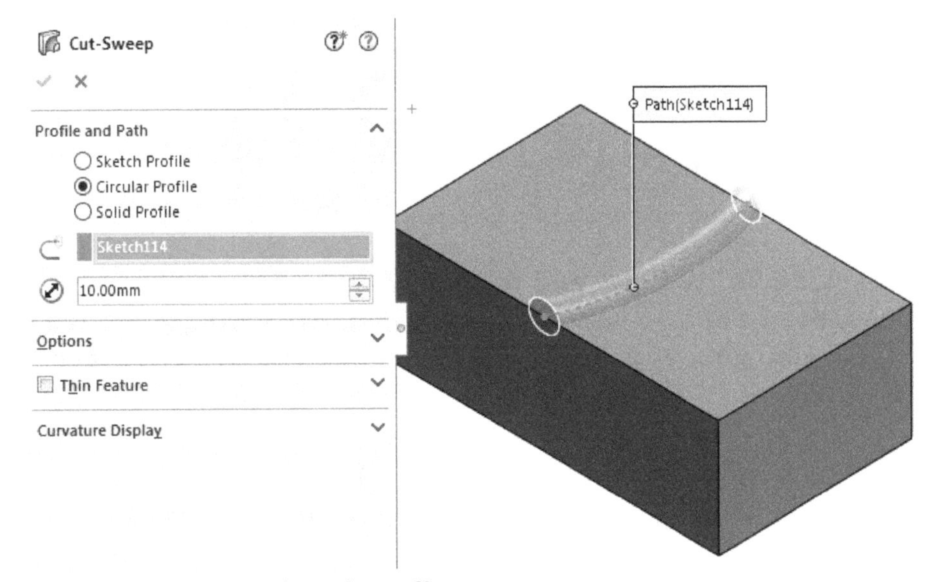

Figure-74. Swept cut with circular profile

- Specify desired value for diameter of circular profile and click on the **OK** button from the **PropertyManager** to create the feature.

Swept Cut with Solid Profile

- Select the **Solid Profile** radio button from the **PropertyManager**. You are asked to select a solid revolve or cylindrical extrude feature as tool body. Note that the solid should be created by using only arc and line, and should not be merged with other bodies in model.
- Select the solid body and then select the path. Note that path should start from or within the tool body; refer to Figure-75.

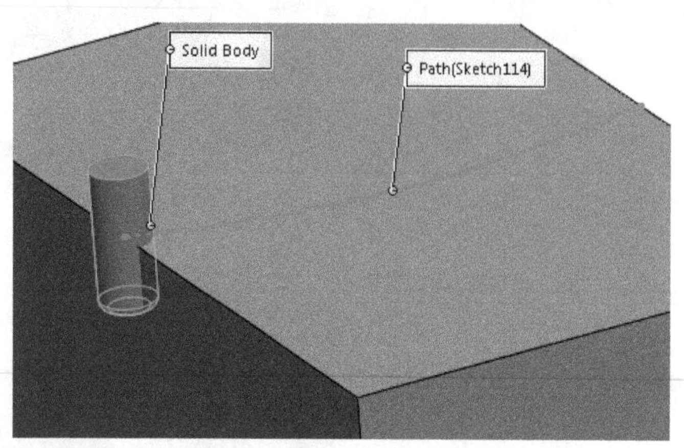

Figure-75. Solid_body_and_path

- Click on the **OK** button from the **PropertyManager** to create the swept cut; refer to Figure-76. This option can be useful in checking the tool impression on a solid body before making any CAM programming.

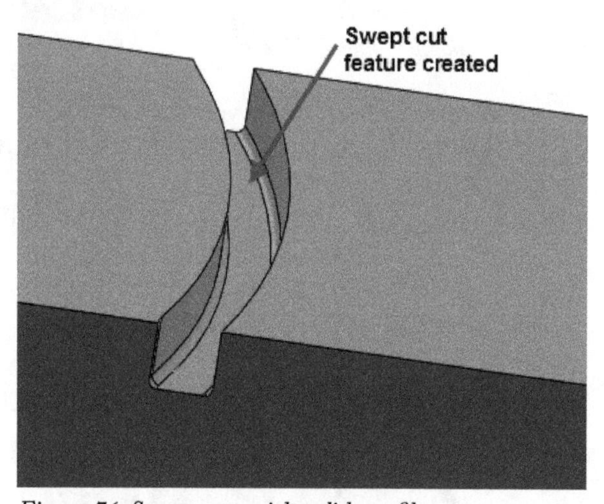

Figure-76. Swept_cut_with_solid_profile

The other tools like Lofted Cut and Boundary Cut work in the same way as their respective Boss/Base tool. I would be very happy if you practice those tools by yourself. If you get any doubt, please let me know at cadcamcaeworks@gmail.com

Hole Wizard

The **Hole Wizard** tool is used to create holes that comply with the real machining tools. SolidWorks has library of standard holes and slots that can be created in the solid model. You can use this standard library or you can create a customized hole/slot by using the tool. The procedure to use this tool is given next.

- Click on the **Hole Wizard** tool from the **Hole Wizard** drop-down in the **Ribbon**. The **Hole Specification PropertyManager** will display as shown in Figure-77.

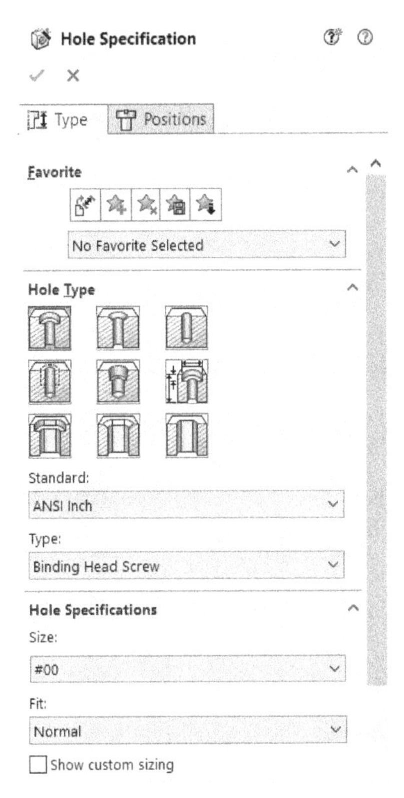

Figure-77. Hole Specification PropertyManager

- The **Favorite** rollout is used to store and reuse specific type of holes that you need again and again. Refer to Figure-78.

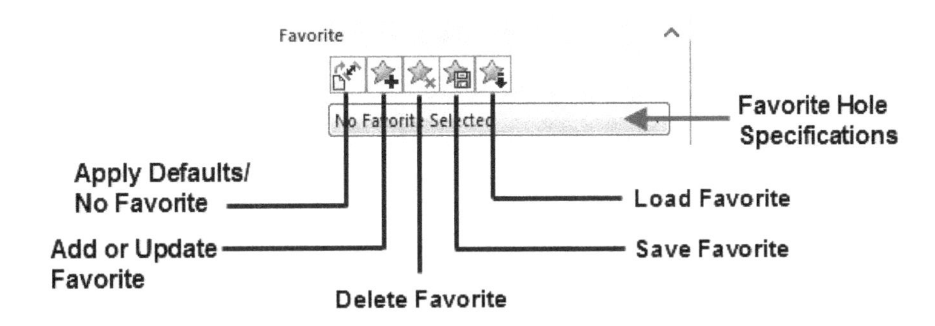

Figure-78. Favorite_rollout

- Click on desired type of hole from the **Hole Type** rollout. The parameters of the selected hole type will display in the **PropertyManager**.
- Select desired hole standard and hole type from the **Standard** and **Type** drop-downs, respectively. The sizes of holes related to selected hole standard and type will display in the **Size** drop-down in the **Hole Specification** rollout.
- Click on the **Show custom sizing** check box if you want to customize the hole.
- Enter desired parameters and select the options as per your need from the other rollouts.
- Click on the **Add or Update Favorite** button from the **Favorite** rollout if you want to use this hole many times in the model.
- Set desired tolerance values in **Tolerance/Precision** rollout for the hole.

- Now, click on the **Positions** tab to specify the position of the hole. The **PropertyManager** will display as shown in Figure-79.

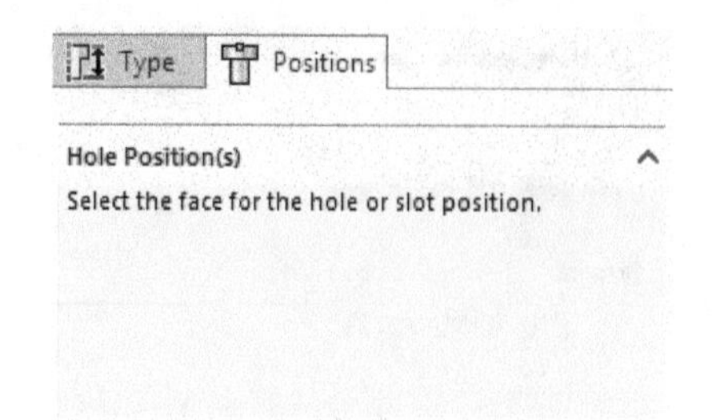

Figure-79. Positions_tab_of_Hole_Specification_PropertyManager

- Click on the face of the solid model where you want to place the hole and then click to specify the position of the hole. Note that later you can position it with dimensions.
- Click on the **OK** button to create the hole.
- On creating the hole, a node of it will be added in the **FeatureManager Design Tree**. Expand this node; refer to Figure-80.

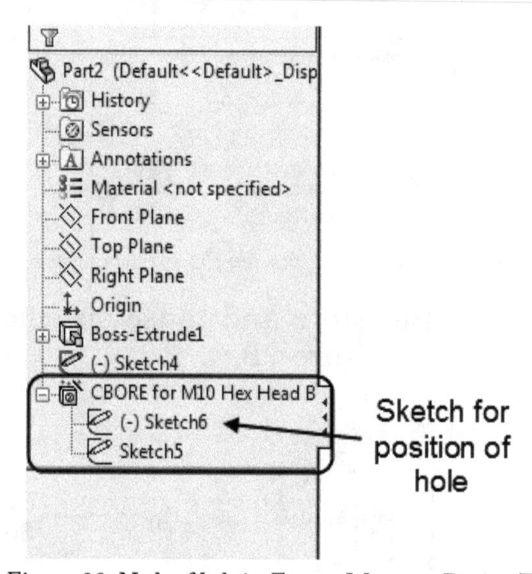

Figure-80. Node of hole in FeatureManager Design Tree

- Right-click on the sketch for position of the hole and select the **Edit Feature** button from the tool box; refer to Figure-81.

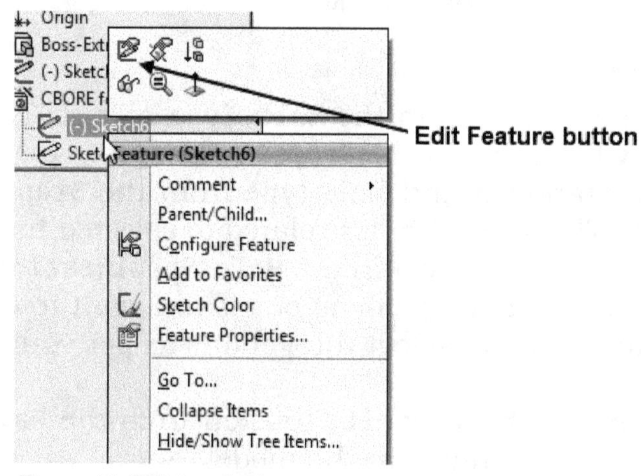

Figure-81. Editing sketch

- The sketching environment will be displayed.

- Here, specify the position of point by dimensioning. Note that here you can create multiple points by using the **Point** tool and the holes will be created on all the points you created. Refer to Figure-82.

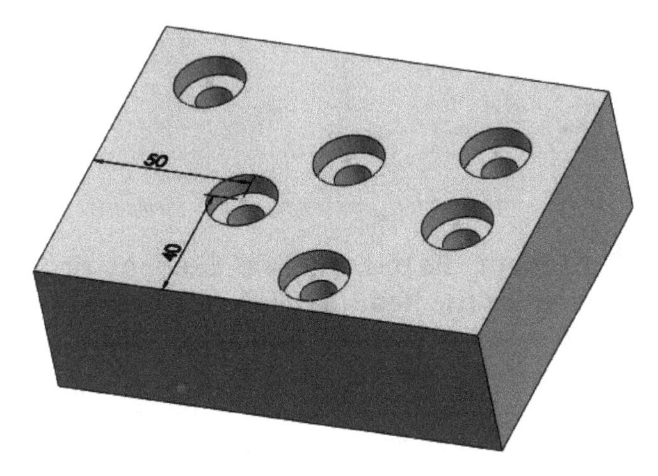

Figure-82. Holes created at sketched points

Advanced Hole

The **Advanced Hole** tool is used to dynamically create holes in solids. This tool was first time introduced in SolidWorks 2017 version. The procedure to use this tool is given next.

- Click on the **Advanced Hole** tool from the **Hole Wizard** drop-down in the **Features CommandManager**. The **Advanced Hole PropertyManager** will be displayed along with the **Near Side** flyout; refer to Figure-83.

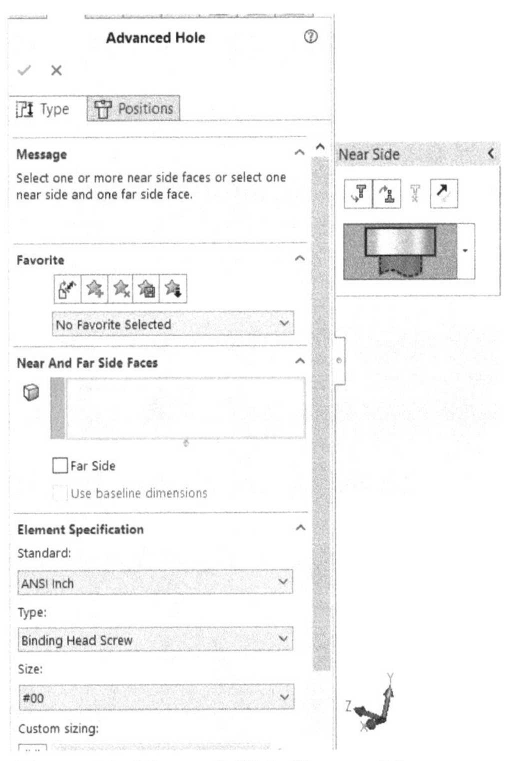

Figure-83. Advanced_Hole_PropertyManager

- Select the face to place the top of hole. Set the parameters for current element of hole in the **PropertyManager**; refer to Figure-84.

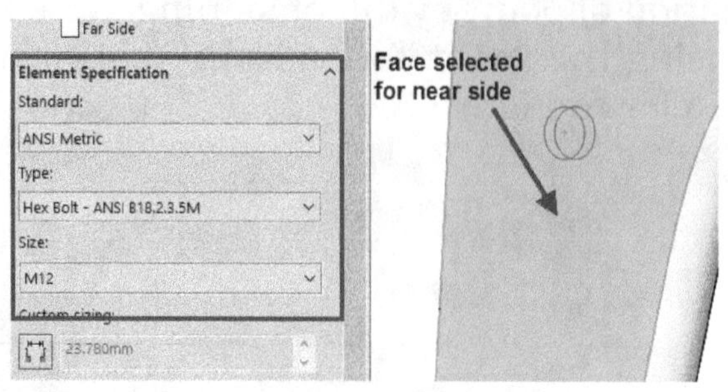

Figure-84. Specifying_parameters_of_hole_element

- Click on the **Insert Element Below Active Element** or **Insert Element Above Active Element** button from the **Near Side** flyout to insert more elements of hole in the near side. A new element will be added to the hole and in the **Near Side** flyout.
- Click on the down arrow next to newly added element and select desired type of element from the drop-down displayed; refer to Figure-85.

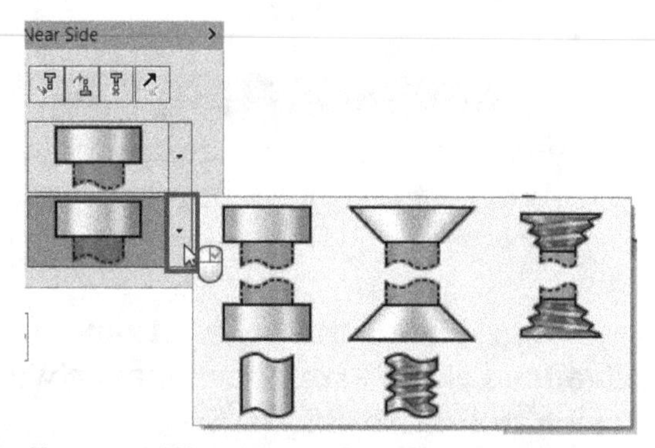

Figure-85. Element types drop-down

- Set the parameters for newly added element in the **PropertyManager** as discussed earlier.
- Select the **Far Side** check box from the **Near and Far Side Faces** rollout in the **PropertyManager** if you want to specify the termination face of hole. The selection box for **Far Side** will be displayed in the rollout.
- Select the face for far side and select desired elements from **Far Side** flyout displayed along the **PropertyManager**; refer to Figure-86.

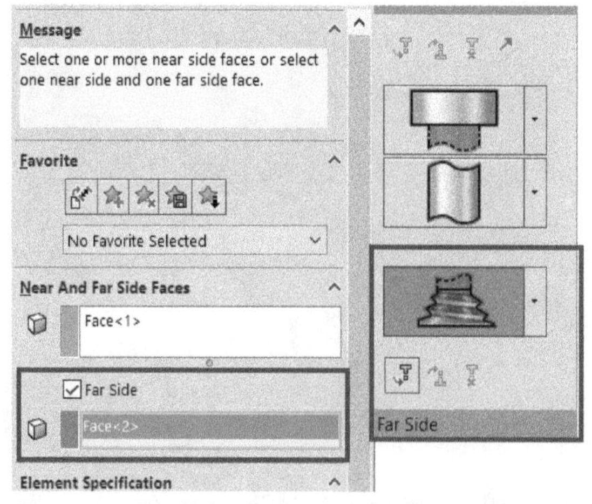

Figure-86. Far Side selection box and flyout

- Set the parameters as discussed earlier

- Click on the **Position** tab in the **PropertyManager** and define the position of the hole.
- Click on the **OK** button to create the hole.

Thread

The **Thread** tool is used to cut helical thread on cylindrical faces. Using this tool, you can save custom threads in library. The procedure to use this tool is discussed next.

- Click on the **Thread** tool from the **Hole Wizard** drop-down in the **Features** tab of the **Ribbon**; refer to Figure-87. The **Thread PropertyManager** will be displayed; refer to Figure-88.

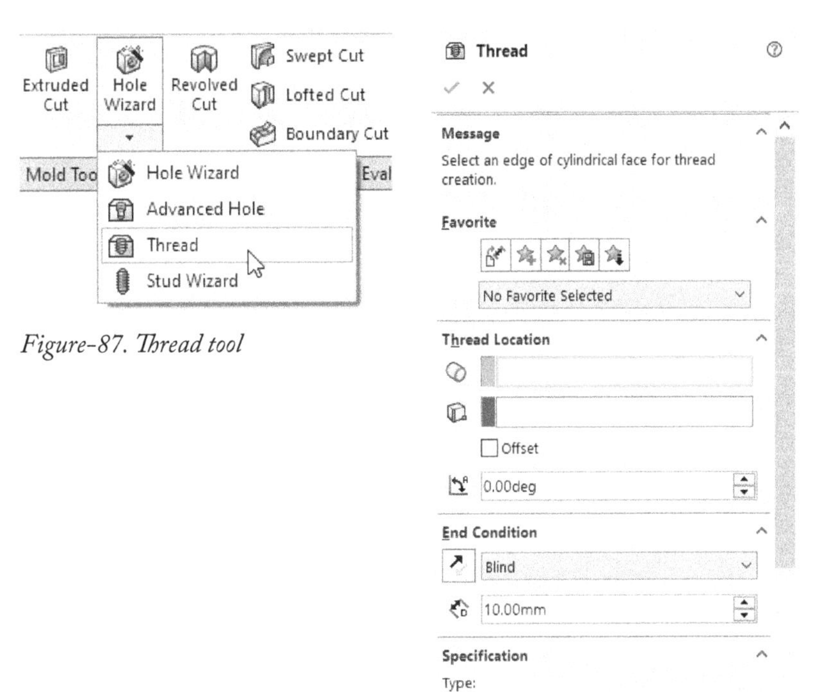

Figure-87. Thread tool

Figure-88. Thread PropertyManager

- Select the round edge of the cylindrical face of hole/boss feature. Preview of the thread will be displayed; refer to Figure-89.

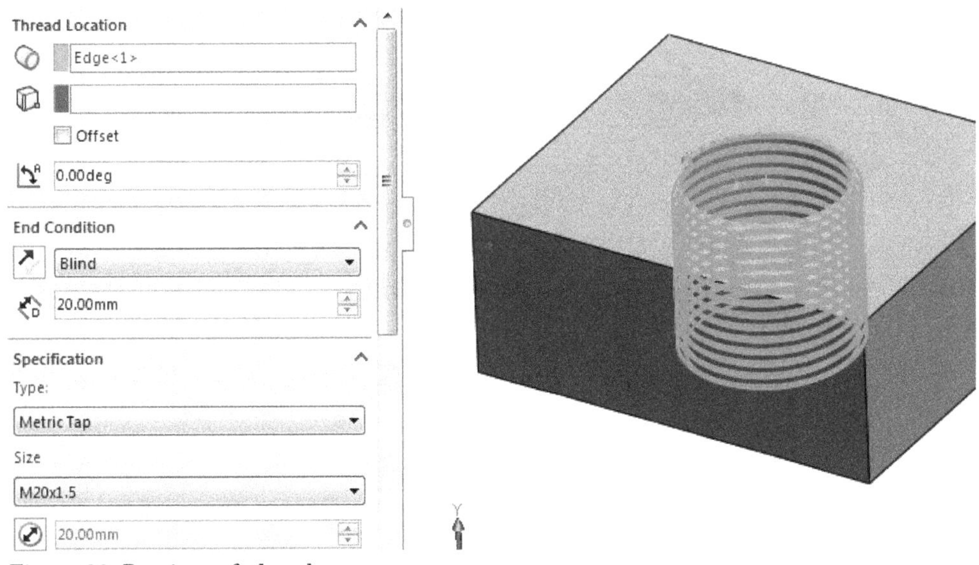

Figure-89. Preview_of_thread

- Select the starting location of the thread after clicking in the **Start Location** selection box. Note that this option is optional and you may not require it in your operation.
- Select the **Offset** check box if you want to start the thread at an offset distance from the starting location. After selecting this check box, an edit box becomes active below it where you can specify the value of offset; refer to Figure-90.

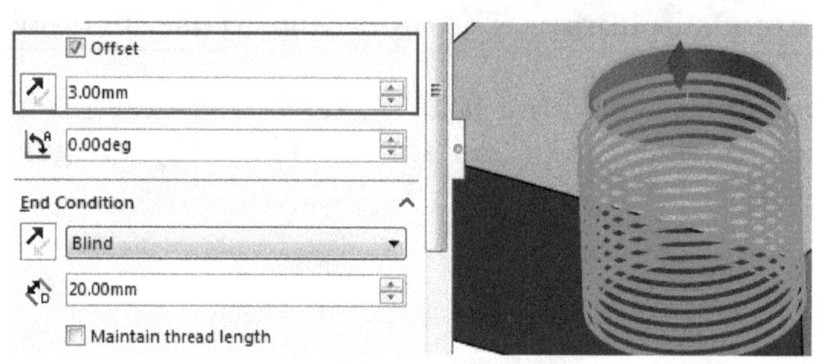

Figure-90. Thread_at_offset_distance

- From the **End Condition** rollout, specify the depth of the thread. By default, the **Blind** option is selected in the drop-down and you need to specify the depth of thread in the edit box below it.
- Select desired threading tool from the **Type** drop-down in the **Specification** rollout of the **PropertyManager**; refer to Figure-91.

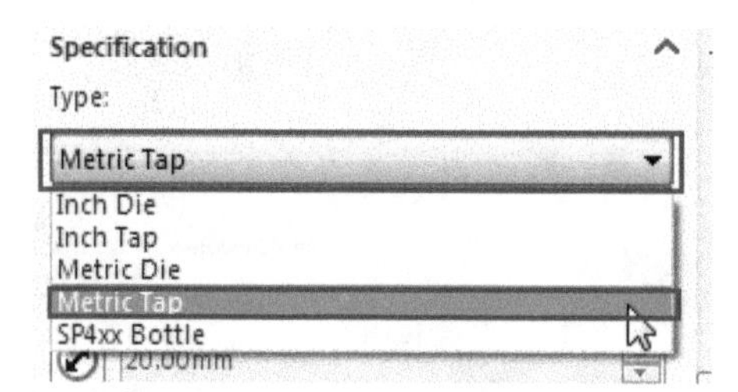

Figure-91. Type_drop-down_for_threading_tools

- Select the thread size from the **Size** drop-down below the **Type** drop-down in the **Specification** rollout. If you want to specify custom size of thread then click on the **Override Diameter** and **Override Pitch** buttons below the drop-down and specify desired values; refer to Figure-92.

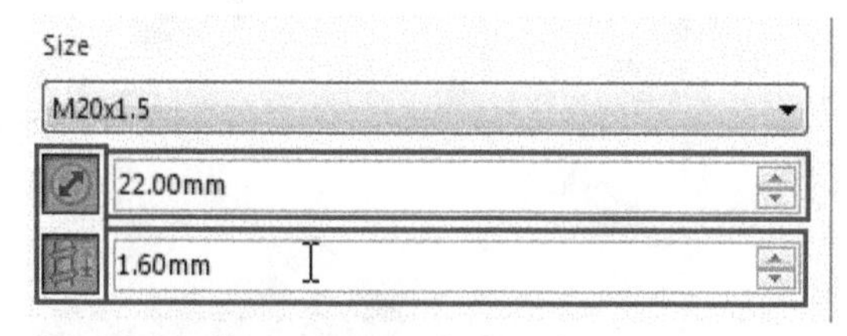

Figure-92. Custom_size_of_thread

- Select desired method from the **Thread Method** area of the rollout. There are two radio buttons available in this area; **Cut thread** and **Extrude thread**. Select the **Cut thread** radio button if you are creating thread inside a hole like in nut. Select the **Extrude thread** radio button if you want to create thread on a boss feature like bolt.

- You can cut/extrude the mirror image of current thread specifications by using the **Mirror Profile** check box. On selecting this check box, the **Mirror horizontally** and **Mirror vertically** radio buttons will become active. Select the **Mirror horizontally** radio button if you want to use horizontal mirror image of the current thread profile or select the **Mirror vertically** radio button if you want to use the vertical mirror image of the current thread profile; refer to Figure-93.

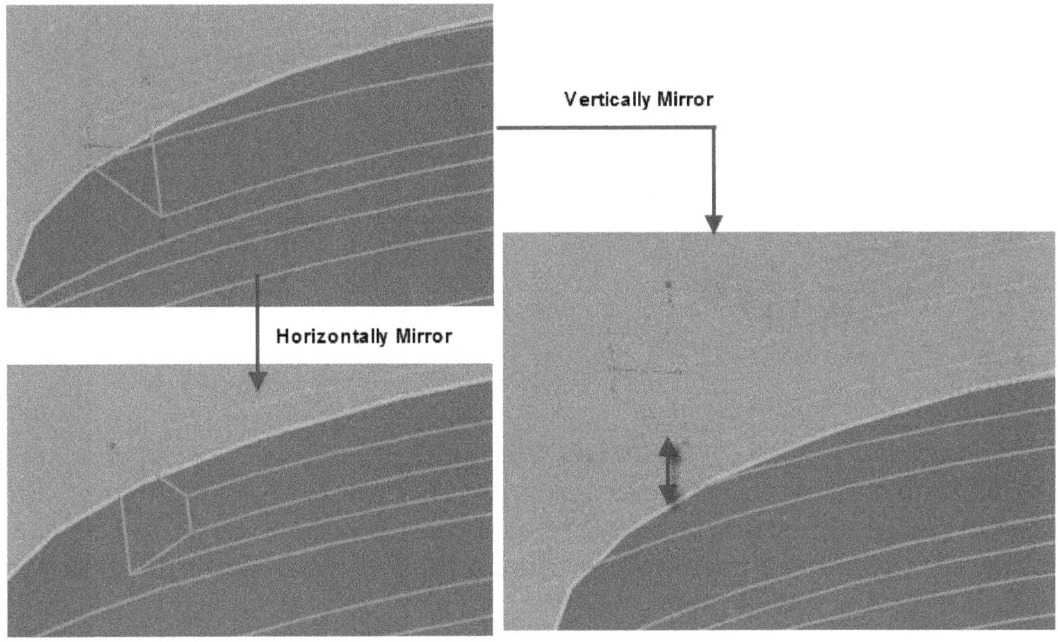

Figure-93. Mirroring_thread_profile

- Specify the thread angle using the **Rotation Angle** edit box.
- From the **Thread Options** rollout, you can select the thread as left handed or right handed.
- If you want to create a multiple start thread then select the **Multiple Start** check box from the **Thread Options** rollout and specify desired number of starts in the spinner below the check box; refer to Figure-94. Note that the thread pitch value must permit multiple starts without causing crossing, self-intersecting threads.

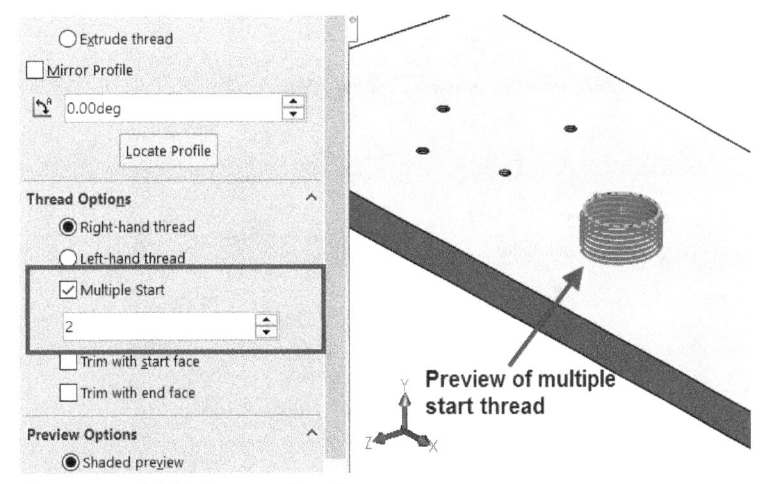

Figure-94. Multiple start thread

- To trim the thread at starting or end, select the **Trim with start face** or **Trim with end face** check box, respectively from the **Thread Options** rollout.
- Check the preview of threads by using desired radio button from the **Preview Options** rollout in the **PropertyManager**.

- Click on the **OK** button from the **PropertyManager** to create the thread; refer to Figure-95.

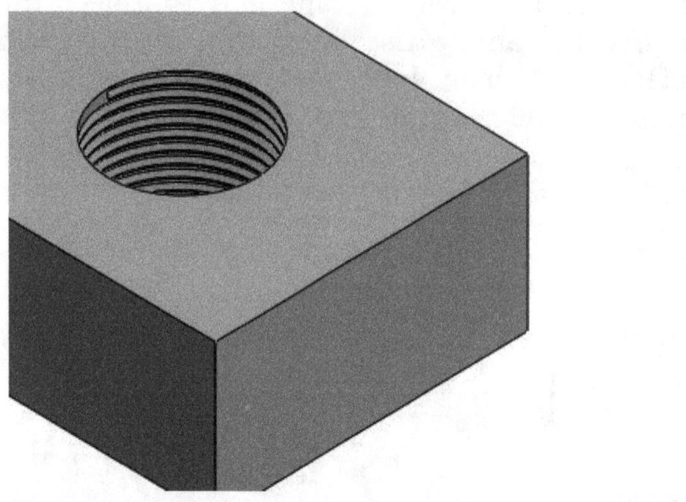
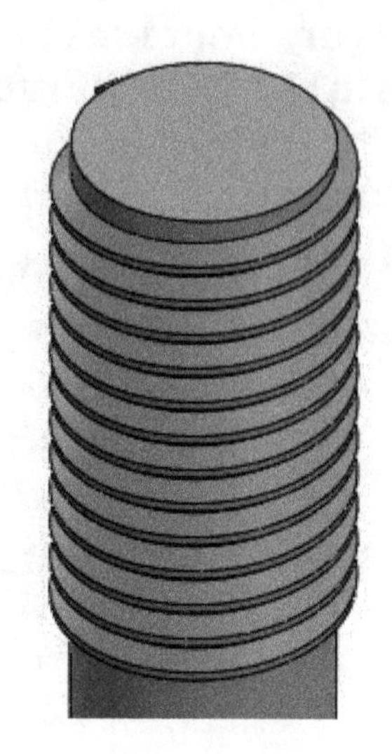

Figure-95. Threaded_solids

Stud Wizard

The **Stud Wizard** tool is used to create external threaded stud features. For using this tool, you define the stud parameters and then position the studs on the model. You can also apply thread parameters to existing cylindrical shafts. The procedure to use this tool is discussed next.

- Click on the **Stud Wizard** tool from **Hole Wizard** drop-down in the **Features** tab of the **Ribbon**. The **Stud Wizard PropertyManager** will be displayed with the **Creates Stud on a Cylindrical Body** option selected; refer to Figure-96. You will be asked to select the edge of cylindrical face to create the stud wizard.

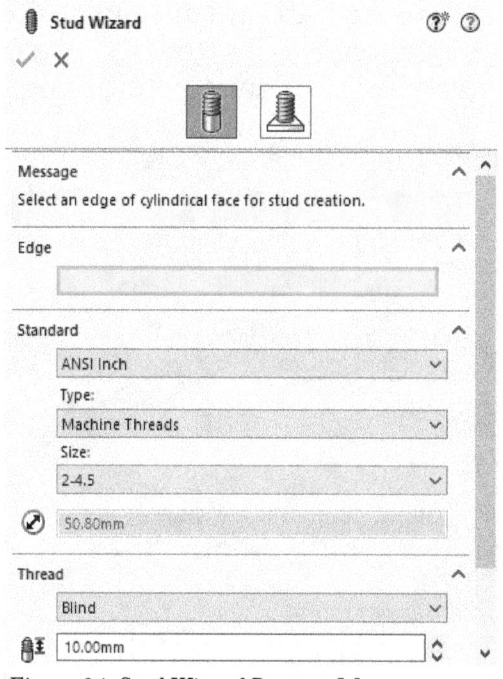

Figure-96. Stud Wizard PropertyManager

- Select the edge of cylindrical face; refer to Figure-97. The preview of stud wizard will be displayed; refer to Figure-98.

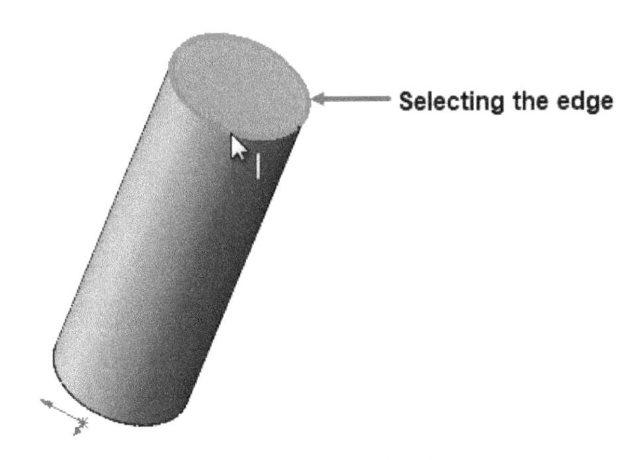

Figure-97. Selecting the edge of cylindrical face

Figure-98. Preview of stud wizard

- Select desired option from **Standard** drop-down in the **Standard** rollout to specify the dimensioning standard, such as ANSI Metric or ISO.
- Select desired type of thread from the **Type** drop-down.
- Select desired size of the thread from **Size** drop-down.
- Specify major diameter of the thread in the **Major Diameter** edit box.
- Specify desired option from **End Condition** drop-down in the **Thread** rollout to extend the thread from the starting edge to the specified end condition.
- Enter desired value to specify the thread length along the stud from the starting edge in the **Thread Depth** edit box.
- Select the **Thread class** check box to specify a thread class fit that measures the looseness or tightness of external mating threads.
- Select the **UnderCut** check box and enter desired values in the **Undercut diameter**, **Undercut depth**, and **Undercut radius** edit boxes to create an undercut at the end of the threaded portion of the stud to provide clearance.

- Click on the **Restore Default Values** button to override the values in the fields and reverts to the default values.
- After specifying desired parameters, click on the **OK** button from the **PropertyManager**. The stud wizard on the edge of cylindrical surface will be created; refer to Figure-99.

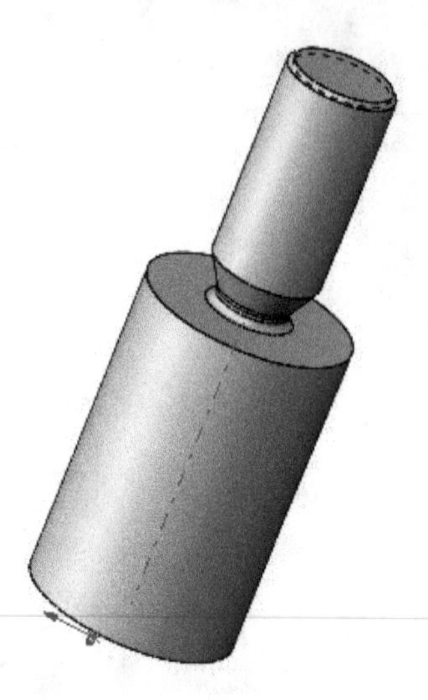

Figure-99. Stud wizard on the edge created

- Click on the **Create Stud on a Surface** option in the **Stud Wizard PropertyManager**. The options in the **PropertyManager** will be displayed as shown in Figure-100.

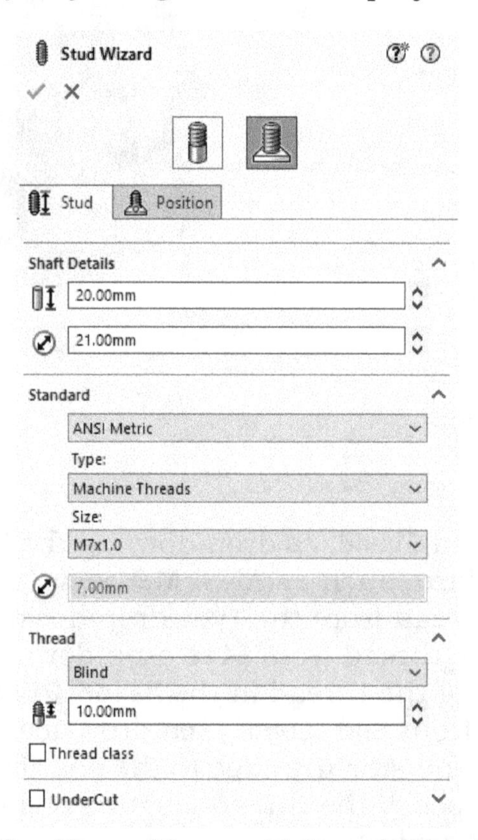

Figure-100. Stud Wizard PropertyManager with Creates Stud on a Surface option selected

- Click on the **Position** tab of the **PropertyManager**. You will be asked to select face of the model for the stud position.

- Click on the face of the model at desired location to specify the position of stud; refer to Figure-101. The preview of stud will be displayed; refer to Figure-102. You will be asked to specify the parameters of stud.

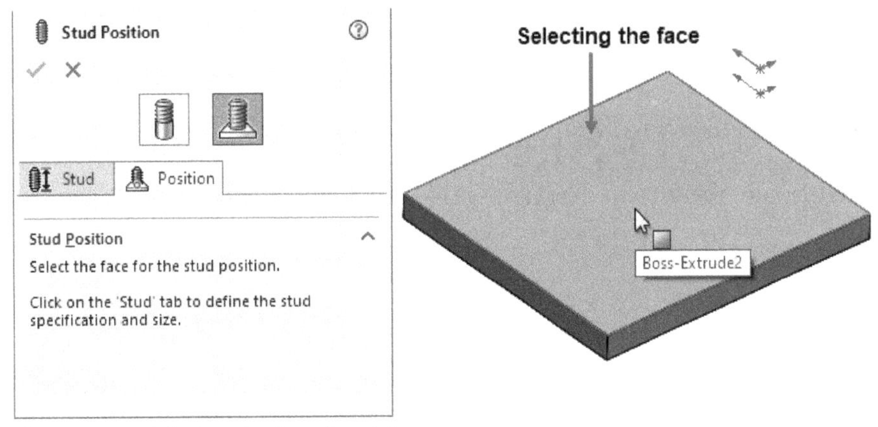

Figure-101. Selecting the face of model

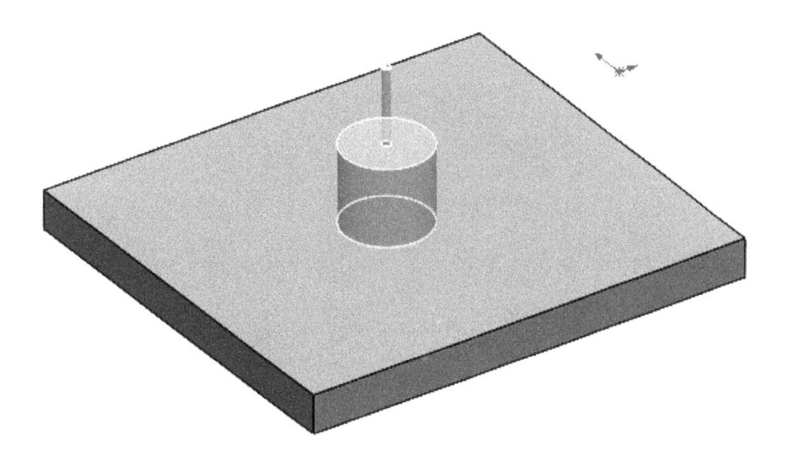

Figure-102. Preview of stud on the surface of model

- Specify desired length and diameter of shaft in the **Shaft Length** and **Shaft Diameter** edit boxes, respectively from **Shaft Details** rollout in the **Stud** tab of the **PropertyManager**.
- Specify the other parameters for the stud as discussed earlier and click on the **OK** button from the **PropertyManager**. The stud on the surface will be created; refer to Figure-103.

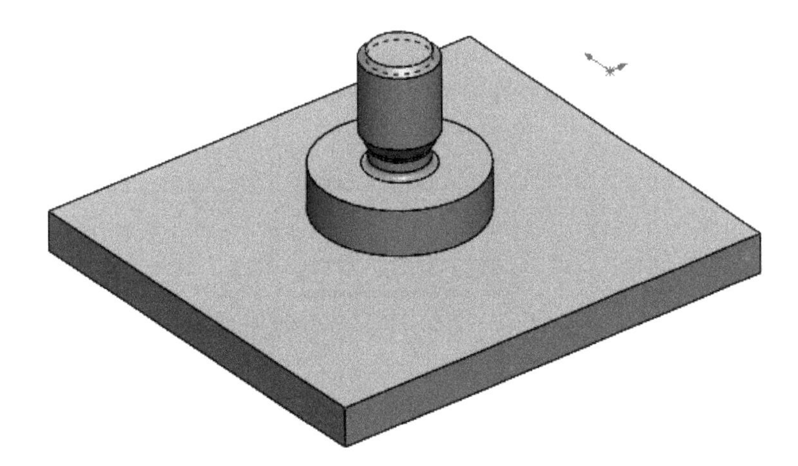

Figure-103. Stud on the surface of model created

Till this chapter, you have learned creation of solid objects and removing material from them. In the next chapter, you will learn the modifying operations that can be performed on solid models. Now, you will practice the 3D sketch that we have discussed in this chapter. Later in the chapters, you will practice on Solid Modeling tools.

Practical on 3D Sketch

Create a 3D sketch as shown in Figure-104.

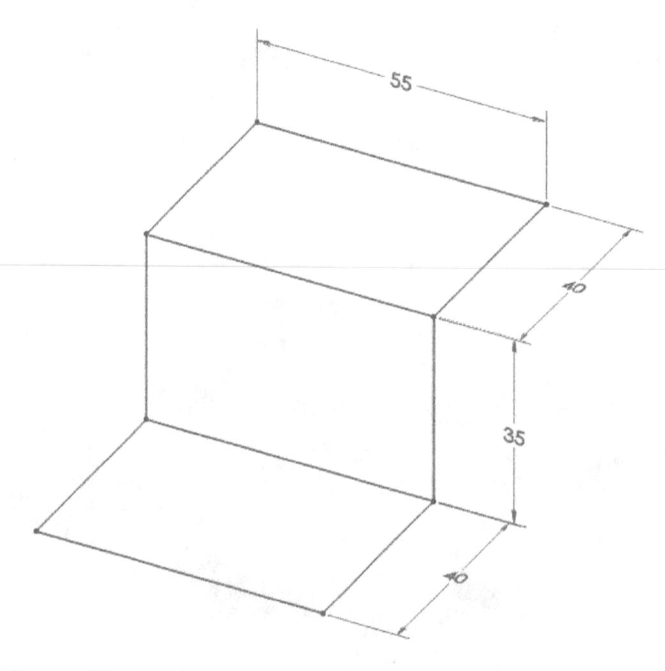

Figure-104. 3D sketch for Practical

Starting Sketch

- Start SolidWorks by double-clicking on the SolidWorks icon from the desktop. (If not started yet.)
- Click on the **New** button from the **Quick Access Toolbar** or press **CTRL+N** from keyboard. The **New SOLIDWORKS Document** dialog box will be displayed.
- Double-click on **Part** icon from the dialog box. The Part environment of SolidWorks will be displayed.
- Click on **Sketch CommandManager** tool from the **Ribbon** and then click on the down arrow below **Sketch** tool. A drop-down will be displayed.
- Click on the **3D Sketch** tool from the drop-down; refer to Figure-105.

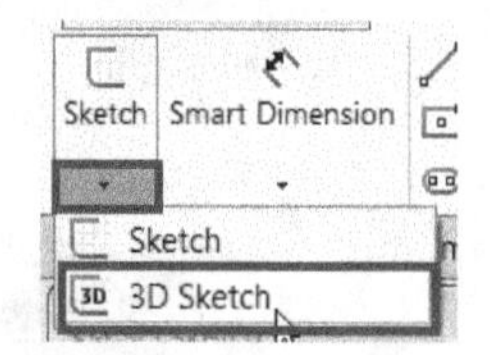

Figure-105. 3D Sketch tool

Creating Lines in 3D

- Click on the **Line** tool from the **Sketch CommandManager** or press **L** from keyboard. The line tool will be activated and coordinate system will be displayed as shown in Figure-106. Also, the name of active plane will be displayed below the cursor.

Figure-106. Coordinate system for 3D sketching

- Click on the coordinate system and press **TAB** key from keyboard till the YZ plane is active.
- Move the cursor along the negative Z axis and enter the length of line as **40** in the input box; refer to Figure-107.

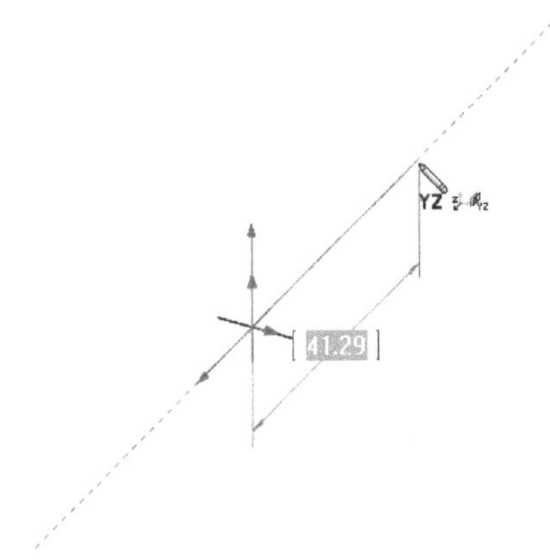

Figure-107. Creating line along Z axis

- Move the cursor vertically upward along Y axis and enter the length as **35**; refer to Figure-108.

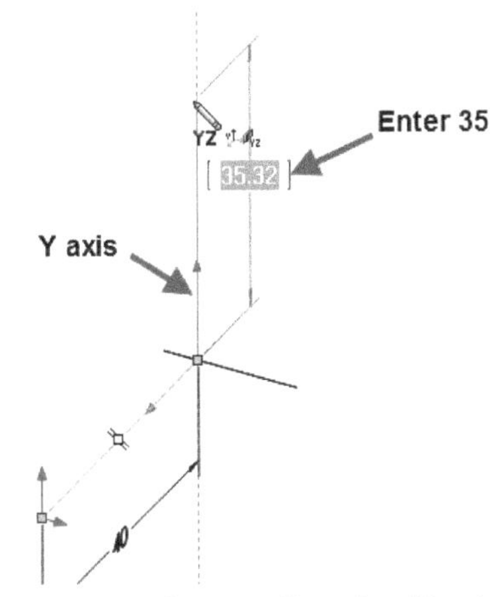

Figure-108. Creating line along Y axis

- Press **TAB** to activate ZX plane and create line of length **40** along negative Z axis; refer to Figure-109.

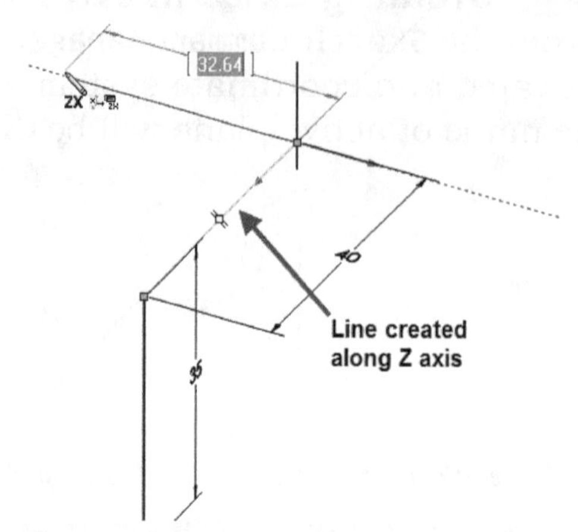

Figure-109. Line created along Z axis

- Move the cursor along X-axis and enter the length as **55** in the input box; refer to Figure-110.

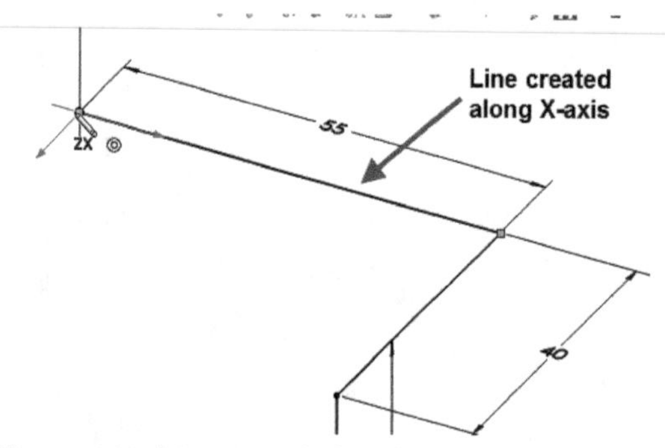

Figure-110. Line created along X axis

- Repeat the same method to create rest of the lines; refer to Figure-111. Press **ESC** to exit the tool.

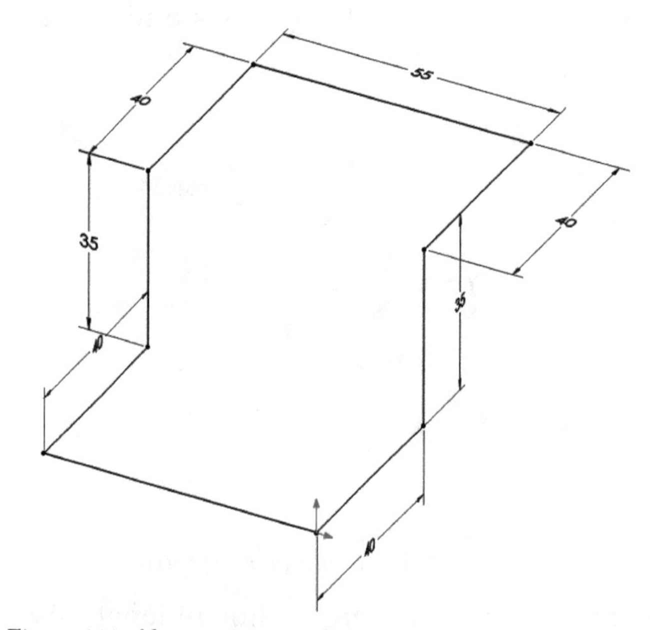

Figure-111. After creating outer lines

- Using the line tool, join the inner points of the sketch created earlier; refer to Figure-112.

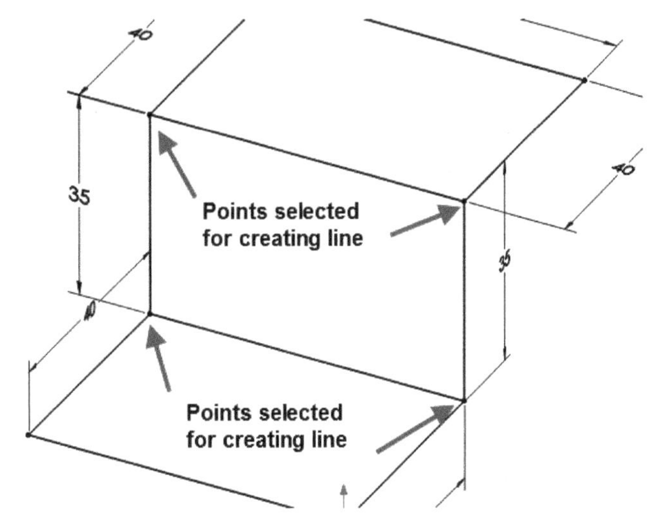

Figure-112. Creating lines to join points

Self Assessment

Q1. A sketch which exists in more than one plane is called ....

Q2. Using the .... key from keyboard, you can change the plane while creating the 3D sketch.

Q3. We can create a box using rectangle as sketch by using the .... tool.

Q4. We can create a cylinder using the rectangle as sketch by using the .... tool.

Q5. The .... tool is used to create a solid volume by moving a sketch along the selected path.

Q6. We can create sweep feature only at one side if the section lies in between the path. (T/F)

Q7. We can create both guide curves and path in same sketch for creating sweep feature. (T/F)

Q8. Using the **Profile Twist** options of the **Swept Boss/Base** tool, you can create springs. (T/F)

Q9. For Circular Profile Sweep feature, the profile of section is by default circle and we are required to specify the diameter of that. (T/F)

Q10. By default,...........(hint: number) planes are available in SolidWorks.

Q11. How many number of maximum references can be provided for creating a reference plane?

a. 2
b. 3
c. 4
d. 5

Q12. After invoking the **Plane PropertyManager** if you select an existing plane then which of the following button is selected by default in the **PropertyManager**?

a. Parallel
b. Perpendicular
c. Coincident
d. Offset distance

Q13. Which of the following combination can be used to create plane at an angle?

a. An Edge and connected planar face
b. An Edge and intersecting plane
c. Two vertices and flat face
d. All of the above.

Q14. Which of the following cannot be selected as origin for creating coordinate system?

a. Vertices of solid
b. Edge of solid
c. Mid Point of an edge
d. Center line of revolve feature

Q15. Which of the following option is available for **Swept cut** tool but not for **Swept Boss/Base** tool?

a. **Circular Profile** radio button
b. **Solid sweep** radio button
c. **Guide Curves** rollout
d. **Mesh preview** check box

FOR STUDENT NOTES

Answer to Self-Assessment:
1. 3D Sketch, **2.** TAB, **3.** Extruded Boss/Base, **4.** Revolve Boss/Base, **5.** Swept Boss/Base, **6.** F, **7.** T, **8.** T, **9.** T, **10.** 3, **11.** b, **12.** d, **13.** d, **14.** d, **15.** b

Chapter 5

Solid Editing and Practical

Topics Covered

The major topics covered in this chapter are:

- ***Fillet/Chamfer tool***
- ***Pattern tools***
- ***Rib tool***
- ***Draft tool***
- ***Shell tool***
- ***Wrap tool, Intersect tool, and Mirror tool***
- ***Equations***
- ***Design Table***
- ***Practical and Practice***

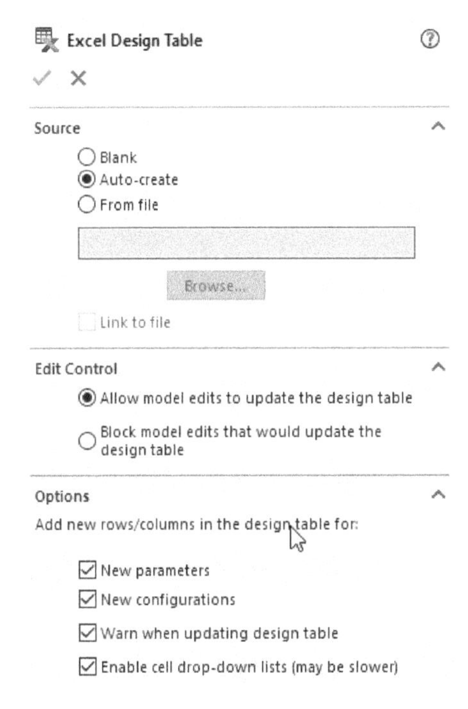

INTRODUCTION

In the previous chapter, we have learned to create solid models and remove material from them. In this chapter, we will learn to edit the models. Like the other tools, SolidWorks has packed all the editing tools into one column. These tools are available in the column next to the column intended for removing material; refer to Figure-1. The tools in this column are explained next.

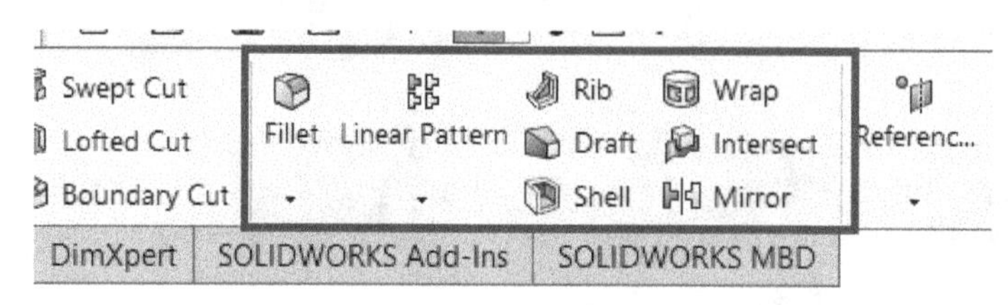

Figure-1. Solid_Editing_tools

FILLET

The **Fillet** tool is used to apply radius at the edges. This tool works in the same way as the **Sketch Fillet** do. It is recommended that you apply the fillets after creating all the featured required in model if possible, because fillets can increase the processing time during modifications. The procedure to use this tool is given next.

Constant Size Fillet

- Click on the **Fillet** tool from the **Fillet** drop-down in the **Ribbon**. The **Fillet PropertyManager** will display as shown in Figure-2.

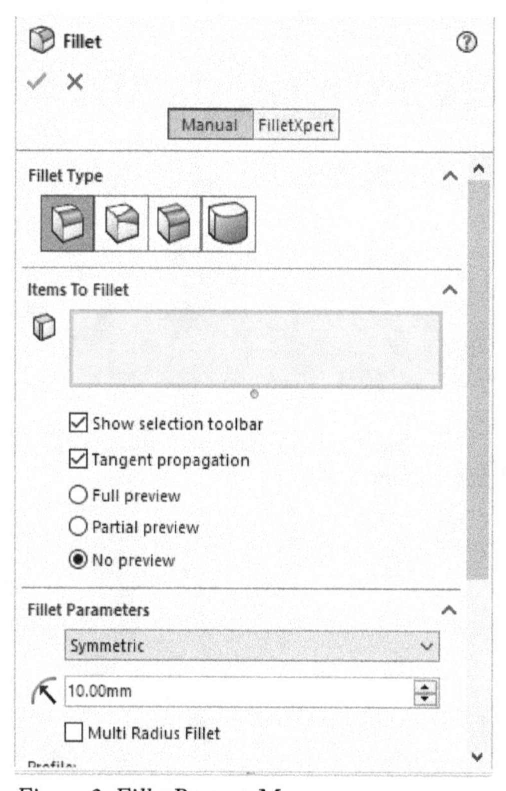

Figure-2. Fillet PropertyManager

- Select the edge on which you want to apply the fillet. In place of selecting edge, you can select the two adjoining faces. The fillet will be created at the intersection of the two faces; refer to Figure-3.

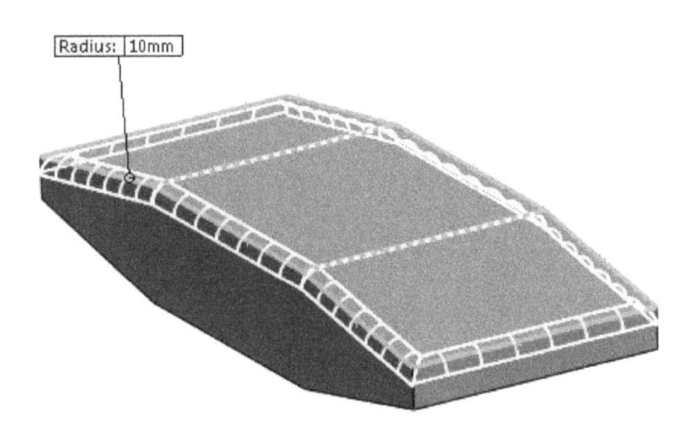

Figure-3. Constant_size_fillet

- In the **Fillet Type** rollout, **Constant Size Fillet** button is selected.
- Specify desired radius for fillet in the **Radius** edit box in **Fillet Parameters** rollout of the **PropertyManager**.
- Select the **Full preview** radio button to check the preview of the fillet.
- You can select desired profile of fillet by using the **Profile** drop-down in **Fillet Parameters** rollout. Preview of fillet on selecting different profile options is given in Figure-4.

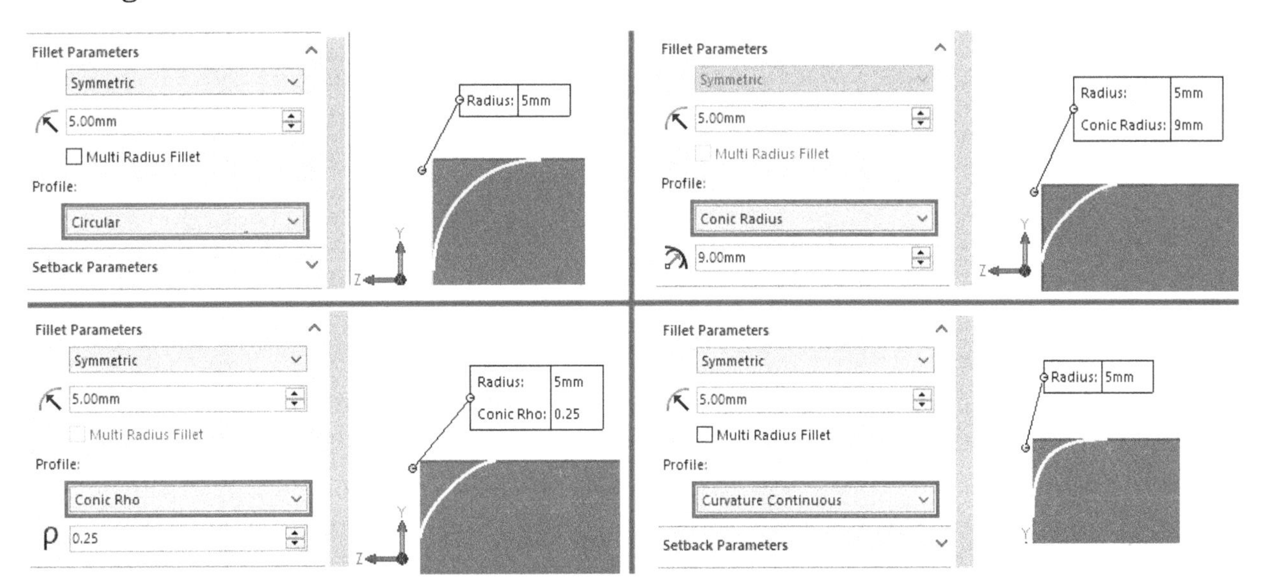

Figure-4. Fillet with different profiles

- Select the **Multi Radius Fillet** check box if you have selected multiple edges and want to specify different radii for each edge.
- You can specify different radius for both sides of round with respect to selected edge by using the **Asymmetric** option from the drop-down in the **Fillet Parameters** rollout.
- You can set the setback parameters at the vertices by using the options in the **Setback Parameters** rollout; refer to Figure-5.

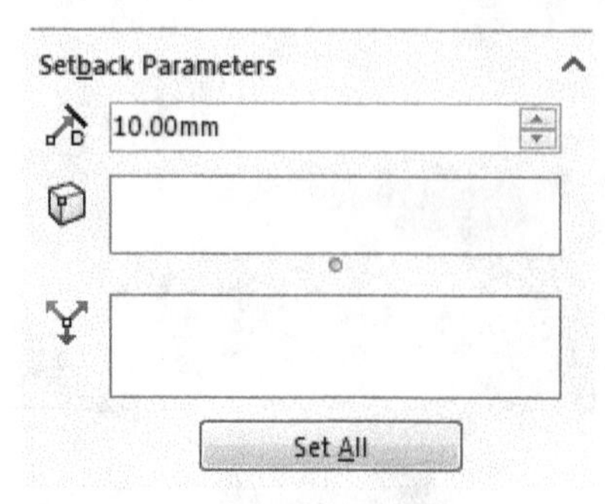

Figure-5. Setback_Parameters_rollout

- Click in the **Setback Vertices** selection box and click on the vertex for which you want to specify setback parameters. Fillet radius at the different joining edges will be displayed in the **Setback Distances** box.
- Select desired direction from the **Setback Distances** box and specify the value of distance by using the **Distance** edit box in the **Setback Parameters** rollout.
- Click on the **OK** button from the **PropertyManager** to create the fillet.

Variable Radius Fillet

You can create fillet with varying radius by selecting the **Variable Size Fillet** button. On selecting this button, the **Variable Radius Parameters** rollout is added in the **Fillet PropertyManager**; refer to Figure-6.

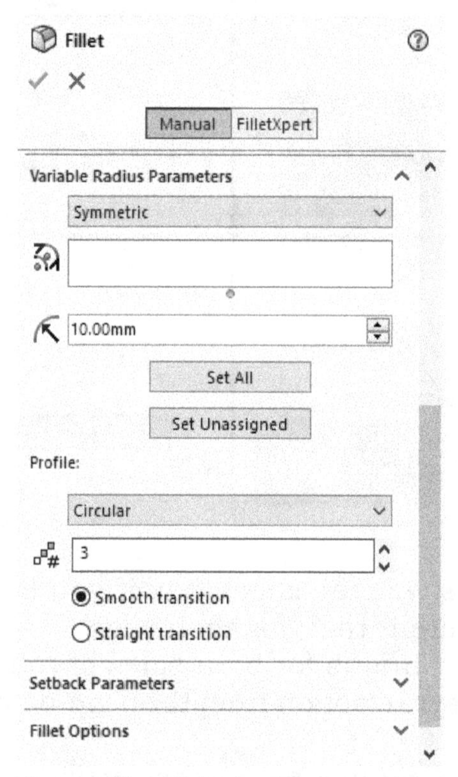

Figure-6. Variable Radius Parameters rollout

- Set the number of points for changing the radius by using the **Number of Instances** spinner in the rollout.
- Click in the **Attached Radii** box for start and end point in the viewport and specify the starting & ending radius.
- Click on the desired point in the preview of fillet and specify the desired radius; refer to Figure-7.

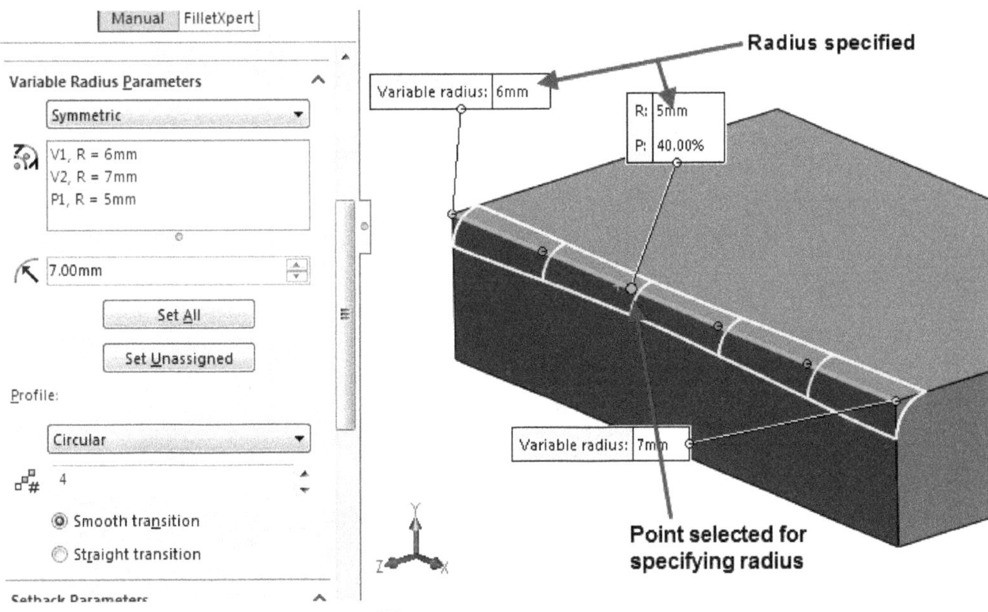

Figure-7. Creating_variable_radius_fillet

- You can change the profile for round by using the **Profile** drop-down in the rollout.

Face fillet

- Select the **Face Fillet** button from the **Fillet Type** rollout to create fillet at the joining edge of two faces.
- On selecting this radio button, the **Items to Fillet** rollout is added in the **Fillet PropertyManager**.
- Click in the first box and select the first face/faces.
- Next, click in the second box and select the second face/faces. Figure-8 shows preview of the fillet.

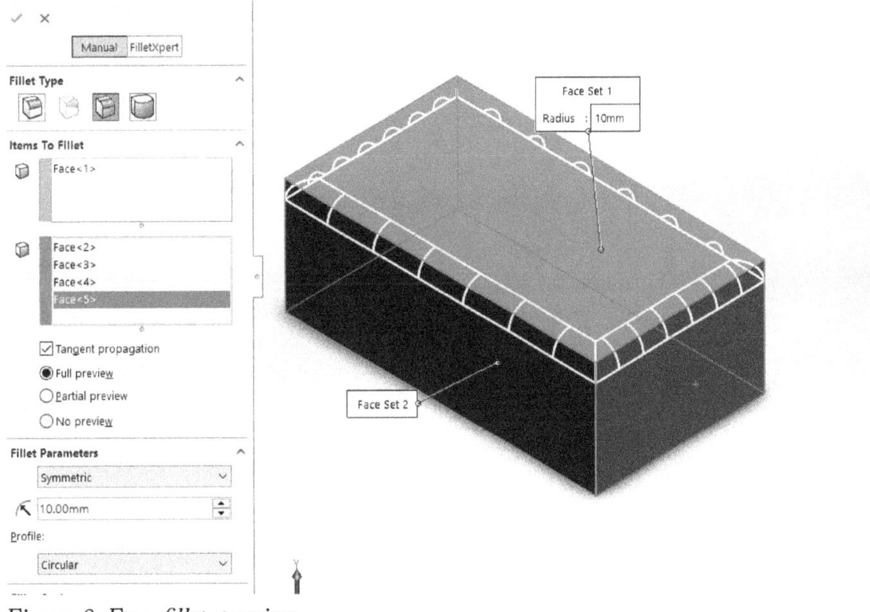

Figure-8. Face_fillet_preview

- Set the other parameters as discussed earlier.

Full round fillet

Select the **Full Round Fillet** button to create a fillet where three faces meet each other. Using this option, you can create dome feature; refer to Figure-9.

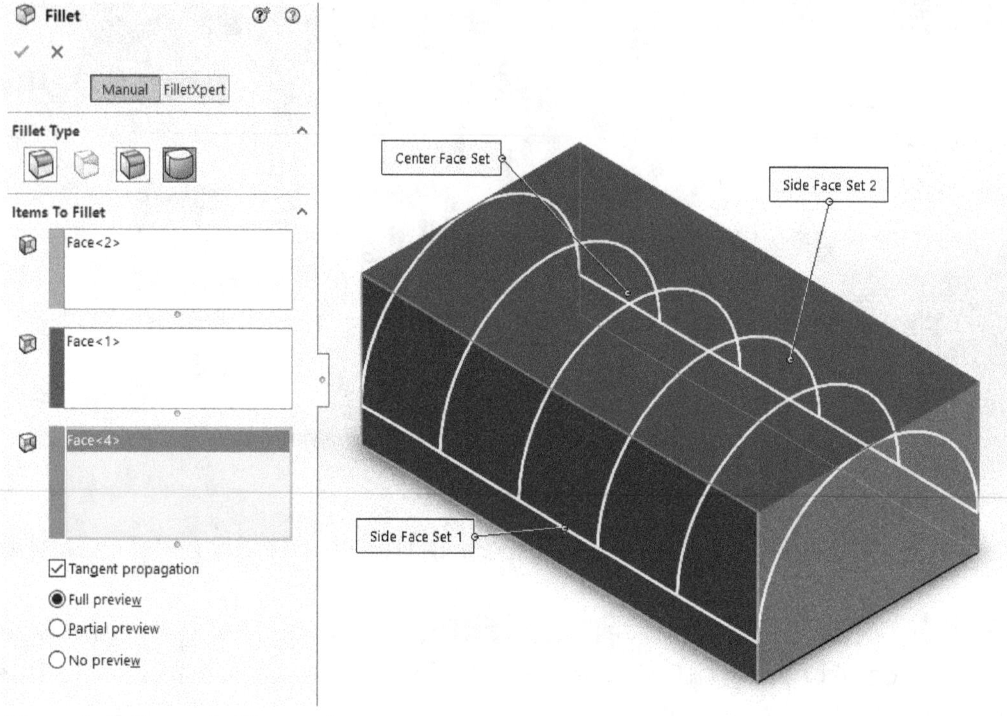

Figure-9. Full_round_fillet_preview

FilletXpert

The **FilletXpert** tool is used to apply different type of fillets in one single mode. The procedure to use this tool is discussed next.

- Click on the **FilletXpert** button from the **Fillet PropertyManager**. The **FilletXpert PropertyManager** will be displayed; refer to Figure-10.

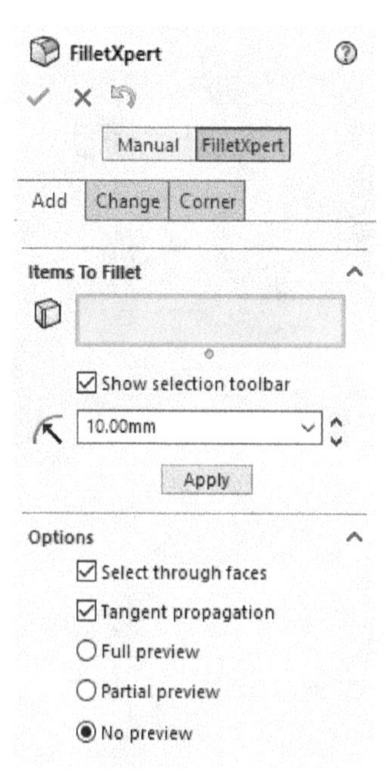

Figure-10. FilletXpert PropertyManager

- Select the face, edge, loop, or feature. Preview of the fillet will be displayed. If you have selected an edge then fillet will be applied to the edge. If you have selected a face then all the edges of the face will be filleted. If you have selected a loop then all the edges connected to the loop will be filleted. If you have selected a feature then fillet will be applied to all the sharp edges of the feature.
- Set desired value of fillet radius in the edit box below **Items To Fillet** selection box.
- Click on the **Apply** button to apply fillet and start with a new set of fillet.

CHAMFER

The **Chamfer** tool is used to bevel the sharp edges of the model. This tool works in the same way as the **Sketch Chamfer** do. The procedure to create chamfer by using this tool is given next.

- Click on the **Chamfer** tool from the **Fillet** drop-down. The **Chamfer PropertyManager** will display as shown in Figure-11.

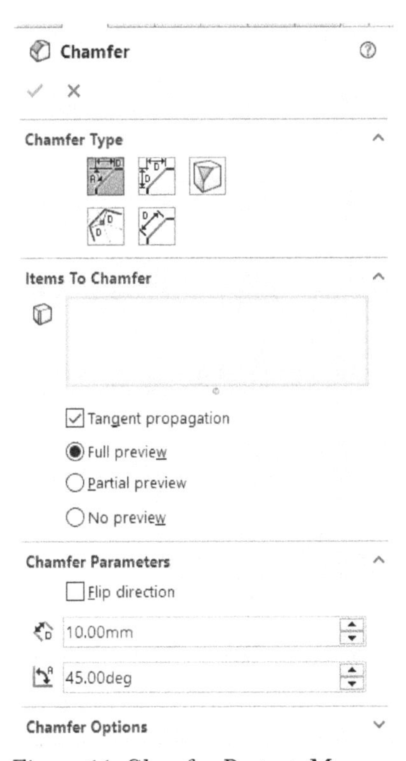

Figure-11. Chamfer_PropertyManager

- Select the edges from the model on which you want to apply chamfer.
- By default, the **Angle Distance** button is selected in the **Chamfer Type** rollout of the **PropertyManager**. Specify the angle and distance parameters in the respective edit boxes in **Chamfer Parameters** rollout. Select the edges to create chamfers.
- If you want to specify distances for both sides of chamfer then select **Distance Distance** button and specify the parameters. Click on the edges to create chamfers.
- If you want to create chamfer at corners then select the **Vertex** button and select the corner of the model.

Figure-12 shows chamfer created at edges and vertices.

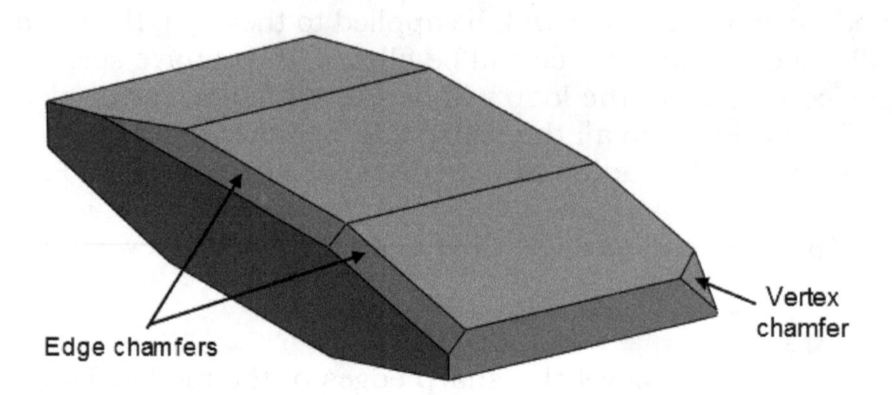

Figure-12. Chamfers_created_on_edges_and_vertex

- Click on the **Offset Face** button from the **Chamfer Type** rollout if you want to create chamfer generated by offset of face adjacent to selected edge; refer to Figure-13. Specify the chamfer distance in the edit box in **PropertyManager** and select the edge/face/feature to create chamfer.

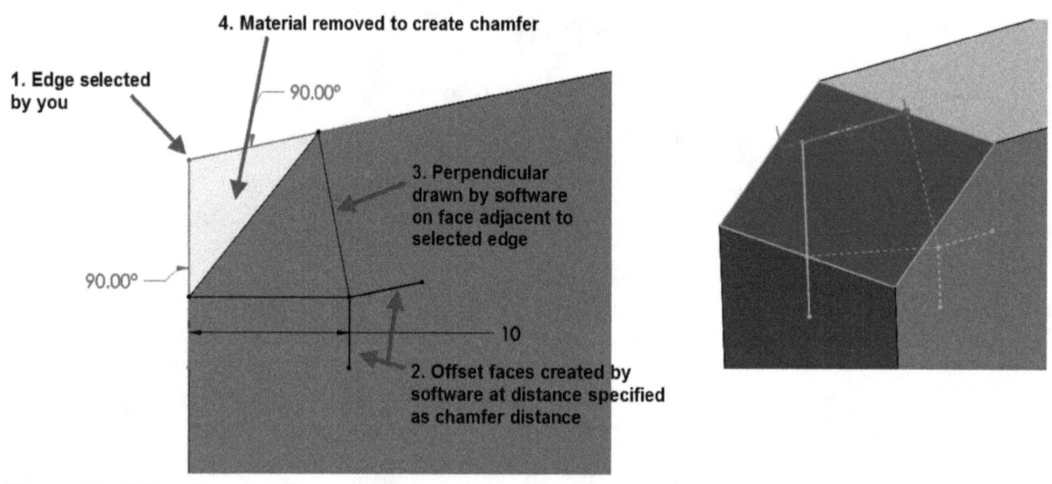

Figure-13. Offset_face_chamfer

- Click on the **Face Face** button from the **Chamfer Type** rollout in the **PropertyManager** if you want to create chamfers on the edges formed by intersection of selected faces. Then, specify the chamfer distance in the edit box in **PropertyManager**. Click on the base face to be used for chamfer. Click in the **Face Set 2** selection box in the **PropertyManager** and then select the other face(s). Preview of chamfer will be displayed; refer to Figure-14.

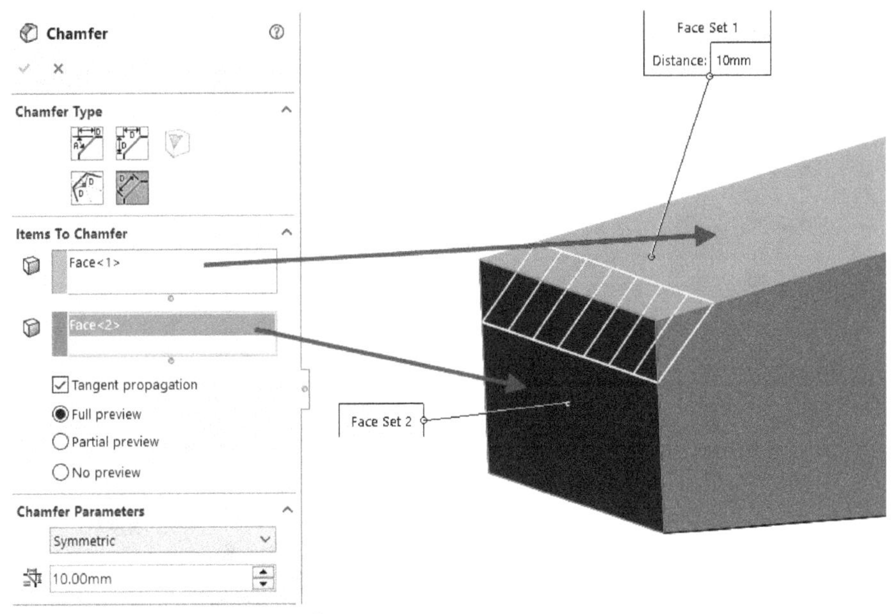

Figure-14. Preview_of_chamfer

- After specifying desired parameters of chamfer, click on the **OK** button from **PropertyManager** to create the chamfer.

LINEAR PATTERN

The **Linear Pattern** tool is used to create linear pattern of solid features in the Modeling environment. This tool is similar to the sketch linear pattern. The procedure to create linear pattern is given next.

- Select all the features that you want to pattern from the **FeatureManager Design Tree** and select the **Linear Pattern** tool from the **Linear Pattern** drop-down. The **Linear Pattern PropertyManager** will be displayed as shown in Figure-15.

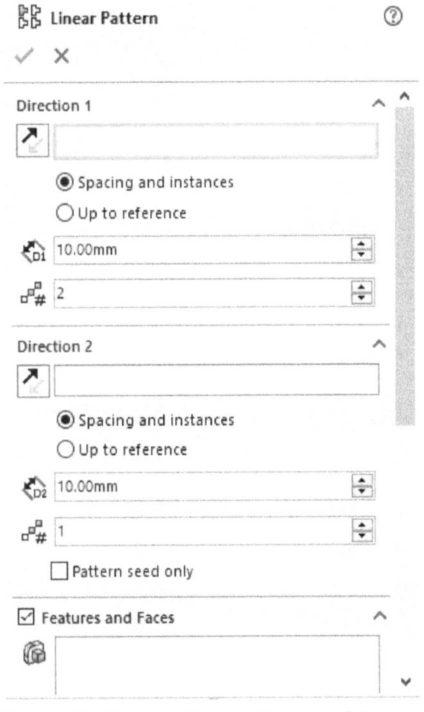

Figure-15. Linear Pattern PropertyManager

- Select the direction reference for **Direction 1** (like edge, face, plane, and so on) and specify the related parameters.
- Similarly, click in the **Direction 2** selection box of **PropertyManager** and specify the reference for **Direction 2**. Specify the related parameters; refer to Figure-16.

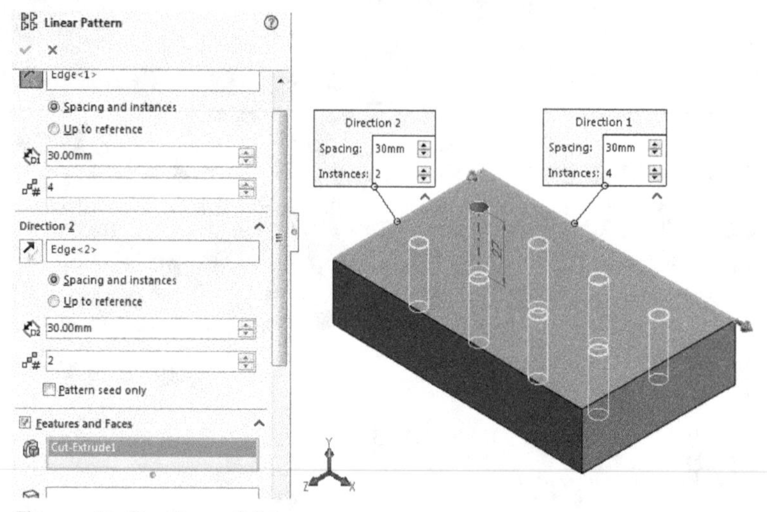

Figure-16. Preview_of_Linear_pattern

- If you want to skip any instance in the pattern then expand the **Instances to Skip** rollout and select the pink dots in the preview to skip them; refer to Figure-17.

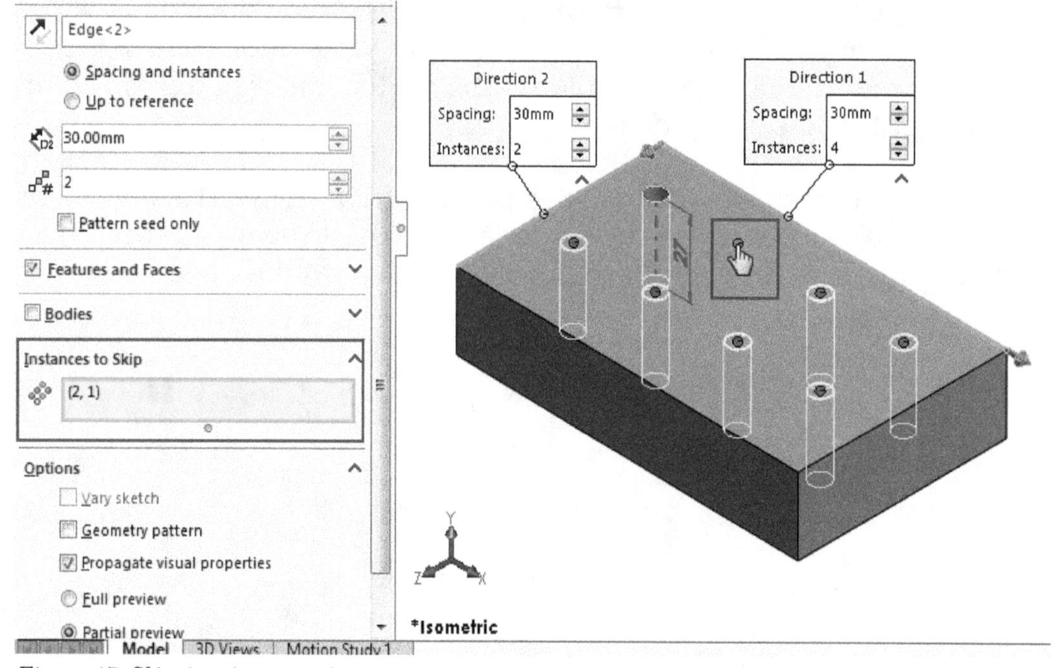

Figure-17. Skipping_instances_in_pattern

- Click on the **OK** button from the **PropertyManager** to create the pattern.

CIRCULAR PATTERN

The **Circular Pattern** tool is used to create multiple instances of selected features around an axis. This tool is similar to the sketch circular pattern. The procedure to create circular pattern is given next.

- Select all the features that you want to pattern and select the **Circular Pattern** tool from the **Linear Pattern** drop-down. The **Circular Pattern PropertyManager** will be displayed as shown in Figure-18.

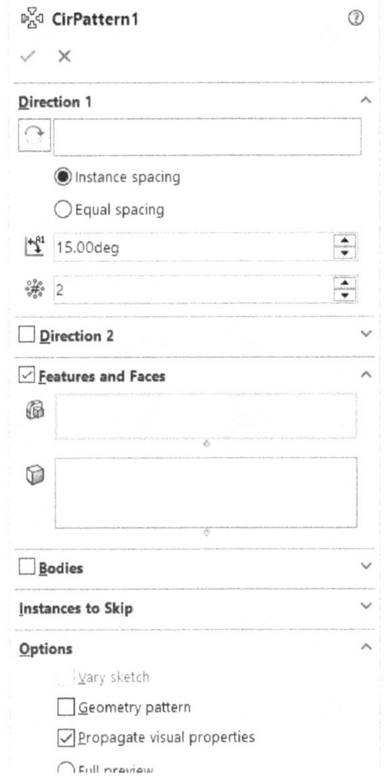

Figure-18. Circular_Pattern_PropertyManager

- Select the edge/axis/circular face(axis of circular face will be automatically selected) about which you want to create the pattern. Preview of the pattern will be displayed.
- Specify the required parameters like angle between two instances, number of instances, and so on for Direction 1 and Direction 2 as discussed under previous tool. The preview of pattern will display; refer to Figure-19.

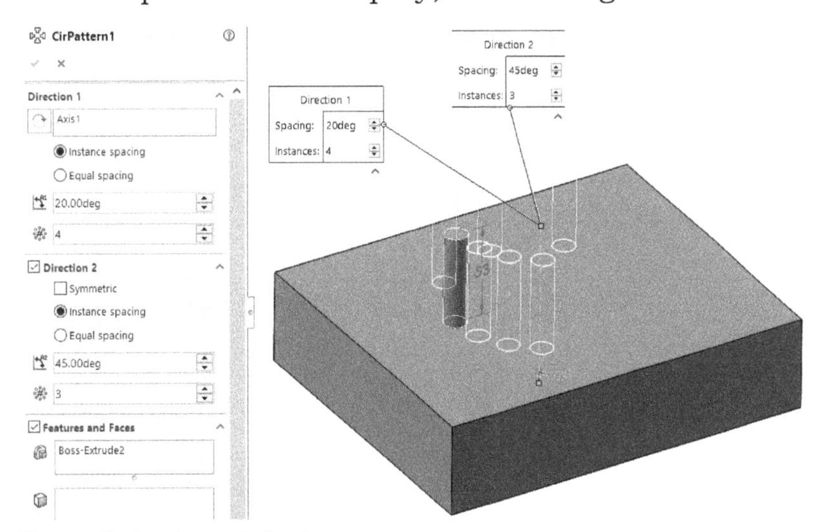

Figure-19. Preview_of_circular_pattern

- Click on the **OK** button to create the pattern.

MIRROR

The **Mirror** tool is used to create mirror copy of the features in the Modeling environment. This tool is similar to the sketch mirror. The procedure to create mirror is given next.

- Select all the features that you want to mirror.
- Click on the **Mirror** tool from the **Linear Pattern** drop-down in the **Ribbon**. The **Mirror PropertyManager** will display as shown in Figure-20.

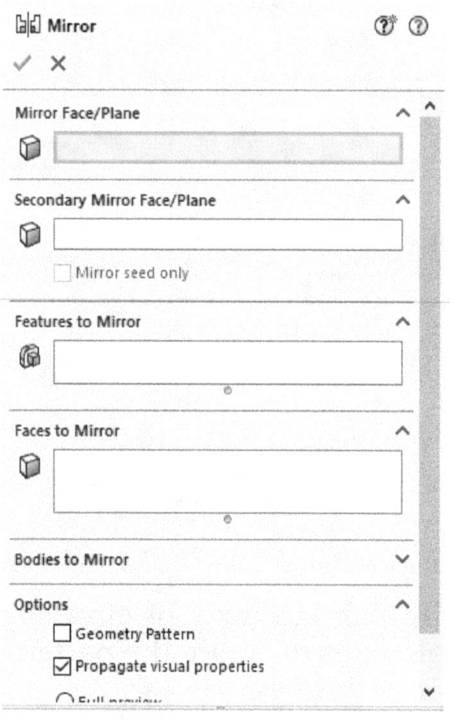

Figure-20. Mirror PropertyManager

- Select the plane or face about which you want to mirror the features for the **Mirror Face/Plane** selection box and select another plane or face for **Secondary Mirror Face/Plane** selection box if you want to mirror the feature on another plane or face also. Preview of the mirror will be displayed; refer to Figure-21.

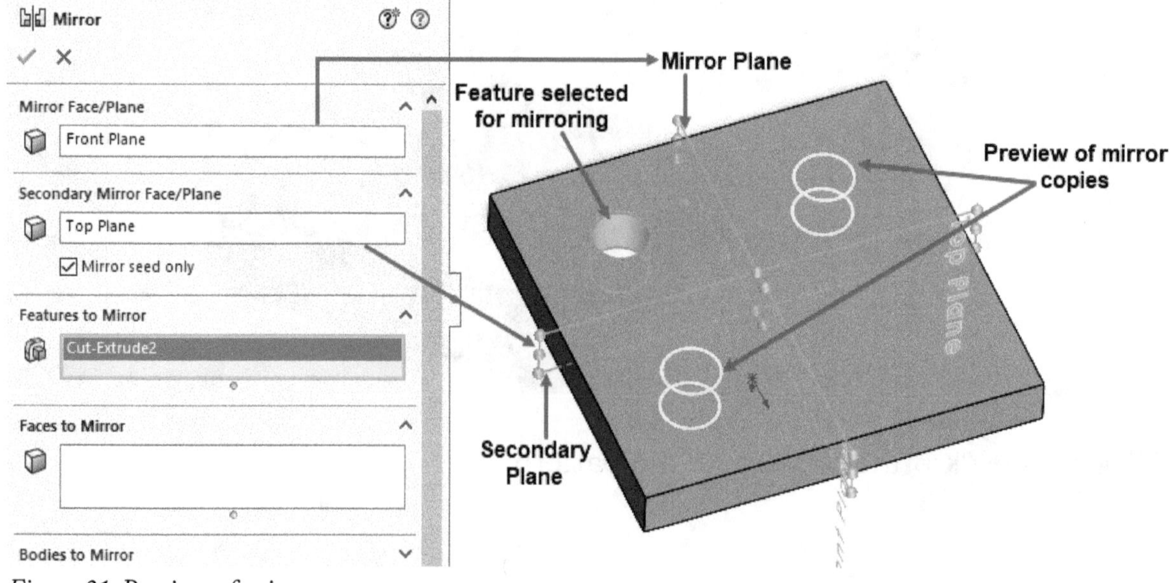

Figure-21. Preview_of_mirror

- The **Mirror seed only** check box will be active only if the secondary plane is perpendicular to the first plane.
- Click on the **OK** button from the **PropertyManager** to create the feature.

If you want to mirror a complete body with respect to mirror plane as in the above figure, then follow the steps given next.

- Click on the **Mirror** tool. The **Mirror PropertyManager** will display.
- Expand the **Bodies to Mirror** rollout and click on the bodies in the viewport that you want to mirror.
- Click in the **Mirror Face/Plane** box in the **Mirror Face/Plane** rollout of the **ProperyManager** and select the mirror plane. Preview of mirror will display; refer to Figure-22. Make sure that you clear the **Merge solids** check box before creating the mirror copy.

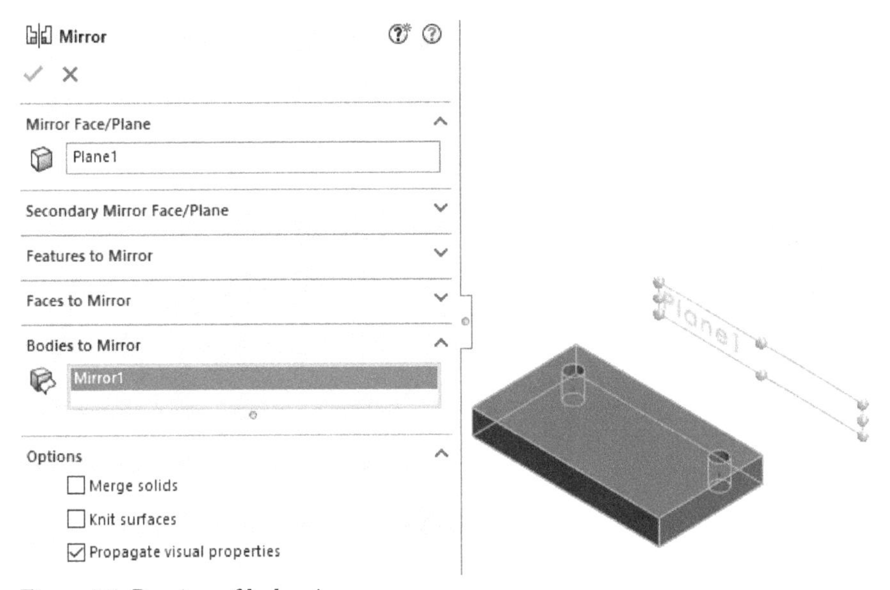

Figure-22. Preview of body mirror

- Click on the **OK** button to create the mirror copy.

CURVE DRIVEN PATTERN

The **Curve Driven Pattern** tool is used to create multiple instances of a solid features along the selected path. The procedure to create curve driven pattern is given next.

- Select all the features that you want to pattern and select the **Curve Driven Pattern** tool from the **Linear Pattern** drop-down. The **Curve Driven Pattern PropertyManager** will display as shown in Figure-23.

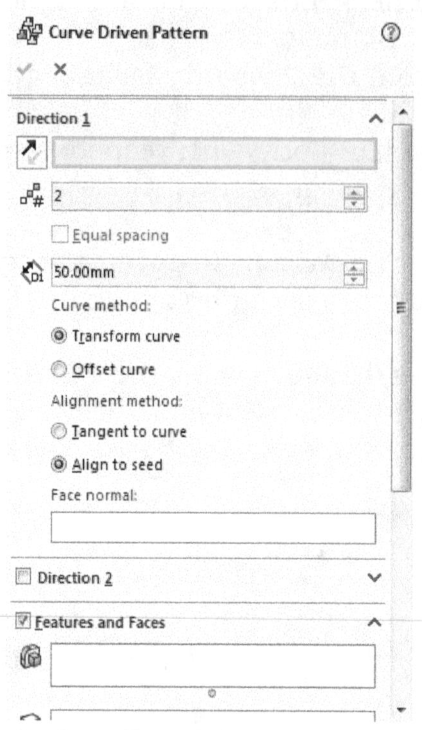

Figure-23. Curve_Driven_Pattern_PropertyManager

- Select the curve along which you want to create the pattern.
- Specify the number of instances and other parameters. Preview of pattern will display; refer to Figure-24.

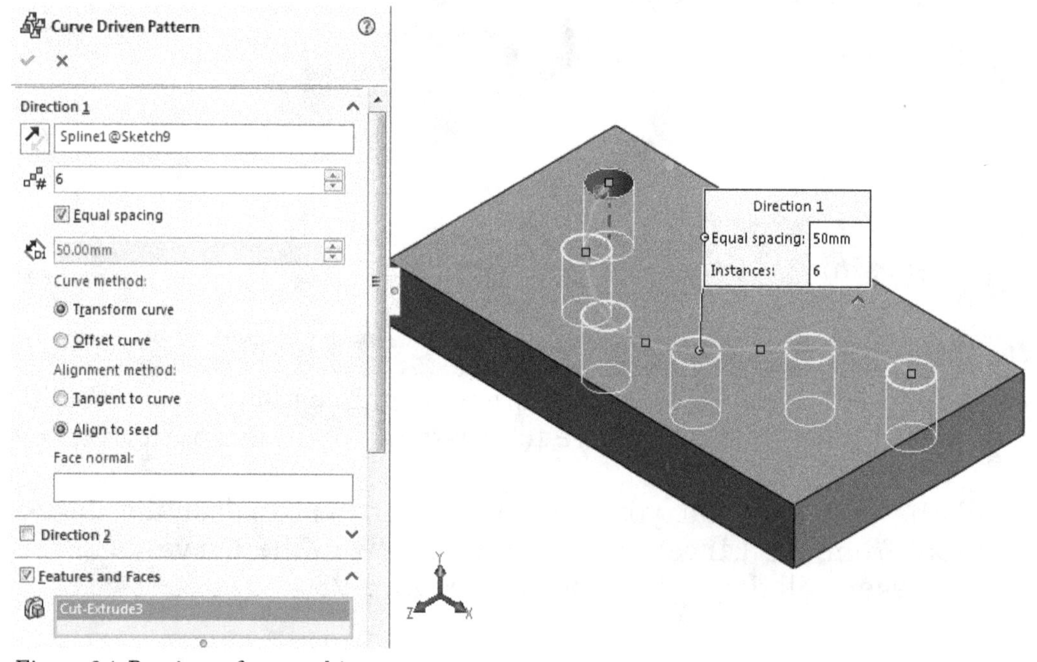

Figure-24. Preview_of_curve_driven_pattern

- Similarly, you can select a curve for direction 2 and specify the related parameters.

SKETCH DRIVEN PATTERN

The **Sketch Driven Pattern** tool is used to create multiple instances of a solid features as per the points specified in the selected sketch. The procedure to create sketch driven pattern is given next.

- Select all the features that you want to pattern and select the **Sketch Driven Pattern** tool from the **Linear Pattern** drop-down. The **Sketch Driven Pattern PropertyManager** will display as shown in Figure-25.

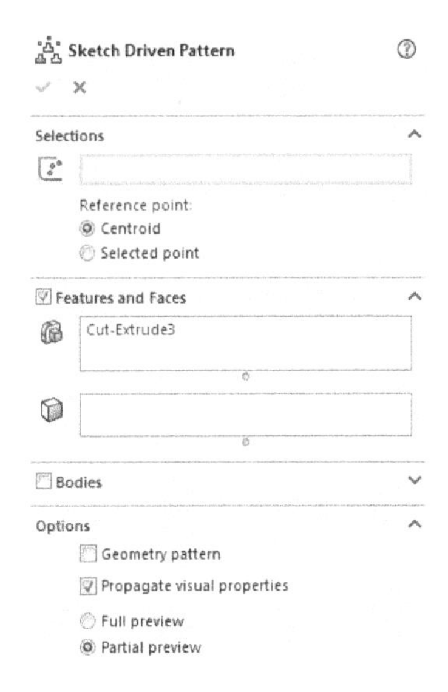

Figure-25. Sketch_Driven_Pattern_PropertyManager

- Select the sketch as reference for pattern. Note that the sketch must have points created in it for referencing pattern instances. Preview of the pattern will be displayed; refer to Figure-26.

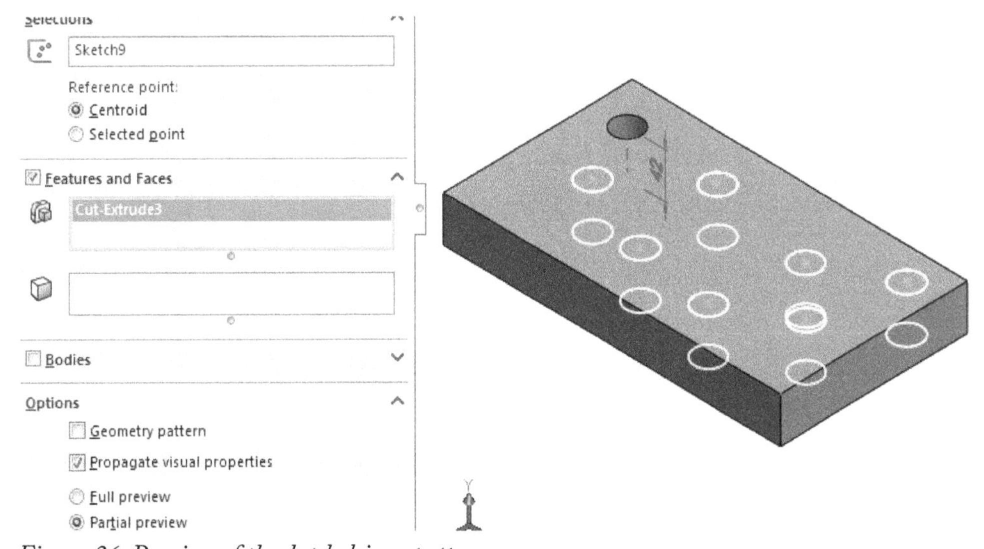

Figure-26. Preview of the sketch driven pattern

- Specify the other parameters as required and then click on the **OK** button from the **PropertyManager** to create the pattern.

TABLE DRIVEN PATTERN

The **Table Driven Pattern** tool is used to create multiple instances of a features as per the coordinates specified in the table. Note that you must have a coordinate system created in the model for referencing the coordinates of the table. The procedure to create table driven pattern is given next.

- Select all the features that you want to pattern and then click on the **Table Driven Pattern** tool from the **Linear Pattern** drop-down in the **Ribbon**. The **Table Driven Pattern** dialog box will be displayed; refer to Figure-27.

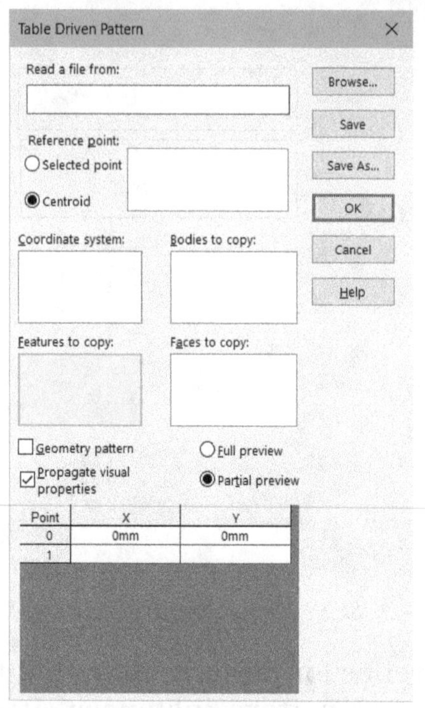

Figure-27. Table Driven Pattern dialog box

- Click in the **Coordinate system** selection box and select the coordinate system. Note that you need to create a coordinate system at desired location before using this tool.
- Now, double-click in the cells of table and specify desired coordinates for various instances of pattern. Preview of the pattern will be displayed; refer to Figure-28.

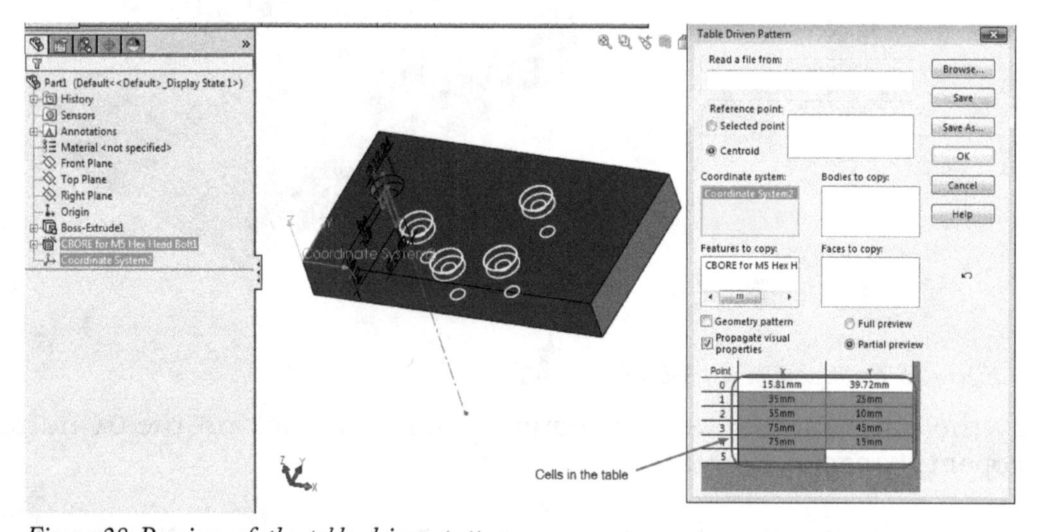

Figure-28. Preview_of_the_table_driven_pattern

- Specify the other parameters as required and then click on the **OK** button from the dialog box to create the pattern.
- If you want to save the current pattern table then click on the **Save** button from the dialog box and save the file at desired location.

- To use an already created pattern table, click on the **Browse** button. The **Open** dialog box will be displayed. Browse to desired file and double-click to use it as pattern table; refer to Figure-29.

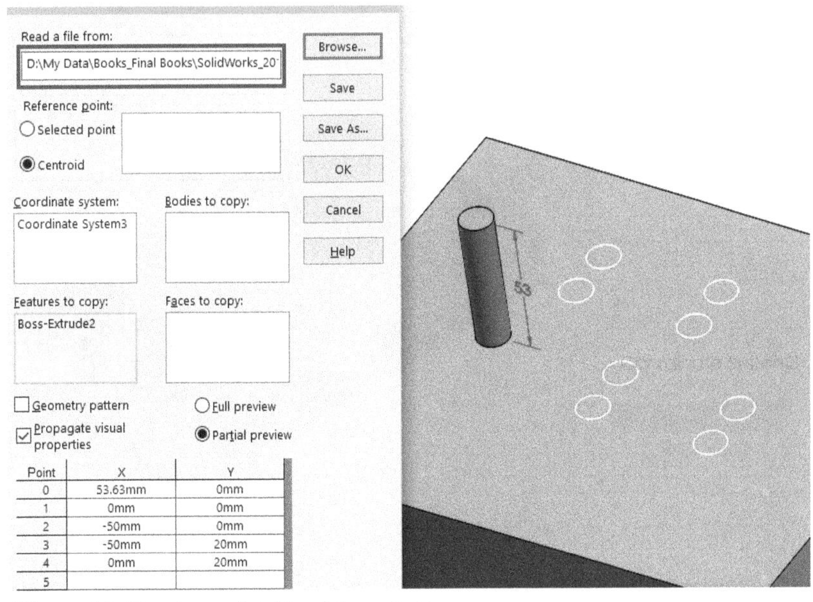

Figure-29. Preview_of_table_driven_pattern

FILL PATTERN

The **Fill Pattern** tool is used to create multiple instances of a feature by filling the selected close region. Note that you must have a closed loop sketch created in the model for referencing the region for fill pattern or you can select a flat face. The procedure to create fill pattern is given next.

- Select all the features that you want to pattern and then click on the **Fill Pattern** tool from the **Linear Pattern** drop-down in the **Ribbon**. The **Fill Pattern PropertyManager** will be displayed; refer to Figure-30.

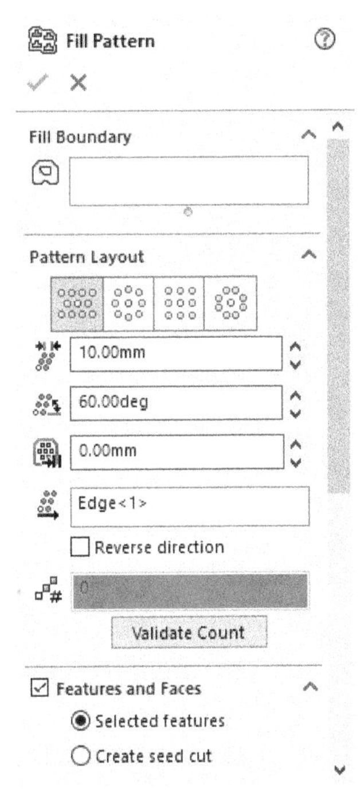

Figure-30. Fill_Pattern_PropertyManager

- Select the sketch for filling with instances of pattern feature. Preview of the pattern will be displayed; refer to Figure-31.

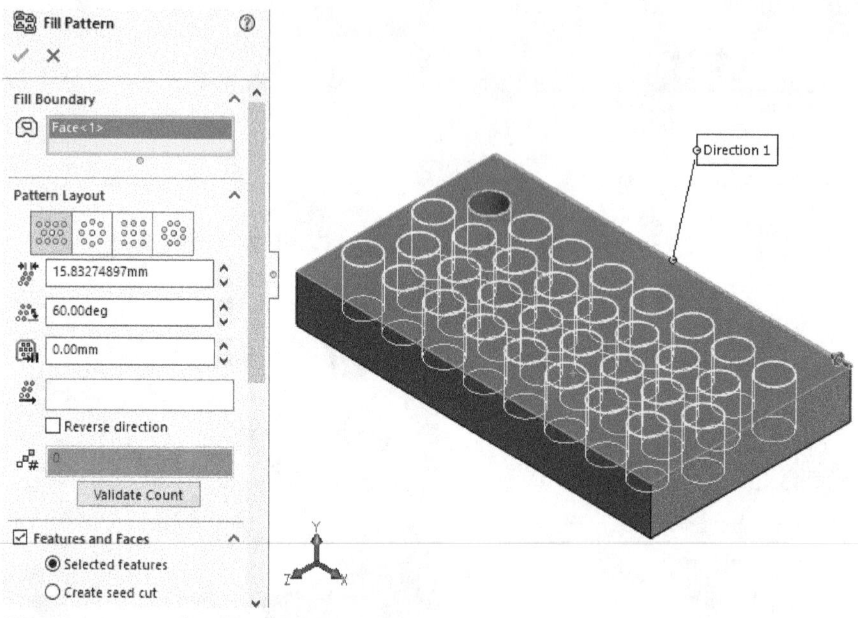

Figure-31. Preview of the fill pattern

- Specify desired parameters like spacing between instances, space between boundary line and instances, and so on.
- Click on the **OK** button from the **PropertyManager** to create the fill pattern.

Using **Fill Pattern**, you can create some predefined cuts in the solid model with the shapes of circle, square, diamond, and polygon. The procedure to use this option is given next.

- Without selecting any feature, click on the **Fill Pattern** tool from the **Linear Pattern** drop-down in the **Ribbon**. The **Fill Pattern PropertyManager** will be displayed.
- Select the **Create seed cut** radio button from the **Features and Faces** rollout of the **PropertyManager** if you want to create predefined shape cuts in fill pattern. Buttons for various shapes will be displayed in the rollout with related parameter options; refer to Figure-32.

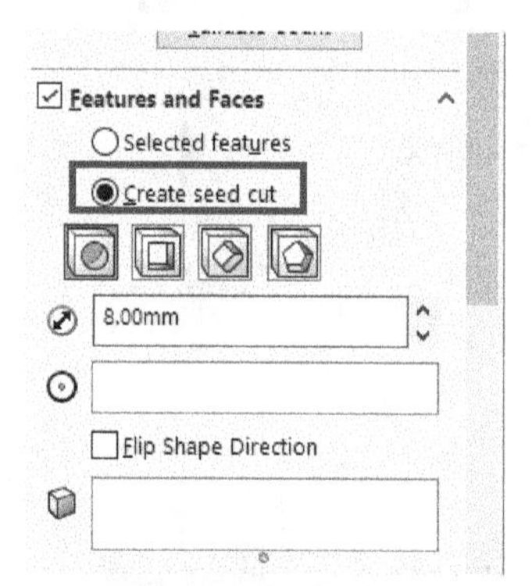

Figure-32. Options_for_creating_seed_cuts

- Select desired shape button from the rollout and specify the related parameters for it.
- If you want to create copies of selected feature then select the **Selected features** radio button and select desired feature from the **FeatureManager Design Tree**.

- Click on the face to apply the cuts. Preview of the fill pattern with seed cuts will be displayed; refer to Figure-33.

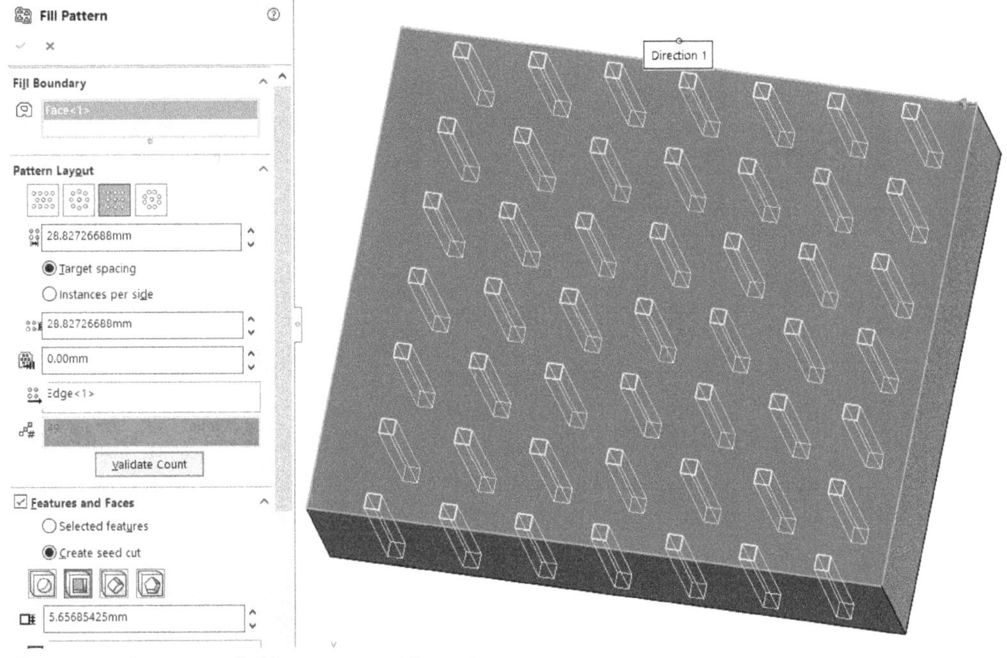

Figure-33. Preview_of_fill_pattern_with_seed_cuts

- Click on the **OK** button from the **PropertyManager** to create the feature.

VARIABLE PATTERN

The **Variable Pattern** tool is used to create pattern of dimensioned features with different parameters. Like, you can create pattern of a rectangular boss feature with different dimensions and at different distances from each other. The procedure to use this tool is given next.

- Click on the **Variable Pattern** tool from the **Linear Pattern** drop-down in the **Ribbon**. The **Variable Pattern PropertyManager** will be displayed; refer to Figure-34.

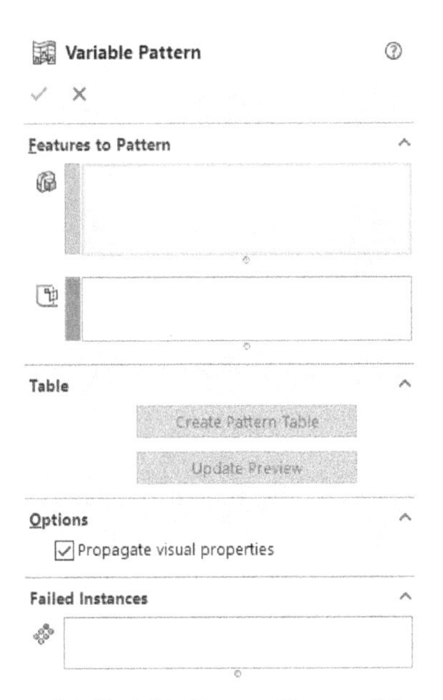

Figure-34. Variable Pattern PropertyManager

- Click on the feature that you want to pattern. The **Create Pattern Table** button will become active in **Table** rollout.
- Click on the **Create Pattern Table** button. The **Pattern Table** dialog box will be displayed; refer to Figure-35.

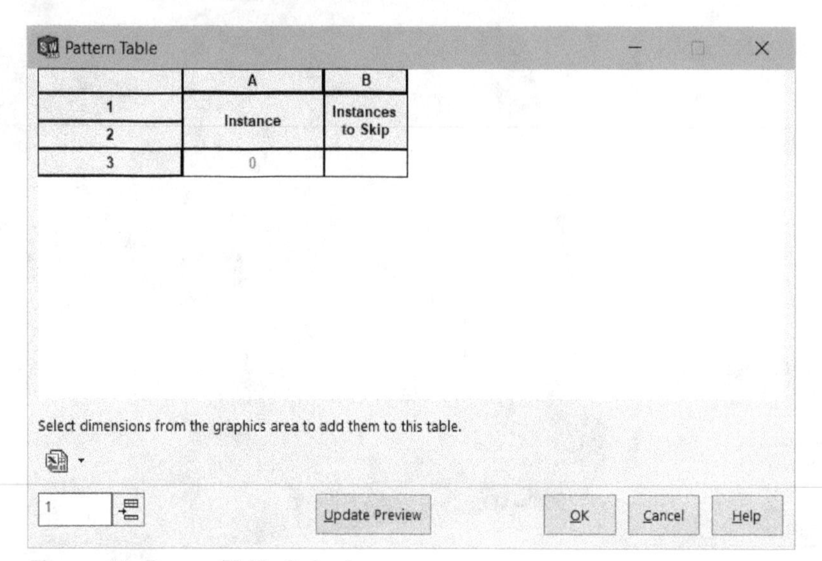

Figure-35. Pattern Table dialog box

- Select the dimensions by which you want to vary the features and distances between instances of pattern directly from the model. The related dimensions will be added as parameters in the table; refer to Figure-36.

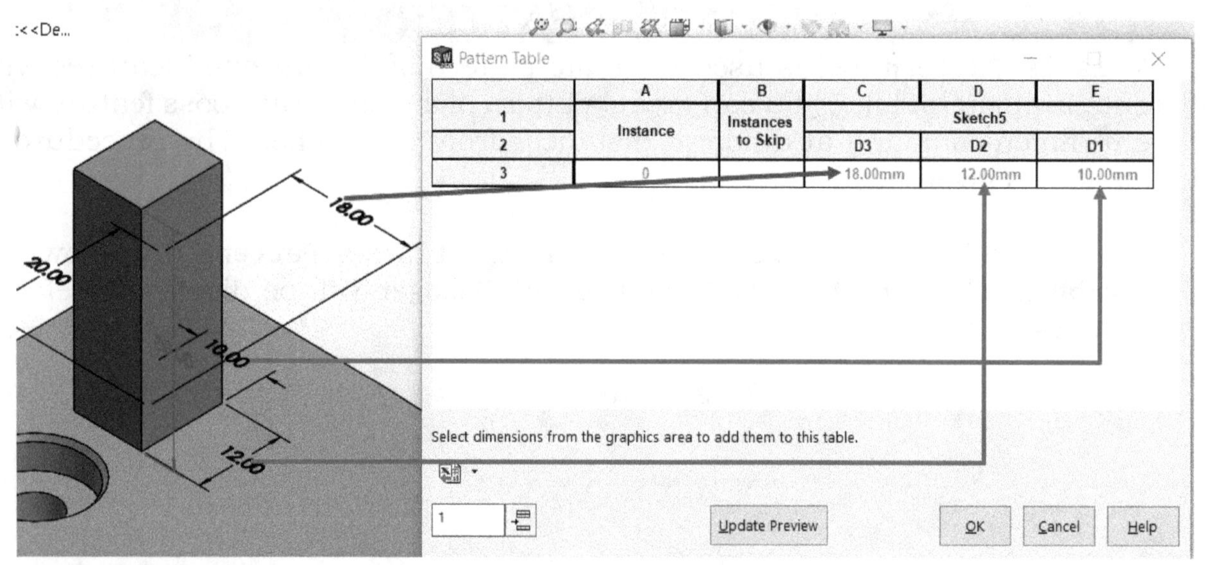

Figure-36. Dimensions added in the table

- Specify desired number of instances to be added in the edit box and click on the **Add instances** button from the bottom of the dialog box. The specified number of instances will be added in the dialog box.
- Modify the parameters as desired and click on the **Update Preview** button. The preview of pattern will be displayed; refer to Figure-37.

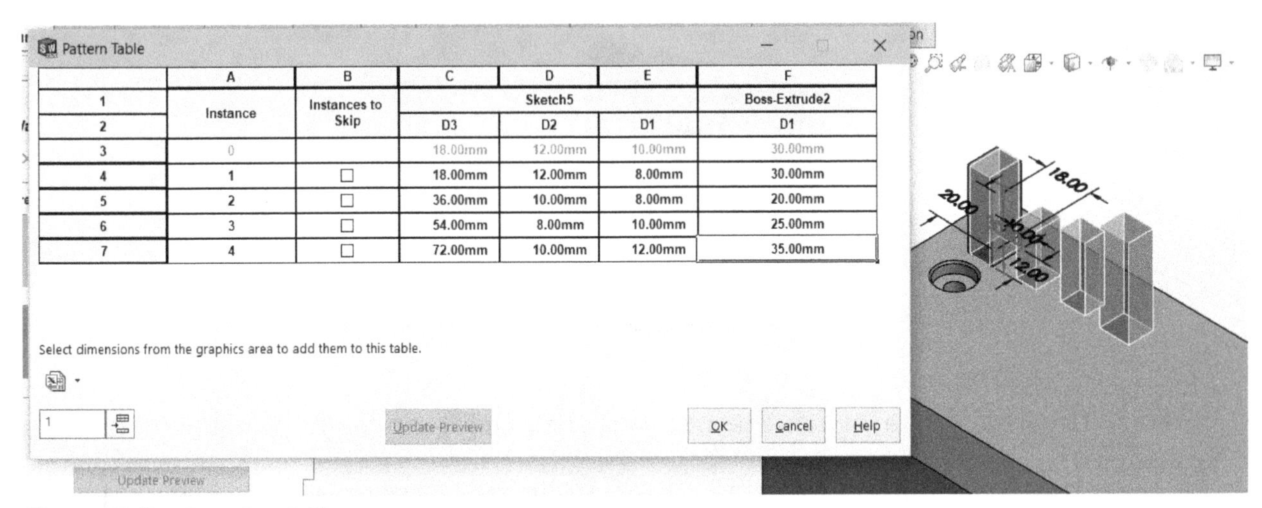

Figure-37. Preview of variable pattern

- Click on the **OK** button from the dialog box to exit.
- Click on the **OK** button from the **PropertyManager** to create the pattern.

RIB

The **Rib** tool is used to create support in the structures to increase their strength. You can find use of rib in various fixtures that are fastened to the wall or in the building columns to support objects. The procedure to create rib feature is given next.

- Click on the **Rib** tool from the **Ribbon**. You are asked to select a sketch or a sketching plane.
- Select the sketch if there is an existing one for rib or select the plane/face and create the sketch. Preview of the rib feature will be displayed. Also, the options in the **PropertyManager** will be modified as per selection; refer to Figure-38.

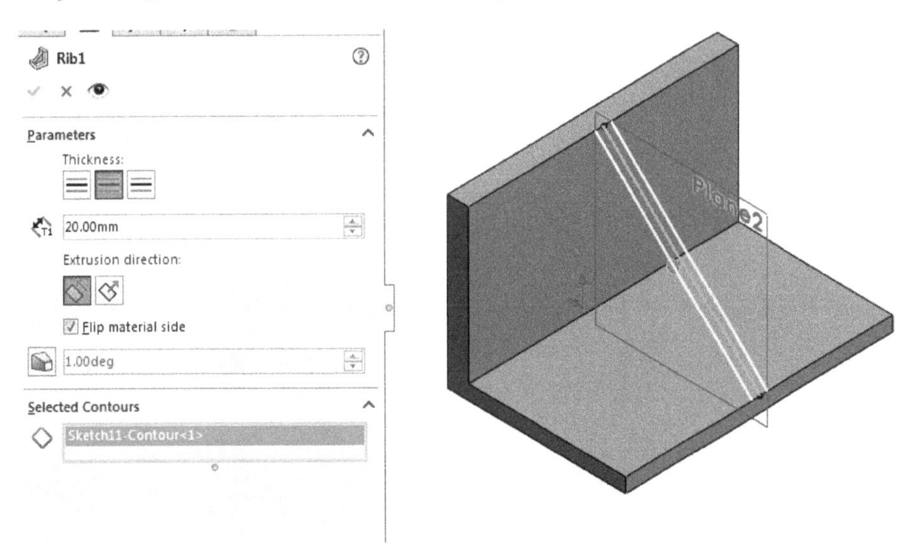

Figure-38. Preview_of_rib_feature

Note that the sketch should be created in such a way that its projection is within solid faces of the model; refer to Figure-39. For creating such sketch, you might need to create reference planes.

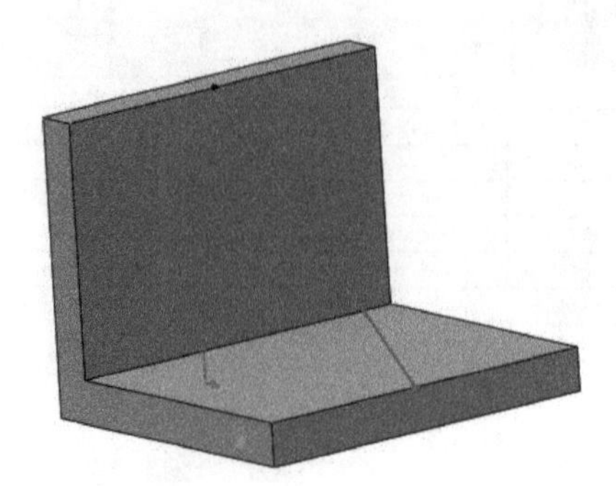

Figure-39. Sketch_for_rib

- Specify the thickness of rib feature and the draft angle in the corresponding spinners.
- You can flip the direction of rib by using the buttons for **Extrusion direction** in the **PropertyManager**. Note that the direction should be in such a way that the rib feature terminates by solid faces; refer to Figure-38.
- You can also change the thickness side by using the three buttons given for **Thickness**.
- Click on the **OK** button to create the rib feature.

DRAFT

The **Draft** tool is used to apply taper to the faces of a solid model. This tool is mainly useful when you are designing components for molding or casting. **Draft** tool applies taper on the faces and this taper allows easy & safe ejection of part from the dies. The procedure to use this tool is given next.

- Click on the **Draft** tool from the **Ribbon**. The **DraftXpert PropertyManager** will display; refer to Figure-40.

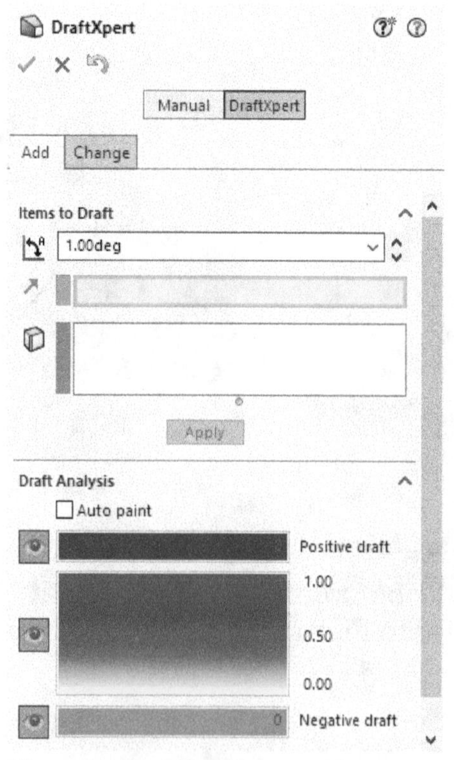

Figure-40. DraftXpert PropertyManager

Using Draft PropertyManager

- Click on the **Manual** button at the top in the **PropertyManager**. The **Draft PropertyManager** will display; refer to Figure-41.

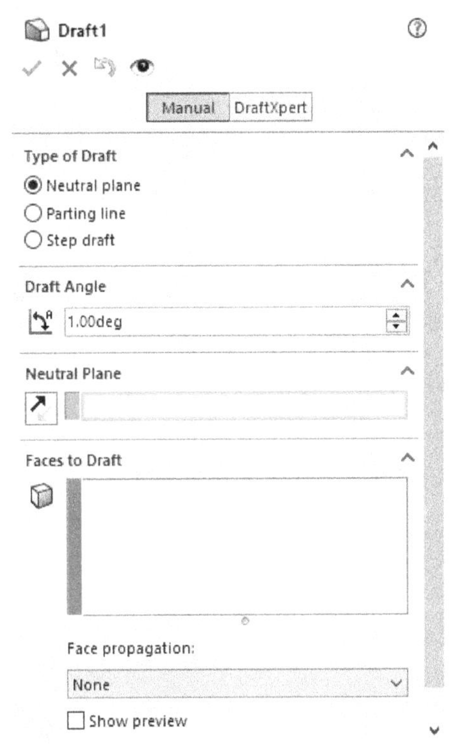

Figure-41. Draft PropertyManager

- Make sure that the **Neutral Plane** radio button is selected and then select a face with respect to which you want to measure all the draft angles. You can flip the direction of draft by using the **Flip** button adjacent to the **Neutral Plane** box.
- On selecting the face, you are asked to select the walls on which you want to apply the draft.
- Change the draft angle by using the spinner in the **Draft Angle** rollout.
- Click on the **Detail Preview** button to check the preview of the draft. Click again on the button to exit the preview; refer to Figure-42.
- Click on **OK** button to create the draft.

We will learn about other options later in the chapter for **Mold Tools**.

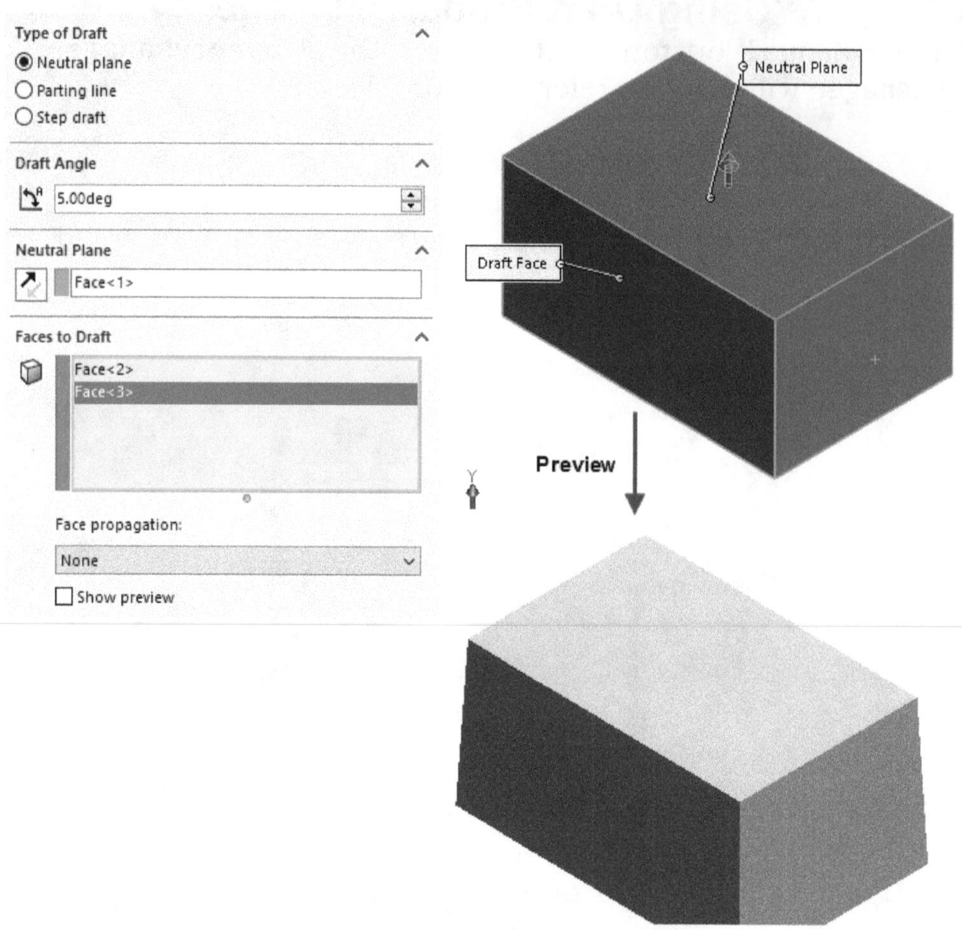

Figure-42. Draft preview

Using DraftXpert PropertyManager

- Click on the **Draft** tool from the **Ribbon** and then click on the **DraftXpert** button if not selected by default. The **DraftXpert PropertyManager** will be displayed.
- Select the face to be used as neutral plane. You will be asked to select the face(s) to which draft is being applied.
- Select the desired face and specify the draft angle in the angle spinner.
- Click on the **Apply** button from the **PropertyManager**. The draft will be created; refer to Figure-43.

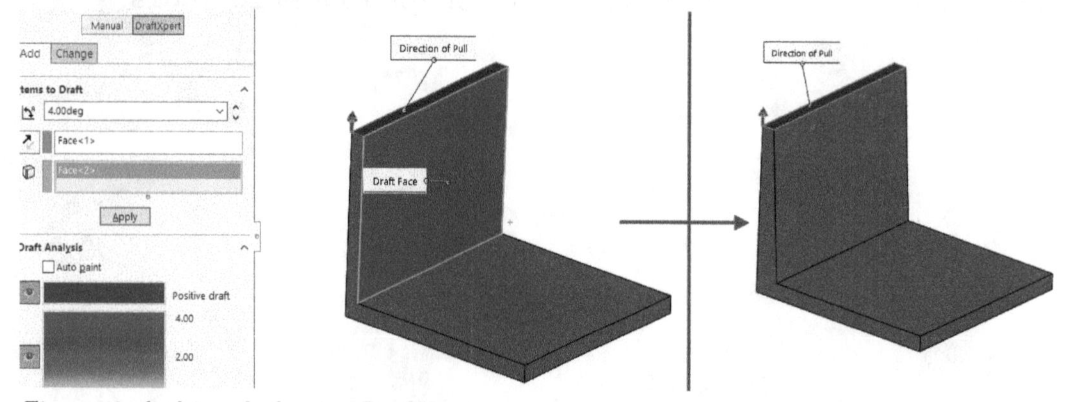

Figure-43. Applying_draft_using_DraftXpert

- Click on the **OK** button from the **PropertyManager** to exit the tool.

SHELL

The **Shell** tool is used to make a solid part hollow and remove one or more faces. The procedure to use this tool is given next.

- Click on the **Shell** tool from the **Ribbon**. The **Shell PropertyManager** will display; refer to Figure-44.

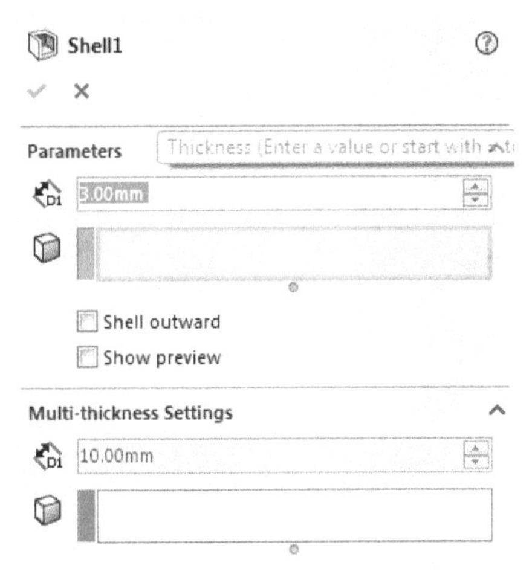

Figure-44. Shell_PropertyManager

- Specify desired thickness in the spinner.
- Click on the **Show preview** check box to display the preview.
- Select a face that you want to remove.
- You can flip the direction of shell by clicking on the **Shell outward** check box.
- Click on the **OK** button to create the feature. Figure-45 shows the preview of the shell feature and output of shell.

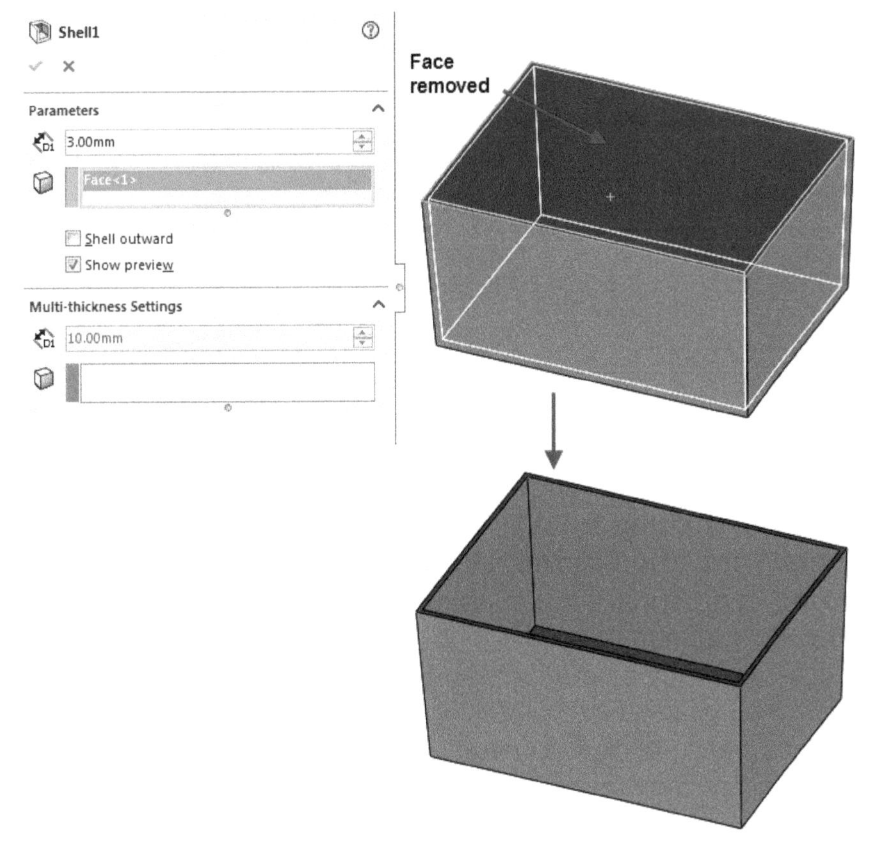

Figure-45. Preview_and_output_of_shell

WRAP

The **Wrap** tool is used to wrap text or any curve/curves on the cylindrical faces. Before using this tool, create a sketch that you want to wrap on a plane parallel to the wrapping face; refer to Figure-46. The steps to create wrap are given next.

Figure-46. Sketch_for_wrap

- Click on the **Wrap** tool and select the sketch from **FeatureManager**. The **Wrap PropertyManager** will display as shown in Figure-47.

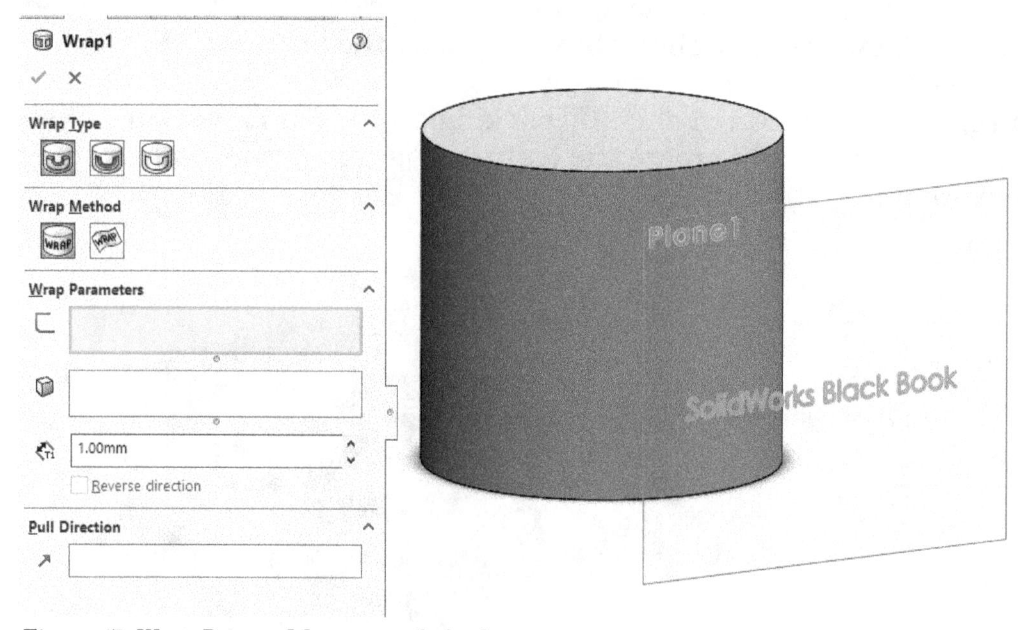

Figure-47. Wrap_PropertyManager_and_sketch

- Select desired button for embossing, engraving (debossing), or scribing from the **Wrap Type** rollout.
- Select desired wrap method from the **Wrap Method** area. There are two buttons in this area; **Analytical** and **Spline Surface**. Select the **Analytical** button if you want to wrap on cylindrical face. Select the **Spline Surface** button if you want to wrap sketch on surface.
- Select the cylindrical face or surface on which you want to emboss or engrave the sketch. Preview will display.
- Set the thickness value in the spinner.
- Click **OK** button to create the feature. Figure-48 shows a sketch embossed on the cylindrical face.

Figure-48. Output_of_wrap

INTERSECT

The **Intersect** tool is used to separate the volume created by intersection of solids, surfaces, or a combination of both. This tool can be very helpful for creating mold tools manually. The steps to use this tool are given next.

- Click on the **Intersect** tool from the **Ribbon**. The **Intersect PropertyManager** will display as shown in Figure-49.

Figure-49. Intersect_PropertyManager

- Select the surfaces/solids from which you want to extract the intersecting volume.
- There are two type of regions formed while intersecting solids; Intersecting regions and internal regions formed during intersection. There are two radio buttons in the **PropertyManager** to find out these regions; **Create intersecting regions** and **Create internal regions**. You can select the **Create both** radio button to include both regions in the list.
- Click on the **Intersect** button from the **PropertyManager**, the preview will be displayed in the viewport. Also, the region that you can exclude will be displayed in the **Regions to Exclude** rollout; refer to Figure-50.

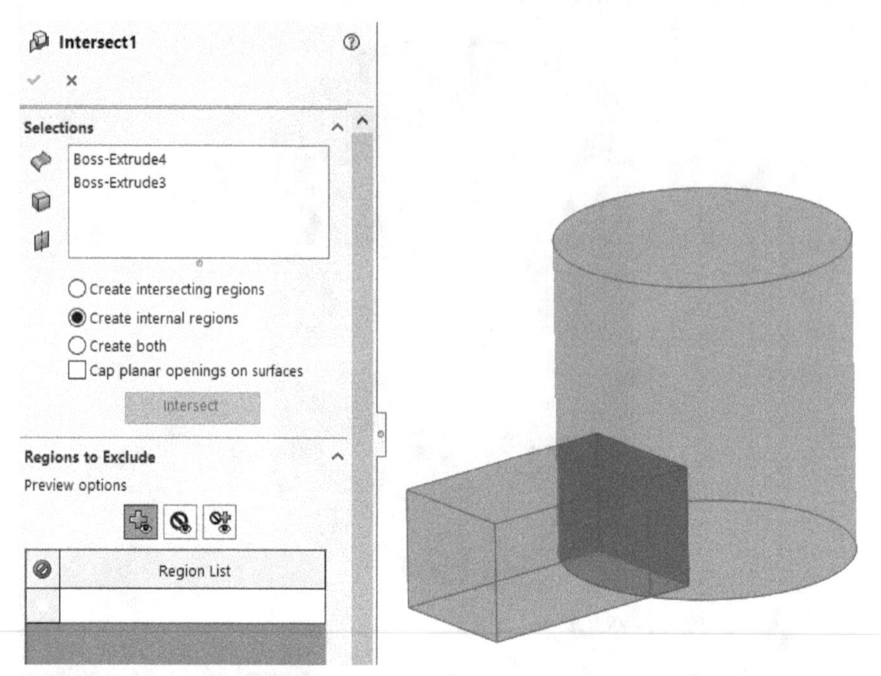

Figure-50. Preview_of_intersection

- Click on the check boxes of feature that you want to exclude from the **Regions to Exclude** rollout or select the portion of model from the viewport.
- Click on the **OK** button to create the intersection portion. Figure-51 shows the output of the intersection.

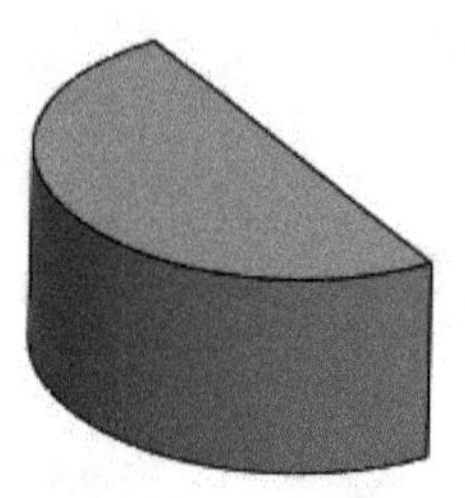

Figure-51. Output of intersection

EQUATIONS

The **Equations** tool is used to manipulate the model based on specified equations for various parameters of part. The procedure to use **Equation** tool is given next.

- Click on the **Equations** tool from the **Tools** menu; refer to Figure-52. The **Equations, Global Variables, and Dimensions** dialog box will be displayed; refer to Figure-53.

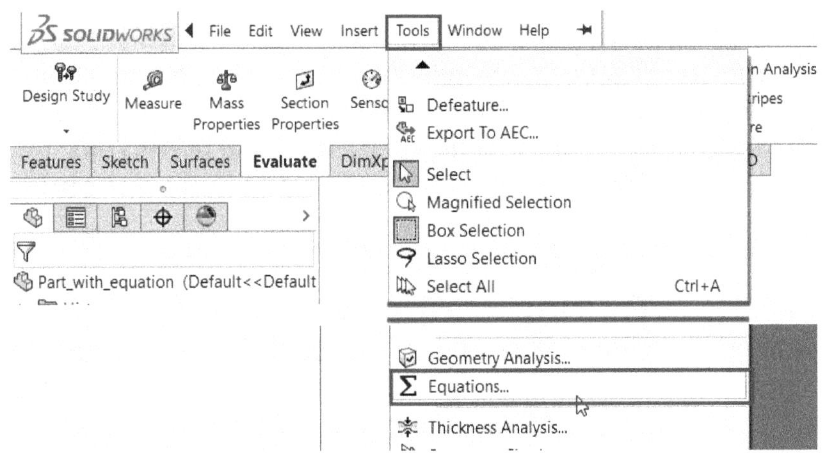

Figure-52. Equations tool

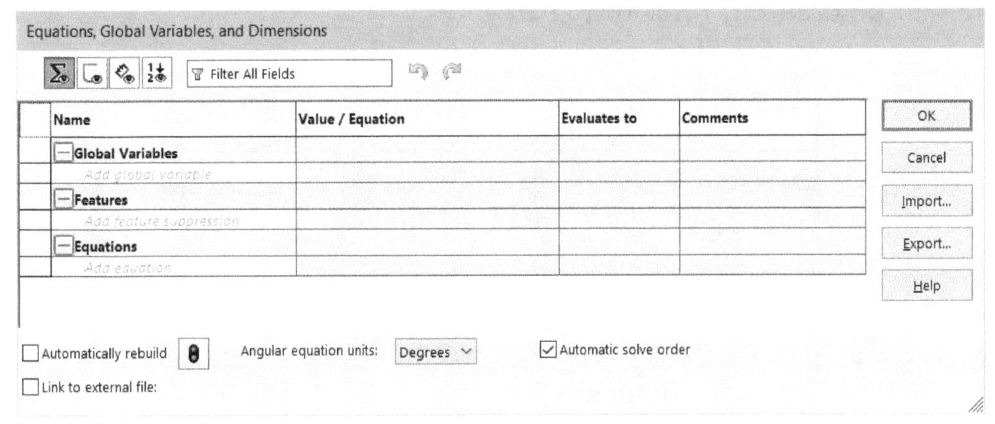

Figure-53. Equations,_Global_Variables,_and_Dimensions_dialog_box

- By default, the **Equation View** button is selected in the dialog box so, you can see the parameters related to equations like global variables, features suppressed/ unsuppressed, and equations. Similarly, select the **Sketch Equation View** button to check equations in various sketches of part.
- Click on the **Dimension View** button from the dialog box as we will be writing equations in this mode. On selecting this button, all the dimensions of the model will be displayed in the dialog box; refer to Figure-54.

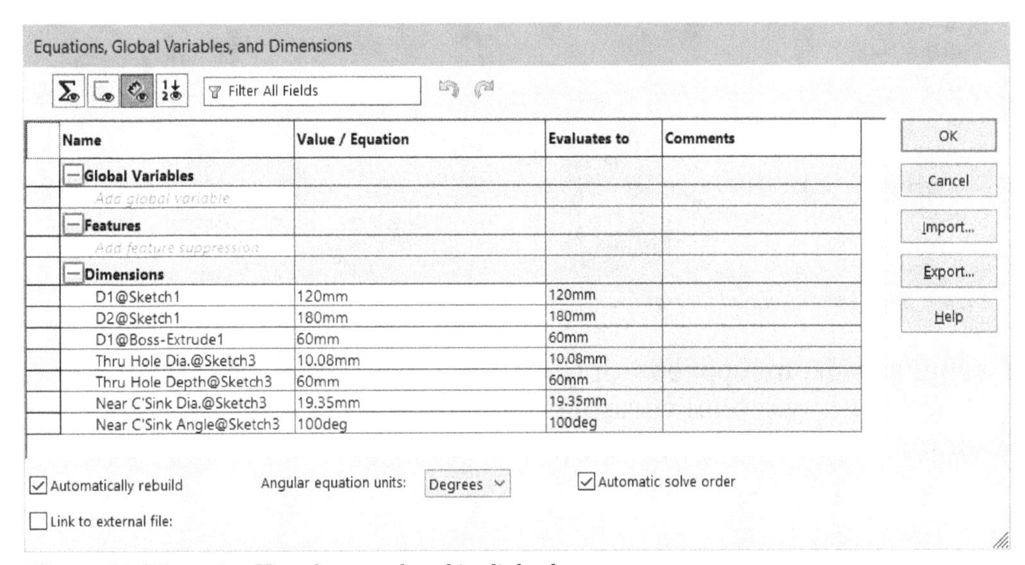

Figure-54. Dimension View button selected in dialog box

The options in this dialog box are divided into three nodes viz. Global Variables, Features, and Dimensions. Global variables are parameters which are user-defined and are used in equations. Like, you can create a global variable scale and use it to scale-up or scale-down the model. In the **Features** node, you can give the name of feature and set it suppressed/unsuppressed. In the **Dimensions** node, you can set the equations for various dimensions. The methods to use options in each node are discussed next.

Adding Global Variables

- Click in the **Add global variable** cell in the **Global Variables** node of the dialog box and type desired name of the variable like scale in the format **"scale"** (note that "" are included). Press **TAB** from keyboard. You will be asked to specify the value of variable.
- Type desired value of variable in the edit box and then press **ENTER**. The global variable will be created; refer to Figure-55.

Name	Value / Equation	Evaluates to	Com
Global Variables			
"scale"	= 2	2	
Add global variable			
Features			
Add feature suppression			
Dimensions			

Figure-55. Global variable created

Suppressing/Un-suppressing Features

- Click in the **Add feature suppression** cell of the **Features** node. You are asked to select the feature to be suppressed/unsuppressed.
- Select the feature from **FeatureManager Design Tree**. A flyout will be displayed in the dialog box with various options; refer to Figure-56.

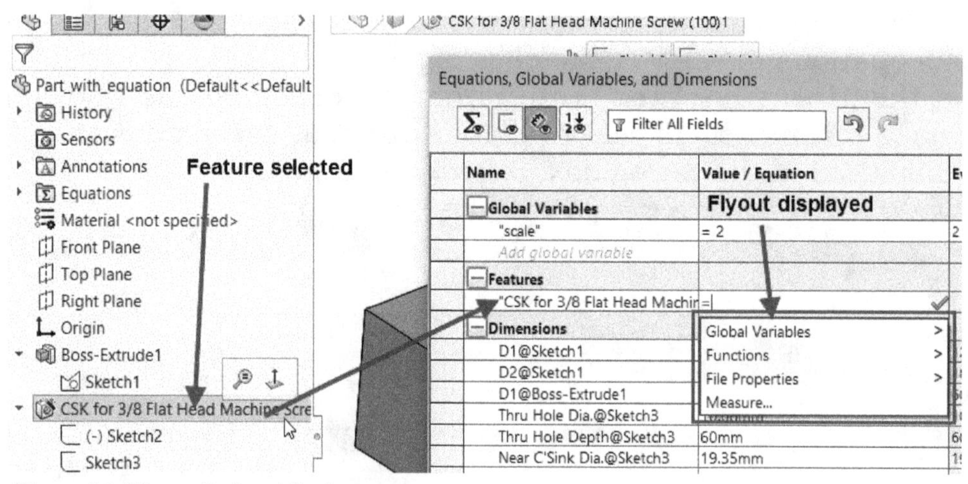

Figure-56. Flyout displayed for features

- Select **suppress** or **unsuppress** option from the **Global Variables** menu in the flyout. Note that any global variable created by you earlier will also be displayed in this menu.

Setting Dimension Equations

- Click in the value area of the dimension that you want to solved by equation and remove its value.
- Type the equation for dimension in the field; refer to Figure-57. Press **ENTER** to apply the equation.

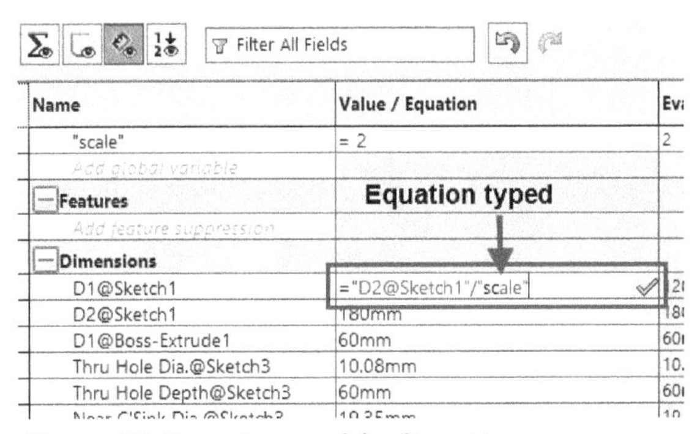

Figure-57. Equation typed for dimention

- Similarly, create other equations and then click on the **Rebuild** button from the dialog box to update the model; refer to Figure-58.

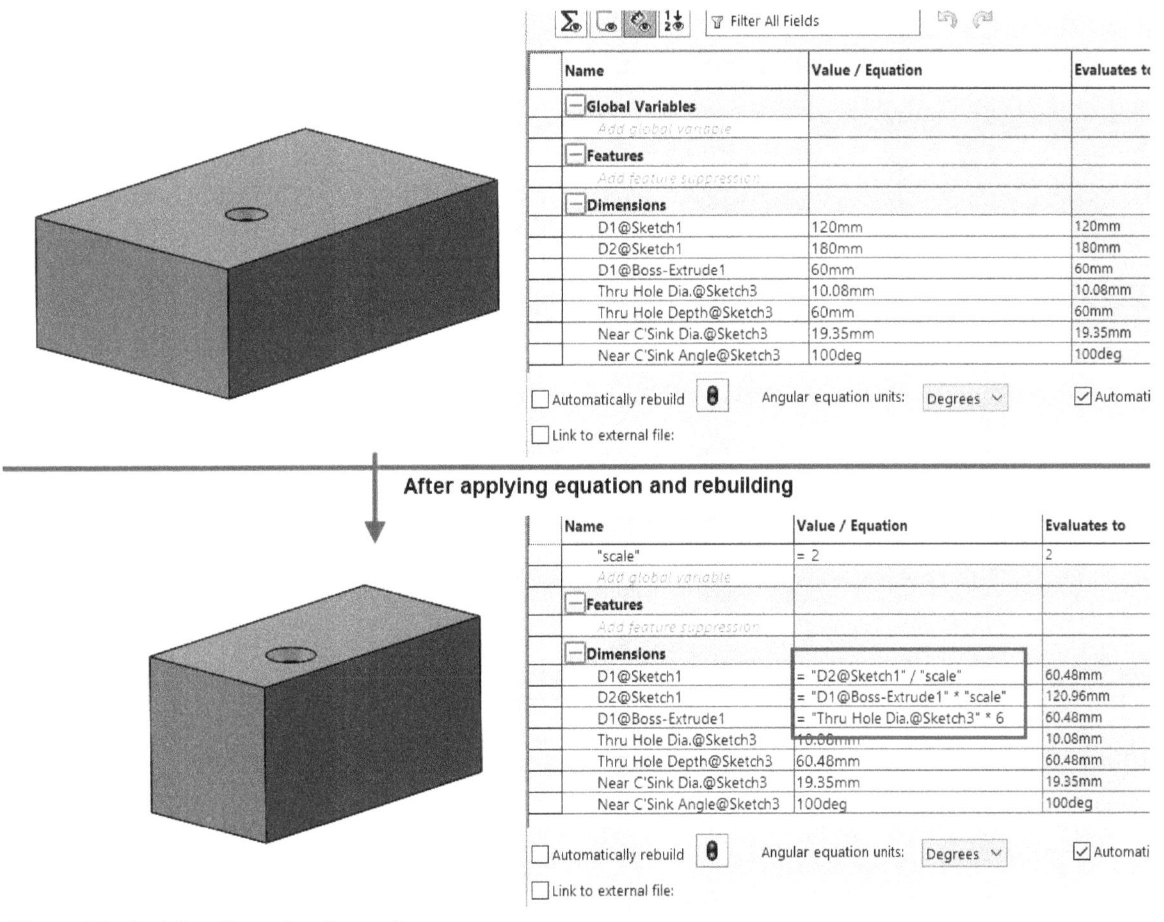

Figure-58. Applying dimensional equations

- Select the **Automatically Rebuild** check box if you want the model to be updated automatically. Click on the **OK** button to apply equations.

DESIGN TABLE

Design table is a very fast way to create multiple configurations of a part. Suppose you want to create multiple sizes of same shaped part like bolt, then you can use the Design table to create multiple parts while modeling the part only one time (Note that the part file is available in the resource kit). The procedure to use the **Excel Design Table** tool is given next with the help of an example shown in Figure-59.

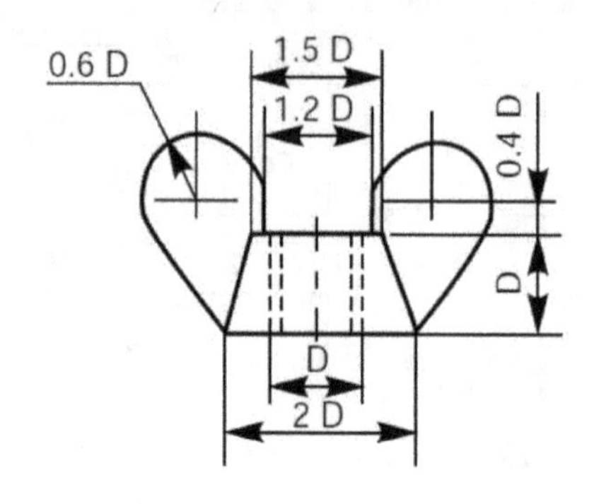

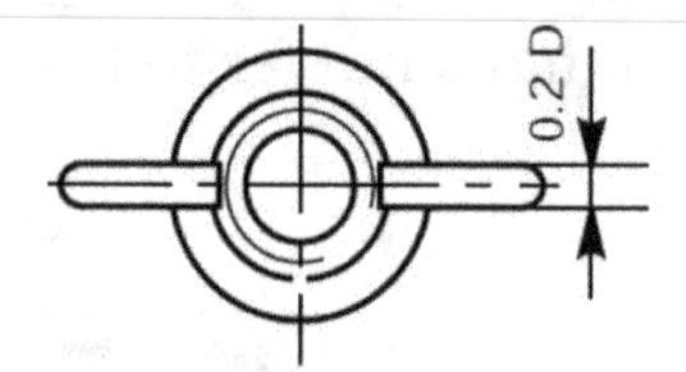

Figure-59. Wing Nut

- Create the model while taking the value of **D** as **10**; refer to Figure-60.

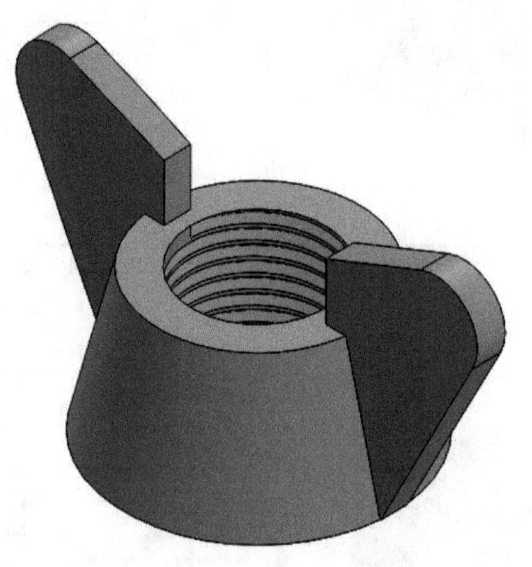

Figure-60. Wing nut created

- Create the equation of part as per the drawing given in Figure-59. The parameters to be specified for equation are given in Figure-61.

Name	Value / Equation	Evaluates to	Com
Global Variables			
Add global variable			
Features			
Add feature suppression			
Dimensions			
D4@Sketch1	= 10	10mm	
D1@Sketch1	= "D4@Sketch1" * 2	20mm	
D2@Sketch1	= "D4@Sketch1"	10mm	
D3@Sketch1	= "D4@Sketch1" * 1.5	15mm	
D1@Revolve1	360deg	360deg	
D8@Thread1	0deg	0deg	
D7@Thread1	1mm	1mm	
D6@Thread1	= "D4@Sketch1"	10mm	
D4@Thread1	= "D4@Sketch1"	10mm	
D1@Thread1	0deg	0deg	
D1@Sketch7	= "D4@Sketch1" * 0.6	6mm	
D2@Sketch7	= "D4@Sketch1" * 0.4	4mm	
D3@Sketch7	= "D4@Sketch1" * 1.2	12mm	
D4@Sketch7	= "D4@Sketch1" * 1.4	14mm	
D1@Boss-Extrude2	= "D4@Sketch1" * 0.2	2mm	

Figure-61. Parameters for wing nut specified for equation

- Click on the **Excel Design Table** tool from the **Insert -> Tables** menu; refer to Figure-62. The **Excel Design Table PropertyManager** will be displayed; refer to Figure-63.

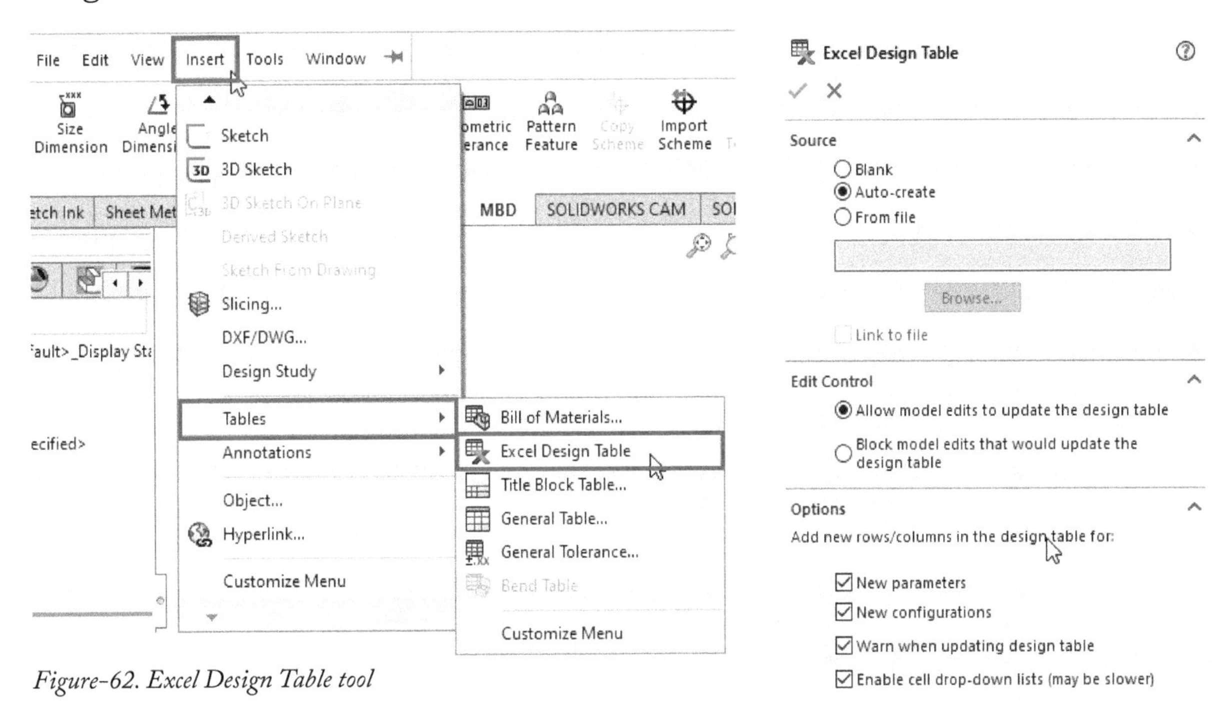

Figure-62. Excel Design Table tool

Figure-63. Excel Design Table PropertyManager

- Click on the **OK** button from the **PropertyManager**. The **Dimensions** selection box will be displayed; refer to Figure-64. Select the dimension which is used in equation to drive other dimensions and click on the **OK** button from the selection box. The design table excel sheet will be displayed; refer to Figure-65.

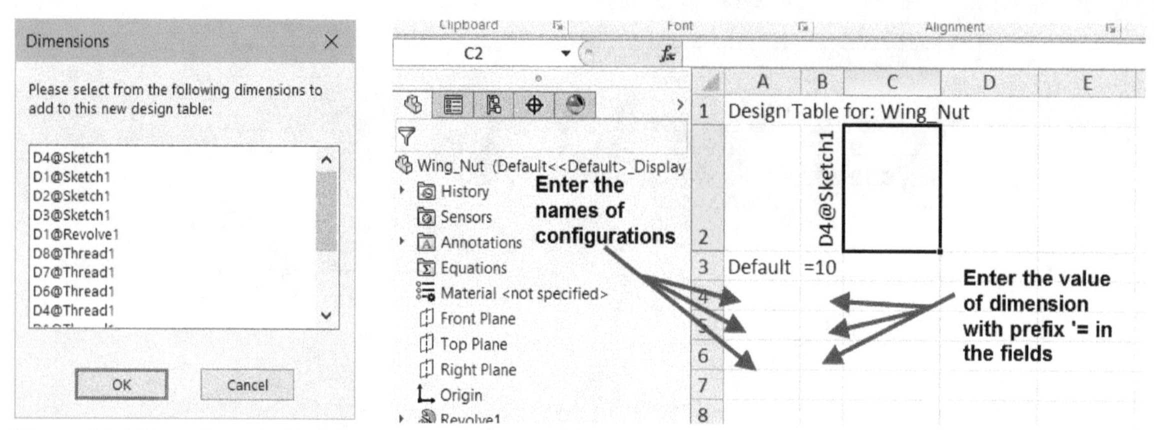

Figure-64. Dimensions selection box *Figure-65. Design table*

- Under the first column, specify the name of configuration like Wing Nut dia 12, Wing Nut dia 15, etc. In the second column, specify the respective dimension value like 12, 15, etc. Refer to Figure-66.

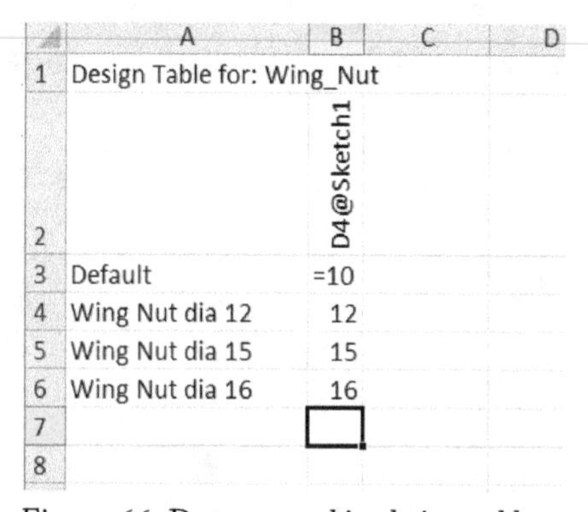

	A	B	C	D
1	Design Table for: Wing_Nut			
2		D4@Sketch1		
3	Default	=10		
4	Wing Nut dia 12	12		
5	Wing Nut dia 15	15		
6	Wing Nut dia 16	16		
7				
8				

Figure-66. Data entered in design table

- Click on the **OK** button displayed in the top right of drawing area (a tick mark). A notification will be displayed telling you that configurations have been generated through design table; refer to Figure-67.
- Click on the **OK** button. The configurations will be created. Double-click on desired configuration in **ConfigurationManager** to check the model; refer to Figure-68.

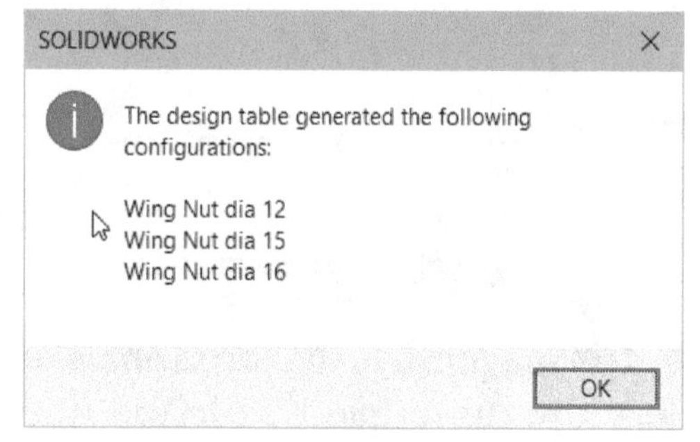

Figure-67. Notification box

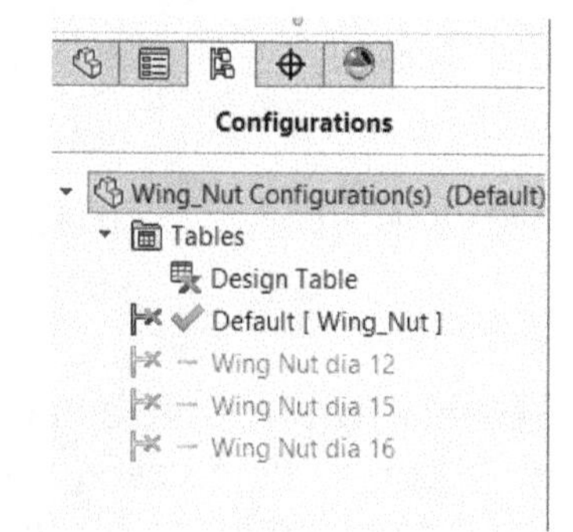

Figure-68. ConfigurationManager

APPLYING MATERIAL TO PART

Material is a very important properties of objects created in any CAD software. Most of the analysis and cost estimates are directly based on the material of part. The procedure to apply material to the part is given next.

- After creating model, click on the **Edit->Appearance->Material** tool from the Menu bar; refer to Figure-69. The **Material** dialog box will be displayed; refer to Figure-70.

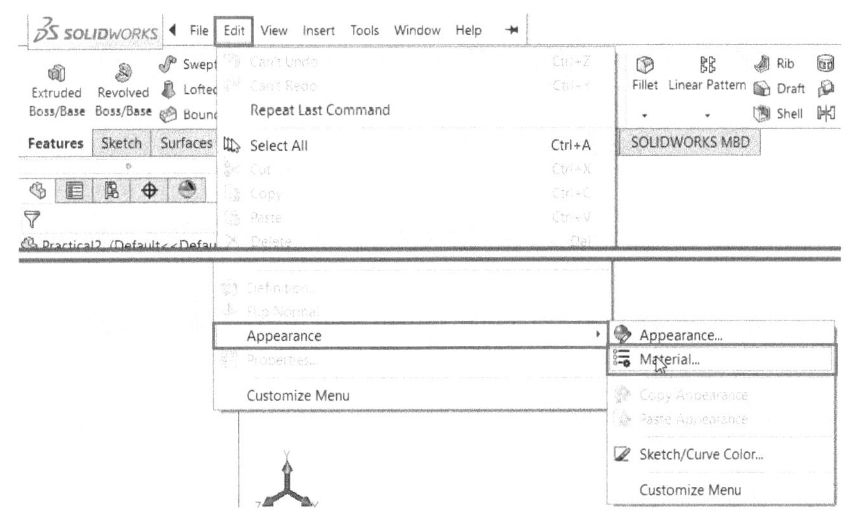

Figure-69. Material tool

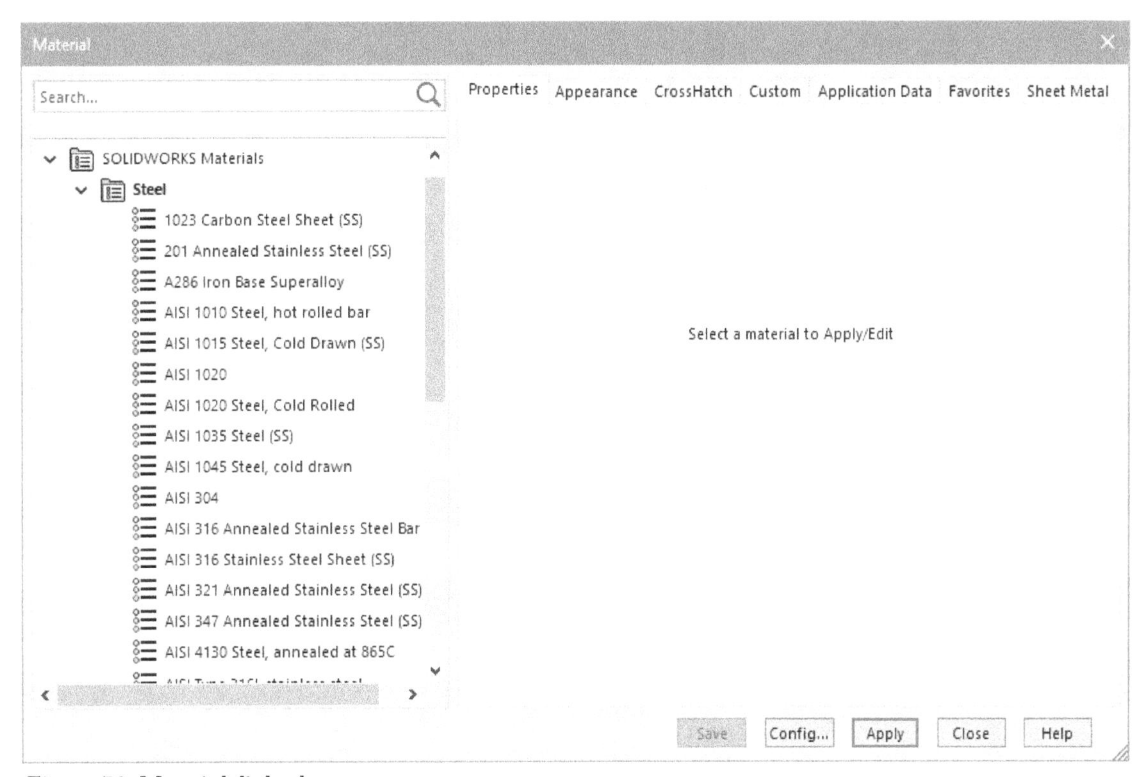

Figure-70. Material dialog box

- Select desired material from the list and click on the **Apply** button to apply it on the current part.
- Click on the **Close** button to exit the dialog box.

MASS PROPERTIES OF PART

Click on the **Mass Properties** tool from the **Evaluate CommandManager** in the **Ribbon**. The mass properties of the part will be displayed; refer to Figure-71.

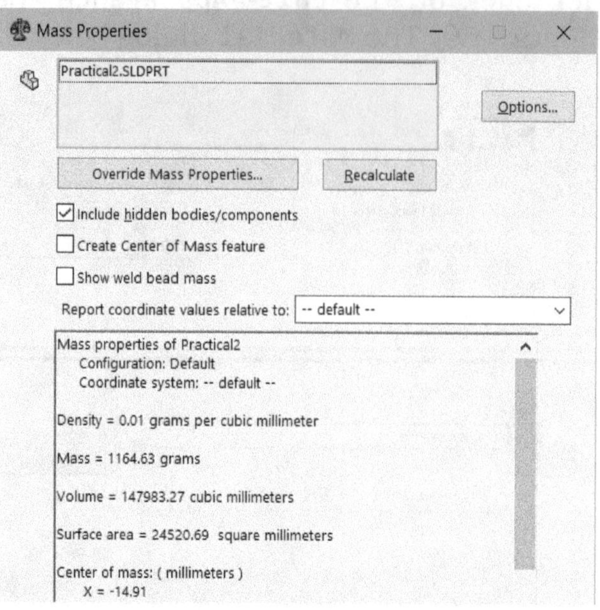

Figure-71. Mass Properties dialog box

Click on the **Close** button to close the dialog box.

Practical 1

Create the model (isometric view) as shown in Figure-72. The views of the model with dimensions are given in Figure-73.

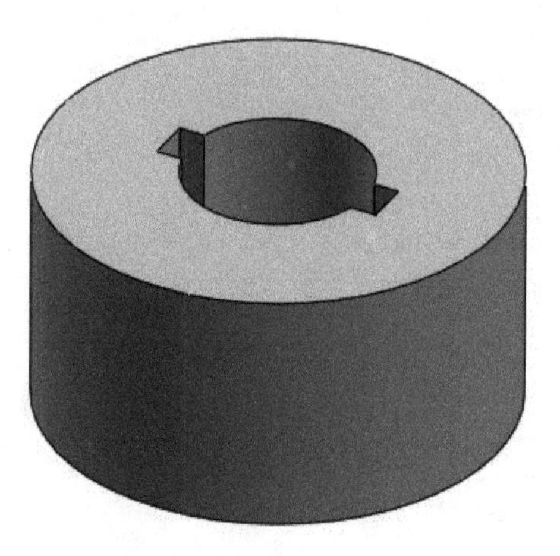

Figure-72. Practical_1_model

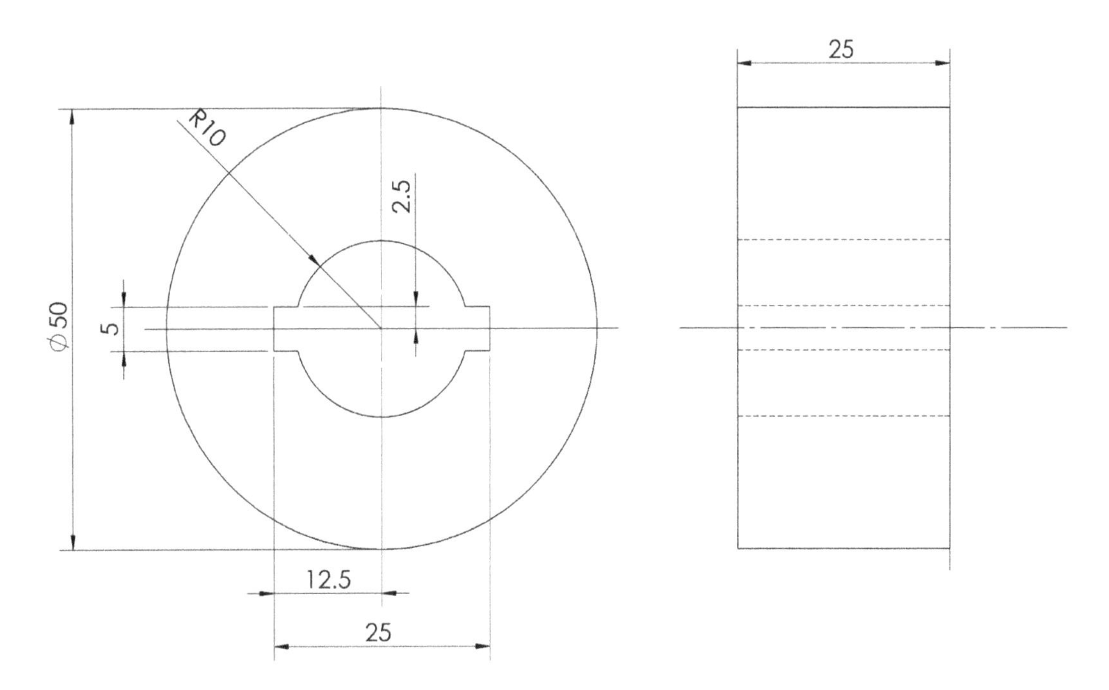

Figure-73. Views_for_Practical_1

Before we start working on the Practical, it is important to understand two terms; first angle projection and third angle projection. These are the standards of placing views in the engineering drawing. The views placed in the above figure are using third angle projection. In first angle projection, the top view of model is placed below the front view and right side view is placed at left of the front view. You will learn more about projection in chapter related to drafting.

Starting SolidWorks Modeling environment and creating Extrude feature

- Double-click on the SolidWorks icon from desktop if you have not started SolidWorks. Start a new part file.
- Click on the **Features CommandManager** in the **Ribbon** if not selected.

From the isometric view as well as from the other views, we can judge that this model can be easily created by extruding the sketch.

- Click on the **Extrude Boss/Base** tool. The **Extrude PropertyManager** will display.
- Select the **Top** plane from the **FeatureManager Design Tree** or from the viewport; refer to Figure-74.

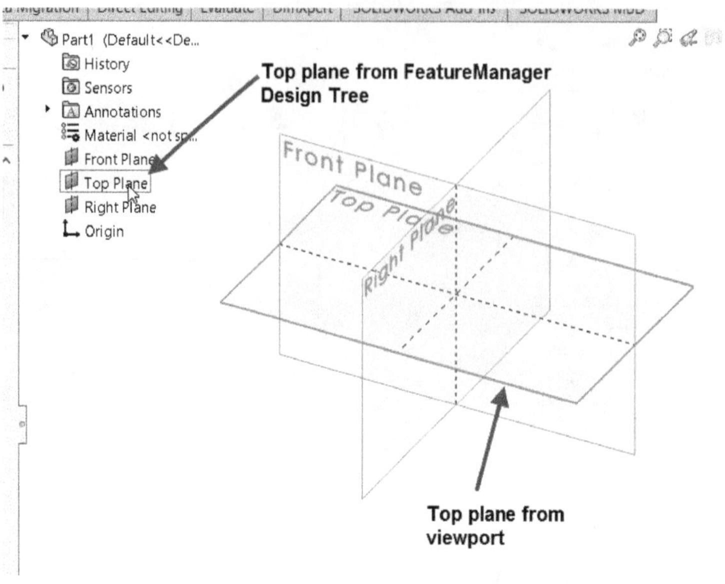

Figure-74. Selecting_Top_plane

We will draw the sketch on top plane to get the isometric view as shown in Figure-72.

- Click on the **Normal To** button from the **View Orientation** drop-down in the **Heads-up View Toolbar** or press **CTRL + 8** if the plane is not parallel to screen. Note that **8** key in shortcut is not from Numpad of keyboard.
- Click on the **Circle** tool from the **Ribbon** and select the **Diameter Dimensions** check box.
- Click at the coordinate system to place center of the circle.
- Drag the cursor and enter the value as **50** in the **Dimension box**.
- Again click at the coordinate system and draw circle of diameter **20**.
- Click on **OK** button from the **Circle PropertyManager**.
- Select the **Center Rectangle** tool from the **Rectangle** drop-down in the **Ribbon** and click at the center of the circles.
- Drag the cursor and specify the dimension of rectangle as **5** and **25** for height and width, respectively. Click **OK** from the **PropertyManager**. The sketch after performing the above steps will display as shown in Figure-75.

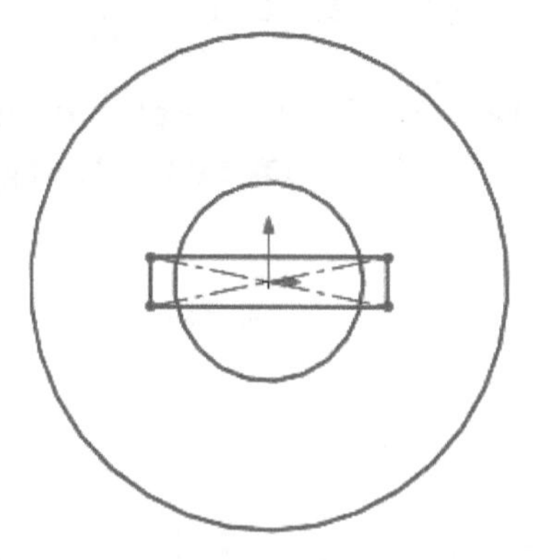

Figure-75. Sketch after creating circles and rectangles

- Select the **Trim Entities** tool from the **Ribbon** and trim the entities in such a way that the sketch is displayed as shown in Figure-76.

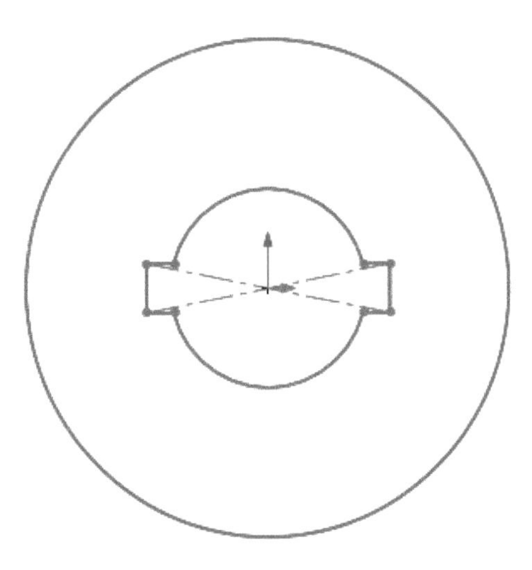

Figure-76. Sketch_after_trimming

- Click on the **Smart Dimension** tool and dimension the sketch. Refer to Figure-77.

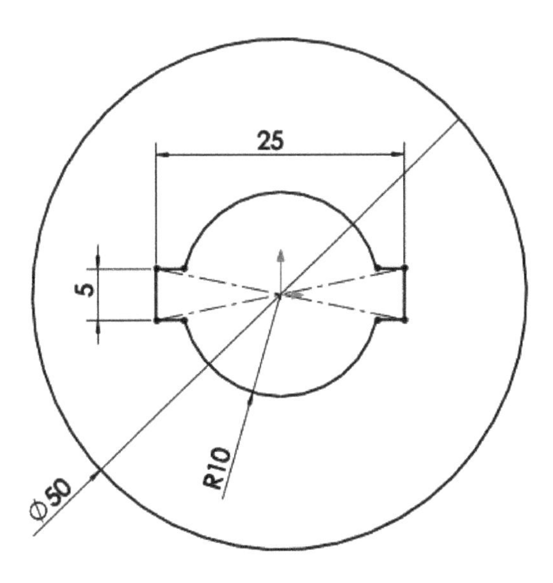

Figure-77. Dimensioned_sketch

- Click on the **Exit Sketch** button. Preview of the Extrude feature will display as shown in Figure-78.

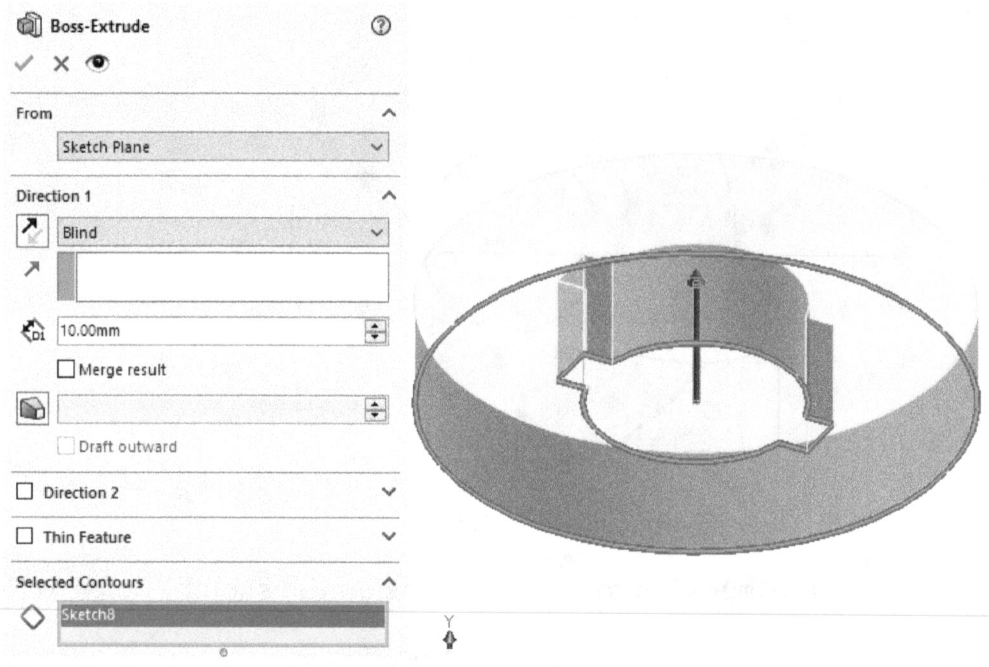

Figure-78. Preview of extrude

- Specify the height of extrusion as **25** and then click on the **OK** button from the **PropertyManager**. The model will be created as in Figure-72.

Practical 2

Create the model (isometric view) as shown in Figure-79. The dimensions and view is given in Figure-80.

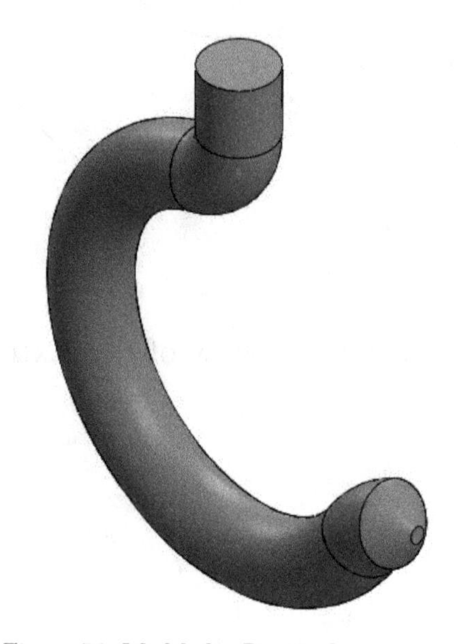

Figure-79. Model_for_Practical

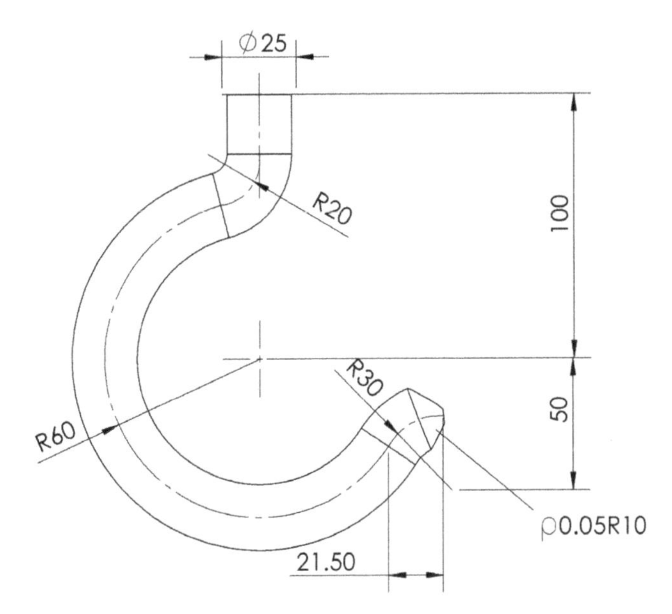

Figure-80. Practical 2 drawing view

Start the SolidWorks if not started and open the modeling environment as explained in previous practical.

You can find out from the model that this hook can be easily created with the help of **Swept Boss/Base** tool. But before we use that tool, we must have sketches for path as well as section. The steps to create them are as follows:

Creating Sketches for Hook

- Start a new sketch on Front Plane.
- Click on the **Circle** tool and create the circle of diameter **120** taking coordinate system as center.
- Draw a straight line starting from top quadrant point and having length of approximately **40**.
- Draw a three point arc in the bottom area of the circle; refer to Figure-81. (It should look like the one given in figure. Accuracy is no required.)

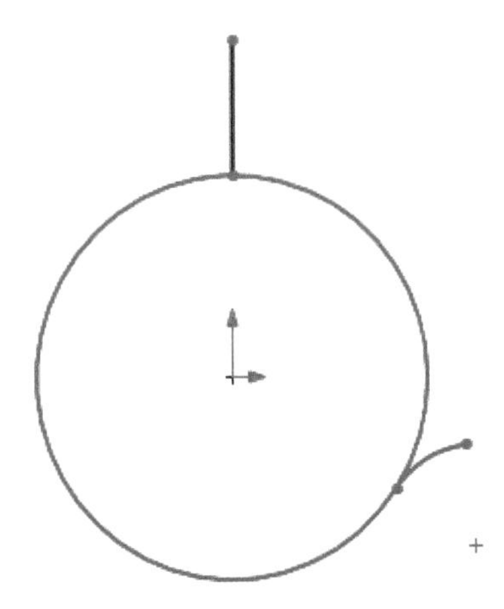

Figure-81. Sketch_after_creating_three_point_arc

- Click on the **Trim Entities** tool and trim the portion between the straight line and the arc; refer to Figure-82.

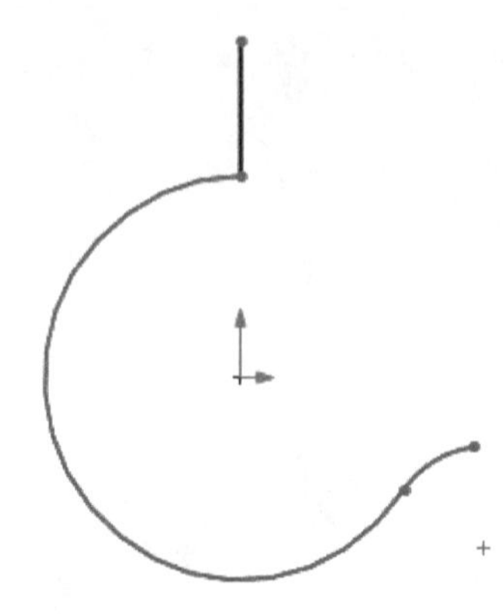

Figure-82. Sketch_after_trimming_circle

- Click on the **Sketch Fillet** tool and apply fillet between the straight line and circle. Specify the fillet radius as **20**.
- Now, dimension the sketch as shown in Figure-83.

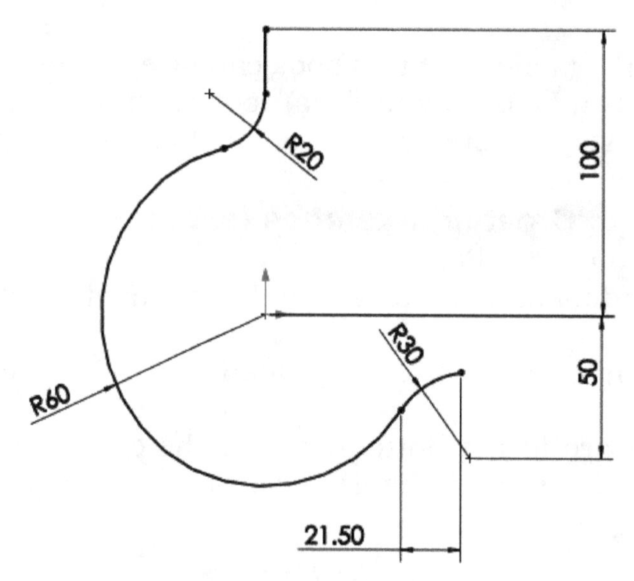

Figure-83. Dimension_sketch_of_hook

- Exit the sketch by clicking **Exit Sketch** button from the **Ribbon**.

In previous steps, we have created the sketch of the path. Now, we will create the section sketch.

- Click on the **Plane** tool from the **Reference** drop-down in the **Features** tab of the **Ribbon**. The **Plane PropertyManager** will display as shown in Figure-84.

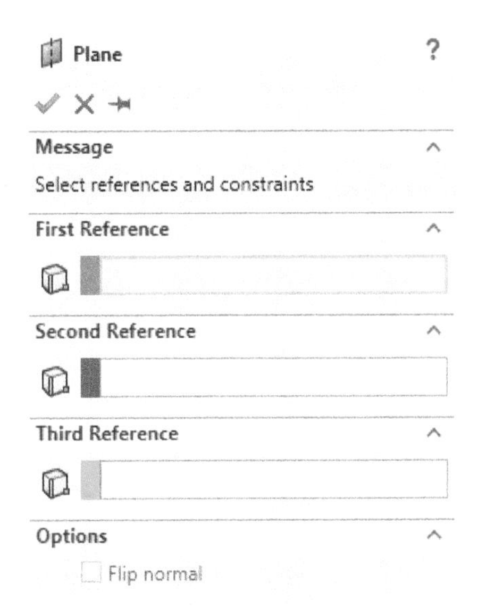

Figure-84. Plane_PropertyManager

- Click on the top point of the line and select the **Top Plane**. Preview of the plane will display; refer to Figure-85.

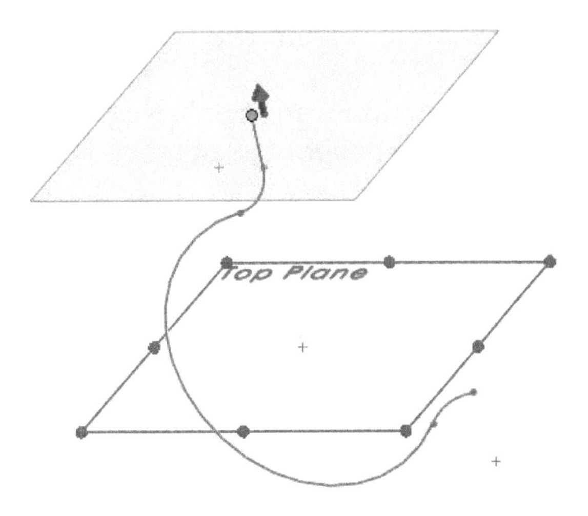

Figure-85. Preview_of_plane

- Click on the **OK** button from the **Plane PropertyManager** to create the plane.
- Select the newly created plane and select the **Sketch** button from the **Sketch** tab of the **Ribbon**. The sketching environment will display.
- Select the **Normal To** button from the **View Orientation** box. The sketching plane will become parallel to the viewport.
- Create a circle of diameter **25** taking end point of line as center; refer to Figure-86.

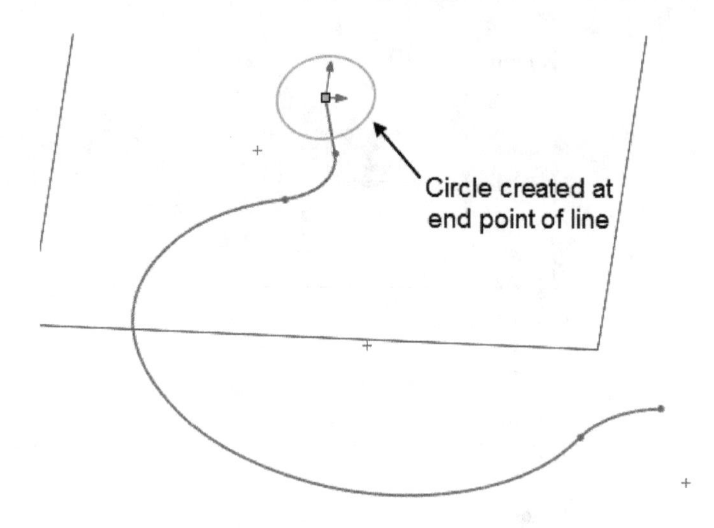

Figure-86. Circle_to_be_created

Creating the Swept Boss/Base feature

Now, we have all the sketches to create the hook. The steps to create the swept boss/ base feature using these sketches is given next.

- Click on the **Swept Boss/Base** tool from the **Ribbon**. The **Sweep PropertyManager** will display.
- Select the circle created for section and then select sketch created for the profile.
- Click on the **OK** button from the **PropertyManager**. The swept base feature will be created; refer to Figure-87.

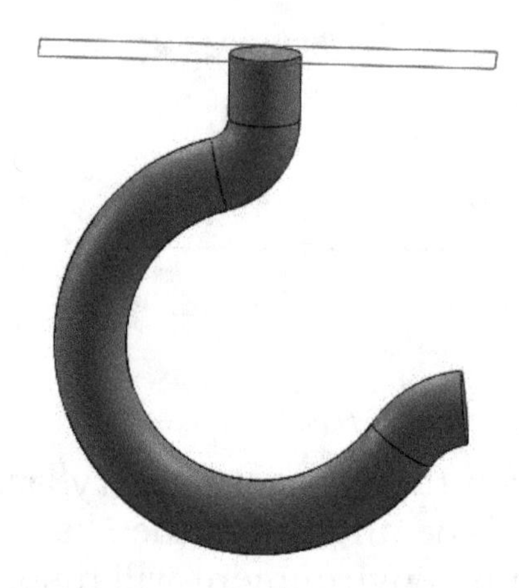

Figure-87. Swept_feature_created

Applying Conic fillet at the end

As we all know hooks doesn't end with sharp edges. So, we need to apply fillet at the end. Steps to do so are given next.

- Click on the **Fillet** tool from the **Fillet** drop-down in the **Ribbon**. The **Fillet PropertyManager** will display.
- Select the edge of the end and specify the parameters in the **PropertyManager** as shown in Figure-88.

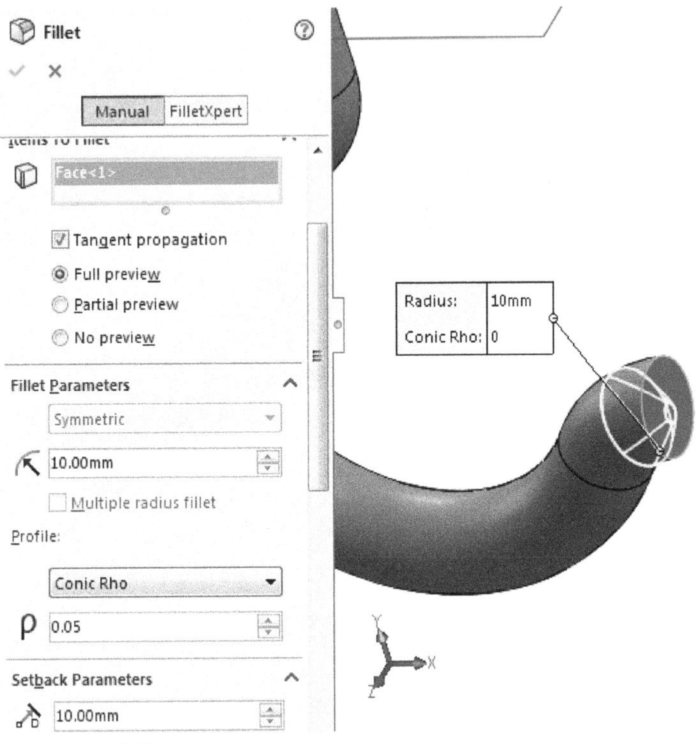

Figure-88. Fillet_at_the_end_edge

- Click on the **OK** button from the **PropertyManager** to create the feature. The model will be displayed as shown in Figure-89.

Figure-89. Final_model

Practical 3

Create the model (isometric view) as shown in Figure-90. The dimensions of the model are given in Figure-91.

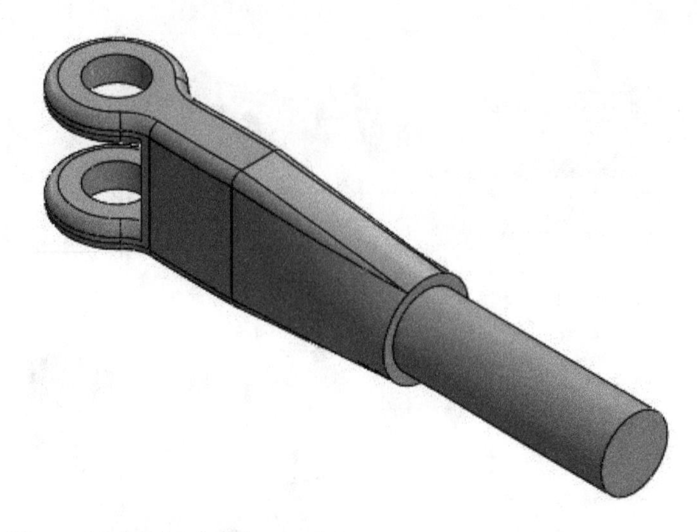

Figure-90. Model_for_Practical_3

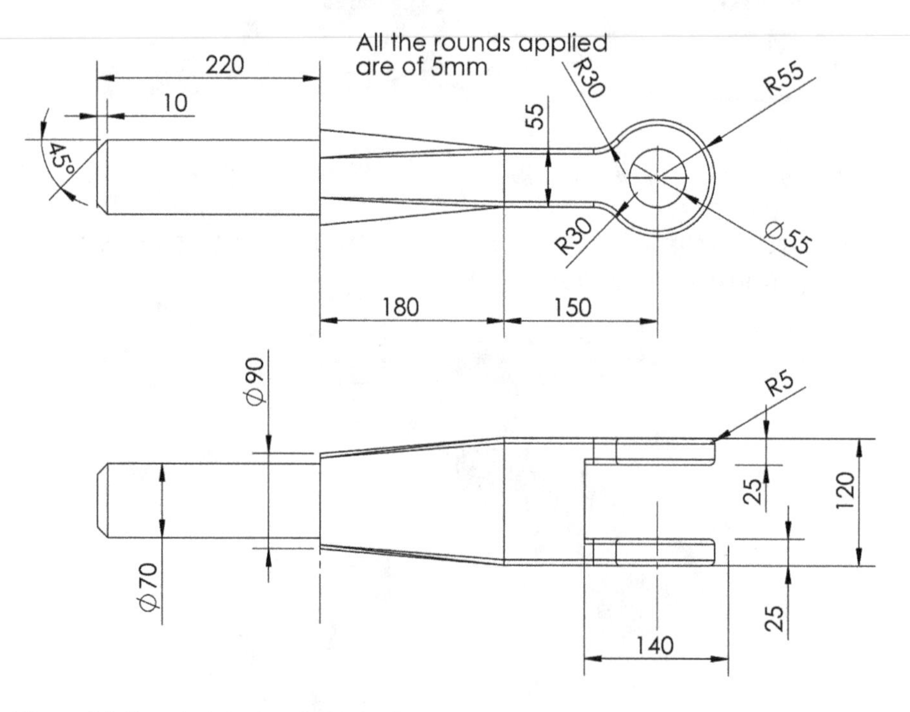

Figure-91. Practical_3_drawing_views

Creating first extrude feature

- Click on the **Extrude Boss/Base** tool from the **Ribbon**. The **Boss-Extrude PropertyManager** will display.
- Select the Top plane from the viewport and create the sketch as shown in Figure-92.

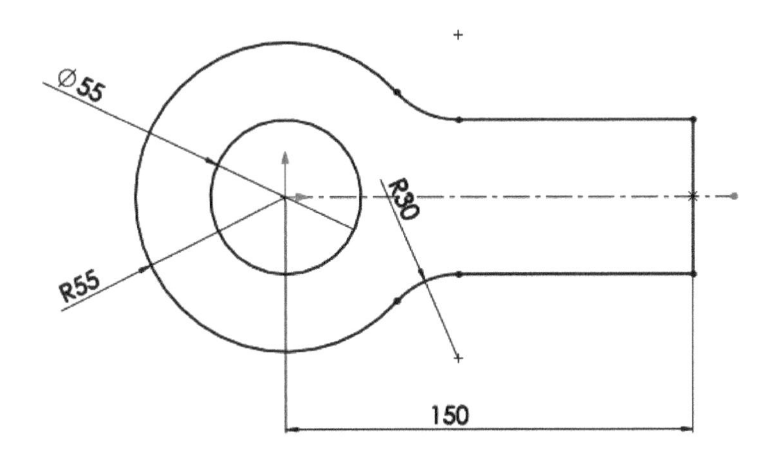

Figure-92. Sketch_created_at_top_plane

- Exit the sketch and extrude it to the height of **120**.

Creating loft feature

Before creating loft feature, we must have at least two sketch sections. First, we will create a plane at a distance of **180** from the vertical flat face of the model and then, we will create sketches and loft feature successively.

- Click on the **Plane** tool from the **Reference Geometry** drop-down and select the vertical flat face of the model.
- Specify the offset distance as **180**; refer to Figure-93.

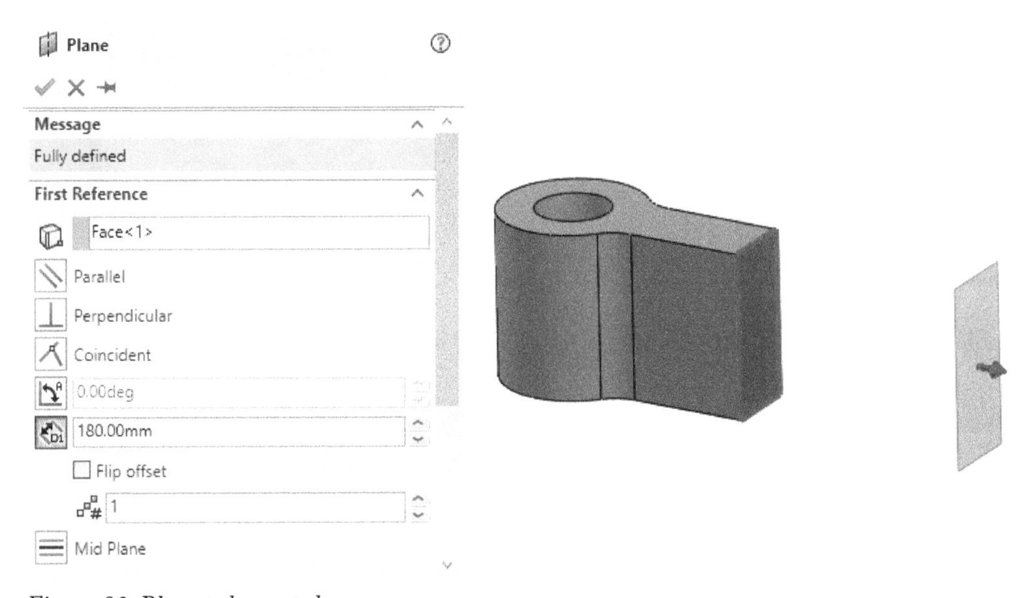

Figure-93. Plane to be created

- Click on **OK** button to create the plane.
- Select this plane and create the sketch as shown in Figure-94.

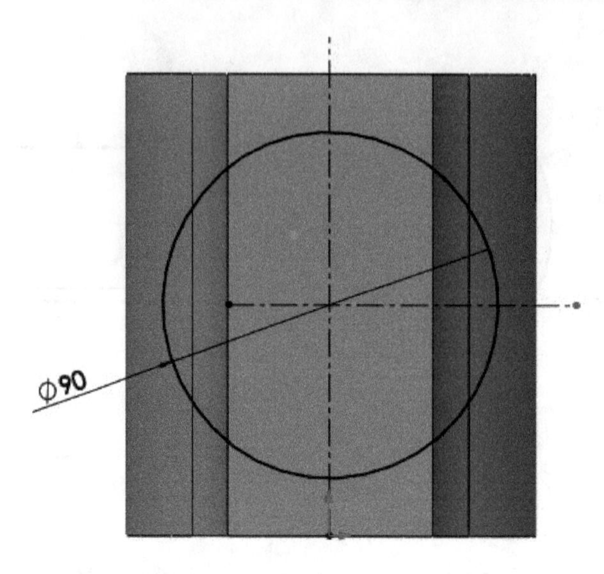

Figure-94. Sketch_to_be_created

- Exit the sketch by clicking **Exit Sketch** button.

Now, we are ready to create the Loft feature.

- Click on the **Lofted Boss/Base** tool from the **Ribbon**. The **Loft PropertyManager** will be displayed.
- Select the flat face of the model which you earlier selected to create plane; refer to Figure-95.

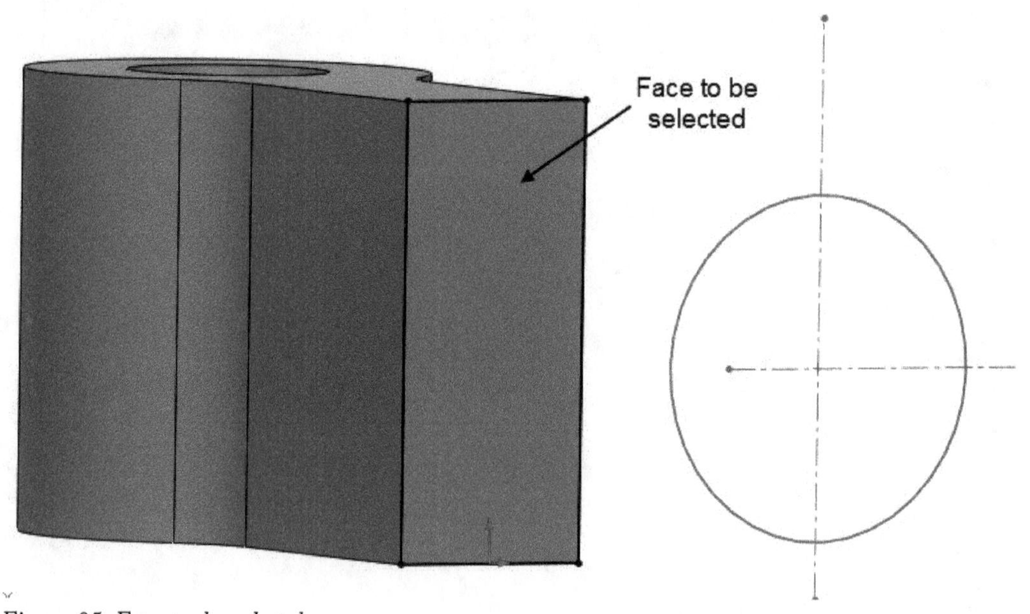

Figure-95. Face_to_be_selected

- Select the sketch created. Preview of the loft feature will be displayed; refer to Figure-96.

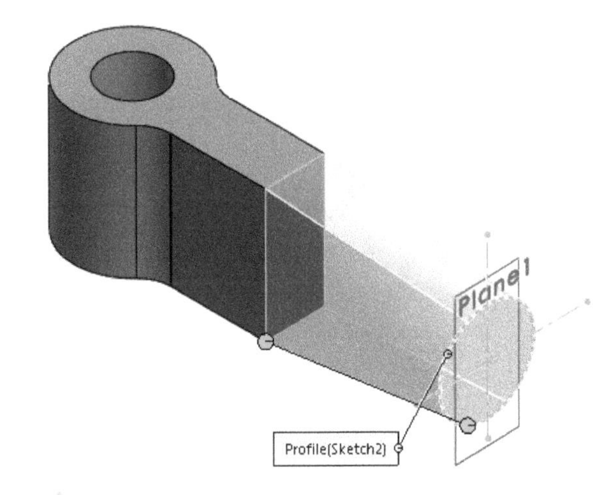

Figure-96. Lofted_feature_to_be_created

- Drag the green handles on the curves to manipulate the shape of lofted feature if you get it twisted.
- Click on the **OK** button from the **PropertyManager**.

Creating extrude feature

- Select the flat face of the lofted feature and click on the **Extruded Boss/Base** tool. The sketching environment will open.
- Create a circle of diameter **70** at the center and exit the sketch. Preview of extruded feature will display.
- Set the extrude depth as **220**. The preview of the model will display as shown in Figure-97.

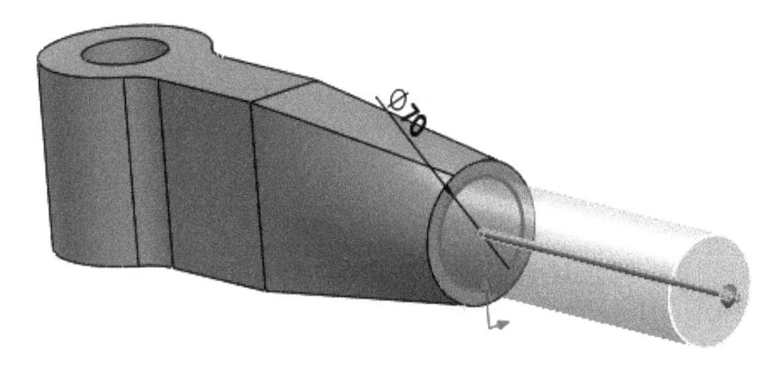

Figure-97. Preview_of_model_after_extrusion

- Click **OK** from the **PropertyManager**.

Now, we need to create the cut feature to give front shape to model.

Creating extrude cut feature

After looking at the model, we can find that the front cut can be easily created by using the **Extruded Cut** tool.

- Click on the **Extruded Cut** tool from the **Ribbon** and select the vertical plane passing through the model i.e. Front plane. The sketching environment will be displayed.
- Create the sketch as shown in Figure-98.

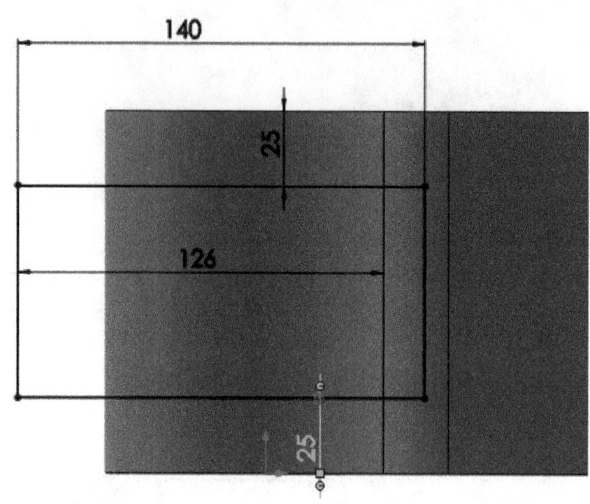

Figure-98. Sketch_for_cut_feature

- Exit the sketch. The preview of extruded cut will display.
- Select the **Mid Plane** option from the **End Condition** drop-down and specify the height as **200**.
- Click on the **OK** button. The cut feature will be created and the model will display as shown in Figure-99.

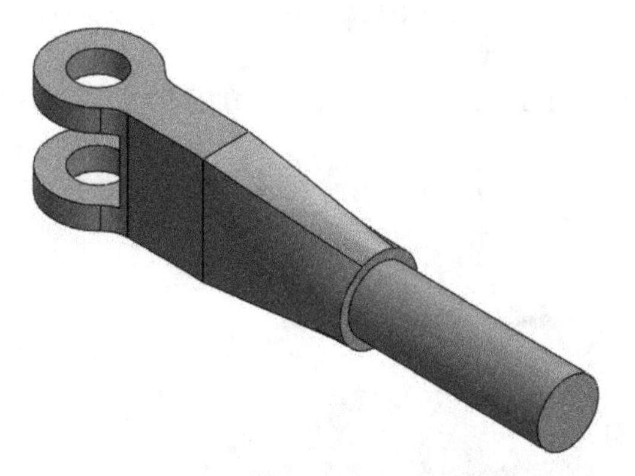

Figure-99. Model_after_cut_feature

Creating Fillets and Chamfers

Click on the **Fillet** tool, specify the radius as **5**, and select all the edges on which you want to create the fillet. Similarly, click on the **Chamfer** tool, specify the parameters as per the drawing and select the edges to apply the chamfer. The model after applying the fillet and chamfers is displayed as shown in Figure-100.

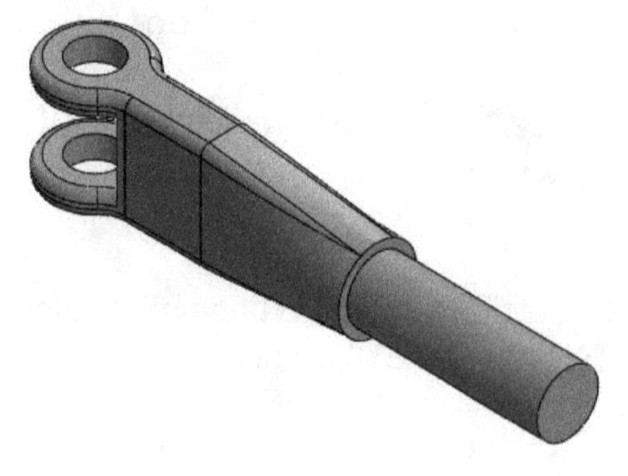

Figure-100. Final model for Practical 3

PRACTICE 1

Create the model as shown in Figure-101. The dimensions are given in Figure-102.

Figure-101. Practice 1

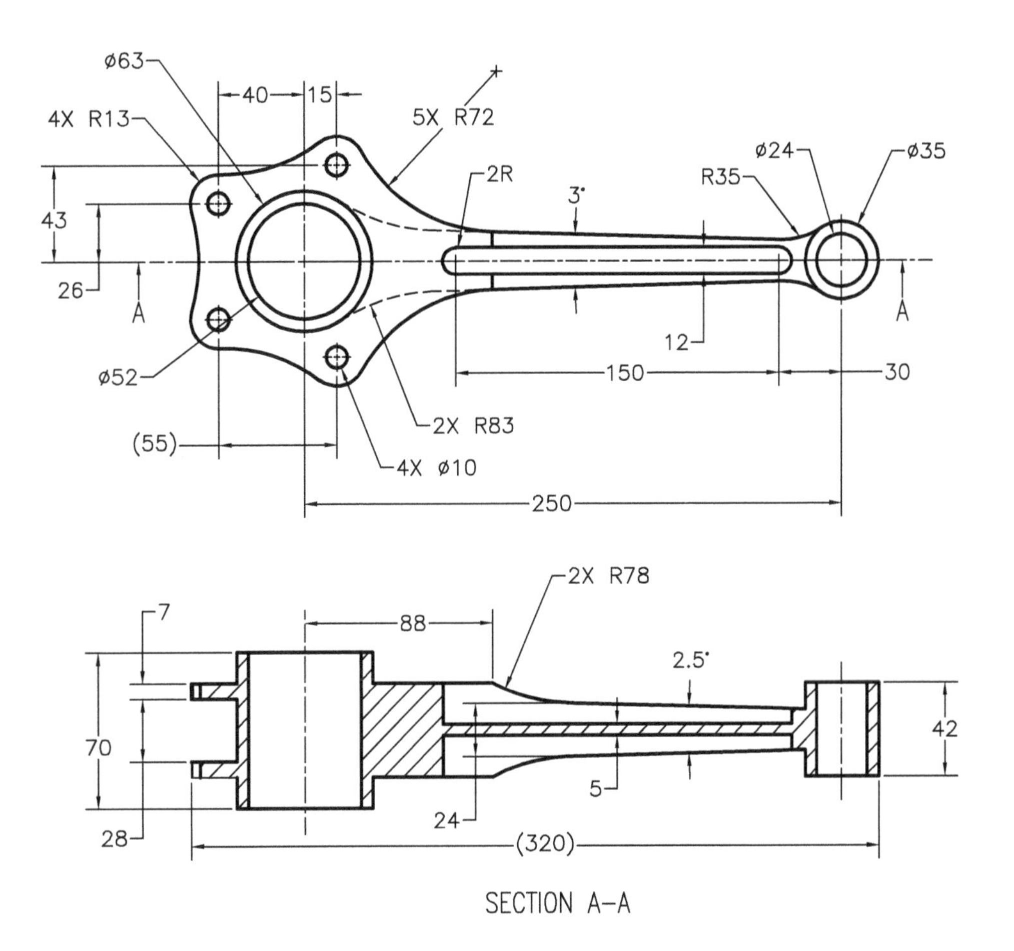

Figure-102. Dimensions of the Practice 1 model

PRACTICE 2

Create the model using the drawings shown in Figure-103.

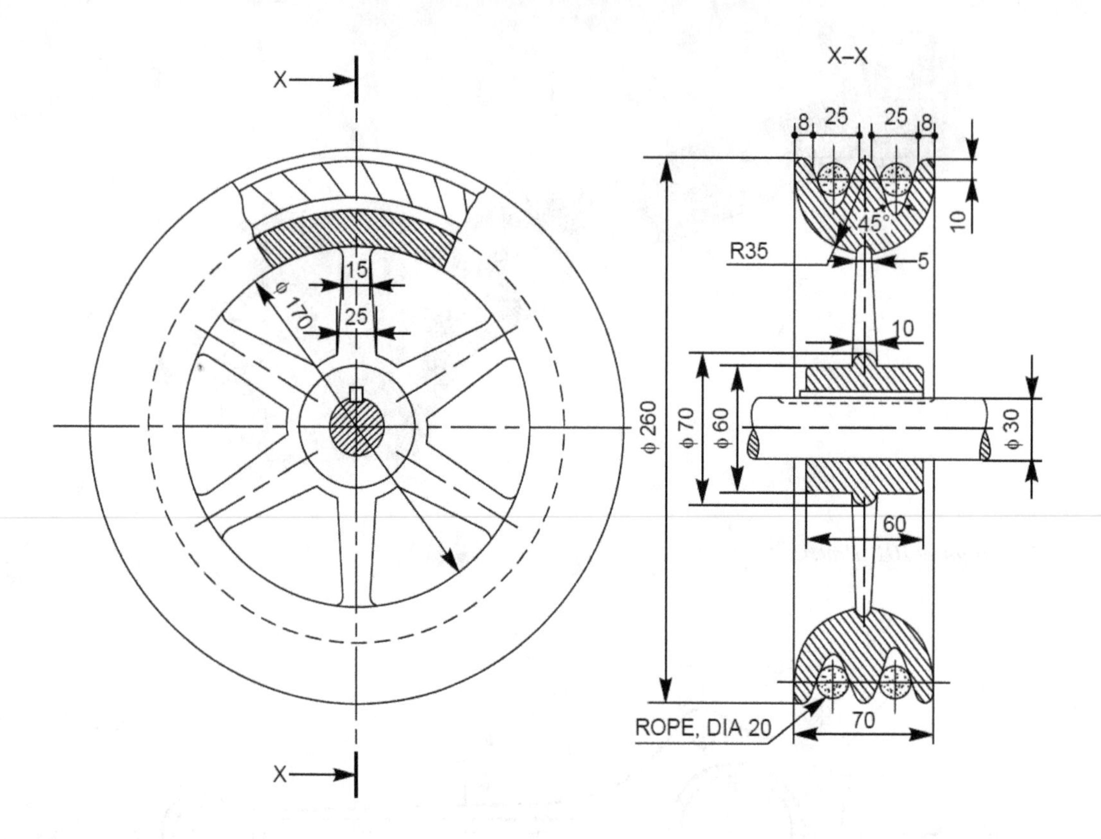

Figure-103. Rope_Pulley

PRACTICE 3

Create the model as shown in Figure-104. Dimensions are given in Figure-105. Assume the missing dimensions.

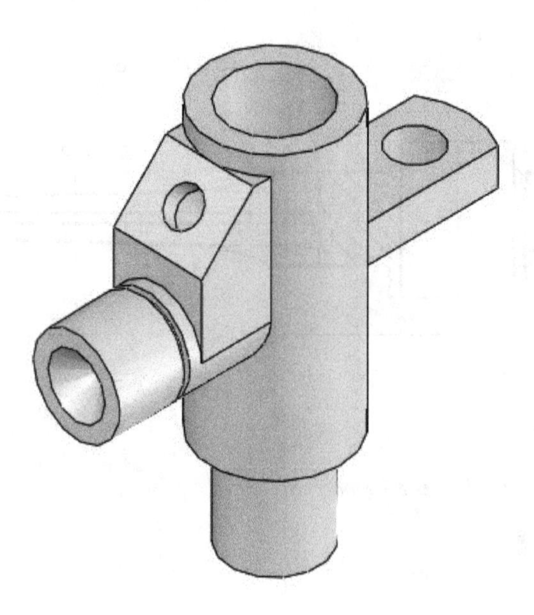

Figure-104. Practice_3_model

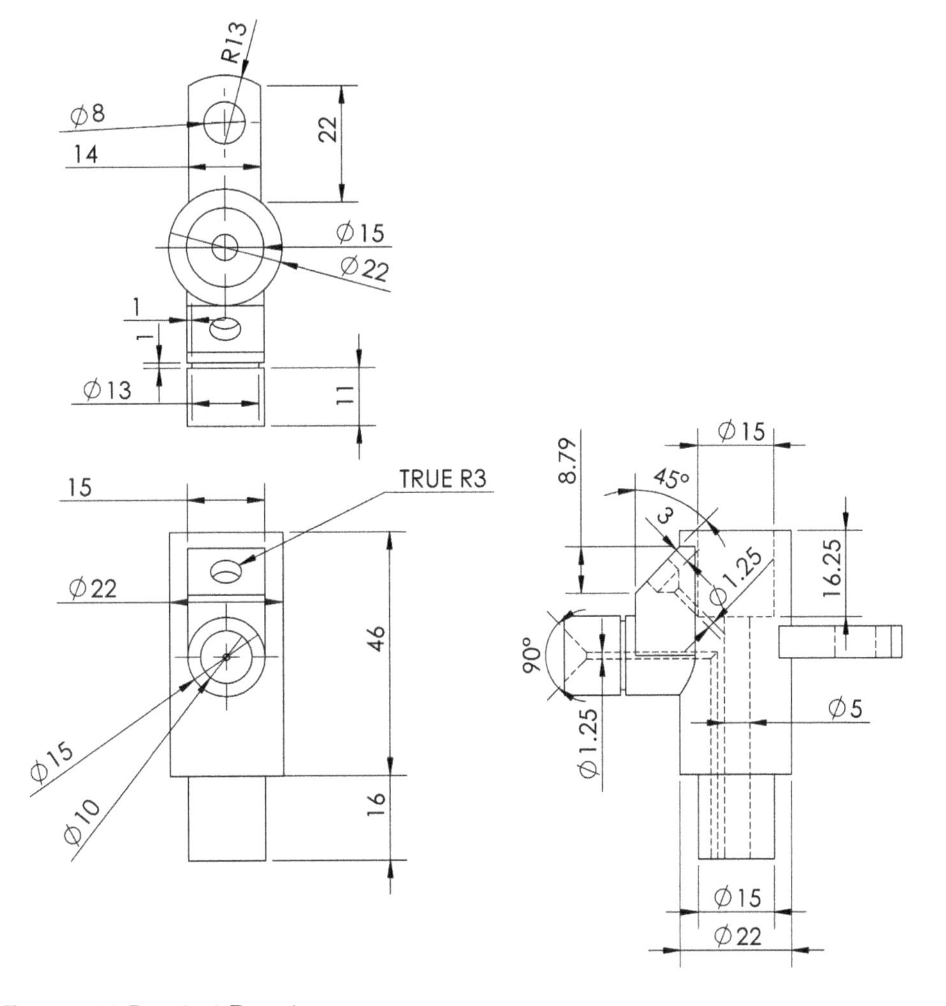

Figure-105. Practice 3 Drawing

PRACTICE 4

Create the model by using the dimensions given in Figure-106.

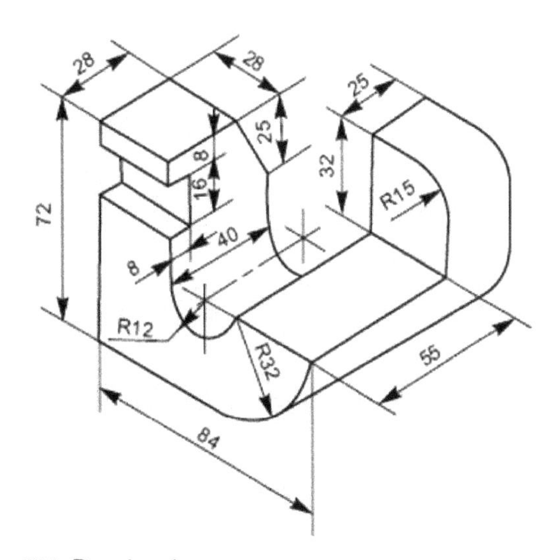

Figure-106. Practice_4

PRACTICE 5

Create the model by using the dimensions given in Figure-107.

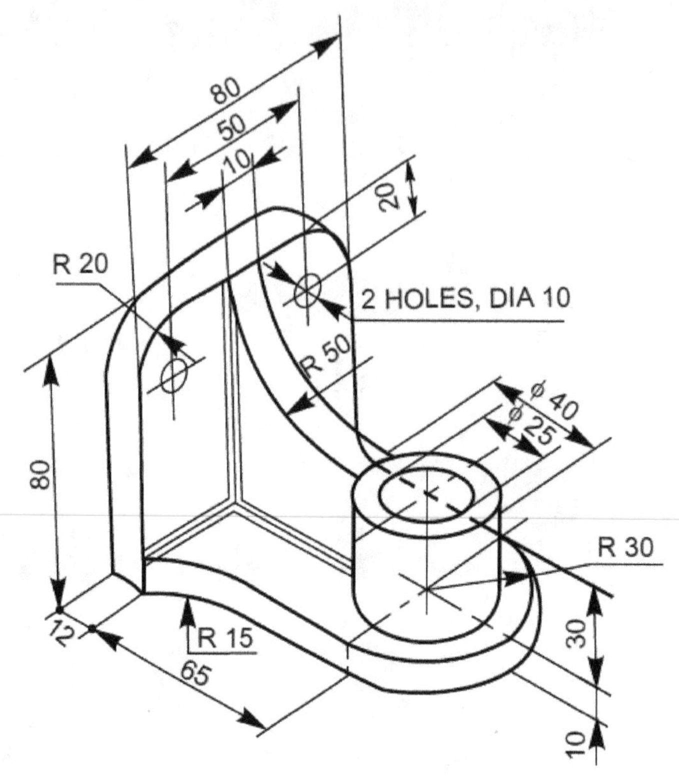

Figure-107. Practice_5

PRACTICE 6

Create the model by using the dimensions given in Figure-108.

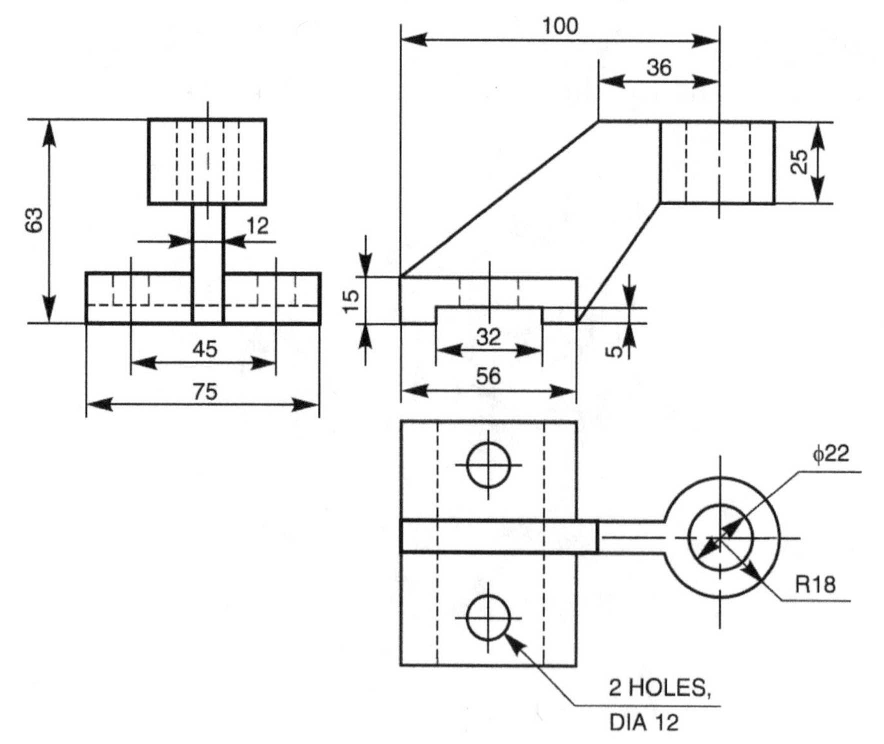

Figure-108. Practice 6

PRACTICE 7

Create a ring nut with value of **D** as **5**,**6**,**8**, and **10** using equation and design table. Dimensions are given in Figure-109.

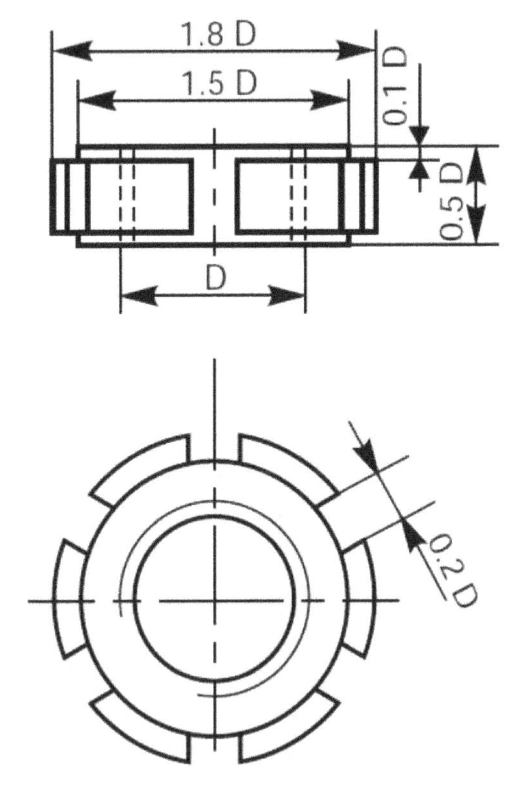

Figure-109. Ring Nut

SELF ASSESSMENT

Q1. The .... tool is used to apply radius at the edges.

Q2. The .... tool is used to apply different type of fillets in one single mode.

Q3. The .... tool is used to bevel the edges of the model.

Q4. The .... tool is used to create multiple instances of a solid features along the selected path.

Q5. The .... tool is used to create multiple instances of a solid features as per the points specified in the selected sketch.

Q6. The .... tool is used to create multiple instances of a features by filling the selected bounded region.

Q7. The .... tool is used to create multiple instances of a features as per the coordinates specified in the table.

Q8. The .... tool is used to create support in the structures to increase their strength.

Q9. The .... tool is used to apply taper to the faces of a solid model.

Q10. The .... tool is used to make a solid part hollow and remove one or more faces.

Answer to Self-Assessment:

1. Fillet, **2.** FilletXpert, **3.** Chamfer, **4.** Curve Driven Pattern, **5.** Sketch Drive Pattern, **6.** Fill Pattern, **7.** Table Driven Pattern, **8.** Rib, **9.** Draft, **10.** Shell

Chapter 6

Advanced Solid Modeling

Topics Covered

The major topics covered in this chapter are:

- ***Dome tool.***
- ***Freeform and Deform tools.***
- ***Indent and Flex tools***
- ***Split tool.***
- ***Move/Copy tool.***
- ***Delete/Keep Body and Convert to Mesh Body tools.***
- ***Importing Bodies***
- ***Fastening Features***
- ***Blocks***
- ***Macros***
- ***Practical and Practice***

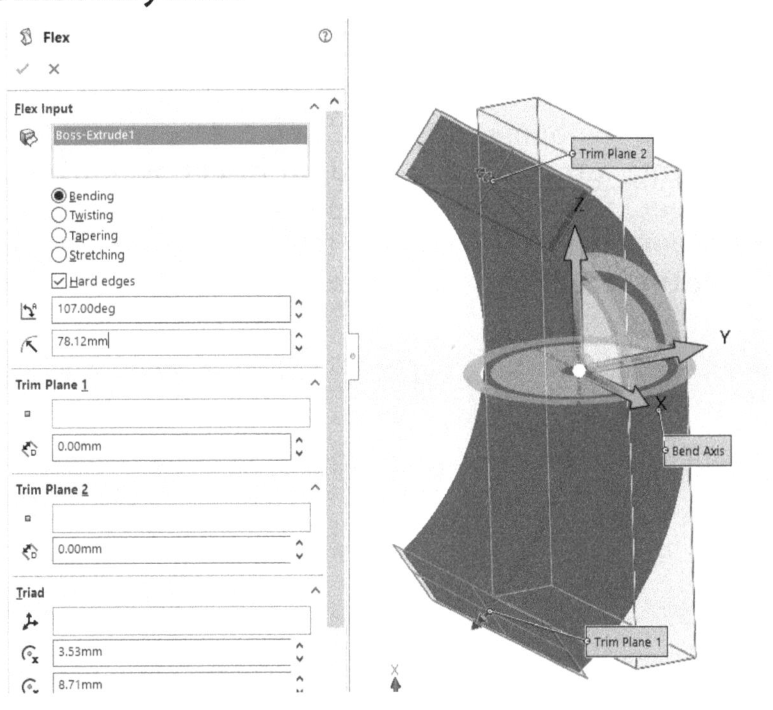

INTRODUCTION

In previous chapter, you have learned about solid editing using various tools like material removal tools, pattern, shell, and so on. In this chapter, we will work with some advanced modeling tools that are used in exceptional cases. Since these tools are not generally used by an average user so they are not found in **Ribbon**. Most of the tools discussed in this chapter will be available in the Menus. We will start with advanced modeling tools in the **Features** cascading menu and then we will move forward to other menus. The tools are discussed next.

CREATING DOME

Although, you can create domes in the modeling by using various surfacing tools but SolidWorks has a direct tool to reduce your efforts. The procedure to create dome is given next.

- Click on the **Dome** tool from the **Features** cascading menu of the **Insert** menu. The **Dome PropertyManager** will be displayed; refer to Figure-1.

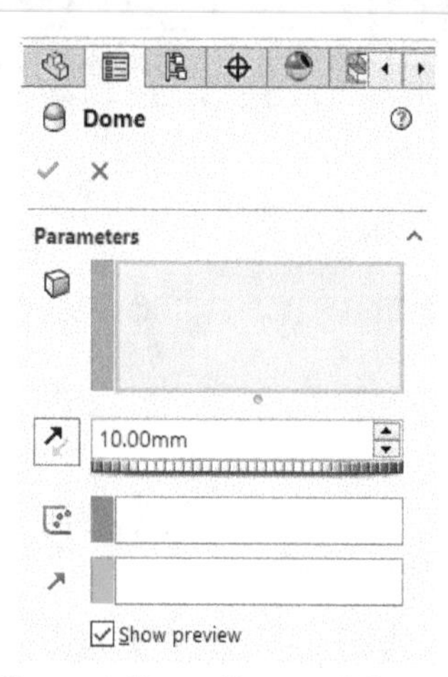

Figure-1. Dome PropertyManager

- Select the flat faces on which you create dome feature. Preview of the features will be displayed; refer to Figure-2.
- Set desired height for dome in the **Distance** edit box. Click on the **Reverse Direction** button if you want to remove material from the model in place of adding using the dome shape.
- If you want to use a custom direction for creating the dome then click in the **Direction** selection box and select desired reference.
- If you want to constrain the dome at some point then click in the **Constraint Point or Sketch** selection box and select desired point; refer to Figure-3.

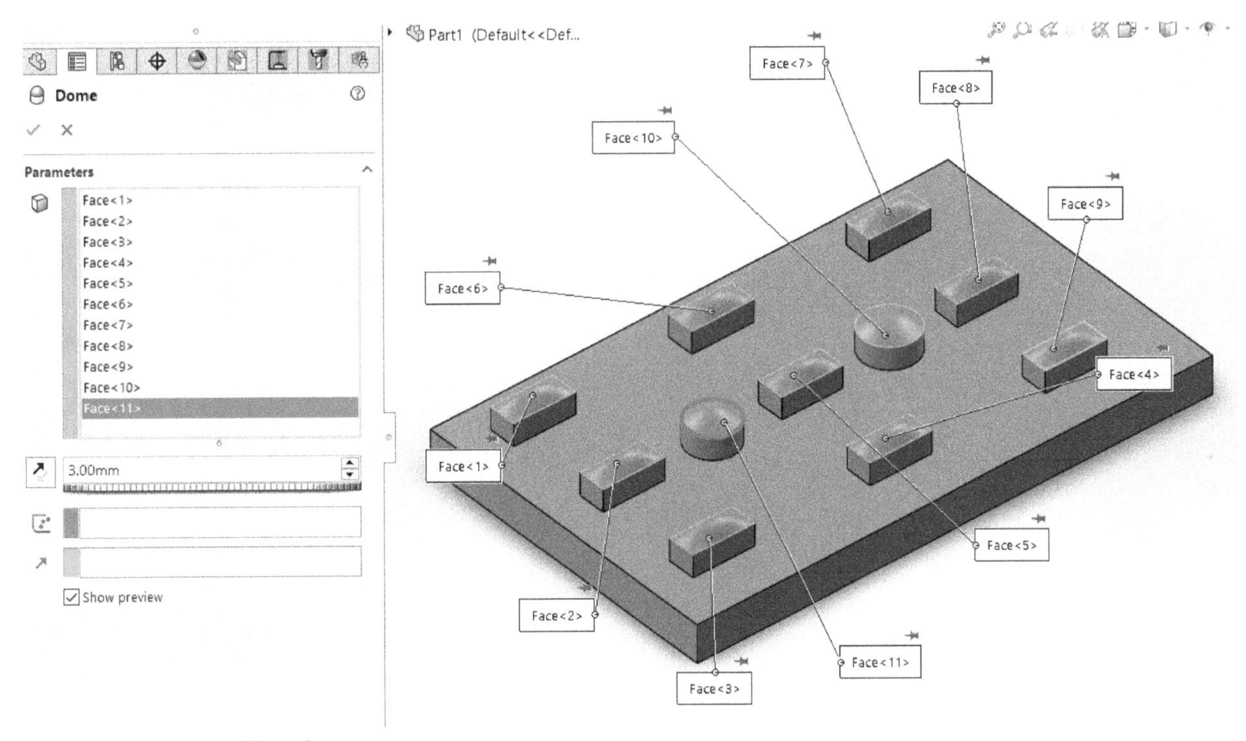

Figure-2. Preview of dome feature

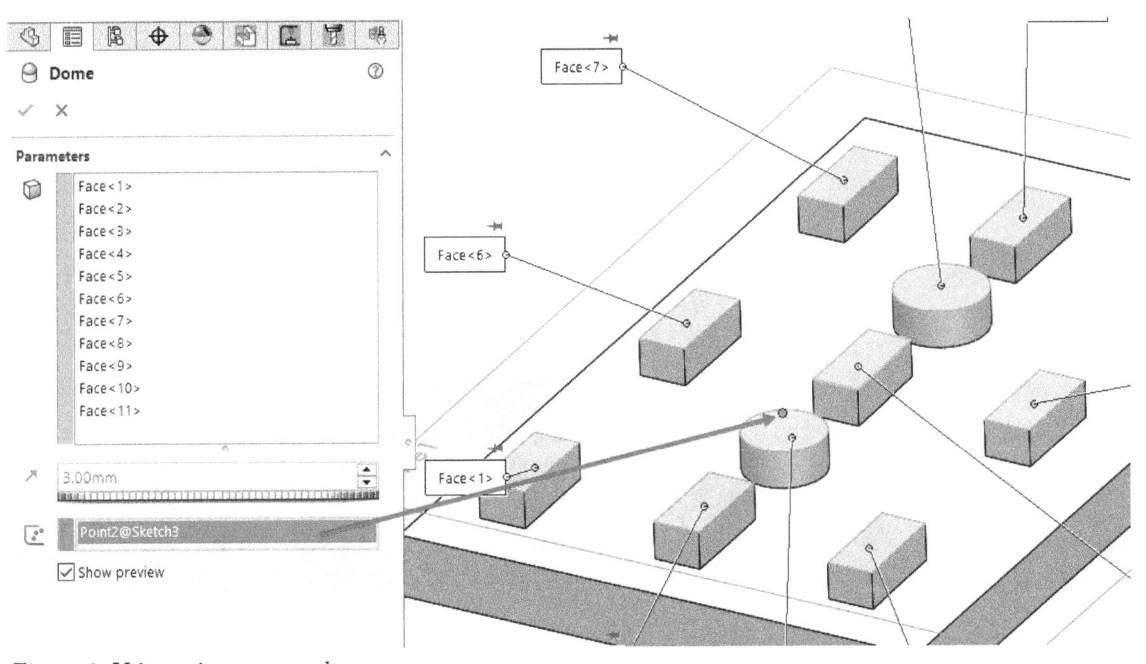

Figure-3. Using point to create dome

- Click on the **OK** button from the **PropertyManager** to create the dome feature.

FREE FORMING SOLID

The **Freeform** tool in SolidWorks is used to modify the shape and size of solid model by using the control points or control polygons. The procedure to use this tool will be discussed in Chapter 9 of Surfacing.

DEFORMING SOLID

The **Deform** tool is used to deform a solid model using different set of instructions. The procedure to use this tool is given next.

- Click on the **Deform** tool from the **Features** cascading menu of the **Insert** menu. The **Deform PropertyManager** will be displayed; refer to Figure-4.

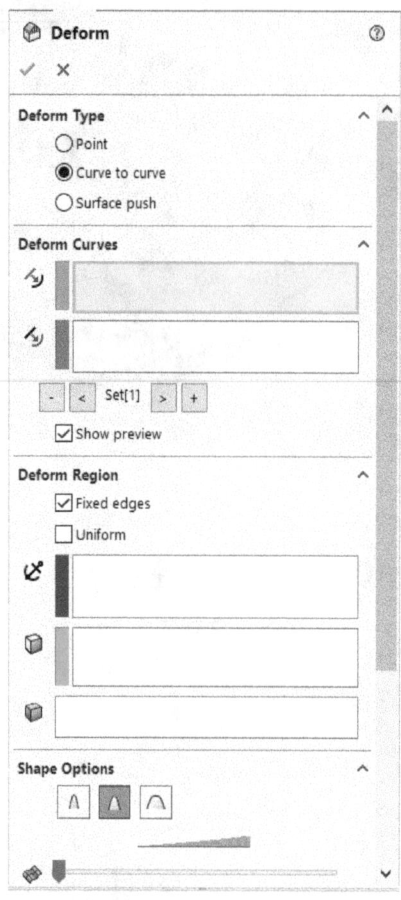

Figure-4. Deform PropertyManager

- Select the **Point** radio button from the **Deform Type** rollout if you want to deform the solid body using a point on it. Select the **Curve to curve** radio button if you want to deform selected edge/curve of the model to match with new one. Select the **Surface Push** radio button if you want to use triad to deforming selected face of the model.

Deforming model using Point Option

- Select the **Point** radio button from the **PropertyManager**. The options will be displayed as shown in Figure-5.
- Select a point on the model that you want to use for deformation. Preview of deformation will be displayed; refer to Figure-6.

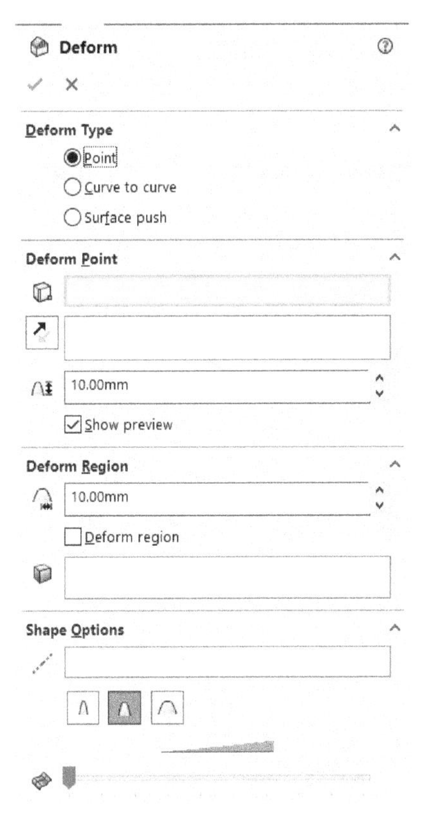

Figure-5. Deform PropertyManager with Point radio button selected

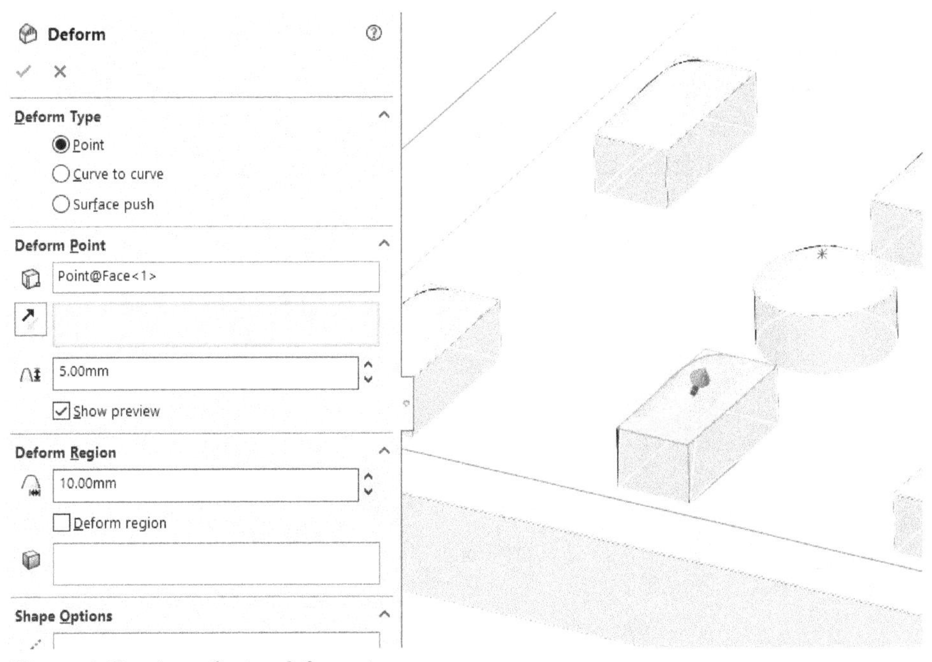

Figure-6. Preview of point deformation

- Set desired value of deformation in the **Deform Distance** edit box of **Deform Point** rollout.
- Set the direction of deformation using the **Deform Direction** selection box as discussed earlier.
- By default, the whole body is deformed on applying this tool but if you want to confine deformation in a specific area then select the **Deform region** check box from the **Deform Region** rollout. Two selection box will be displayed below the check box.

- Click in the green selection box and select the faces that you want to be deformed.
- Using the options in **Shape Options** rollout, you can define the shape of deformation.
- After setting desired values, click on the **OK** button.

Deforming Model using Curve to Curve Option

- Select the **Curve to curve** radio button from the **Deform Type** rollout. The **PropertyManager** will be displayed as shown in Figure-7.

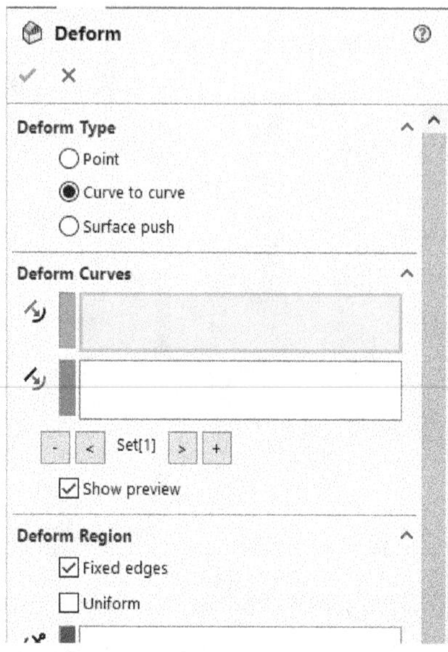

Figure-7. Deform PropertyManager with Curve to Curve options

- Select the edge(s) that you want to be deformed.
- Click in the **Target Curves** selection box (pink colored selection box) from the **Deform Curves** rollout. You will be asked to select the target curve. Select a curve. The model will be deformed accordingly; refer to Figure-8.

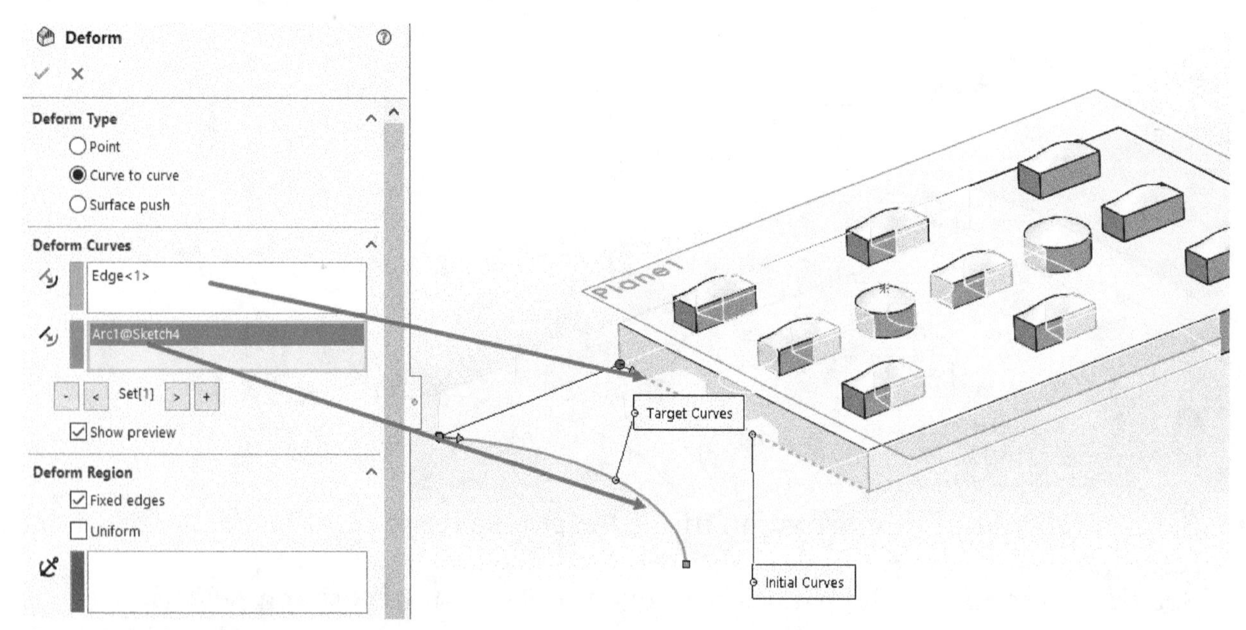

Figure-8. Deformed by curves

- To deform another set of curves, click on the **+** button in the **Deform Curves** rollout.
- After performing desired deformation, click on the **OK** button from the **PropertyManager**.

Deforming Model using Surface push Option

- Select the **Surface push** radio button from the **Deform Type** rollout. Rest of the steps are similar to the previous two options.

CREATING INDENT ON SOLID MODEL

The **Indent** tool is used to create indent on the solid based on the selected solid face or surface. The procedure to use this tool is given next.

- Click on the **Indent** tool from the **Features** cascading menu of the **Insert** menu. The **Indent PropertyManager** will be displayed; refer to Figure-9.

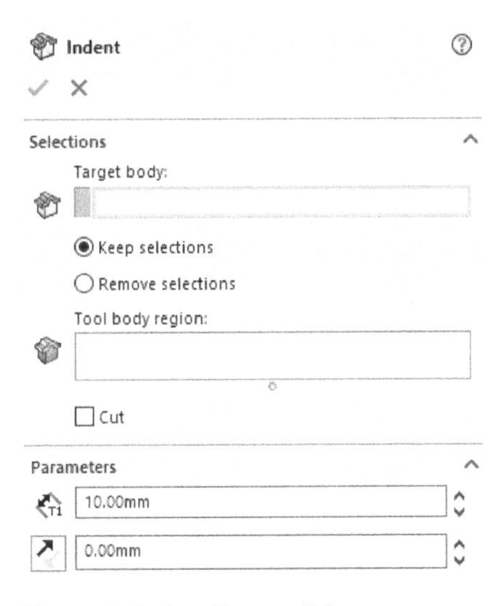

Figure-9. Indent PropertyManager

- Select the body on which you want to make indent of the tool.
- Click in the **Tool body region** selection box and select the body to be used as tool for indent.
- Select the **Keep selections** radio button if you want the target body and tool body after performing indent operation.
- Select the **Remove selections** radio button if you want the target body and tool body after performing indent operation.
- By default, the indent is added to the selected body but if you want to remove the material from the target body then select the **Cut** check box. Preview of the indent will be displayed; refer to Figure-10.
- Click on the **OK** button to perform the operation. Note that you may need to hide the tool body manually to check the indent. To do so, select the tool body from the **FeatureManager Design Tree** and right-click on it. A shortcut menu will be displayed. Click on the **Hide** button from the shortcut menu.

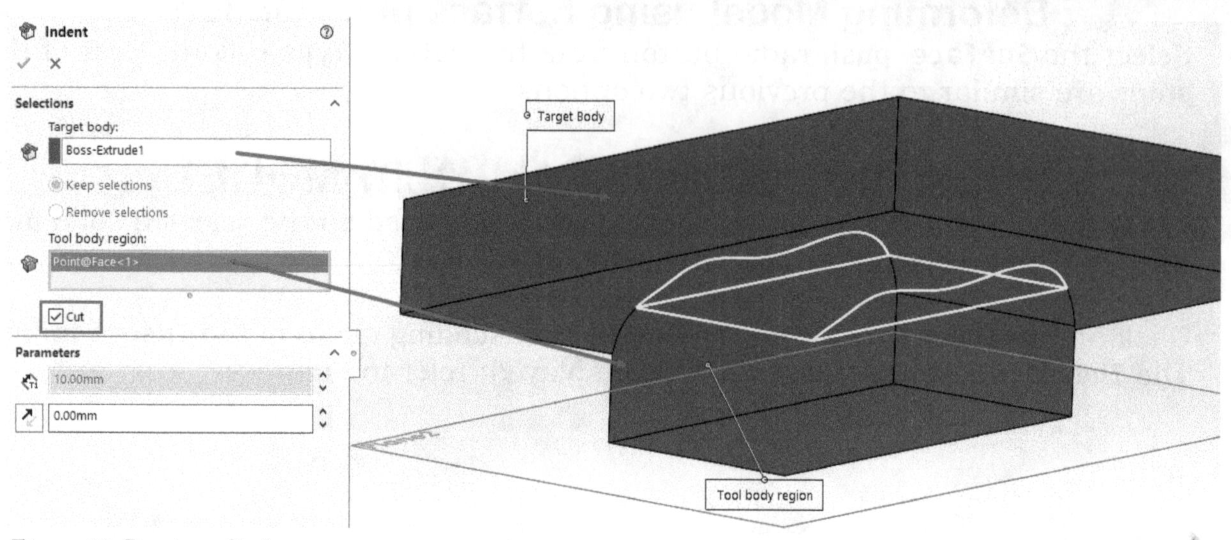

Figure-10. Preview of indent

APPLYING FLEX TRANSFORMATION

The **Flex** tool is used to dynamically transform a solid body using the control handles. The procedure to use this tool is given next.

- Click on the **Flex** tool from the **Features** cascading menu of the **Insert** menu. The **Flex PropertyManager** will be displayed; refer to Figure-11.

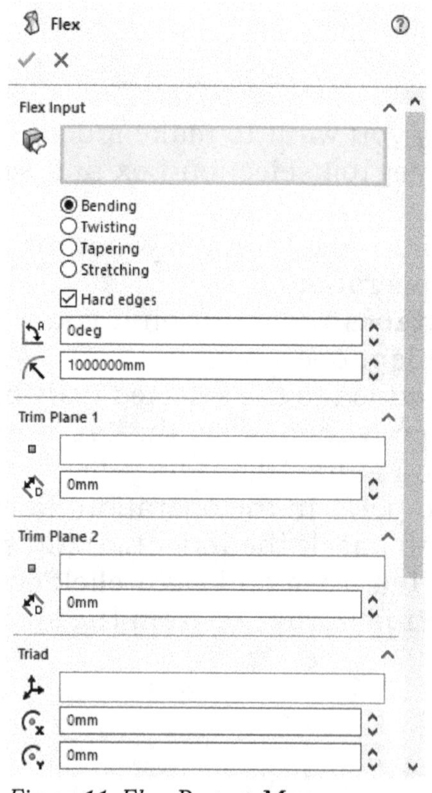

Figure-11. Flex_PropertyManager

- Select the body on which you want to perform flex operations.
- Click on desired radio button from the **Flex Input** rollout to perform bending, twisting, tapering, or stretching. The procedure to perform various operations is discussed next.

Bending

- Select the **Bending** radio button. The model will be displayed as shown in Figure-12. Note that the model can be bent only about the bend axis.

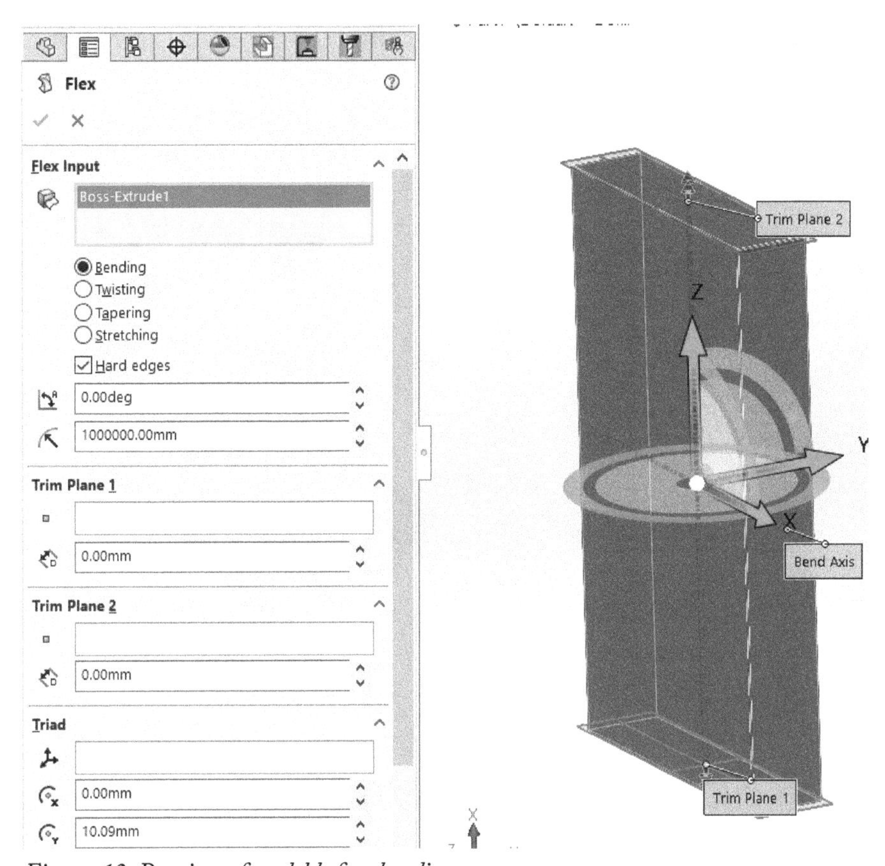

Figure-12. Preview of model before bending

- By default, the bend axis is at the center of the model. If you want to change the location of the bend axis then you need to move and orient the triad as required. Set the desired values in the edit boxes of **Triad** rollout or you can use the handles of triad.
- Specify the desired value of angle in the **Angle** edit box of the **Flex Input** rollout. Preview of bending will be displayed; refer to Figure-13.
- Click on the **OK** button to apply bending.

Similarly, you can use the **Twisting**, **Tapering**, and **Stretching** radio buttons to perform twisting, tapering and stretching of the solid bodies.

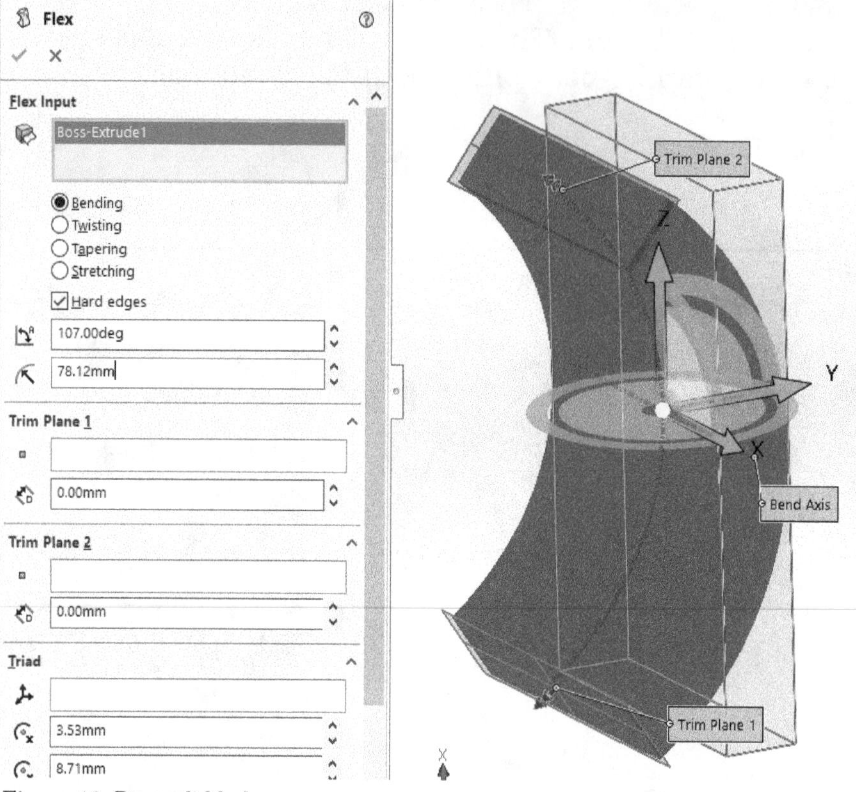

Figure-13. Bent solid body

COMBINING BODIES

The **Combine** tool is used to add, subtract, or create common body by intersection of two bodies. The procedure to use this tool is given next.

- Click on the **Combine** tool from the **Features** cascading menu of the **Insert** menu. The **Combine PropertyManager** will be displayed; refer to Figure-14.

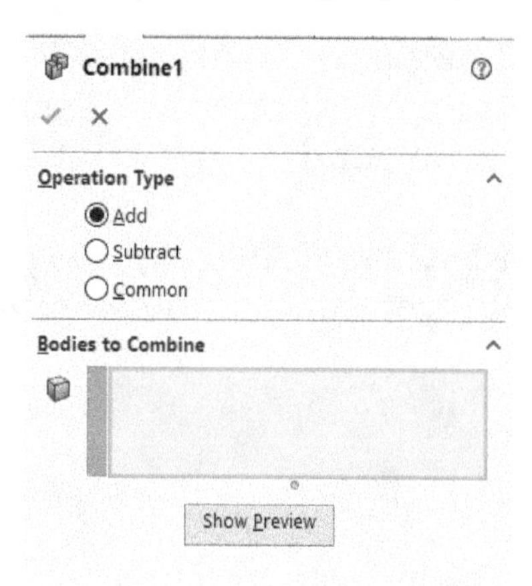

Figure-14. Combine PropertyManager

- Select desired option from the **Operation Type** rollout.
- Select the **Add** radio button and then select the two bodies to combine them. Click on the **Show Preview** button to check the preview of combination.
- Select the **Subtract** radio button if you want to subtract one body from another intersecting body. The **PropertyManager** will be displayed as shown in Figure-15.

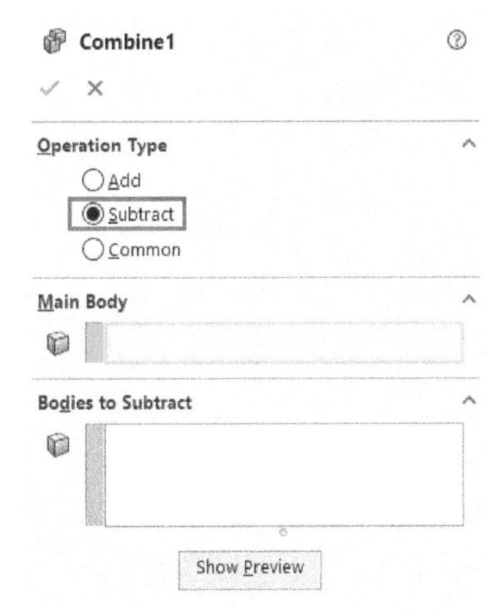

Figure-15. Combine PropertyManager with Subtract radio button

- Click in the **Main Body** selection box and select the body from which you want to remove the other body. Click in the **Bodies to Subtract** selection box and select the bodies to be subtracted from the main body. Click on the **Show Preview** button from the **PropertyManager** to check the preview.
- Select the **Common** radio button if you want to create intersection body of the selected bodies. After selecting this radio button, select the bodies to be combined. Click on the **Show Preview** button to check preview.
- After setting desired parameters, click on the **OK** button.

SPLITTING A SOLID BODY

The splitting of solid body is generally used in preparing mold. You will know more about molding later in this book. The procedure to perform splitting is discussed next.

- Click on the **Split** tool from the **Features** cascading menu in the **Insert** menu. The **Split PropertyManager** will be displayed; refer to Figure-16.

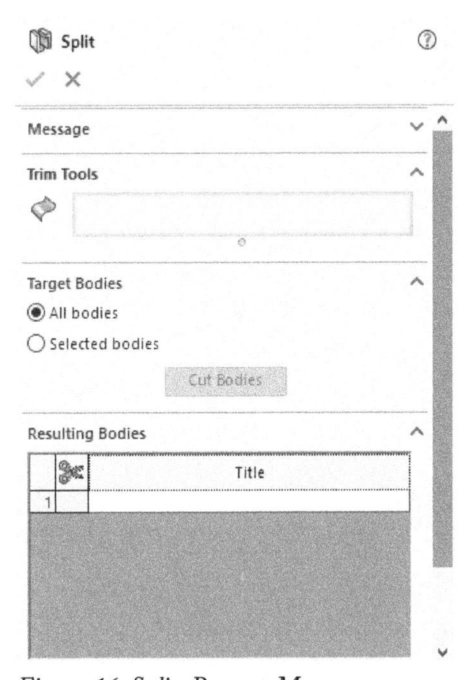

Figure-16. Split_PropertyManager

- Select the plane or surface by which you want to split the solid.
- Select desired radio button from the **Target Bodies** rollout and click on the **Cut Bodies** button from the **PropertyManager**. Based on your selection, the split-up bodies will be displayed in the preview and **PropertyManager**; refer to Figure-17.

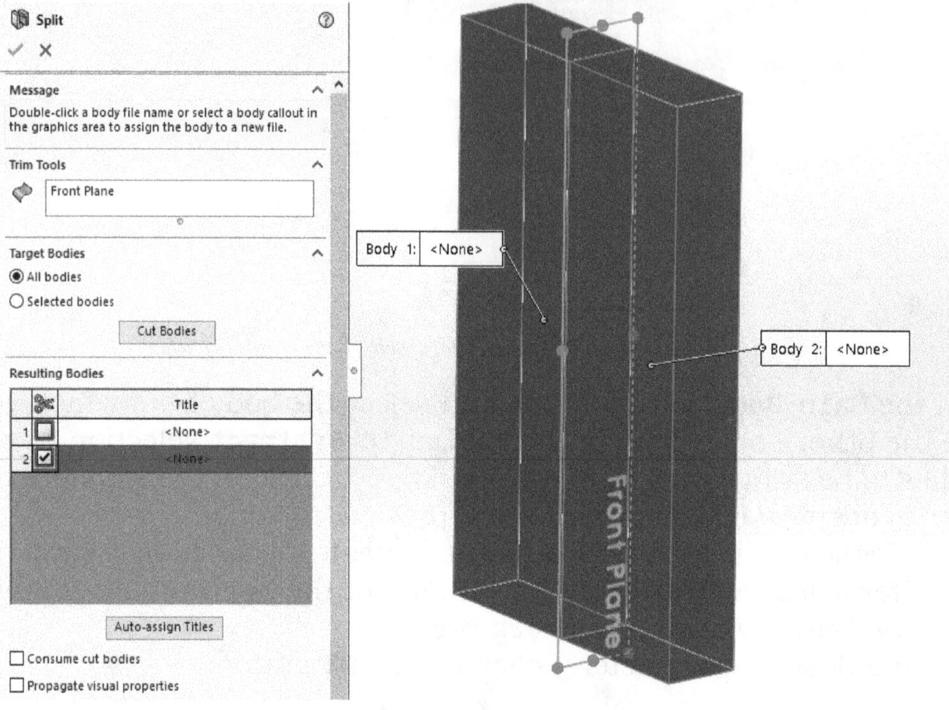

Figure-17. Preview of split-up bodies

- Click on the **OK** button from the **PropertyManager** to create split-up bodies. Note that the bodies can be found in the **Solid Bodies** folder of the **FeatureManager Design Tree**.

MOVING AND COPYING OBJECTS

The **Move/Copy** tool is used to create copies of the selected body or move the bodies using constraint. Using this tool, you can also move the bodies dynamically using a triad. The procedure to use this tool is given next.

- Click on the **Move/Copy** tool from the **Features** cascading menu of the **Insert** menu. The **Move/Copy Body PropertyManager** will be displayed; refer to Figure-18.
- Select the body/bodies that you want to move or copy.
- Click on desired constraint button from the **Mate Settings** rollout.
- Now, click in the selection box of **Mate Settings** rollout and select the faces on which you want to apply the constraints.

Or

- Click on the **Translate/Rotate** button at the bottom in the **PropertyManager**. The **PropertyManager** will be displayed as shown in Figure-19.

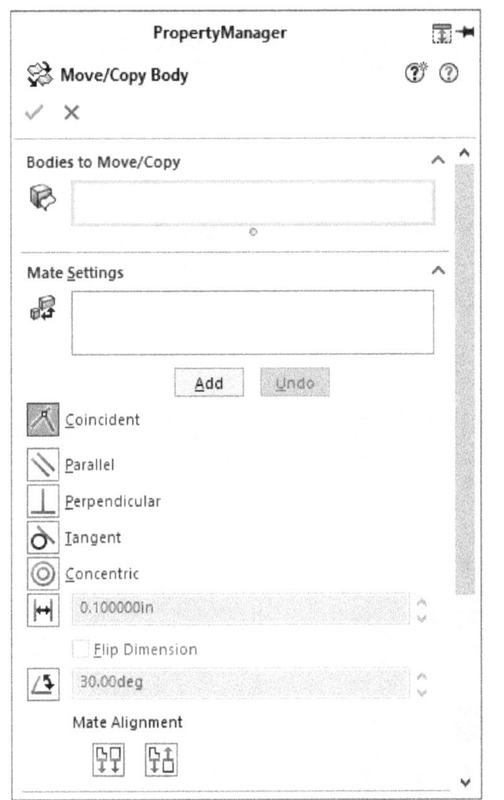

Figure-18. Move_Copy_Body_PropertyManager

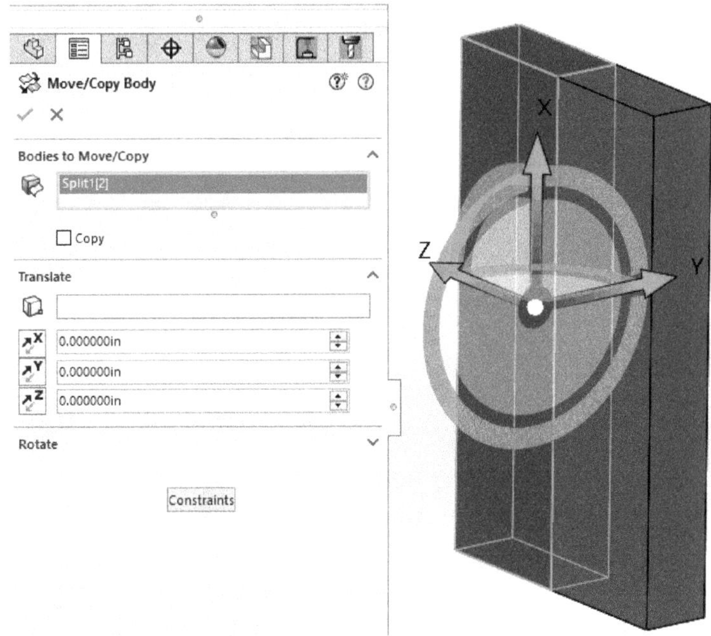

Figure-19. Using Translate option for moving

- Select the **Copy** check box if you want to create a copy of the body while moving it.
- Enter desired values in the edit boxes of the **Translate** rollout or drag the center point of the body at desired location; refer to Figure-20.

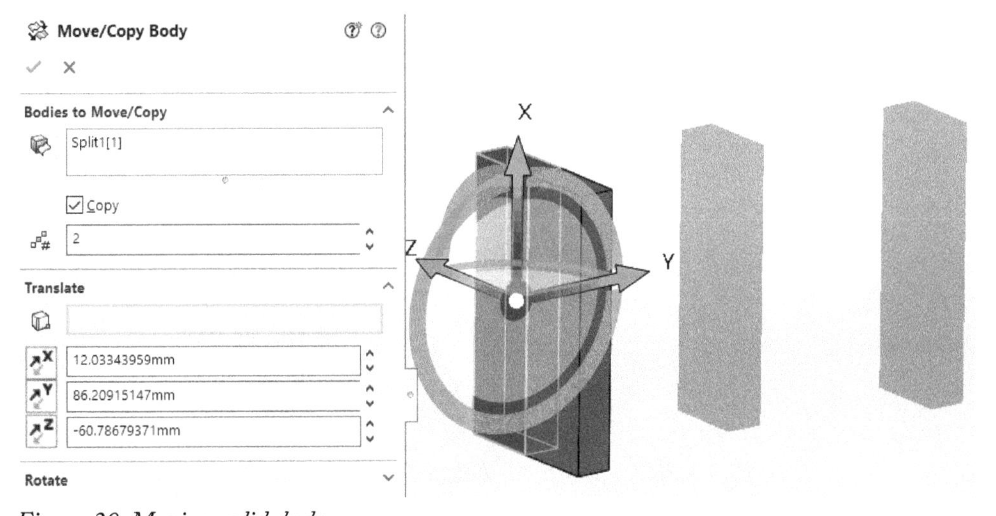

Figure-20. Moving_solid_body

- Click on the **OK** button from the **PropertyManager** to apply the changes.

DELETING BODIES

The **Delete/Keep Body** tool is used to delete selected bodies. The procedure to do so is given next.

- Click on the **Delete/Keep Body** tool from the **Features** cascading menu of the **Insert** menu. The **Delete/Keep Body PropertyManager** will be displayed; refer to Figure-21.

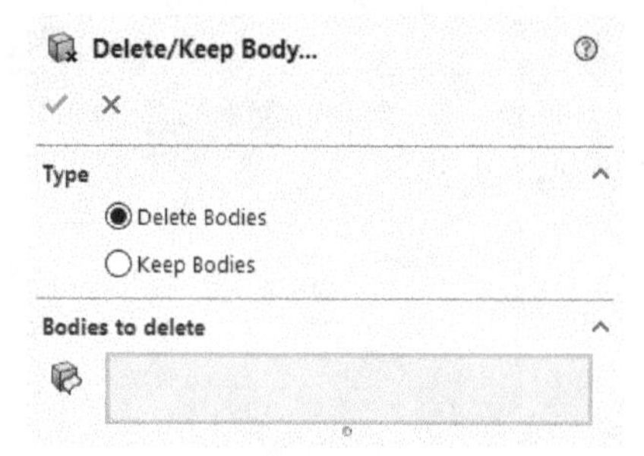

Figure-21. Delete/Keep Body PropertyManager

- Select the **Delete Bodies** radio button if you want the selected bodies to be deleted and rest of the bodies will be kept. Select the **Keep Bodies** radio button if you want the selected body to be kept and rest of the bodies to be deleted.
- Select desired bodies to be deleted or kept.
- Click on the **OK** button to apply the changes.

CONVERTING TO MESH BODIES

A mesh is construction of polygons. The procedure to convert model into mesh is given next.

- Click on the **Convert to Mesh Body** tool from the **Features** cascading menu of the **Insert** menu. The **Convert to Mesh Body PropertyManager** will be displayed; refer to Figure-22.

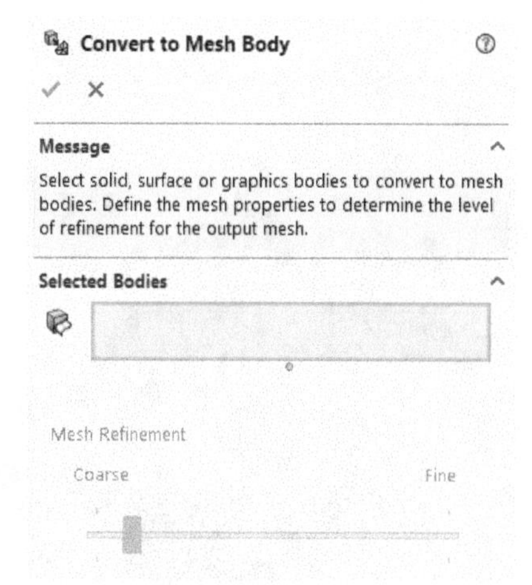

Figure-22. Convert to Mesh Body PropertyManager

- Select the body that you want to convert into mesh. The body will be displayed as mesh; refer to Figure-23.

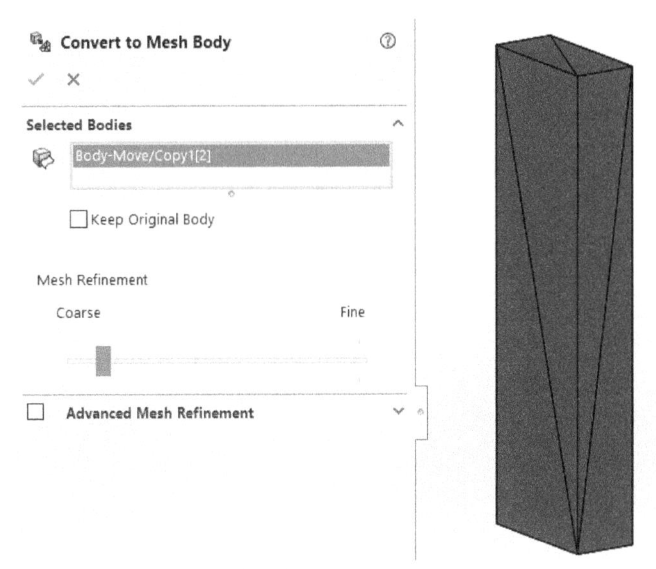

Figure-23. Preview of mesh

- Move the slider towards **Coarse** or **Fine** to refine the mesh.
- If you want to define the element size and other parameters manually then select the **Advanced Mesh Refinement** check box and then select the **Define Maximum Element Size** check box. The options will be displayed as shown in Figure-24.

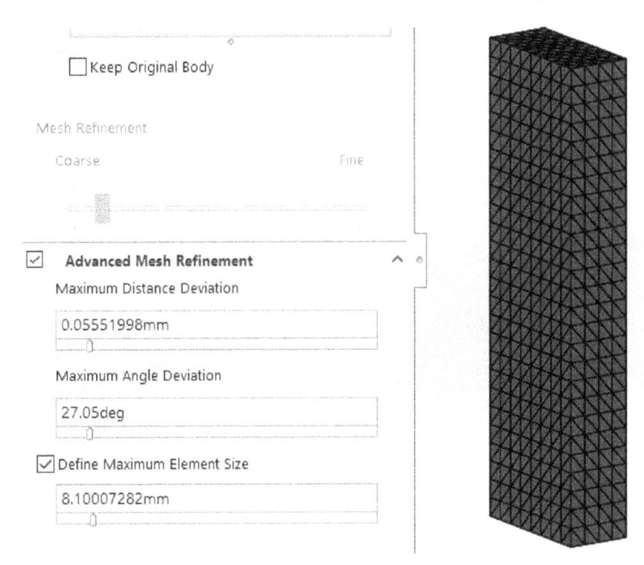

Figure-24. Options for advanced refinement

- Set desired element size and click on the **OK** button to create mesh.

APPLYING 3D TEXTURE

The **3D Texture** tool is used to convert solid or surface body into 3D textured graphic body. Note that you need to apply 3D texture from Miscellaneous category of appearances on the body before using this tool.

- Click on the **3D Texture** tool from the **Features** cascading menu of the **Insert** menu. The **3D Texture PropertyManager** will be displayed; refer to Figure-25.

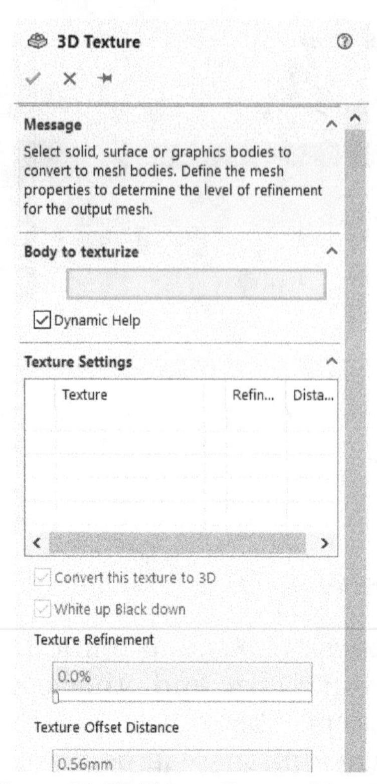

Figure-25. 3D Texture PropertyManager

- Select the body on which 3D texture is to be applied. Preview of texture will be displayed; refer to Figure-26.

Figure-26. Preview of 3D textured body

- Set desired parameters in the **Texture Settings** rollout and click on the **OK** button from the **PropertyManager** to apply texture.

IMPORTING MODEL IN PART ENVIRONMENT

The procedure to import model in Part environment is given next.

- Click on the **Imported** tool from the **Features** cascading menu in the **Insert** menu. The **Open** dialog box will be displayed.
- Select desired model file and click on the **Open** button. The part model will be imported at the default origin; refer to Figure-27.

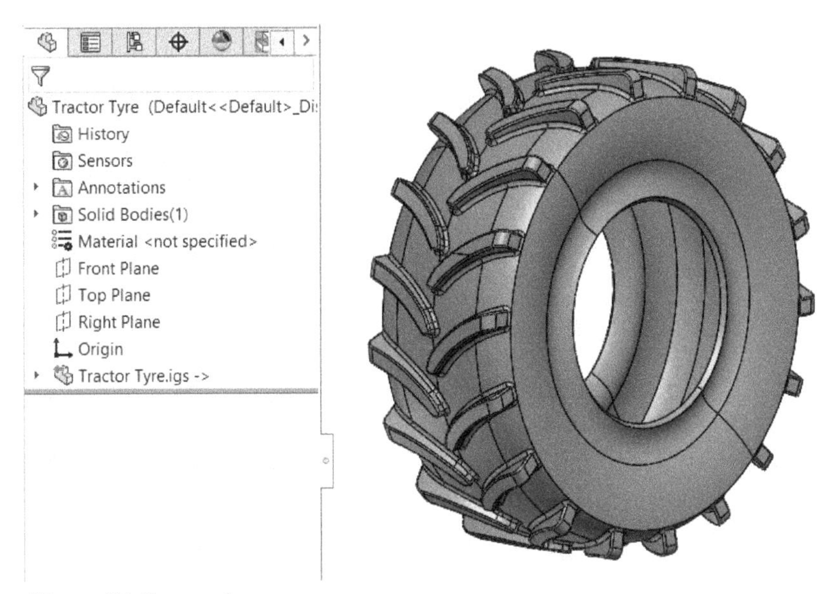

Figure-27. Imported part

FASTENING FEATURES

The Fastening features of SolidWorks are useful when creating plastic molded parts for assembly. These features include mounting boss, snap hook, snap hook groove, and so on. The procedures to create these features are given next.

Creating Mounting Boss

The Mounting boss is used to assemble two components while providing location stability. The procedure to create mounting boss is given next.

- Click on the **Mounting Boss** tool from the **Fastening Feature** cascading menu of the **Insert** menu. The **Mounting Boss PropertyManager** will be displayed; refer to Figure-28.
- Select the face which you want to create the mounting boss. Preview of the mounting boss will be displayed.
- Click in the **Select Direction** selection box and select the reference for defining the direction of mounting boss.
- If you want to place the mounting boss on a circular edge then click in the **Circular Edge** selection box in the **Position** rollout and select the round edge on which you want the mounting boss to be created.

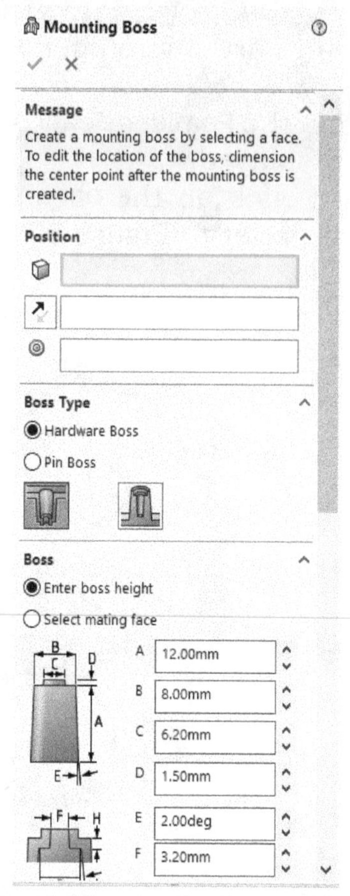

Figure-28. Mounting Boss PropertyManager

- Select desired type of boss from the **Boss Type** rollout and specify desired dimensions of the boss.
- After setting desired parameters, click on the **OK** button to create mounting boss.

Creating Snap Hook

The **Snap Hook** tool is used to create snap hook for plastic parts. The procedure to do so is given next.

- Click on the **Snap Hook** tool from the **Fastening Features** cascading menu of the **Insert** menu. The **Snap Hook PropertyManager** will be displayed; refer to Figure-29.
- Select the edge on which you want to create snap hook. Preview of the snap hook will be displayed; refer to Figure-30.
- Click in the pink selection box and select desired reference for vertical direction. Select the **Reverse direction** check box to flip the direction.
- Click in the violet colored selection box and select the reference to define horizontal direction of the snap hook. The snap hook will be displayed as in Figure-31.
- Select the **Select mating face** radio button and select desired face if you want the height of snap hook to be decided based on the selected face. Otherwise, select the **Enter body height** radio button and specify the snap hook height in the respective edit box.
- Set the other parameters as required and click on the **OK** button from the **PropertyManager** to create the snap hook.

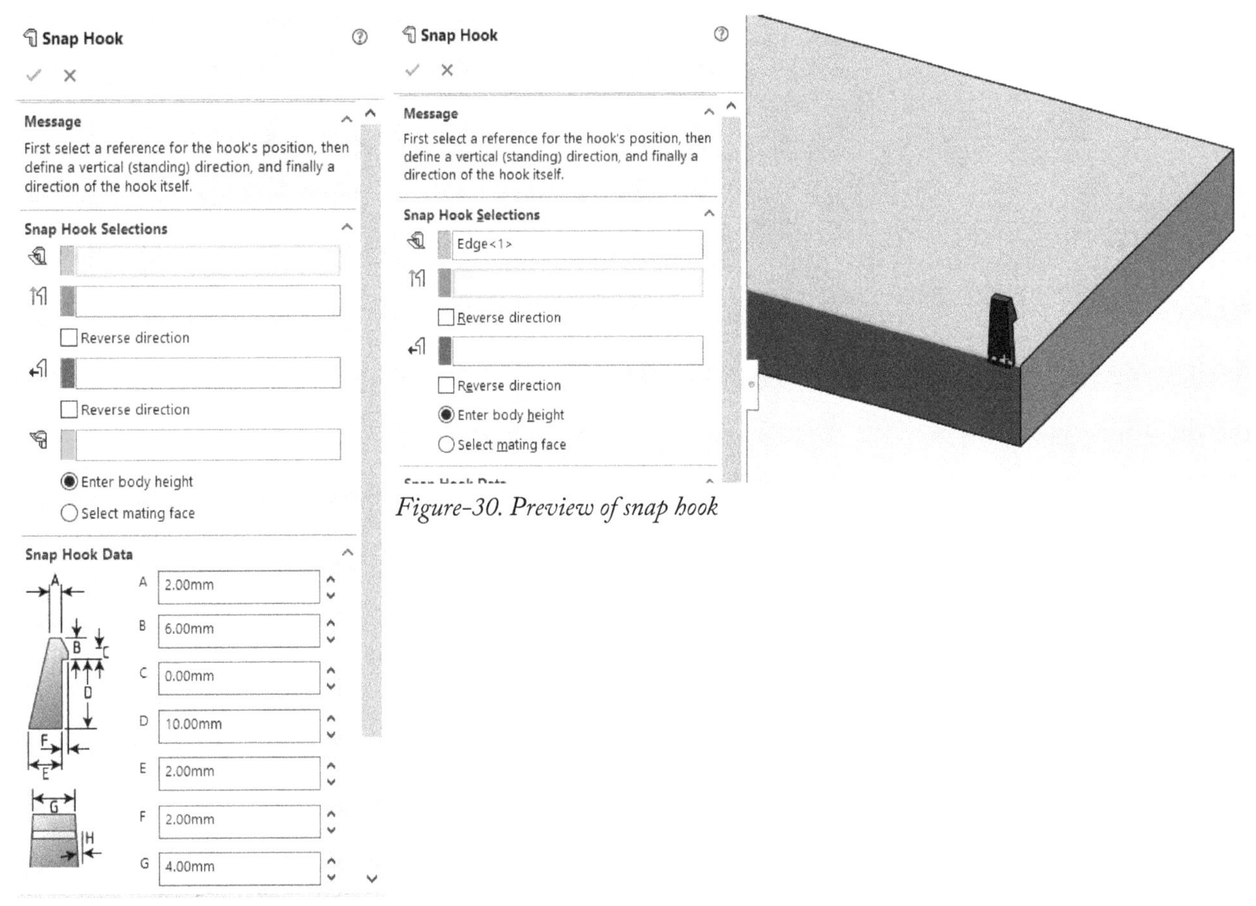

Figure-29. Snap Hook Property-Manager

Figure-30. Preview of snap hook

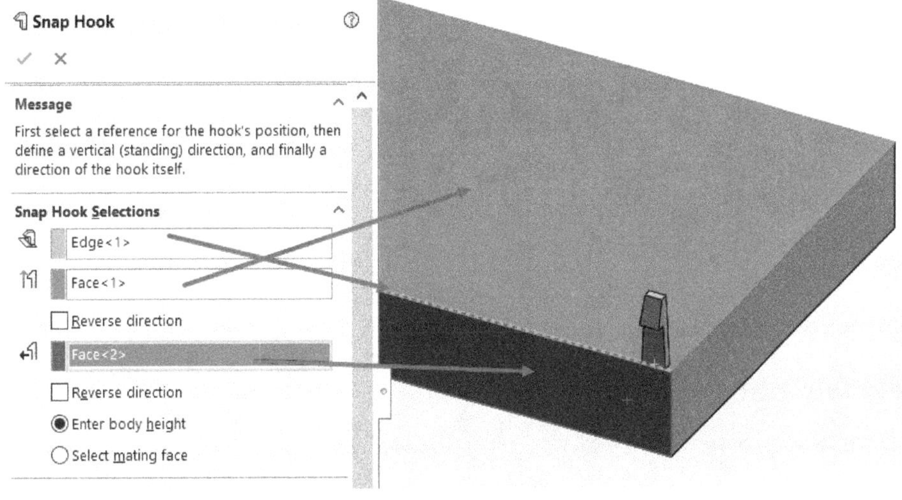

Figure-31. Snap hook placement

Creating Snap Hook Groove

The **Snap Hook Groove** tool is used to create groove for snap hook. The procedure to create snap hook groove is given next.

- Click on the **Snap Hook Groove** tool from the **Fastening Feature** cascading menu of the **Insert** menu. The **Snap Hook Groove PropertyManager** will be displayed; refer to Figure-32.

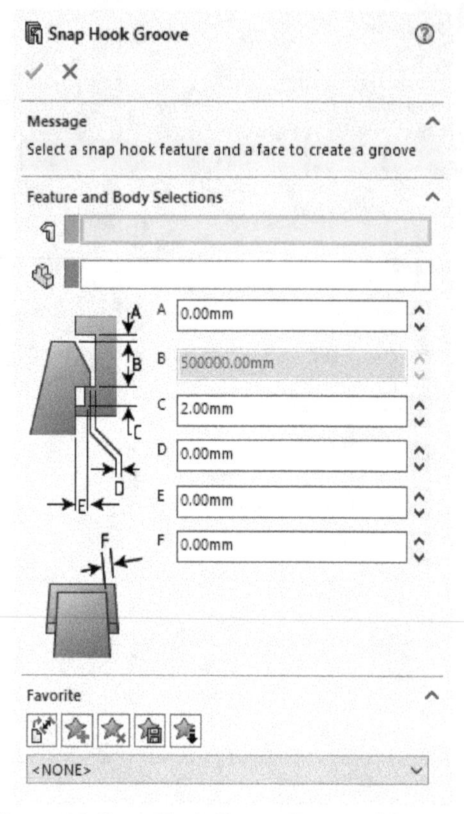

Figure-32. Snap Hook Groove PropertyManager

- Select the snap hook for which you want to create the groove. You will be asked to select the body on which the groove is to be created.
- Select the body on which you want to create groove. Note that this body should be different from the one on which the snap hook is created and it must intersect with the snap hook; refer to Figure-33. On selecting the body, preview of the snap hook groove will be displayed; refer to Figure-34.

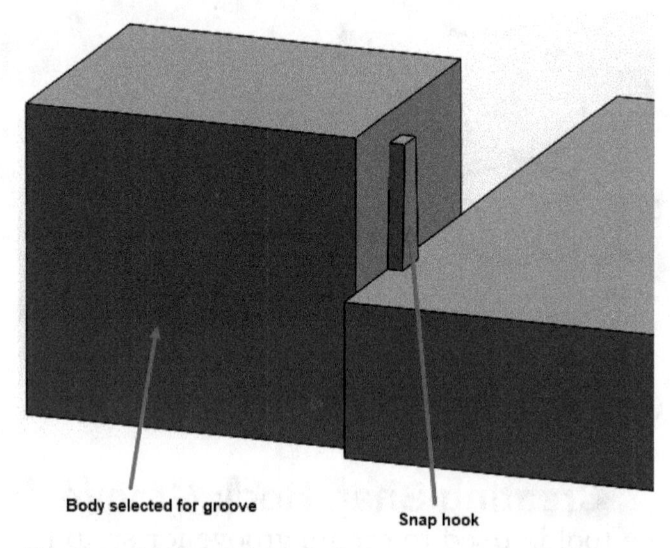

Figure-33. Selection for snap hook groove

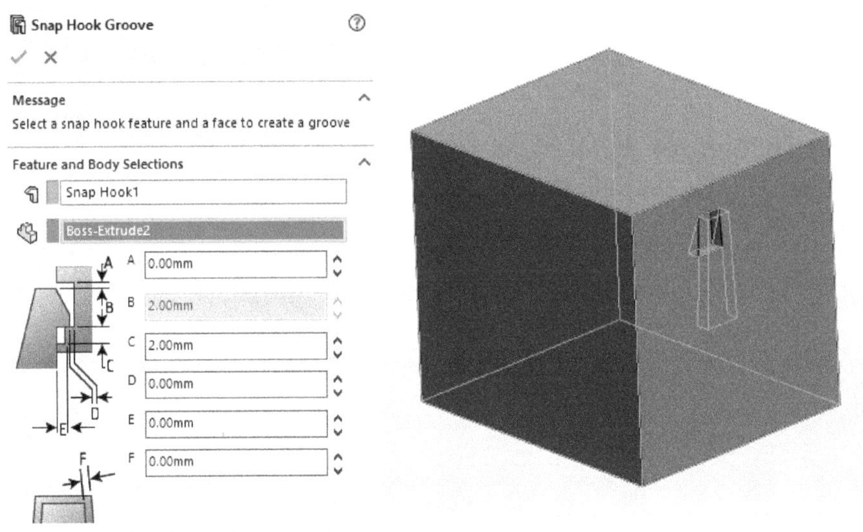

Figure-34. Preview of snap hook groove

- Set the parameters as required in the **PropertyManager** and click on the **OK** button to create the groove.

Creating Vent

Vent is used to provide cut for air flow. The procedure to create vent is given next.

- Click on the **Vent** tool from the **Fastening Feature** cascading menu of the **Insert** menu. The **Vent PropertyManager** will be displayed; refer to Figure-35.

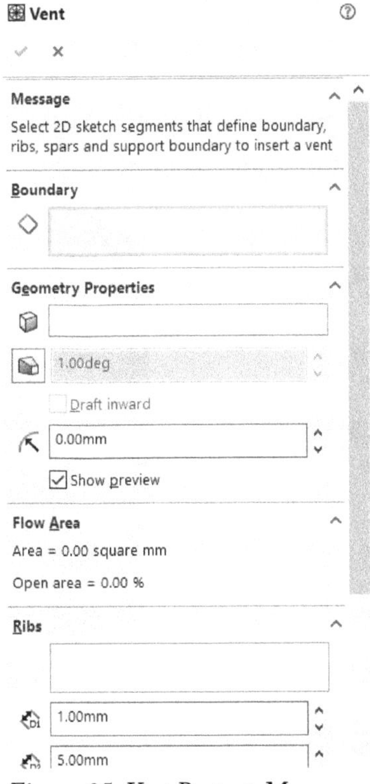

Figure-35. Vent PropertyManager

- The procedure to create vent will be discussed later in this book.

Creating Lip and Groove

Lip and Groove tool is used to guide assembly of plastic parts. The procedure to create lip or groove is given next.

- Click on the **Lip/Groove** tool from the **Fastening Feature** cascading menu of the **Insert** menu. The **Lip/Groove PropertyManager** will be displayed; refer to Figure-36.

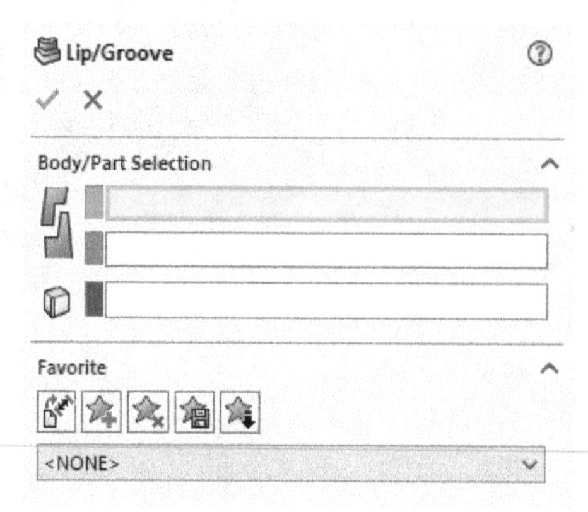

Figure-36. Lip Groove PropertyManager

- Select the body on which you want to create groove. You will be asked to select the body on which the lip is to be created.
- Select the other body. You will be asked to select the reference plane/face/edge/axis for direction of lip and groove mate.
- Select desired reference for direction; refer to Figure-37. The body selected for lip will hide automatically and you will be asked to select the face of other body to create groove.

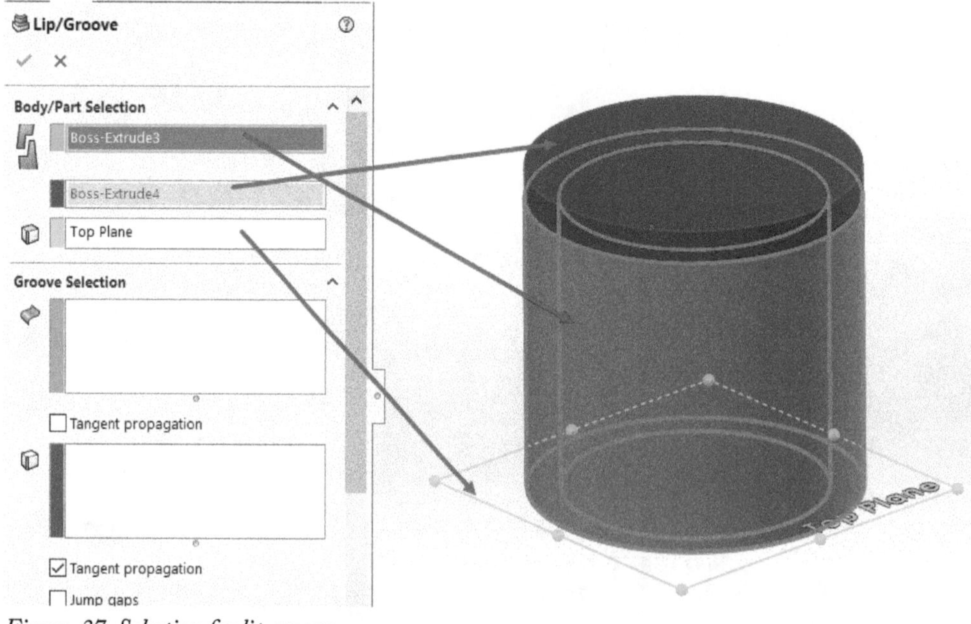

Figure-37. Selection for lip groove

- Select the face where this body meets the other.
- Click in the next selection box of the **Groove Selection** rollout. You will be asked to select the edge from where the groove should start.
- Select desired edge. Preview of the groove will be displayed; refer to Figure-38.

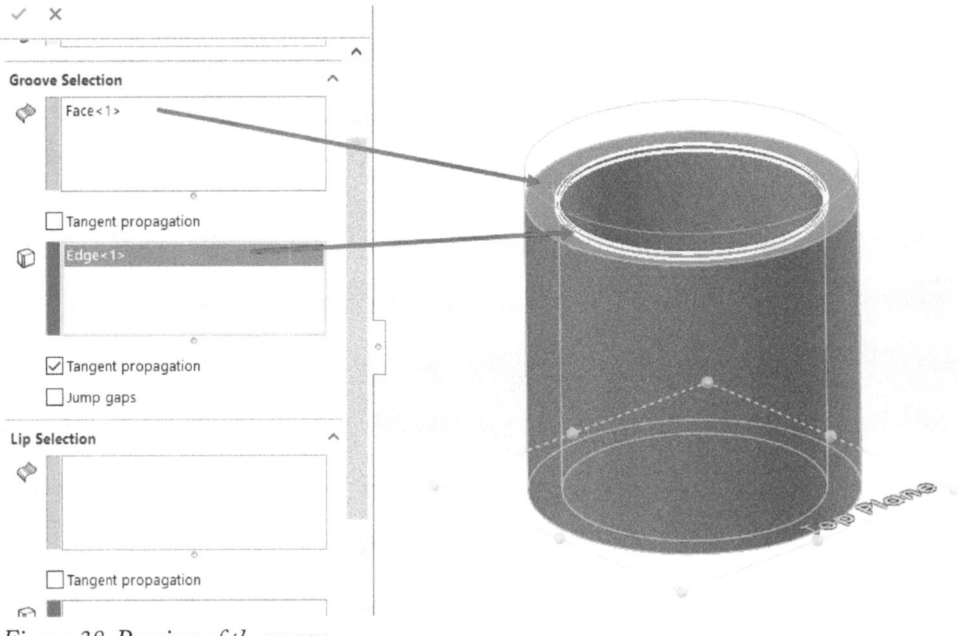

Figure-38. Preview of the groove

- Note that if there is a gap in the groove and you want to create groove through this gap then select the **Jump Gaps** check box. Preview of the groove will be displayed accordingly; refer to Figure-39.

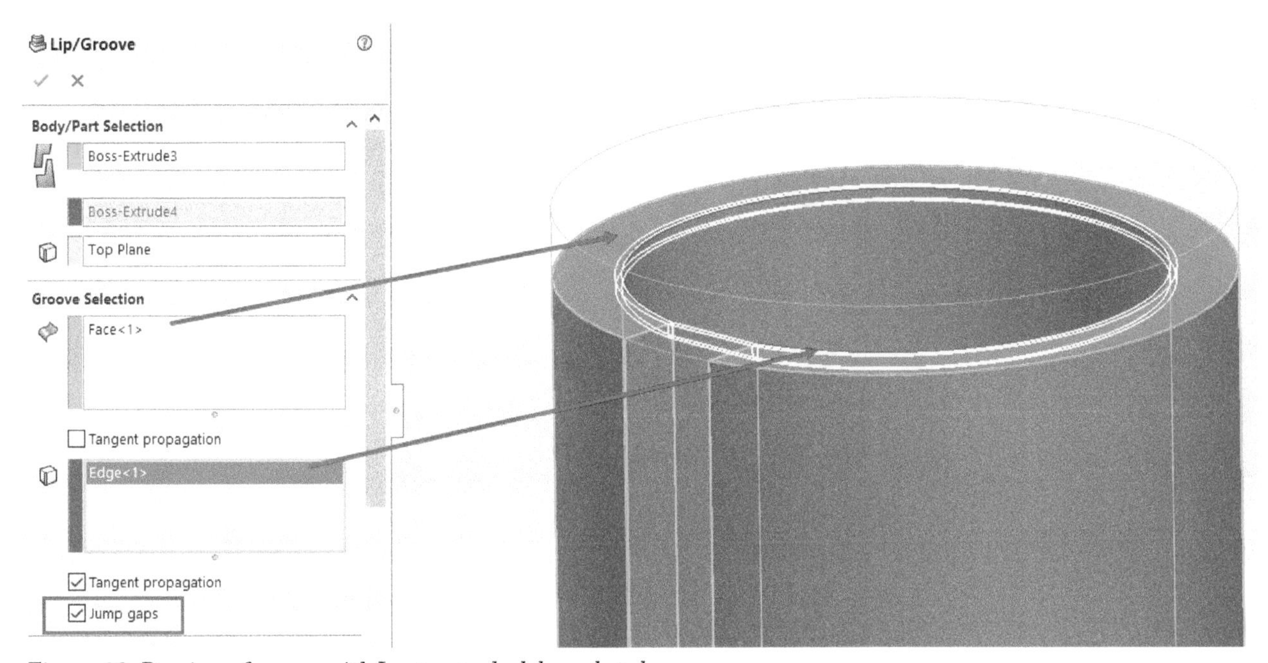

Figure-39. Preview of groove with Jump gaps check box selected

- Click in the first selection box of the **Lip Selection** rollout. The body selected for lip will be displayed and the other body will hide automatically. Select the face mating to the other body.
- Click in the next edit box of the **Lip Selection** rollout. You will be asked to select the edge on which the lip feature is to be created.
- Select desired edge. Preview of the lip feature will be displayed; refer to Figure-40.

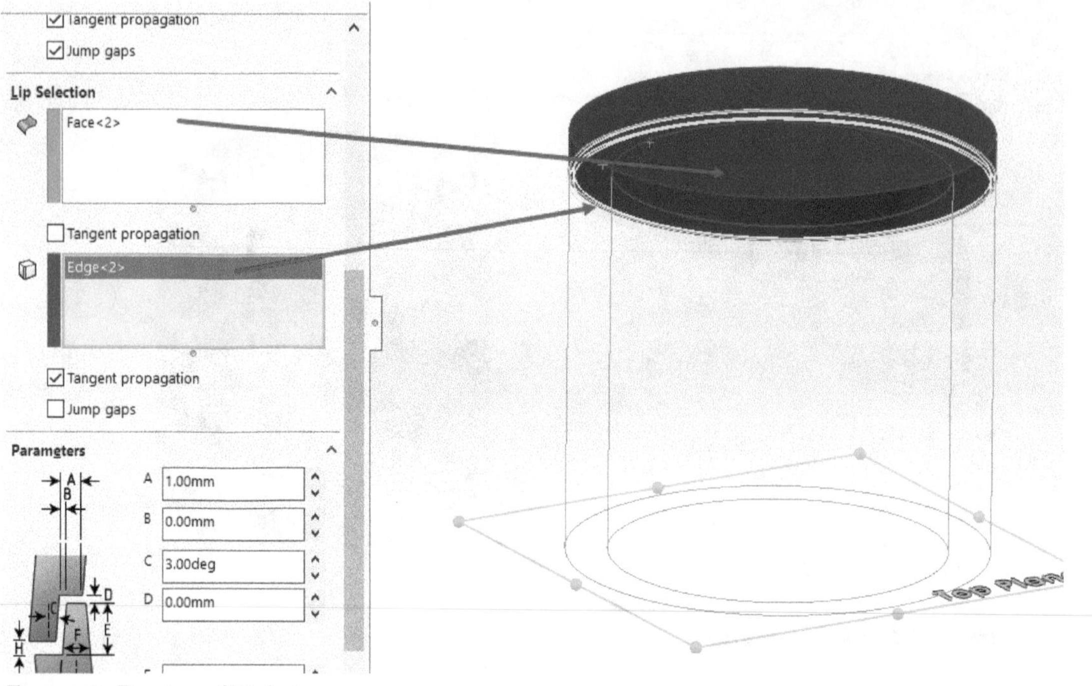

Figure-40. Preview of lip feature

- Set desired parameters in the **PropertyManager** and click on the **OK** button to create the feature.

BLOCK DESIGNING

The block designing is a powerful method to group the objects and perform various manipulations. The tools to create block in Part environment are available in the **Blocks** cascading menu of the **Tools** menu; refer to Figure-41.

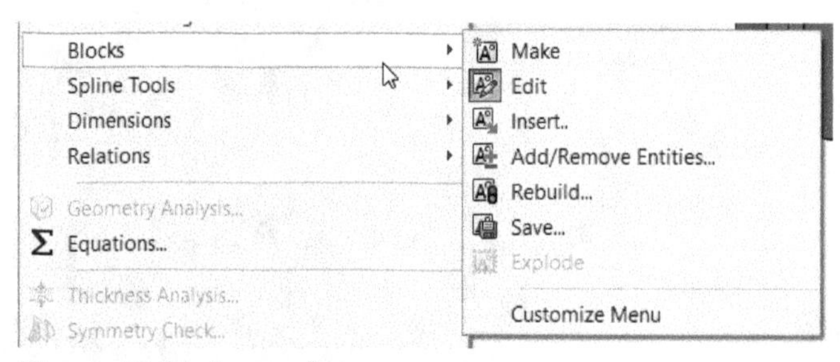

Figure-41. Blocks cascading menu

Note that the tools in this menu are active when you are editing or creating a block.

Creating Block Using Sketch

The blocks are created in sketching environment. The procedure to do so is given next.

- Create a sketch of the model that you want to use as block; refer to Figure-42.
- Make sure you are still in the sketching environment and then select the **Make** tool from the **Blocks** cascading menu of the **Tools** menu. The **Make Block PropertyManager** will be displayed; refer to Figure-43.

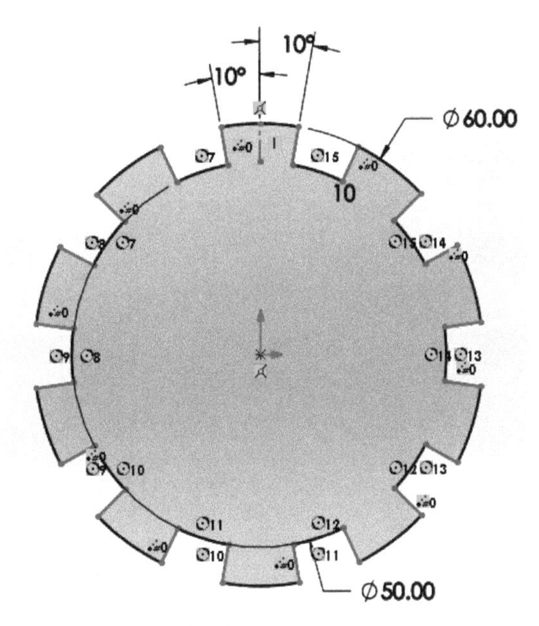

Figure-42. Sketch for block

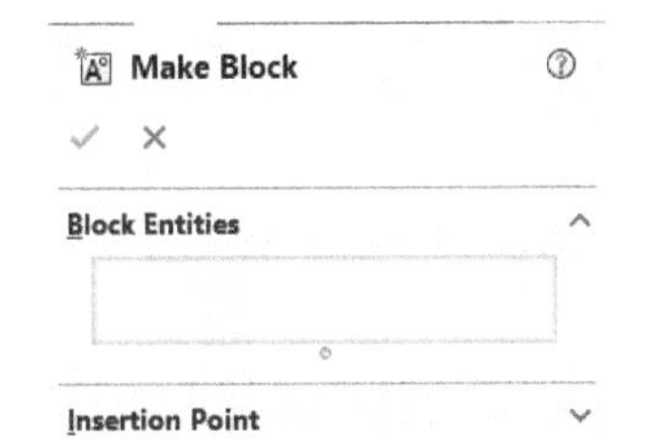

Figure-43. Make Block PropertyManager

- Select the entities of the sketch to be included in the block.
- To define the insertion point for the block, expand the **Insertion Point** rollout and drag the blue colored coordinate system at desired location; refer to Figure-44.

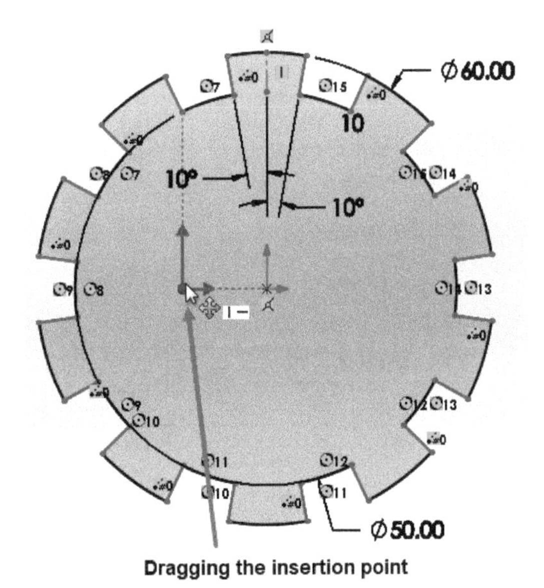

Figure-44. Dragging the insertion point

- Click on the **OK** button from the **PropertyManager** to convert the sketch into block; refer to Figure-45.

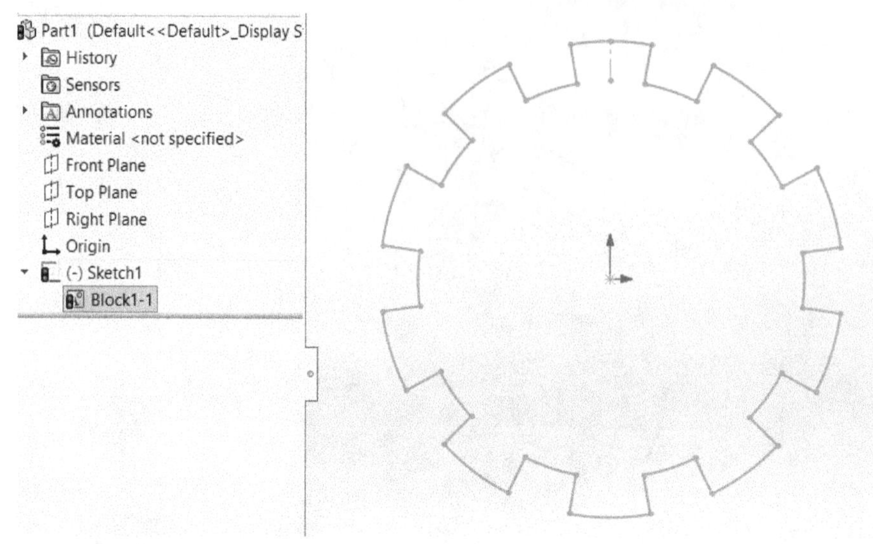

Figure-45. Sketch converted to block

Saving the Block

After making the block as discussed earlier, click on the **Save** tool from the **Blocks** cascading menu of the **Tools** menu. (Make sure you do not leave the sketching environment). The **Save As** dialog box will be displayed. Specify desired name and location for the file and click on the **Save** button to save the block.

Editing a Block

Editing a block involves sketching tools and block toolbar. The procedure to edit a block is given next.

- To edit a block, select it from the **FeatureManager Design Tree** and right-click on it. A shortcut menu will be displayed; refer to Figure-46.

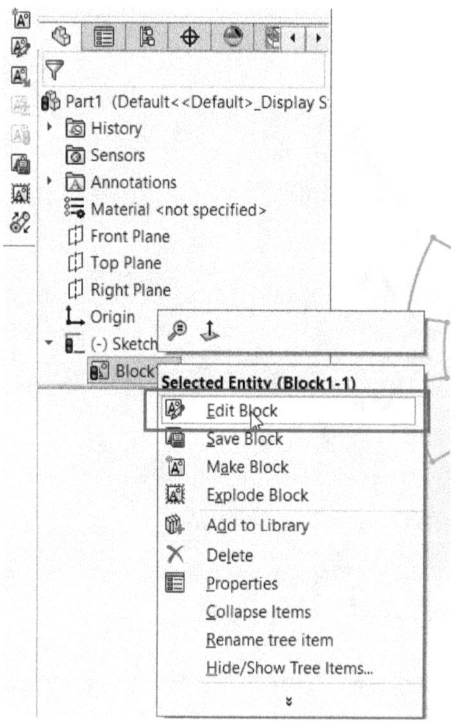

Figure-46. Edit block option

- Click on the **Edit Block** option from the shortcut menu. The **Sketching** environment will be displayed with **Blocks** tool bar; refer to Figure-47.

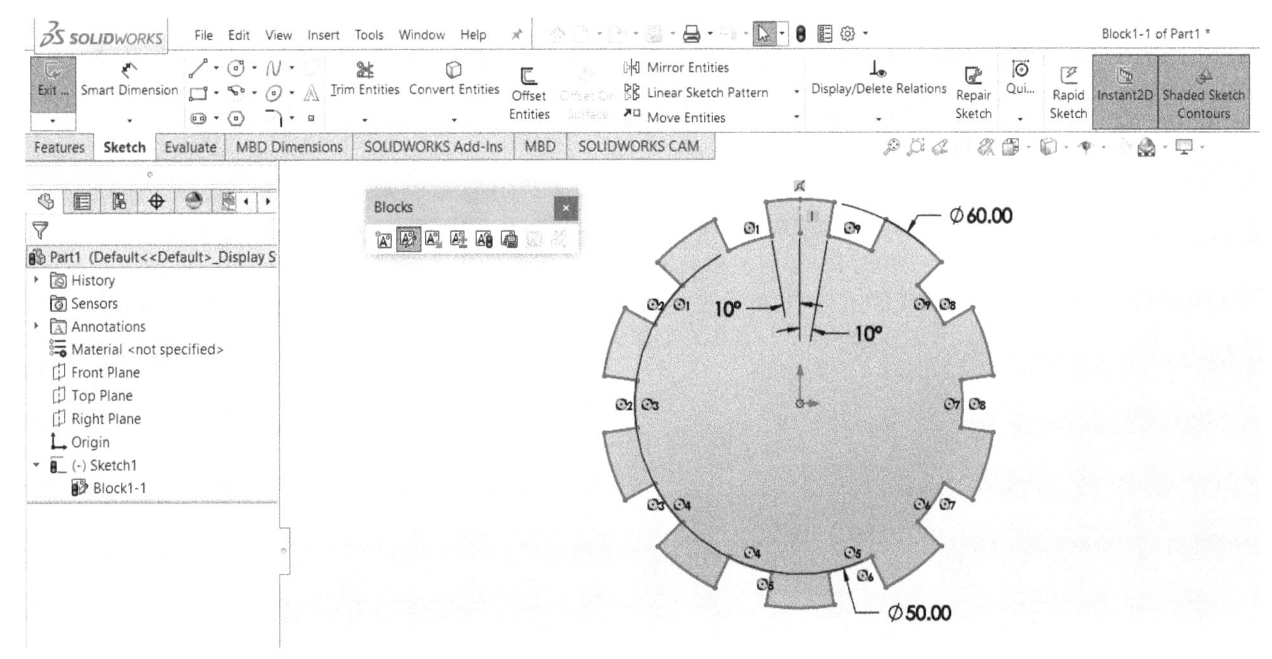

Figure-47. Editing a block

- Edit the block as required by using the sketching tools.
- If you want to add/remove anything from the block then click on the **Add/Remove** button from the **Blocks** toolbar. The **Add/Remove Entities PropertyManager** will be displayed; refer to Figure-48.

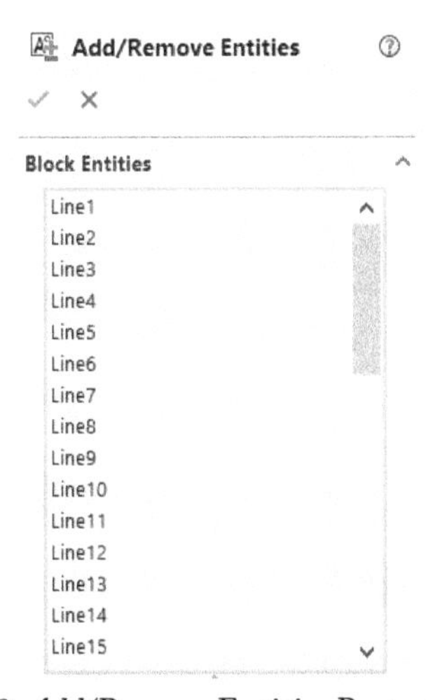

Figure-48. Add/Remove Entities PropertyManager

- Select the entities that you want to be added in the block from the drawing area.
- To remove an entity from the block, select it from the list. Right-click on it and select the **Delete** option.
- Click on the **OK** button from the **PropertyManager** to apply the changes.
- Now, click on the **Rebuild** button tool from the **Blocks** toolbar to update the block.
- Click on the **Save Block** tool from the toolbar to save the block.

Exploding Block

The **Explode Block** tool is used to break the block into sketch entities. To do so, select the block from the **FeatureManager Design Tree** and right-click on it. A shortcut menu will be displayed. Click on the **Explode Block** tool from the shortcut menu. The block will be converted to a sketch.

MACROS

Macro is a set of instructions to be run by computer automatically once giving a call. Macros can reduce the modeling time at a greater extent if used efficiently. The tools to create and manage macros are available in the **Macro** cascading menu of the **Tools** menu; refer to Figure-49. These tools are discussed next.

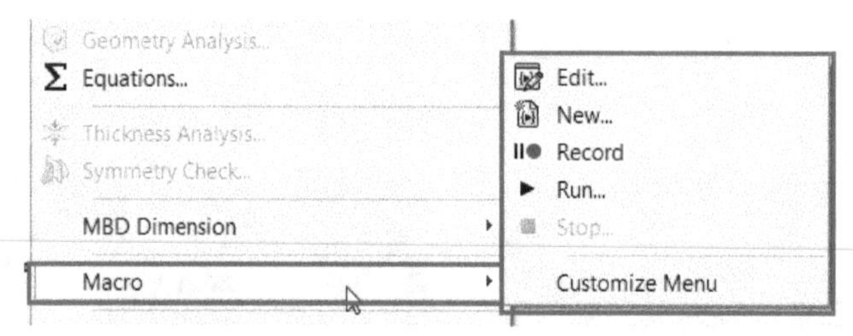

Figure-49. Macro cascading menu

Creating Macros

The procedure to create macro is given next.

- Click on the **New** tool from the **Macro** cascading menu of the **Tools** menu. The **Save As** dialog box will be displayed; refer to Figure-50.

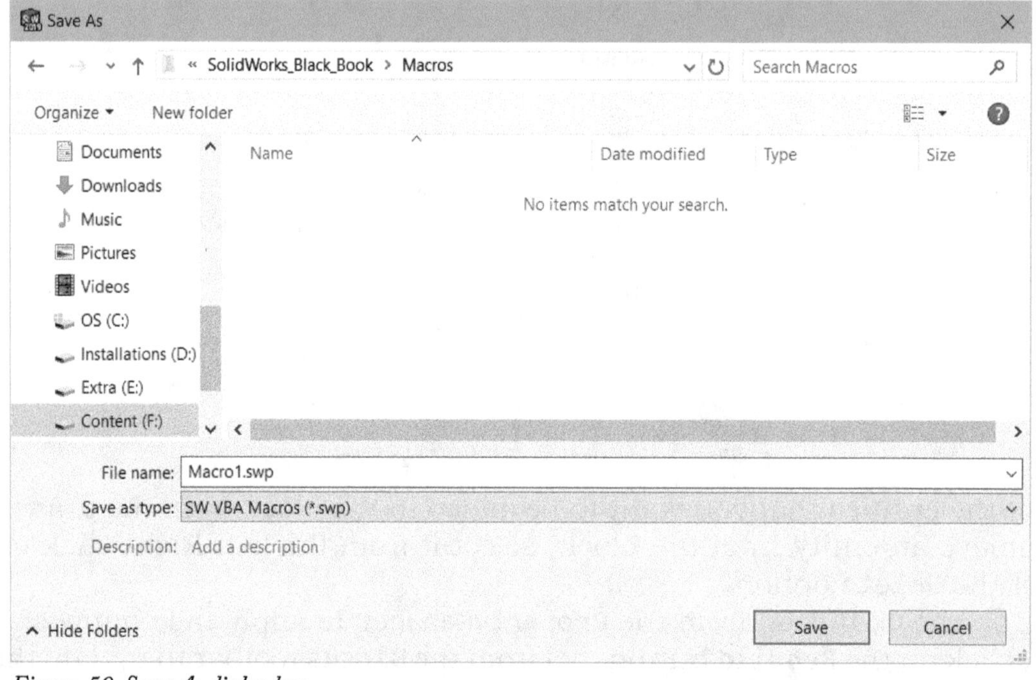

Figure-50. Save As dialog box

- Specify desired name and save the file at desired location by clicking the **Save** button. The **Microsoft Visual Basic** interface will be displayed with the macro file open for editing; refer to Figure-51.

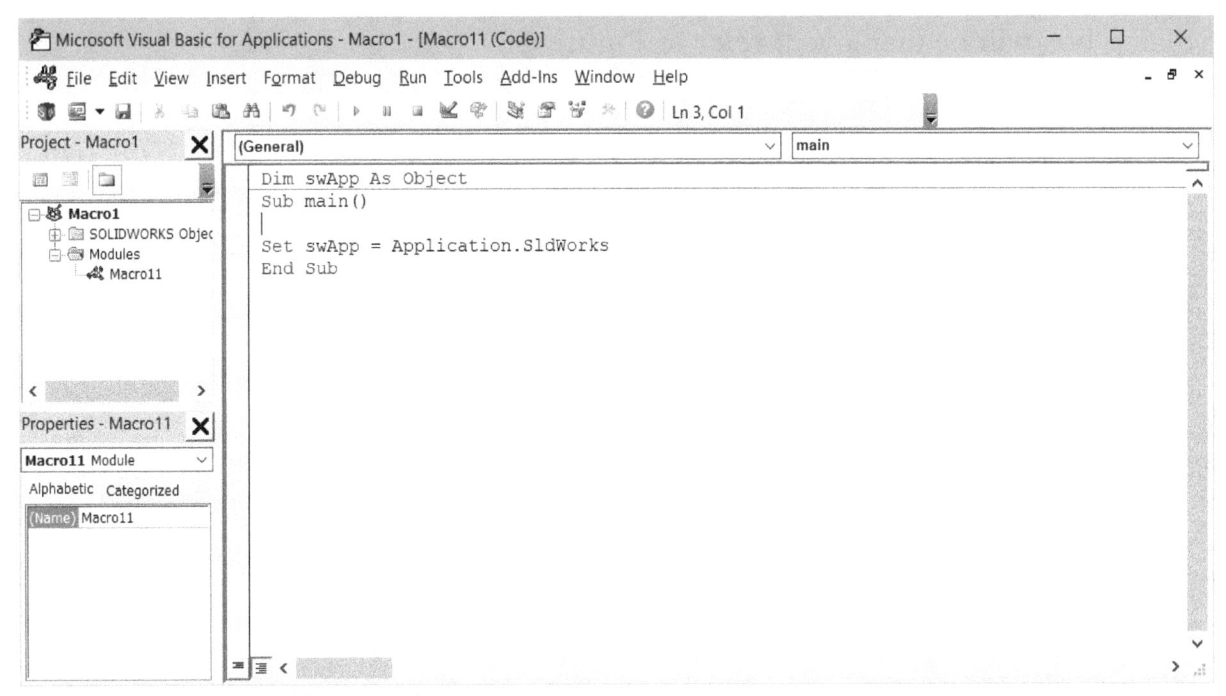

Figure-51. Microsoft Visual Basic interface

- Create desired program using visual basic codes and API of SolidWorks.
- After creating the program, save the file and close the application window.

Note that you need to have working knowledge of Microsoft VBA to create codes for SolidWorks macros. Apart from this hard way, there is an easier way to create macro by recording it while working in SolidWorks.

Recording Macros in SolidWorks

The **Record** tool in **Macro** cascading menu is used to record macros. The procedure is discussed next.

- Click on the **Record** tool from the **Macro** cascading menu of the **Tools** menu. The **Macro** toolbar will be added above the **Ribbon**; refer to Figure-52.

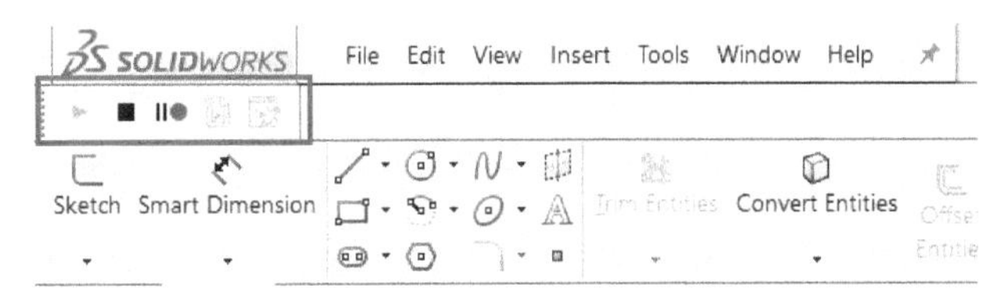

Figure-52. Macro toolbar

- Perform the steps as required and then click on the **Stop** button. The **Save As** dialog box will be displayed.
- Specify desired name and click on the **Save** button. The macro will be created.

Running a Macro

The **Run** tool in the **Macro** cascading menu is used to run recorded macros. The procedure is given next.

- Click on the **Run** tool from the **Macro** cascading menu of the **Tools** menu. The **Open** dialog box will be displayed; refer to Figure-53.

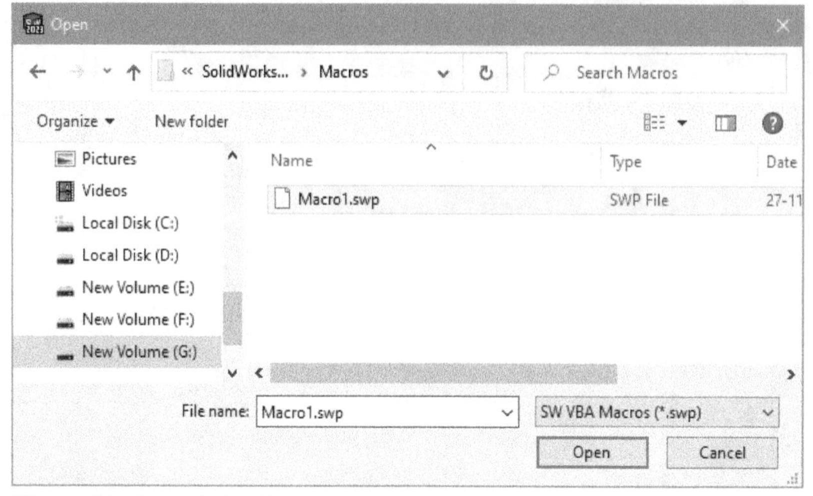

Figure-53. Open_dialog_box

- Double-click on the macro that you want to run. The macro will run automatically.

Editing Macro

The **Edit** tool in the **Macro** cascading menu is used to edit earlier created macro. The procedure to edit the macro is given next.

- Click on the **Edit** tool from the **Macro** cascading menu of the **Tools** menu. The **Open** dialog box will be displayed.
- Double-click on the macro file that you want to edit. The Microsoft VBA application will be displayed with the file opened; refer to Figure-54.

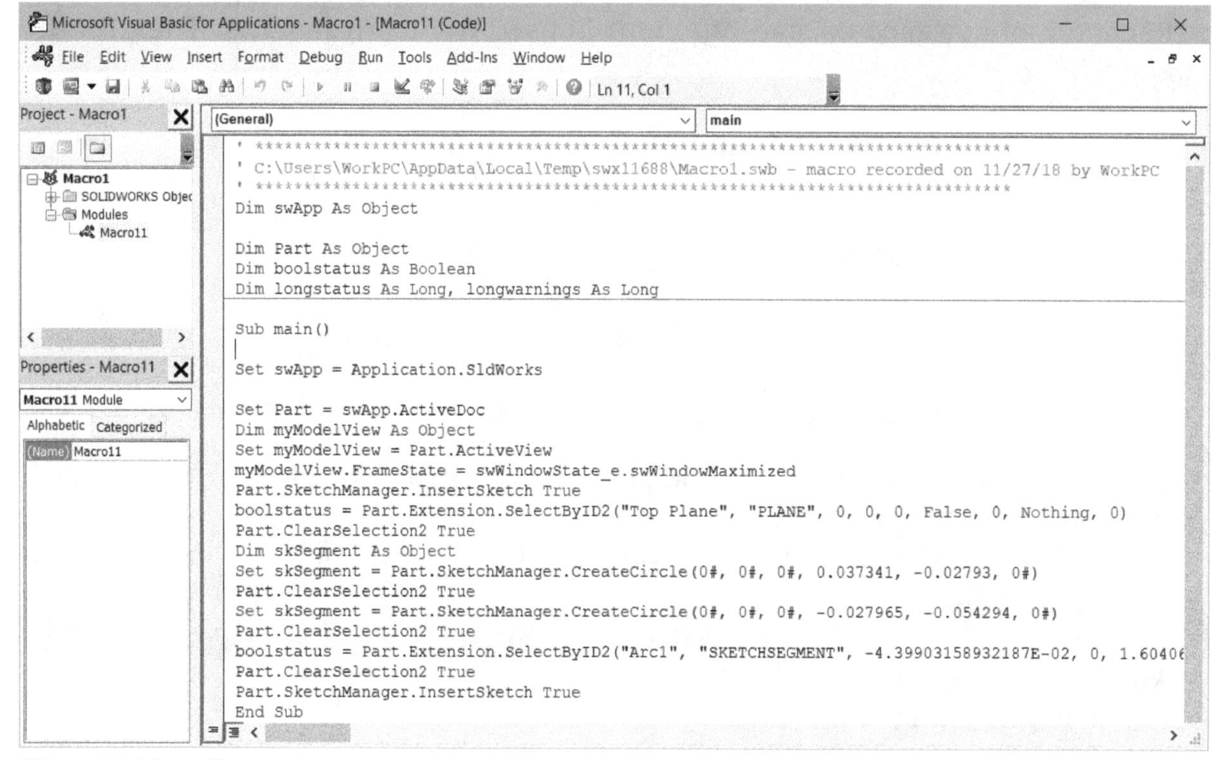

Figure-54. Macro file opened

- Edit the macro and save the file.

MESH MODELING

The Mesh models are bodies created by triangular polygons called facets. Each facet has three vertices and three edges. The tools to create and manage mesh models are available in the **Mesh Modeling CommandManager**; refer to Figure-55.

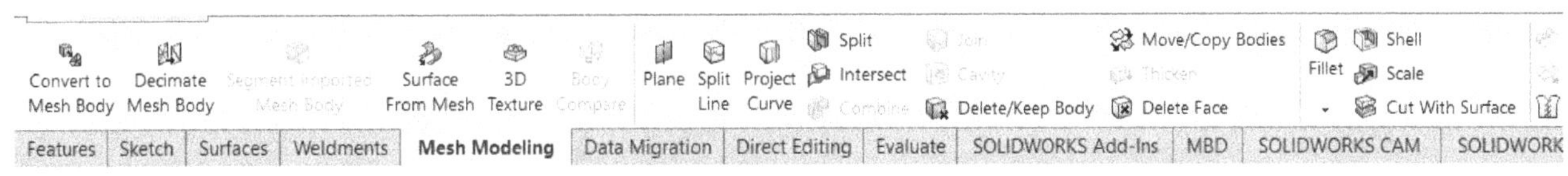

Figure-55. Mesh Modeling CommandManager

Most of the tools of this **CommandManager** have been discussed earlier. Rest of the tools are discussed next.

Decimating Mesh Body

The **Decimate Mesh Body** tool is used to reduce the number of facets in mesh and simply the model. This tool is active only when a mesh body is present in the drawing area. This tool does not support BREP mesh bodies and **Convert to Mesh Body** tool discussed earlier converts solid into BREP mesh body. So, you will need a graphic mesh body created in other mesh modeling software like Autodesk Maya, Rhino, and so on. The procedure to use this tool is given next.

- Click on the **Decimate Mesh Body** tool from the **Mesh Modeling CommandManager**. The **Decimate Mesh Body PropertyManager** will be displayed; refer to Figure-56.

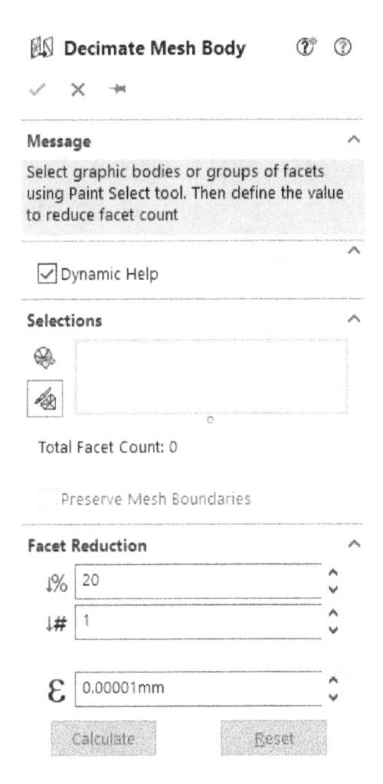

Figure-56. Decimate Mesh Body PropertyManager

- If you want to select full body then click on the mesh body imported. The body will get selected. If you want to decimate specific portion of mesh body then click on the **Paint Select** button from the **Selections** rollout of the **PropertyManager**. The cursor mark will change to circle and a toolbox will be displayed for defining size of selection circle.

- Set desired parameters in the toolbox and drag the cursor over the portion of mesh body to select desired facets; refer to Figure-57.

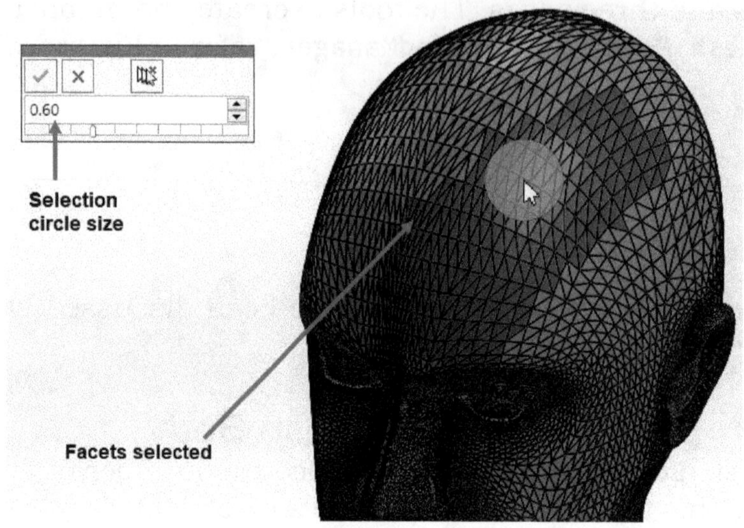

Figure-57. Selecting facets for decimating

- After selecting facets, click on the **OK** button from the toolbox.
- Set desired values for reduction in facets by percentage or number of faces.
- Set desired value of tolerance upto which the facet can deviate from the original boundaries.
- Click on the **Calculate** button to display preview of mesh body decimation. If the changes are not as desired then click on the **Reset** button and set the parameters as desired.
- Click on the **OK** button from the **PropertyManager** to apply changes.

Segmenting Mesh Body

The **Segment Imported Mesh Body** tool is used to create segments of the mesh body based on the orientation and grouping of facets. To use this tool, you need to import a graphic mesh and convert it to mesh body by using **Convert to Mesh Body** tool then only this tool will be active in **CommandManager**. The procedure to use this tool is given next.

- Select the imported mesh body and click on the **Segment Imported Mesh Body** tool from the **Mesh Modeling CommandManager**. The **Segment Mesh PropertyManager** will be displayed; refer to Figure-58.

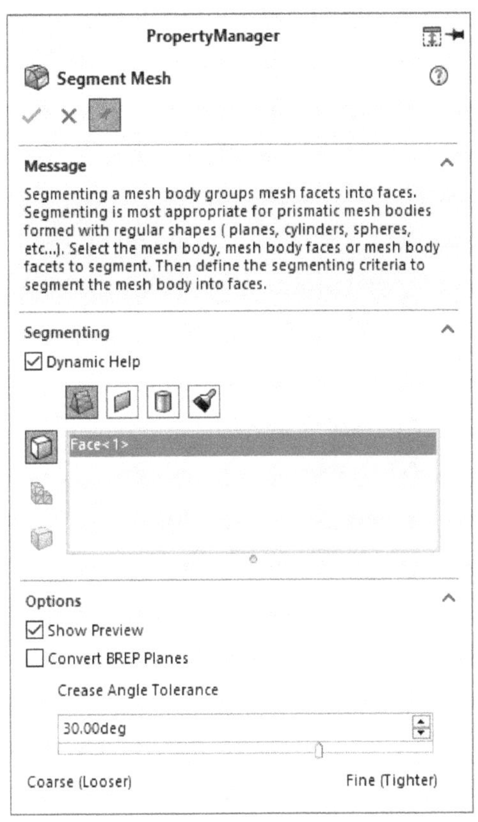

Figure-58. Segment_Mesh_PropertyManager

- Select desired button from the **Segmenting** rollout to define what type of segments are to be created.
- Select the **Show Preview** check box to display the preview of segmenting; refer to Figure-59.

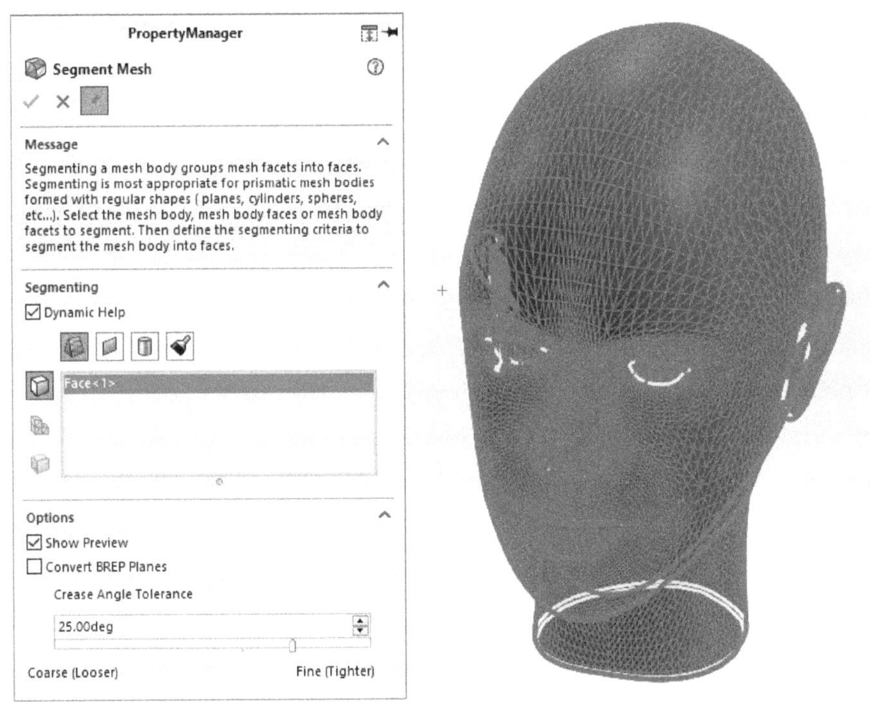

Figure-59. Segmenting_preview

- Set desired value of tolerance using the slider.

- Click on the **OK** button from the **PropertyManager**.
- After creating the segment, click on the **Close** button to exit the **PropertyManager**.

Comparing Bodies

The **Body Compare** tool is used to compare two mesh bodies. The procedure to use this tool is given next.

- Click on the **Body Compare** tool from the **Mesh Modeling CommandManager** in the **Ribbon**. The **Body Compare PropertyManager** will be displayed; refer to Figure-60.

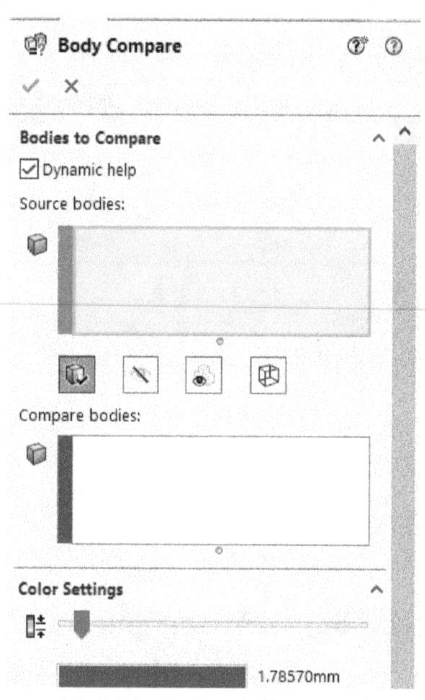

Figure-60. Body Compare PropertyManager

- Select the source body from the drawing area and then click in the **Compare bodies** selection box. You will be asked to select the other body with which the source body is to be compared.
- Set desired color parameters for the faces in the **Color Settings** rollout.
- Click on the **OK** button from the **PropertyManager** to display comparison. Click again on the **Body Compare** button to exit comparison.

PRACTICE 1

In this practical, we will create a sketch for the drawing given in Figure-61.

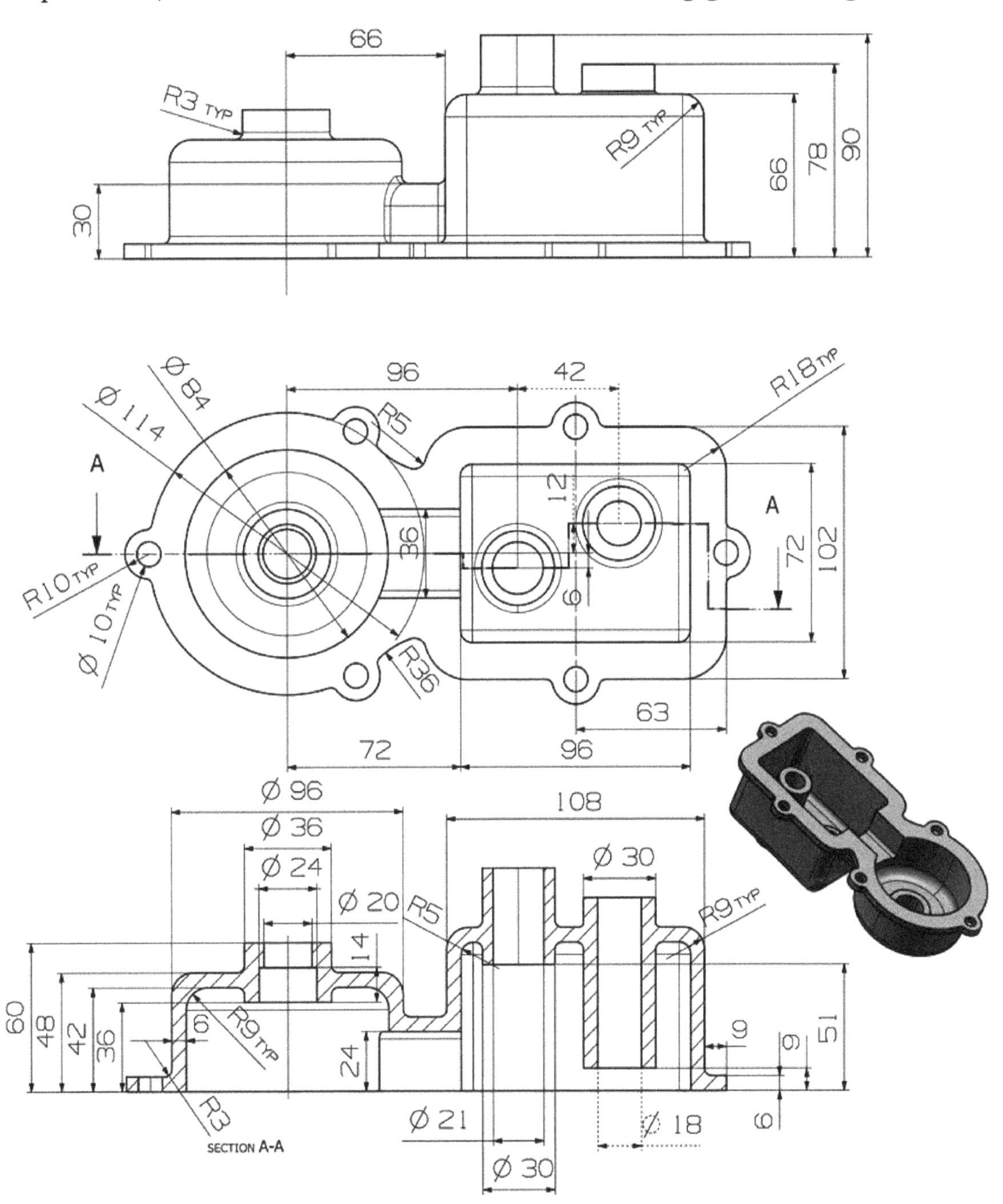

Figure-61. Practice 1

SELF ASSESSMENT

Q1. The **Dome** tool is available in the .... cascading menu of **Insert** menu.

Q2. The .... tool is used to modify shape and size of solid model by using control points or control polygons.

Q3. The .... tool is used to deform a solid model using different set of instructions.

Q4. The **Indent** tool is used to create indent on the solid based on the selected solid face or surface. (T/F)

Q5. The .... tool is used to split solid into pieces.

Q6. The **Snap Hook** tool is used to create groove of snap hook. (T/F)

Q7. The **Explode Block** tool is used to delete selected block. (T/F)

Answer to Self-Assessment:

1. Features, **2.** Freeform, **3.** Deform, **4.** T, **5.** Split, **6.** F **7.** F

Chapter 7

Assembly and Motion Study

Topics Covered

The major topics covered in this chapter are:

- ***Inserting Components in Assembly.***
- ***Assembly Constraints.***
- ***Reference and Assembly features.***
- ***Exploded View.***
- ***Bill of Material.***
- ***Motion Study.***

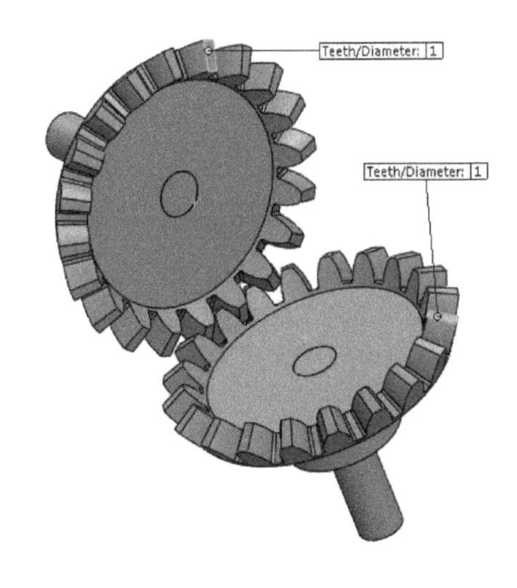

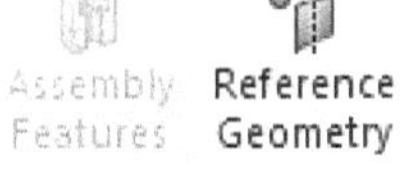

ASSEMBLY

In engineer's language, assembly is the combination of two or more components and these components are constrained to each other in a specified manner called assembly constraints. In SolidWorks, Assembly Design is a separate environment. To start the Assembly Design, click on the **New** button from the **Menu Bar**. The **New SOLIDWORKS Document** dialog box will display; refer to Figure-1. Double-click on the **Assembly** button from the dialog box. The Assembly Design environment will display as shown in Figure-2. The tools related to assembly are available in the **Ribbon**. Note that in the left of the screen, the **Begin Assembly PropertyManager** is displayed. The **Open** dialog box is also displayed prompting you to select file(s) to be inserted in the assembly.

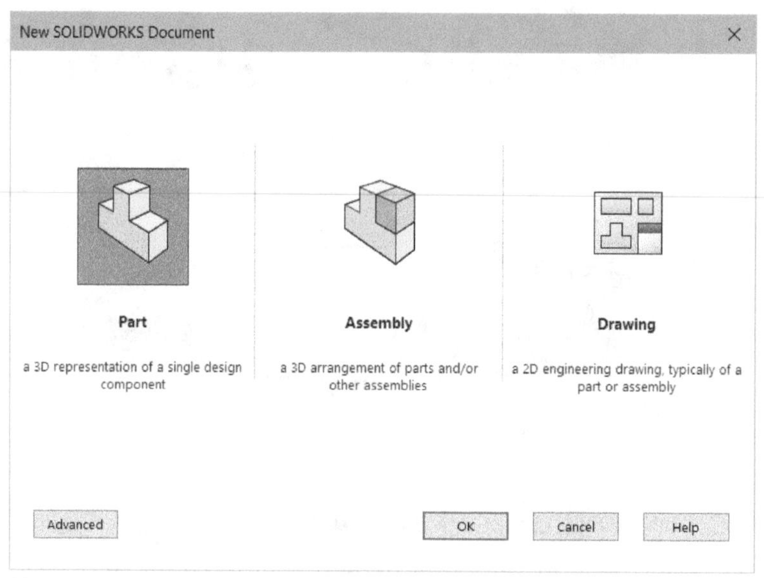

Figure-1. New_SOLIDWORKS_Document_dialog_box

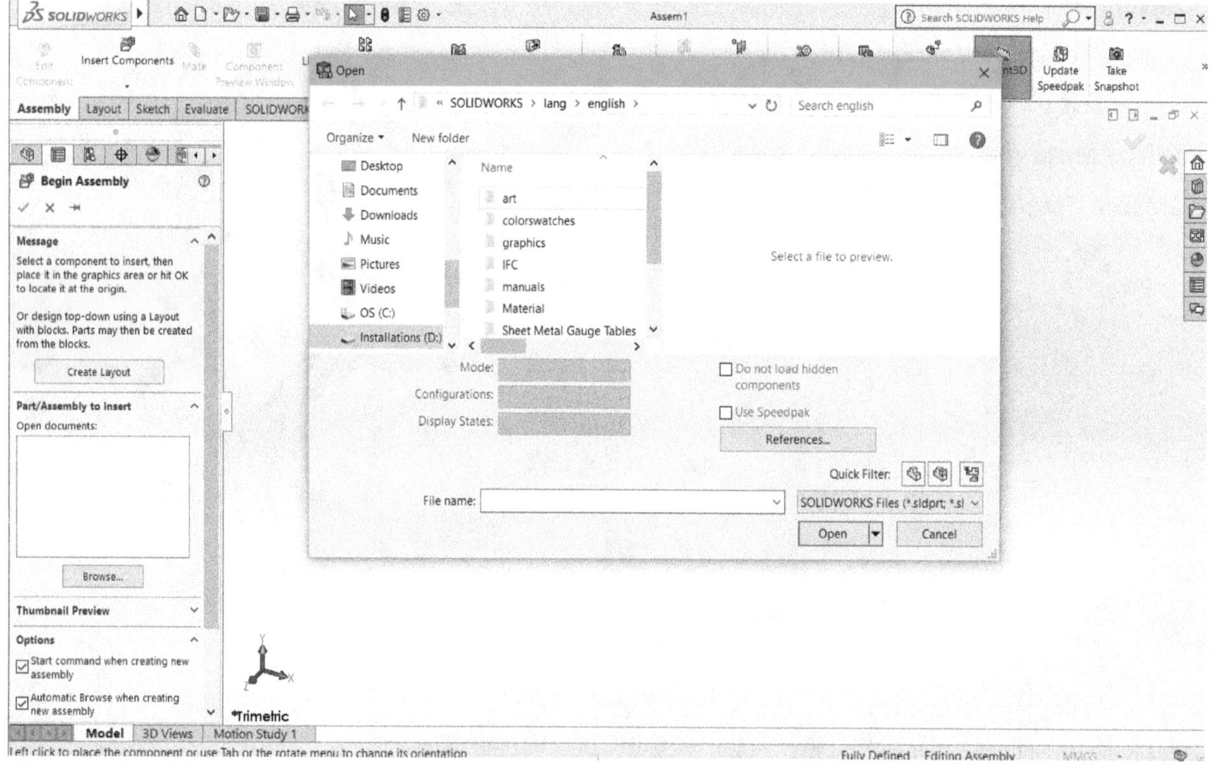

Figure-2. Assembly Design environment

INSERTING BASE COMPONENT

- After starting Assembly environment, select desired part file from the **Open** dialog box and click on the **Open** button. The part will get attached to the cursor; refer to Figure-3.

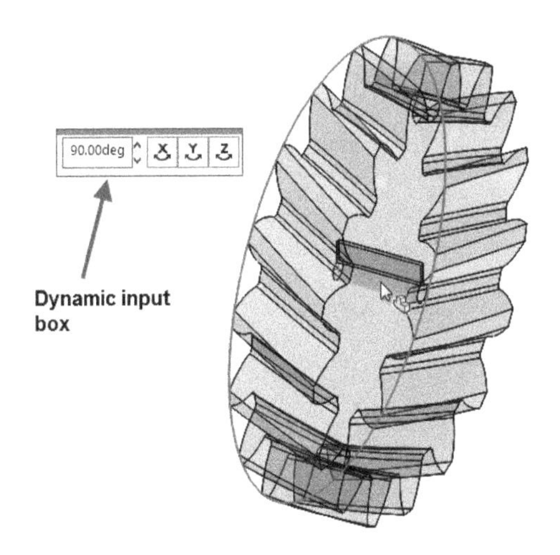

Figure-3. Part attached to cursor

- Specify the rotation angle value in the edit box and click on desired direction button from the dynamic input box to rotate the part.
- Click on the **OK** button from the **PropertyManager** to place the part. Note that the base component of assembly is inserted as fixed in SolidWorks and **(f)** is displayed next to its name in the **FeatureManager Design Tree**; refer to Figure-4.

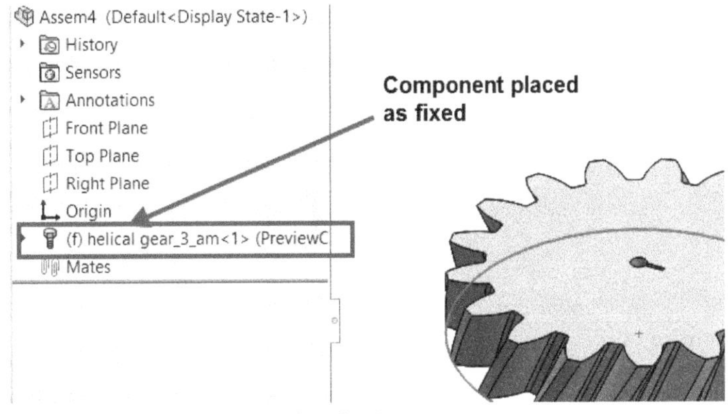

Figure-4. Component placed as fixed

INSERTING COMPONENTS IN ASSEMBLY

Once you have inserted the base component in the assembly, follow the steps given next to insert more components.

- Click on the **Insert Components** tool from the **Assembly CommandManager** in the **Ribbon**. The **Open** dialog box will display; refer to Figure-5.
- Browse to the location of your desired part file and double-click on the file to add it. The component will be attached to the cursor; refer to Figure-6. You can rotate the component by using the **Rotate** options in dynamic input box.
- Click in the viewport to place the component. The component will be inserted in the assembly.

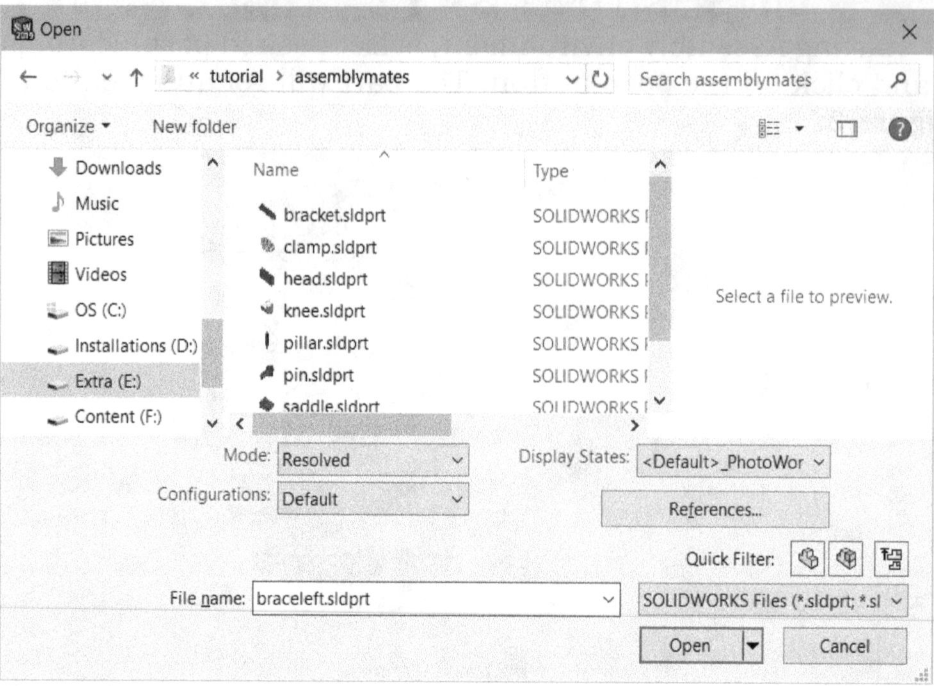

Figure-5. Open dialog box

Figure-6. Component attached to cursor

- The component will be placed as floating and can move anywhere by dragging. Also **(-)** mark will be added before its name in the **FeatureManager Design Tree** displayed at the left.
- You can use the **Rotate** menu to rotate the part before inserting.
- Similarly, you can insert more components as per your requirement in the assembly.
- If you want to reinsert a component then hold the **CTRL** key from keyboard and drag desired component already existing in the display area. One more instance of component will be inserted; refer to Figure-7.

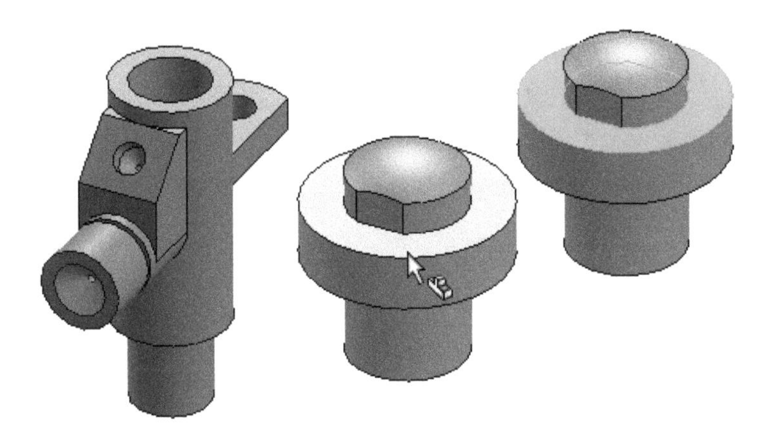

Figure-7. Reinserting a component

- From SolidWorks 2016 onwards, you can also choose desired configuration of the part being inserted in assembly. To do so, click on the **Configuration** drop-down in the **Insert Component PropertyManager** and select desired configuration; refer to Figure-8.

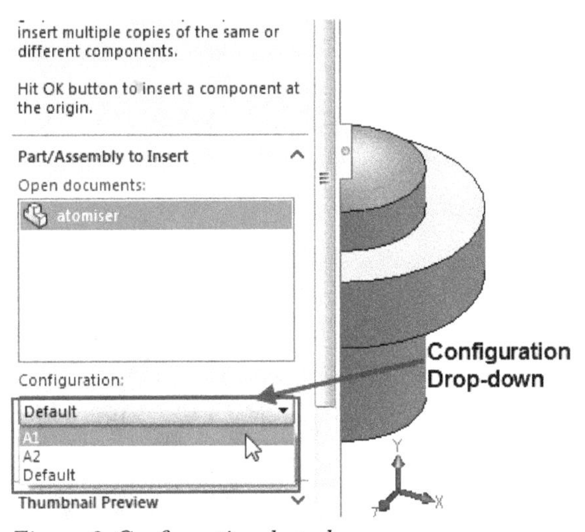

Figure-8. Configuration drop-down

After adding all the components in assembly, the next task is to apply proper constraints to them. The next section explains the use of assembly constraints (also called **Mates** in the language of SolidWorks).

ASSEMBLY CONSTRAINTS (MATES)

The options to apply assembly constraints are available in the **Mate PropertyManager**. The use of options in this **PropertyManager** is discussed next.

- After inserting desired components, click on the **Mate** tool from the **Ribbon**. The **Mate PropertyManager** will display as shown in Figure-9.
- From SolidWorks 2016 onwards, you can make the first selection for mate as transparent for easy identification. To do so, scroll down in **Mate PropertyManager** and select the **Make first selection transparent** check box; refer to Figure-10.

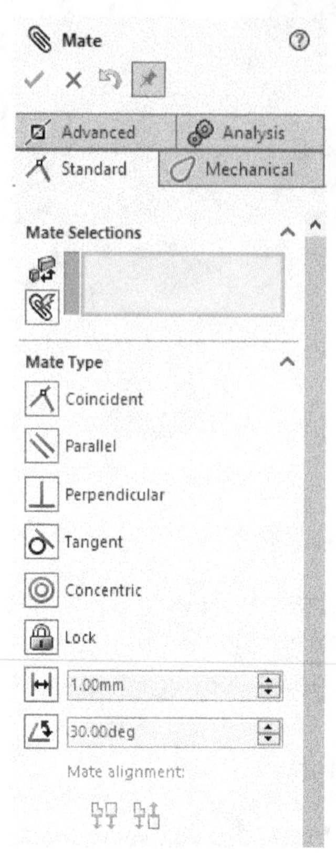

Figure-9. Mate PropertyManager

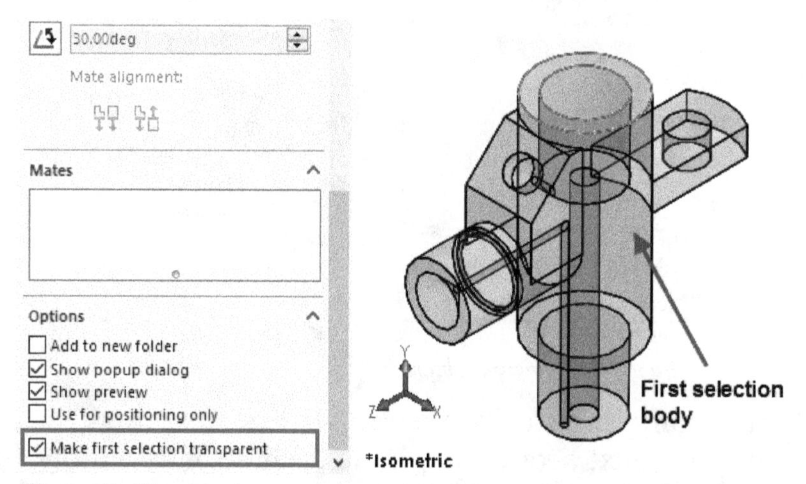

Figure-10. First selection transparency option

- If you want to position the components by using the mates but do not want to apply the mates then select the **Use for positioning only** check box from the **Options** rollout in the **PropertyManager**.

The buttons used for applying constraints are discussed next.

Coincident

The **Coincident** button is used to make two components coincide at the selected references. The steps to use this constraint are given next.

- Click on the **Coincident** button from the **PropertyManager**. You are asked to select two references for this mate.

- Select two axes/faces/planes/curves that you want to make coincident. The preview of constraint will display with a pop-up toolbar; refer to Figure-11.

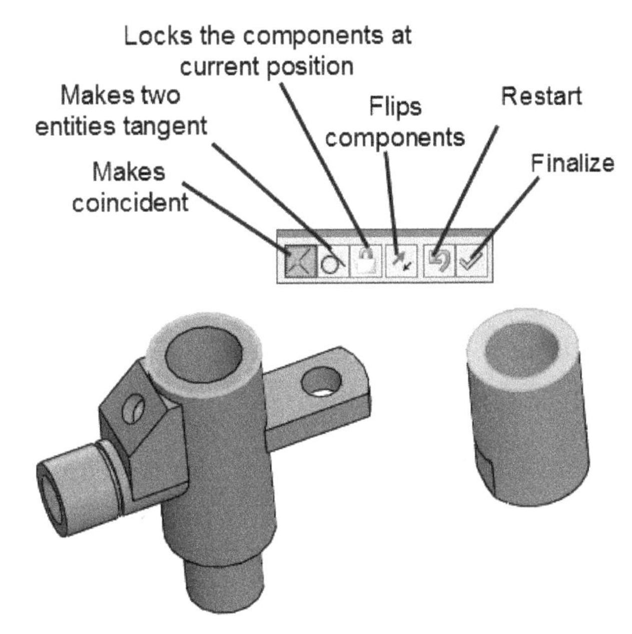

Figure-11. Coincident constraint preview

- Click on the **Flip** button from pop-up toolbar to change the orientation of components; refer to Figure-12.

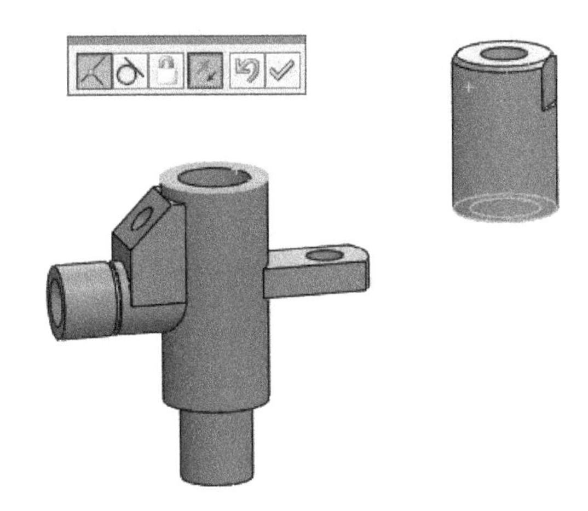

Figure-12. Components after flipping

Parallel

The **Parallel** button is used to make two components parallel with respect to each other. The steps to use this button are given next.

- Click on the **Parallel** button and select the two faces/axes/planes. The components will be parallel with respect to these references. Refer to Figure-13.

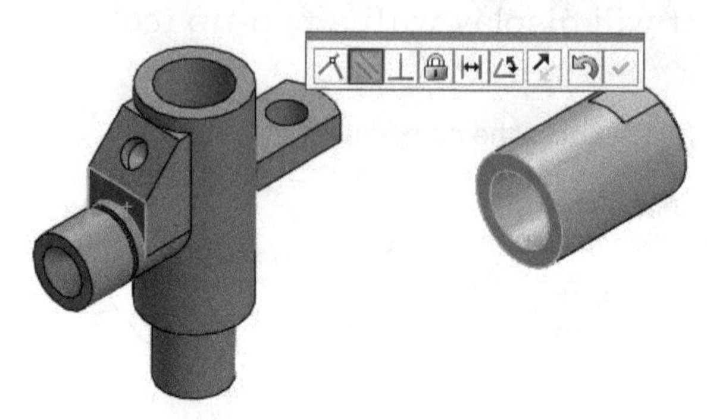

Figure-13. Parallel constraints

Note that the tools displayed in the pop-up toolbar are also available in the **PropertyManager** and work in the same way. The pop-up toolbar gives us facility to change the constraint type after the command is activated.

Perpendicular

The **Perpendicular** button is used to make two components perpendicular to each other. The steps are given next.

- Click on the **Perpendicular** button and select two faces/axes/planes that you want to make perpendicular to each other. The preview will be displayed; refer to Figure-14.

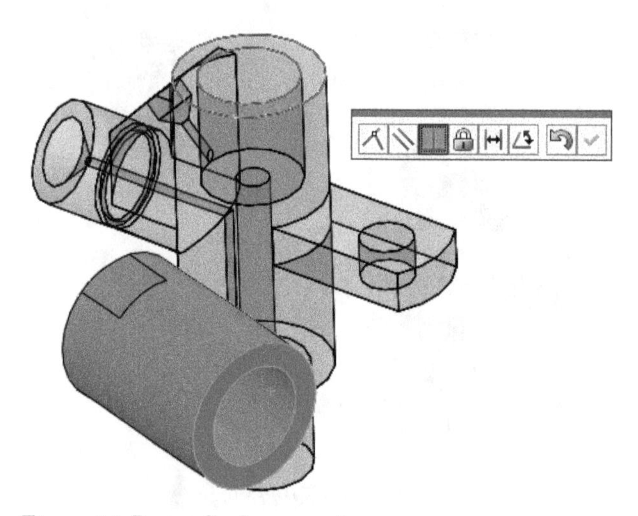

Figure-14. Perpendicular constraint

Tangent

The **Tangent** button is used to make two components tangent to each other. The steps are given next.

- Click on the **Tangent** button and select two faces that you want to make tangent. The preview of tangent constraint will display as shown in Figure-15.

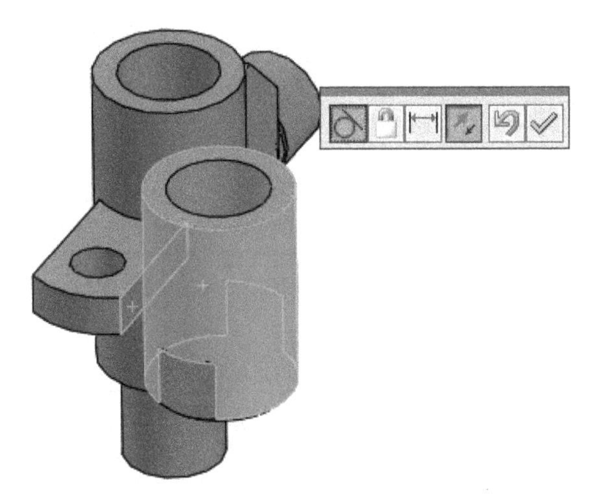

Figure-15. Tangent constraint

Concentric

The **Concentric** button is used to make two round components share the same center axis. The steps are given next.

- Click on the **Concentric** button and select two round faces that you want to make concentric. The preview of tangent constraint will display as shown in Figure-16.

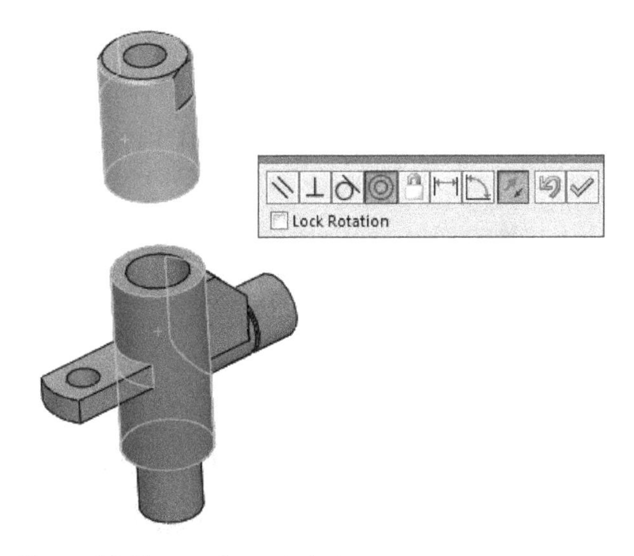

Figure-16. Concentric constraint

Note that you can lock the rotation of selected components by selecting the **Lock Rotation** check box in the **PropertyManager**/pop-up toolbar.

Lock

The **Lock** button is used to lock the component at its current position. The steps to do so are given next.

- Click on the **Lock** button and select the components you want to fix. The preview of components will display in blue color.
- Click **OK** to fix the component. Now, drag one of component to check the effect.

Note that you cannot move the first component of assembly by default. To move the first component, right-click on its name in the **FeatureManager Design Tree** and select the **Float** option from the menu displayed; refer to Figure-17. Now, you will be able to move the component.

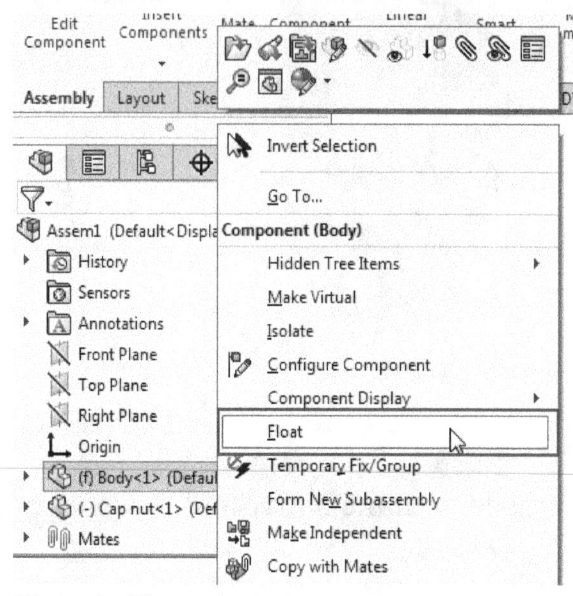

Figure-17. Float option in shortcut menu

Distance

The **Distance** button is used to set distance between two selected faces. The steps are given next.

- Click on the **Distance** button and select two flat faces that you want to use for setting distance. The preview of distance constraint will display; refer to Figure-18.

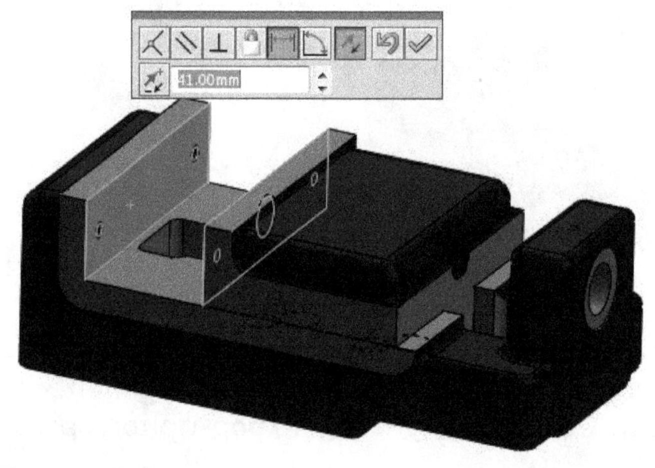

Figure-18. Distance constraints

- Set desired distance in the pop-up toolbar and select the **Flip** button if required.

Angle

The **Angle** button is used to set angle between two selected faces. The steps are given next.

- Click on the **Angle** button and select two flat faces that you want to use for setting angle. The preview of angle constraint will display; refer to Figure-19.

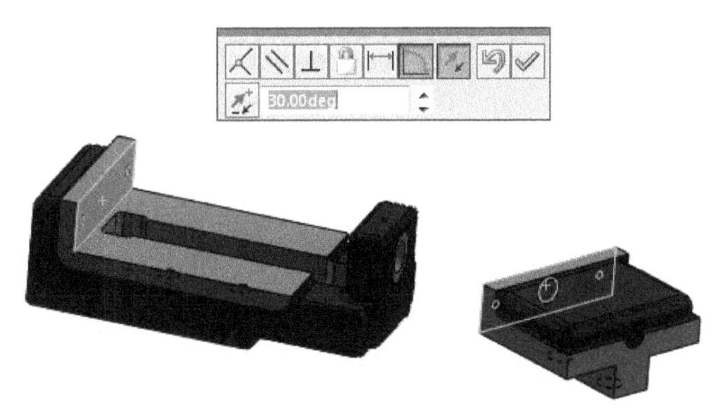

Figure-19. Angle constraint

- Set desired angle in the pop-up toolbar and select the **Flip** button if required.

Note that if you have applied any wrong constraint and want to delete it then click on **+** sign next to **Mates** in the **FeatureManager Design Tree**, select desired constraint from the list and press **DEL** button from keyboard.

Till this point, we have learned **Standard Mates** that are used for rigid assemblies. Now, we will move to advanced constraints (Mates) that play key role in motion study. Note that you might need to delete all the standard mates to apply advanced and mechanical mates for checking motion.

ADVANCED MATES

Expand the **Advanced Mates** rollout from the **FeatureManager** to display the advanced mates; refer to Figure-20. The mates in this rollout are explained next.

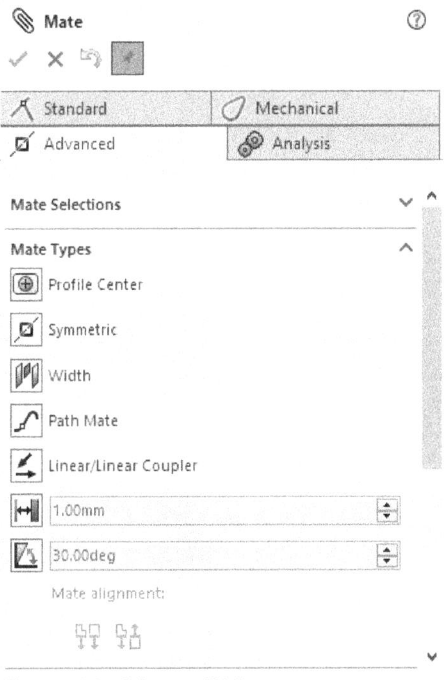

Figure-20. Advanced Mates

Profile Center

The **Profile Center** button is used to align the two components at a common center of the faces selected for mate. The steps to use this mate are given next.

- Click on the **Profile Center** button and select the flat faces of the components. The faces will be aligned at their profile centers; refer to Figure-21.

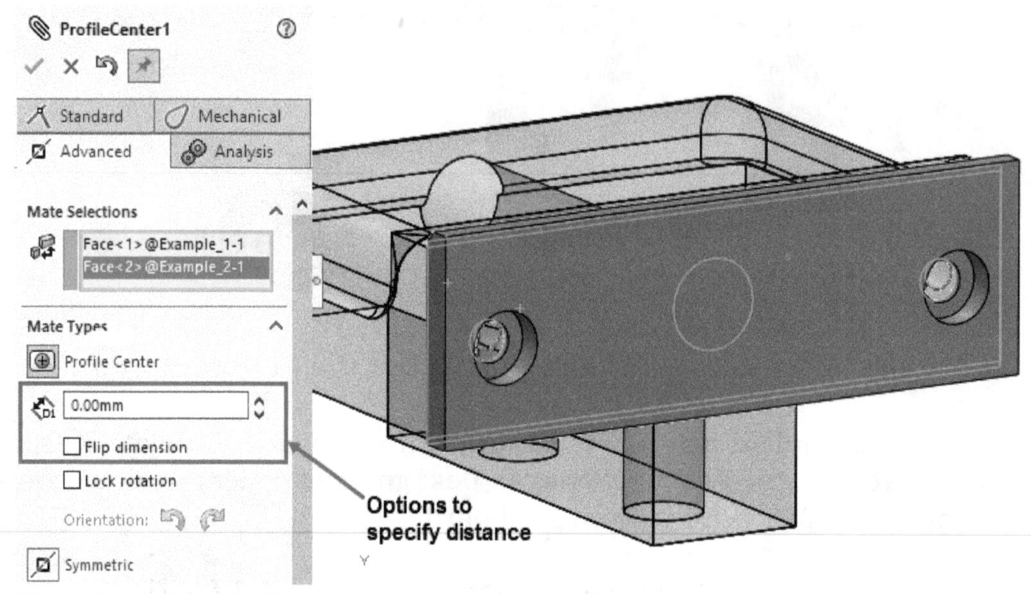

Figure-21. Applying Profile center mate

- Specify desired distance between the faces by using the edit box displayed below the **Profile Center** button in the **PropertyManager**.

Symmetric

The **Symmetric** button is used to make faces of components symmetric with respect to a reference. In other words, the distance by which one component move will be the same as distance moved by other symmetric component. The steps are given next.

- Click on the **Symmetric** button and select two flat faces that you want to make symmetric.
- Click in the **Symmetry plane** selection box in the **Mate Selections** rollout and select the plane about which you want to make the components symmetric. The preview of applied mate will display; refer to Figure-22.

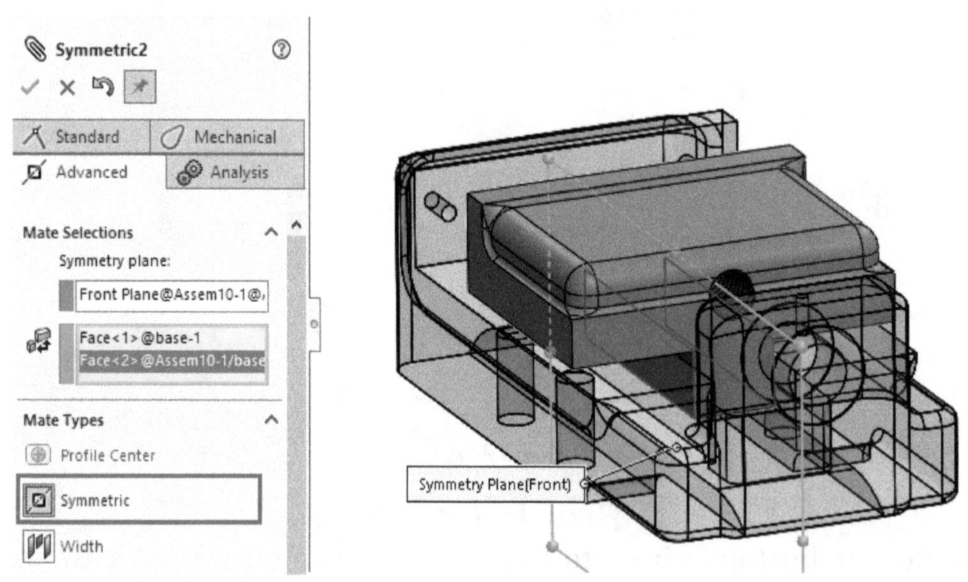

Figure-22. Symmetry constraint

Width

The **Width** button is used to fit a component in the selected width reference. The steps to use this mate are given next.

- Click on the **Width** button and select two flat faces in the assembly that define the width reference.
- Now, select the two flat faces of your components to define total width of your component. The preview of mate will display; refer to Figure-23.

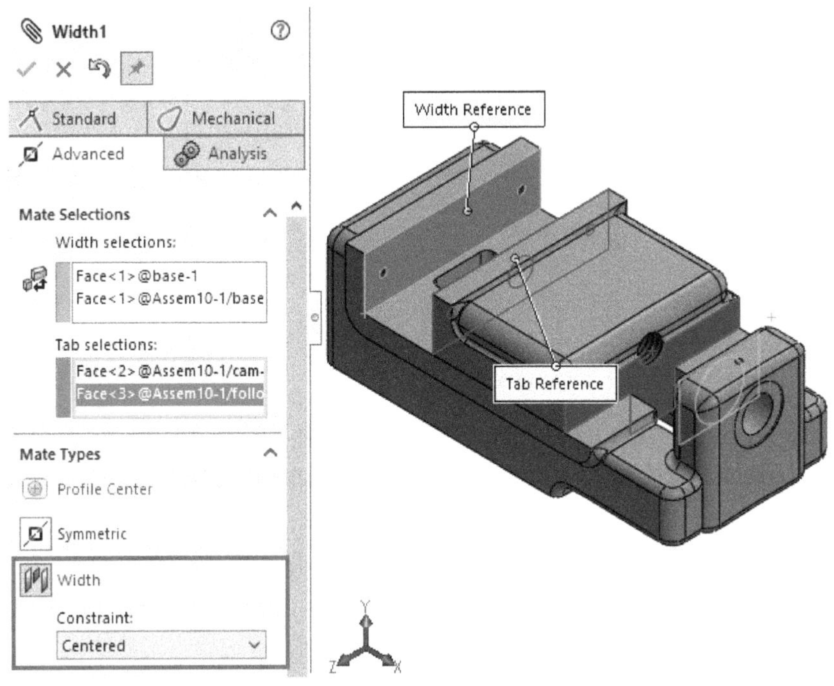

Figure-23. Width mate

Note that if the width of references is more than the limits of component then the component will be inserted in the middle of the width references.

Path Mate

The **Path Mate** button is used to make pointed component following the specified path. The steps to use this mate are given next.

- Click on the **Path Mate** button and select the vertex of the component that you want to make follower.
- Select sketch of the path that you want to make as guide for the follower. The preview of path mate will display; refer to Figure-24.

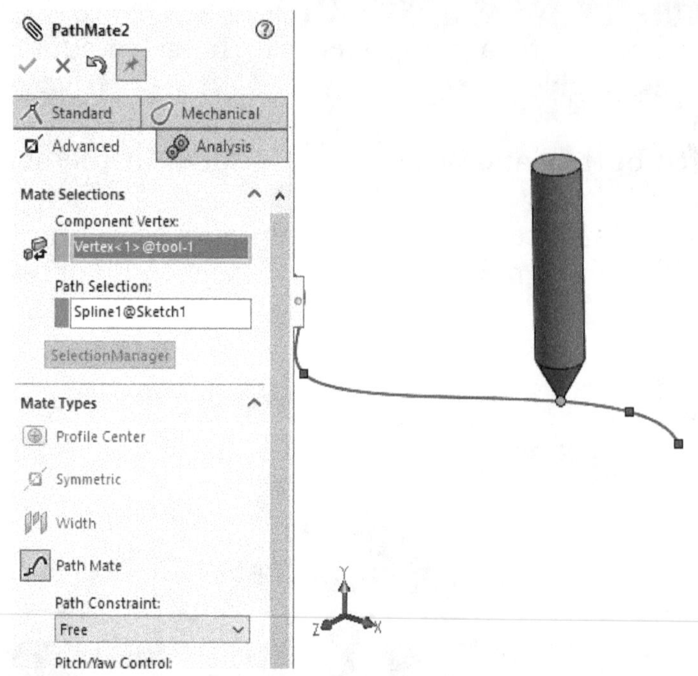

Figure-24. Path mate constraint

- You can change the options related to movement of object by using the drop-down and edit boxes displayed below **Path Mate** in **PropertyManager**; refer to Figure-25.

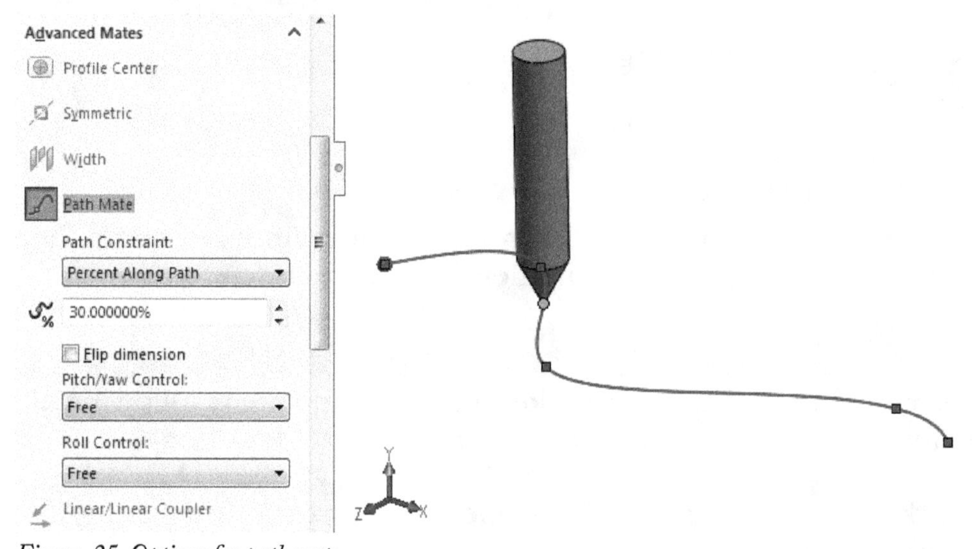

Figure-25. Options for path mate

- If you want to move the component at certain distance or path length percentage then select the respective option from the **Path Constraint** drop-down in the **PropertyManager** and specify the related value; refer to Figure-26.

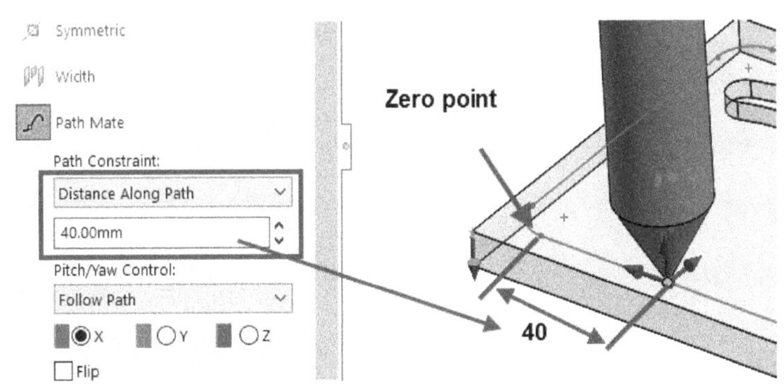

Figure-26. Using Path Constraint drop-down

- By default, the vertex in path mate follows the path but the part is free to rotate about the vertex. To fix the orientation of part, select the **Up Vector** option from the **Roll Control** drop-down in the **PropertyManager**. The options for up vector will be displayed as shown in Figure-27.

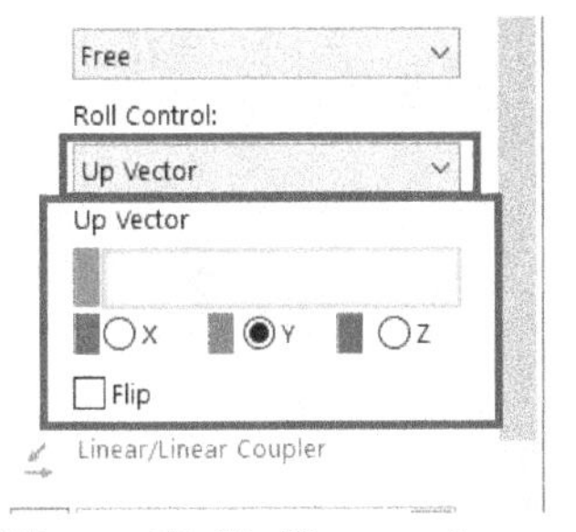

Figure-27. Up Vector options

- Select an edge/plane/axis from the assembly to reference up vector. Select desired direction radio button from the **PropertyManager**. The follower will be aligned accordingly; refer to Figure-28.

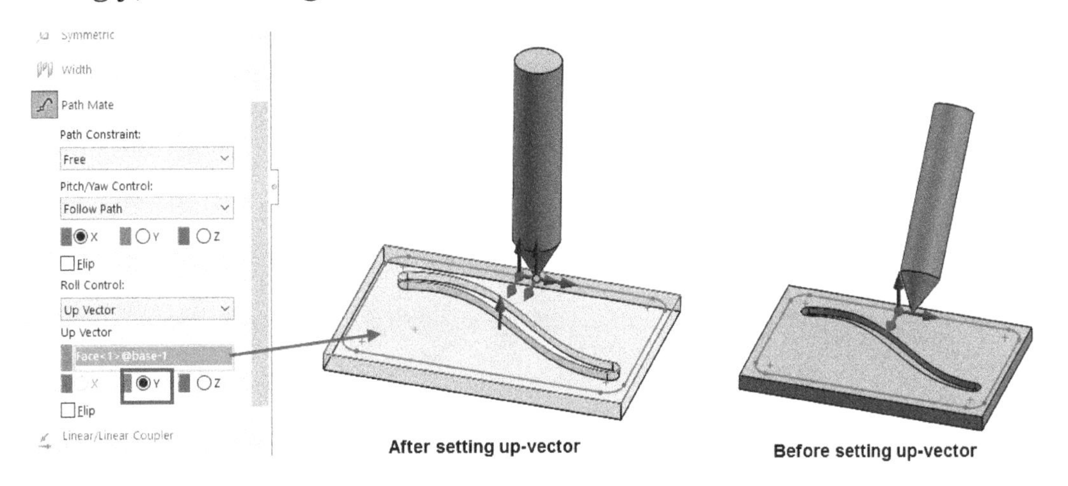

Figure-28. Setting Up Vector for follower part

Linear/Linear Coupler

The **Linear/Linear Coupler** button is used to make two components move with respect to each other by a specified ratio. The steps to use this mate are given next.

- Click on the **Linear/Linear Coupler** button and select the faces of the component between which you want to apply the mate. Preview of mate will display; refer to Figure-29.

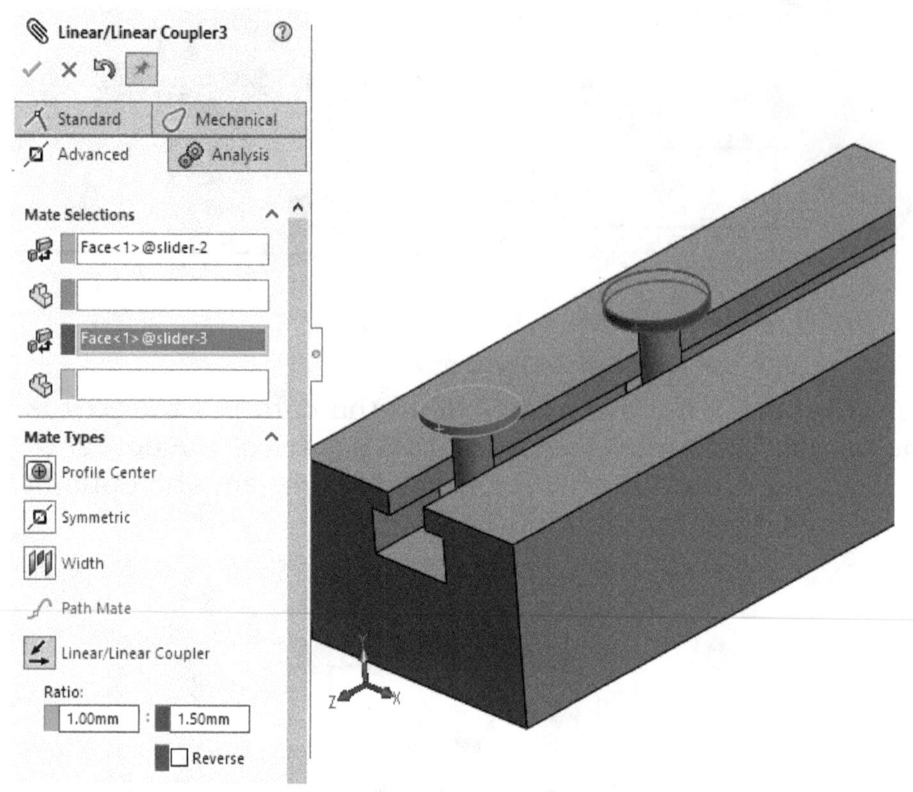

Figure-29. Linear coupler mate

- Set desired ratio between the components. Click **OK** to apply the mate.
- To check the motion, drag one of the component in linear direction. The other component will move automatically.

Advanced Distance

The **Advanced Distance** button is used to apply maximum and minimum movement limit of a component. The steps to apply this mate are given next.

- Click on the **Distance** button from the **Advanced Mates** tab. The options in the rollout will display as shown in Figure-30.
- Select the face/plane of component and then select the face/plane of reference to limit the movement. Preview of mate will display as shown in Figure-31.

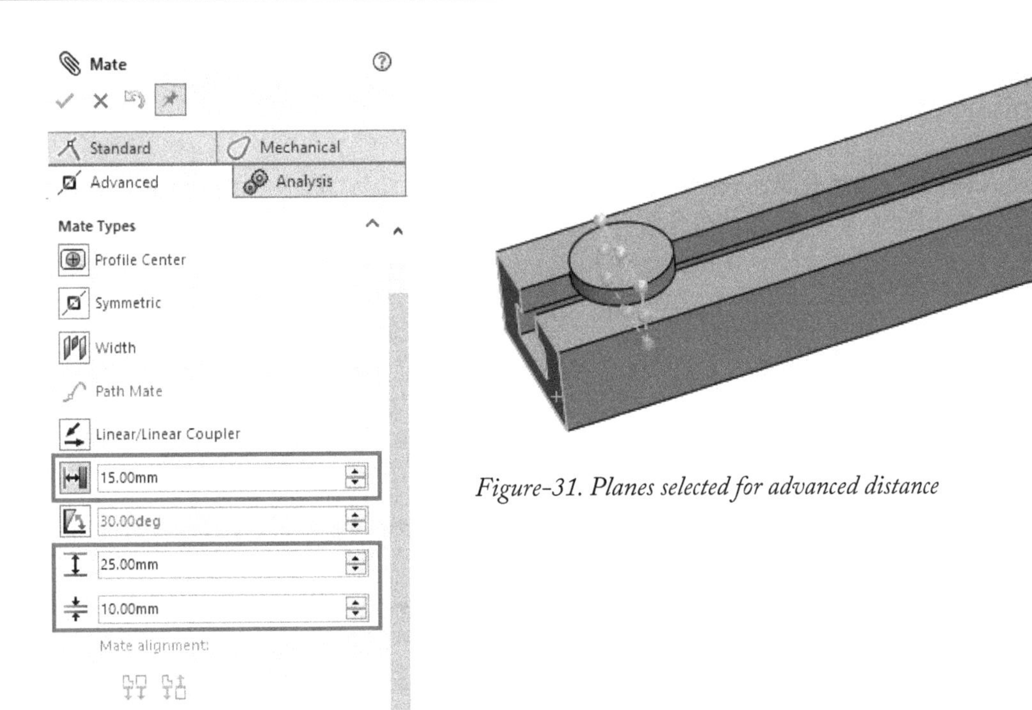

Figure-31. Planes selected for advanced distance

Figure-30. Advanced distance mate options

Advanced Angle

The **Advanced Angle** button is used to apply maximum and minimum rotation limit of a component. This mate works in the way similar to **Advanced Distance** mate.

Till this point, we have learned Standard Mates and Advanced Mates. There are also a few mates that represent mechanical motion in assembly. These constraints are grouped in a tab named **Mechanical Mates**. To use these mates, click on the **Mechanical Mates** tab; refer to Figure-32. The buttons in this tab are explained next.

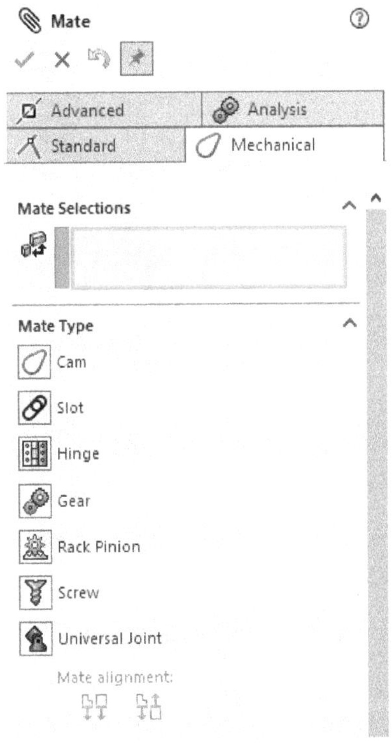

Figure-32. Mechanical Mates

Cam

The **Cam** button is used to create cam-follower mate between two entities. The steps given next explain the procedure.

- Click on the **Cam** button from the **Mechanical Mates** tab. The selection boxes will be displayed as shown in Figure-33.

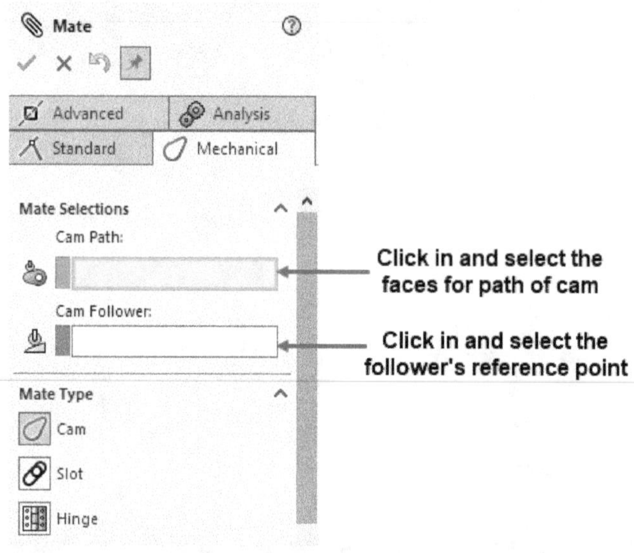

Figure-33. Cam_Mate

- Follow the steps in the above figure. Figure-34 shows an example of cam mate.

Note that you need to limit the motion of follower so that it move only in vertical direction.

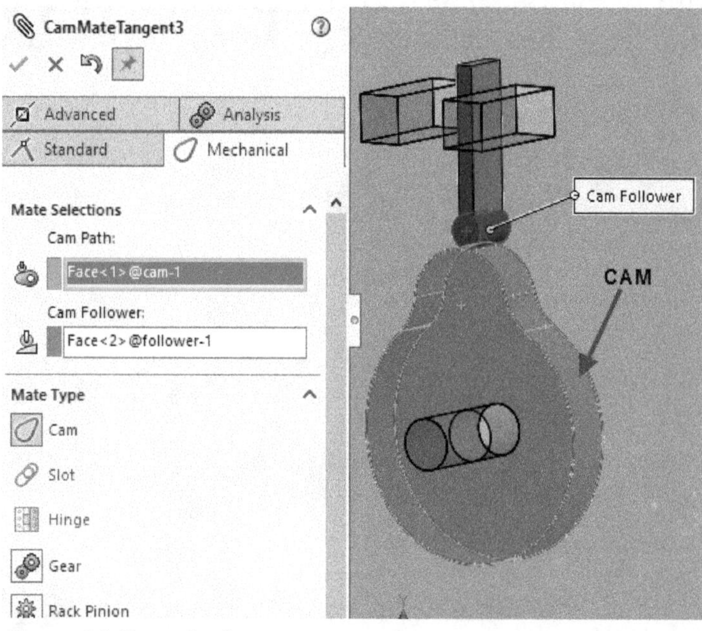

Figure-34. Example of cam mate

Slot

The **Slot** button is used to create slot-follower mate between two entities. After applying this mate, the component will move only in the limits of slot. The steps given next explain the procedure.

- Click on the **Slot** button from the **Mechanical Mates** tab.
- Select the round face of slot and the follower, the mate will be applied.
- You can set the starting position of follower by using the options in the **Constraint** drop-down displayed below the **Slot** mate in **PropertyManager**. Figure-35 shows an example of slot mate.

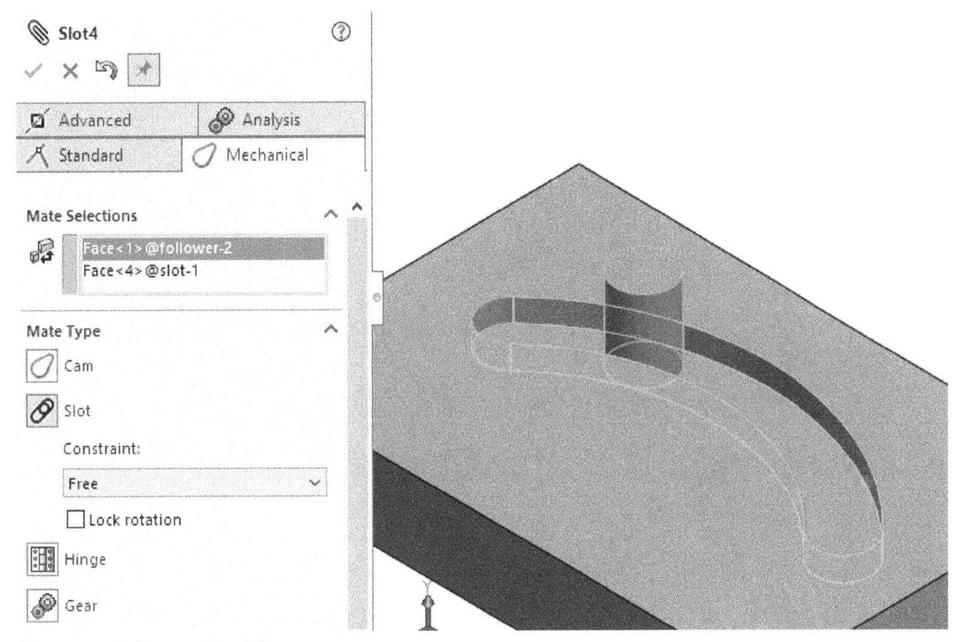

Figure-35. Example of slot mate

Hinge

The **Hinge** button is used to make the two parts behave as hinged. The procedure to use this button is given next.

- Click on the **Hinge** button from the **Mechanical Mates** tab. The selection boxes in the **Mate Selections** rollout are displayed as shown in Figure-36.

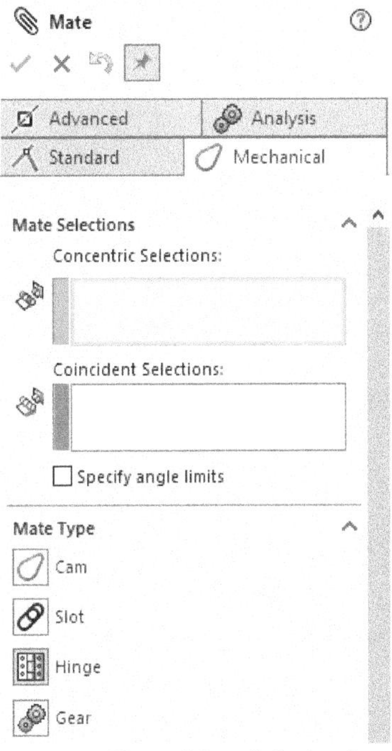

Figure-36. Hinge Mate Selection boxes

- Select the two round faces that you want to be concentric.
- Select the two flat faces that define the angular motion limit of the hinge.
- Select the **Specify angle limits** check box and specify the limits as explained in advanced mates. Figure-37 shows a hinge mate being applied to the assembly.

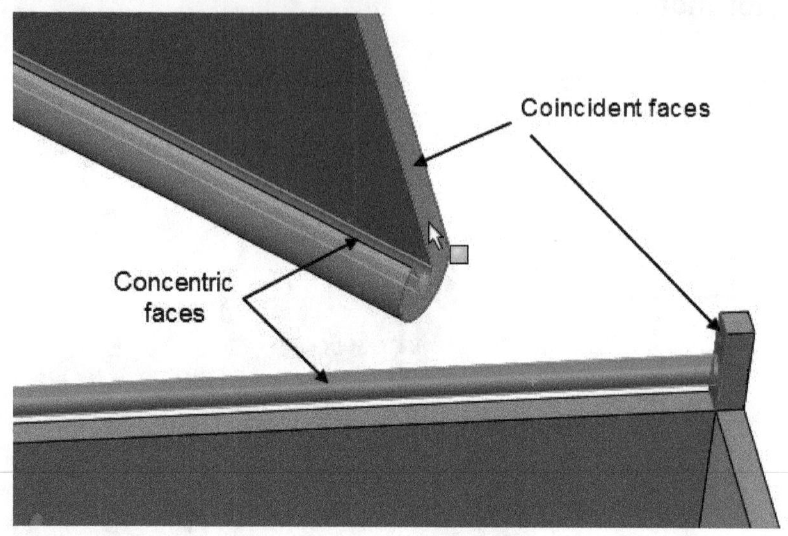

Figure-37. Example of hinge mate

Gear

The **Gear** button is used to create a joint between two gears. The procedure to use this button is given next.

- Click on the **Gear** button from the **Mechanical Mates** tab.
- Select the flat faces of the gears. The **Gear Mate PropertyManager** will display as shown in Figure-38. Also, the gears will display as shown in Figure-39.

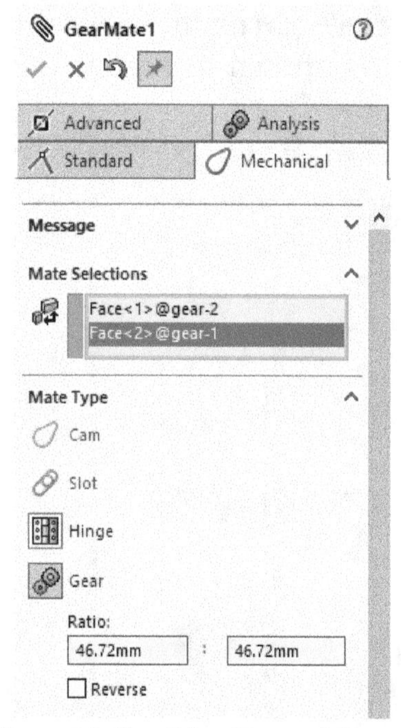

Figure-38. Gear Mate PropertyManager

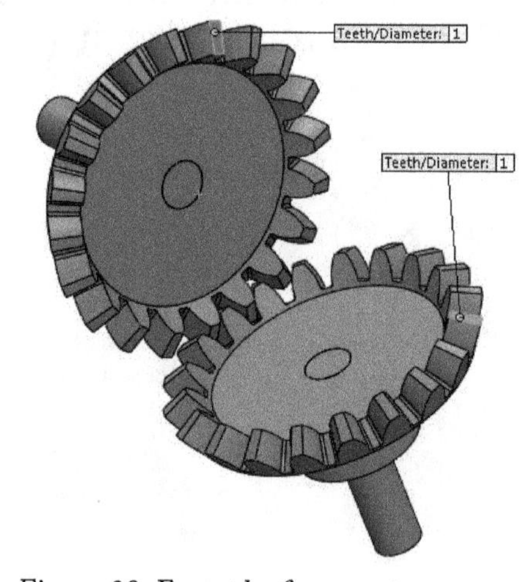

Figure-39. Example of gear mate

Rack Pinion

The **Rack Pinion** button is used to create joint between a rack and a pinion. The procedure to use this button is given next.

- Click on the **Rack Pinion** button from the **Mechanical Mates** tab.
- Select edge of the rack and flat face of the pinion gear. The rack and pinion mate will be created; refer to Figure-40. Specify desired ratio in the **PropertyManager**.

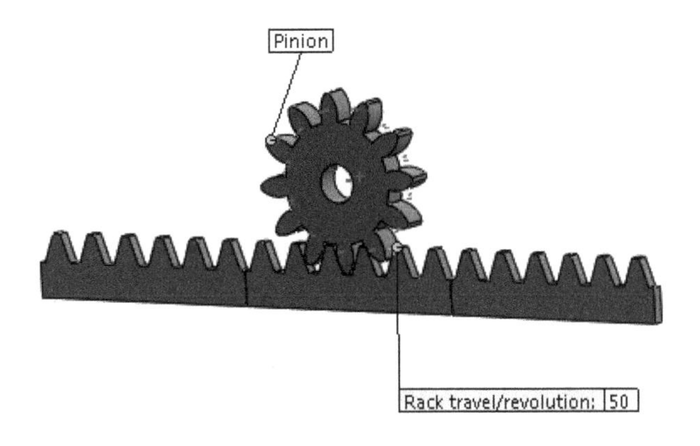

Figure-40. Rack pinion mate

Screw

The **Screw** button is used to create screw joint between a cylinder and a flat face. The procedure to use this button is given next.

- Click on the **Screw** button from the **Mechanical Mates** tab.
- Select the cylindrical face of first component and flat face of other component. The preview will be displayed; refer to Figure-41.

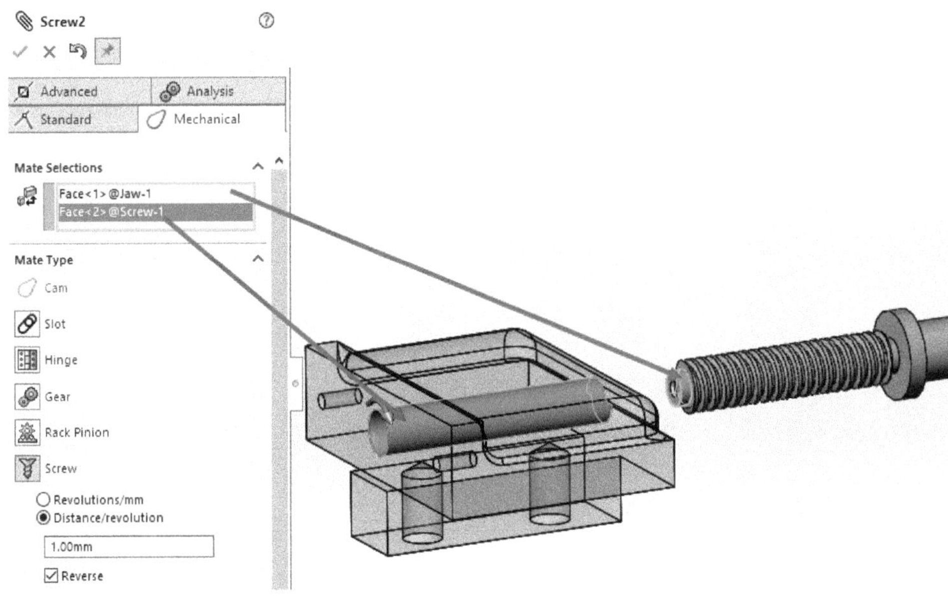

Figure-41. Screw Mate

- Specify desired value in the box displayed below **Screw** button in the **PropertyManager**.

Universal Joint

The **Universal Joint** button is used to create universal joint between two components. After clicking this button, select the two round faces. The universal joint will be created; refer to Figure-42.

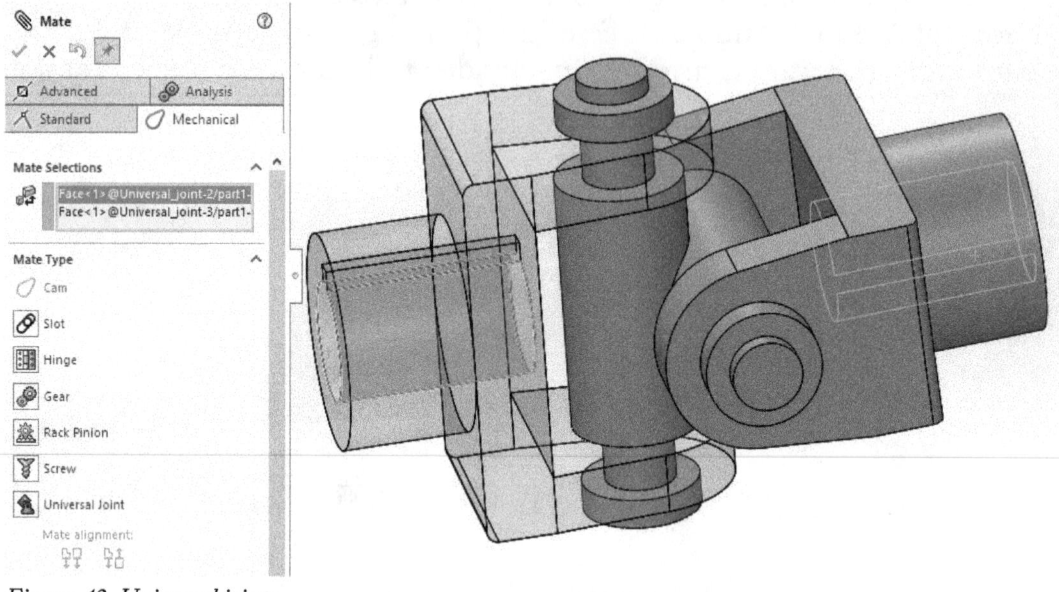

Figure-42. Universal joint

EXPLODED VIEW

After creating assembly, we are required to display the components of assembly expanded in the way they are assembled. Follow the steps given below to create the exploded view of assembly.

- Select all the components that you want to explode from the main assembly and click on the **Exploded View** tool from the **Ribbon**. The **Explode PropertyManager** will be displayed; refer to Figure-43.

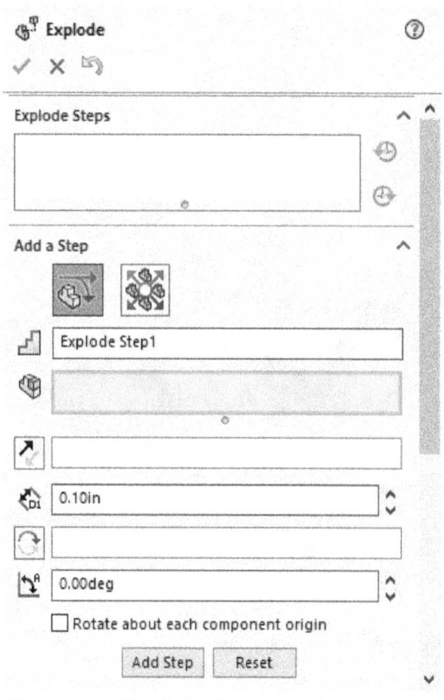

Figure-43. Explode_PropertyManager

- Select the **Auto-space components** check box from the **Options** rollout in the **Explode PropertyManager** and drag the components in desired direction using the arrow handles displayed. Using the slider below the **Auto-space components** check box, you can set distance between components of assemblies.
- Click on the **OK** button from the **PropertyManager**. The components will be exploded automatically.

Or

- Click on the **Exploded View** button and select the component you want to move. A triad will display on the component.
- Drag the component using the handles displayed to desired distance or specify desired distance and rotation in the edit boxes available in the **Settings** rollout of the **PropertyManager**.
- Repeat the above steps until you explode all the components.
- Click on the **OK** button from the **PropertyManager**. The components will be exploded.

Explode Lines

To display the explode lines, click on the **Explode Line Sketch** button from the **Exploded View** drop-down in **Ribbon**. The **Route Line PropertyManager** will display; refer to Figure-44.

Figure-44. Route_Line_PropertyManager

- Click on the assembly reference of the component and then the corresponding assembly reference of the base component. The explode line will be created; refer to Figure-45.

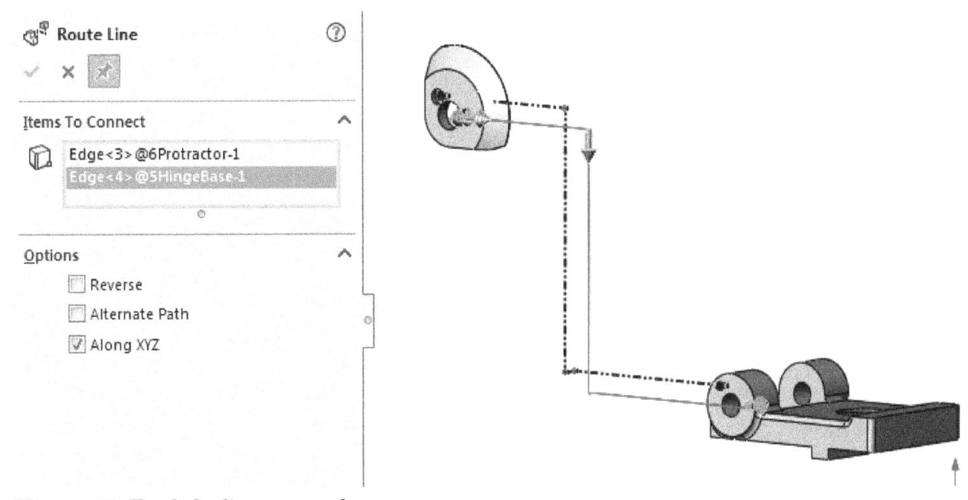

Figure-45. Explode_lines_created

Similarly, you can use the **Insert/Edit Smart Explode Lines** tool from the **Exploded View** drop-down in the **Ribbon**. Note that explode lines are created automatically by using this tool.

BILL OF MATERIALS

Bill of materials is used to list the total components of the assembly in the form of a table. To create bill of materials, follow the steps given below.

- Click on the **Bill of Materials** button from the **Ribbon**. The **Bill of Materials PropertyManager** will display; refer to Figure-46.

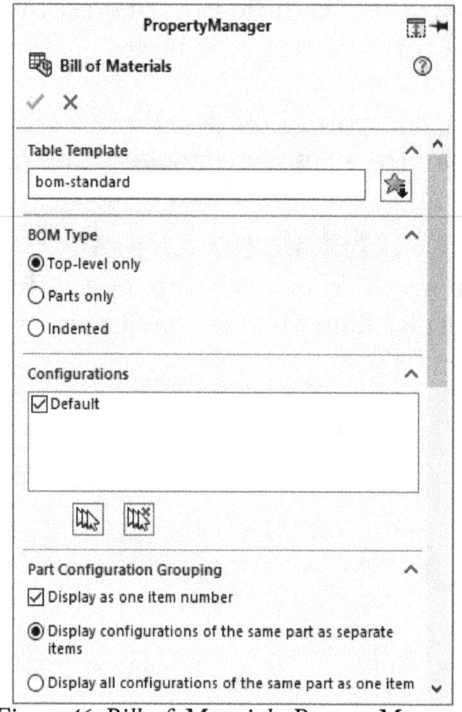

Figure-46. Bill_of_Materials_PropertyManager

- Click on the button to display templates. The **Open** dialog box will be displayed with default templates of bill of materials; refer to Figure-47.

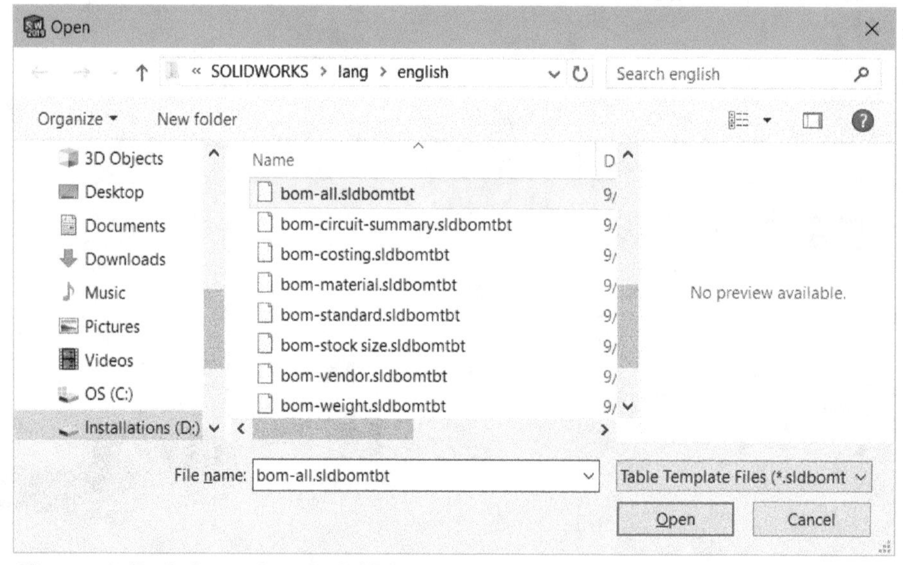

Figure-47. Default templates for BOM

- Select desired template from the dialog box and click on the **Open** button. The template will be activated.
- Set desired parameters in the **PropertyManager** like BOM type, configurations, and so on.
- Click on the **OK** button from the **PropertyManager**. The **Select Annotation View** dialog box will be displayed and you will be asked to place the table of bill of materials.
- Select desired radio button and click **OK** button from the **Select Annotation View** dialog box.
- Click in the viewport to place the table; refer to Figure-48.

ITEM NO.	PART NUMBER	DESCRIPTION	WEIGHT	QTY.
1	Body			1
2	Washer			1
3	Valve spindle			1
4	Spring			1
5	Spring cap nut			1
6	Pre adjusting nut			1
7	Nozzle valve			1
8	Nozzle body			1
9	Cap nut			2
10	atomiser			1

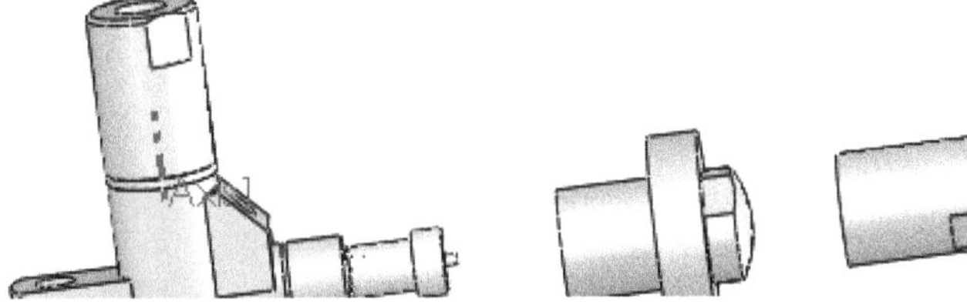

Figure-48. Bill of materials

MATE CONTROLLER

The **Mate Controller** tool was added in SolidWorks 2016. Using this tool, you can manage the distance and angle mates applied in the assembly. The procedure to use this tool is given next.

- Click on the **Mate Controller** tool from the **Insert** menu; refer to Figure-49. The **Mate Controller PropertyManager** will be displayed; refer to Figure-50.

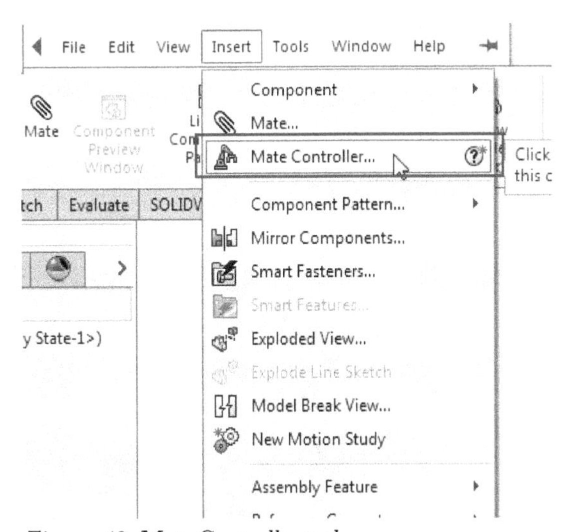

Figure-49. Mate Controller tool

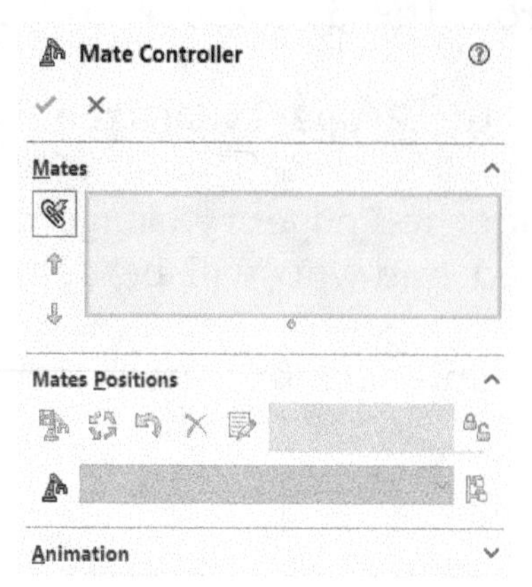

Figure-50. Mate Controller PropertyManager

- Click on the **Collect All Supported Mates** button from the **Mates** rollout of the **PropertyManager**. The mates that can be controlled with their angular or distance dimension will be displayed in the selection box; refer to Figure-51.

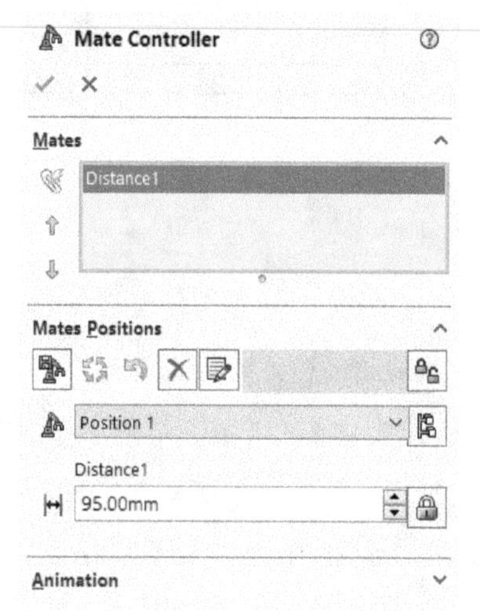

Figure-51. Mates selected automatically

- Set desired dimension spinners available in the **Mates Positions** rollout.
- You can save the positions by using the **Add Position** button from the **Mates Positions** rollout. On doing so, the **Name Position** dialog box will be displayed; refer to Figure-52.

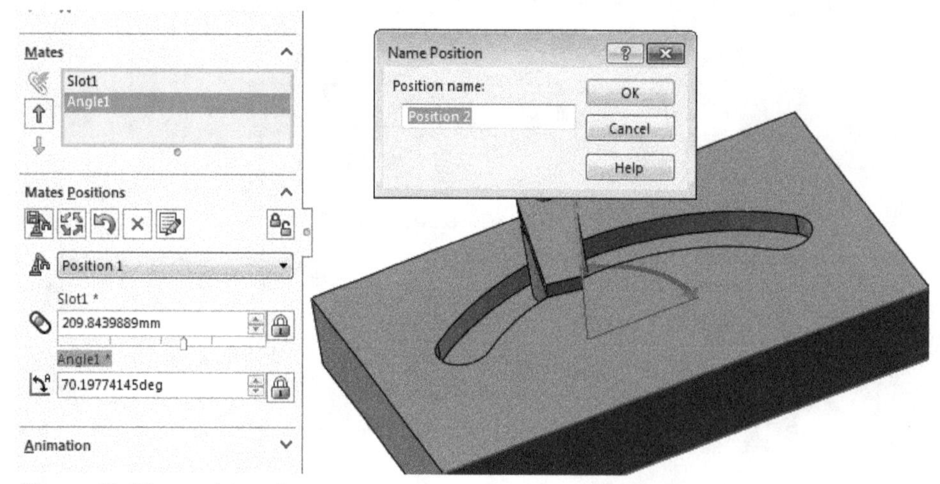

Figure-52. New position of mates

- Specify desired name and click on the **OK** button from the dialog box.
- Click on the **Add Configuration** button to make a new configuration of assembly at current position.
- You can select desired position from the drop-down in the **Mates Positions** rollout; refer to Figure-53.

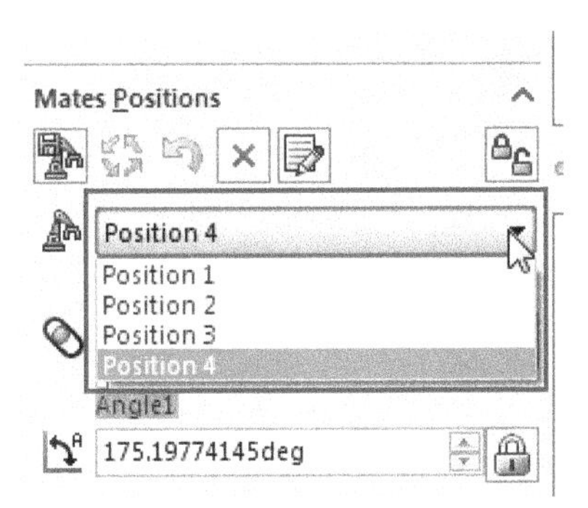

Figure-53. Positions rollout

- To animate the position change, expand the **Animation** rollout in the **PropertyManager** and click on the **Calculation Animation** button from it; refer to Figure-54.

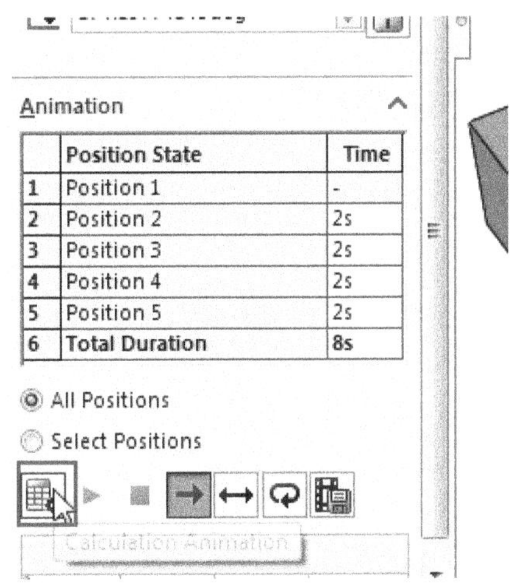

Figure-54. Animation rollout

- After calculation, you can use the **Play** button from the **Animation** rollout to play the animation.
- Click on the **OK** button from the **PropertyManager** to apply the mate control. The Mate Controller feature will be added in the **FeatureManager Design Tree**.

MOTION STUDY

Motion study is used to check the motion of components with respect to each other after applying the driving force the components. This feature of SolidWorks help to understand the mechanism of our assembly in real world conditions. To start the motion study, click on the **Motion Study 'x'** tab displayed at the bottom bar of the viewport. Note that here **'x'** is the sequence number. After clicking on this tab, the interface is displayed as shown in Figure-55.

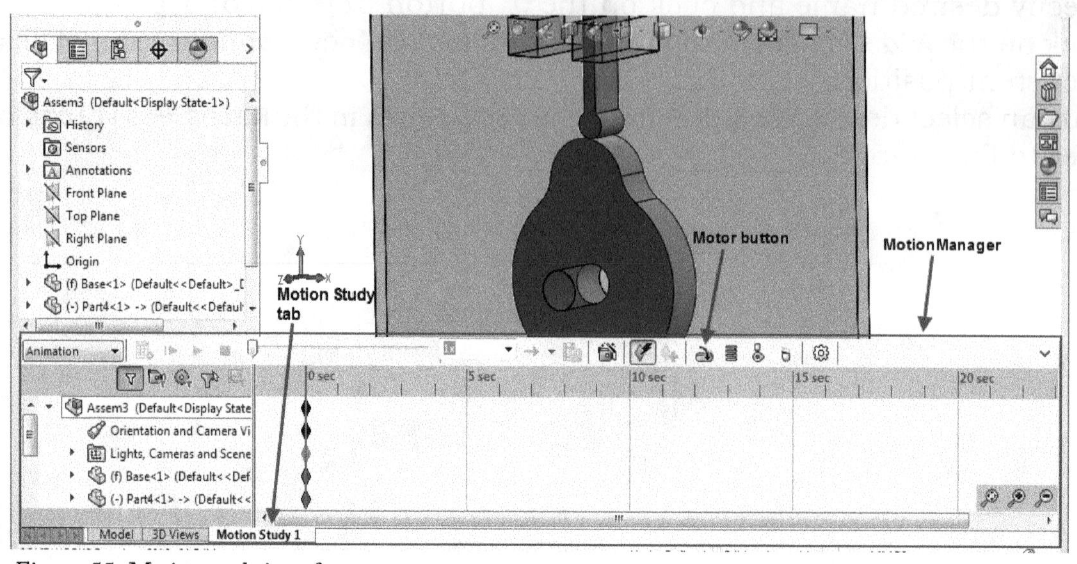

Figure-55. Motion study interface

Now, we need to apply some driving force to one component so that we can check the motion of other parts. In this case, we are using the cam-follower mechanism as displayed above. We need to add the rotary motion to the cam to check motion of follower. The steps to do so are given next.

- Click on the **Motor** button shown in the above figure. The **Motor PropertyManager** will display; refer to Figure-56.

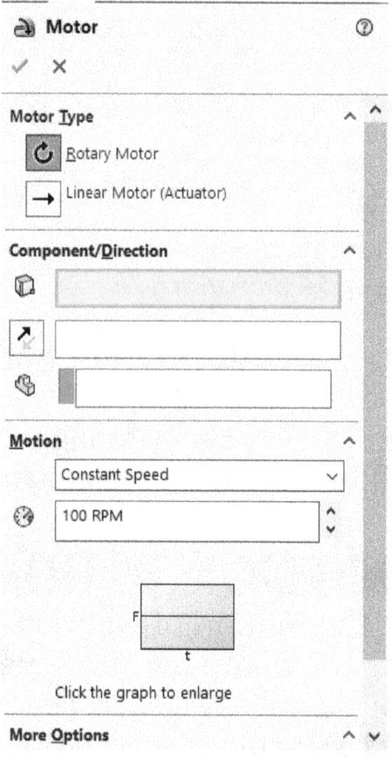

Figure-56. Motor PropertyManager

- Click on the circular face of the part to apply the rotary motion; refer to Figure-57.

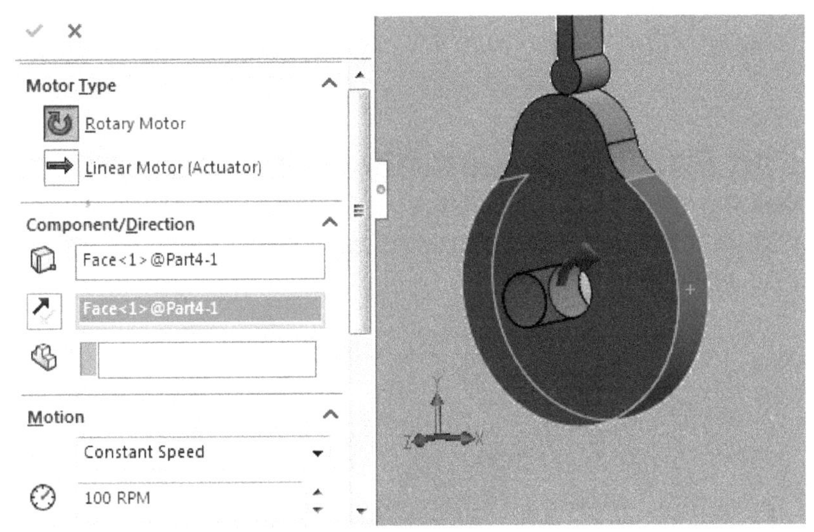

Figure-57. Circular face selected for rotary motion

- Specify desired options for speed in the **Motion** rollout and click **OK** to apply the motion. The motor will be added in the **MotionManager**.

Note that in the same way you can apply linear motion to a flat face by selecting **Linear Motor (Actuator)** button from the **Motor PropertyManager** in place of **Rotary Motor** button.

If you want to start a new motion study, then click on the **New Motion Study** tool from the **Assembly** tab in the **Ribbon**. A new motion study tab will be added in the **MotionManager**.

Playing Motion Study

To animate the motion study, click on the **Calculate** ( ) button from the **MotionManager** tool bar. The animation of mechanism will display in the viewport. To increase the time of animation, drag the key point of assembly to desired point; refer to Figure-58. Click on the **Calculate** button again to run the animation.

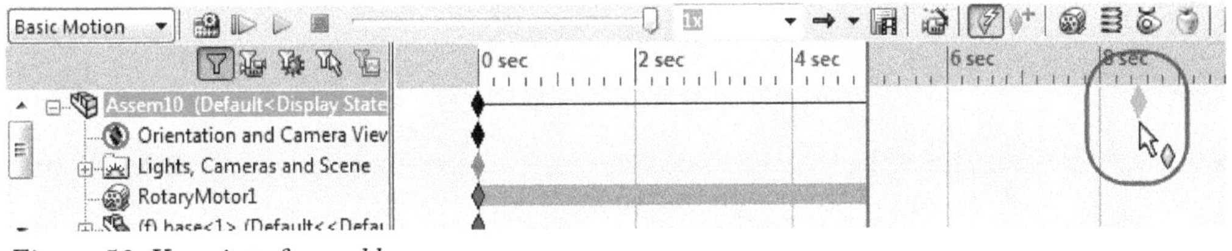

Figure-58. Keypoint of assembly

You can save the motion in movie format by selecting the **Save** ( )button from the **MotionManager** tool bar.

BOTTOM UP APPROACH AND TOP DOWN APPROACH

There are two ways of creating assembly in most of the CAD packages; Bottom Up approach and Top Down approach. In Bottom Up approach, all the components are created separately in Part environment and then assembled in Assembly environment by inserting them one by one.

In Top Down approach, we create all the parts in assembly environment and apply the mates on the spot. The procedure to create parts in assembly environment is given next.

Creating Parts in Assembly (Top Down Approach)

- Click on the **New Part** tool from the **Insert Components** drop-down in the **Assembly** tab of **Ribbon**; refer to Figure-59. You are asked to select a face or plane on which the part is to be positioned.

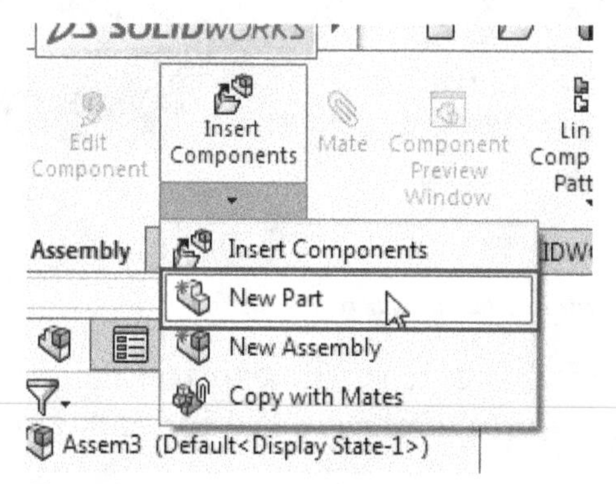

Figure-59. New Part tool

- Select desired plane from the **FeatureManager Design Tree**. The sketching environment will be displayed; refer to Figure-60.

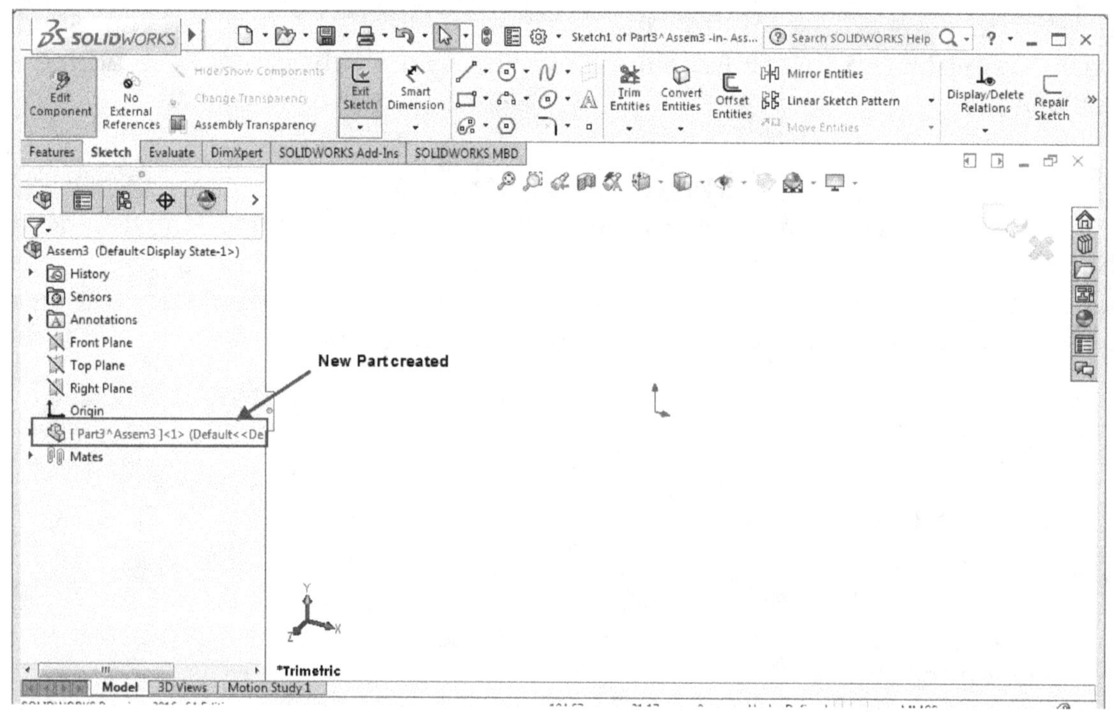

Figure-60. Sketching environment for assembly part

- Create desired sketch and perform operations using the tools in **Features** tab in the **Ribbon**; refer to Figure-61.

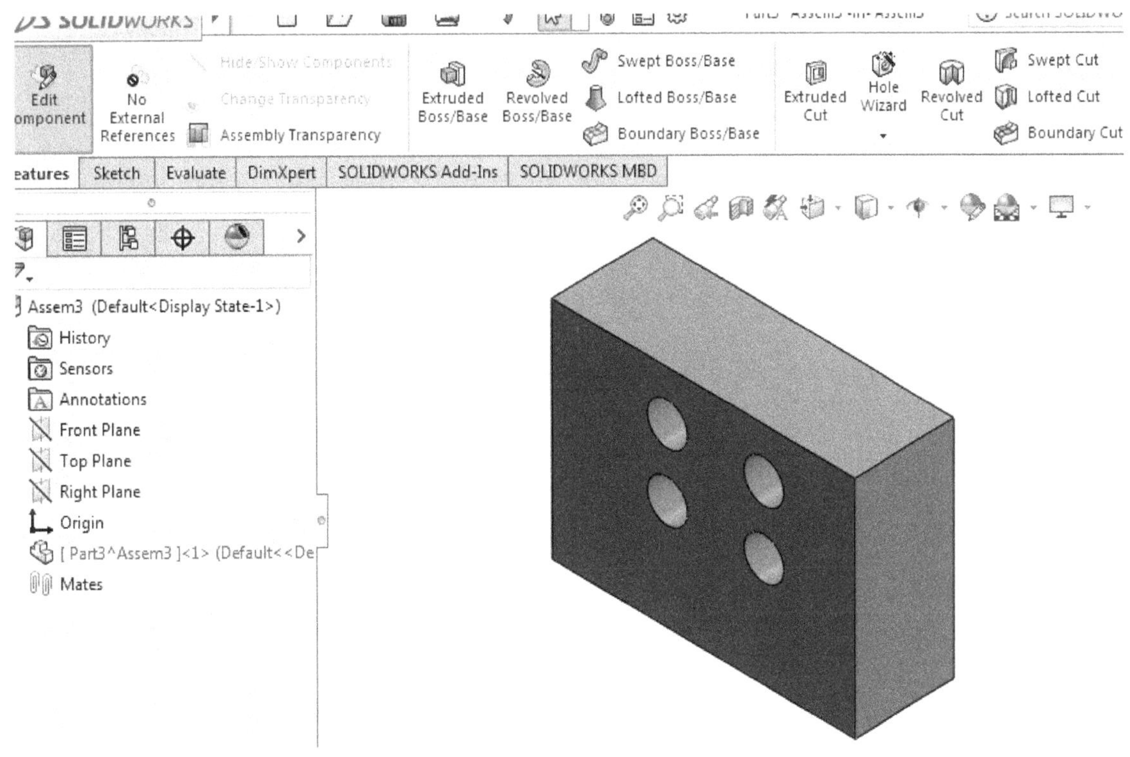

Figure-61. Part created

- Once you have created all the features of part, click on the **Exit editing component** button at the top-right in the viewport; refer to Figure-62.

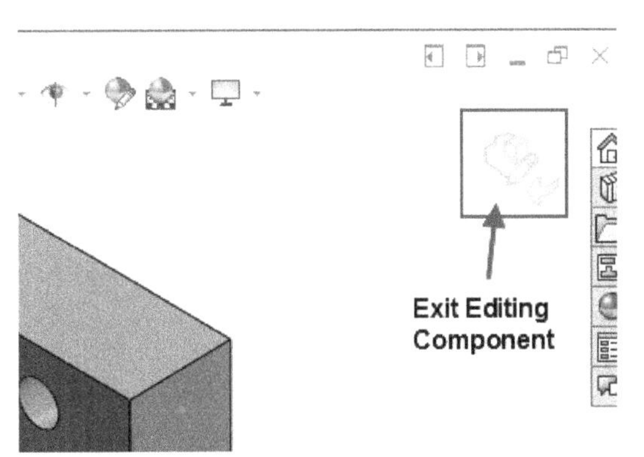

Figure-62. Exit Editing Component

- Similarly, you can create other components in assembly and apply the mates as discussed earlier.

Till this point, we have used various assembly tools. In the next chapter, we will apply these tools to perform practical and work out with practice questions.

SELF ASSESSMENT

Q1. Assembly is the combination of two or more components in any manner. (T/F)

Q2. Assembly Design is a separate environment in SolidWorks to perform operations related to assembly of components. (T/F)

Q3. The first component inserted in assembly is fixed by default. (T/F)

Q4. You can reinsert a component by holding the **CTRL** key from keyboard while dragging desired component. (T/F)

Q5. You can make the first component transparent while applying mates in SolidWorks. (T/F)

Q6. Which of the following tool can be used to manage positions of various distance and angle constraints in assembly?

a. Edit Component
b. Mate
c. Mate Controller
d. Explode View

Q7. You can insert an already created assembly in the new assembly file. (T/F)

Answer to Self-Assessment:
1. F, **2.** T, **3.** T, **4.** T, **5.** T, **6.** c, **7.** T

Chapter 8

Advanced Assembly Practical and Practice

Topics Covered

The major topics covered in this chapter are:

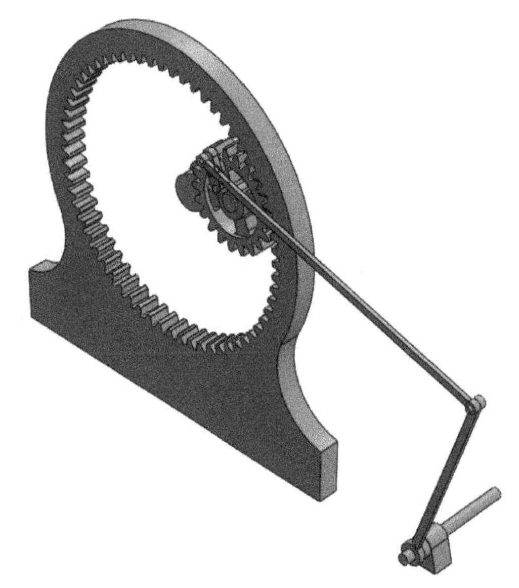

- ***Hole Series, Weld Bead, Chain/Belts***
- ***Smart Fasteners***
- ***Toolbox Utilities***
- ***Magnetic Mate***
- ***Asset Publisher***
- ***Configurations***
- ***Assembly Practical 1***
- ***Assembly Practical 2***
- ***Assembly and Motion Practical 3***
- ***Practice Exercises.***

Note: Before starting this chapter, mail us at cadcamcaeworks@gmail.com to get the part files required to complete this chapter.

INTRODUCTION

In this chapter, you will learn about some advanced tools related to assemblies. You will also learn about handling large assemblies in SolidWorks. After that, you will work on practical and practice files.

ASSEMBLY EDITING TOOLS

Most of the tools that are used for editing assembly are similar to the tools discussed for Solid model editing. The editing tools are available in the **Assembly Features** drop-down of the **Assembly CommandManager** in the **Ribbon**; refer to Figure-1. The editing tools which were not discussed earlier are discussed next.

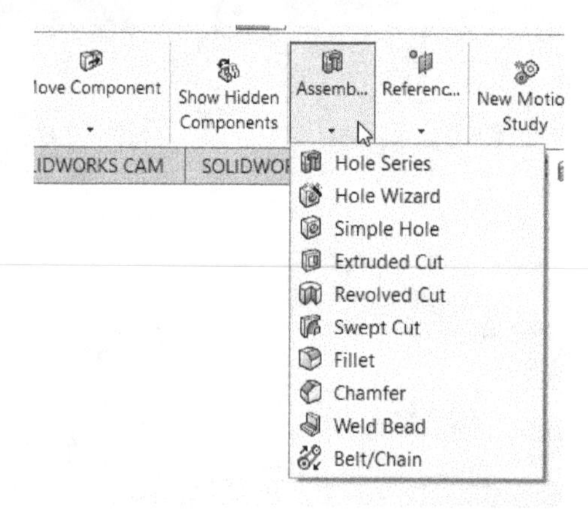

Figure-1. Assembly_Features_drop-down

Creating Hole Series

The **Hole Series** tool is used to create a series of holes in the components assembled together. If you want to create single hole passing through many components in the assembly then this is the tool you should use. The procedure to use this tool is given next.

- Click on the **Hole Series** tool from the **Assembly Features** drop-down in the **Assembly CommandManager**. The **Hole Position PropertyManager** will be displayed; refer to Figure-2. Also, you will be asked to select a location for placing the new hole.

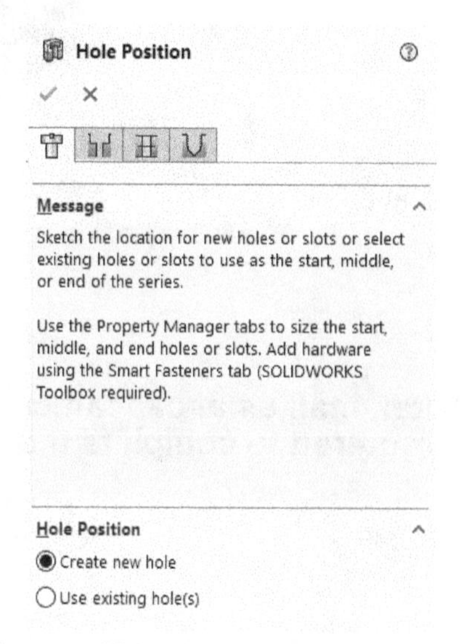

Figure-2. Hole_Position_PropertyManager

- Click at desired location on the part. Preview of the hole will be displayed; refer to Figure-3. Click at more positions to place more holes.

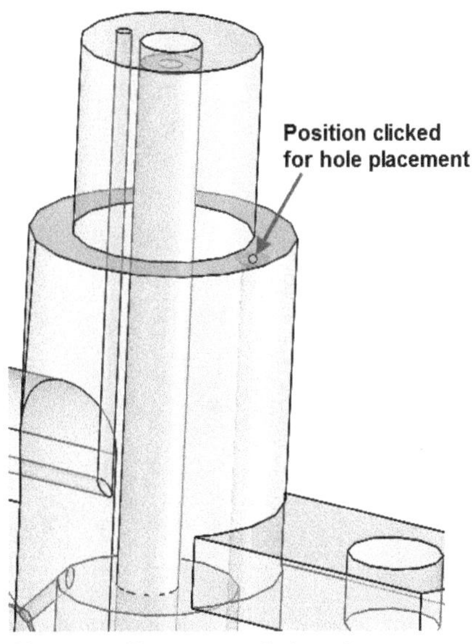

Figure-3. Positioning hole series

- Every hole series has three sections; start hole, middle hole, and end hole. These sections can be edited individually. Click on the **First Part** tab in the **PropertyManager** to display options related to start hole; refer to Figure-4. Set the options as required in the **PropertyManager**.
- Similarly, click on the **Middle Parts** tab and **Last Part** tab to modify the middle and last holes in the hole series. The **PropertyManager** on selecting these tabs are shown in Figure-5.

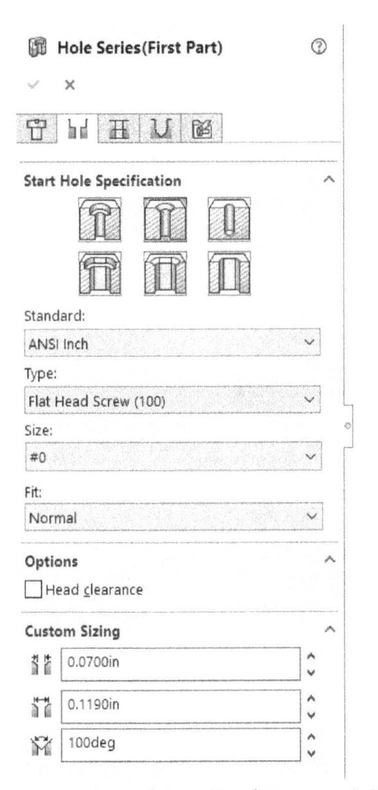

Figure-4. Hole Series(First Part) PropertyManager

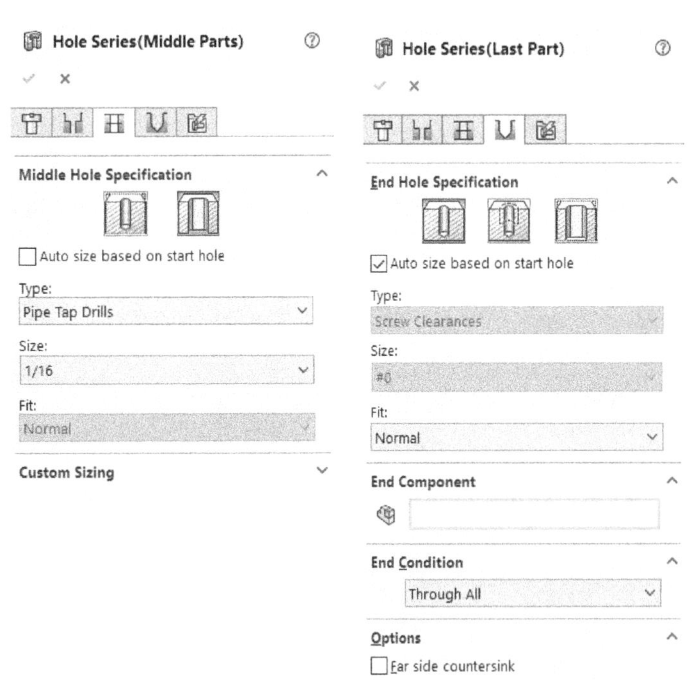

Figure-5. Hole Series PropertyManager for Middle Parts and Last Part tab

- After specifying desired parameters, click on the **OK** button from the **PropertyManager** to create the hole series.

Creating Weld Bead in Assembly

- Click on the **Weld Bead** tool from the **Assembly Features** drop-down in the **Assembly CommandManager**. The **Weld Bead PropertyManager** will be displayed; refer to Figure-6.

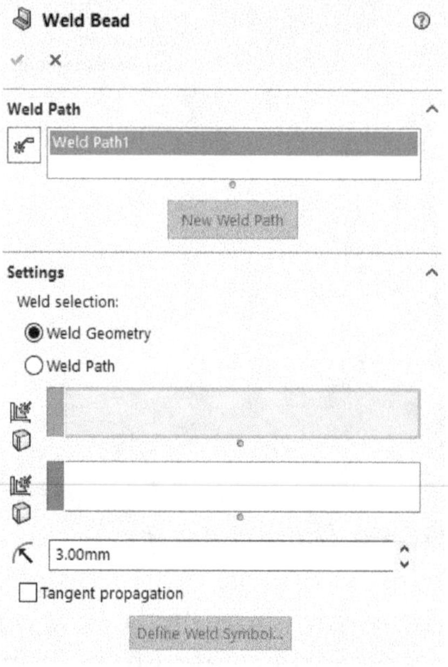

Figure-6. Weld Bead PropertyManager

- Select first reference face/edge for welding.
- Click in the second selection box in the **Settings** rollout of the **PropertyManager** and select the second face/edge. Preview of the weld bead will be displayed; refer to Figure-7.

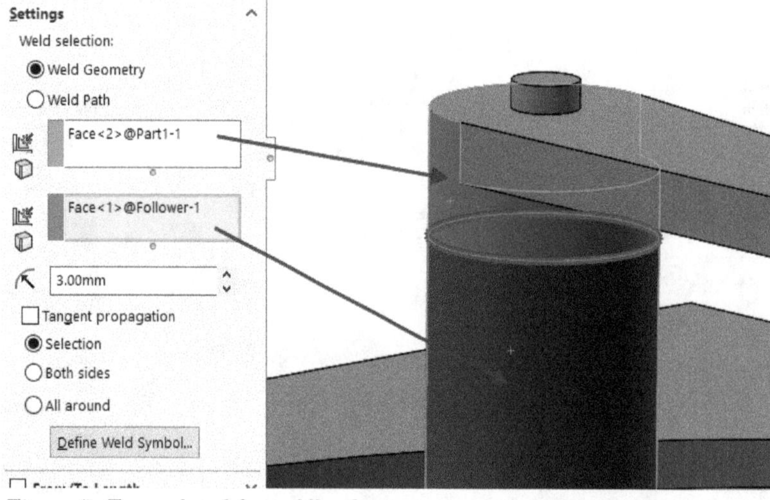

Figure-7. Faces selected for weldbead

- If you want to select the weld path in place or references then select the **Weld Path** radio button from the **Settings** rollout of the **PropertyManager** and select the path from drawing area.
- Set the other parameters of weld bead as required.
- Click on the **Define Weld Symbol** from the **PropertyManager** and set the weld symbol as required. Note that you will learn more about weld bead in chapter related to welding in this book.
- Click on the **OK** button from the **PropertyManager** to create the weld bead.

Creating Belt/Chain in Assembly

The **Belt/Chain** tool in the **Assembly Feature** drop-down is used to create representation of a real belt/chain used for motion transfer. The procedure to use this tool is given next.

- Click on the **Belt/Chain** tool from the **Assembly Features** panel in the **Assembly CommandManager** of the **Ribbon**. The **Belt/Chain PropertyManager** will be displayed; refer to Figure-8.

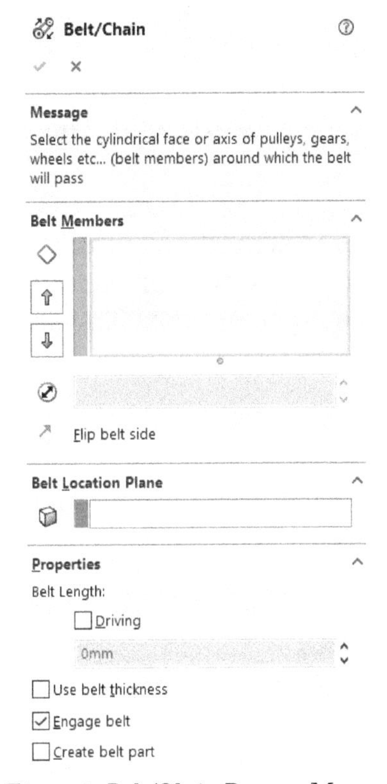

Figure-8. Belt/Chain PropertyManager

- Select the round edges of the pulley around which belt is to be created. Preview of the belt will be displayed; refer to Figure-9.

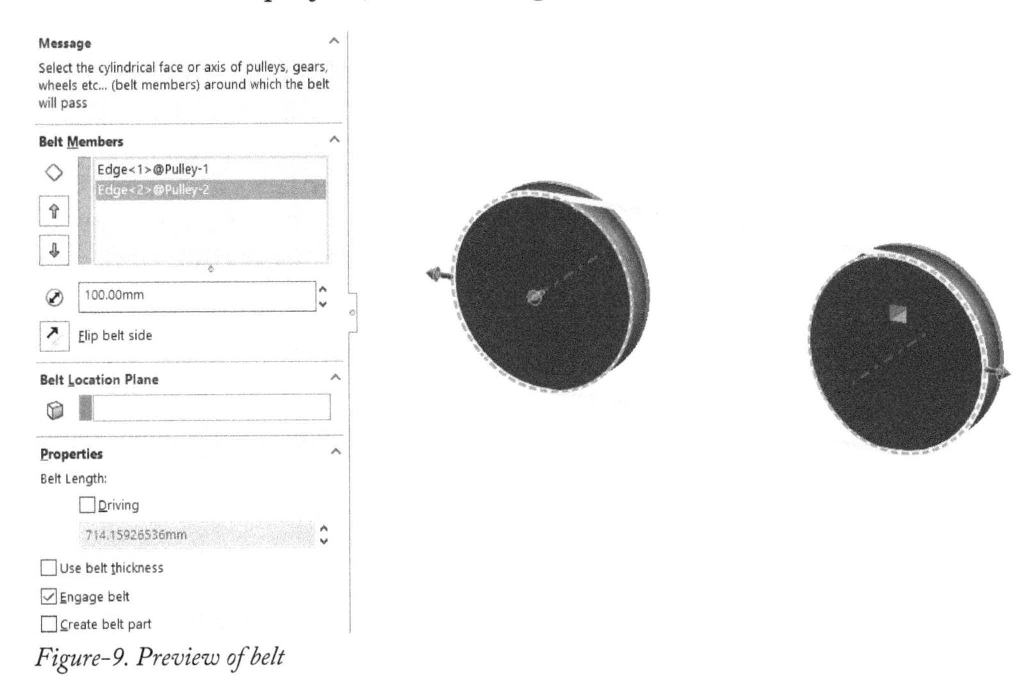

Figure-9. Preview of belt

- Select the **Engage belt** check box in the **Properties** rollout of **PropertyManager** if you want both the pulleys to behave like they are engaged by a belt. Now, if you rotate one pulley by dragging then other pulley will also rotate by same amount.
- Select the **Create belt part** check box if you want to create belt as a part for the assembly.
- Click on the **OK** button from the **PropertyManager** to create belt feature/part.

SMART FASTENERS

The **Smart Fasteners** tool is used to add fasteners to the assembly. This tool is available only when SOLIDWORKS Toolbox is installed and is active. To activate the SOLIDWORKS Toolbox, click on **SOLIDWORKS Toolbox** tool from the **SOLIDWORKS Add-Ins CommandManager** in the **Ribbon**. The procedure to use Smart Fasteners is given next.

- Click on the **Smart Fasteners** tool from the **Assembly CommandManager** in the **Ribbon**. A message box may be displayed which says smart fasteners can take longer time if there are unresolved components or there are too many holes in the assembly.
- Click on the **OK** button from the message box. The **Smart Fasteners PropertyManager** will be displayed; refer to Figure-10.

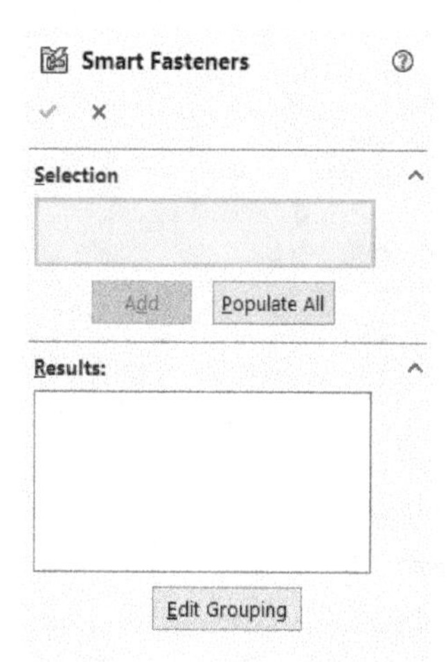

Figure-10. Smart Fasteners PropertyManager

- Select the hole to which you want to apply fastener or click on the **Populate All** button from the **Selection** rollout of the **PropertyManager**. If you have selected holes manually then click on the **Add** button from the rollout. The preview of fasteners will be displayed with updated **PropertyManager**; refer to Figure-11.
- Right-click in the **Fastener** selection box in the **Series Components** rollout of the **PropertyManager**; refer to Figure-12 and select the **Change Fastener Type** option from the shortcut menu displayed. The **Smart Fastener** dialog box will be displayed; refer to Figure-13.

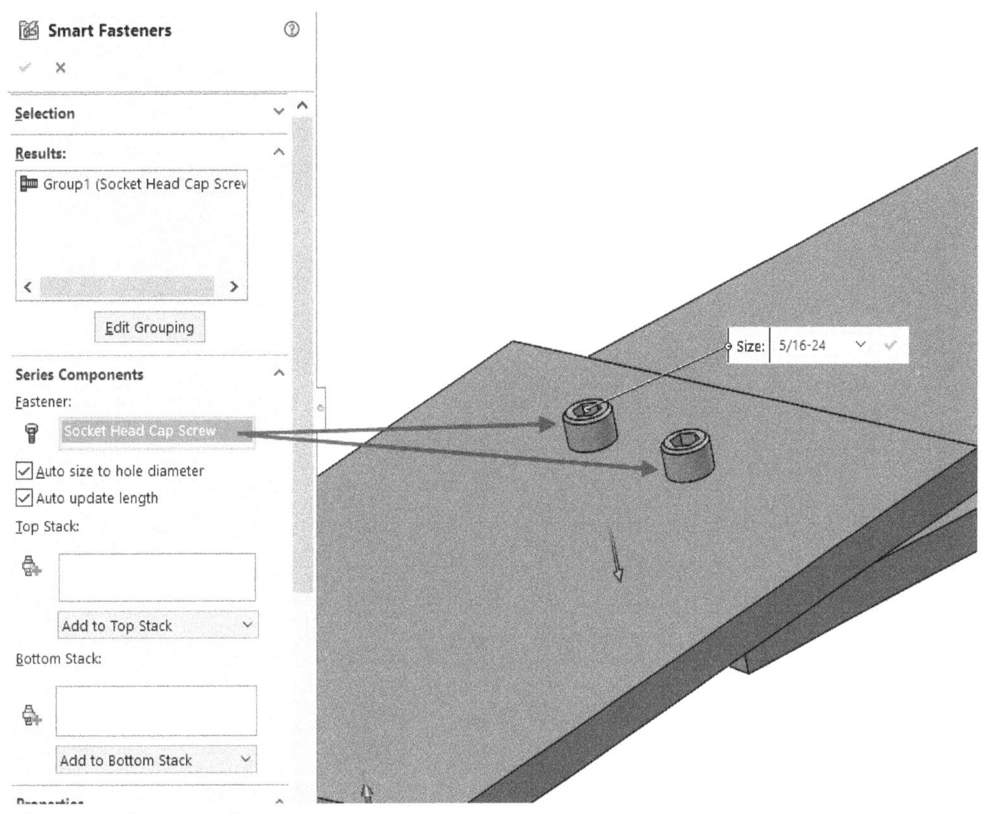

Figure-11. Preview of fasteners

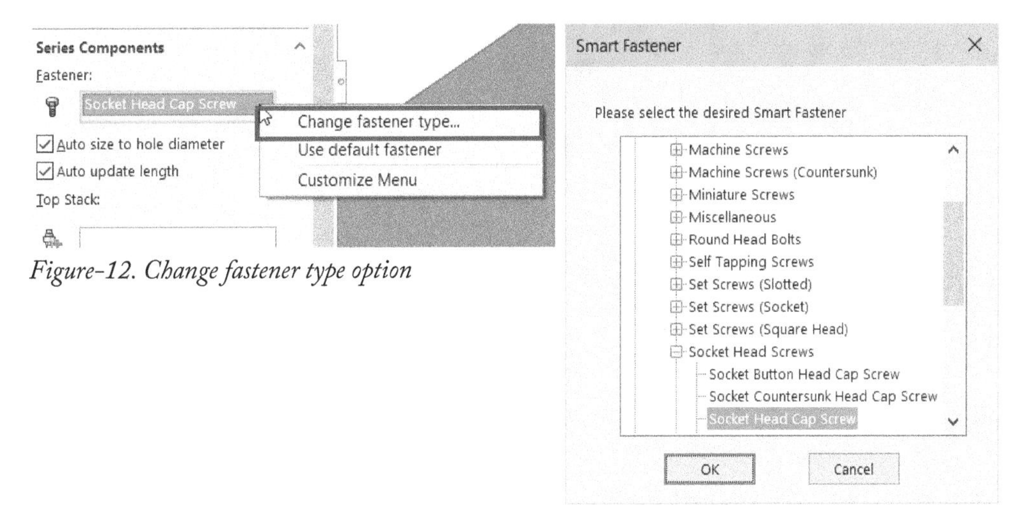

Figure-12. Change fastener type option

Figure-13. Smart Fastener dialog box

- Select desired fastener type and click on the **OK** button from the dialog box. The fastener in preview and **Fastener** selection box will get changed accordingly.
- By default, the **Auto size to hole diameter** and **Auto update length** check boxes are selected in the **PropertyManager** so length and size of fastener is calculated automatically. If you want to manually specify the size and length then clear the respective check boxes and specify the parameters in the **Properties** rollout of the **PropertyManager**.
- To add components like washers to the top of fastener, click in the **Add to Top Stack** drop-down. A list of options will be displayed; refer to Figure-14. Select desired component for stacking at top. If you want to add more components in top stack then select the component again to list.

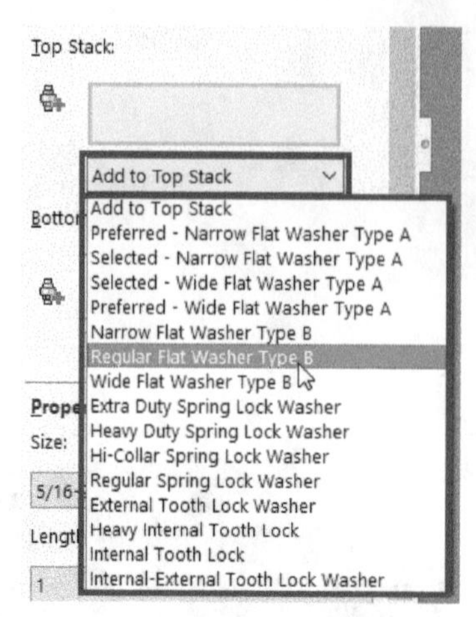

Figure-14. Add to Top Stack drop-down

- To change the parameters of the top stack component, select it from the **Top Stack** selection box and specify the parameters in the **Properties** rollout of the **PropertyManager**. You can also specify desired parameter in the dynamic input box displayed in the preview; refer to Figure-15.

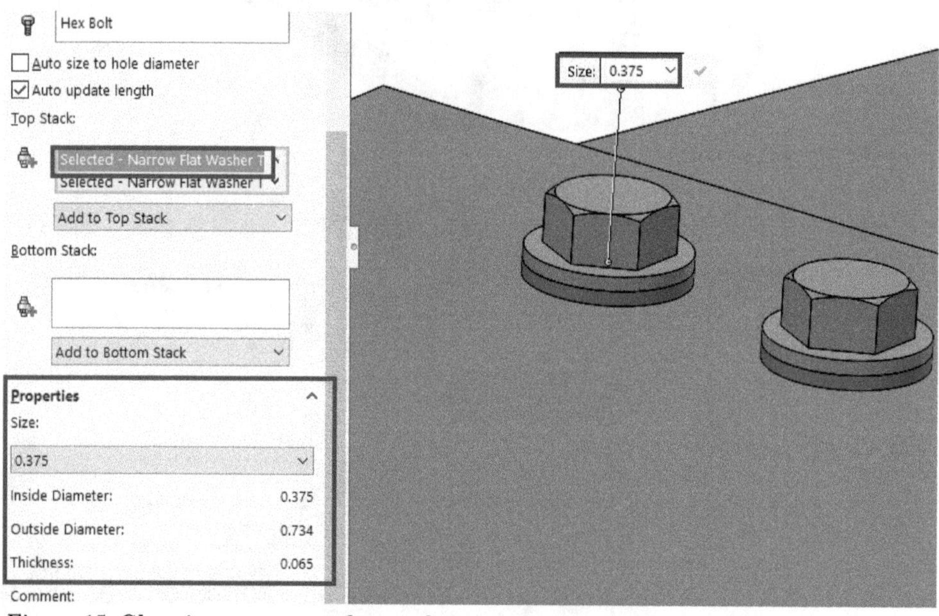

Figure-15. Changing parameters of top stack component

- Similarly, you can add components like nuts in the Bottom stack.
- After adding desired fasteners, click on the **OK** button from the **PropertyManager** to create fastener components.

TOOLBOX

There are various tools in the toolbox to increase the productivity of software in design work like Structural Steel Calculator, Beam Calculator, Cam Calculator, etc. These tools are available in the **SOLIDWORKS Add-Ins CommandManager** if you have activated the **SOLIDWORKS Toolbox** button from the **CommandManager**. **Note that you can use most of these tools in Part environment only so you need to create a new file in Part environment to use these tools.** Various tools in the SOLIDWORKS Toolbox are discussed next.

Structural Steel Beam Calculator

The **Structural Steel Calculator** tool is used to create section of the structural steel based on the inputs provided by you. The procedure to use this tool is given next.

- Click on the **Structural Steel Beam Calculator** tool from the **SOLIDWORKS Add-Ins CommandManager**. The **Structural Steel** dialog box will be displayed; refer to Figure-16.

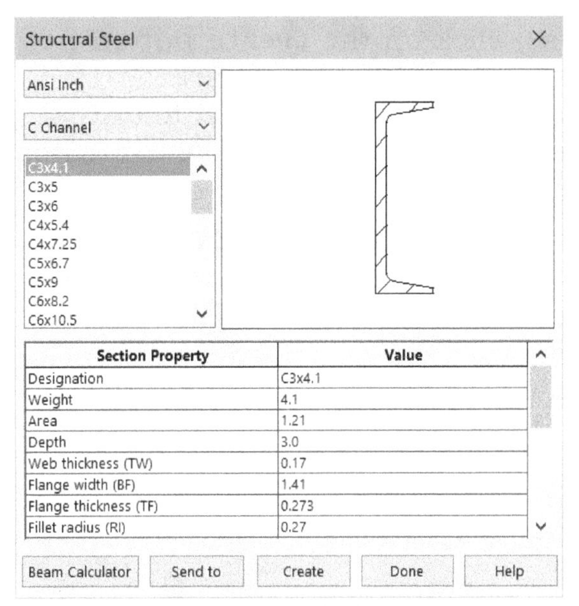

Figure-16. Structural Steel dialog box

- Select desired beam standard and structural beam section from the respective drop-downs in the dialog box.
- Select desired structural component from the list box. If you know the engineering aspects of using the current steel structure then click on the **Create** button to create the section of structural member.
- If you want to test the current section under load then click on the **Beam Calculator** button from the dialog box. The **Beam Calculator** dialog box will be displayed; refer to Figure-17.

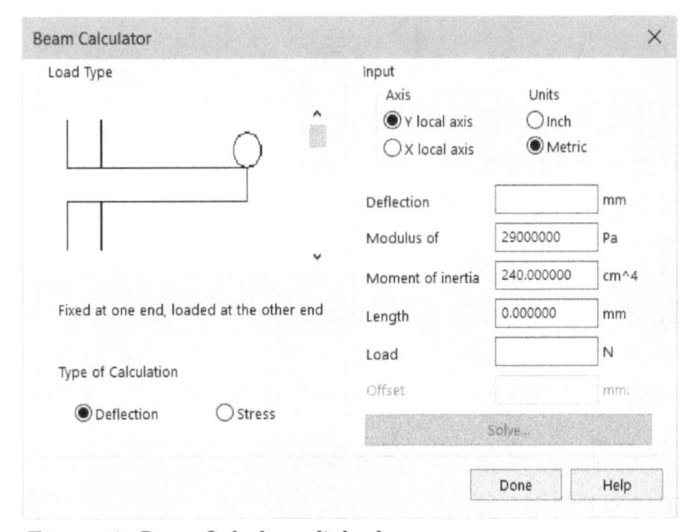

Figure-17. Beam Calculator dialog box

- Select desired load type from the **Load Type** list box by scrolling down in the dialog box.
- Select the parameter for which calculation is to be done from the **Type of Calculation** area of the dialog box.
- Set the value of load in the **Load** edit box and click on the **Solve** button. The deflection or Stress will be displayed in the respective edit box of the dialog box.
- Click on the **Done** button to exit the dialog box.
- If the deflection or stress is more than what is expected then perform the changes in the part. Otherwise, click on the **Create** button from the **Structural Steel** dialog box to create the sketch section of part.
- Click on the **Done** button from the dialog box to exit.
- Now, you can use the **Extrude Boss/Base** tool to create the beam.

Creating Grooves

The **Grooves** tool in the **SOLIDWORKS Add-Ins CommandManager** is used to create O-ring and retaining grooves on the solid. The procedure to use this tool is given next.

- Click on the **Grooves** tool from the **SOLIDWORKS Add-Ins CommandManager** in the **Ribbon**. The **Grooves** dialog box will be displayed; refer to Figure-18.

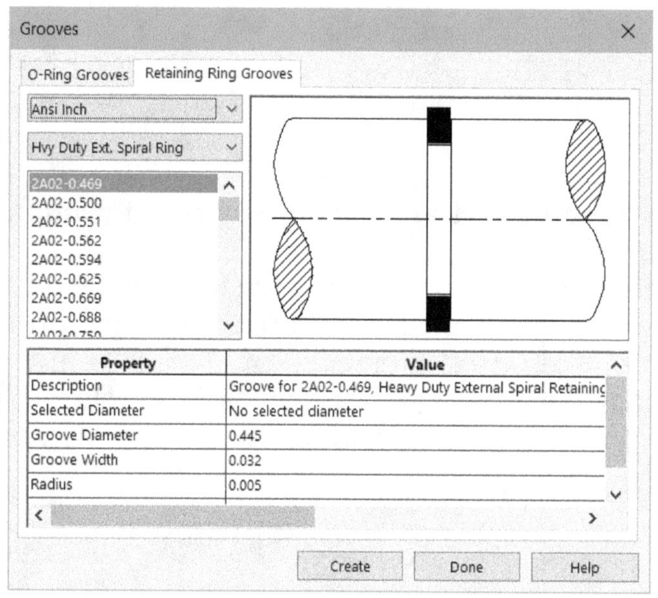

Figure-18. Grooves dialog box

- Click on the **O-Ring Grooves** tab if you want to create a O-ring groove in the solid or click on the **Retaining Ring Grooves** tab if you want to create a retaining groove in the solid.
- Select the standard for rings from the drop-down in the dialog box and then select the groove type from the drop-down below it.
- Select the size of groove from the list box. Details of the groove will be displayed in the **Property** box and preview will be displayed in the **Preview** area of the dialog box.
- Click on the solid revolve feature at desired location to create groove. A sketch will be created with a point at specified location; refer to Figure-19.

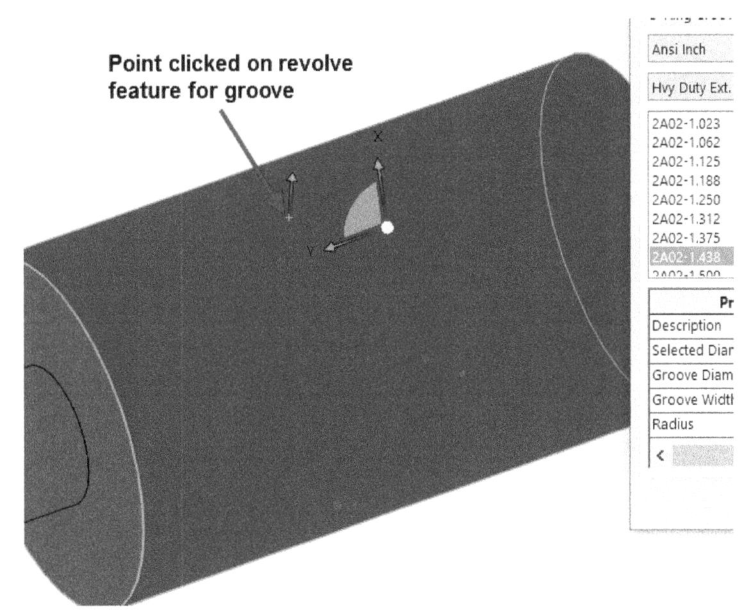

Figure-19. Point selected for groove feature

- Click on the **Create** button from the dialog box. The groove will be created; refer to Figure-20.

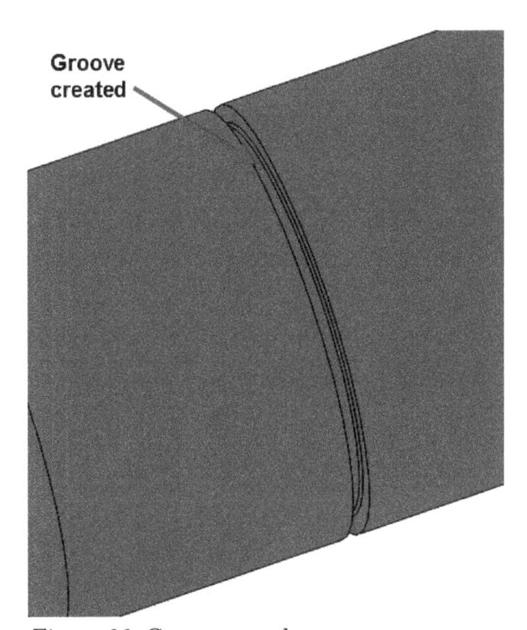

Figure-20. Groove created

- Click on the **Done** button from the dialog box to exit.

Creating Cams

The **Cams** tool is used to create circular and linear cams. The procedure to use this tools is given next.

- Click on the **Cams** tool from the **SOLIDWORKS Add-Ins CommandManager** in the **Ribbon**. The **Cam** dialog box will be displayed; refer to Figure-21.

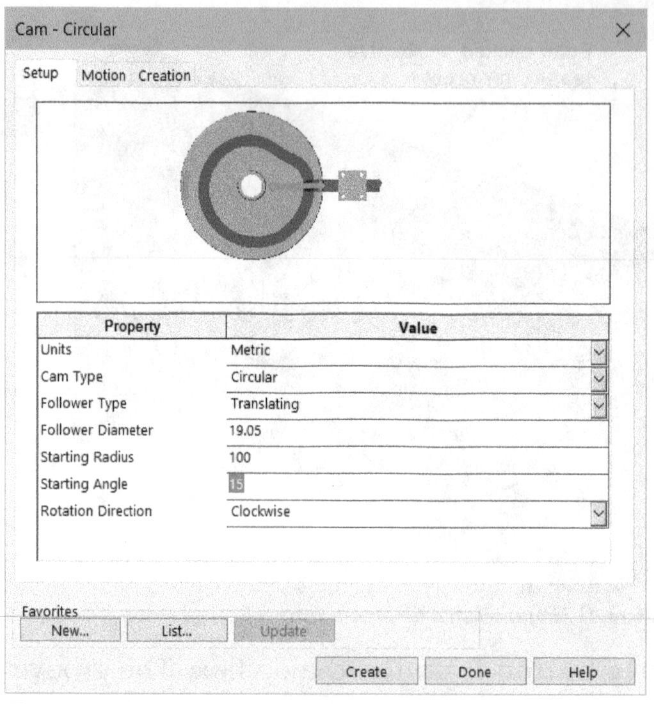

Figure-21. Cam-Circular dialog box

- Set desired unit in the **Units** field of the table.

Circular Cam

- Select the **Circular** option from the **Cam Type** field in the table.
- Set desired parameters in the table.
- Click on the **Motion** tab and using **Add** button, create the motion of Cam; refer to Figure-22.

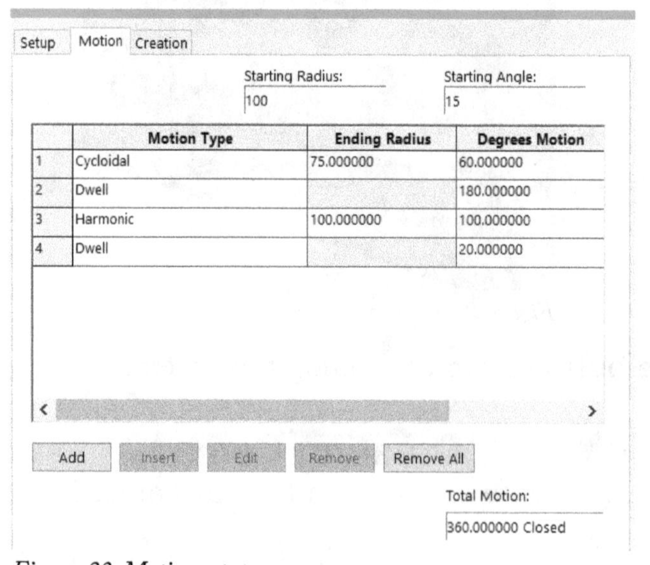

Figure-22. Motion setup

- Similarly, click on the **Creation** tab and set the physical parameters of the cam; refer to Figure-23.

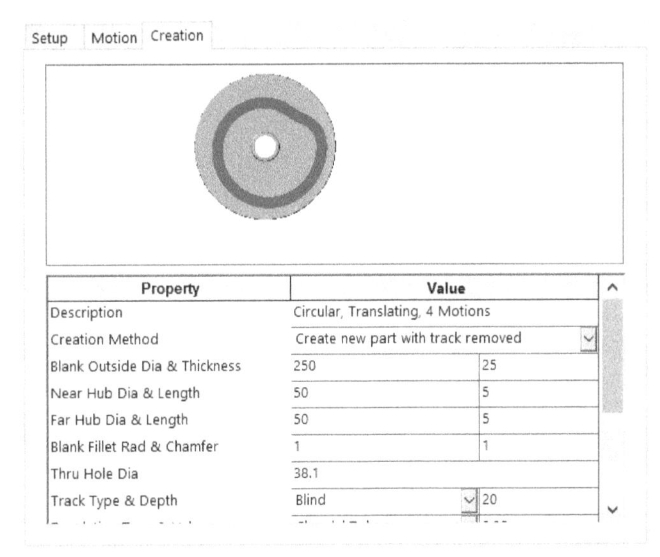

Figure-23. Physical parameters of cam in Creation tab

- Click on the **Create** button from the dialog box to create cam; refer to Figure-24.

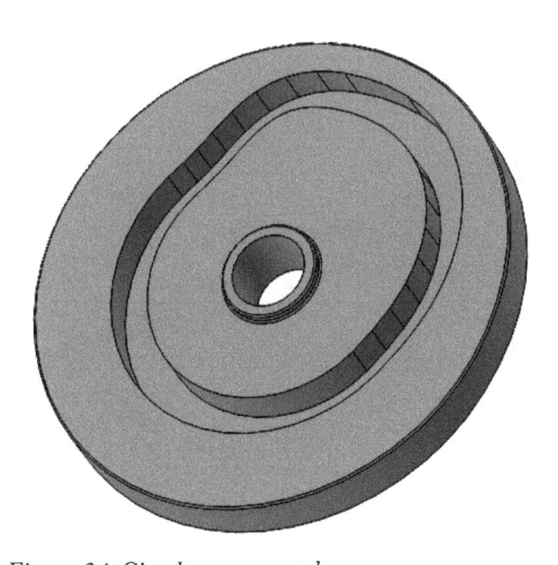

Figure-24. Circular cam created

- Click on the **Done** button to exit the dialog box.

Creating Linear Cam

- Select the **Linear** option from the **Cam Type** drop-down in the table. The **Cam-Linear** dialog box will be displayed; refer to Figure-25.
- Set the parameters as discussed earlier and click on the **Create** button. The cam will be created; refer to Figure-26.

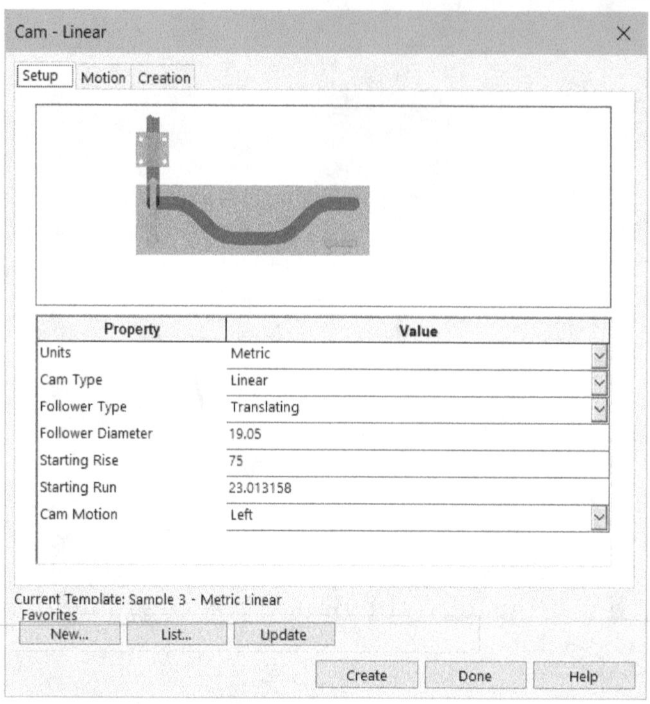

Figure-25. Cam-Linear dialog box

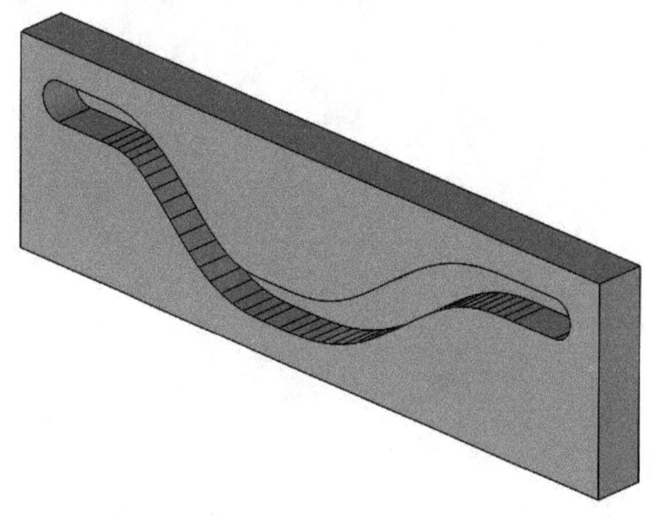

Figure-26. Linear cam created

- To add the current cam in the favorite list, click on the **New** button at the bottom of the dialog box and specify the name of the cam in the **New Favorite Name** dialog box displayed. Select the **Template** check box to make it as template.
- Click on the **OK** button from the dialog box to save it as favorite.
- To load previously saved cam, click on the **List** button from the dialog box and select desired cam from the **Favorite** dialog box displayed. Click on the **Load** button from the dialog box.
- Click on the **Done** button from the dialog box to exit the tool.

Beam Calculator and Bearing Calculator

The **Bearing Calculator** tool is used to calculate the load bearing capacity of bearing and life time of bearing. You can use the tool in the same way as discussed for **Structural Steel Beam Calculator** tool. Similarly, you can use the **Beam Calculator** tool to calculate load capacity of beams.

MAGNETIC MATES

Magnetic mates are used to assemble the components automatically based on magnetic snap points defined on them. The magnetics mates are specially beneficial for piping and ducting work. It can also be used in large plant layouts where applying mates on many instances of same component becomes very time consuming. To use magnetic mates, first we need to prepare components for that. The tool used for preparing component is discussed next.

Asset Publisher

The **Asset Publisher** tool is used to create connection points on components so that they can snap to each other in assembly. The procedure to use this tool is given next.

- Open the part file which you want to use in magnetic mates like file of pipe, duct, etc.
- Click on the **Asset Publisher** tool from the **Tools** menu; refer to Figure-27. The **Asset Publisher PropertyManager** will be displayed; refer to Figure-28. Also, you will be asked to select a flat face to be attached to ground plane while inserting this component in assembly.

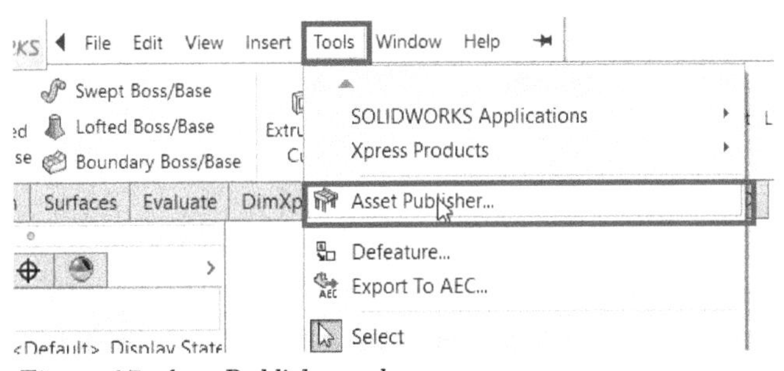

Figure-27. Asset Publisher tool

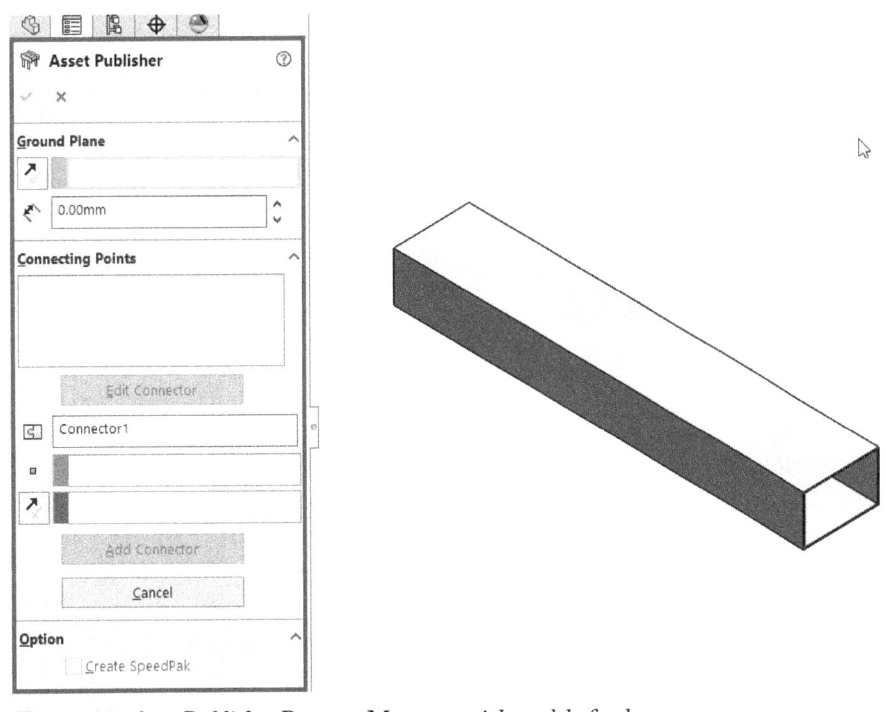

Figure-28. Asset Publisher PropertyManager with model of a duct

- Select the flat face of the component; refer to Figure-29. If you want the part to be above or below the ground plane then specify desired distance value in the edit box in **Ground Plane** rollout of the **PropertyManager**. You can reverse the direction of placement by using the **Reverse Direction** button in the rollout.

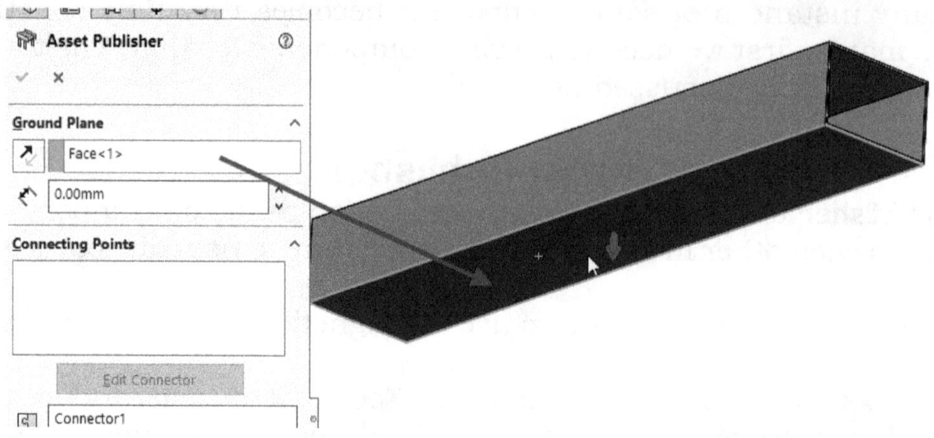

Figure-29. Face of duct selected for ground plane

- Now, click on the edge/point on the part that you want to use as connector point and then select a face to specify direction of connection; refer to Figure-30.

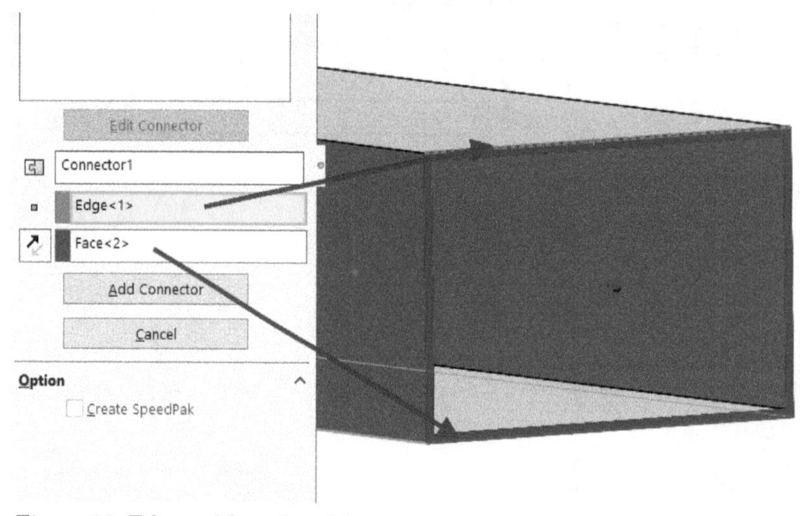

Figure-30. Edge and face selected for connection point

- Click on the **Add Connector** button to add it in **Connecting Points** list box of **PropertyManager**.
- Repeat the previous two steps to create more connection points on the part.
- After specifying desired connection points, click on the **OK** button from the **PropertyManager**.

In the same way, you can prepare other parts for magnetic mates. Now, we have a part with connection points ready for magnetic mates. But, what if we need different lengths/sizes of same component. To enable fast conversion of size of same part, we use configurations. The procedure to create configuration is given next.

Creating Configurations

- Click on the **ConfigurationManager** tab in the **Manager** pane at the left of application window. The **ConfigurationManager** will be displayed; refer to Figure-31.

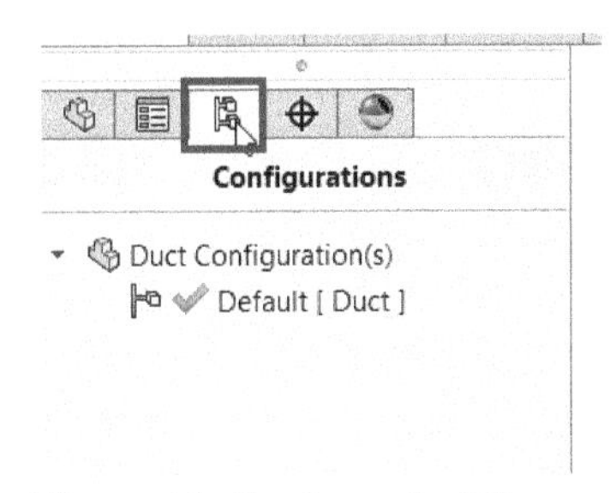

Figure-31. ConfigurationManager

- Right-click on the part name in the **ConfigurationManager**. A shortcut menu will be displayed.
- Select the **Add Configuration** option from the shortcut menu; refer to Figure-32. The **Add Configuration PropertyManager** will be displayed; refer to Figure-33.

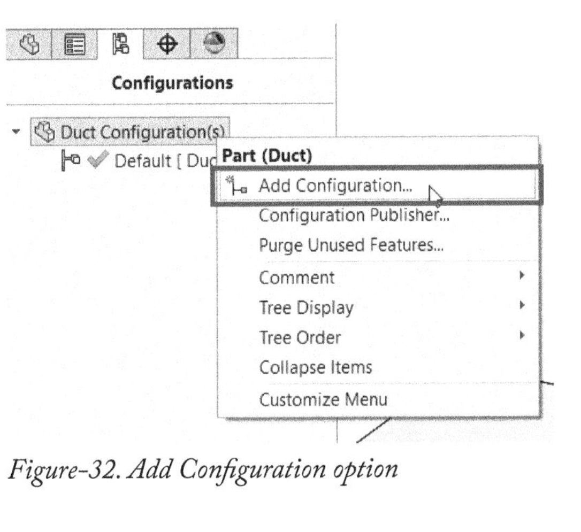

Figure-32. Add Configuration option

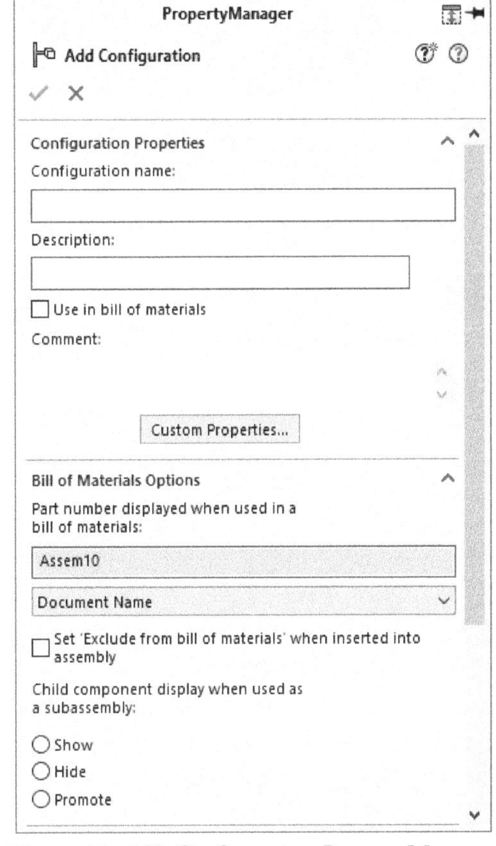

Figure-33. Add_Configuration_PropertyManager

- Specify desired name of the configuration like “length 2k” in the **Configuration name** edit box of **PropertyManager**.
- Specify desired description of configuration in the **Description** edit box of **PropertyManager**.
- Set the other options as required and then click on the **OK** button. Now, the newly created configuration is active.
- Change the parameters of part like length of extrusion, width in sketch, etc. by using **FeatureManager Design Tree**.
- After performing changes, save the file.
- Repeat the procedure to create more configurations.
- You can check different sizes of the part by double-clicking on the respective configurations in the **ConfigurationManager**.

Note that you can create the configurations of assembly in the same way.

Creating Assembly with Magnetic Mates

- Start a new assembly as discussed earlier.
- Press **ESC** if the **Open** dialog box is displayed automatically and exit the **Begin Assembly PropertyManager** displayed automatically on starting new assembly.
- Click on the **Ground Plane** tool from the **Insert->Reference Geometry** menu; refer to Figure-34. The **Ground Plane PropertyManager** will be displayed; refer to Figure-35.

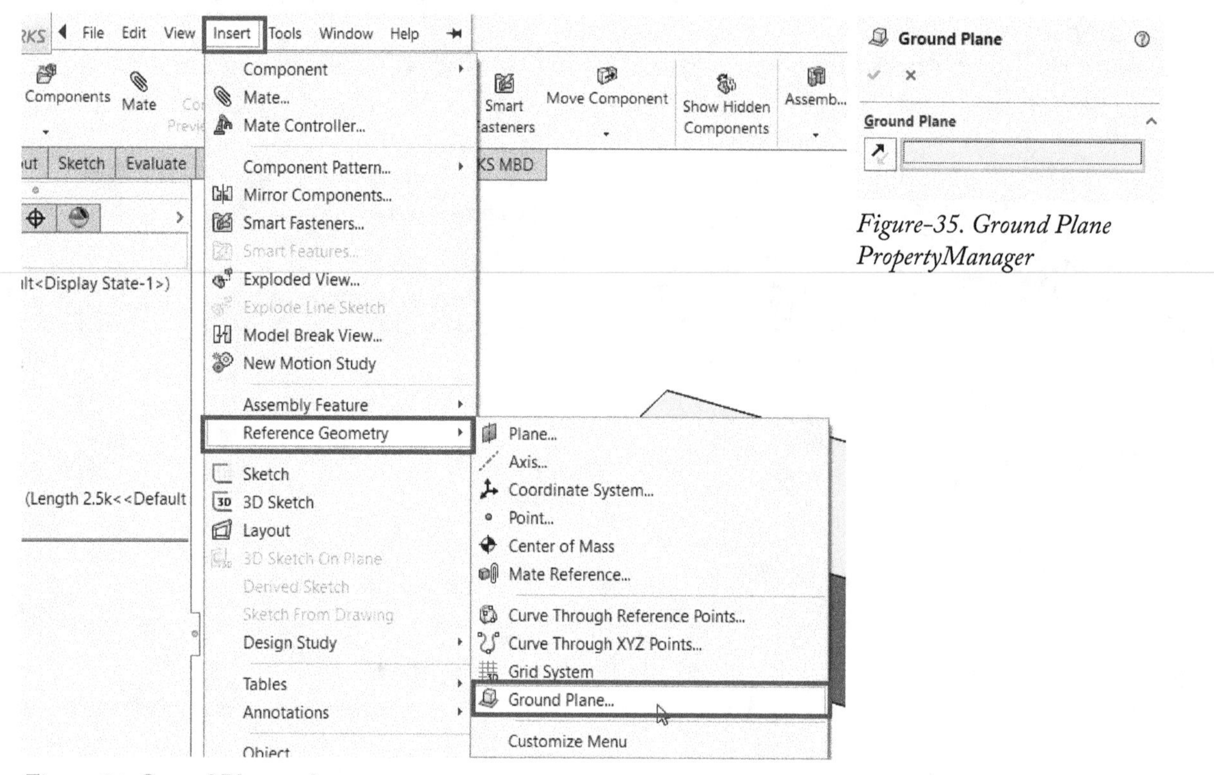

Figure-35. Ground Plane PropertyManager

Figure-34. Ground Plane tool

- Select the plane/face that you want to use as ground plane for magnetic mate components.
- Click on the **OK** button from the **PropertyManager**.
- Toggle **Magnetic Mate ON/OFF** tool in the **Tools** menu to **ON** so that magnetic mate can be applied.
- Click on the **Insert Components** tool from the **Assembly CommandManager** and double-click on the part (in **Open** dialog box displayed) that you have prepared for magnetic mate earlier. The part will get attached to cursor; refer to Figure-36. Note that the pink dots on the part are connection points created earlier using the **Asset Publisher** tool.
- Select desired configuration of part from the **Configuration** drop-down in the **Insert Component PropertyManager**; refer to Figure-37.
- Click on the **OK** button from the **PropertyManager**. The part will be place automatically based on the ground plane specified in part and assembly.
- Again, click on the **Insert Components** tool and double-click on the second part to be inserted. Now, if you move the part near the other part, the connections point will snap each other and purple line of connection will be displayed; refer to Figure-38.

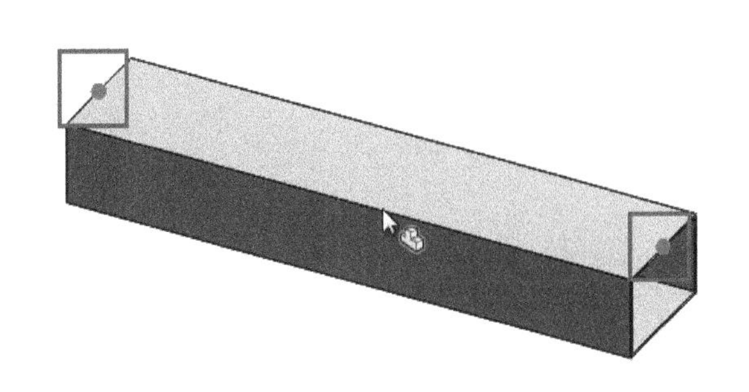

Figure-36. Part attached to cursor with magnetic connectors

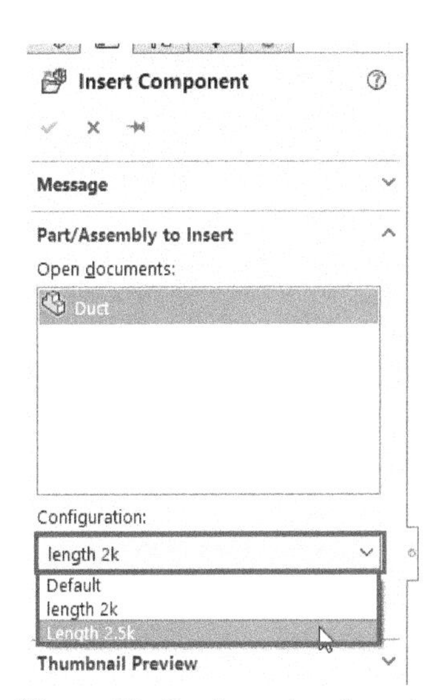

Figure-37. Configuration drop-down

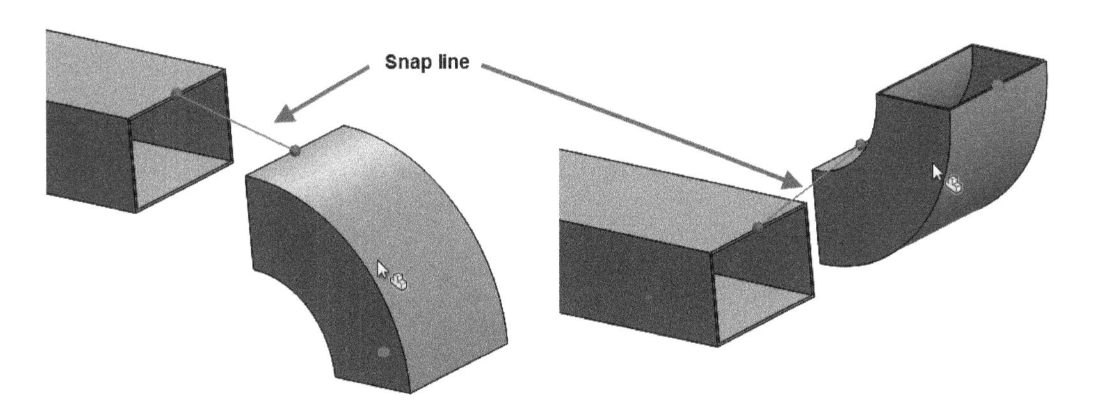

Figure-38. Autosnapping of magnetic connectors

- Click in the viewport when desired connection points are snapping. The parts will be assembled by magnetic mate; refer to Figure-39. Note that you can use the **Rotate** toolbar to rotate the part if it is not positioned as required.

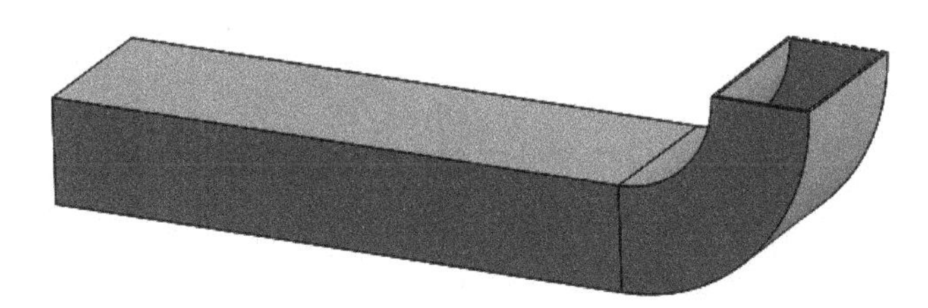

Figure-39. Part connected with magnetic mate

- In the same way, you can insert the other components with connection points.

Note that you should not specify the ground plane for components when the same component is to be assembled in different orientations. Like, we have not specified ground plane for Duct Turn component in example discussed here. The part file for this example are available in the Magnetic Mates folder of Chapter 8 resources in resource kit. You can check the effect of specifying ground plane Duct turn component by yourself.

Working with Large Assemblies: If you are working with a large assembly then you can select the **Large Assembly Settings** button to activate large assembly mode.

The other tools in **Ribbon** have already been discussed.

Practical 1

Assemble the parts of Fuel Injection Nozzle as shown in Figure-40. The exploded view of assembly is displayed as shown in Figure-41.

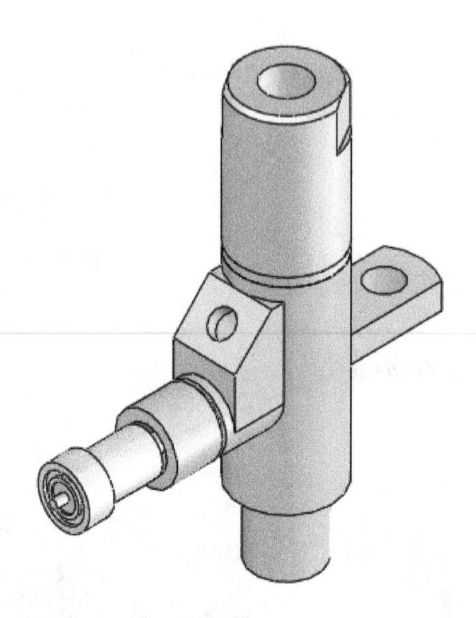

Figure-40. Assemble view of nozzle

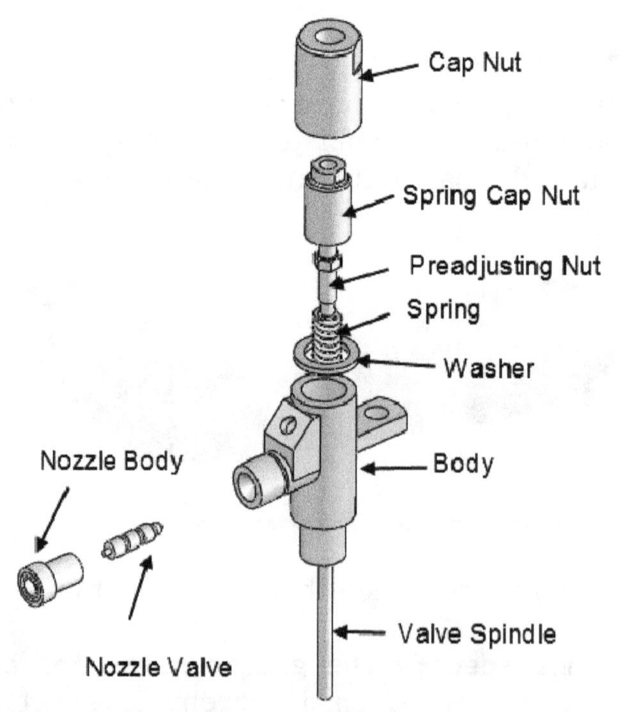

Figure-41. Exploded view of assembly

All the part files can be downloaded from the provided link. In this practical, we will use parts of **Fuel injection nozzle** folder from downloaded files/folders in resource kit.

The steps to assemble these parts are given next.

Inserting Body

- Start the assembly environment by selecting **Assembly** button from the **New SolidWorks Document** dialog box. The **Open** dialog box will display and you are asked to insert first part.
- Double-click on the Body part in the **Fuel injection nozzle** folder. The part will attach to the cursor.
- Click in the viewport to place the part.

Inserting and constraining Washer

- Click on the **Insert Components** button from the **Ribbon**. And then click on the **Browse** button from the **PropertyManager**. The **Open** dialog box will display.
- Double click on the **Washer** part in the **Fuel injection nozzle** folder.
- The part will attach to the cursor.
- Click in the viewport to place the part.
- Click on the **Mate** button from the **Ribbon**. The **Mate PropertyManager** will display.
- Select the **Coincident** button from the **PropertyManager** and select the round edges as shown in Figure-42.

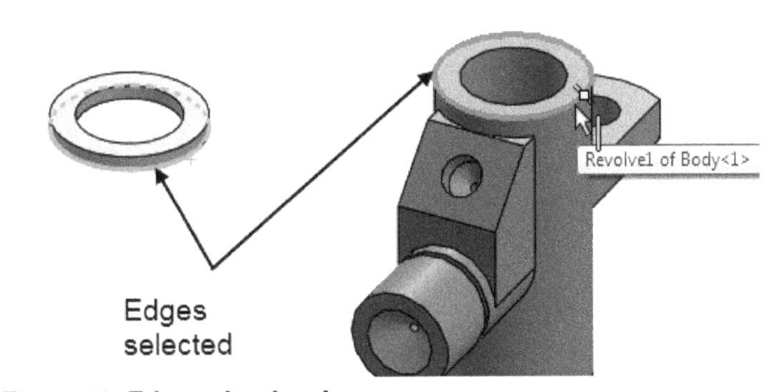

Figure-42. Edges to be selected

Inserting and constraining Spring

- Click on the **Insert Components** button from the **Ribbon**. And then click on the **Browse** button from the **PropertyManager**. The **Open** dialog box will display.
- Double click on the **Spring** part in the **Fuel injection nozzle** folder.
- The part will attach to the cursor.
- Click in the viewport to place the part.
- Click on the **Mate** button from the **Ribbon**. The **Mate PropertyManager** will display.
- Select the **Coincident** button from the **PropertyManager** and select the flat faces as shown in Figure-43.
- Click on the **Flip Mate Alignment** button( ) to align the spring properly.

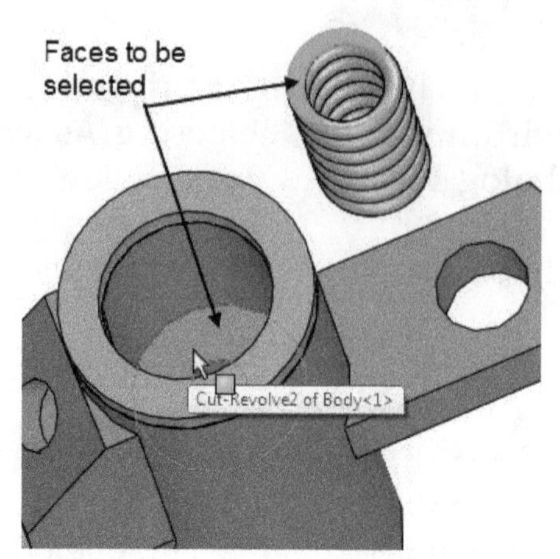

Figure-43. Faces to be selected

- Click on the **OK** button from the **PropertyManager** to apply the mate.
- Click on the **View Temporary Axes** button and **View Axes** button from the **Hide/Show Items** drop-down to display axes; refer to Figure-44.

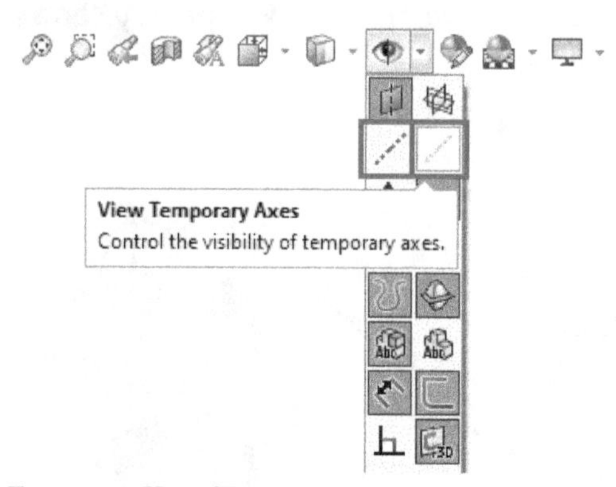

Figure-44. View Temporary Axes button

- Click on the **Coincident** button again from the **PropertyManager** and select the axes as shown in Figure-45.

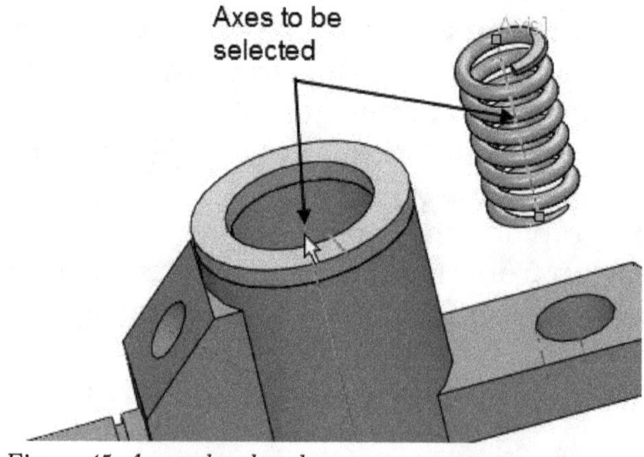

Figure-45. Axes to be selected

- Click **OK** button from the **PropertyManager**. The mate will be applied.

Inserting and constraining Pre adjusting Nut

- Click on the **Insert Components** button from the **Ribbon** and then click on the **Browse** button from the **PropertyManager**. The **Open** dialog box will display.

- Double-click on the **Spring** part in the **Fuel injection nozzle** folder.
- The part will attach to the cursor.
- Click in the viewport to place the part.
- Click on the **Mate** button from the **Ribbon**. The **Mate PropertyManager** will display.
- Select the **Coincident** button from the **PropertyManager** and select the flat faces as shown in Figure-46.

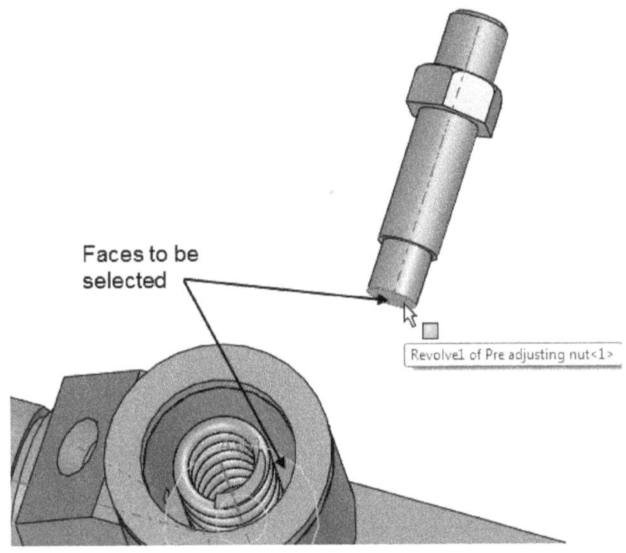

Figure-46. Faces to be coincident

- Click **OK** from the **PropertyManager** to apply the mate.
- Click on the **Coincident** button again from the **PropertyManager** and select the center axes as shown in Figure-47.

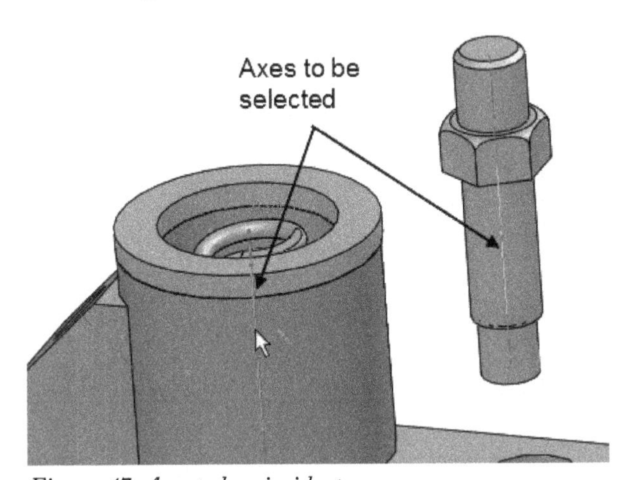

Figure-47. Axes to be coincident

- Click on the **OK** button from the **PropertyManager** to apply the mate.

Inserting and Constraining other components

In the same way, you can assemble the other components of the injection nozzle; refer to Figure-40 and Figure-41.

Practical 2

Assemble the parts of handle as shown in Figure-48. The exploded view of assembly is displayed as shown in Figure-49.

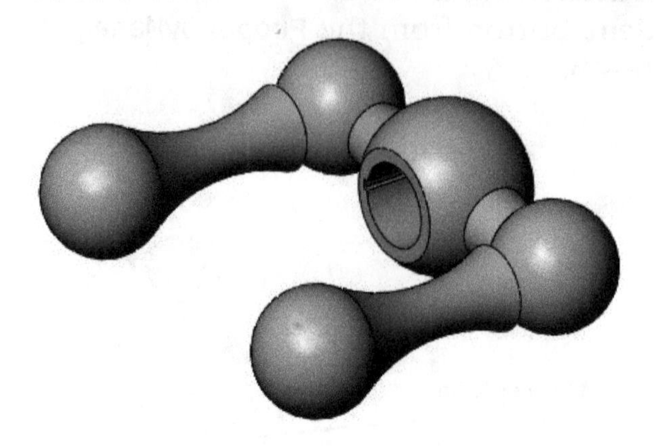

Figure-48. Handle assembled

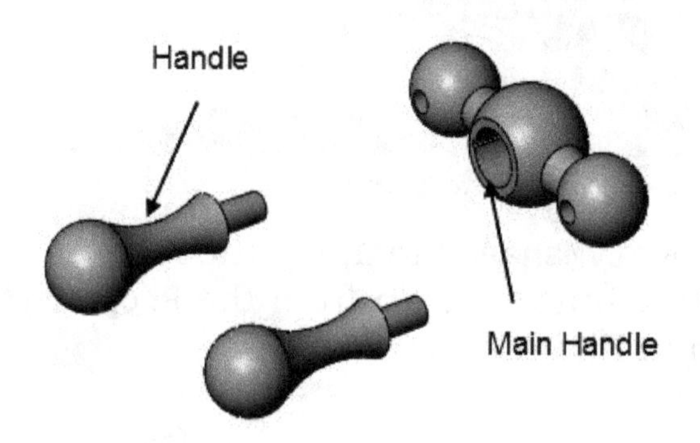

Figure-49. Exploded view of handle

All the part files can be downloaded from the provided link. In this practical, we will use parts of **Handle assembly** folder from downloaded files/folders.

The steps to assemble these parts are given next.

Inserting Main Handle

- Start the assembly environment by selecting **Assembly** button from the **New SOLIDWORKS Document** dialog box. The **Open** dialog box will display. You will be asked to insert first part.
- Double-click on the **Main Handle** part in the **Handle assembly** folder. The part will attach to the cursor.
- Click in the viewport to place the part.

Inserting and constraining Handle

- Click on the **Insert Components** button from the **Ribbon** and then click on the **Browse** button from the **PropertyManager**. The **Open** dialog box will display.
- Double-click on the **Handle** part in the **Handle assembly** folder. The part will attach to the cursor.
- Click in the viewport to place the part.
- Click on the **Mate** button from the **Ribbon**. The **Mate PropertyManager** will display.

- Select the **Concentric** button from the **PropertyManager** and select the round faces as shown in Figure-50.

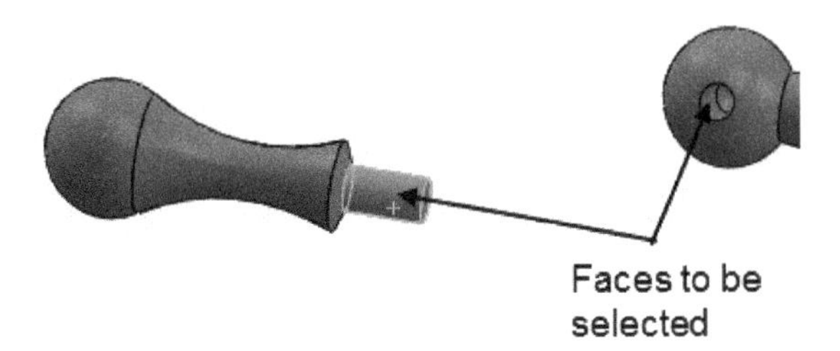

Figure-50. Faces selected for concentric mate

- Click **OK** from the **PropertyManager** to apply the mate.
- Display the planes by selecting the **View Planes** button from the **Hide/Show Items** drop-down in the **Heads-up View** toolbar; refer to Figure-51.

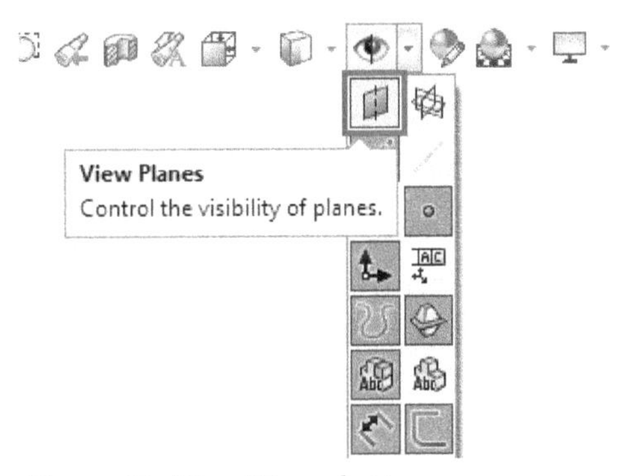

Figure-51. View Planes button

- Click on the **Coincident** button from the **Mate PropertyManager** and select the plane and edge as shown in Figure-52.

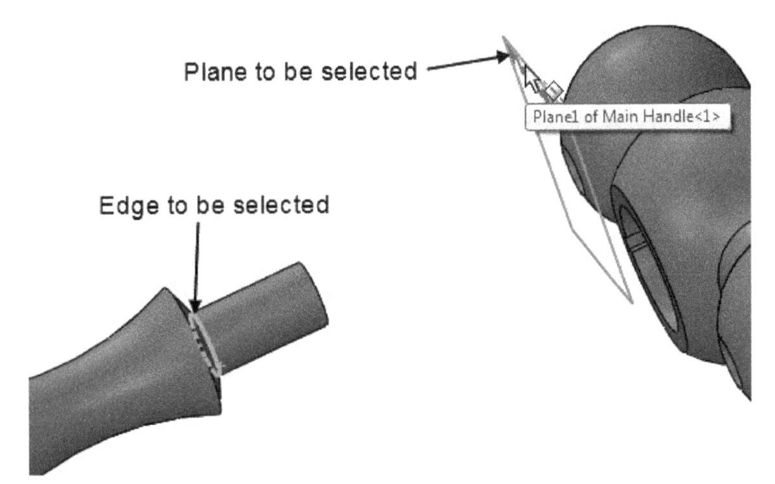

Figure-52. Selection for coincident mate

- Click on the **OK** button from the **PropertyManager** to apply the mate.

The assembled handle will displayed as shown in Figure-53.

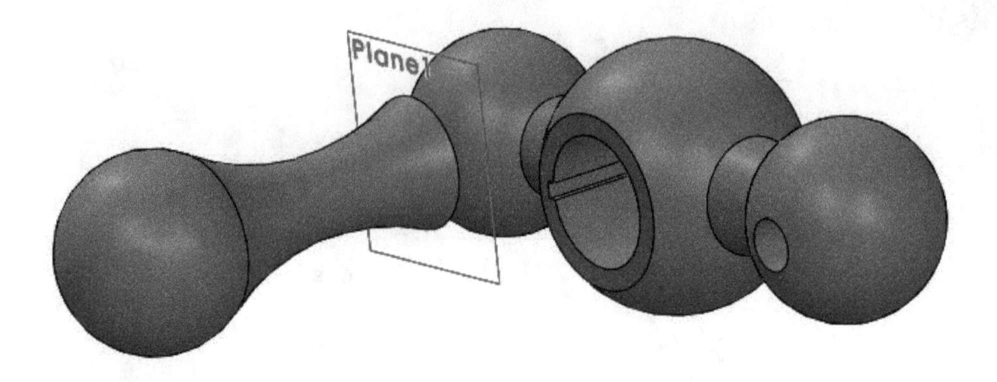

Figure-53. Assembled handle

Similarly, assemble the other handle so that the model displays as shown in Figure-48.

Practical 3

Assemble the parts of Epicyclic Gear Mechanism as shown in Figure-54. The exploded view if assembly is displayed as shown in Figure-55. Also, apply the simulation to mechanism as shown in video file provided in the resources.

Figure-54. Epicyclic gear mechanism assembly

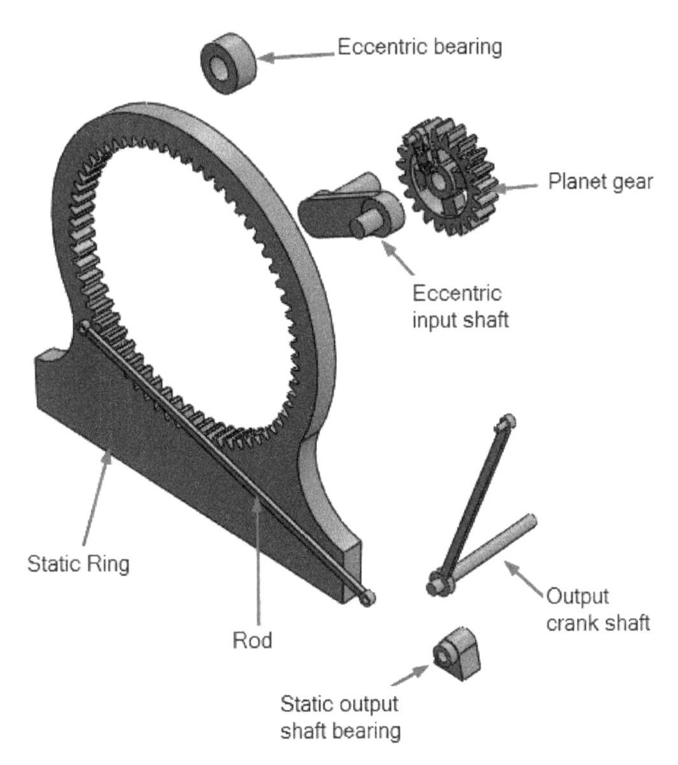

Figure-55. Exploded view of epicyclic gear mechanism

All the part files can be downloaded from the provided link. In this practical, we will use parts of **Epicyclic Gear Mechanism** folder from downloaded files/folders.

The steps to assemble these parts are given next.

Inserting Static Ring

- Start the assembly environment by selecting Assembly button from the **New SOLIDWORKS Document** dialog box. The **Open** dialog box will display. You will be asked to insert first part.
- Double-click on the Static Ring part in the **Epicyclic Gear Mechanism** folder. The part will attach to the cursor.
- Click in the viewport to place the part. The part will be fixed at the specified position.

Inserting and constraining Washer

- Click on the **Insert Components** button from the **Ribbon**. And then click on the **Browse** button from the **PropertyManager**. The **Open** dialog box will display.
- Double click on the **Eccentric Input shaft** part in the **Epicyclic Gear Mechanism** folder.
- The part will attach to the cursor.
- Click anywhere in the viewport to place the component.
- Select the round face of **Input shaft** and **Static ring**, and select the **Concentric** button from the **Mates** toolbar displayed; refer to Figure-56.

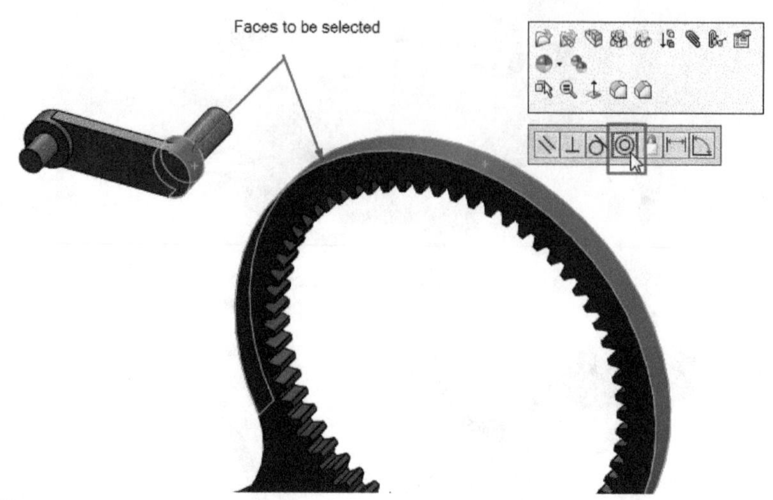

Figure-56. Applying concentric mate

- Select the flat face of Input shaft and Static ring as shown in and apply the **Coincident** mate from the **Mates** toolbar.

Inserting and constraining Eccentric static bearing

- Click on the **Insert Components** button from the **Ribbon**. And then click on the **Browse** button from the **PropertyManager**. The **Open** dialog box will display.
- Double click on the **Static Eccentric Bearing** part in the **Epicyclic Gear Mechanism** folder.
- The part will attach to the cursor.
- Click anywhere in the viewport to place the component.
- Select the round face of **Input shaft** and **Static Eccentric Bearing**, and select the **Concentric** button from the **Mates** toolbar displayed; refer to Figure-57.

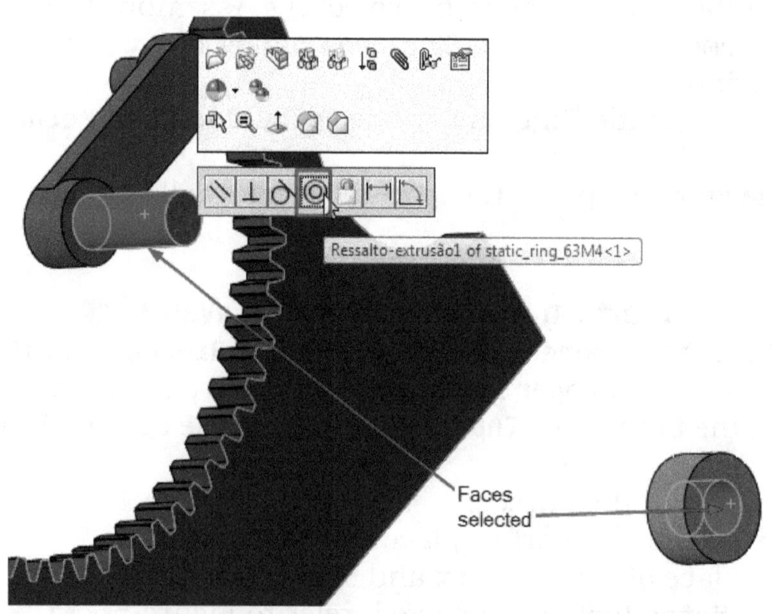

Figure-57. Applying concentric mate on bearing

- Similarly, make the flat faces of the bearing and shaft coincident; refer to Figure-58.

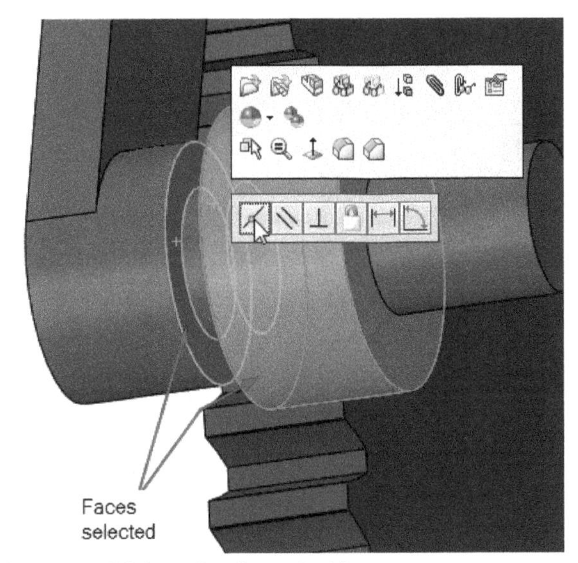

Figure-58. Making flat faces coincident

Inserting and constraining Planet gear

- Click on the **Insert Components** button from the **Ribbon**. And then click on the **Browse** button from the **PropertyManager**. The **Open** dialog box will display.
- Double-click on the **Planet Gear** part in the **Epicyclic Gear Mechanism** folder. The part will attach to the cursor.
- Click anywhere in the viewport to place the component.
- Select the round face of **Input shaft** and **Plant Gear**, and select the **Concentric** button from the **Mates** toolbar displayed; refer to Figure-59.

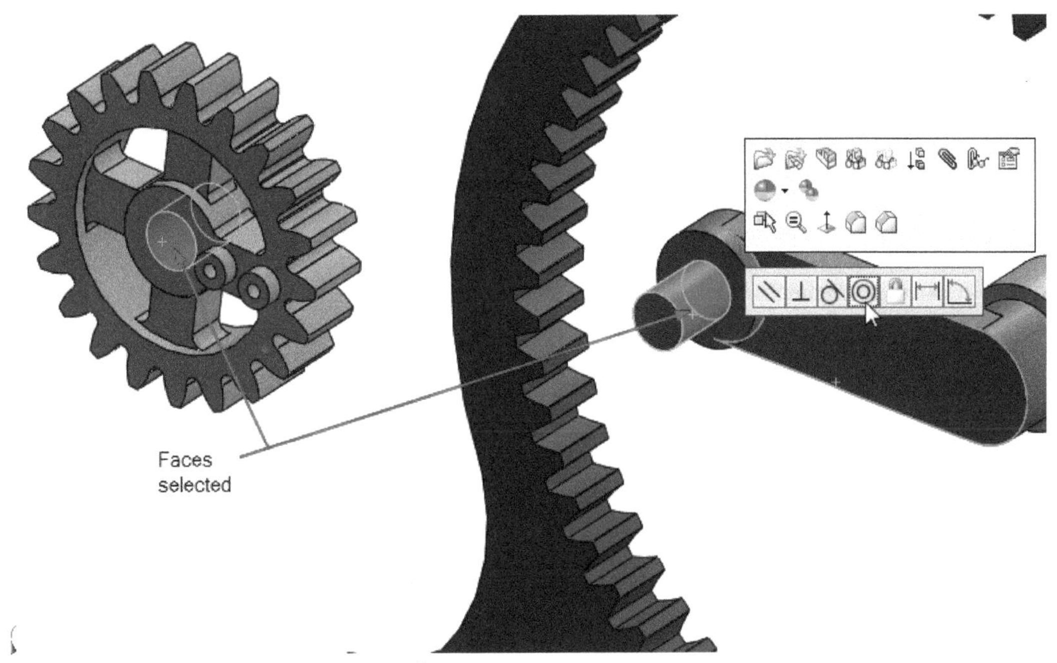

Figure-59. Applying concentric mate on planet gear

- Similarly, applying the coincident mate on the flat face of gear and Input shaft; refer to Figure-60.

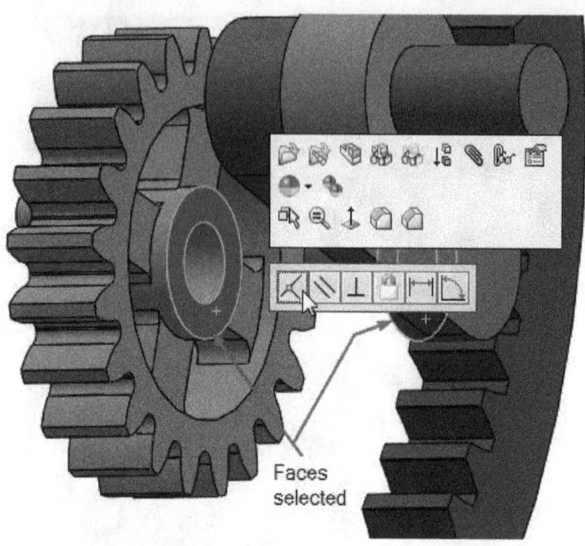

Figure-60. Making flat faces of gear and shaft coincident

Inserting and constraining other components

In the same way, insert and constrain the other components refer to the figures given next for reference.

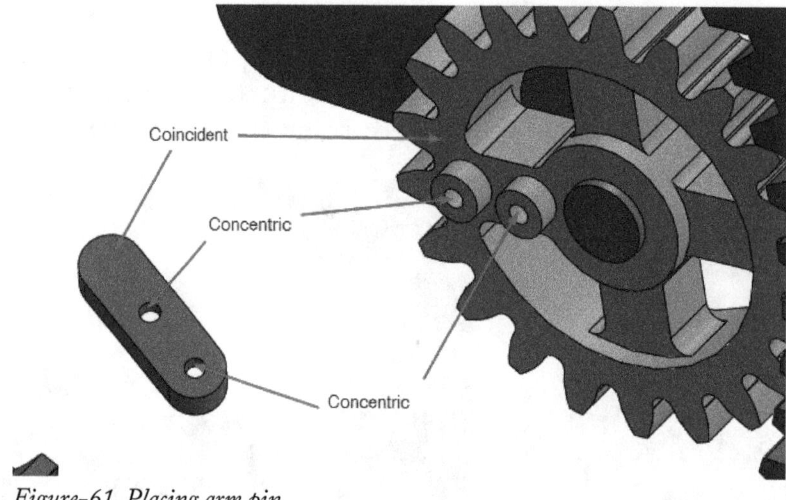

Figure-61. Placing arm pin

Figure-62. Placing the rod

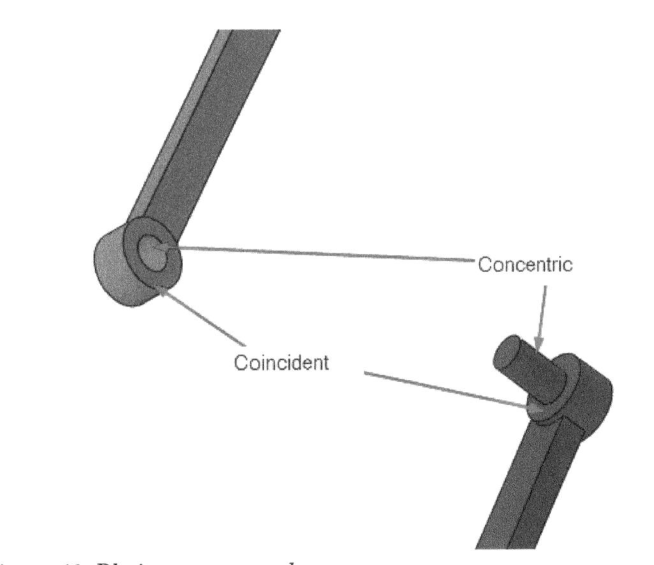

Figure-63. Placing output crank

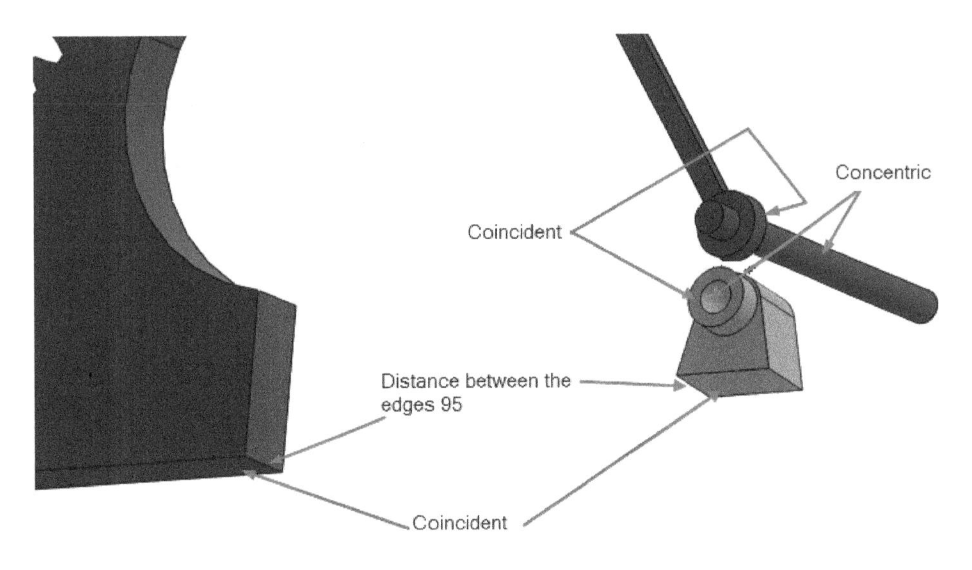

Figure-64. Placing the output shaft bearing

Applying relation between gears

- After placing all the components and applying the constraints, click on the **Mate** button from the **Ribbon**. The **Mate PropertyManager** will be displayed. Select the **Gearmate** button from the **Mechanical Mates** tab in the **PropertyManager**; refer to Figure-65.

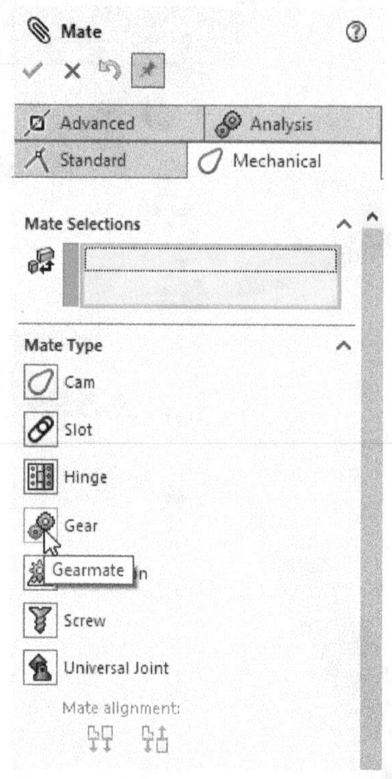

Figure-65. Gear mate

- Select the two edges of gear and static ring; refer to Figure-66.

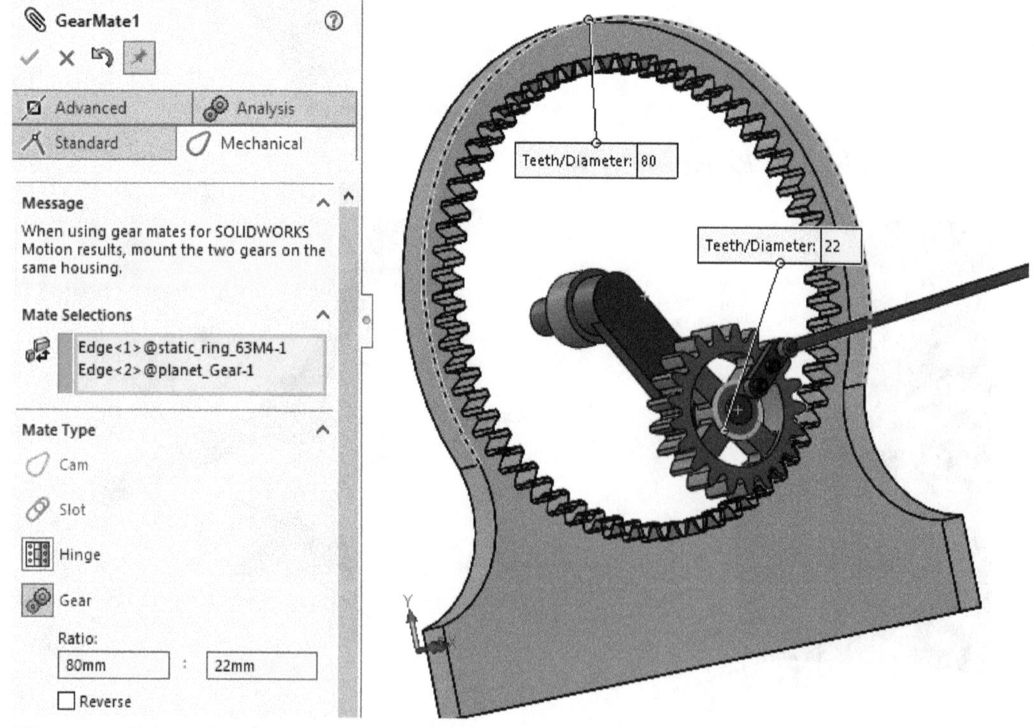

Figure-66. Edges selected for gear mate

- Specify the gear ratio as **80:22** in the **Mechanical Mates** rollout in the **PropertyManager** and click on the **OK** button from it twice.

Creating Motion Study

- Click on the **Motion Study 1** tab from the bottom bar. The Motion Study interface will be displayed; refer to Figure-67.

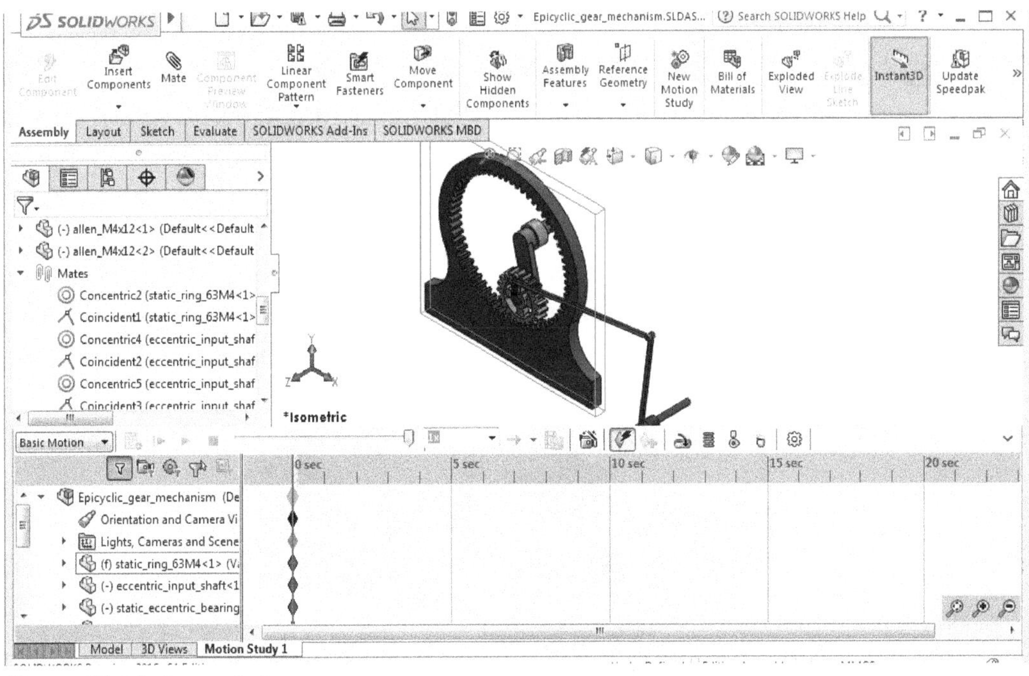

Figure-67. Motion study for gear

- Click on the **Motor** button from the Motion Study interface.
- Select the face shown in Figure-68. Specify the RPM as **10** and click on the **OK** button from the **Motor PropertyManager**.

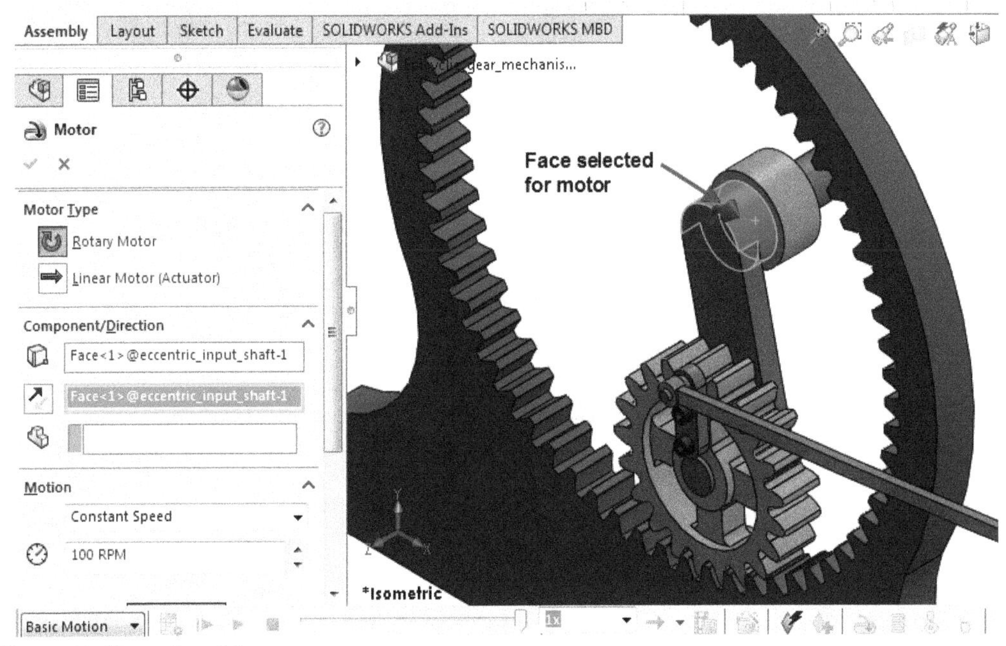

Figure-68. Face selected for motor

- Select the **Basic Motion** option from the drop-down; refer to Figure-69 and then click on the **Calculate** button to check the motion.

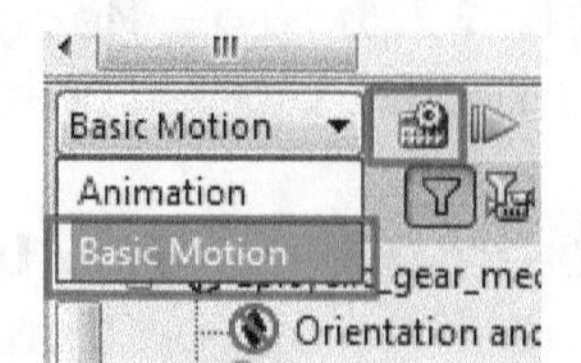

Figure-69. Basic Motion option

PRACTICE 1

Assemble the model as shown in Figure-70. The exploded view is given in Figure-71.

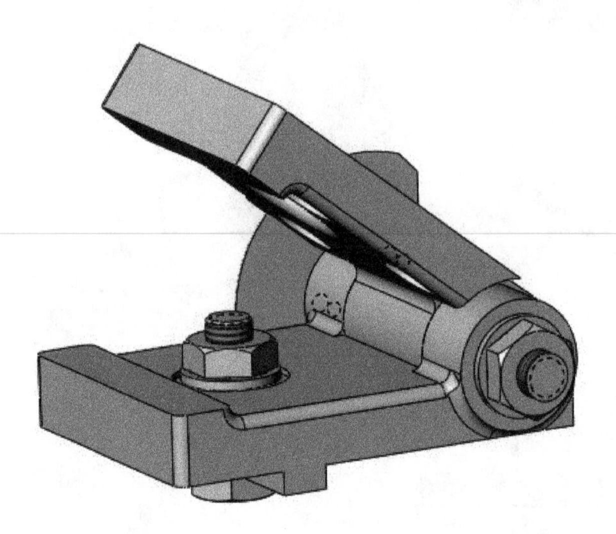
Figure-70. Bottom assembly

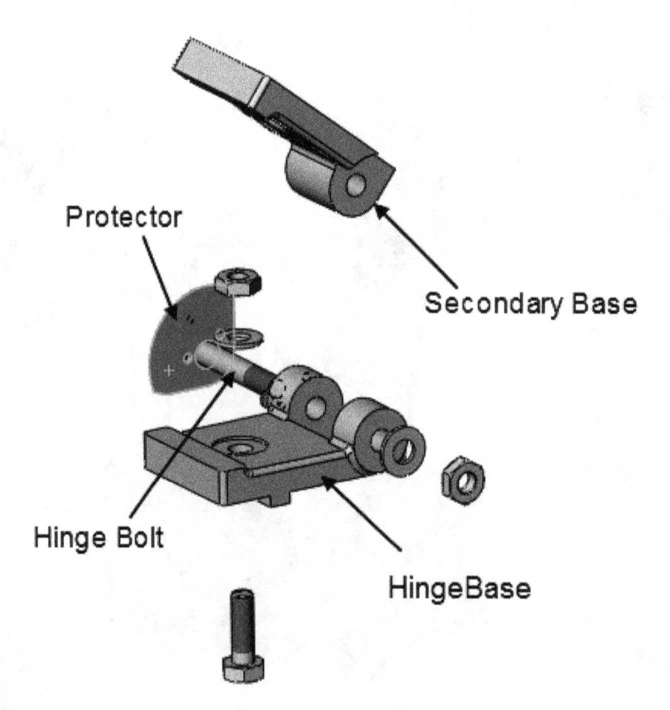

Figure-71. Exploded view of bottom assembly

Note that you need to import nuts and bolts from the toolbox provided by SolidWorks; refer to Figure-72. The toolbox is the standard library of components provided by SolidWorks. This library contains nuts, bolts, bearings, transmission parts, washer and a lot more with inch as well as mm specifications.

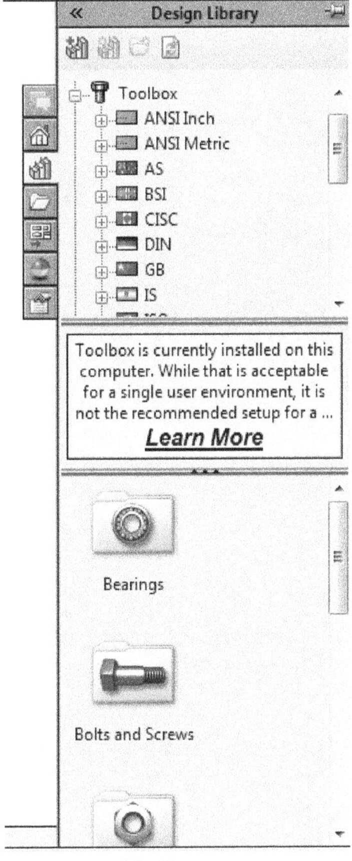

Figure-72. Toolbox

PRACTICE 2

Assemble the model as shown in Figure-73. The exploded view is given in Figure-74.

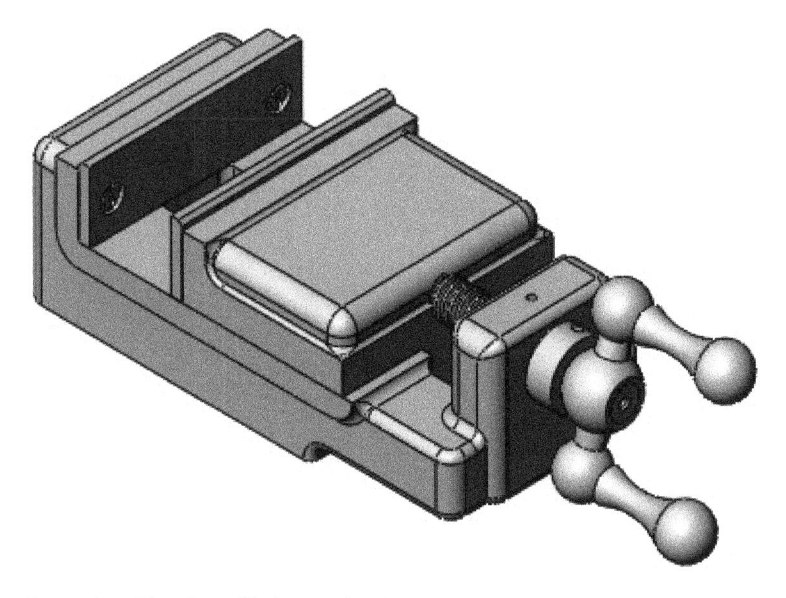

Figure-73. Top Assembly isometric view

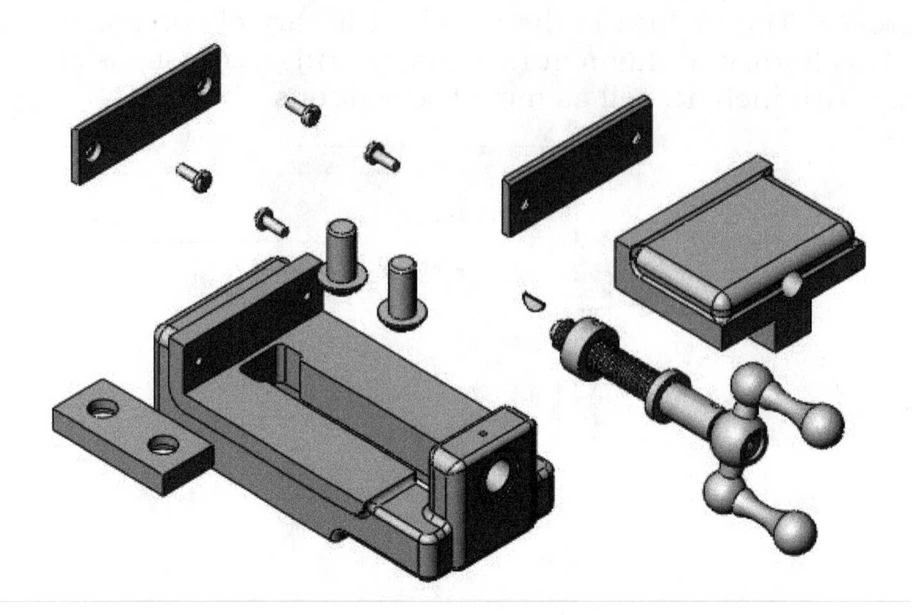

Figure-74. Exploded top assembly

PRACTICE 3

Assemble the Top Assembly and Bottom Assembly created in Practice 1 and Practice 2 for the assembly as shown in Figure-75.

Note that in this practice, you will insert the sub-assemblies in place of inserting single parts.

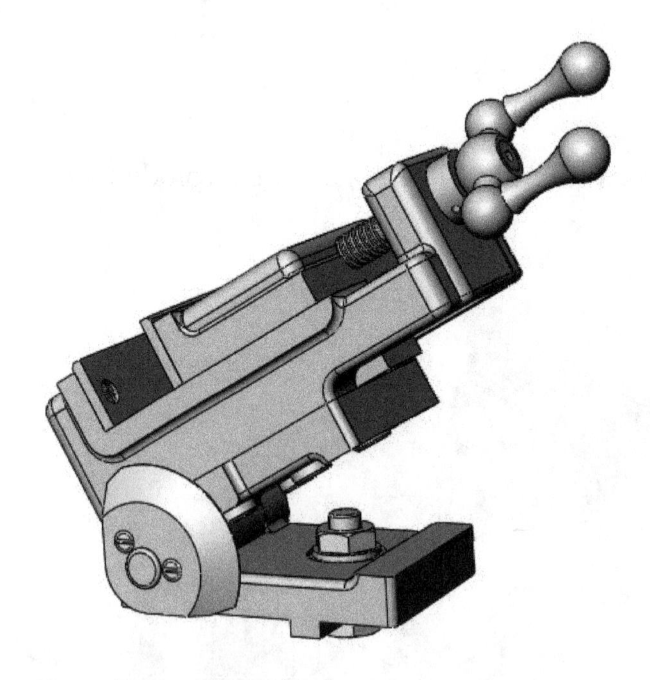

Figure-75. Assembly for practice 3

To get more parts for practice, write us at cadcamcaeworks@gmail.com

SELF ASSESSMENT

Q1. Which tool you should use to create a hole passing through many components in the assembly?

Q2. What is the use of **Smart Fasteners** tool and how you can activate it?

Q3. What are the **Magnetic Mates**?

Q4. If you want to rotate both the pulleys at the same amount by dragging only single pulley, then which of the following check boxes you should select in the **Belt/Chain PropertyManager**?

a) Create belt part
b) Engage belt
c) Driving
d) None of the above

Q5. The **Grooves** tool is used to create .... and .... on the solid.

FOR STUDENT NOTES

Answer to Self-Assessment:
1. Hole Series, **4.** (b), **5.** O-Ring grooves and Retaining Ring grooves

Chapter 9

Surfacing and Practice

Topics Covered

The major topics covered in this chapter are:

- ***Surfacing Introduction.***
- ***Surfacing tools similar to solid creation tools.***
- ***Special tools for surfacing.***
- ***Surface editing tools.***
- ***Surface to solid conversion.***
- ***Practical and Practice.***

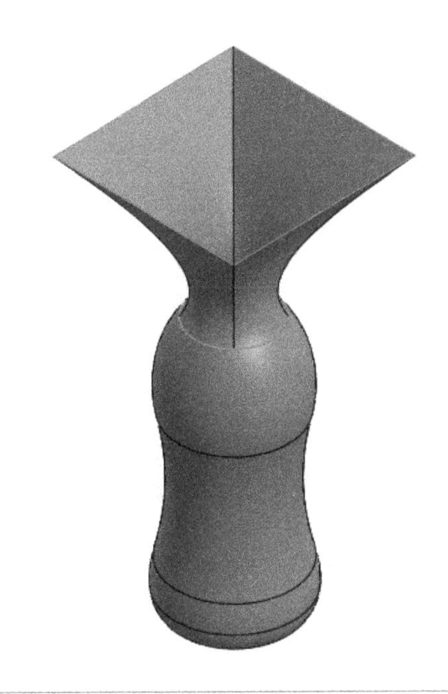

Extruded Surface Revolved Surface Swept Surface Lofted Surface Boundary Surface Filled Surface Freeform
Planar Surface Offset Surface Ruled Surface Surface Flatten Fillet Delete Face Replace Face Knit Surface Cut With Surface Referenc... Curves
Features Sketch Sketch Ink Surfaces Sheet Metal Mold Tools Markup Evaluate MBD Dimensions SOLIDWORKS Add-Ins MBD SOLIDWORKS CAM SOLIDWORKS

SURFACING

Surfacing is a separate world in the field of CAD. The complicated shapes which are difficult for solid modeling are most of the time easy for surfacing. Basic tools of surfacing are very similar to the solid creation tools discussed earlier like, extrude, revolve, sweep, and so on. But there are some other tools that allow to modify the part shape freely in 3D. To start surfacing in SolidWorks, click on the **Surfaces** tab in the **Ribbon** of Part environment. If the tab is not available by default, then right-click on any of the tab in the **Ribbon** and select the **Surfaces** option from the menu. The interface will display as shown in Figure-1. All the tools used for surfacing in SolidWorks are explained as follow:

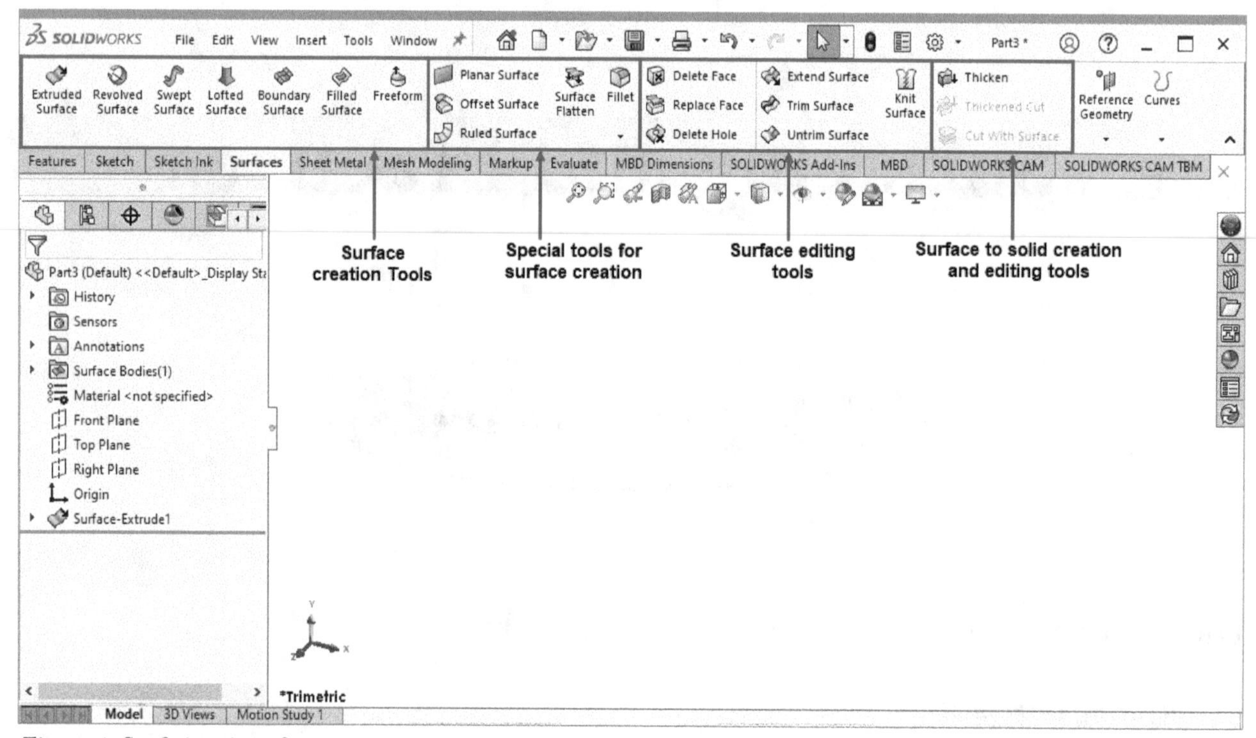

Figure-1. Surfacing_interface

SURFACING TOOLS SIMILAR TO SOLID CREATION TOOLS

The list of these tools is given next.

- **Extruded Surface** similar to **Extruded Boss/Base**
- **Revolved Surface** similar to **Revolved Boss/Base**
- **Swept Surface** similar to **Swept Boss/Base**
- **Lofted Surface** similar to **Lofted Boss/Base**
- **Boundary Surface** similar to **Boundary Boss/Base**

These tools are one by one discussed next.

Extruded Surface

The **Extrude Surface** tool is used to extrude a close or open sketch to the specified height to form a surface. The steps to use this tool are given next.

- Click on the **Extruded Surface** tool from the **Ribbon**. You are asked to select a plane to draw sketch.
- Draw open or close sketch of the surface on desired plane.
- Exit the sketch. Preview of the surface will display; refer to Figure-2.

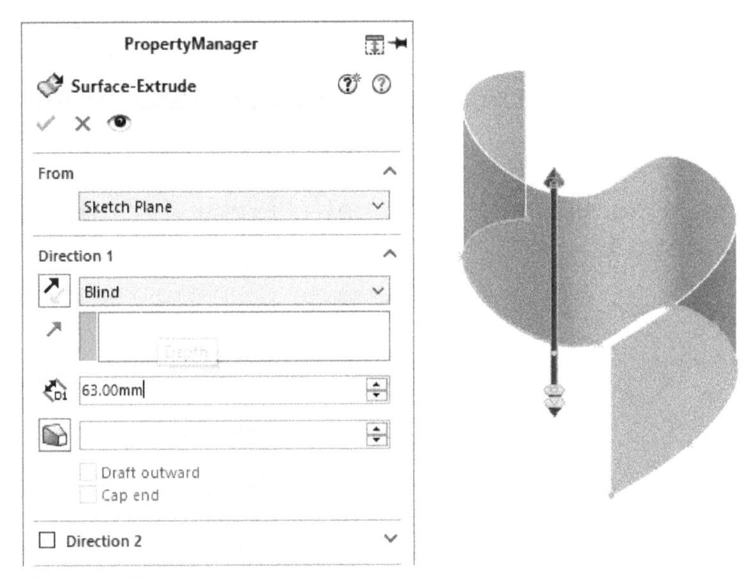

Figure-2. Extruded_surface

- If you draw a close sketch, then you can select the **Cap end** check box to close the end edges.
- The other options are similar to **Extruded Boss/Base** tool.
- Click on the **OK** button from the **PropertyManager** to create extruded surface.

Revolved Surface

The **Revolved Surface** tool is used to revolve a close or open sketch to specified angle with respect to selected reference to form a surface. The steps to use this tool are given next.

- Click on the **Revolved Surface** tool from the **Ribbon**. You are asked to select a plane to draw sketch.
- Draw open or close sketch of the surface on desired plane. Make sure that you create a center line for revolving the sketch.
- Exit the sketch. Preview of the surface will display; refer to Figure-3. Options in the **PropertyManager** have been discussed already.

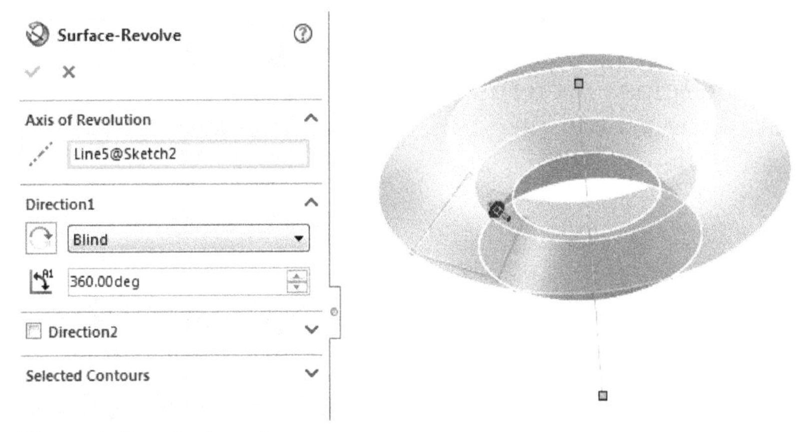

Figure-3. Revolved_surface

- Click on the **OK** button from the **PropertyManager** to create the revolved surface.

Swept Surface

The **Swept Surface** tool is used to sweep a section along the selected path to form a surface. The steps to use this tool are given next.

- Click on the **Swept Surface** tool from the **Ribbon**. **Surface-Sweep PropertyManager** will display.
- Select the open or close section sketch.
- Select the open or close sketch for the path. Preview of the Swept surface will display; refer to Figure-4.

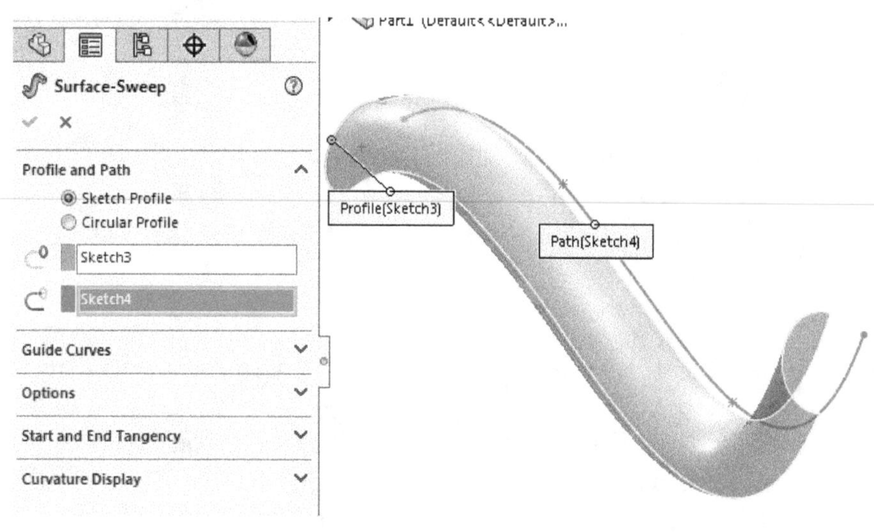

Figure-4. Swept surface

- The options in the **PropertyManager** are same as for **Swept Boss/Base** tool.
- Click on the **OK** button from the **PropertyManager** to create the surface.

Note that for creating **Swept Boss/Base** feature as well as for **Swept surface**, you can select edge of an existing model as a path if required.

Lofted Surface

The **Lofted Surface** tool is used to join two or more open/close sections to form surface. The steps to use this tool are given next.

- Click on the **Lofted Surface** tool from the **Ribbon**. **Surface-Loft PropertyManager** will display.
- Select the first open/close section sketch.
- Select the second open/close section sketch. Preview of the lofted surface will display; refer to Figure-5.

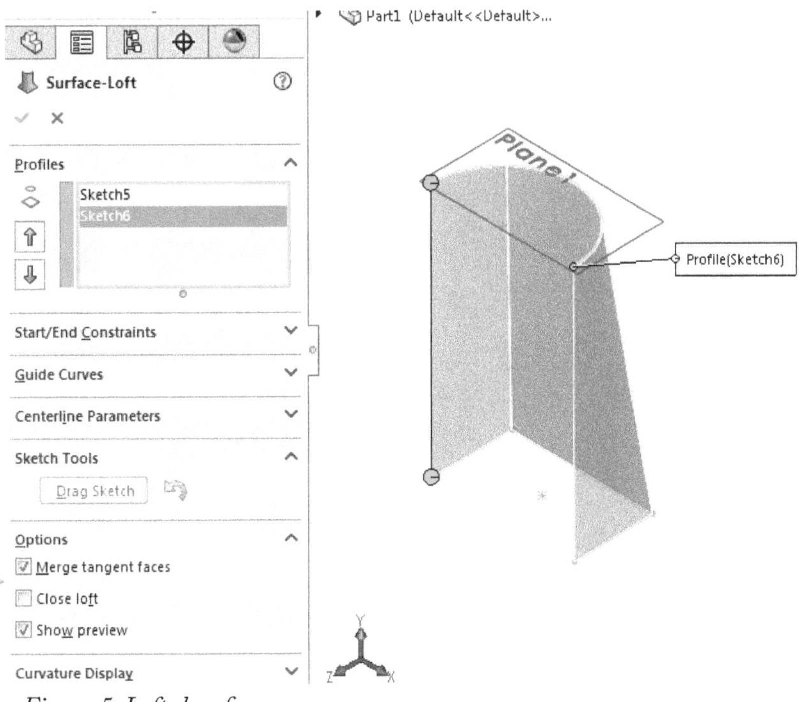

Figure-5. Lofted surface

- Click **OK** button to create the surface.

Note that if you are selecting first section as open sketch then the second section should also be open sketch. Same with the close sketches.

Boundary Surface

The **Boundary Surface** tool is used to join two open/close sections to form surface. The steps to use this tool are given next. This tool works in the same way as the lofted surface.

Note that using the Lofted Surface tool and Boundary Surface tool, you can join two or more surfaces, edges of solids and face of solids. Refer to Figure-6 and Figure-7.

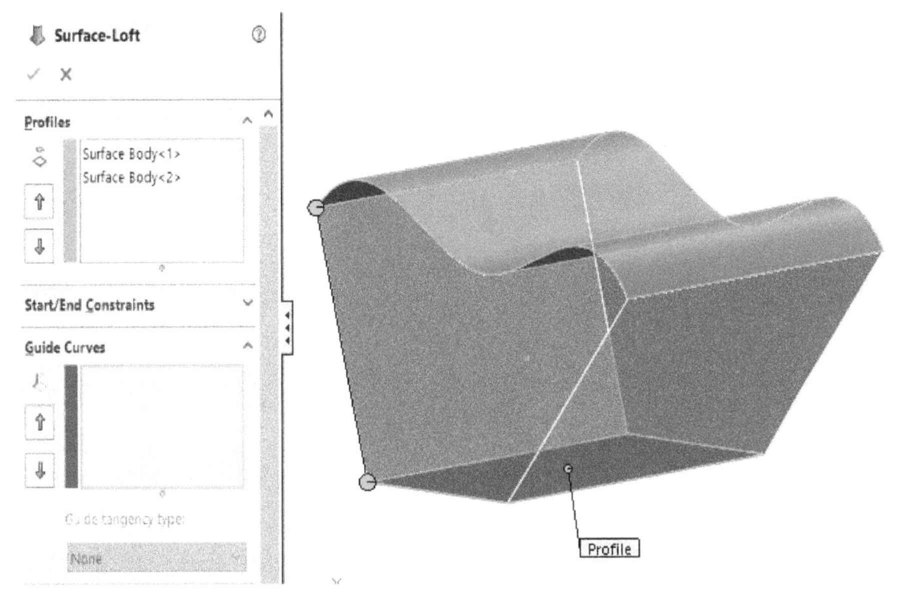

Figure-6. Surfaces joined by loft

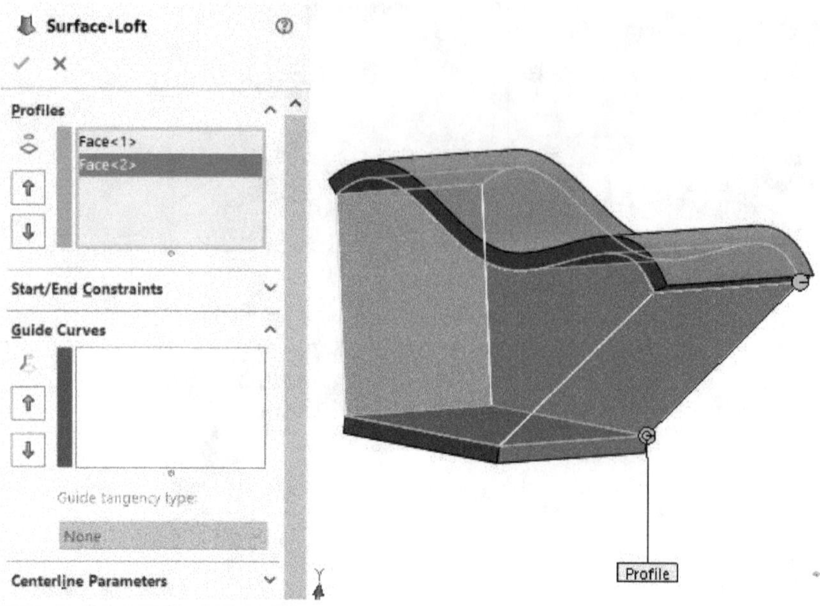

Figure-7. Faces of solids joined by loft

Filled Surface

The **Filled Surface** tool is use to fill gap by selecting the close boundary. This close boundary can be made by creating sketches or it can be formed by intersection of solids/surfaces. The steps to use this tool are given next.

- Click on the **Filled Surface** tool from the **Ribbon**. The **Fill Surface PropertyManager** will display; refer to Figure-8.

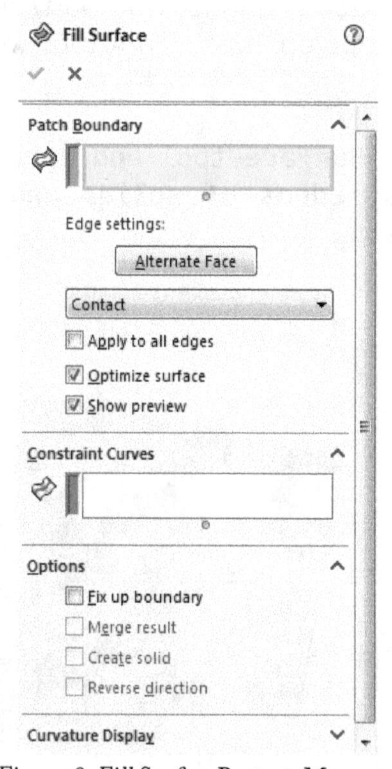

Figure-8. Fill Surface PropertyManager

- Select the boundary of area that you want to fill using the surface. Preview of the surface will display; refer to Figure-9.

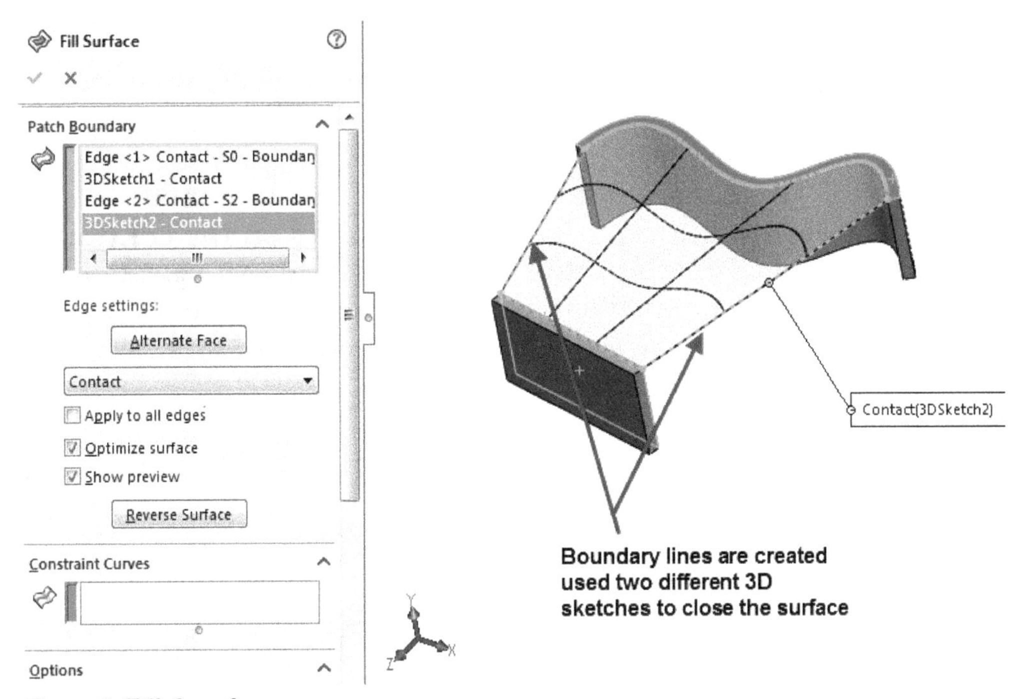

Figure-9. Filled_surface

This tool is the most used one for closing surfaces. Using this tool, you can also close circular holes in surfaces/solids. Note that when you need to fill holes in model for creating parting surface for mold design, this is the tool to be used most.

Freeform

The **Freeform** tool is used to freely deform solid faces/surfaces. The steps to use this tool are given next.

- Click on the **Freeform** tool from the **Ribbon**. The **Freeform PropertyManager** will display; refer to Figure-10.

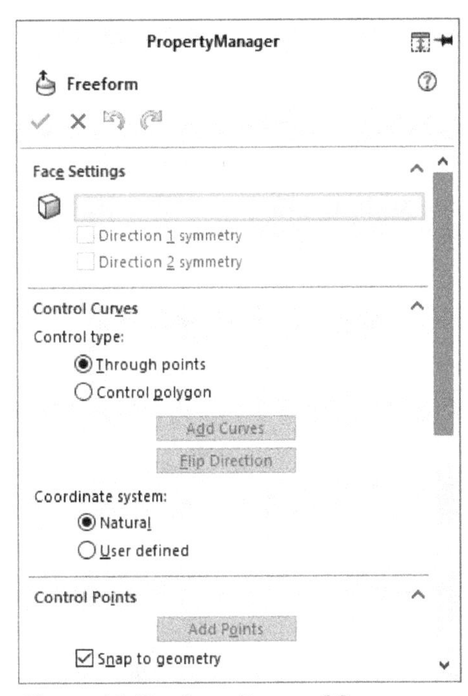

Figure-10. Freeform_PropertyManager

- Select the surface/face you want to deform. Mesh on curves will display on the surface.
- Click on the **Add Curves** button and click at desired position over the surface to set the highlighted line as control curve. You can select more than one curve for controlling the surface shape.
- Click again on the **Add Curves** button to exit the selection mode.
- Click on the **Add Points** button to add control points on the control curve.
- Click again on the **Add Points** button or right-click to exit selection mode.
- Drag the point you have created earlier to change the shape of the surface/face. Refer to Figure-11.

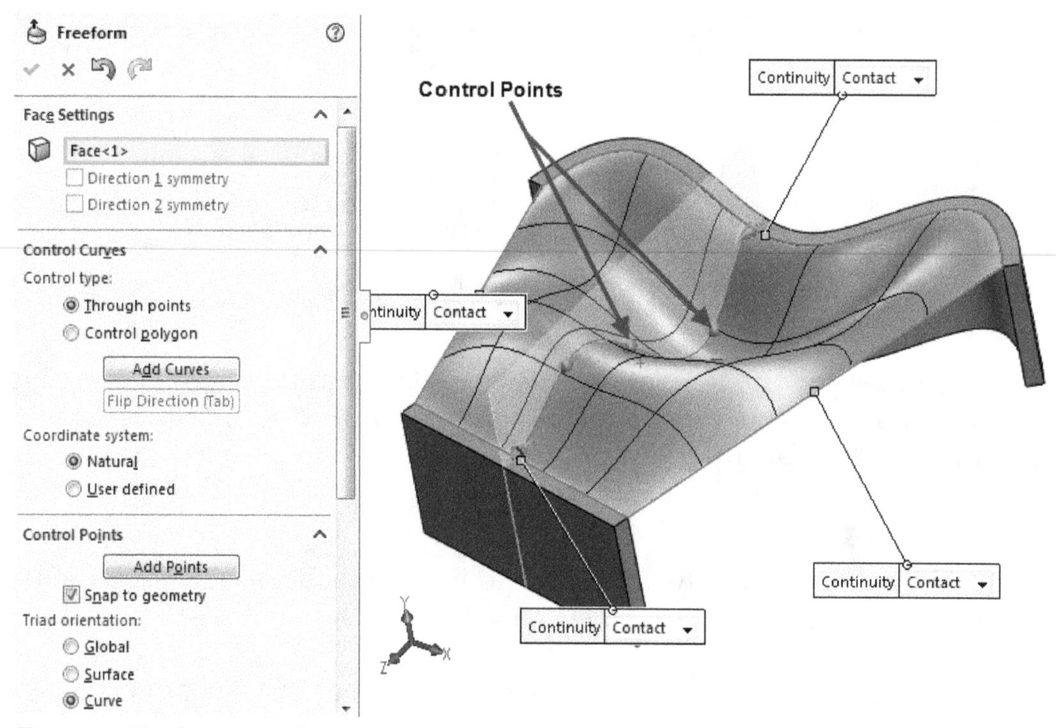

Figure-11. Freeforming_surface

- Click on the **OK** button from the **PropertyManager** to apply the modification.

SPECIAL SURFACING TOOLS

Earlier, you have learned the tools that were similar to the Solid creation tools. Now, we will discuss about the special tools provided in SolidWorks to create surfaces. These tools are discussed next.

Planar Surface

The **Planar Surface** tool is used to create a surface joining two or more edges or a sketch that is in the same plane. The steps to use this tool are given next.

- Click on the **Planar Surface** tool from the **Ribbon**. The **Planar Surface PropertyManager** will display; refer to Figure-12.

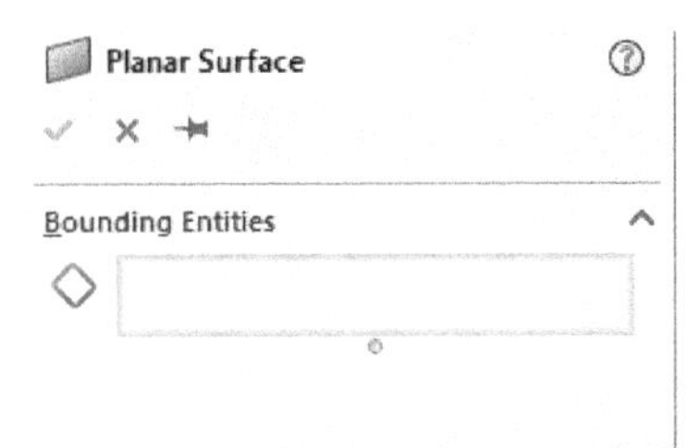

Figure-12. Planar Surface PropertyManager

- Select the edges that form a planar surface. The preview will be displayed; refer to Figure-13.

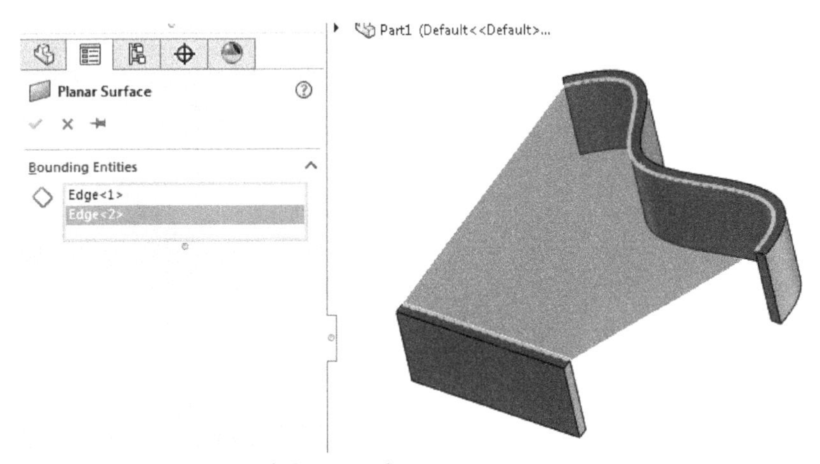

Figure-13. Preview of planar surface

- Click on the **OK** button from the **PropertyManager** to create the surface.

Offset Surface

The **Offset Surface** tool is used to create a surface at an offset distance from the selected face/surface. The steps to create offset surface are given next.

- Click on the **Offset Surface** tool from the **Ribbon**. The **Offset Surface PropertyManager** will display; refer to Figure-14.
- Select a surface/face by which you want to create the offset surface. Preview of the surface will be displayed; refer to Figure-15.

Figure-14. Offset Surface PropertyManager

Figure-15. Preview of offset surface

- Specify desired distance in the spinner and click on the **OK** button to create the offset surface.

Ruled Surface

The **Ruled Surface** tool is used to create a combination of surface adjoining to each other. This type of surface becomes very important while creating parting surface for molding/casting. The steps to use this tool are given next.

- Click on the **Ruled Surface** tool from the **Ribbon**. The **Ruled Surface PropertyManager** will display; refer to Figure-16.
- Select the edges using which you want to create the ruled surface. Preview of surface will display; refer to Figure-17.

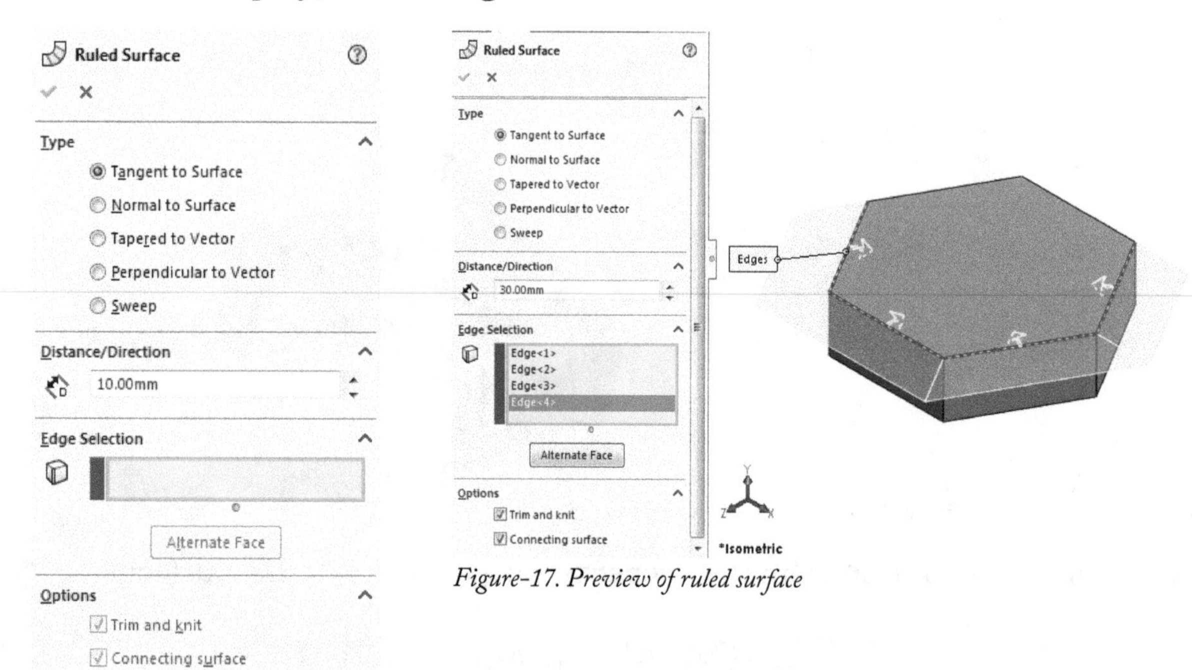

Figure-17. Preview of ruled surface

Figure-16. Ruled_Surface_PropertyManager

- Increase the length of surfaces by using the spinner in the **Distance/Direction** rollout.
- If you select the **Normal to Surface** radio button, then the surfaces will be displayed as shown in Figure-18.

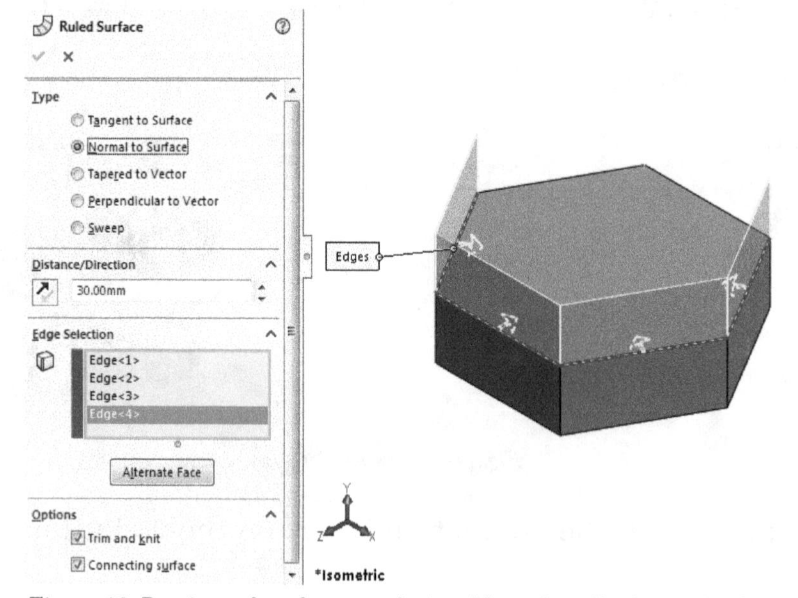

Figure-18. Preview_of_surface_on_selecting_Normal_to_Surface_radio_button

- If you want to create surface tapered to the selected vector, then select the **Tapered to Vector** radio button and then select the edge to specify vector. Also, set the angle in the **Angle** spinner. Preview of the surface will be displayed; refer to Figure-19.

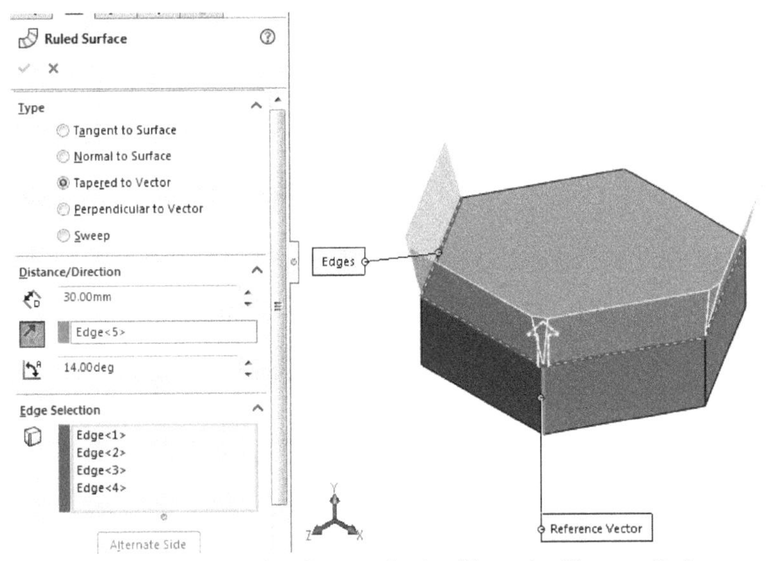

Figure-19. Preview of surface on selecting Tapered to Vector radio button

- Select the **Perpendicular to Vector** radio button to create the surfaces perpendicular to selected vector. Refer to Figure-20.

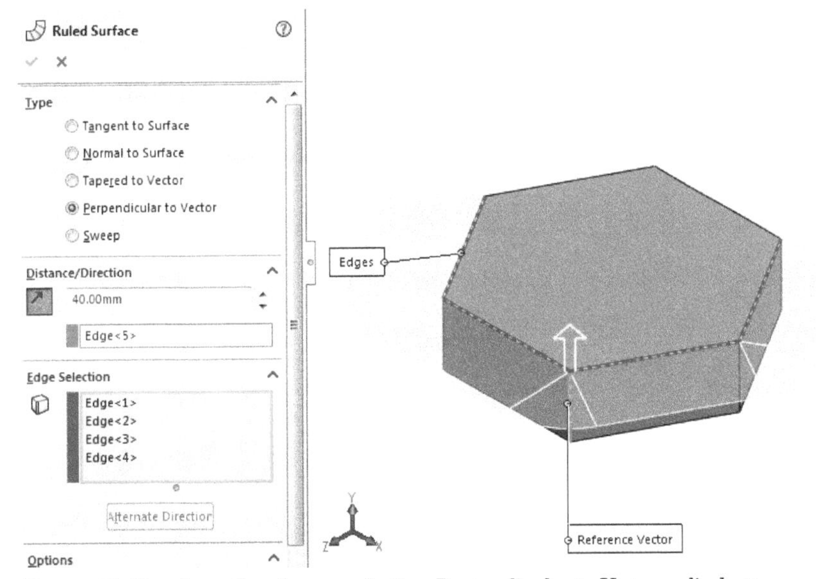

Figure-20. Preview of surface on selecting Perpendicular to Vector radio button

- Select the **Sweep** radio button to sweep the surface along the selected vector. Preview will be displayed; refer to Figure-21.
- After selecting desired option, click on the **OK** button to create the surfaces.

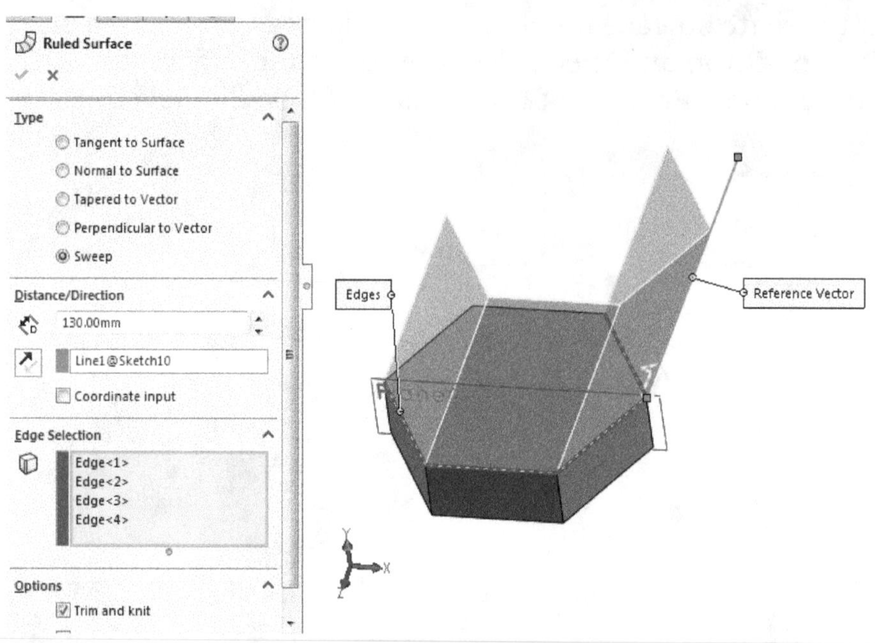

Figure-21. Preview of surface on selecting Sweep radio button

Surface Flatten tool

The **Surface Flatten** tool as the name suggests is used to make a flat surface by flattening various interconnected surfaces. Note that before using this tool, there must be surfaces in the viewport. The procedure to use this tool is given next.

- Click on the **Surface Flatten** tool from the **Ribbon**. The **Flatten PropertyManager** will be displayed; refer to Figure-22.
- Select the surface/surfaces that you want to flatten; refer to Figure-23.

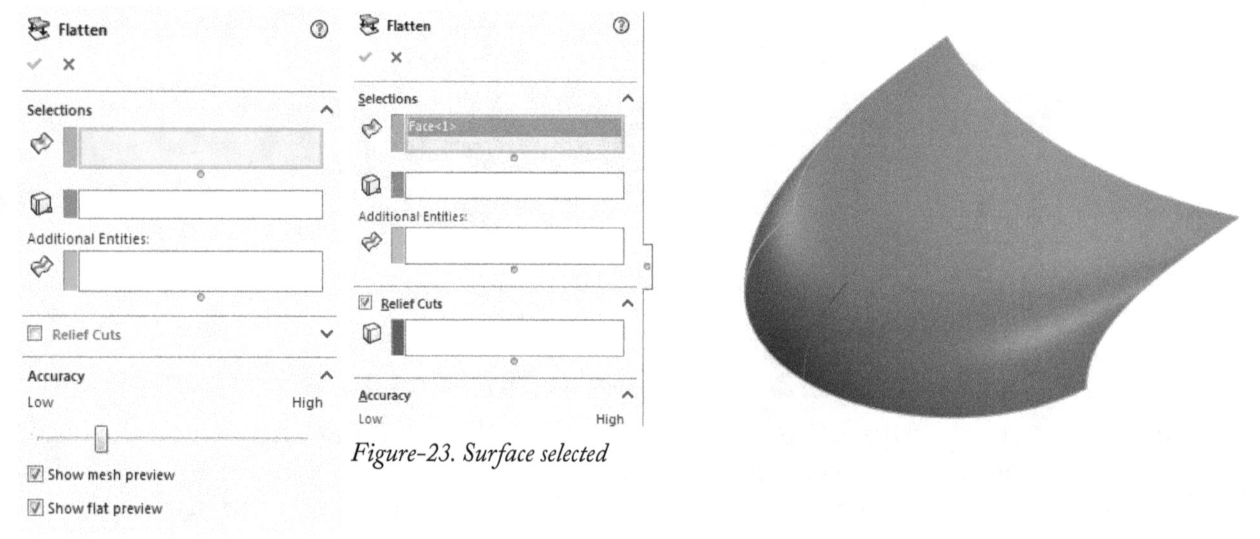

Figure-22. Flatten PropertyManager

Figure-23. Surface selected

- Click in the **Vertex/Point** collector to select the fixed reference.
- Select the corner vertex of the surface. Preview of the surface will be displayed; refer to Figure-24.

Figure-24. Preview of flattened surface

- If you want to apply relief in the surface, then select the **Relief Cuts** check box and click in the selection box in **Relief Cuts** rollout. You will be asked to select a curve.
- Click on the curve that you want to use for relief cut; refer to Figure-25.

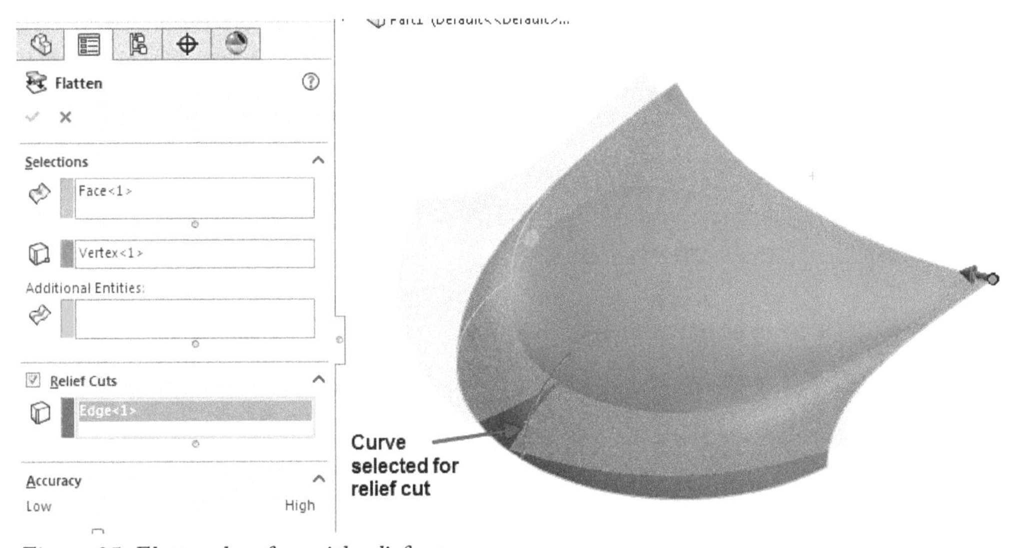

Figure-25. Flattened surface with relief cut

- Click on the **OK** button from the **PropertyManager** to create the flattened surface. Note that the parent surface will not be deleted by this operation.

You can use the **Fillet** and **Chamfer** tools in the same way as discussed earlier.

SURFACE EDITING TOOLS

The tools in this category are used to edit the surfaces. For example, trimming the surface, deleting some portion of the surface, knitting two surfaces for applying fillet. These tools are explained next.

Delete Face

The **Delete Face** tool is used to remove a face/surface from the model. The procedure to use this tool are given next.

- Click on the **Delete Face** tool from the **Ribbon**. The **Delete Face PropertyManager** will display as shown in Figure-26.
- Select the face that you want to remove.

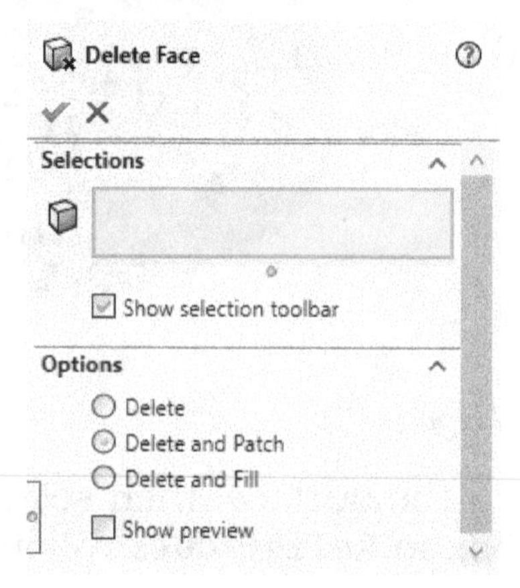

Figure-26. Delete Face PropertyManager

- Select desired radio button, if you want to patch or fill the gap created by deleting the surface.
- Click on the **OK** button from the **PropertyManager**. The face will be deleted. Note that using this tool, you cannot delete a single face surface body.

Replace Face

The **Replace Face** tool is used to extend the selected face to the replacement face. The procedure to use this is given next.

- Click on the **Replace Face** tool. The **Replace Face PropertyManager** will display; refer to Figure-27.

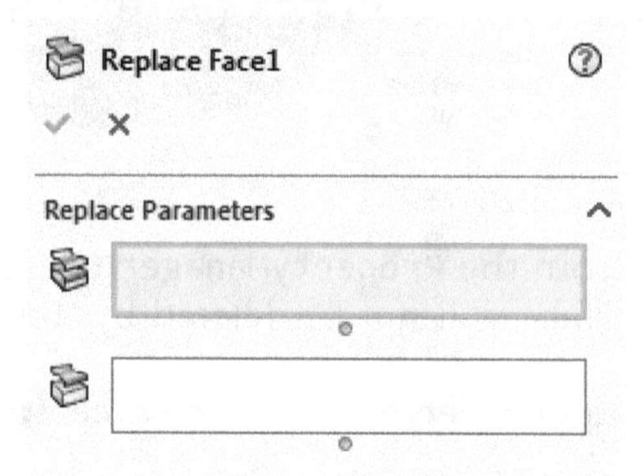

Figure-27. Replace Face PropertyManager

- Select the face/faces that you want to replace.
- Click in the next selection box in the **PropertyManager** and select the surface by which you want to replace the face.
- Click on the **OK** button to apply the replacement; refer to Figure-28.

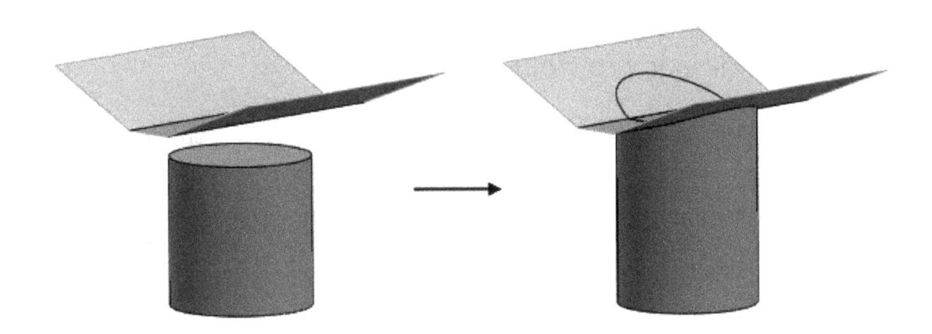

Figure-28. Output of replace face

Delete Hole

The **Delete Hole** tool is used to remove holes created on the surface by using a closed profile. The procedure to use this tool is given next.

- Click on the **Delete Hole** tool from the **Ribbon**. The **Delete Face PropertyManager** will be displayed; refer to Figure-29.

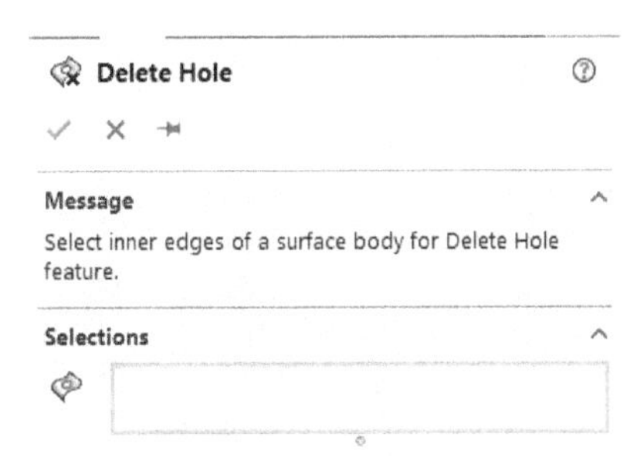

Figure-29. Delete Hole PropertyManager

- Select the edges of holes/cuts to be removed from the surface. Preview of filled surface will be displayed; refer to Figure-30.

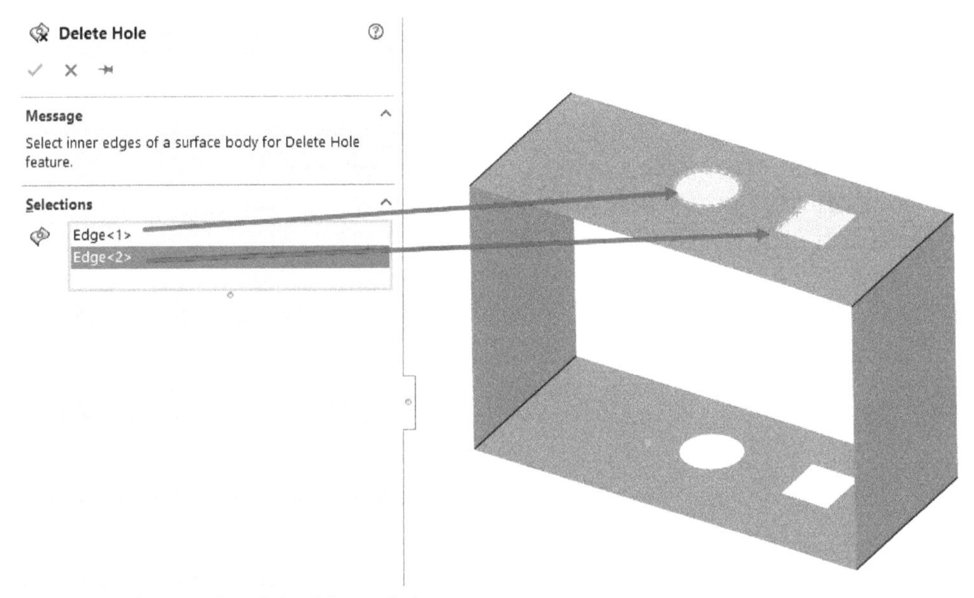

Figure-30. Edges selected for deleting hole

- Click on the **OK** button from **PropertyManager** to delete holes.

Extend Surface

The **Extend Surface** tool is used to extend selected surface by specified value. The procedure to use this tool is given next.

- Click on the **Extend Surface** tool. The **Extend Surface PropertyManager** will display; refer to Figure-31.
- Select the edge of the surface that you want to extend. Preview of extension will display; refer to Figure-32.

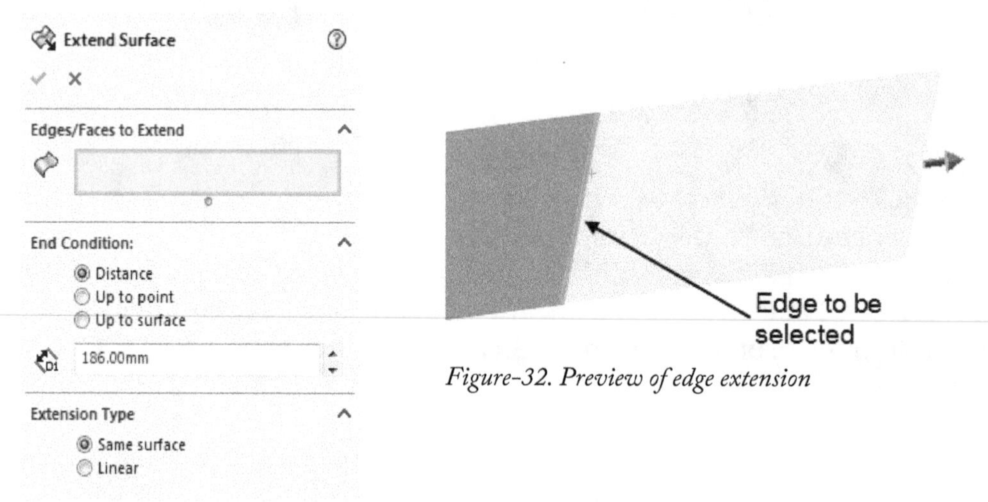

Figure-32. Preview of edge extension

Figure-31. Extend Surface PropertyManager

- Specify the length of extension in the spinner. If you want to extend the surface to a selected point/surface then select the respective radio button and then the reference. The surface will extend to that reference.

Trim Surface

The **Trim Surface** tool is used to remove desired portion of the surface by using sketch/other surface. The procedure to use this tool is given next.

- Click on the **Trim Surface** tool. The **Trim Surface PropertyManager** will display; refer to Figure-33.
- Select the sketch or surface that you want to use as trimming tool. You are asked to select the portion that you want to keep; refer to Figure-34.
- Select the portion that you want to keep. The surface on the other side of trimming surface or sketch will be removed.

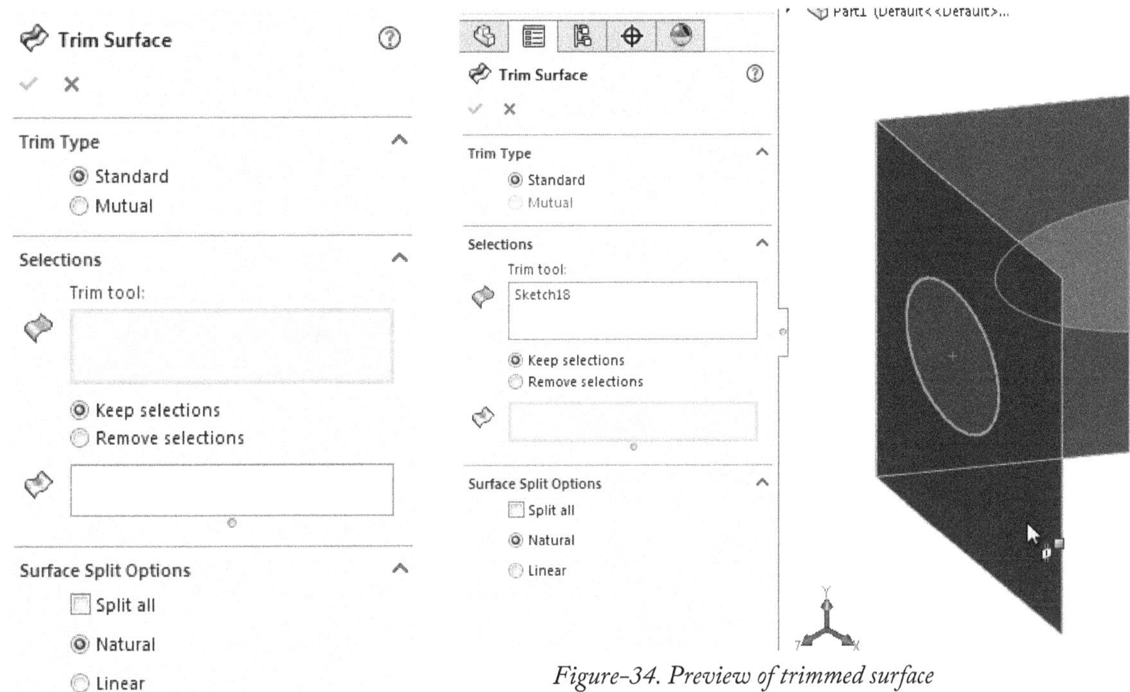

Figure-33. Trim Surface PropertyManager

Figure-34. Preview of trimmed surface

Untrim Surface

The **Untrim** tool is used to undo the trimmed surfaces. The following steps explain the procedure to untrim the surface.

- Select the **Untrim Surface** tool from the **Ribbon**. The **Untrim Surface PropertyManager** will display; refer to Figure-35.

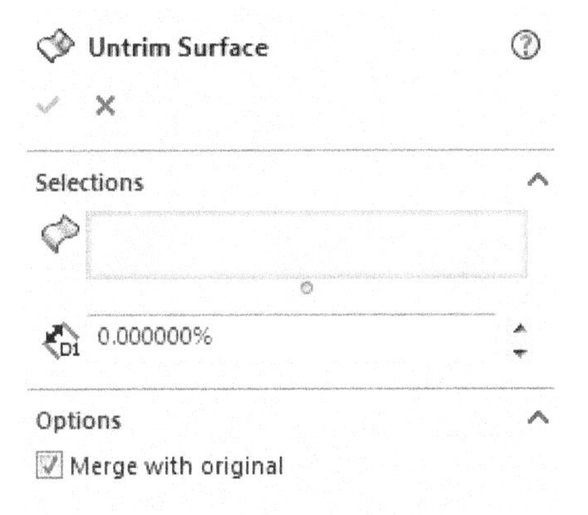

Figure-35. Untrim Surface PropertyManager

- Select the surface on which the trimming operations are to be performed.
- Click on the **OK** button from the **PropertyManager**. The surface will be untrimmed and all the gaps will be patched.

Knit Surface

The **Knit Surface** tool is used to combine two or more surfaces at their common edges. If the surface form a close boundary then this tool turn the surfaces in to a solid. The steps to use this tool are given next.

- Click on the **Knit Surface** tool from the **Ribbon**. The **Knit Surface PropertyManager** will display; refer to Figure-36.

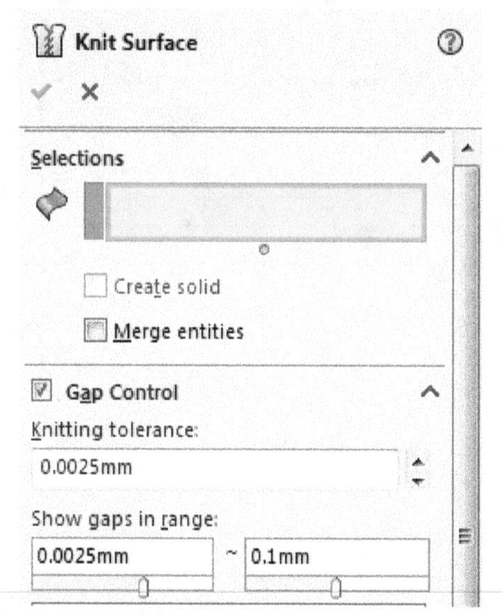

Figure-36. Knit Surface PropertyManager

- Select the surfaces that you want to knit together.
- Click **OK** button. The surfaces will be combined into a single entity. All the knit surfaces are added in the **Surface Bodies** folder; refer to Figure-37.

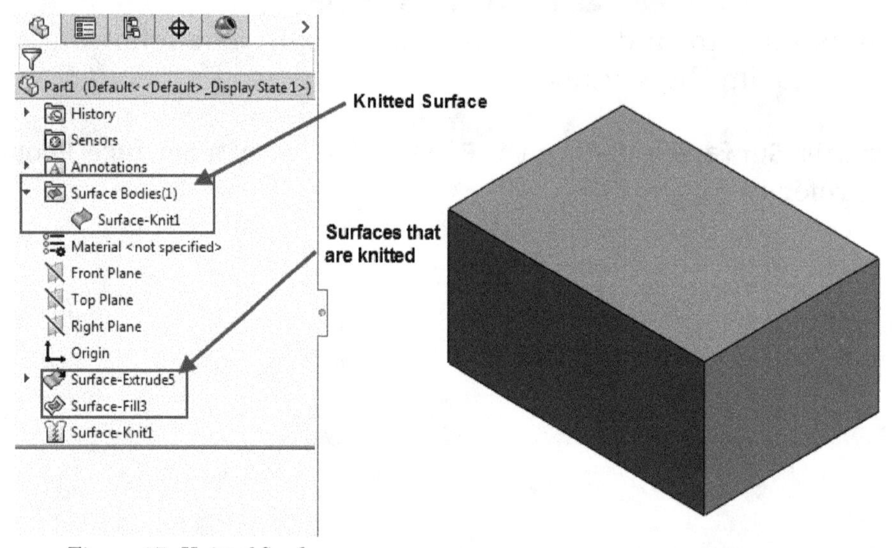

Figure-37. Knitted Surfaces

Note that the knitted surfaces are used as parting surfacing in molding and casting. Also, if the surfaces form a closed boundary then select the **Create solid** check box to create solid from surface.

Thicken

The **Thicken** tool is used to add thickness to the surface. The procedure to use this tool is given next.

- Click on the **Thicken** tool from the **Ribbon**. The **Thicken PropertyManager** will display; refer to Figure-38.

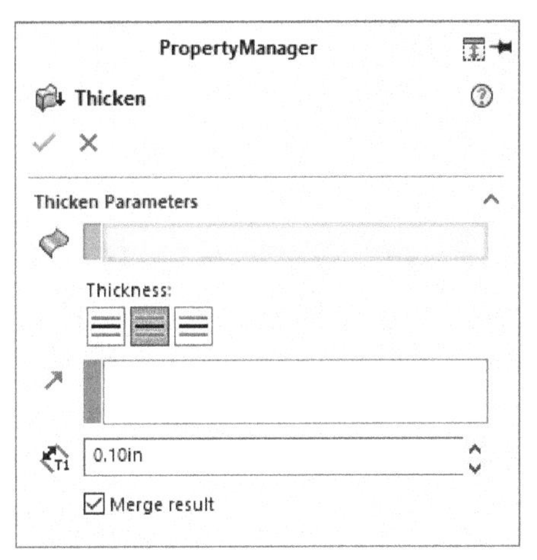

Figure-38. Thicken_PropertyManager

- Select the surface to which you want to add thickness. The preview of thickened surface will display; refer to Figure-39.

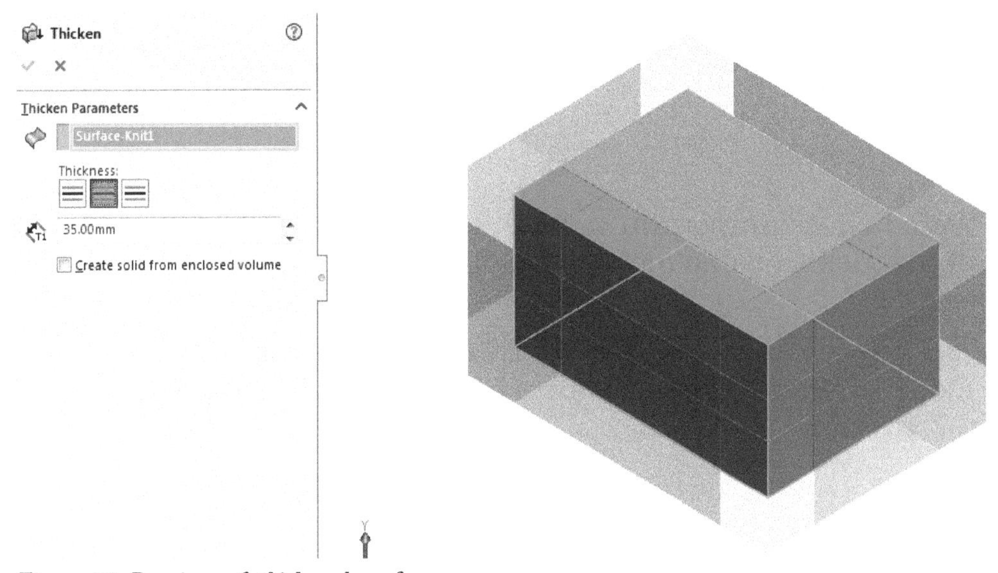

Figure-39. Preview_of_thickened_surface

- Specify desired thickness in the spinner and change the side of thickness application by selecting desired button.

Thickened Cut

The **Thickened Cut** tool is used to remove material by thickening the surface. This tool works in the same way as **Thicken** tool but it removes material in place of adding.

Cut with Surface

The **Cut with Surface** tool is used to cut solids by using the surface. The steps to use this tool are given next.

- Click on the **Cut with Surface** tool from the **Ribbon**. The **SurfaceCut PropertyManager** will display; refer to Figure-40.

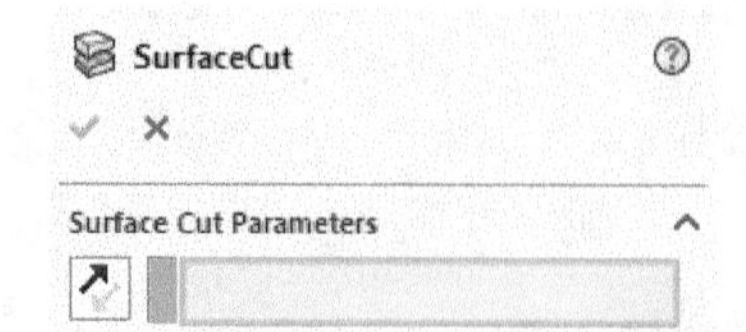

Figure-40. SurfaceCut PropertyManager

- Select the surface which you want to use for cutting the solids.
- The solids in the direction of arrow will be removed. Click on the arrow to flip the direction.
- Click on the **OK** button to cut the solids; refer to Figure-41.

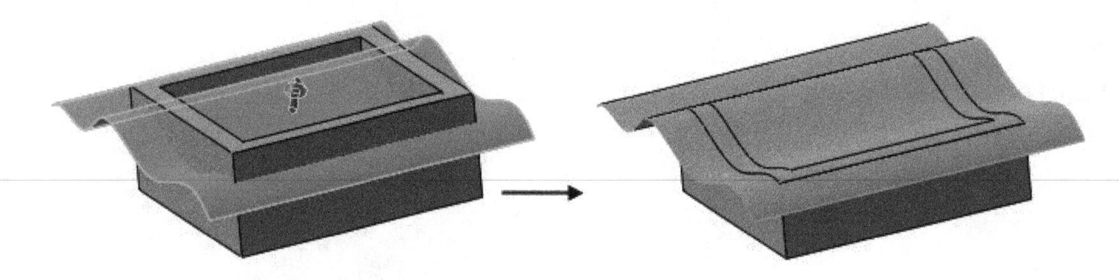

Figure-41. Solid cut by surface

We have covered all the important tools that are used for surfacing in SolidWorks. Now, we will practice on some models to apply these tools.

Practical 1

Create the model of helmet glass as shown in Figure-42. The dimensions of the model are given in Figure-43.

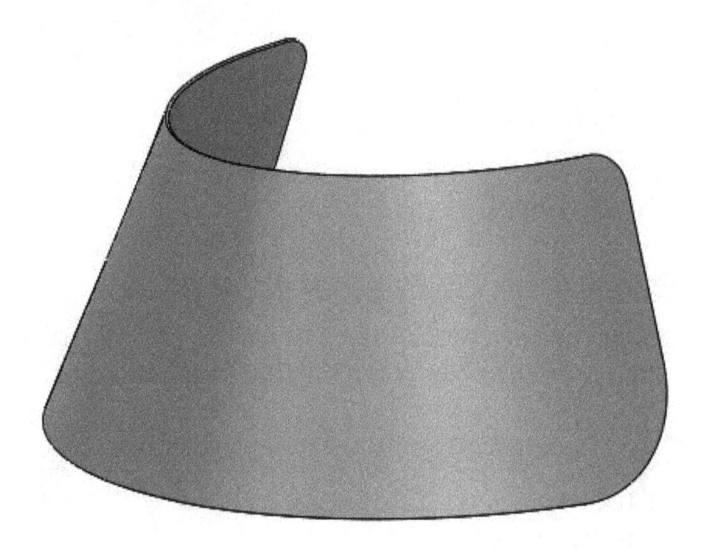

Figure-42. Practical1 model

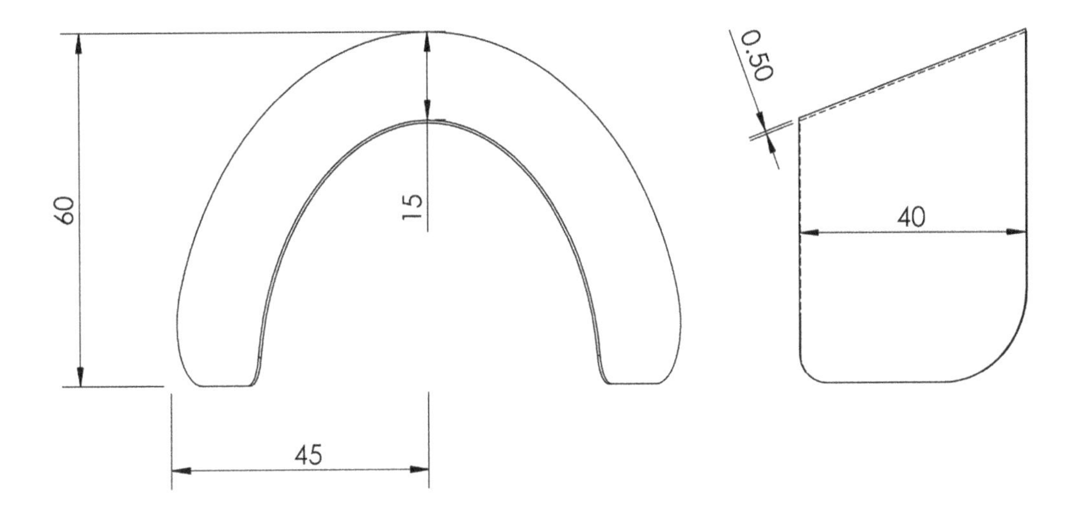

Figure-43. Practical1 drawing

The model displayed is having very low thickness and its having complex 3D shape. So, it is a good idea to use surfacing in this case.

We can create this model by lofted surface easily. For that we need to have two sketches.

Creating first sketch

- Click on the **Sketch** tab and select the **Sketch** button. You are asked to select sketching plane.
- Click on the **Top** plane. The sketching environment will display.
- Click on the **Ellipse** tool from the **Ribbon** and draw an ellipse as shown in Figure-44.
- Draw center line passing through coordinate system and trim the bottom portion of the ellipse; refer to Figure-45.

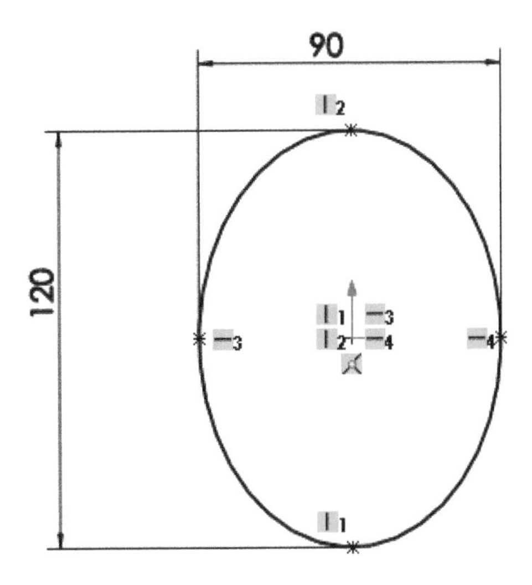

Figure-44. Ellipse to be drawn

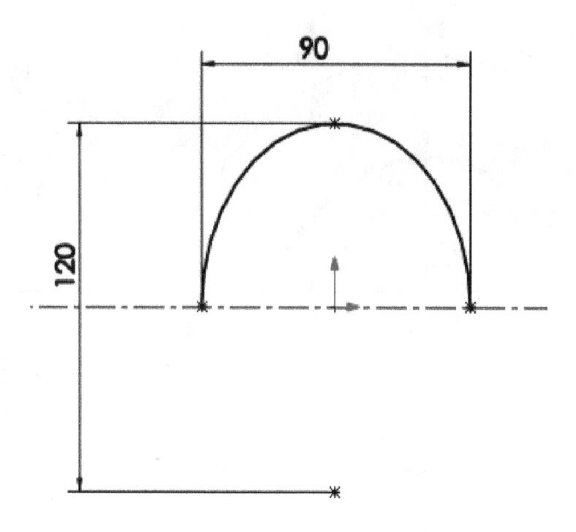

Figure-45. Trimmed ellipse

- Click on the **Exit Sketch** button to exit the sketch.

Creating second sketch

- Click on the **Plane** tool from **Reference Geometry** drop-down in **Surfaces** tab of the **Ribbon** and create a plane at an offset distance of **40** above the **Top** plane; refer to Figure-46.

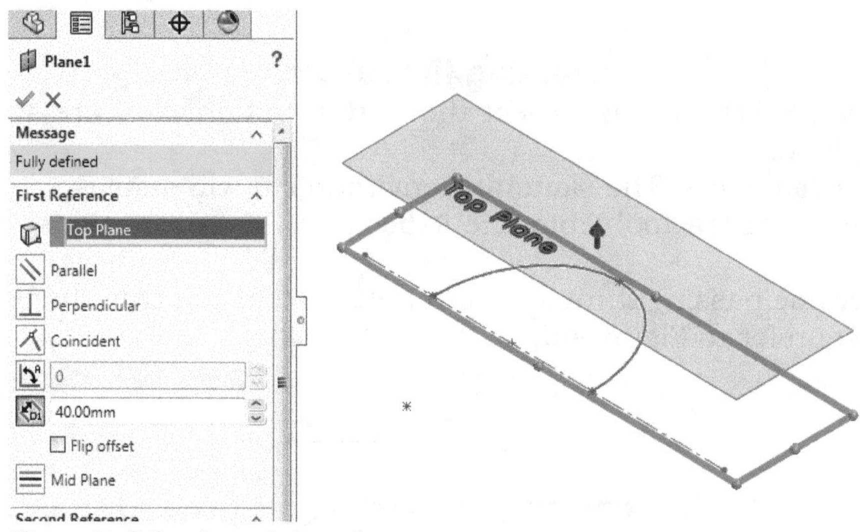

Figure-46. Offset plane to be created

- Click on the **Sketch** button from the **Sketch** tab and select the **Top** plane as sketching plane. Press **CTRL+8** if sketching plane is not parallel.
- Click on the **Offset Entities** tool from the **Ribbon** and selected the earlier created sketch.
- Click on the **Reverse** check box and specify the value as **15** in the spinner; refer to Figure-47.

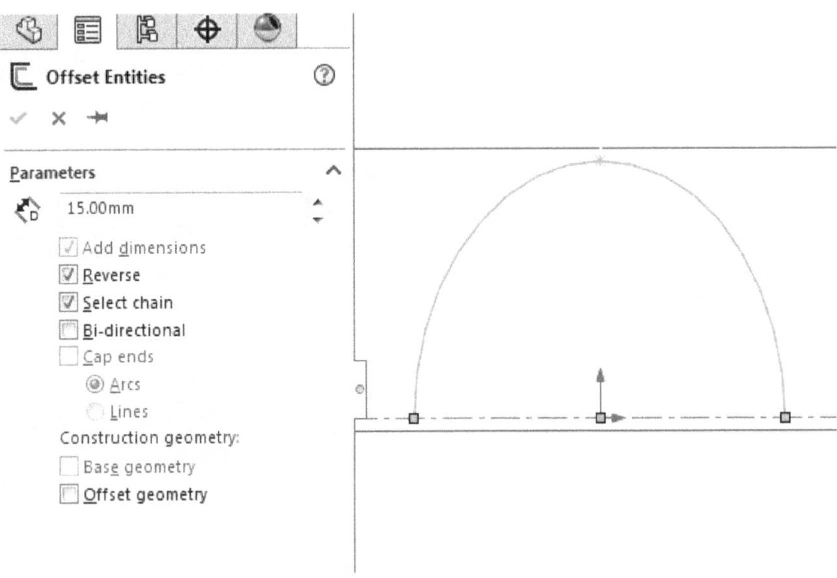

Figure-47. Preview of offset entities

- Click on the **OK** button to create the offset curve.
- Click on the **Exit Sketch** button.

Creating lofted surface

- Click on the **Lofted Surface** tool from **Surfaces** tab of the **Ribbon**. The **Surface-Loft PropertyManager** will display.
- Select the two sketches one by one. Preview of the surface will display; refer to Figure-48.

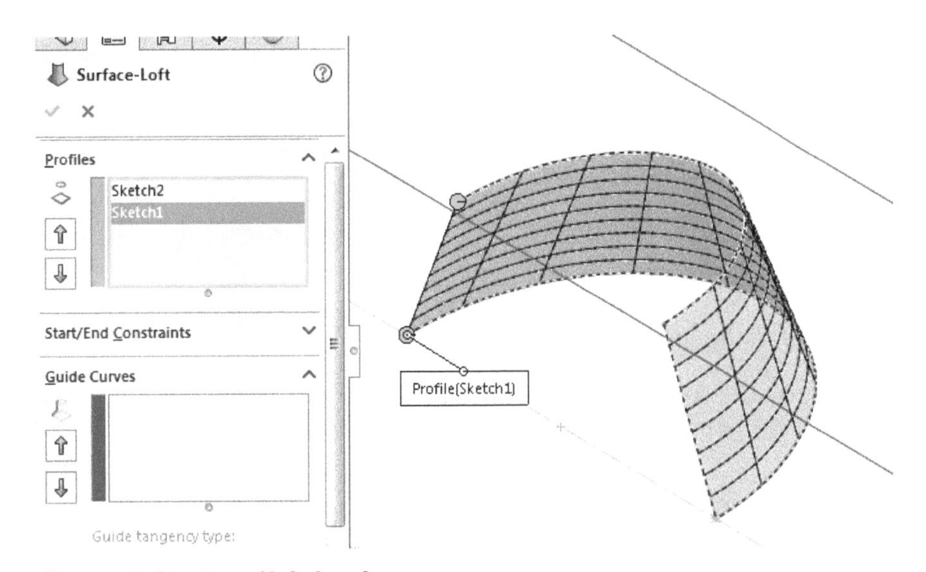

Figure-48. Preview of lofted surface

- Click on the **OK** button to create the surface.
- To hide the plane, select it and right-click on it. Select the **Hide** button from the shortcut menu box.

Thickening surface and applying fillet

- Click on the **Thicken** tool from the **Ribbon** and select the surface. Preview of thickened surface will display.
- Enter the thickness value as **0.5**.

- Click **OK** to create the solid.
- Click on the **Fillet** tool and apply suitable fillets at the small edges on the corners. Refer to Figure-42.

Practical 2

Create the model of flower vase as shown in Figure-49. The dimensions of the model are given in Figure-50.

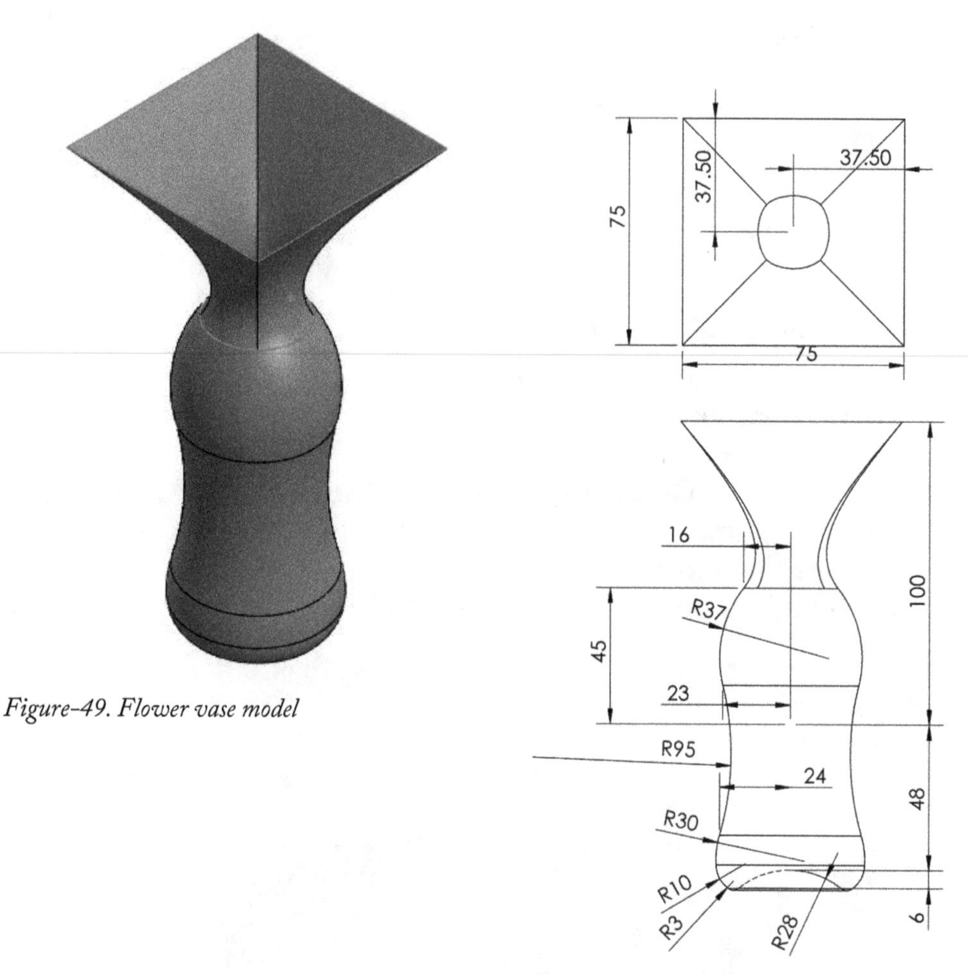

Figure-49. Flower vase model

Figure-50. Practical 2 Drawing

The model displayed is having no thickness and its having complex 3D shape. So, we will be using surfacing to create this model.

We can create this model in two steps. In step 1, we will create a revolved surface and in step 2 we will connect the revolved surface to a rectangle by using **Lofted Surface** tool. The procedure to create this surface model is given next.

Creating Revolved Surface

- Click on the **Sketch** tab and select the **Sketch** button. You are asked to select sketching plane.
- Click on the **FRONT** plane. The sketching environment will display.
- Using the **3 Point Arc** tool create the sketch as shown in Figure-51. Make sure you create a centerline as shown in the figure.

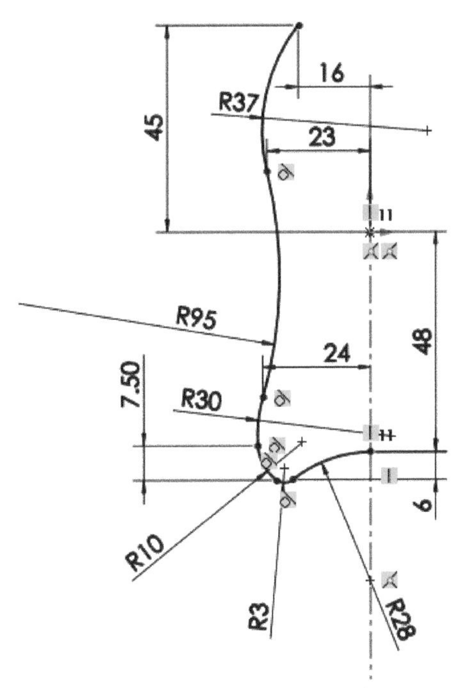

Figure-51. Sketch for revolved surface

- Click on the **Exit Sketch** button and Select the **Revolved Surface** tool from the **Surfaces** tab in the **Ribbon**.
- Select the sketch created recently. Preview of the revolved surface will be displayed; refer to Figure-52.

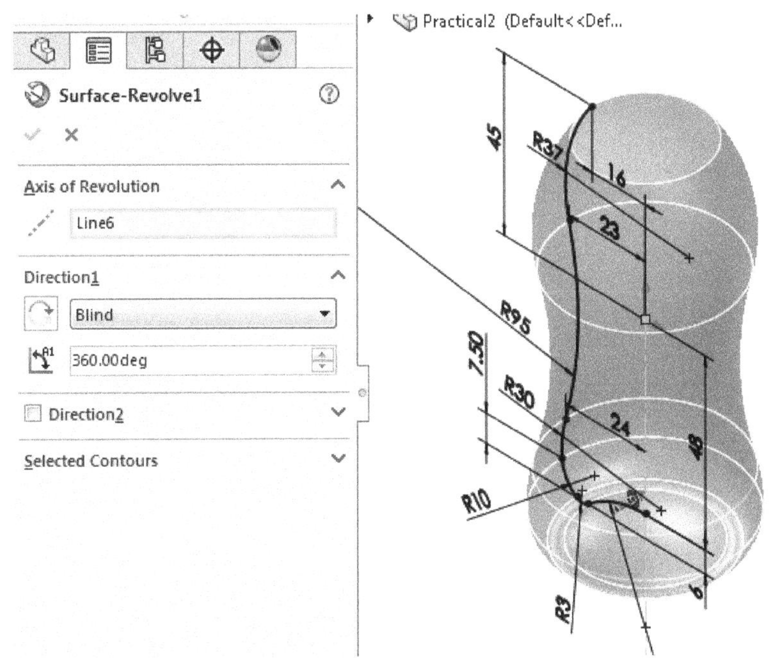

Figure-52. Preview of revolved surface

- Click on the **OK** button from the **PropertyManager** displayed to create the surface.

Creating Lofted Surface

- Click on the **Plane** tool from the **Reference Geometry** drop-down in the **Ribbon** and create an offset plane at distance of **100** from the TOP plane; refer to Figure-53.
- Create a rectangle on the selected plane; refer to Figure-54.

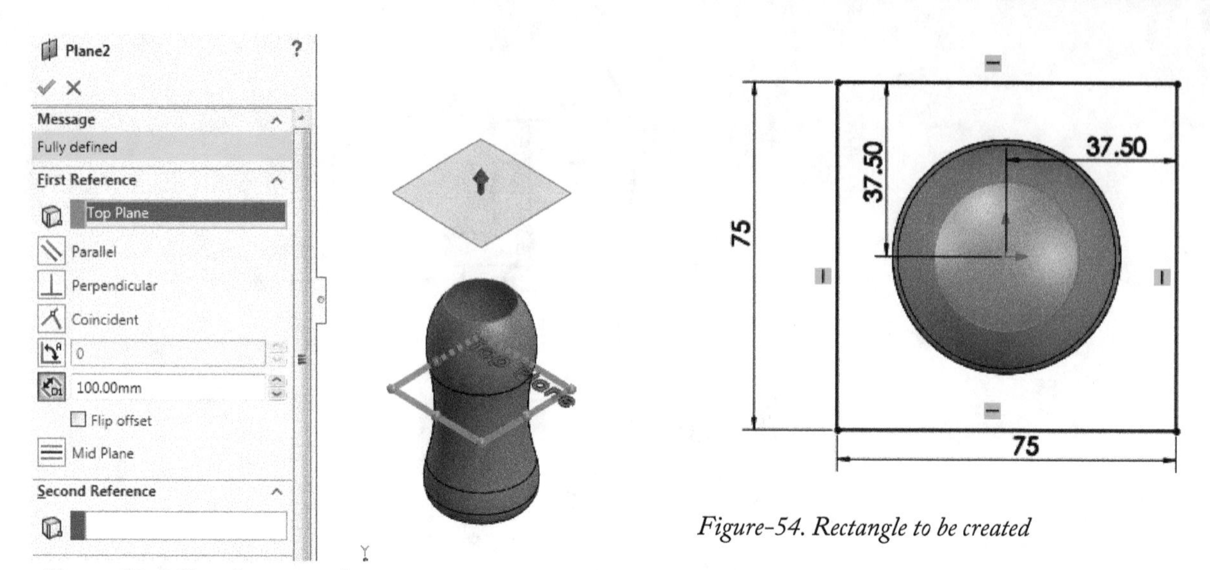

Figure-53. Offset plane created

Figure-54. Rectangle to be created

- Click on the **Lofted Surface** tool from the **Ribbon**. The **Surface-Loft PropertyManager** will be displayed.
- Select the recently created sketch and edge of the surface. Preview of the lofted surface will be displayed; refer to Figure-55.

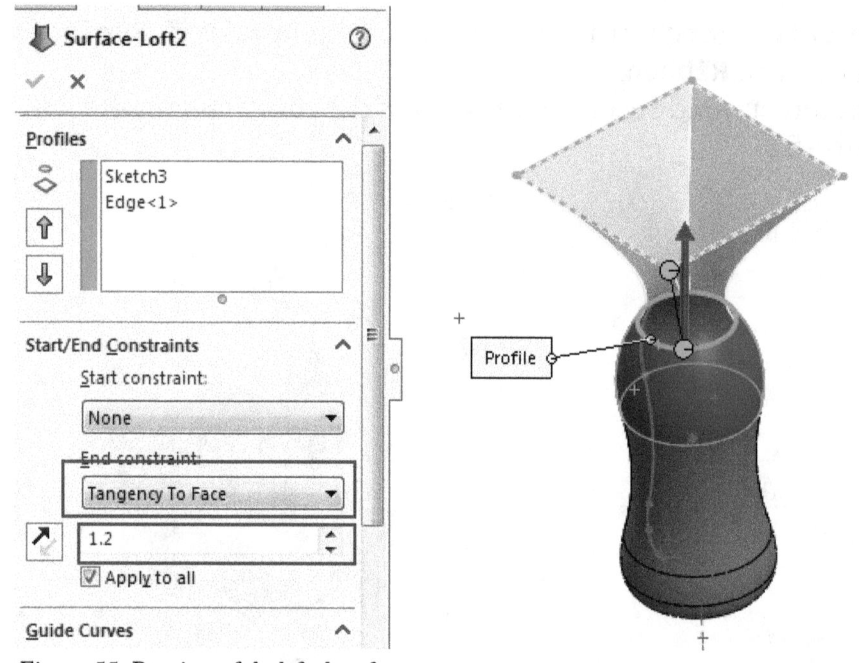

Figure-55. Preview of the lofted surface

- Make sure you set the same values as highlighted in red boxes in the Figure-55. Click on the **OK** button from the **PropertyManager** to create the feature.

PRACTICE 1

Create the surface model of tank as shown in Figure-56. The dimensions of the model are given in Figure-57.

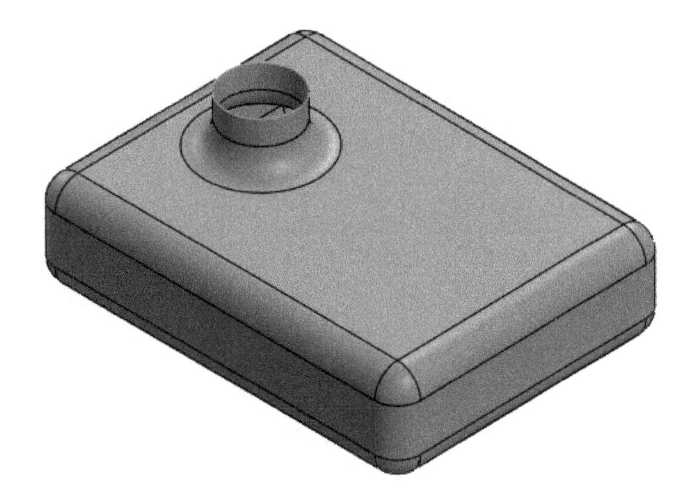

Figure-56. Practice1 model

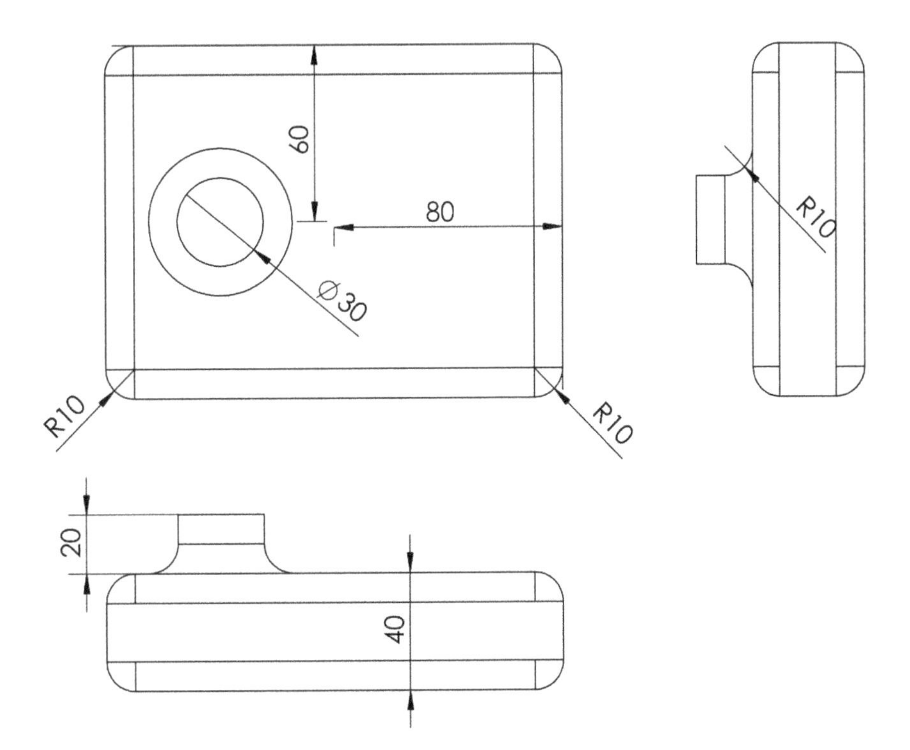

Figure-57. Practice1 drawing

PRACTICE 2

Create the surface model of car bumper as shown in Figure-58. The dimensions of the model are given in Figure-59. **Assume the missing dimensions.**

Figure-58. Practice 2 model

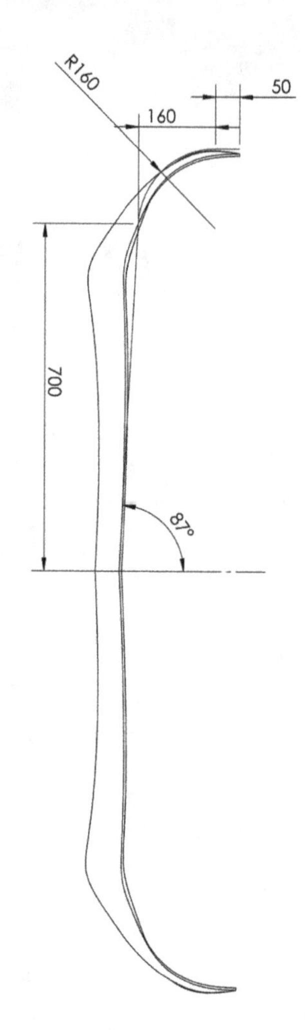

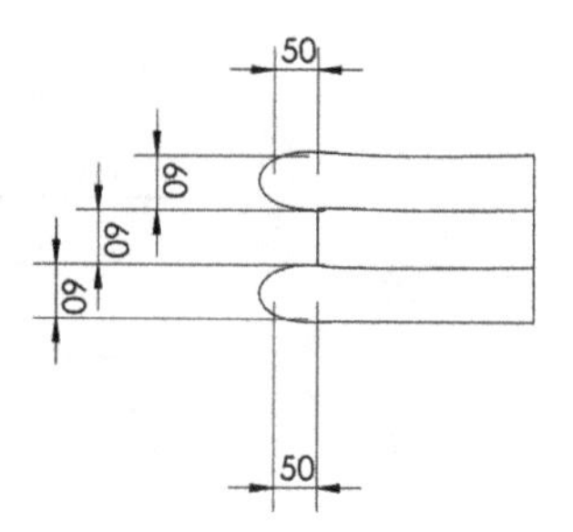

Figure-59. car bumper

SELF ASSESSMENT

Q1. We can create an extruded surface by using an open section. (T/F)

Q2. We do not need any centerline for creating revolved surface. (T/F)

Q3. We can create a swept surface by using open section and closed path. (T/F)

Q4. The **Freeform** tool can not deform face of a solid body.(T/F)

Q5. The **Ruled Surface** tool is used to create a combination of surfaces adjoining to each other. (T/F)

Q6. Which of the following tool can be used to create solid from the surface?

a. **Knit Surface**
b. **Untrim Surface**
c. **Trim Surface**
d. **Extend Surface**

Q7. Which of the following tool is used to remove desired portion of the surface by using sketch/other surface.

a. **Thickened Cut**
b. **Trim Surface**
c. **Cut with Surface**
d. **Delete Face**

Q8. The .... tool is used to extend the selected face to the replacement face.

Q9. The .... tool is used to undo the trimmed surfaces.

Q10. The .... tool is used to combine two or more surfaces at their common edges.

FOR STUDENT NOTES

Answer to Self-Assessment:
1. T, **2.** F, **3.** T, **4.** F, **5.** T, **6.** a, **7.** b, **8.** Replace Face, **9.** Untrim Surface, **10.** Knit Surface

Chapter 10

Drawing and Practice

Topics Covered

The major topics covered in this chapter are:

- ***Drawing Introduction.***
- ***Drawing Sheet Selection.***
- ***Adding Views to sheet.***
- ***Annotating Views.***
- ***Exploded View and Bill of Material.***
- ***Balloons and Title Block***
- ***Practice.***

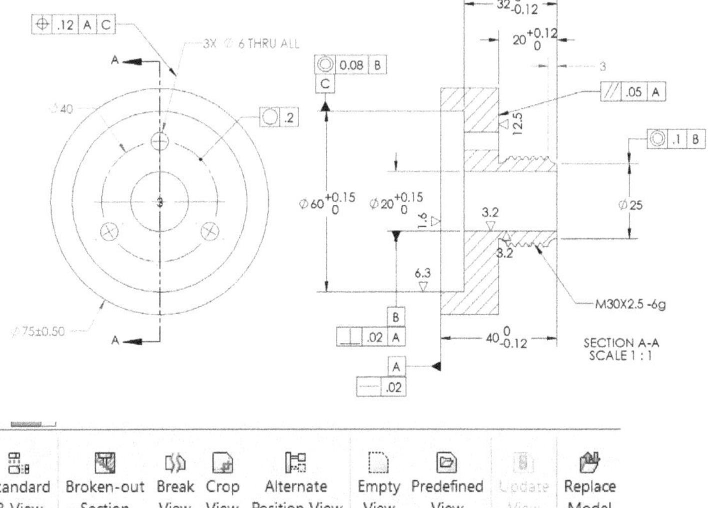

INTRODUCTION

Drawing is the engineering representation of a model on the paper. For manufacturing a model in real world, we need some means by which we can tell the manufacturer what to manufacture. For this purpose, we create drawings from the models. These drawings have information like dimensions, material, tolerances, objective, precautions, and so on. In SolidWorks, we create drawings by using the Drawing environment. To activate this environment, click on the **New** button from the **Menu Bar**. The **New SOLIDWORKS Document** dialog box will display. Double-click on the **Drawing** button; refer to Figure-1. The Drawing environment will open.

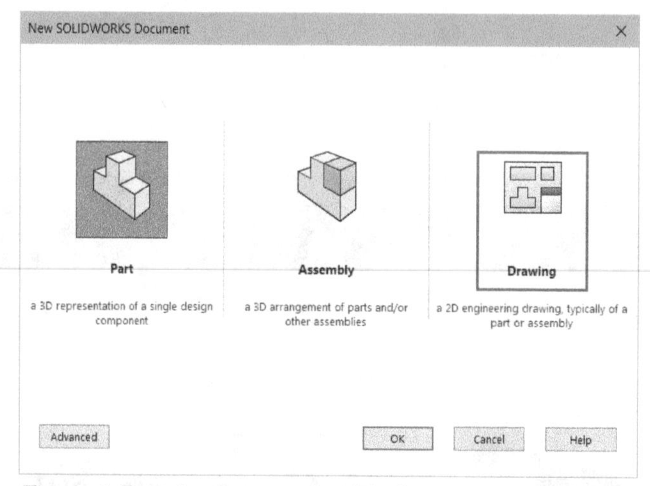

Figure-1. Drawing button to be selected

DRAWING SHEET SELECTION

On starting the Drawing environment, the interface will be displayed as shown in Figure-2. Also, the **Sheet Format/Size** dialog box will display. Using this dialog box, you can set the size of sheet for placing the drawing views.

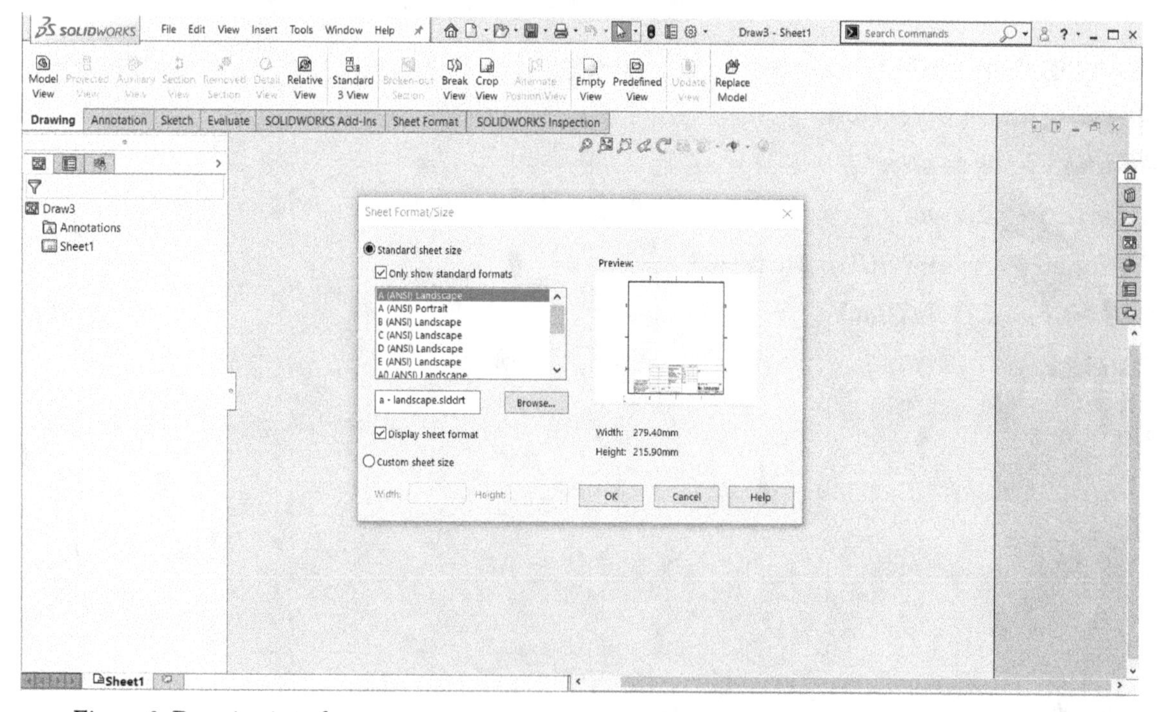

Figure-2. Drawing interface

Select desired size from the list of sheet sizes and click on the **OK** button from the list. The **Model View PropertyManager** will be displayed. Close the **PropertyManager** by clicking on **Cancel** button. (We will discuss about **PropertyManager** later in this chapter). If you want to change the sheet size, follow the steps given below.

- Right-click on the **Sheet** tab from the bottom bar displayed below drawing area. Shortcut menu will display; refer to Figure-3.

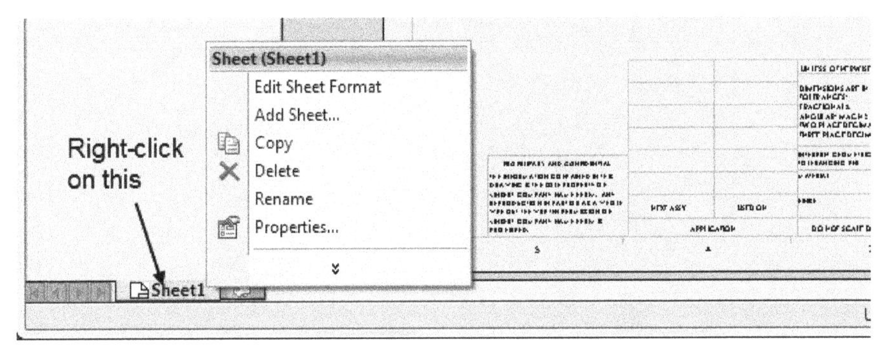

Figure-3. Shortcut menu displayed

- Select the **Properties** button from the menu. The **Sheet Properties** dialog box will display; refer to Figure-4.

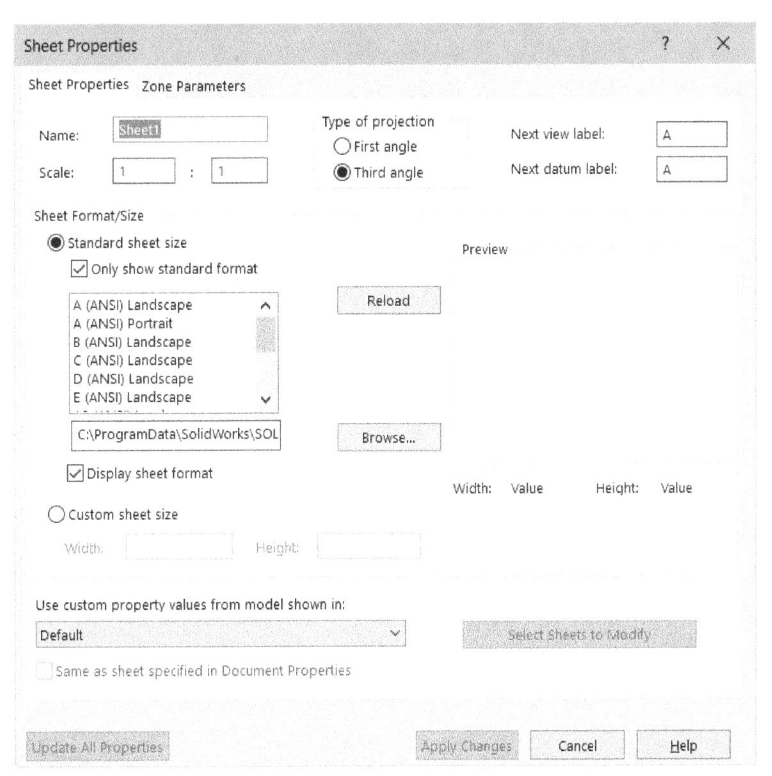

Figure-4. Sheet Properties dialog box

- Select desired sheet size from the list in this dialog box. You can also set the scale for view and projection type (First angle or Third angle) by using the options in this dialog box.

First Angle Projection and Third Angle Projection

Figure-5 shows an object with different view directions say, a, b, c, d, e, and f.

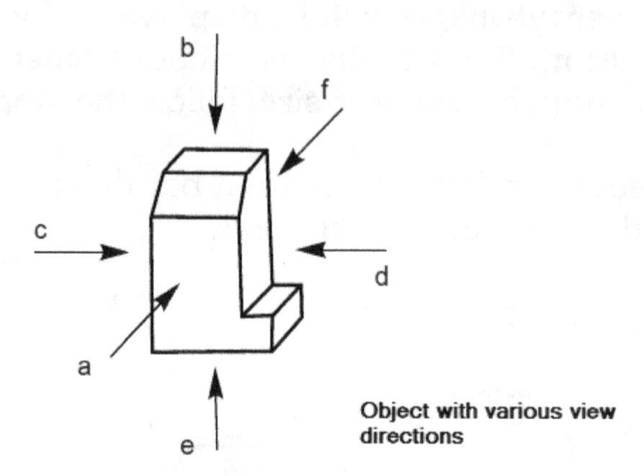

Figure-5. Object with view directions

Here,
1. View in the direction a = view from the front
2. View in the direction b = view from top
3. View in the direction c = view from the left
4. View in the direction d = view from the right
5. View in the direction e = view from bottom
6. View in the direction f = view from the back

In First Angle projection, these views are arranged as shown in Figure-6. In Third Angle projection, these views are arranged as shown in Figure-7.

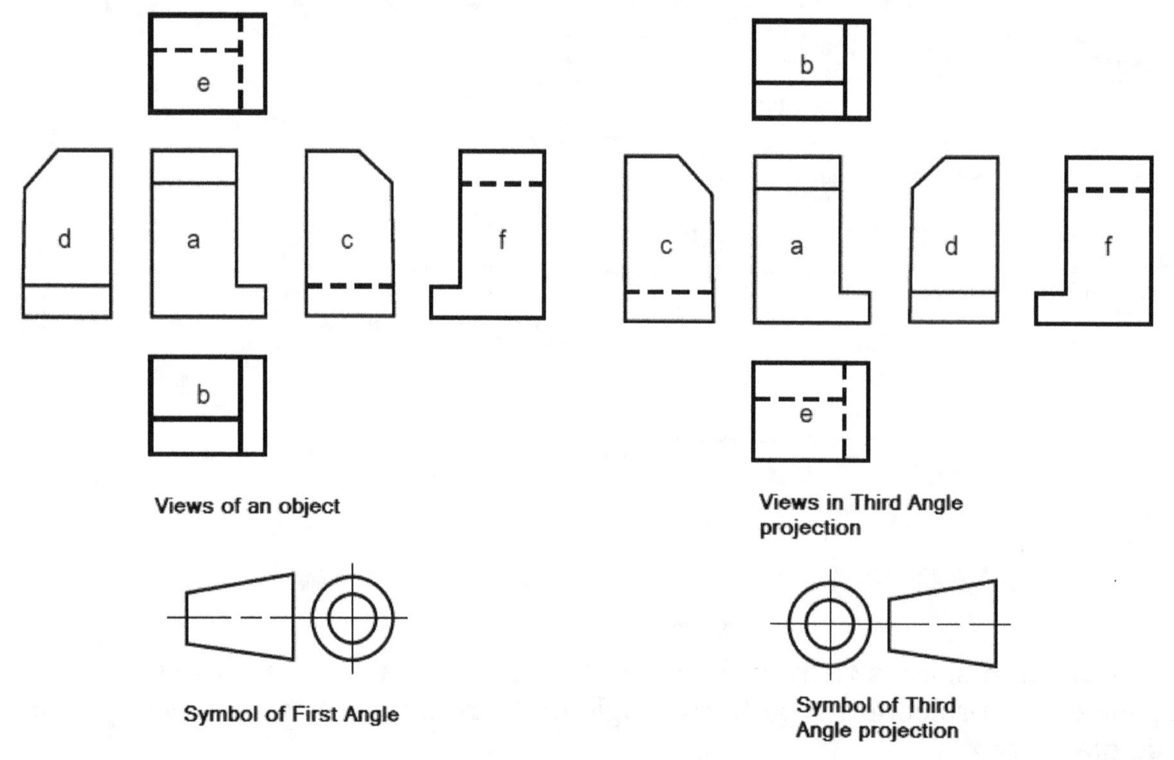

Figure-6. Views in First Angle projection

Figure-7. Views in Third Angle projection

- You can set you custom dimensions by selecting the **Custom sheet size** radio button.
- Set desired parameters in the **Zone Parameters** tab and click on the **Apply Changes** button to accept the sheet size.

After setting the sheet size, the next step is to add various views of the model in the sheet.

ADDING VIEWS TO SHEET

The tools to add views are available in the **Drawing** tab of the **Ribbon**. These tools are discussed next.

Model View

The **Model View** tool is used to add base view to the sheet. To add these views, follow the steps given next.

- Click on the **Model View** button from the **Ribbon**. The **Model View PropertyManager** will display; refer to Figure-8.
- Click on the **Browse** button from the **PropertyManager**. You will be asked to select part/assembly model.
- Double-click on the file of model for which you want to create the views. The view of model will be attached to cursor and the options in **Model View PropertyManager** will be modified; refer to Figure-9.
- Select desired configuration of part/assembly from the **Reference Configuration** drop-down at the top in the **PropertyManager**.
- If you want to place 3 primary views (Top view, Front view, and Right view) of model then select the **Create multiple views** check box and select respective buttons for views from the **Orientation** rollout. Preview of views will be displayed; refer to Figure-10.
- Click on the **OK** button from the **PropertyManager**. The views will be created.

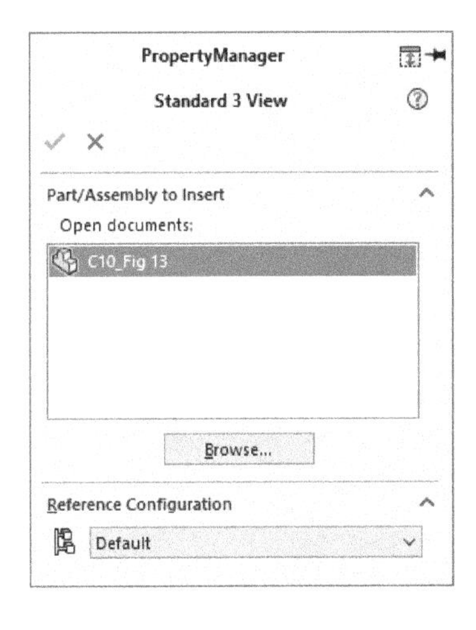

Figure-8. Model View PropertyManager

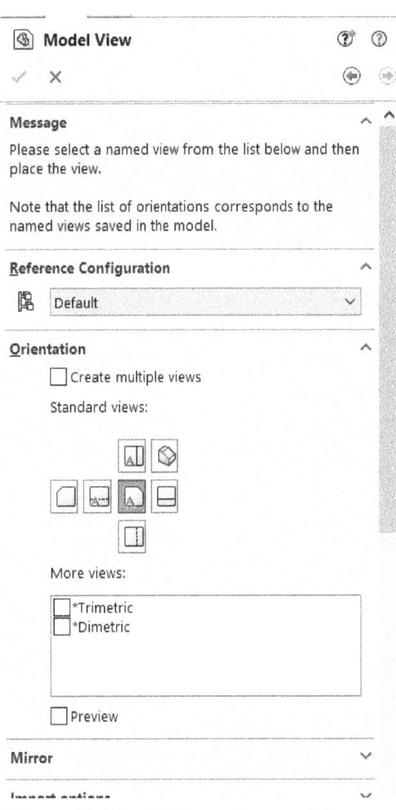

Figure-9. Model View PropertyManager

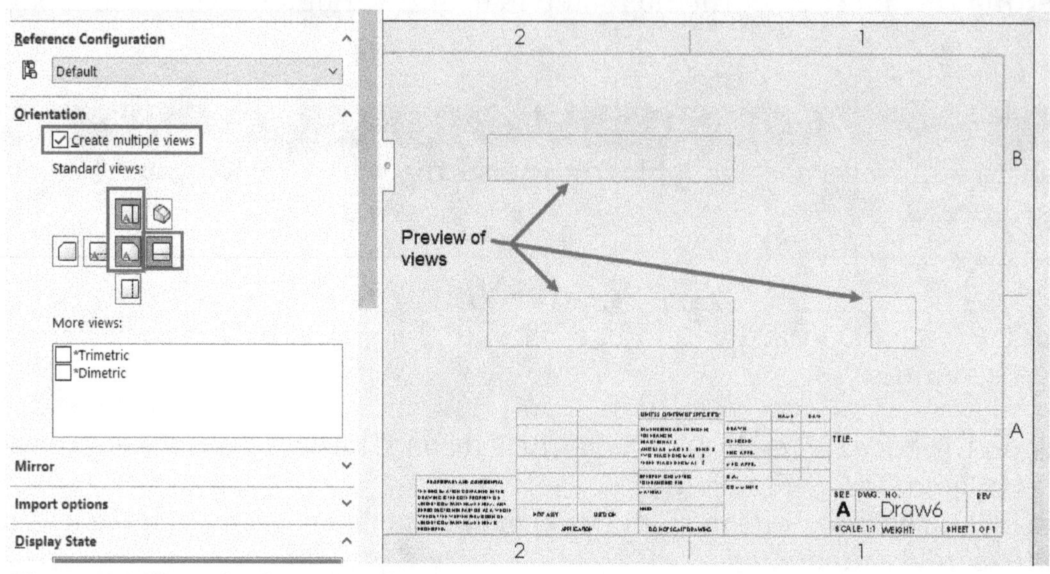

Figure-10. 3 Views placed automatically

To change the properties of any of the view placed, select it from the drawing. The **Drawing View PropertyManager** will display. If you select the main view then the **PropertyManager** will display as shown in Figure-11. If you select any projected view then the **PropertyManager** will display as shown in Figure-12.

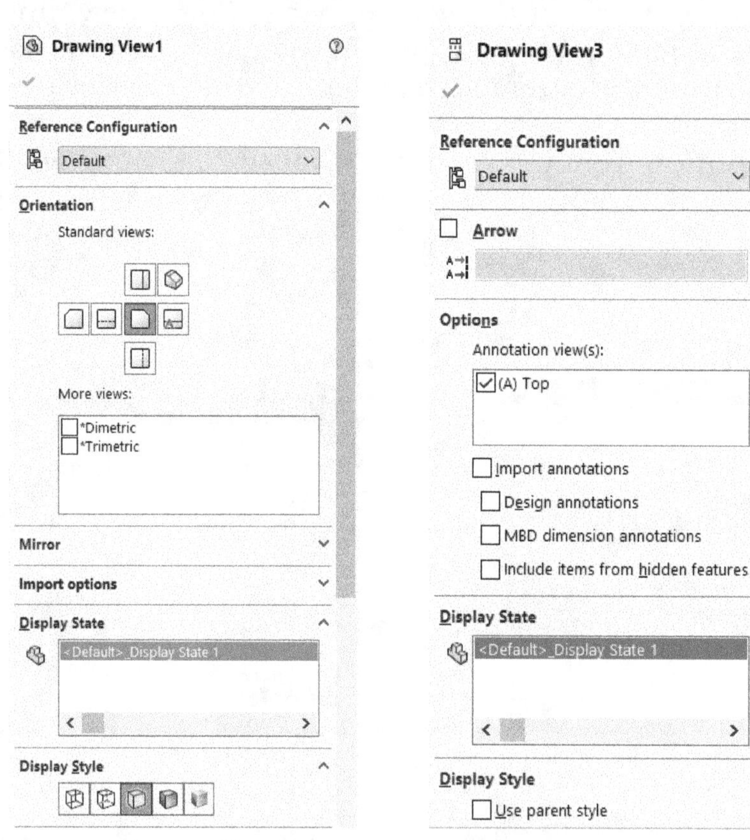

Figure-11. Drawing View PropertyManager on selecting main view

Figure-12. Drawing View PropertyManager on selecting projected view

The options in the **PropertyManager** are discussed next.

Reference Configuration rollout

Select desired configuration of part for which you want to create the drawing views from the drop-down in this rollout.

Orientation rollout

Select desired button to change the view orientation of the model. You can display dimetric or trimetric orientation by selecting the respective check box.

Arrow rollout

Select the check box for **Arrow** rollout of projected view. Preview of projection arrows will be displayed. Specify desired alphabet for label in the **Label** edit box; refer to Figure-13.

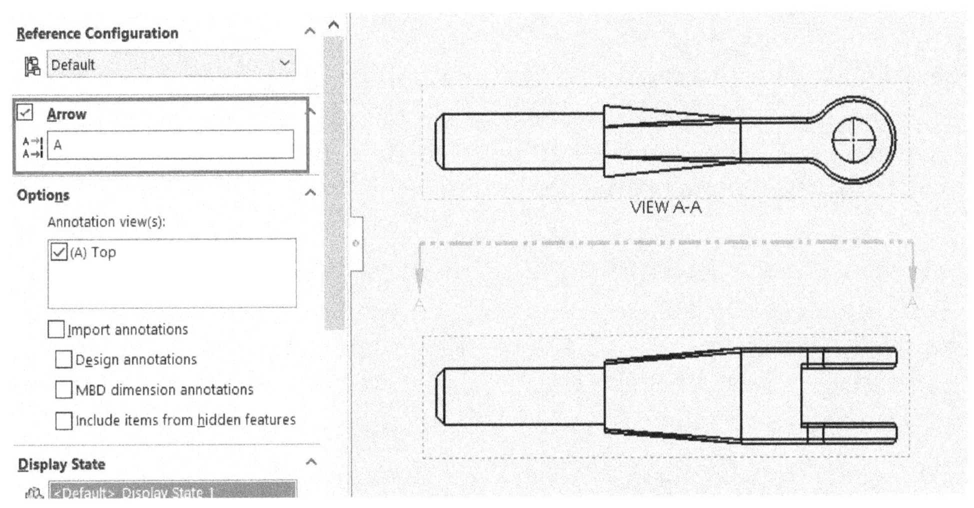

Figure-13. Adding projection arrows

Mirror rollout

The rollout is available for main view only. Select the **Mirror view** check box in this rollout and select the **Horizontal** or **Vertical** radio button to mirror the selected view horizontally or vertically.

Import Options rollout

Options in this rollout are used to import information from the Part environment. If you want to use dimensions or other annotations created in Part environment, then select the **Import annotations** check box. The other check boxes below it will become active. Similarly, you can select other check boxes to include design annotations, DimXpert annotations, hidden features from Part environment, and 3D view annotations.

Display Style rollout

Options in this rollout are used to change the display of model view. If you want to display hidden lines then select the **Hidden Lines Visible** button from the rollout. Similarly, you can select other buttons as per your requirement.

Scale rollout

Options in this rollout are used to change the size of view by using the scale value. To set desired scale, click on the **Use custom scale** radio button from the rollout and enter desired ratio in the edit box below it or select desired ratio from the drop-down below the radio buttons.

Dimension Type rollout

Select the **Projected** radio button from the rollout if you want to dimension projected areas of model. Select the **True** radio button if you want to create actual dimension.

Cosmetic Thread Display rollout

Select desired radio button from the rollout to define quality of cosmetic thread. Note that selecting the **High quality** radio button will increase the processing time.

Select the **Exclude from automatic update** check box from the **Automatic View Update** rollout if you want to manually update the drawing view after changes have been made in the part or assembly environment. After setting parameters, click on the **OK** button from the **PropertyManager**. The views will be created.

Drawing View Properties

Select the drawing view whose properties are to be modified from the **FeatureManager Design Tree**, right-click on it, and select the **Properties** option from the shortcut menu. The **Drawing View Properties** dialog box is displayed; refer to Figure-14. Set desired parameters in various tabs of the dialog box and click on the **OK** button.

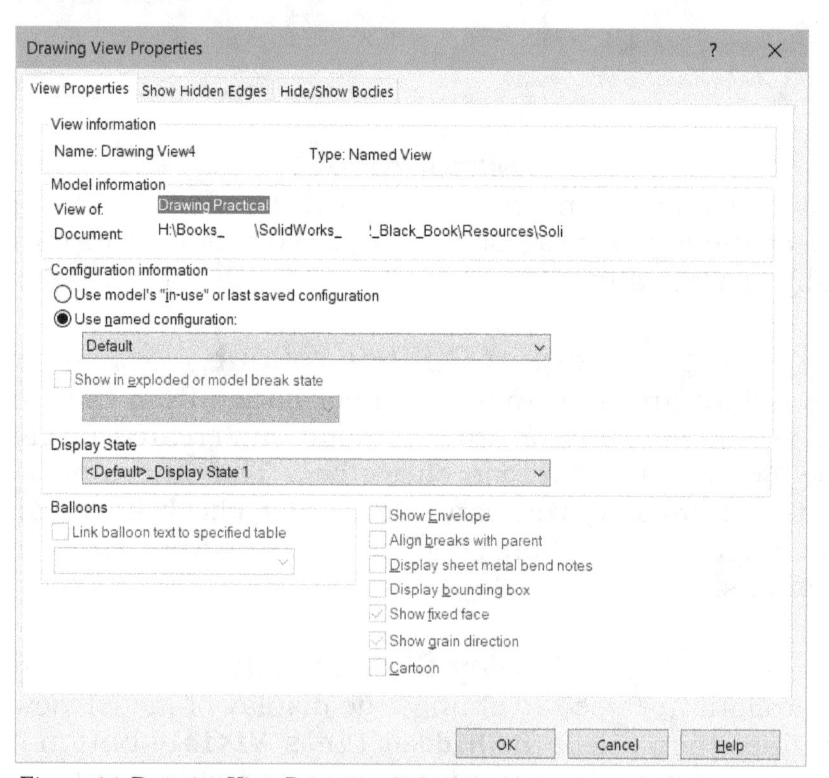

Figure-14. Drawing View Properties dialog box

Note that if you want to delete any view/views from the sheet, then select it and press DEL key from keyboard.

Projected View

The **Projected View** tool is used to create projected view of the selected view in the sheet. The procedure to use this tool is given next.

- Select desired view and click on the **Projected View** tool from the **Ribbon**. The **Projected View PropertyManager** will be displayed with preview of projected view attached; refer to Figure-15.

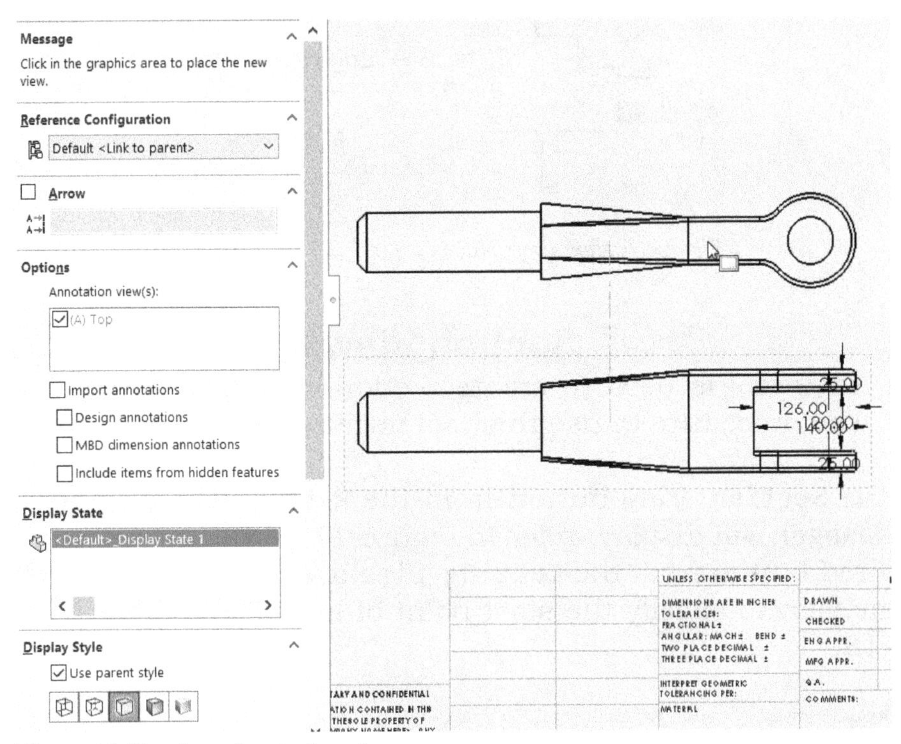

Figure-15. Preview of projection view

- Set desired parameters in the **PropertyManager** as discussed earlier.
- Click at desired location in the sheet to place the projection of selected view.

Auxiliary View

The **Auxiliary View** tool is used to create projected view from a view by making selected edge of view parallel to screen. The procedure to use this tool is given next.

- Click on the **Auxiliary View** tool from the **Ribbon**. The **Auxiliary View PropertyManager** will display and you will be asked to select an axis, edge, or sketch line to generate the view.
- Select the axis/edge/sketch line. The preview of view will be displayed; refer to Figure-16.
- Set desired parameters in the **PropertyManager** like arrow, display style, scale, dimension type and so on.
- Click at desired location to place the view.

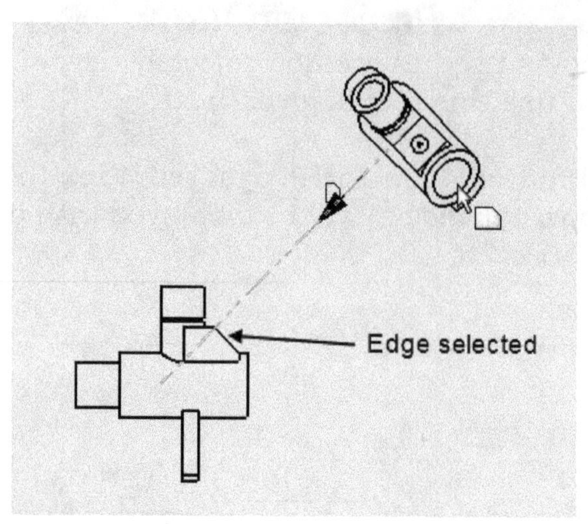

Figure-16. Auxiliary view

Section View

The **Section View** tool is used to create section view by cutting the model using section lines. The procedure to use this tool is given next.

- Click on the **Section View** button from the **Ribbon**. The **Section View Assist PropertyManager** will display; refer to Figure-17.
- Select desired button from the **Cutting Line** area of the **PropertyManager**.
- Click in the view to specify the start point of line. The toolbar will display; refer to Figure-18.

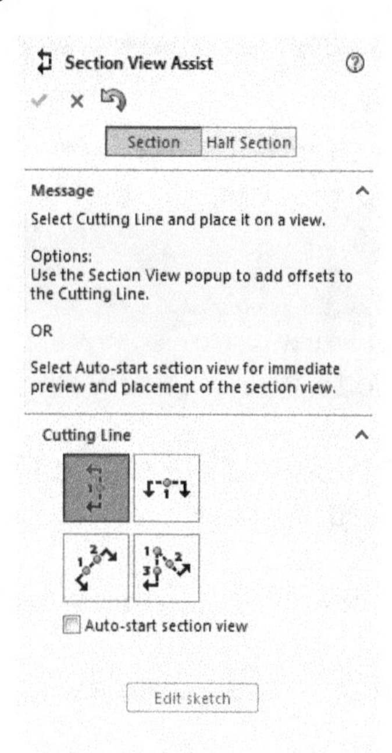

Figure-17. Section View Assist PropertyManager

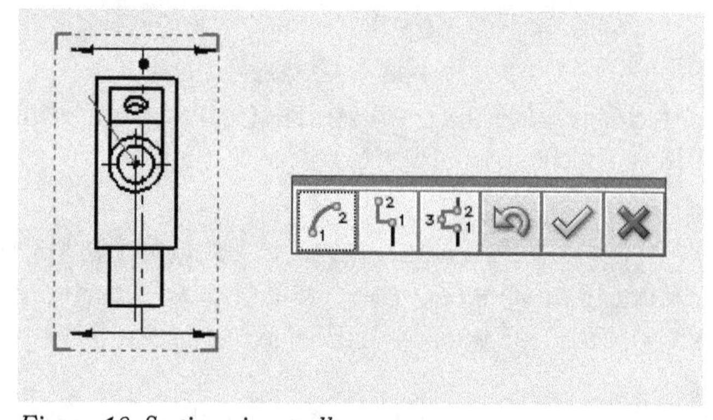

Figure-18. Section view toolbar

- Select desired button from toolbar to change the style of cutting line and specify the points.
- Click on the **OK** button from the toolbar, the preview will display; refer to Figure-19.
- Click to place the view. The **Section View PropertyManager** will display; refer to Figure-20.

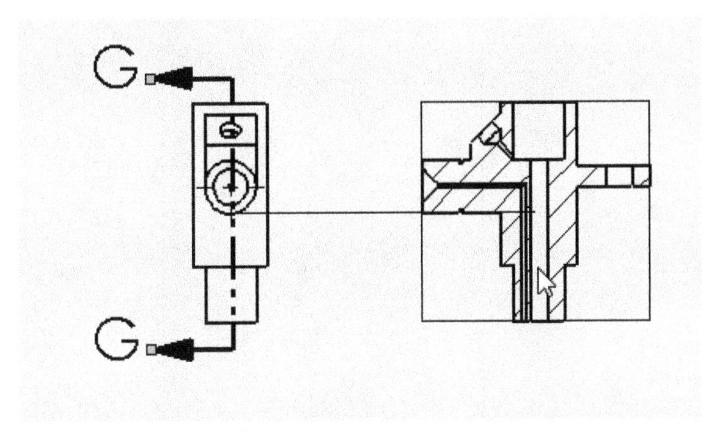

Figure-19. Preview of section view

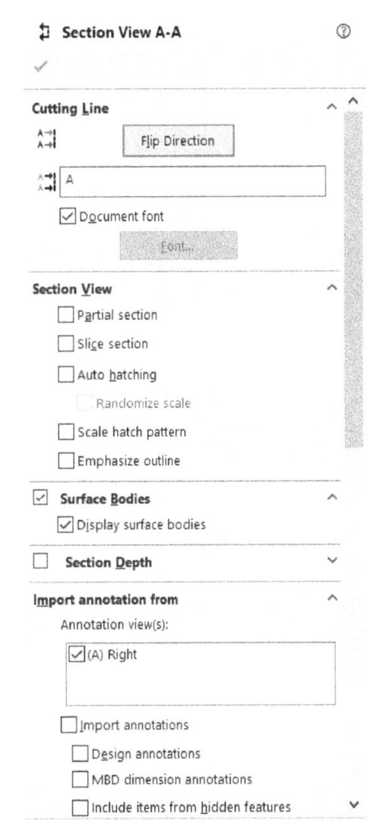

Figure-20. Section View PropertyManager

Section View rollout

- Select the **Partial Section** check box if you want to section partial area of the model. Note that the section depends on section line and you can drag the end point of section line to change it for partial view; refer to Figure-21. (After changing section line, click **OK** from **PropertyManager** and then click on the **Rebuild** tool from the **Menu**.)

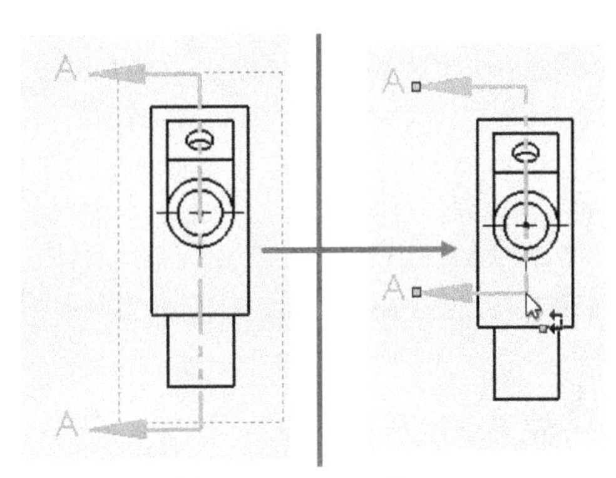

Figure-21. Changing section line

- Select the **Slice section** check box to display only the section which overlaps with section plane; refer to Figure-22.

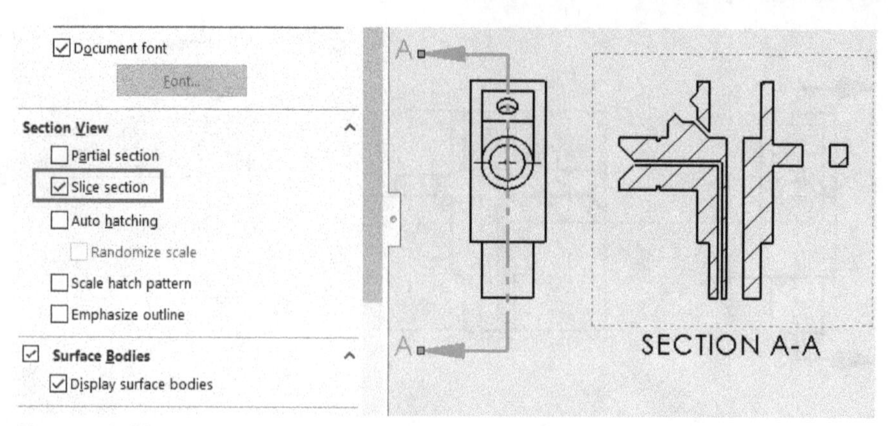

Figure-22. Slice section

- Select the **Auto hatching** check box reverses hatch pattern when two components are in contact in the section view; refer to Figure-23.

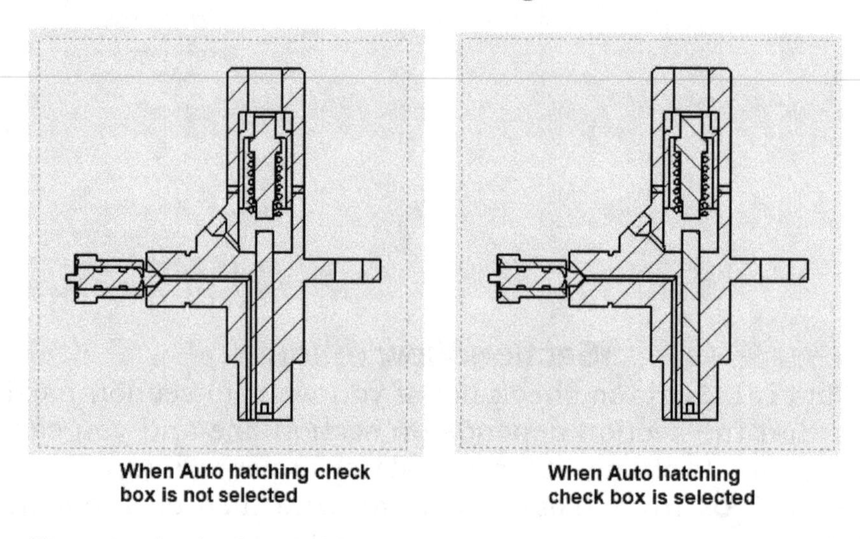

Figure-23. Auto hatching check box

- Select the **Randomize scale** check box if you want to change scale of hatch when two or more components are with same material.
- Select the Scale hatch pattern check box to scale the hatch pattern as per sheet size.
- Select the **Emphasize outline** check box in the **Section View** rollout of **PropertyManager** to create sections compliant to ISO 128-50 which has bold lines for component boundary.

Surface Bodies rollout

- Select the **Surface Bodies** check box if you want to include surface bodies in the section view. Select the **Display surface bodies** check box to display surfaces in the view.

Section Depth rollout

- Select the **Section Depth** check box to define the depth for section view. The options of rollout will be displayed as shown in Figure-24.
- Specify desired depth value for section in the **Depth** edit box or drag the section depth line using pink arrow on it.

- You can also use a reference point or edge to define section depth. Click in the **Depth Reference** selection box and select desired edge/point on the model.
- After setting desired parameters, click on the **Preview** button to check the preview.

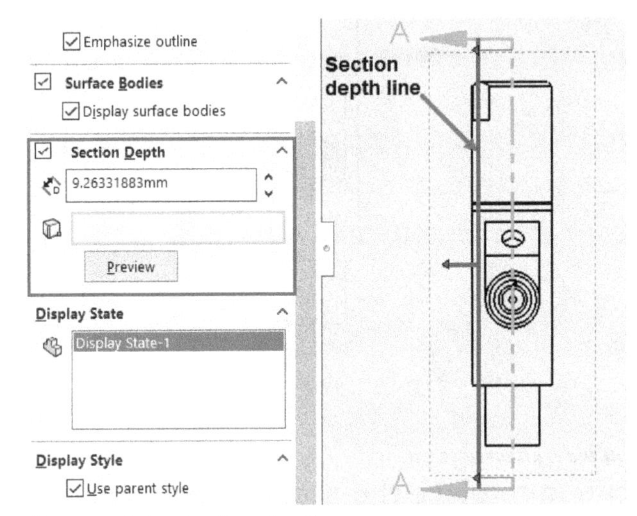

Figure-24. Section Depth rollout

- You can set the other parameters in the **PropertyManager** as discussed earlier.

Half Section

- After clicking on the **Section View** tool from the **Ribbon**, click on the **Half Section** button from the **Section View Assist PropertyManager** to create half section of the model; refer to Figure-25.

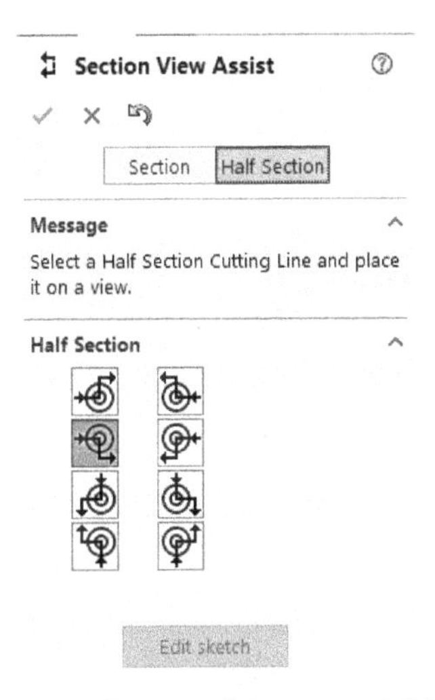

Figure-25. Section View Property Manager with Half Section option

- Select desired button from the **Half Section** rollout and click in the view at desired location to create half section view. The **Section View** dialog box will be displayed. Set desired options and click on the **OK** button. The section view will get attached to the cursor; refer to Figure-26.

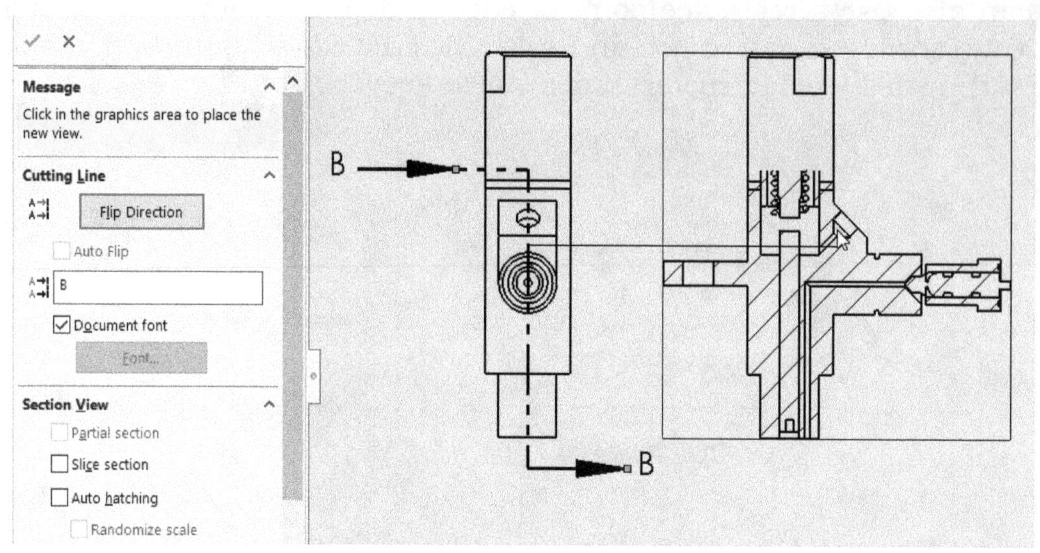

Figure-26. Half section view attached to cursor

- Click at desired location to place the section view and set desired parameters in the **PropertyManager**.
- Click on **OK** button from the **PropertyManager** to create the section view.

Removed Section

The **Removed Section** tool is used to create a section between two selected edges. The procedure to use this tool is given next.

- Click on the **Removed Section** tool from the **Drawing CommandManager**. The **Removed Section PropertyManager** will be displayed; refer to Figure-27.

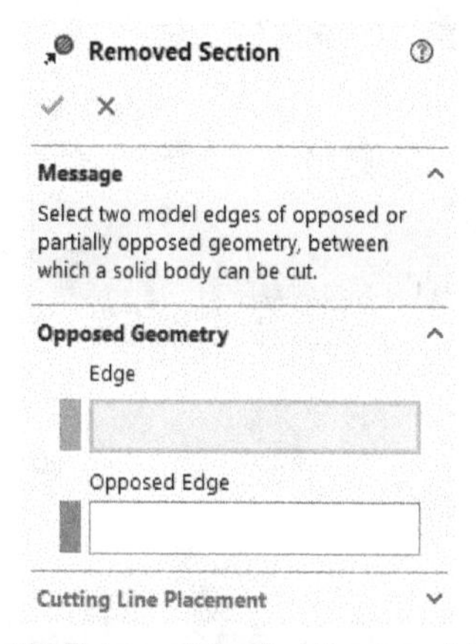

Figure-27. Removed Section PropertyManager

- Select the two edges to mark section to be removed; refer to Figure-28. A section line will get attached to cursor.

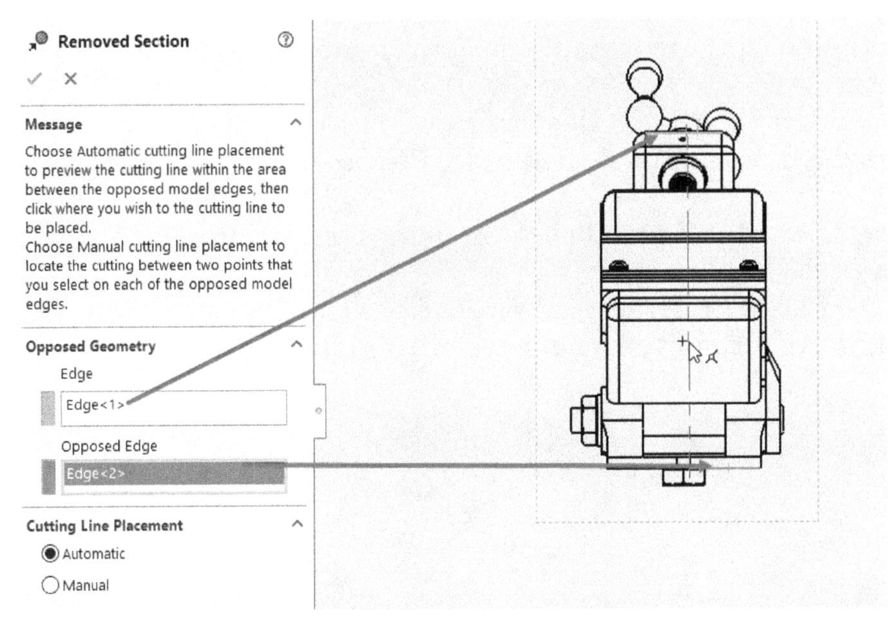

Figure-28. Edges selected for remove section

- Click at desired location to place the section line. The updated **Removed Section PropertyManager** will be displayed; refer to Figure-29.

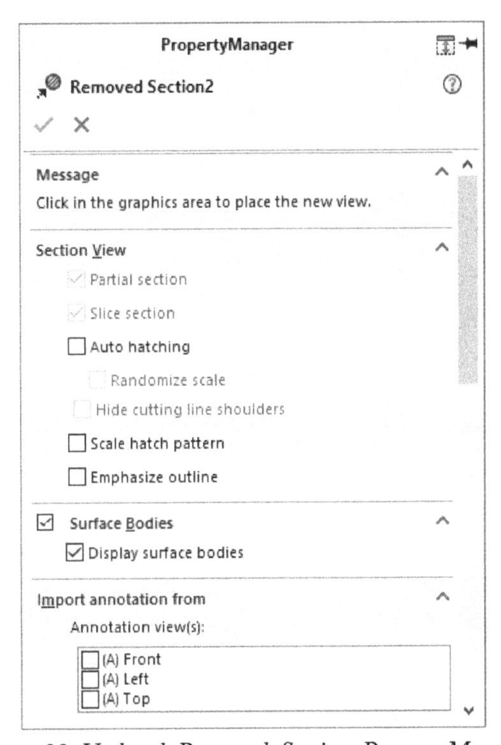

Figure-29. Updated_Removed_Section_PropertyManager

- Select desired check boxes and click on the **OK** button. The view will get attached to cursor.
- Set desired parameters in the **PropertyManager** as discussed earlier and click at desired location.
- Click on the **OK** button from the **PropertyManager**.

Detail View

The **Detail View** tool is used to create detailed view of a specific portion of the model. The procedure to create detail view is given next.

- Click on the **Detail View** button from the **Ribbon**. The **Detail View PropertyManager** will display. Also, you are asked to specify center point of circle for creating detail view circle.
- Click in the drawing view at desired location to specify center of the detail circle.
- Click to specify the radius of the circle. Preview of detail view will be attached to the cursor.
- Change the scale of the detail view by using the options in the **PropertyManager**; refer to Figure-30.
- Select desired style of detail view circle from the **Style** drop-down in the **Detail Circle** rollout of **PropertyManager**; refer to Figure-31.

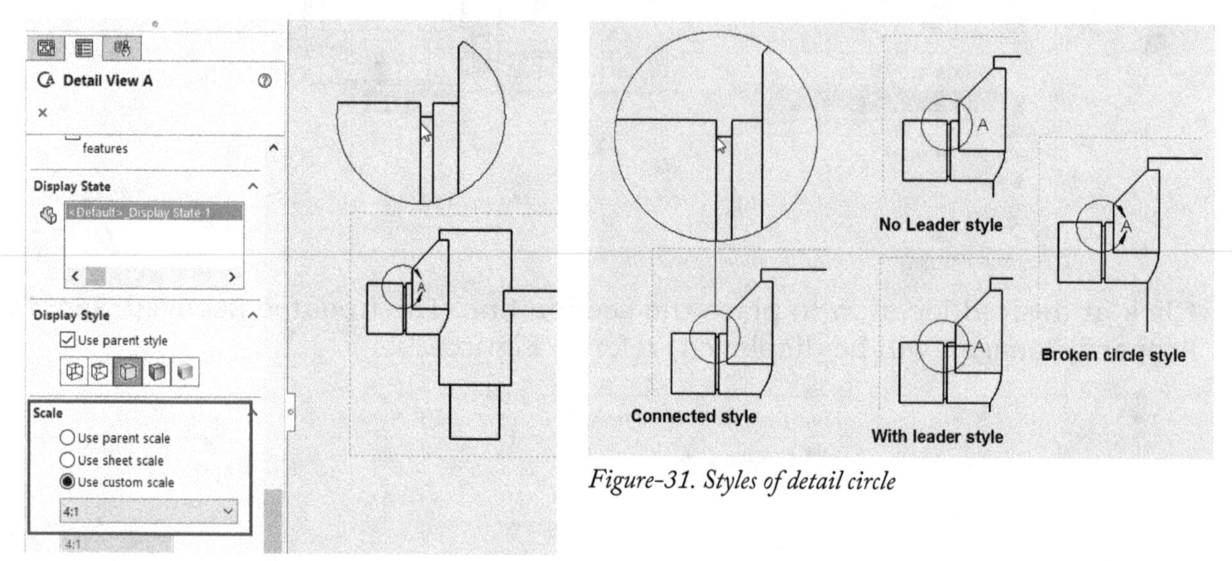

Figure-30. Setting scale of detail view

Figure-31. Styles of detail circle

- Select desired check box from the **Detail View** rollout of the **PropertyManager** to change the outline of detail view.
- Click to place the detail view. Press **ESC** to exit the tool.

Relative View

The **Relative View** tool is used to create view of the component using specified faces for orientation. The procedure to use this tool is given next.

- Click on the **Relative View** tool from the **Drawing CommandManager** in the **Ribbon**. The **Relative View PropertyManager** will be displayed.
- Select desired radio button from the **PropertyManager** to define quality of thread in view.
- Click on the view to define faces. The model will open in part/assembly environment of SolidWorks and the **Relative View PropertyManager** will be displayed; refer to Figure-32.
- Select desired options from the **First Orientation** and **Second Orientation** drop-downs.
- Click in the selection boxes for orientation and select respective faces.
- Click on the **OK** button from the **PropertyManager**. The view will get attached to the cursor.
- Click at desired location to place the view and set desired parameter.

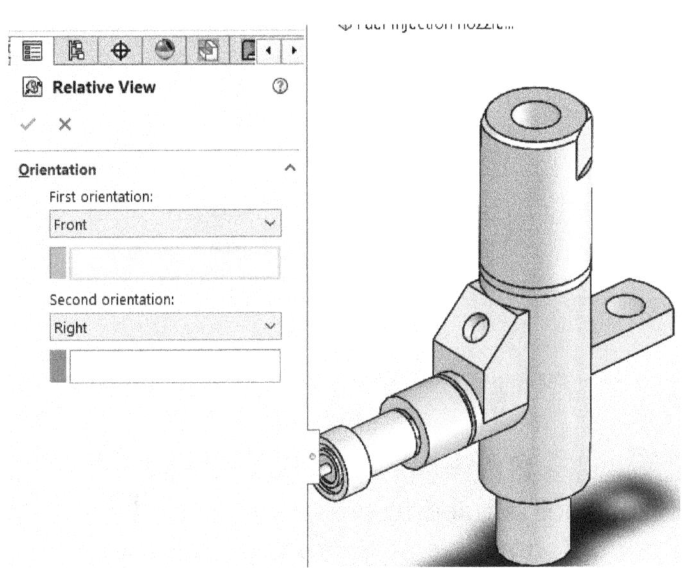

Figure-32. Relative View PropertyManager

Standard 3 View

The **Standard 3 View** tool is used to add 3 standard views (Top view, Front view, and Right view) to the sheet. To add these views, follow the steps given next.

- Click on the **Standard 3 View** button from the **Ribbon**. The **Standard 3 View PropertyManager** will display; refer to Figure-33.
- Click on the **Browse** button from the **PropertyManager**. You will be asked to select part/assembly model.
- Double-click on the file of model for which you want to create the views. Click on the **OK** button from the **PropertyManager**. Views of the model will be placed automatically. Figure-34 shows the views placed in Third angle projection.
- You can drag the view by selecting any curve of model to place at your desired location. The other views will be shifted accordingly; refer to Figure-35.

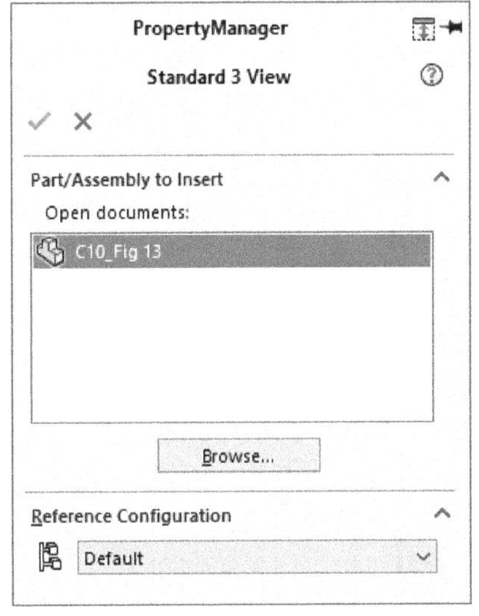

Figure-33. Standard_3_View_Property-Manager

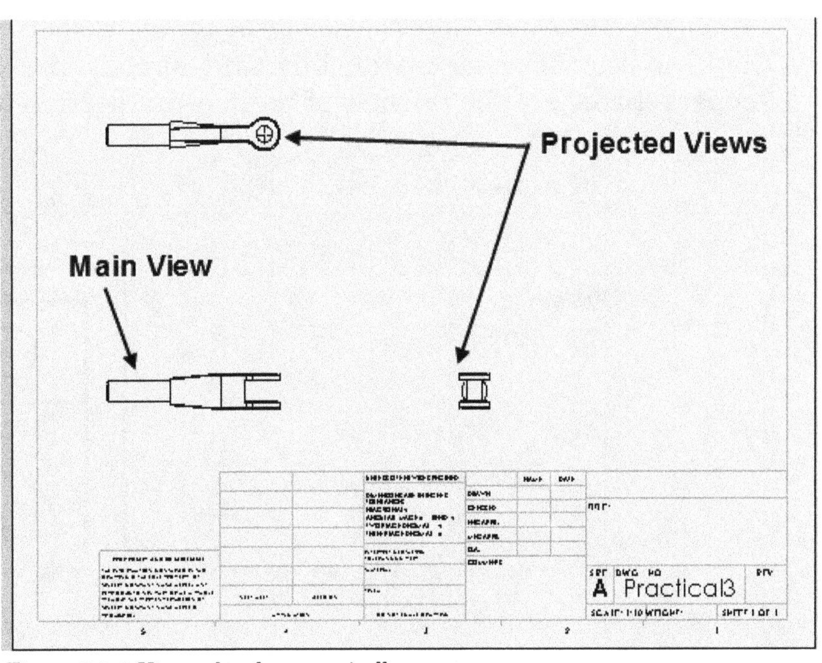

Figure-34. 3 Views placed automatically

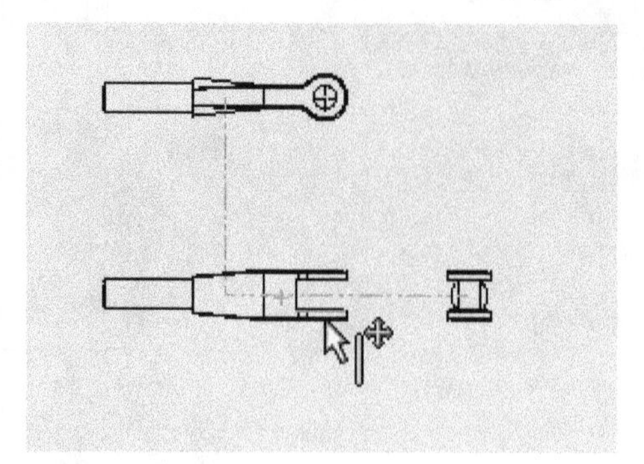

Figure-35. Moving views

Broken-out Section

The **Broken-out Section** tool is used to create section of the selected view to display inner detail of the model. Follow the steps to use this tool.

- Click on the **Broken-out Section** tool from the **Ribbon**. You will be asked to specify the start point of the section spline.
- Click in the view to start spline and create a close spline; refer to Figure-36.

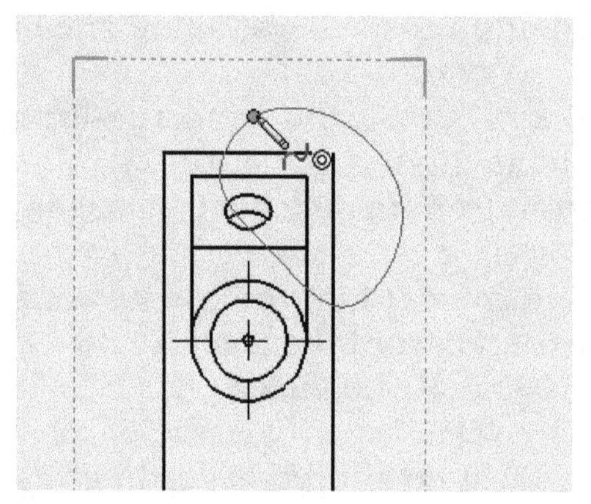

Figure-36. Spline for broken-out section

- Click on the **Preview** check box and specify the depth in the spinner in the **PropertyManager**. The preview of broken-out section will display; refer to Figure-37.

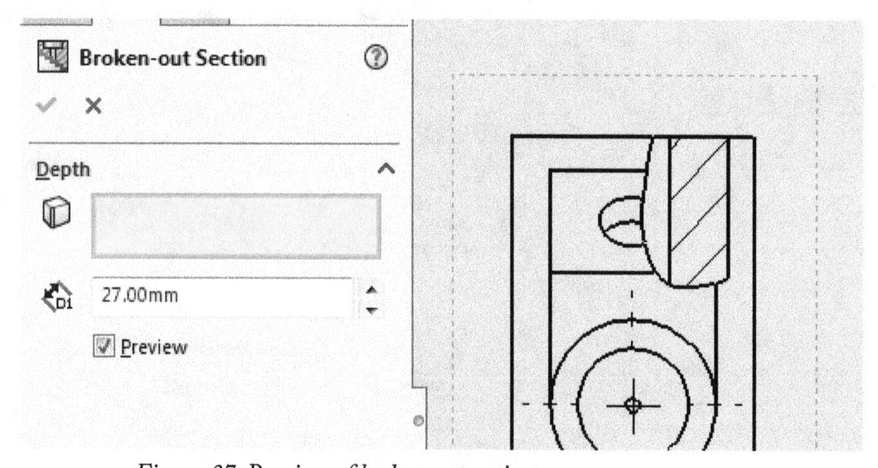

Figure-37. Preview of broken-out section

- Click on the **OK** button to create the section.

Break

The **Break** tool is used to represent very long objects in the drawing by breaking them at specific span. The procedure to use this tool is given next.

- Click on **Break** tool from the **Ribbon**.
- Click on the view that you want to break. The break line will attach to the cursor; refer to Figure-38.
- Set desired gap and break line style in the **PropertyManager**, and click to specify starting point of break span.
- Click to specify the end of break span. The selected span will be removed and the broken view will display with specified gap; refer to Figure-39.

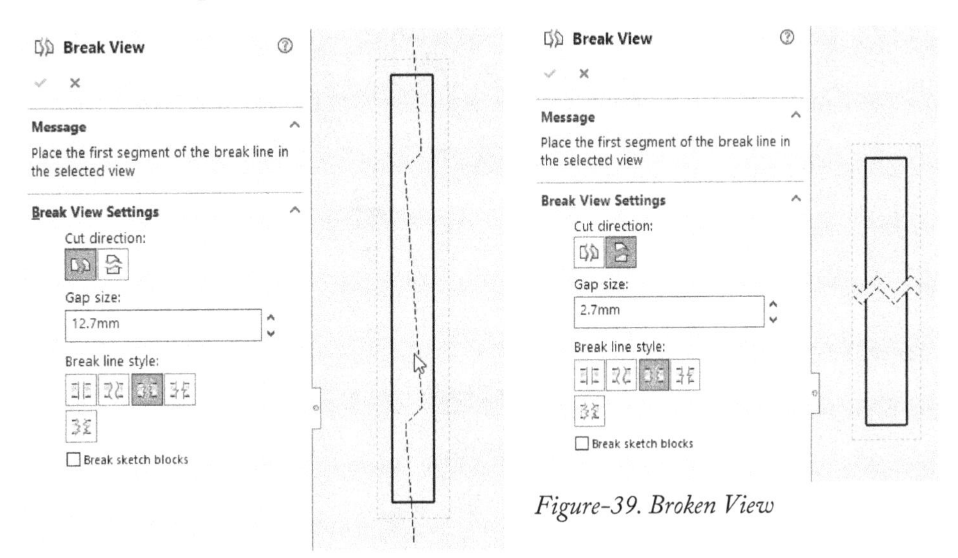

Figure-38. Breakline attached to cursor

Figure-39. Broken View

- Click on the **OK** button to create the view.

Crop View

The **Crop View** tool is used to crop a selected view. The procedure to create crop view is given next.

- Click on the **Circle** tool from the **Sketch CommandManager** and draw circle of desired diameter in the view; refer to Figure-40. Only the region inside the circle will remain after cropping the view.
- Click on the **Crop View** tool from the **Drawing CommandManager** in the **Ribbon**. The view will be cropped accordingly; refer to Figure-41.

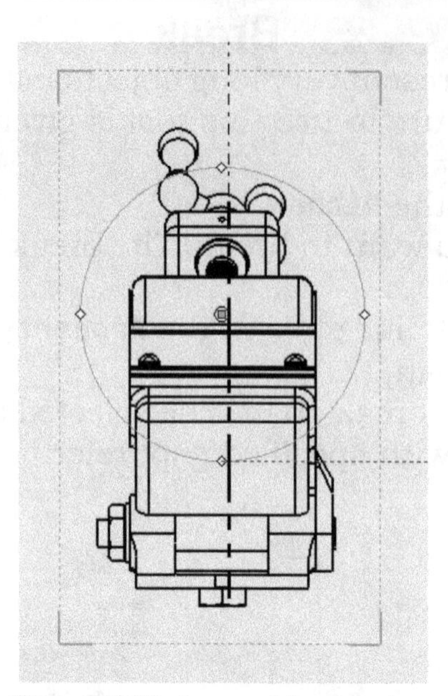

Figure-40. Circle created for crop view

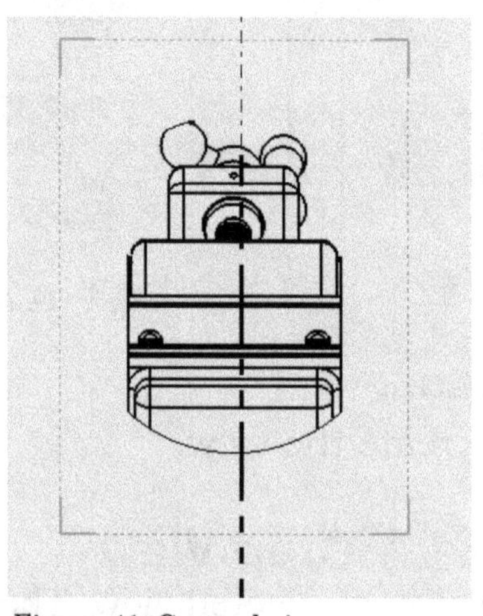

Figure-41. Cropped view

Creating Alternate Position View

The **Alternate Position View** tool is used to superimpose different positions of objects of an assembly in the same view. The procedure to create alternate position view is given next.

- Click on the view for which you want to create alternate position and click on the **Alternate Position View** tool from the **Drawing CommandManager**. The **Alternate Position PropertyManager** will be displayed; refer to Figure-42.

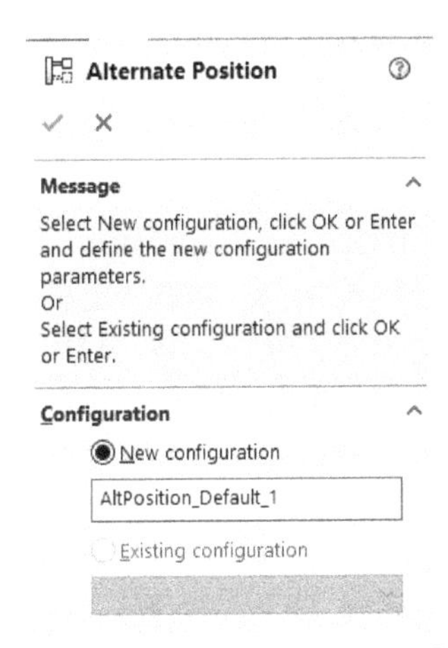

Figure-42. Alternate Position PropertyManager

- Set desired name for configuration in the edit box under **Configuration** rollout and click on the **OK** button from the **PropertyManager**. The assembly model will be displayed with **SmartMates PropertyManager**; refer to Figure-43.

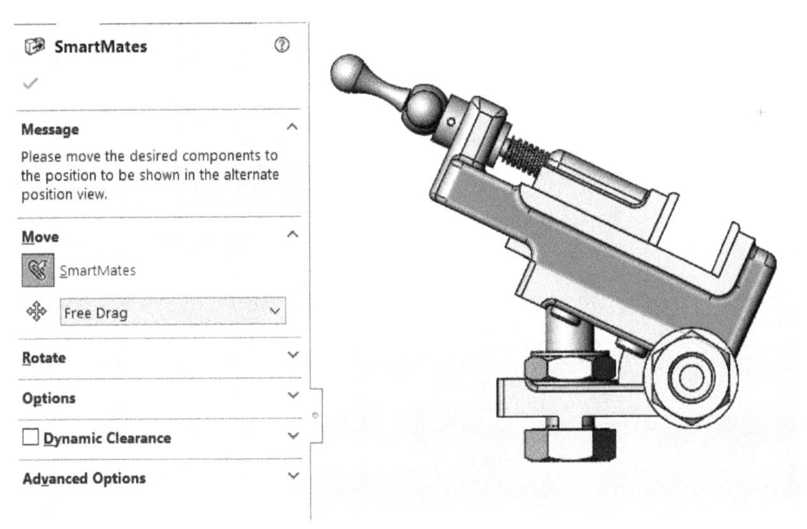

Figure-43. Model with SmartMates PropertyManager

- Move the components of assembly as desired by dragging and click on the **OK** button. The alternate position view will be displayed; refer to Figure-44.

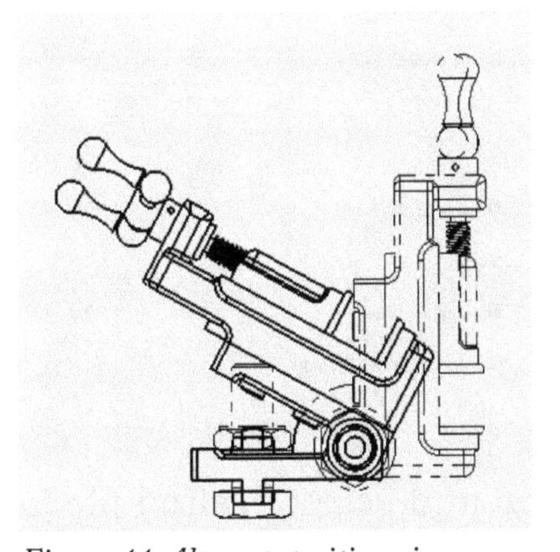

Figure-44. Alternate position view

Creating Empty View

The **Empty View** tool is used to create an empty view boundary in which you can create sketch using the sketching tools. The procedure to use this tool is given next.

- Click on the **Empty View** tool from the **Ribbon**. The empty view will get attached to the cursor.
- Click at desired location to place the view. The **Drawing View PropertyManager** will be displayed as discussed earlier.
- Click on the **OK** button from the **PropertyManager** to create the view. Create desired sketch in the view boundary using the tools in the **Sketch CommandManager**.

Creating Predefined View

The **Predefined View** tool is used to place desired orthogonal, projected, or named view of the model. The procedure to use this tool is given next.

- Click on the **Predefined View** tool from the **Drawing CommandManager**. The view will get attached to cursor.
- Click at desired location to place the view. The **Drawing View PropertyManager** will be displayed; refer to Figure-45.

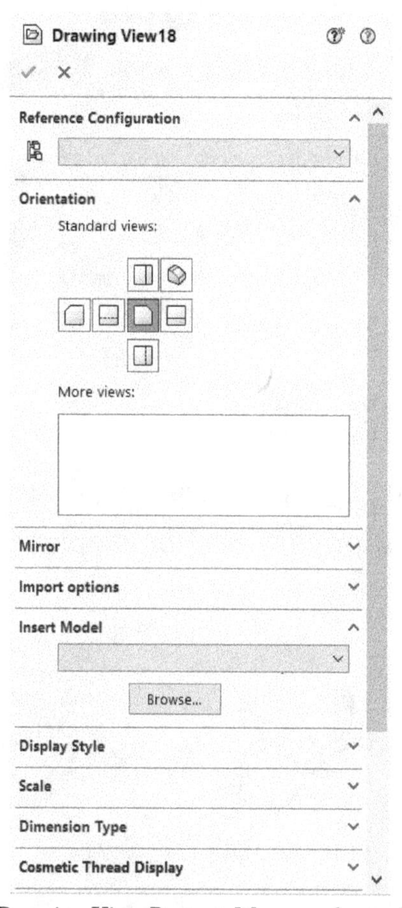

Figure-45. Drawing View PropertyManager for predefined view

- Click on the **Browse** button and select desired file for drawing view.
- Set the other parameters in **PropertyManager** as discussed earlier and click on the **OK** button.

Replacing Model

The **Replace Model** tool is used to replace the model in selected view. The procedure to use this tool is given next.

- Click on the **Replace Model** tool from the **Drawing CommandManager** in the **Ribbon**. The **Replace Model PropertyManager** will be displayed; refer to Figure-46.

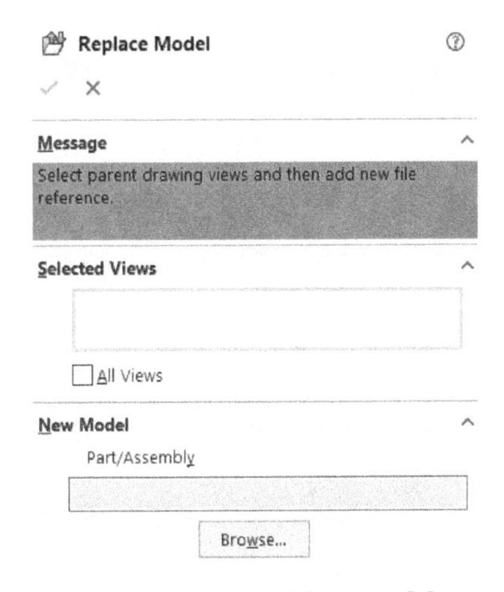

Figure-46. Replace Model PropertyManager

- Select desired view from the drawing area from which you want to replace the model and click on the **Browse** button. The **Open** dialog box will be displayed.
- Select desired model and click on the **Open** button.
- Click on the **OK** button from the **PropertyManager**. The model will be replaced.

ADDING ANNOTATIONS TO VIEW

Till this point, we have learned to place views but without dimensions and annotations these views are of no use for manufacturers. So, we will now add annotations to views. The tools to apply annotations are available in the **Annotation** tab of the **Ribbon**; refer to Figure-47.

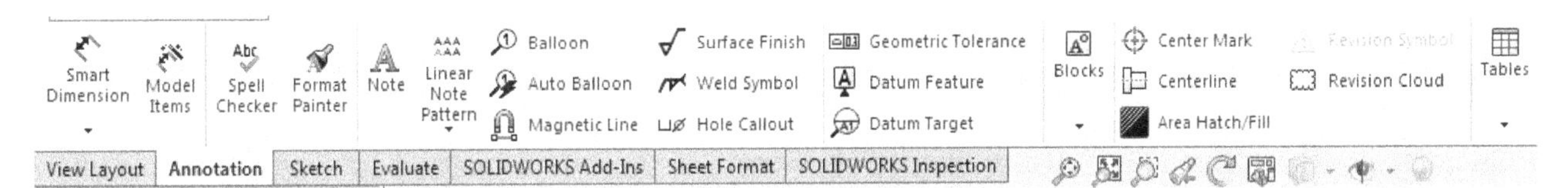

Figure-47. Annotation_tab_of_ribbon

Smart Dimension

The **Smart Dimension** tool is used to dimension each entity of the model on paper. This tool and most of the other tools in the **Smart Dimension** drop-down work in the same way as they do in Part environment.

Angular Running Dimension

The **Angular Running Dimension** tool is used to create angular dimension similar to ordinate dimensioning. The procedure to use this tool is given next.

- Click on the **Angular Running Dimension** tool from the **Smart Dimension** drop-down in the **Annotation CommandManager** of **Ribbon**. You will be asked to select an edge and vertex to define zero angle position.
- Select desired edge and vertex to define 0 degree position; refer to Figure-48.

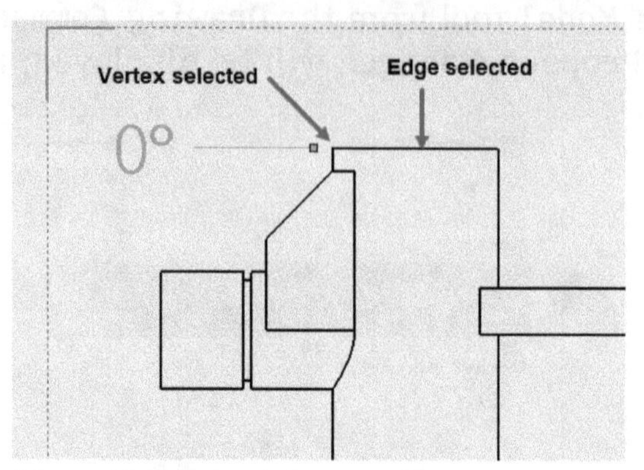

Figure-48. Defining 0 degree position

- Click on the next edge to be dimensioned; refer to Figure-49.

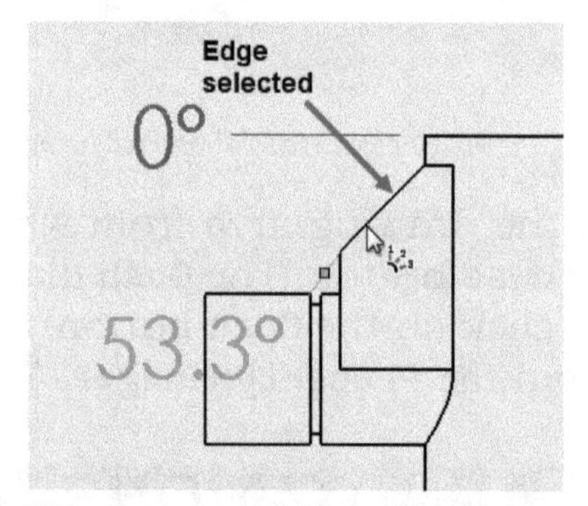

Figure-49. Angular dimension generated

- Press **Esc** to exit the tool.

Chamfer Dimension

The **Chamfer Dimension** tool is used to dimension chamfers in the drawing views. The procedure to use this tool is given next.

- Click on the **Chamfer Dimension** tool from the **Smart Dimension** drop-down in the **Annotation CommandManager** of the **Ribbon**. You will be asked to select the edge to be chamfered.
- Click on the chamfered edge. You will be asked to select the reference edge.
- Click on the adjoining reference edge. The dimension will get attached to cursor; refer to Figure-50.

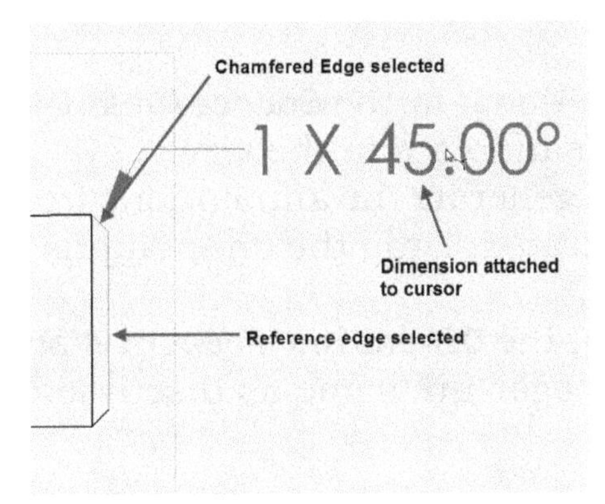

Figure-50. Chamfer dimension attached to cursor

- Click at desired location to place the chamfer dimension. The **Dimension PropertyManager** will be displayed. The options in this **PropertyManager** are discussed later in this chapter. For the time being press **ESC** to exit the **PropertyManager**.

Model Items

The **Model Items** tool is used to import all the dimensions/annotations applied to the model in Part modeling environment. To import the dimensions/annotations, follow the steps given next.

- Click on the **Model Items** tool from the **Ribbon**. The **Model Items PropertyManager** will display; refer to Figure-51.

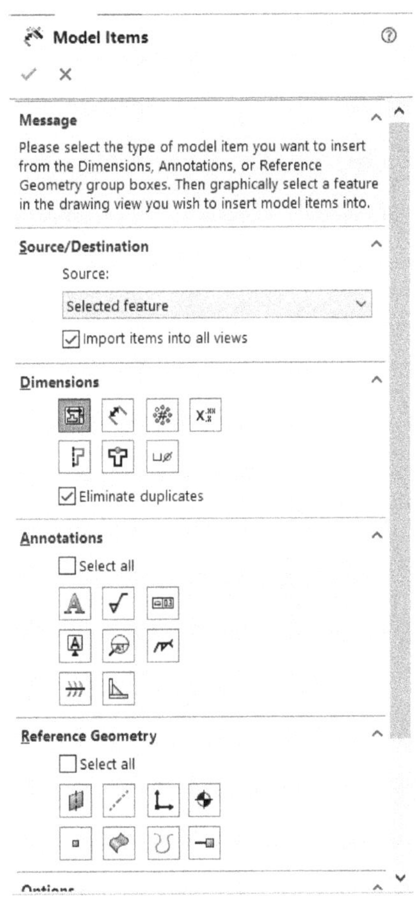

Figure-51. Model Items PropertyManager

- Select desired buttons from the **PropertyManager** to import the respective annotations in the view.
- Click in the **Source** drop-down in the **Source/Destination** rollout to specify the entities that you want to annotate in the view.
- Click on the **OK** button to generate the annotations from the model. The automatic annotations will be generated. Drag the annotations to place them properly.

Select one of the dimension, the **Dimension PropertyManager** will be displayed. The options in this **PropertyManager** are same as discussed in **Advanced Dimensioning Chapter**.

Note

The **Note** tool is used to specify extra information in the drawing that are not mentioned in the dimensions. For example, if you want to say "All dimensions are in mm" then this is the tool to do so. The steps to use this tool are given next.

- Click on the **Note** tool from the **Ribbon**. The **Note PropertyManager** will display and note box will get attached to the cursor; refer to Figure-52.

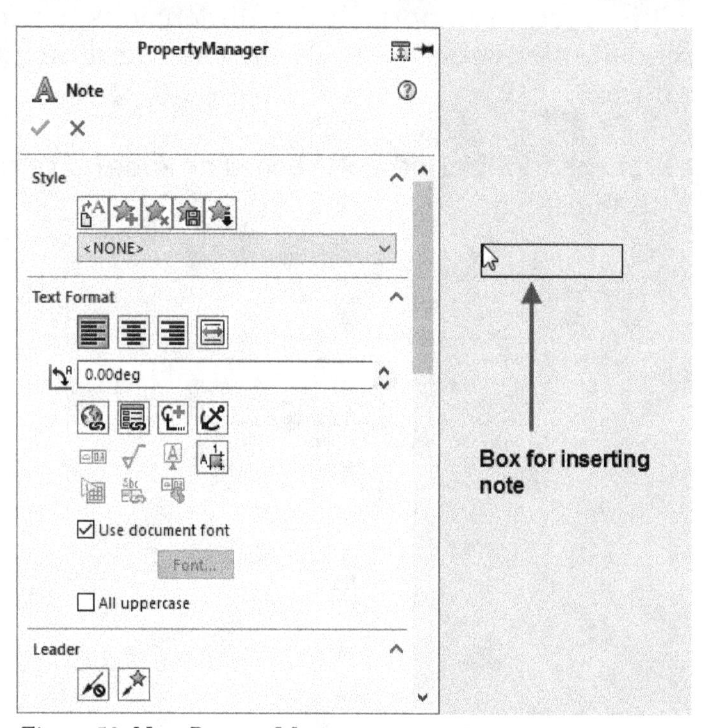

Figure-52. Note_PropertyManager

- If you move the cursor over any entity in the view, then the leader will be added before the note box; refer to Figure-53.

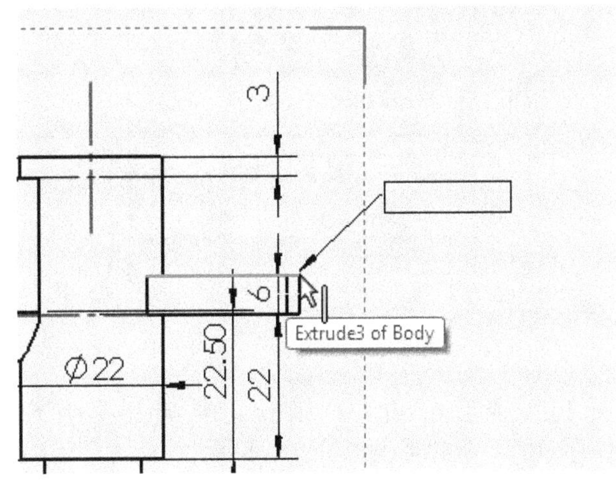

Figure-53. Note box with leader

- Click to place the box. If leader is attached then click again to place the note box.
- On clicking, the editing mode of note will activate. Enter desired text in the box. Apply desired formatting by using the options displayed.
- Click on the **OK** button to create the note.

Flag Notes

The Flag Notes are used to cross-reference listed notes to specific area or feature in the drawing. In other words, flag notes are numbered list of items referenced to different area of drawing. The procedure to create flag notes is given next.

- Click on the **Note** tool from the **Ribbon** and click in the drawing area to specify insertion point.
- Click on the **Number** button from the toolbar displayed; refer to Figure-54.

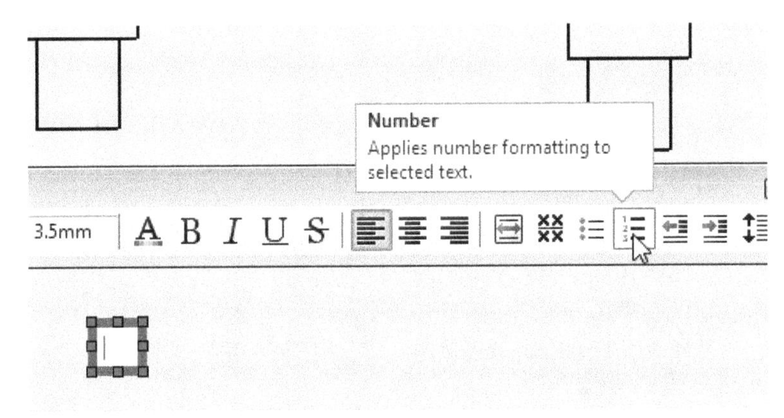

Figure-54. Number button in toolbar

- Type desired note in the text box in the form of numbered list; refer to Figure-55.

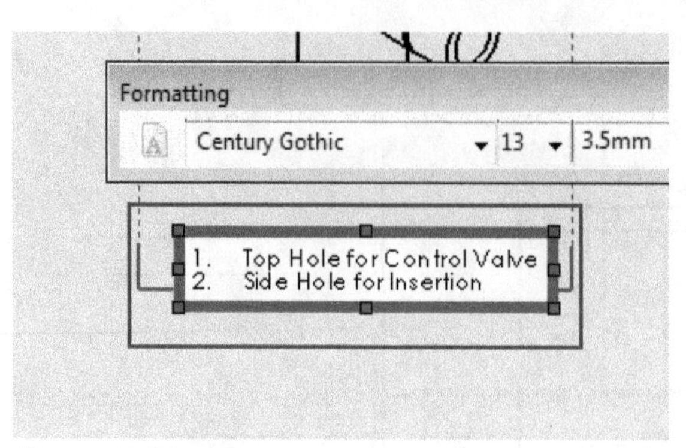

Figure-55. List of notes created

- Click on the number in the list which is to be added in the Flag Note Bank and select the **Add to Flag Note Bank** check box from the **PropertyManager** displayed; refer to Figure-56.
- Similarly, you can add other numbered notes in the Flag Note Bank.
- Click on the **OK** button from the **Note PropertyManager** after adding all the notes in the Flag Note Bank.

Now, we will add balloons in the drawing as per the notes. The procedure to add balloons is given next.

- Click on the **Balloon** tool from the **Annotation** tab in the **Ribbon**. The **Balloon PropertyManager** will be displayed; refer to Figure-57.

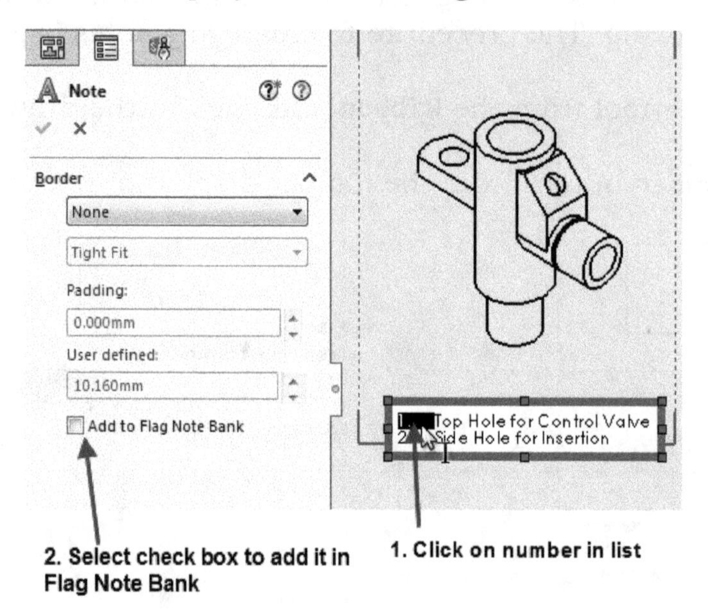

Figure-56. Adding notes flag note bank

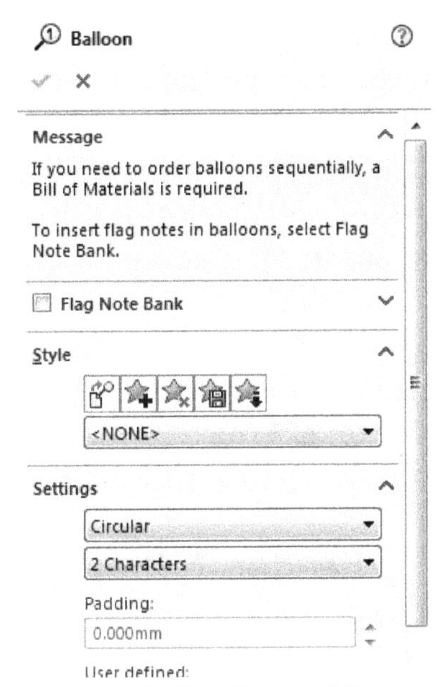

Figure-57. Balloon PropertyManager

- Select the **Flag Note Bank** check box. The notes earlier saved will be displayed in the list; refer to Figure-58.

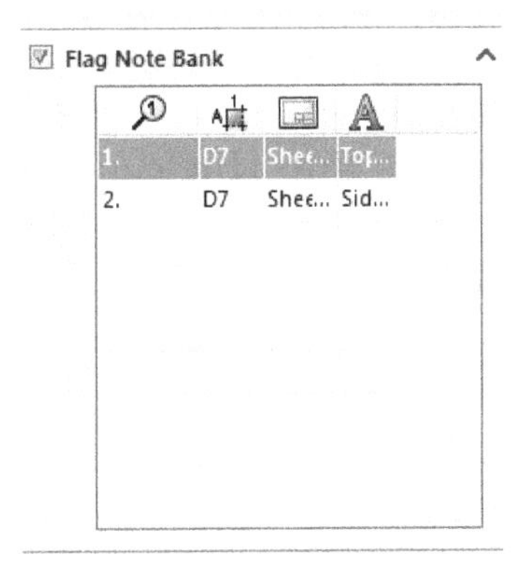

Figure-58. Flag Note Bank

- Select desired note from the list. A balloon will get attached to the cursor.
- Click on desired reference in the drawing and place the balloon; refer to Figure-59.

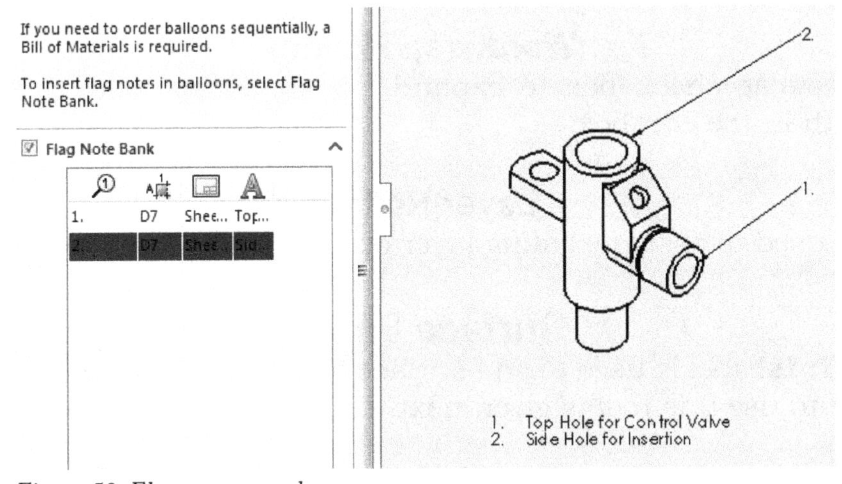

Figure-59. Flag notes created

- Click on the **OK** button from the **Balloon PropertyManager** to exit.

Note that using the options in the **Note PropertyManager**, you can generate all type of geometric annotations like, Datum reference, surface finish symbol, Geometric Tolerance box and so on. The options in the **Note PropertyManager** are discussed next.

Style Rollout

You can save a note as favorite using the options in the **Style** rollout. The options available in this rollout are the same as discussed earlier.

Text Format Rollout

The **Text Format** rollout is used to set the format of the text such as font, size, justification, and rotation of the text. You can also add symbols and hyperlinks to the text using the options available in this rollout.

Leader Rollout

The options in the **Leader** rollout are used to define the style of arrows and leaders that are displayed in the notes.

Leader Style Rollout

The options in this rollout are used to define the style and thickness of the leader. By default, the **Use document display** check box is selected. So, the leader will be displayed with the default style and thickness. On clearing this check box, the **Leader Style** and **Leader Thickness** drop-down lists will be enabled. Using these drop-down lists, you can specify different styles and thickness for the leader.

Border Rollout

The options in the **Border** rollout are used to define the border in which the note text will be displayed. You can assign various types of borders from the **Style** drop-down list. The **Size** drop-down list available in this rollout is used to define the size of the border in which the text will be placed.

Parameters Rollout

The **Parameters** rollout is used to specify the X and Y coordinate values of the note center.

Wordwrap Rollout

Select the **Wordwrap** check box to expand the **Wordwrap** rollout. Specify desired wordwrap width in the edit box.

Layer Rollout

This rollout is used to assign existing layer or create new layer to the notes.

Surface Finish

The **Surface Finish** tool is used to add surface finish symbol to the surface of a part. The procedure to use this tool is given next.

- Click on the **Surface Finish** tool from the **Annotation CommandManager** in the **Ribbon**. The surface finish symbol will get attached to cursor and the **Surface Finish PropertyManager** will be displayed; refer to Figure-60.

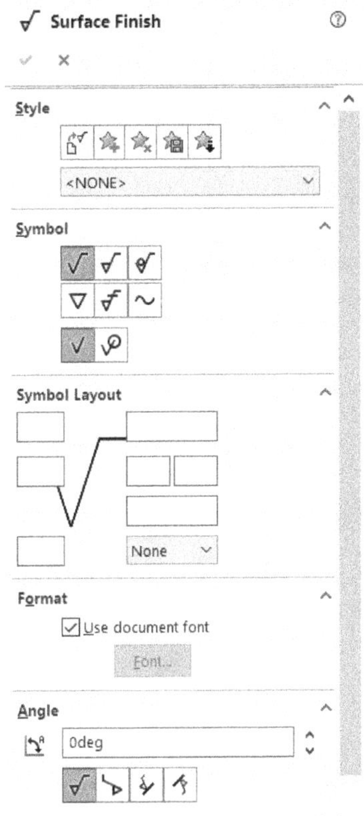

Figure-60. Surface Finish PropertyManager

- Select desired symbol button from the **Symbol** rollout of the **PropertyManager**.
- Specify the respective parameters in the **Symbol Layout** rollout.
- Set the orientation of symbol from the **Angle** rollout.
- Click at desired location in drawing view to place the symbol.
- Click on the **OK** button from the **PropertyManager** to exit the tool.

Surface Texture Symbols or Surface Roughness Symbols

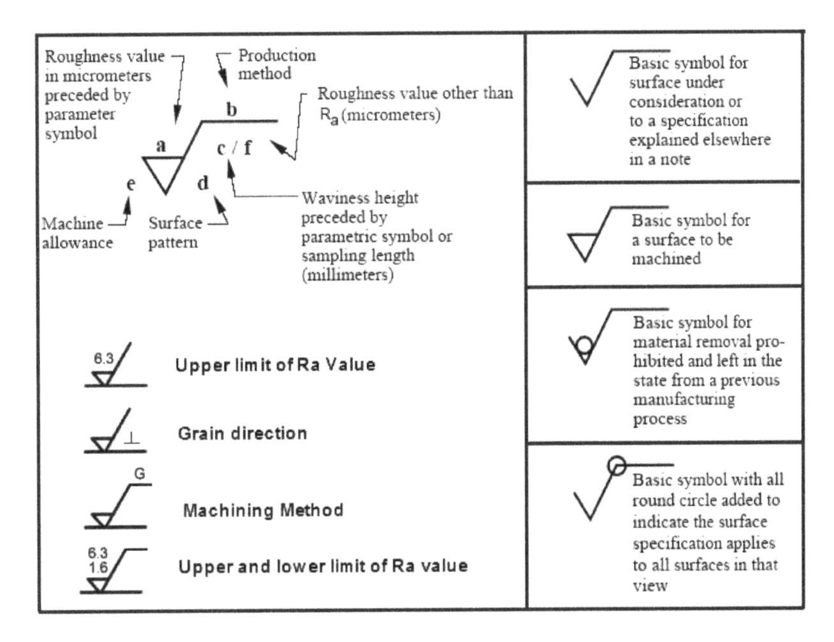

Datum Feature

The **Datum Feature** tool is used to add reference for measuring all the geometric tolerances. For example, if you want to check perpendicularity of a face then you need to give a reference with respect to which the perpendicularity will be measured. The steps to create datum features are given next.

- Click on the **Datum Feature** button from the **Ribbon**. The datum symbol will attach to the cursor and the **Datum Features PropertyManager** will display; refer to Figure-61.

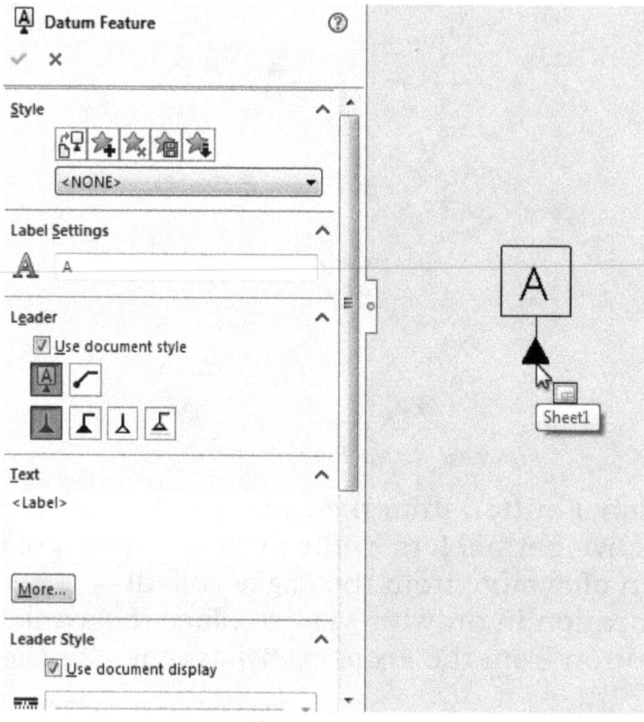

Figure-61. Datum Feature PropertyManager

- Click on the reference to place the leader's start point. Move the cursor to specify length of the leader.
- Click to place the label of the datum feature. Next, datum feature box will attach to the cursor. Press **ESC** to exit the tool. Refer to Figure-62 for datum feature placement.

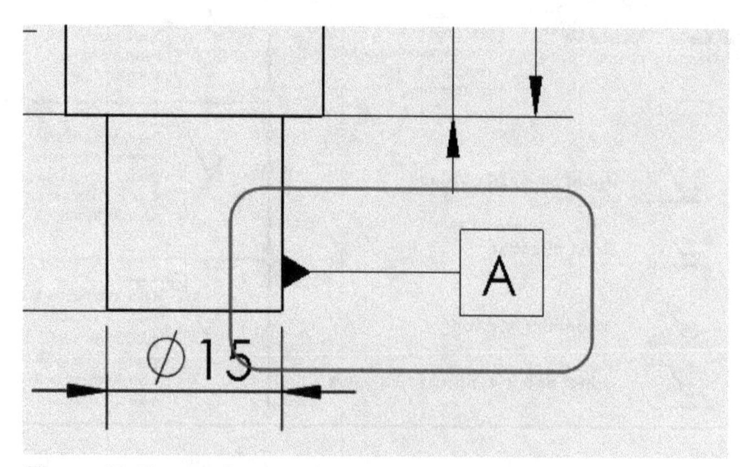

Figure-62. Datum_feature_placement

Figure-63 shows the break-up of a datum feature symbol.

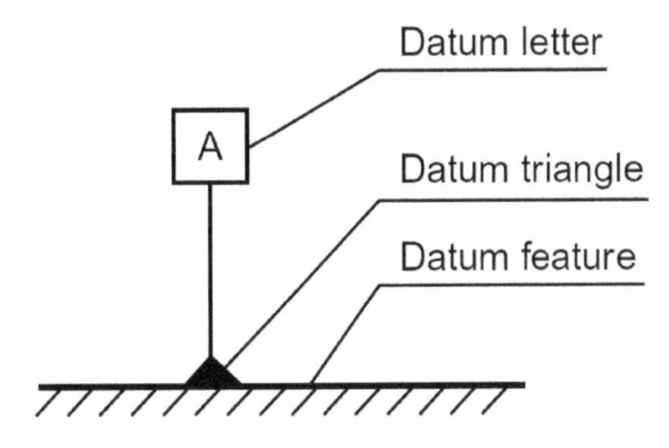

Figure-63. Datum_feature_symbol

Datum Target

Datum targets are circular frames divided in two parts by a horizontal line. The lower half represents the datum feature, and the upper half is for additional information, such as dimensions of the datum target area; refer to Figure-64.

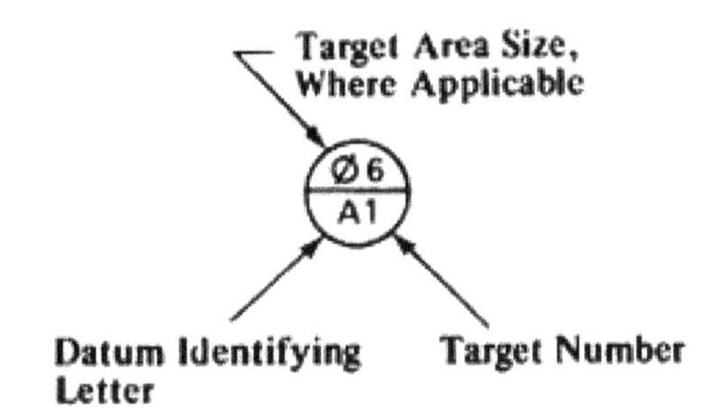

Figure-64. Datum target symbol

The **Datum Target** tool is used to add datum target to the drawing. The steps to do so are given next.

- Click on the **Datum Target** button from the **Ribbon**. The **Datum Target PropertyManager** will display; refer to Figure-65.

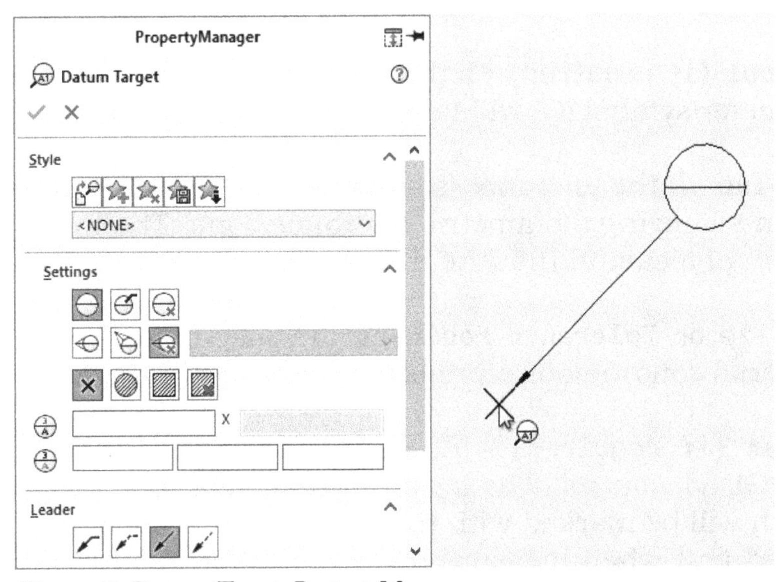

Figure-65. Datum_Target_PropertyManager

- Specify desired value and click in the drawing view to place the leader start point.
- Stretch the symbol and click to place at desired position.

Geometric Tolerance

The feature control frame is also known as GD&T box in laymen's language. The method to insert Feature Control Frame in drawing is same as discussed for Surface Finish symbol. In GD&T, a feature control frame is required to describe the conditions and tolerances of a geometric control on a part's feature. The feature control frame consists of four pieces of information:

1. GD&T symbol or control symbol
2. Tolerance zone type and dimensions
3. Tolerance zone modifiers: features of size, projections, and so on
4. Datum references (if required by the GD&T symbol)

This information provides everything you need to know about geometry of part like what geometrical tolerance needs to be on the part and how to measure or determine if the part is in specification; refer to Figure-66. The common elements of feature control frame are discussed next.

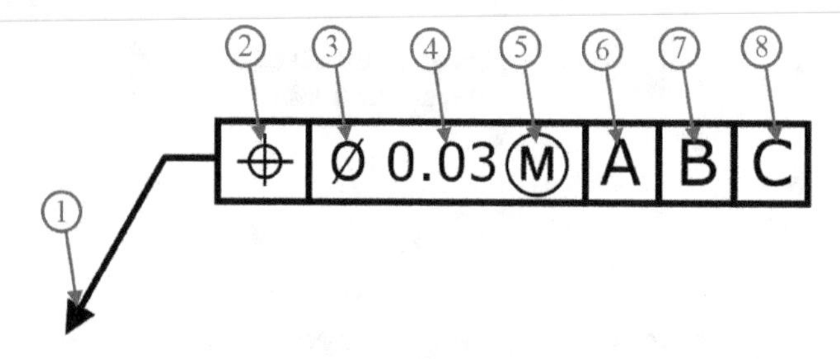

Figure-66. Feature control frame

1. **Leader Arrow** – This arrow points to the feature that the geometric control is placed on. If the arrow points to a surface than the surface is controlled by the GD&T. If it points to a diametric dimension, then the axis is controlled by GD&T. The arrow is optional but helps clarify the feature being controlled.

2. **Geometric Symbol** – This is where your geometric control is specified.

3. **Diameter Symbol (if required)** – If the geometric control is a diametrical tolerance then the diameter symbol (Ø) will be in front of the tolerance value.

4. **Tolerance Value** – If the tolerance is a diameter you will see the Ø symbol next to the dimension signifying a diametric tolerance zone. The tolerance of the GD&T is in same unit of measure that the drawing is written in.

5. **Feature of Size or Tolerance Modifiers (if required)** – This is where you call out max material condition or a projected tolerance in the feature control frame.

6. **Primary Datum (if required)** – If a datum is required, this is the main datum used for the GD&T control. The letter corresponds to a feature somewhere on the part which will be marked with the same letter. This is the datum that must be constrained first when measuring the part. Note: The order of the datum is important for measurement of the part. The primary datum is usually held in three places to fix 3 degrees of freedom

7. **Secondary Datum (if required)** – If a secondary datum is required, it will be to the right of the primary datum. This letter corresponds to a feature somewhere on the part which will be marked with the same letter. During measurement, this is the datum fixated after the primary datum.

8. **Tertiary Datum (if required)** – If a third datum is required, it will be to the right of the secondary datum. This letter corresponds to a feature somewhere on the part which will be marked with the same letter. During measurement, this is the datum fixated last.

Reading Feature Control Frame

The feature control frame forms a kind of sentence when you read it. Below is how you would read the frame in order to describe the feature.

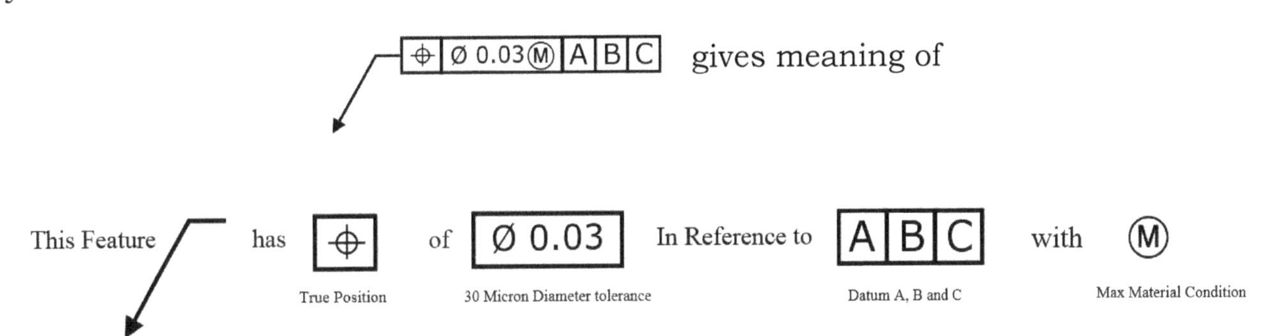

Meaning of various geometric symbols are given in Figure-67.

SYMBOL	CHARACTERISTICS	CATEGORY
⏤	Straightness	Form
⏥	Flatness	
○	Circulatity	
⌭	Cylindricity	
⌒	Profile of a Line	Profile
⌓	Profile of Surface	
∠	Angularity	Orientation
⊥	Perpendicularity	
//	Parallelism	
⌖	Position	Location
◎	Concentricity	
⌯	Symmetry	
↗	Circular Runout	Runout
⌰	Total Runout	

Figure-67. Geometric Symbols

Figure-68 and Figure-69 show the use of geometric tolerances in real-drawing.

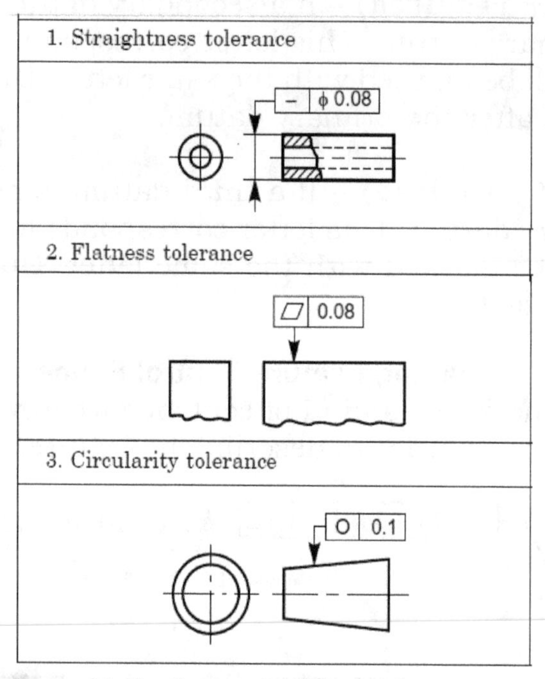

Figure-68. Use of geometric tolerance 1

Note that in applying most of the Geometrical tolerances, you need to define a datum plane like in Perpendicularity, Parallelism and so on.

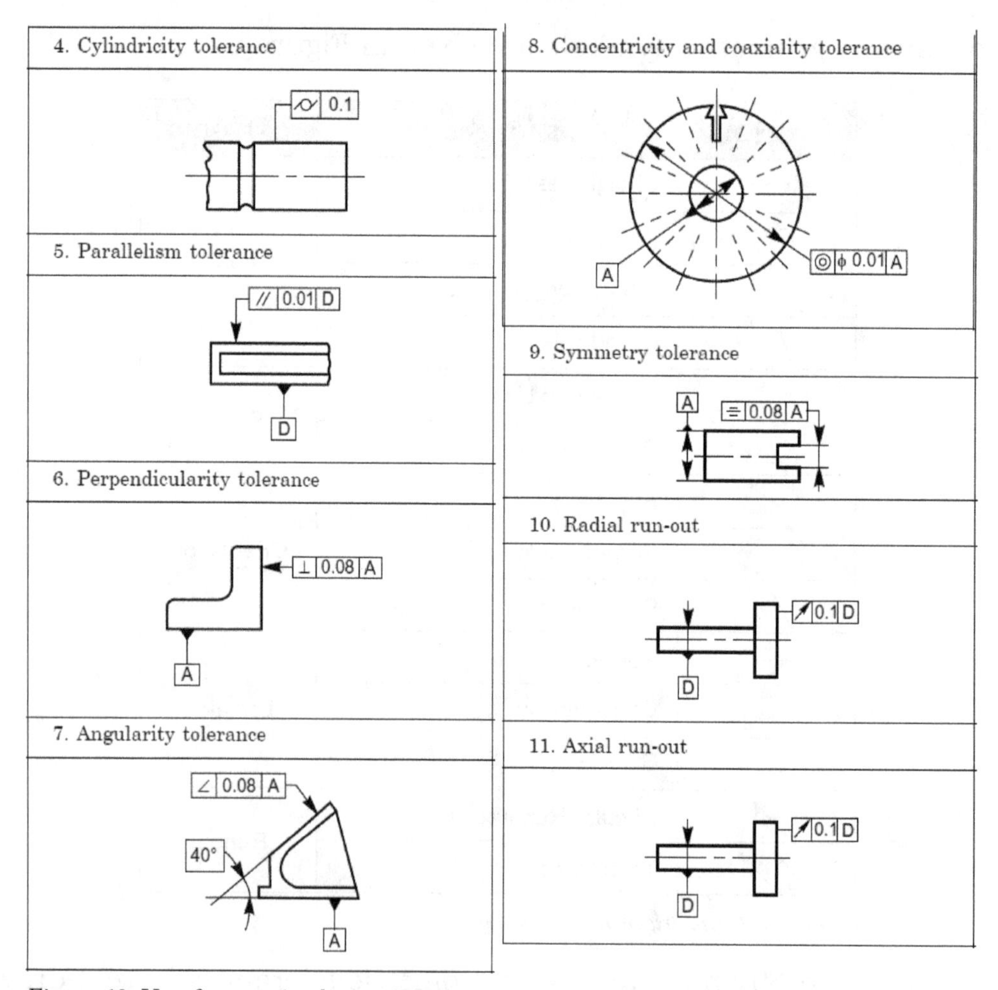

Figure-69. Use of geometric tolerance 2

There are a few dimensioning symbols also used in geometric dimensioning and tolerances, which are given in Figure-70.

Symbol	Meaning
Ⓛ	LMC – Least Material Condition
Ⓜ	MMC – Maximum Material Condition
Ⓣ	Tangent Plane
Ⓟ	Projected Tolerance Zone
Ⓕ	Free State
Ø	Diameter
R	Radius
SR	Spherical Radius
SØ	Spherical Diameter
CR	Controlled Radius
(ST)	Statistical Tolerance
[77]	Basic Dimension
(77)	Reference Dimension
5X	Places

Symbol	Meaning
←⌖	Dimension Origin
⌴	Counterbore
⌵	Countersink
↧	Depth
⌀	All Around
↔	Between
×	Target Point
⌲	Conical Taper
⌳	Slope
□	Square

Figure-70. Dimensioning symbols

The steps to create geometric tolerance box are given next.

- Click on the **Geometric Tolerance** button from the **Annotations** tab in the **Ribbon**. The **Geometric Tolerance PropertyManager** will be displayed; refer to Figure-71.

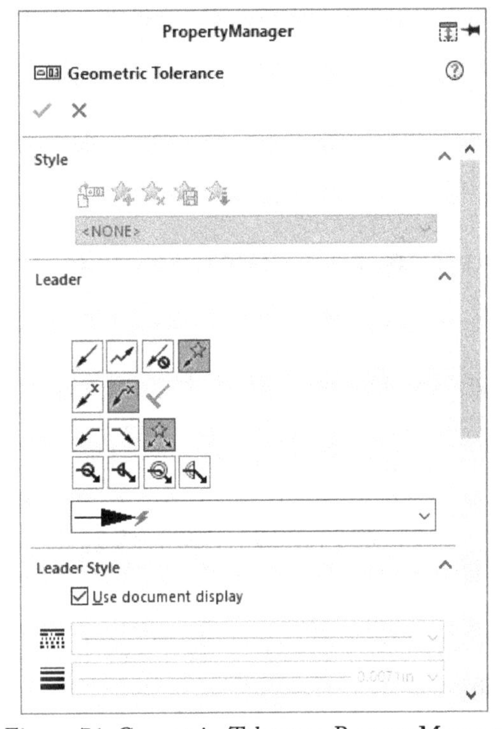

Figure-71. Geometric_Tolerance_PropertyManager

- Click on the geometry to which you want to specify the tolerance. The tolerance box will be attached to cursor; refer to Figure-72.

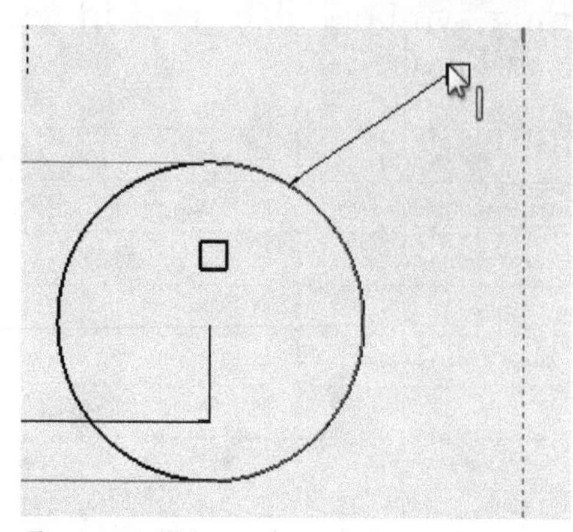

Figure-72. Tolerance_attached_to_cursor

- Move the cursor away and click at the location on the graphics area where you want to place the symbol. The **Tolerance** dialog box will be displayed; refer to Figure-73.
- Select desired tolerance symbol from the dialog box. The updated **Tolerance** dialog box will be displayed; refer to Figure-74.

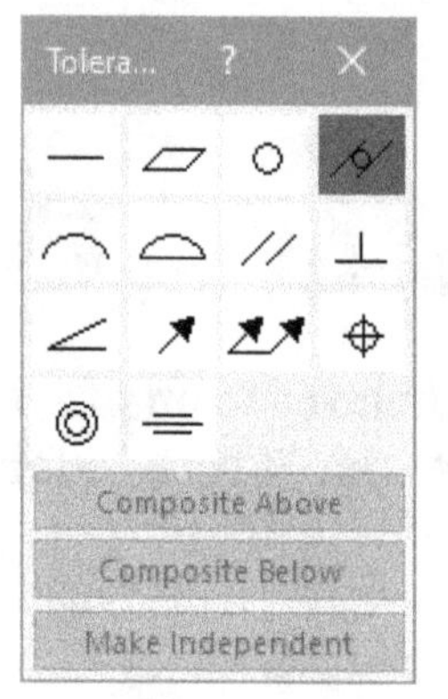

Figure-73. Tolerance_dialog_box

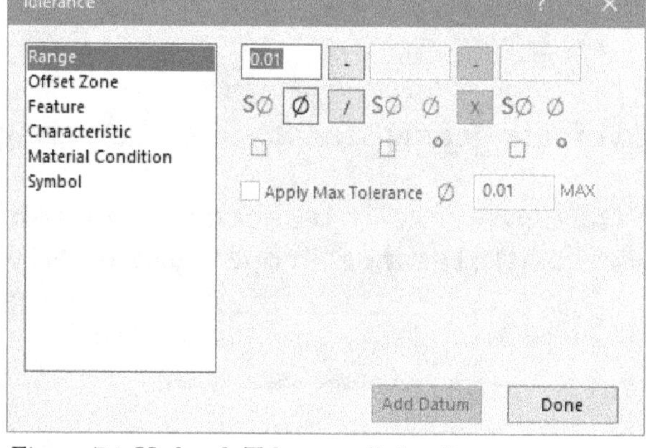

Figure-74. Updated_Tolerance_dialog_box

- Specify desired parameters in the dialog box and **PropertyManager** and click on the **OK** button to create the symbol.
- Click on the **Help** button from the **Properties** dialog box to know more about the box.

Weld Symbol and Hole Callout

The weld symbols are placed in the same way as you place geometric tolerances. Click on the tool, define the parameters and place the symbol by clicking.

To add the hole callout, click on the **Hole Callout** tool and select hole or slot. The callout will be generated automatically. Click to place the callout.

In the same way, you can annotate center line and center mark by selecting the **Centerline** and **Center Mark** tool.

Now, we will add exploded view of the assembly and then we will add bill of material and balloons.

GENERATING EXPLODED VIEW OF ASSEMBLY

To generate exploded view, you must explode the assembly first in the assembly environment. After exploding, save the assembly and then follow the steps to generate exploded view.

- Start a new drawing and click on the **Browse** button from the **Model View PropertyManager**.
- Double-click on the assembly file that you saved earlier.
- Click on the **Show in exploded or model break state** check box and select the exploded view from the drop-down. Note that to use exploded state view, you must have saved a configuration of assembly in exploded state. The topic has been discuss in previous chapters.
- Select the **Isometric** button from the **Orientation** rollout and click in the drawing area to place the assembly; refer to Figure-75.

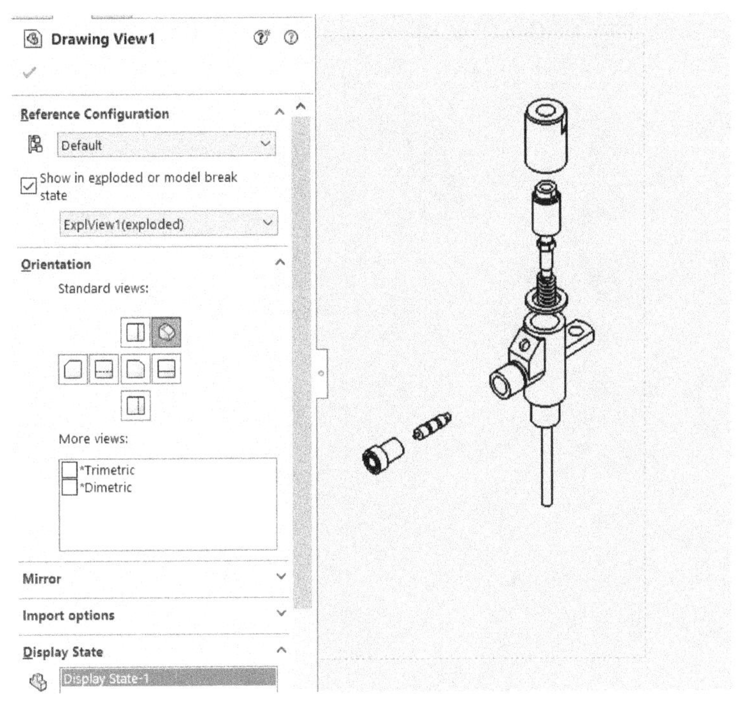

Figure-75. Exploded view of assembly

- You can change the scale as per requirement by using the open in the **PropertyManager**.
- Click on the **OK** button to create the view.

Generating Bill of Material

- Click on the **Annotation** tab and click on the down arrow of **Tables** at the right corner; refer to Figure-76.
- Click on the **Bill of Materials** button. You will be asked to select the view.
- Select the exploded view. The **Bill of Materials PropertyManager** will display.

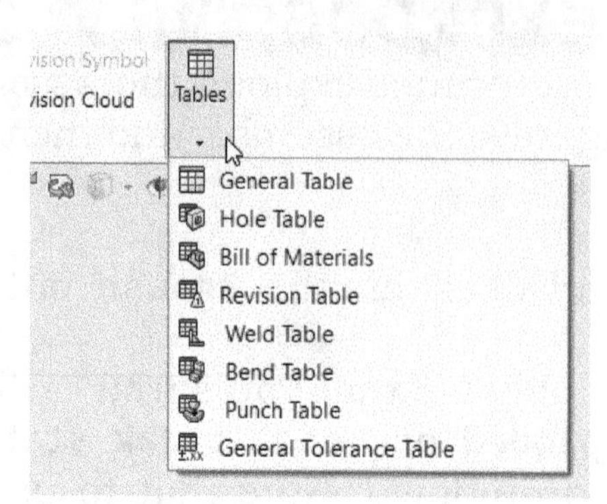

Figure-76. Tables drop-down

- Click on the **OK** button from the **PropertyManager**. The Bill of Materials table will attach to the cursor.
- Click to place the table; refer to Figure-77.

Note that in the same way, you can use the other table options in the **Tables** drop-down.

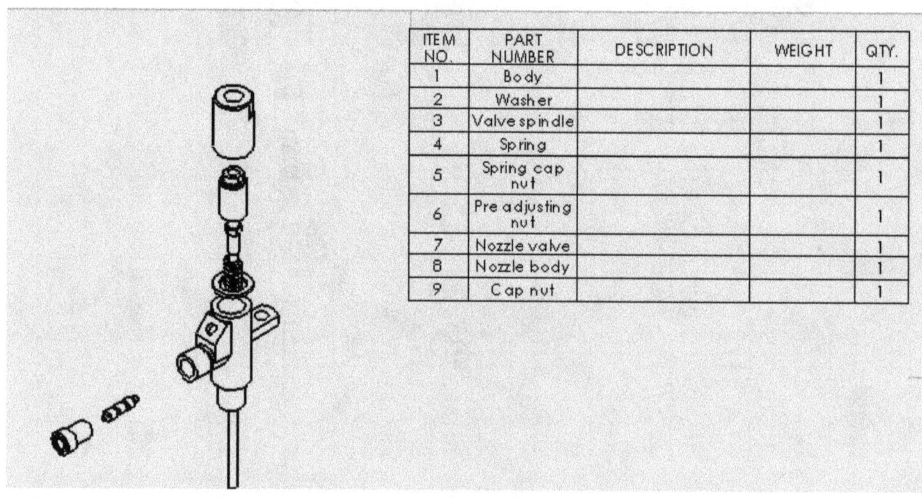

ITEM NO.	PART NUMBER	DESCRIPTION	WEIGHT	QTY.
1	Body			1
2	Washer			1
3	Valve spindle			1
4	Spring			1
5	Spring cap nut			1
6	Pre adjusting nut			1
7	Nozzle valve			1
8	Nozzle body			1
9	Cap nut			1

Figure-77. Bill of Materials

Generating Balloons for Bill of Material

- Click on the **Auto Balloon** button from the **Ribbon** and click on the view. The balloons will be generated automatically.
- Click on the **OK** button from the **PropertyManager** to generate the balloons.
- Drag the balloons to desired positions.

You can use the **Sketch** tab and use the sketcher tools to create custom entities in the drawing.

EDITING TITLE BLOCK

You can edit the title block to specify the information related to designer, part name, versions and so on. To edit the title block, follow the steps given next.

- Right-click on the **Sheet Format** from the **FeatureManager Design Tree**. The shortcut menu will display; refer to Figure-78.

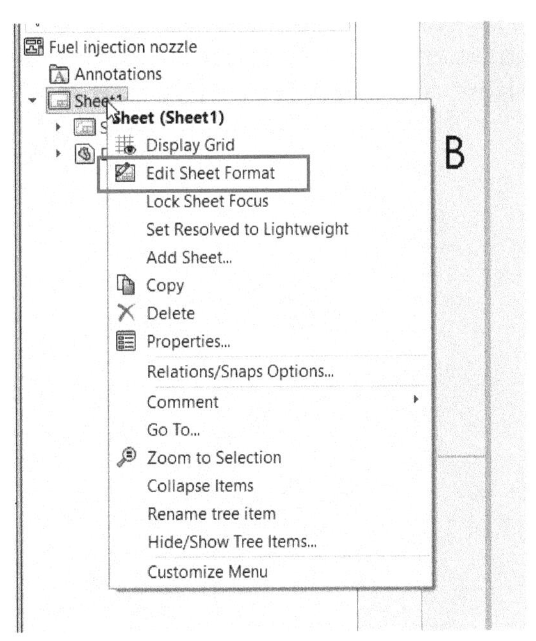

Figure-78. Right click shortcut menu

- Click on the **Edit Sheet Format** button, the title block will become edit able; refer to Figure-79. You can also use the **Edit Sheet Format** tool from the **Sheet Format CommandManager** to edit the title block.
- Double-click in desired field to specify desired information.

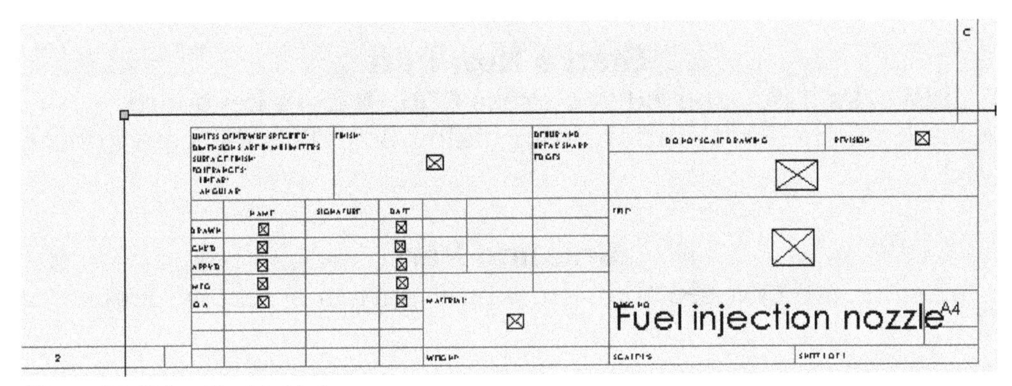

Figure-79. Edit-able title block

- After specifying the information, click on the **OK** button from the **PropertyManager**.
- Similarly, you can edit the existing information by double-clicking on them.
- Click on the **Return** button at the top-right corner of the graphic window to return in Drawing mode; refer to Figure-80.

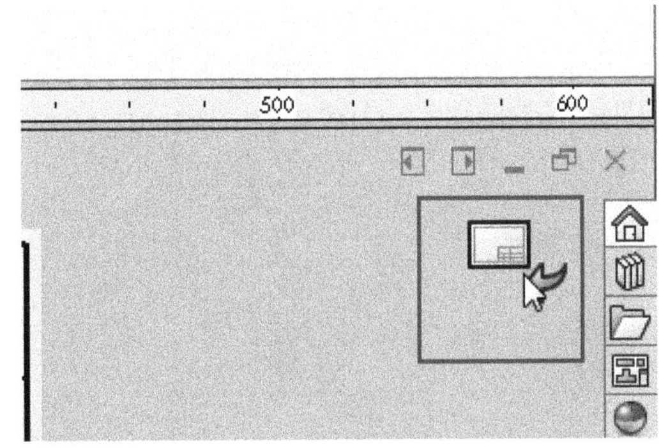

Figure-80. Return to drawing

Practical

In this practical, you will first create the model of part as per the production drawing given in Figure-81 and then you will create the same production drawing of part using the model.

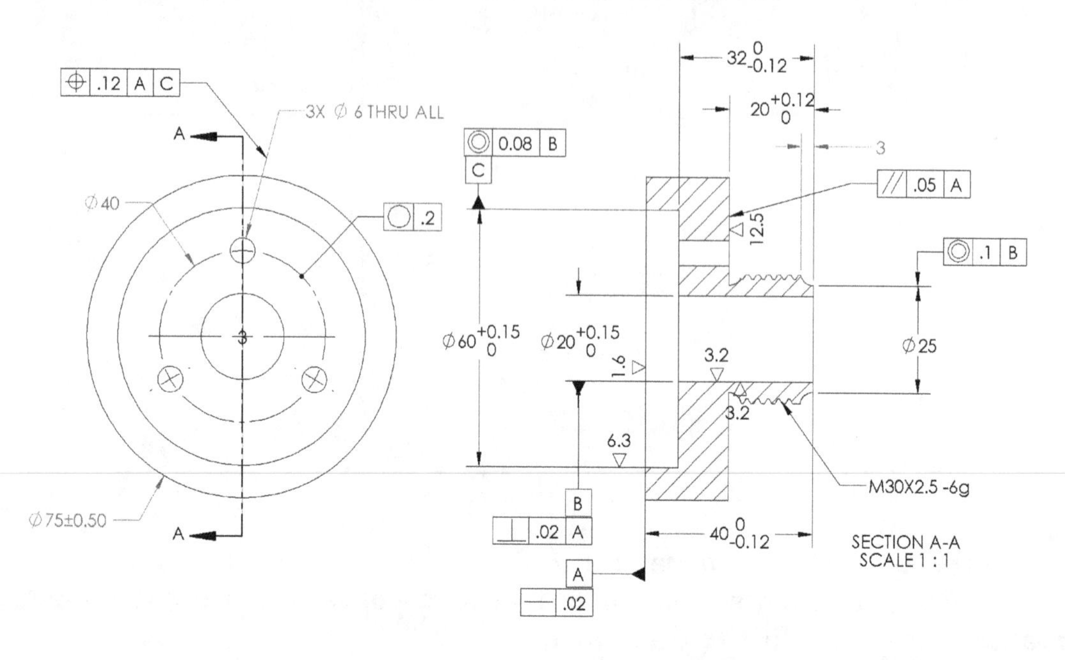

Figure-81. Production drawing for practical

Start a New Part

- Start SolidWorks if not started yet. Press **CTRL+N** from keyboard.
- Double-click on the **Part** button from dialog box. The Part environment will be displayed.

Creating Part

From the drawing, we can see that the whole part is a revolve feature with holes created on it.

- Click on the **Revolved Base/Boss** tool from the **Features CommandManager**. You will be asked to select the sketching plane.
- Click on the **Front Plane** to create sketch on front plane.
- Draw the sketch as shown in Figure-82 while taking the base dimensions from the drawing.
- Select the **75** dimension from the sketch. The **Dimension PropertyManager** will be displayed.
- Click on the **Diameter** button from the **Dimension Text** rollout of the **PropertyManager**. A diameter symbol will be added to the dimension text; refer to Figure-83.

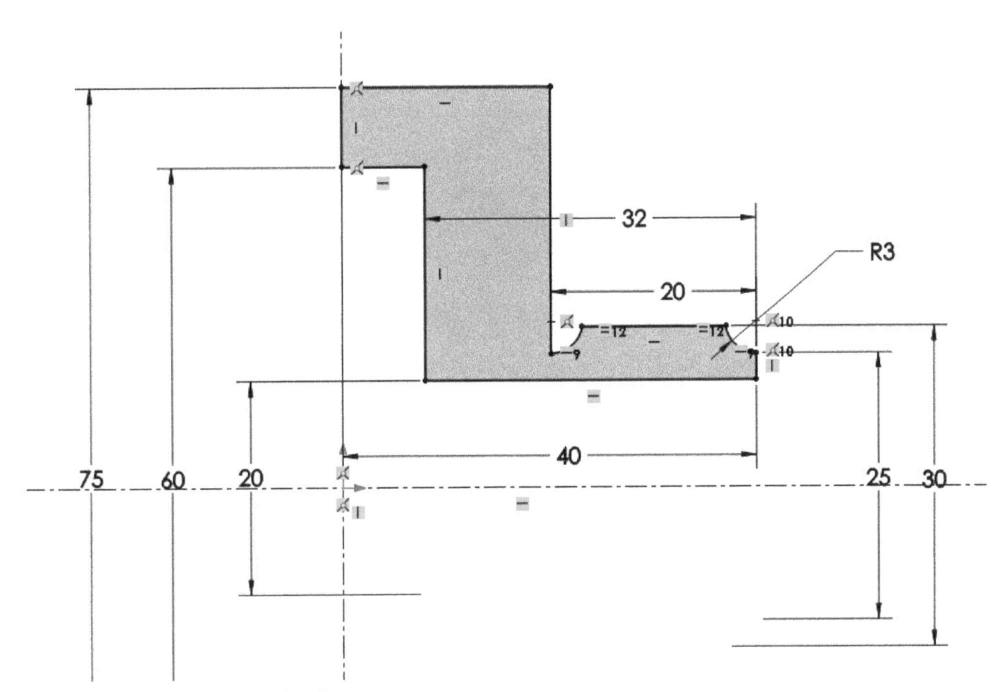

Figure-82. Sketch for revolve feature

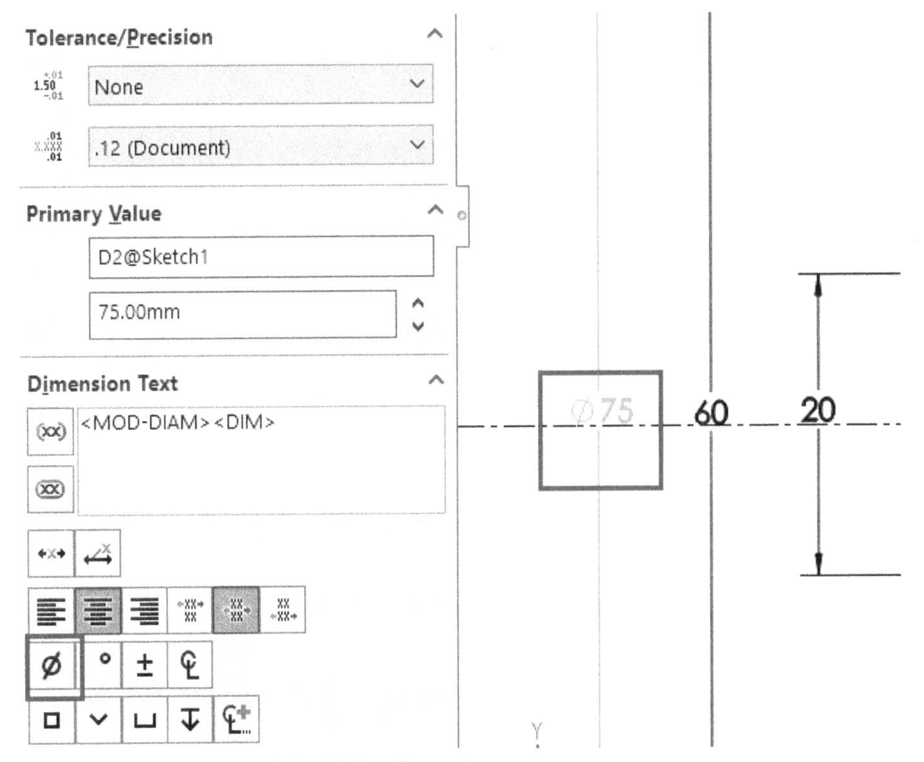

Figure-83. Diameter symbol added to dimension text

- Select the **Symmetric** option from the **Tolerance Type** drop-down in the **Tolerance/ Precision** rollout of **PropertyManager** and specify the value of tolerance as **0.50** in the respective edit box. The dimension will be displayed as shown in Figure-84.
- Click on the **OK** button from the **PropertyManager**.
- Select the other dimensions one by one and repeat the procedure to specify diameter symbol and tolerances as required. After applying symbols and tolerances, the sketch should be displayed as shown in Figure-85.

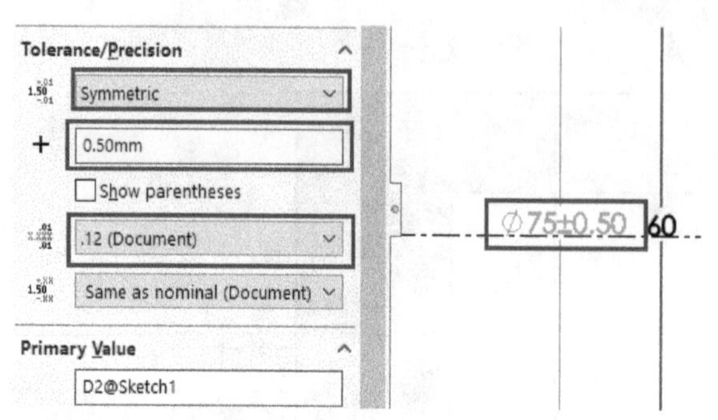

Figure-84. Tolerance applied to dimension

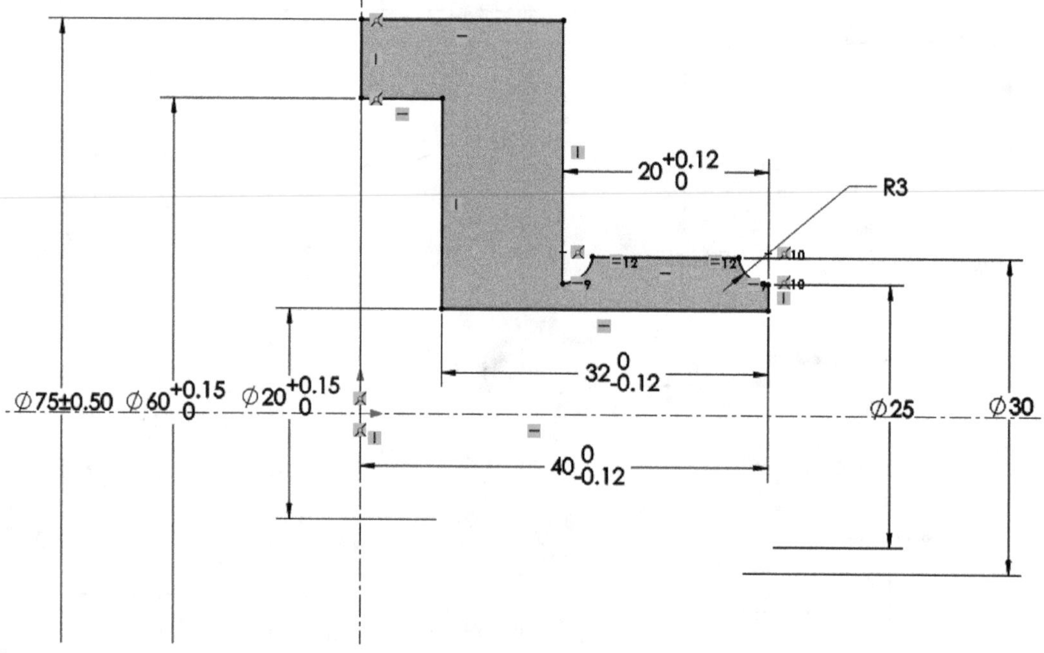

Figure-85. Sketch after applying tolerances and symbols

- Click on the **Exit Sketch** button from the **CommandManager**. You will be asked to select the axis of revolution.
- Select the centerline created in sketch. Make sure the revolution angle is set to **360** degree.
- Click on the **OK** button from the **Revolve PropertyManager**. The revolve feature will be created; refer to Figure-86.
- Click on the **Hole Wizard** tool and create a drill hole of diameter **6** at the top quadrant point of a construction circle **Ø40** on the face revolve feature; refer to Figure-87.

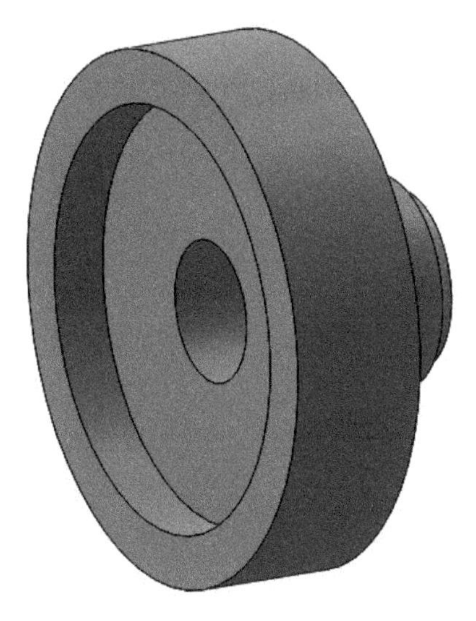

Figure-86. Revolve feature created

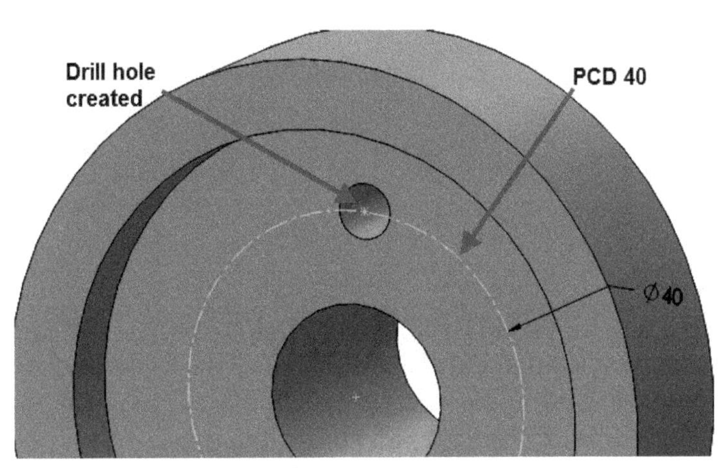

Figure-87. Drill hole created

- Create the circular pattern of hole equidistant on PCD with 3 instances; refer to Figure-88.

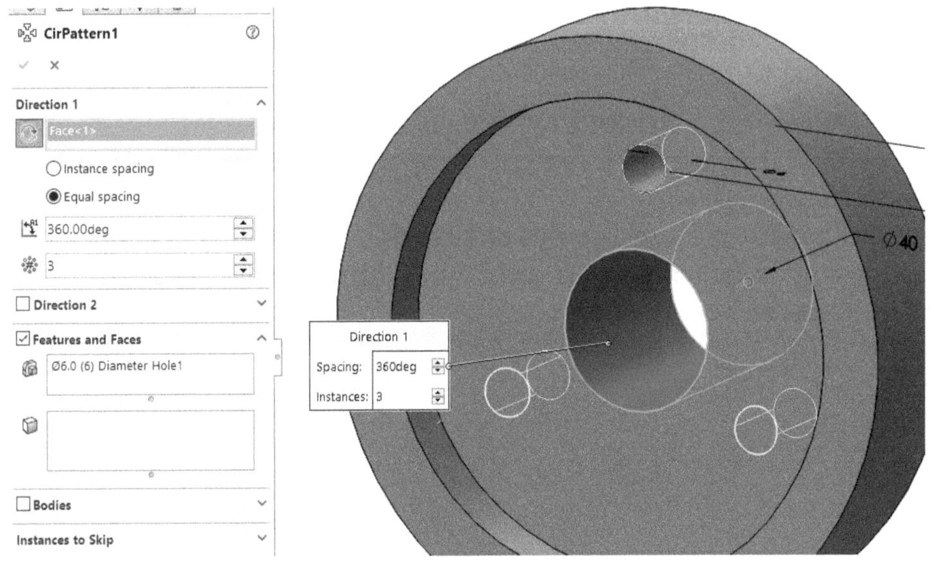

Figure-88. Preview of circular pattern of hole

- Click on the **Thread** tool from the **Hole Wizard** drop-down in the **Features CommandManager** and create the **Metric Die** thread on **M30x2.0**; refer to Figure-89.
- Save the part with the name **Drawing Practical**.

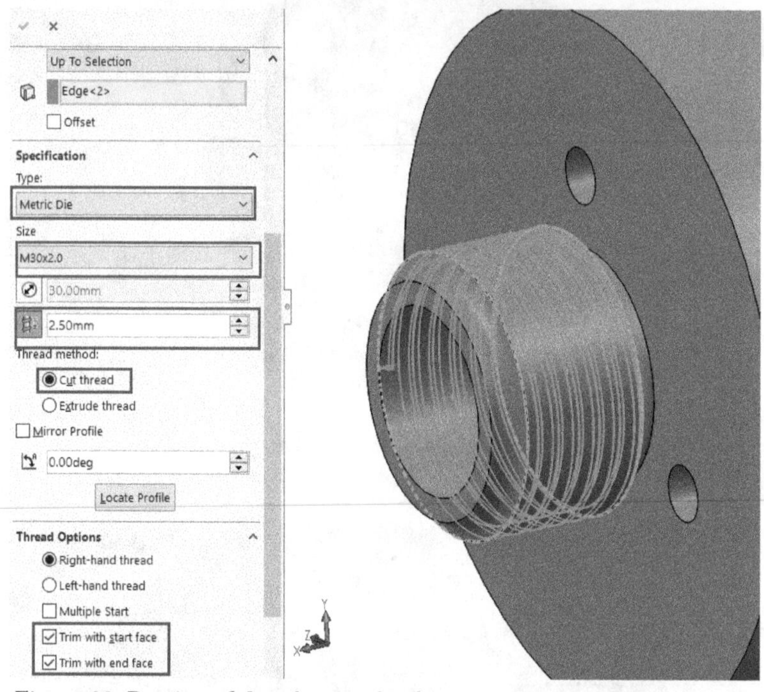

Figure-89. Preview of thread on revolve feature

Starting Drawing File

- Press **CTRL+N** from keyboard. Double-click on the **Drawing** button from the **New SOLIDWORKS Document** dialog box displayed. The Drawing environment will be displayed along with **Sheet Format/Size** dialog box.
- Select the **A4 (ANSI) Landscape** option the list box and click on the **OK** button. (Remember that in First Chapter of the book, we have selected **ANSI** as default drafting standard so we are displayed only ANSI templates of sheet. Click on the **Browse** button from the dialog box to access other templates.)
- On clicking the **OK** button from the **Sheet Format/Size** dialog box; the **Model View PropertyManager** is displayed with open part selected in the box.

Inserting the Base View

- Click on the **Next** button from the **PropertyManager**. The front view of model will get attached to cursor.
- Select the **Left** button from the **Orientation** rollout of the **PropertyManager** and place the view near left edge of the drawing sheet; refer to Figure-90.
- Press **ESC** to exit the tool.

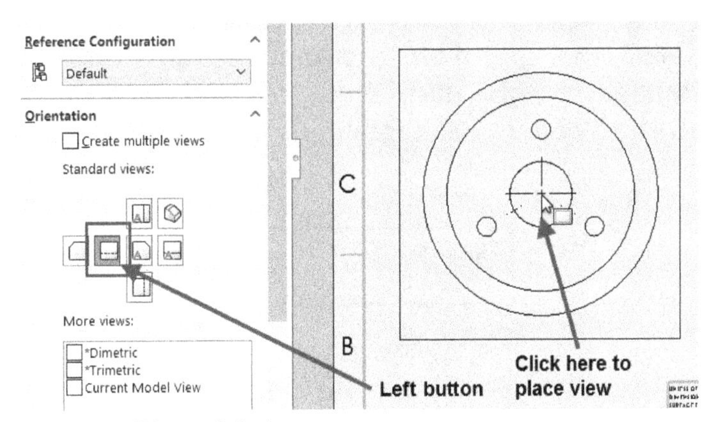

Figure-90. Placing left view

Creating Section View

- Click on the **Section View** tool from the **Drawing CommandManager** in the **Ribbon**. The section line will get attached to cursor.
- Click in the center of the view. A toolbar will be displayed; refer to Figure-91.

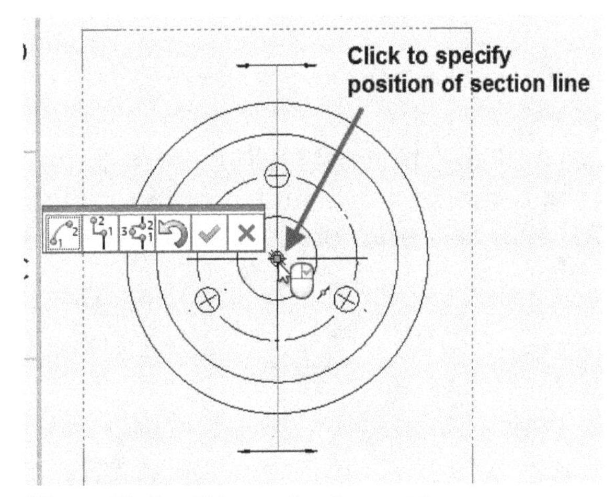

Figure-91. Specifying section line position

- Right-click to accept the section line. The section view will get attached to cursor.
- Place the view at a suitable distance from base view so that all annotations to be applied later can accommodate; refer to Figure-92. Press **ESC** to exit the tool.

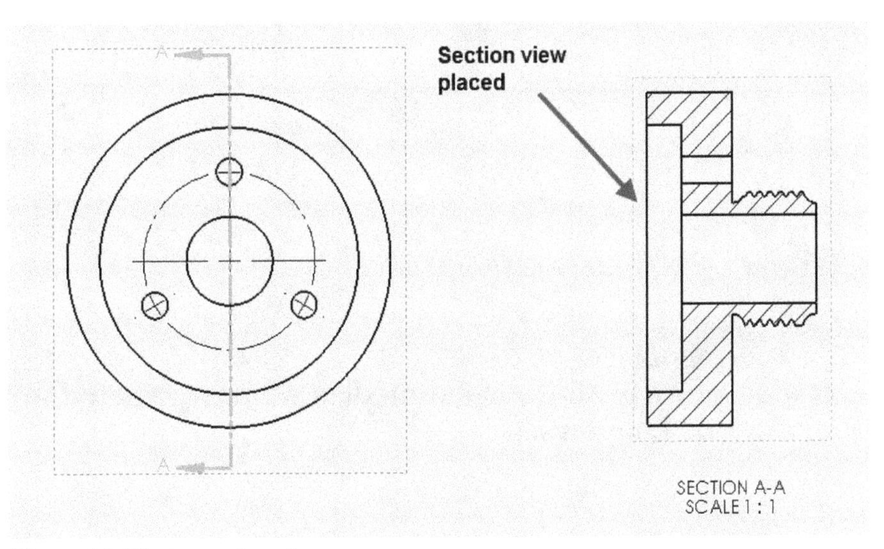

Figure-92. Placing section view

Applying Annotations

- Click on the **Model Items** tool from the **Annotation CommandManager**. The **Model Items PropertyManager** will be displayed.
- Set the options in the **PropertyManager** as shown in Figure-93.

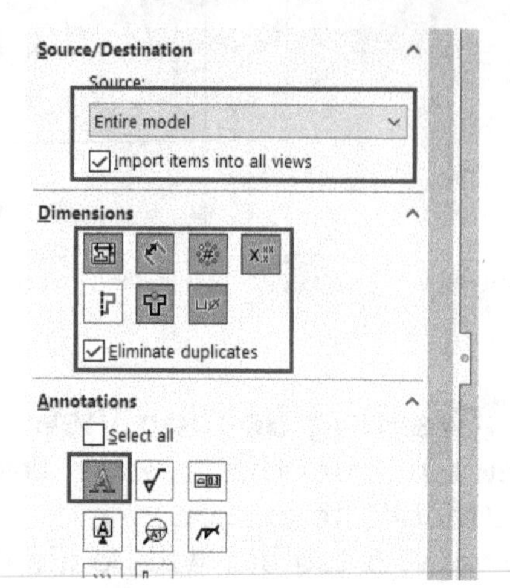

Figure-93. Options selected for model item annotations

- Click on the **OK** button from the **PropertyManager**. The dimensions will be placed in the drawing views.
- Drag the dimensions to suitable places. Delete extra dimensions and use the **Smart Dimension** tool to create dimensions which are left; refer to Figure-94.
- Add the **Geometric Tolerance** and **Datum Feature** symbols in the drawing as per Figure-81.

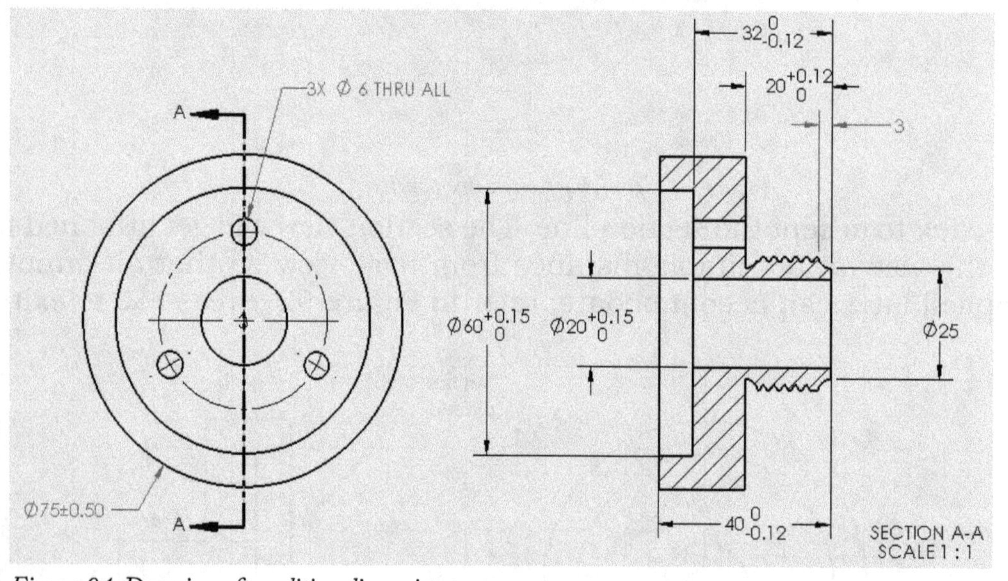

Figure-94. Drawing after editing dimensions

PRACTICE 1

Generate the drawing views of all the solid models we have created in Practices and Practicals in the previous chapters.

PRACTICE 2

Generate the exploded views, bill of materials and balloons from all the assembly models we have worked on till this chapter.

PRACTICE 3

Create the model and drawing given in Figure-95.

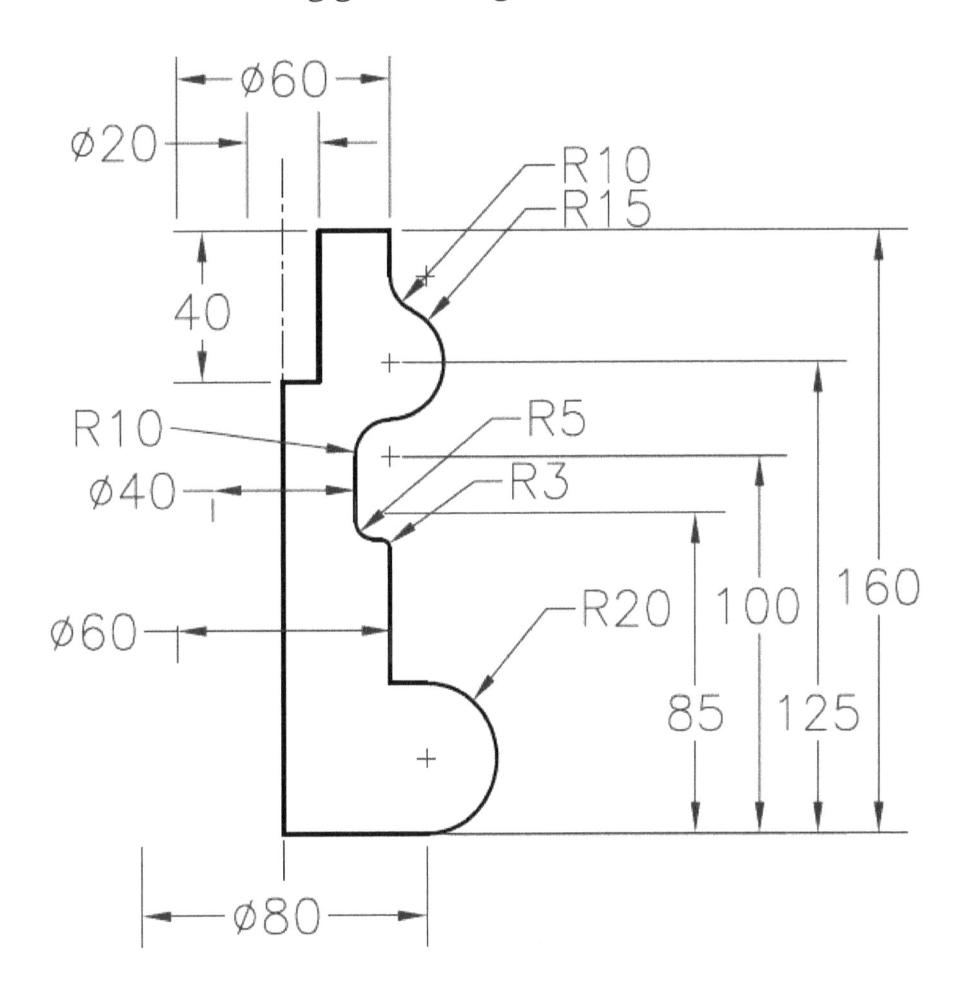

Figure-95. Practice 3

PRACTICE 4

Create the model and drawing given in Figure-96.

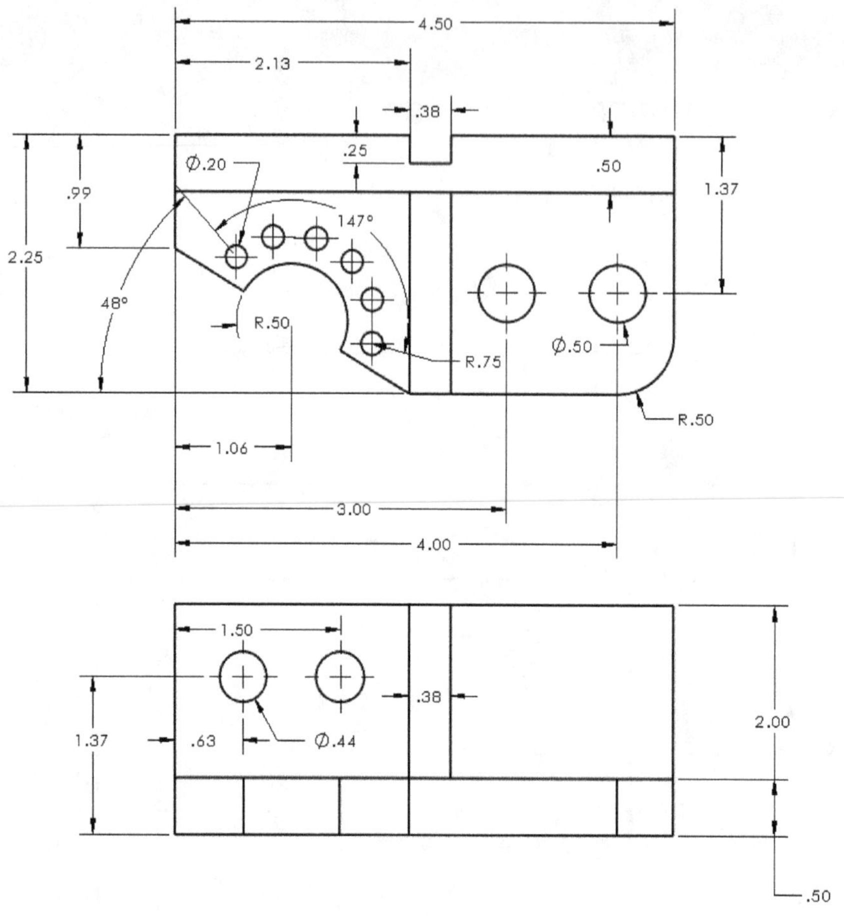

Figure-96. Practice 4

Note that the drawing given in this book are for practice purpose only.

SELF ASSESSMENT

Q1. In which dialog box, we can set the projection type of views being inserted in drawing?

a. **Sheet Format/Size** dialog box
b. **Sheet Properties** dialog box
c. **New SolidWorks Document** dialog box
d. **Geometric Tolerance** dialog box

Q2. In which of the following tab, the tools to add views in the drawing are available?

a. **Drawing** tab
b. **Annotation** tab
c. **Evaluate** tab
d. **Sheet Format** tab
Q3. We can insert as many model views of a part as required. (T/F)

Q4. We can insert the model views of different parts in same drawing by using the **Model View** tool. (T/F)

Q5. The **Auxiliary View** tool is used to create projected view from selected view by making selected edge of view parallel to screen. (T/F)

Q6. The............ tool is used to create section in the selected view to display inner detail of the model.

Q7. The..........tool is used to represent very long objects in the drawing by breaking them at specific span.

Q8. ....is defined as the maximum permissible overall variation of form or position of a feature.

FOR STUDENT NOTES

Answer to Self-Assessment:

1. b, **2.** a, **3.** T, **4.** T, **5.** T, **6.** Broken-out Section, **7.** Break View, **8.** Geometric Tolerance

Chapter 11

AnalysisXpress

Topics Covered

The major topics covered in this chapter are:

- ***Perform Simulation Xpress Analysis.***
- ***Perform Flow Xpress Analysis.***
- ***Perform DFM Xpress Analysis.***
- ***Perform Costing of manufacturing process***
- ***Perform Sustainability analysis of part***

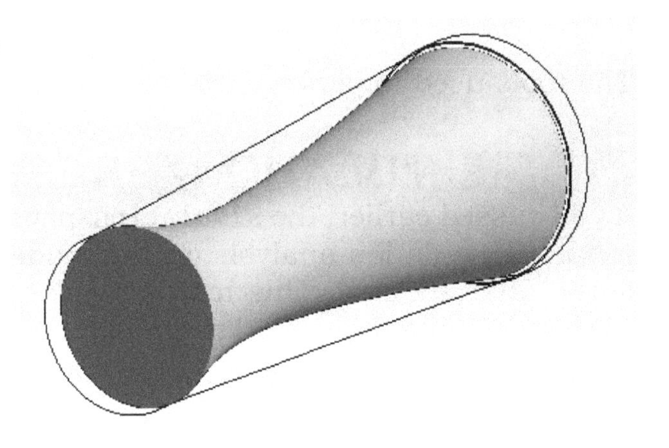

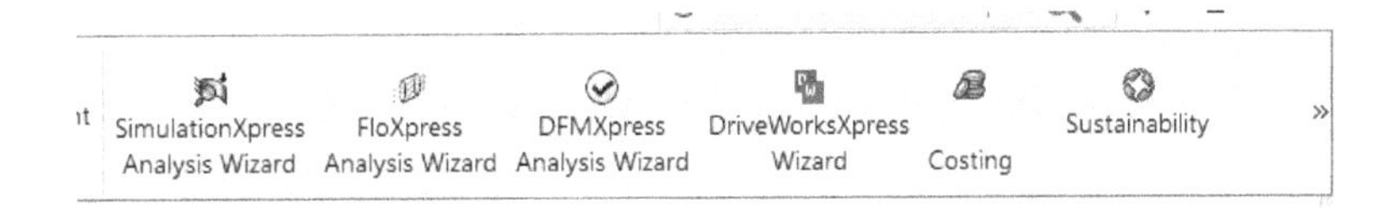

INTRODUCTION

Analysis Xpress is the combination of tools available in SolidWorks to perform some very useful analyses quickly. Various types of analyses available in SolidWorks are SimulationXpress Analysis, FlowXpress Analysis, and so on. These analyses are performed for the following functions:

SimulationXpress Analysis: This analysis is used to check whether the component will fail on the specified force/pressure conditions or not. You can generate a report and perform the optimization.

FloXpress Analysis: This analysis is used to check the flow of a fluid through the designed passage.

DFMXpress Analysis: This analysis is used to check whether the created component is manufacturable or not.

SustainabilityXpress: This analysis is used to check the impact of environment on the component.

Part Reviewer: This analysis is used to check how the part was created in SolidWorks. This analysis becomes very useful to find error in modeling in SolidWorks.

Costing: This analysis is used to check the manufacturing cost of model created in SolidWorks. Note that the cost is automatically modified if you change the model.

Along with the above analysis tools, there is one more tool named DriveWorksXpress. This tool is used to increase the speed of designing by making the products formulae based. For example, you can create a formula for Nuts or Bolts and then specify the driving dimensions to create multiple instances of Nuts/Bolts with different sizes.

Note that you need to enter the product codes before using the features discussed in this chapter.

The tools used to perform the named above analyses are discussed next.

SIMULATIONXPRESS ANALYSIS WIZARD

As discussed earlier, the **SimulationXpress Analysis Wizard** tool is used to perform a quick simulation analysis on the model. You can use this tool to perform linear static analysis only. This tool is available in the **Evaluate CommandManager** of the **Ribbon**; refer to Figure-1. The procedure to use this tool is given next.

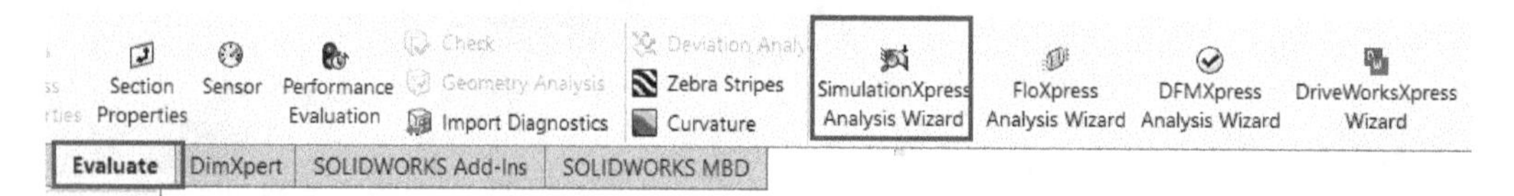

Figure-1. SimulationXpress_Tool

- Click on the **SimulationXpress Analysis Wizard** tool to start SimulationXpress Analysis. The **SOLIDWORKS SimulationXpress** pane will be displayed on the right in the program window; refer to Figure-2. **Make sure that you have opened the solid model for which you want to perform the analysis.**
- Click on the **Options** button in the right of the application window; refer to Figure-3. On doing so, the **SimulationXpress Options** dialog box will be displayed as shown in Figure-4.

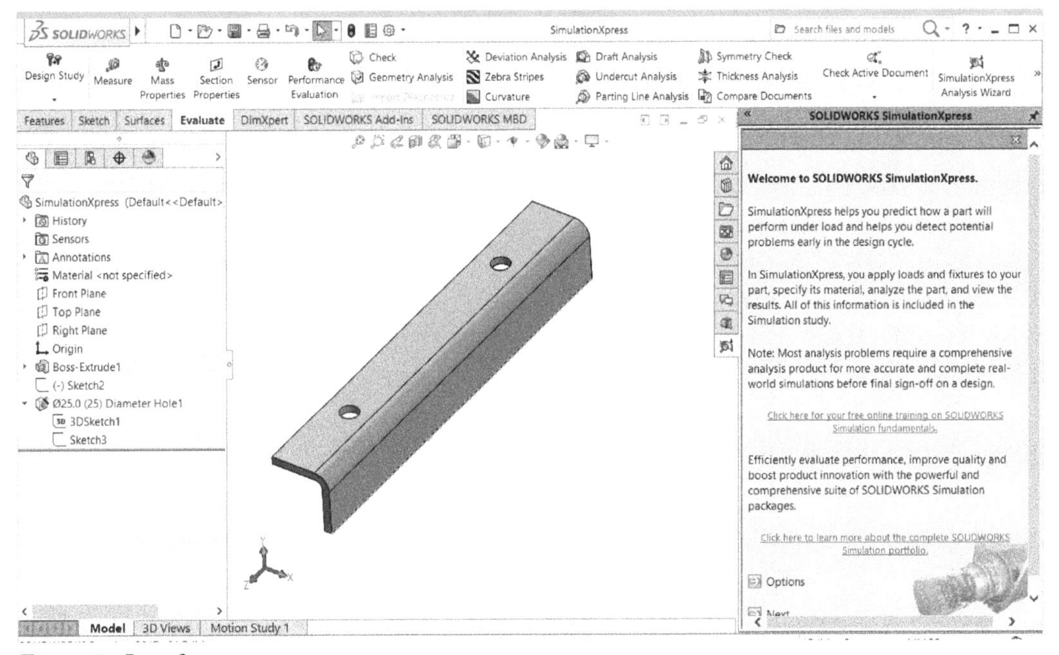

Figure-2. Interface

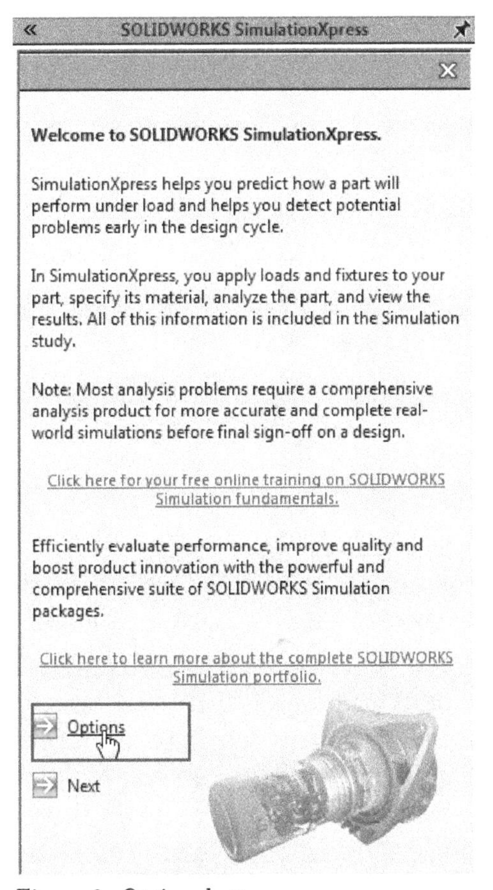

Figure-3. Options button

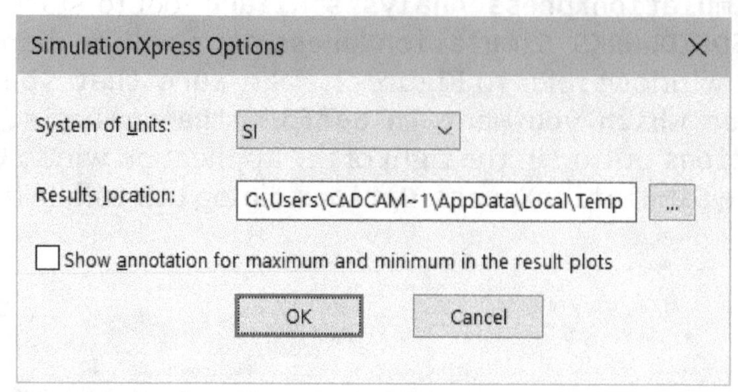

Figure-4. SimulationXpress Options dialog box

The options in this dialog box are used to set the unit system of the analysis and save directory for the analysis report.

- Select the **SI** option from the **System of units** drop-down and change the location of result report to the desired one by using the Browse button. After setting all the parameters, select the **OK** button from the dialog box.

Fixture Setting

- Select the **Next** button from the **SolidWorks SimulationXpress** task pane, the **Fixtures page** of SimulationXpress will be displayed as shown in Figure-5.

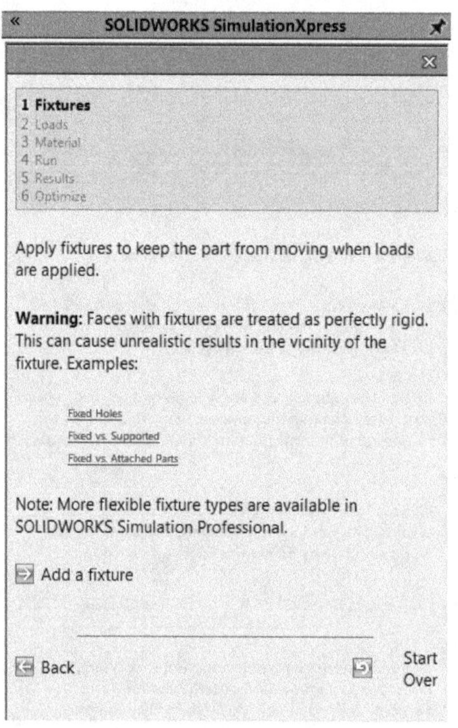

Figure-5. Fixtures page of SimulationXpress

- Click on the **Add a fixture** button to fix a face. On clicking the **Add a fixture** button, the **Fixture PropertyManager** will be displayed; refer to Figure-6.
- Select the face of the model that you want to be fixed; refer to Figure-7.
- Click on the **OK** button from the **Fixture PropertyManager** and then click on the **Next** link button from the **SimulationXpress** displayed in the right. On doing so, the **Loads** page will be displayed in the right; refer to Figure-8.

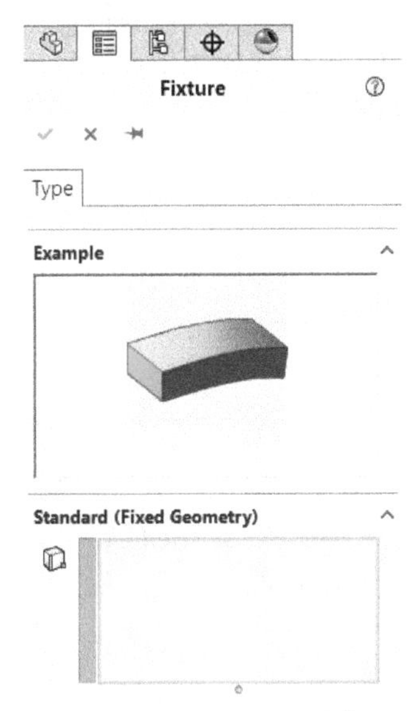

Figure-6. Fixture PropertyManager

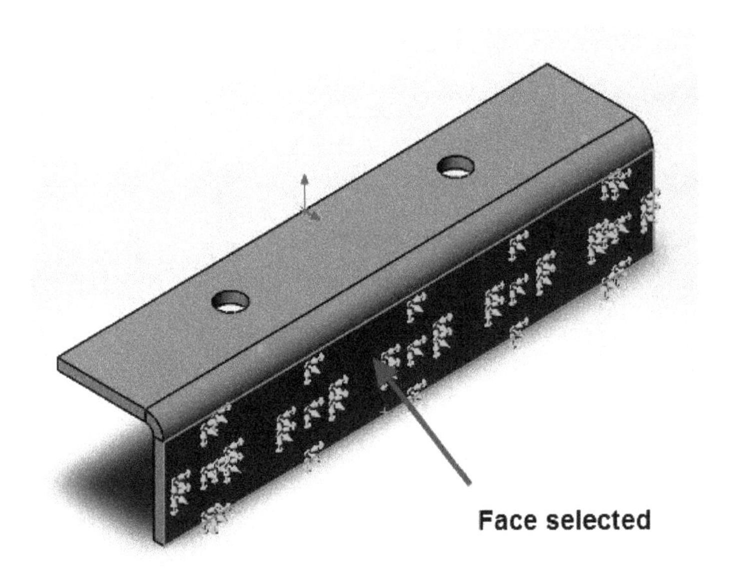

Figure-7. Fixed face

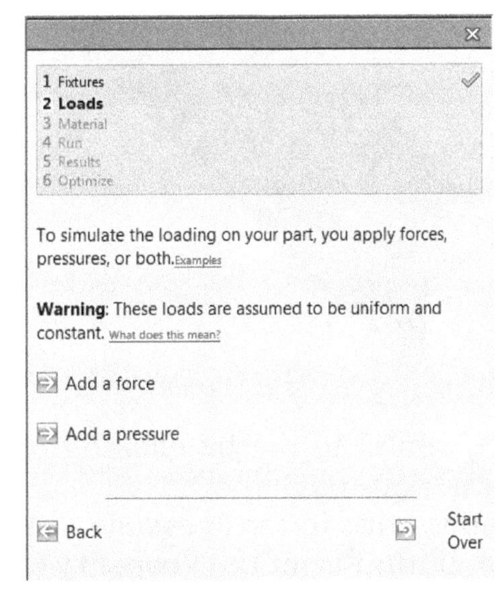

Figure-8. Loads page

Load Setting

- Click on **Add a force** or **Add a pressure** link button from the **SimulationXpress**. In our case, the **Add a pressure** link button is selected. On doing so, the **Pressure PropertyManager** will be displayed as shown in Figure-9.

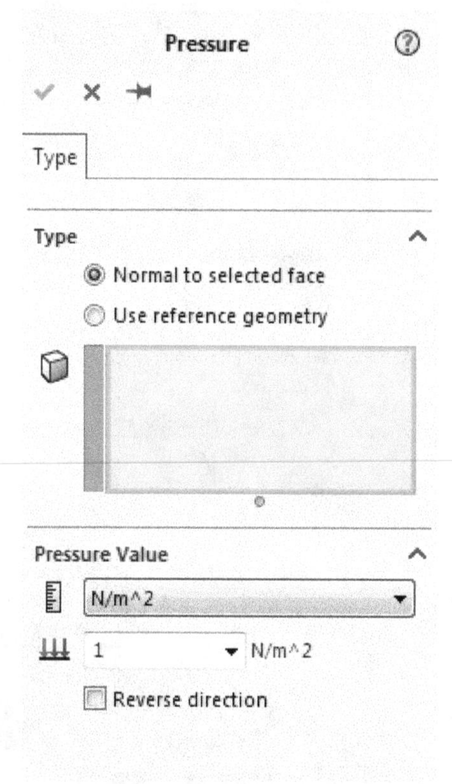

Figure-9. Pressure PropertyManager

- Select the face of the model on which you want to apply the load; refer to Figure-10.

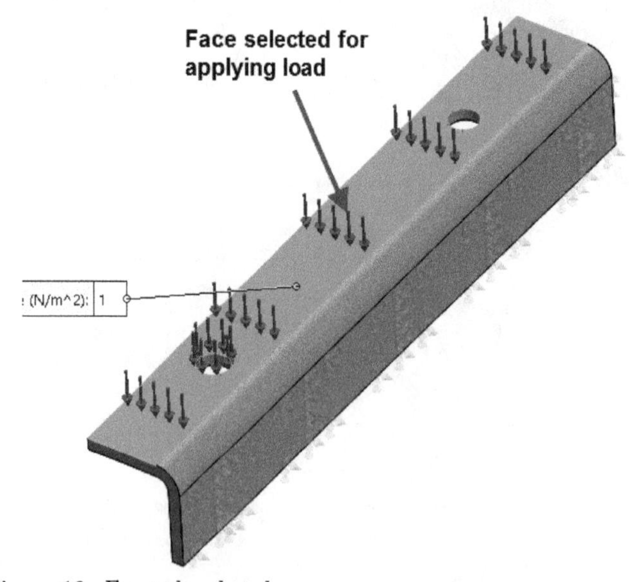

Figure-10. Face to be selected

- Change the unit from N/mm^2 to psi by clicking on the **Unit** drop-down and selecting the **psi** option, if required.
- Specify the desired pressure value in the **Pressure Value** edit box. After specifying the value, click on the **OK** button from the **PropertyManager**.

You can change the direction of pressure by selecting the Reverse direction check box from the FeatureManager.

- You can add more loads by using the **Add a force** or **Add a pressure** link button. After specifying all desired loads, click on the **Next** link button from the **SimulationXpress**. On doing so, the **Material** page will be displayed; refer to Figure-11.

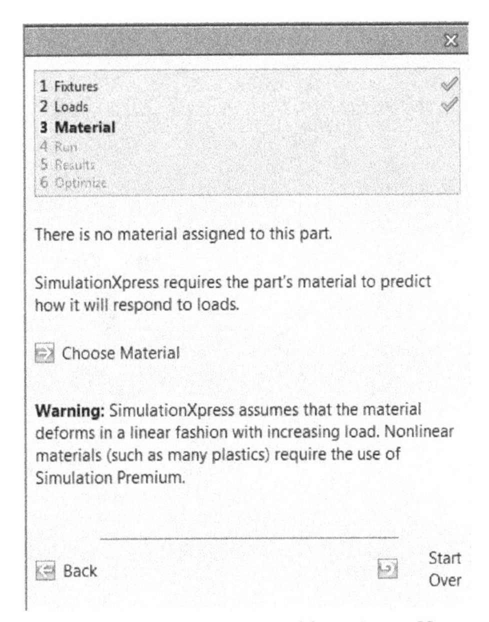

Figure-11. Material page of SimulationXpress

Material Setting

- Check the Warning message in the task pane carefully (This is the conditions of using this analysis) and then click on the **Choose Material** link button from the **SimulationXpress**, the **Material** dialog box will be displayed as shown in Figure-12.
- Note that some of the properties of the material are highlighted in red color. These highlighted properties are the driving properties for the analysis. Select the desired material from the list in the left and then select the **Apply** button from the dialog box.
- Click on the **Close** button to exit the dialog box. The material will be applied and its properties will be displayed in the **SimulationXpress task pane**.
- Select the **Next** button from the **SimulationXpress** Task Pane. On selecting the button, the **Run** page will be displayed; refer to Figure-13.

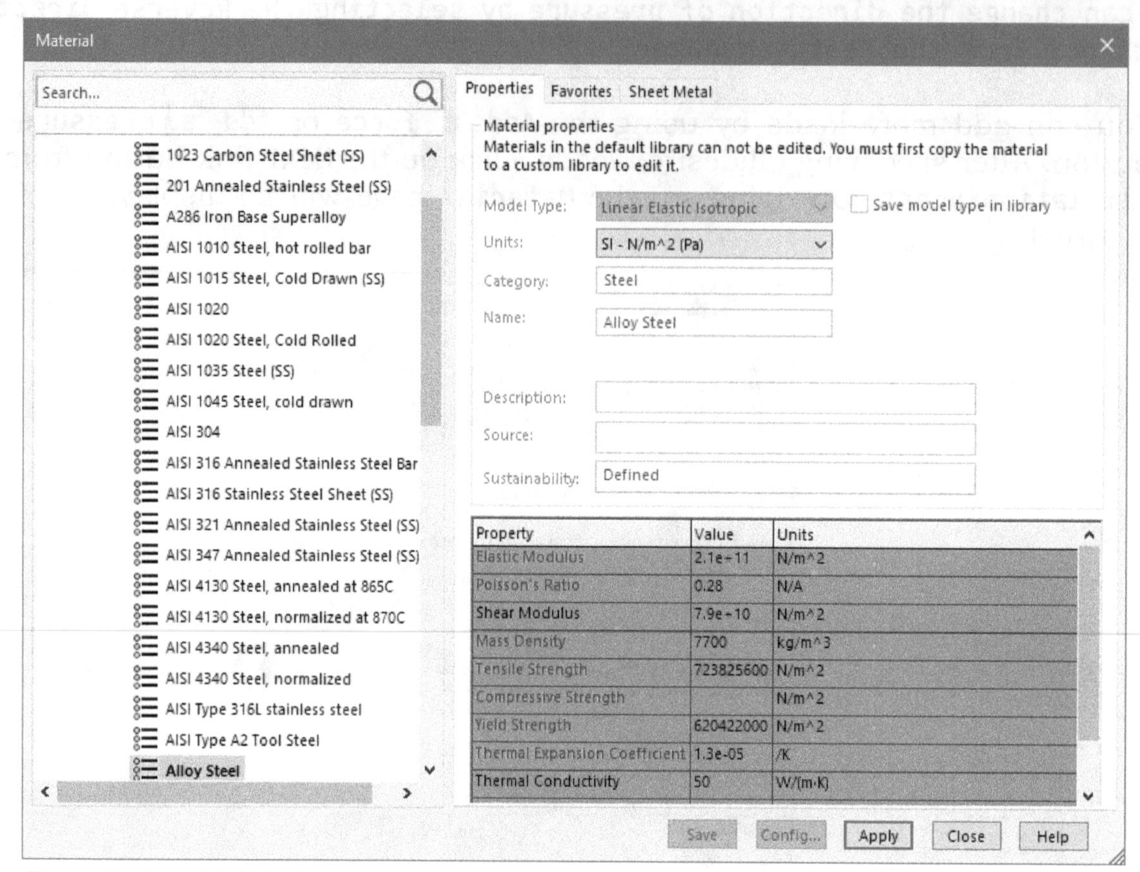

Figure-12. Material dialog box

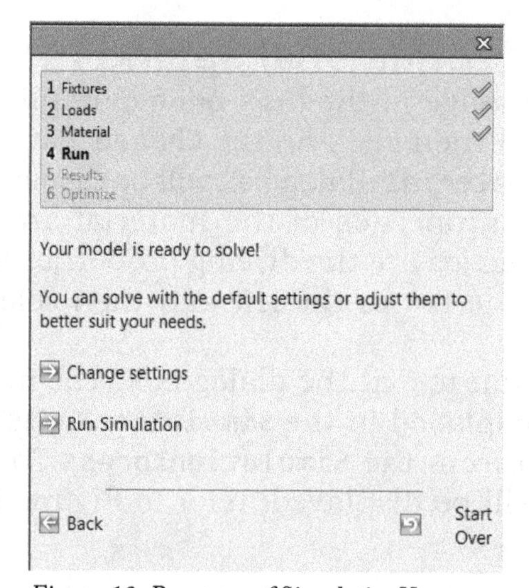

Figure-13. Run page of SimulationXpress

Changing Mesh Density

- Click on the **Change settings** link button from the **Run** page and then click on the **Change mesh density** link button from the page. The **Mesh PropertyManager** will be displayed; refer to Figure-14.

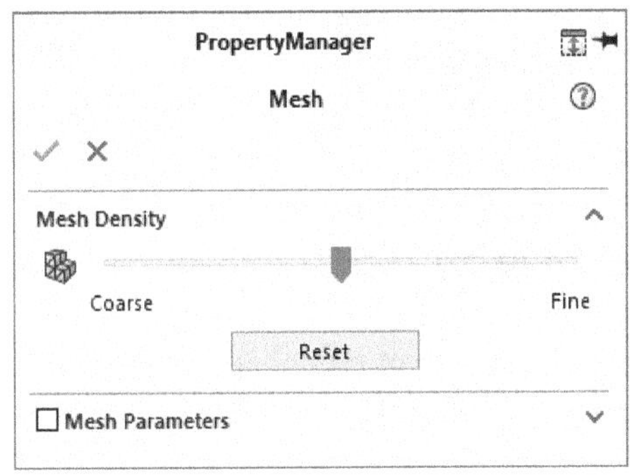

Figure-14. Mesh_PropertyManager

- Click on the **Mesh Parameters** check box in the **PropertyManager** and set the desired value of Global mesh element size as well as the tolerance in the respective edit boxes.

Or

- Drag the **Mesh Density** slider to desired location to make the mesh coarse or fine. Note that the finer you make the mesh, the longer system will take to solve simulation.
- Click on the **OK** button from the **PropertyManager** to apply the changes.
- Click on the **Next** button from the **Task Pane** to return to **Run** page.

Running Simulation

- Click on the **Run Simulation** link button from the **SimulationXpress** to check the output. The result will be displayed in the Modeling area and Results page will be displayed as shown in Figure-15.

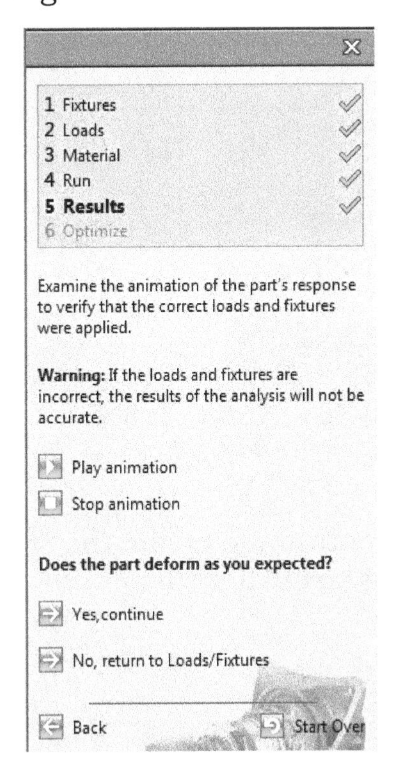

Figure-15. Results_page_of_SimulationXpress

Results

- Select the **Play animation** and **Stop animation** button to start and stop the simulation. If you agree with the result, you need to select the **Yes,continue** link button and if you disagree then select the **No, return to Loads/Fixtures** link button to change the parameters of analysis.
- On selecting the **Yes,continue** button, the modified **Results** page will be displayed as shown in Figure-16.
- You can display von Mises stress or displacement by using the respective button from the SimulationXpress Task Pane. Also, the suggested FOS will also be displayed in the SimulationXpress.
- After checking the results, select the **Done viewing results** link button; the modified results page will be displayed as shown in Figure-17. Now, you can generate an eDrawing or you can generate a report by selecting the respective button from the SimulationXpress.

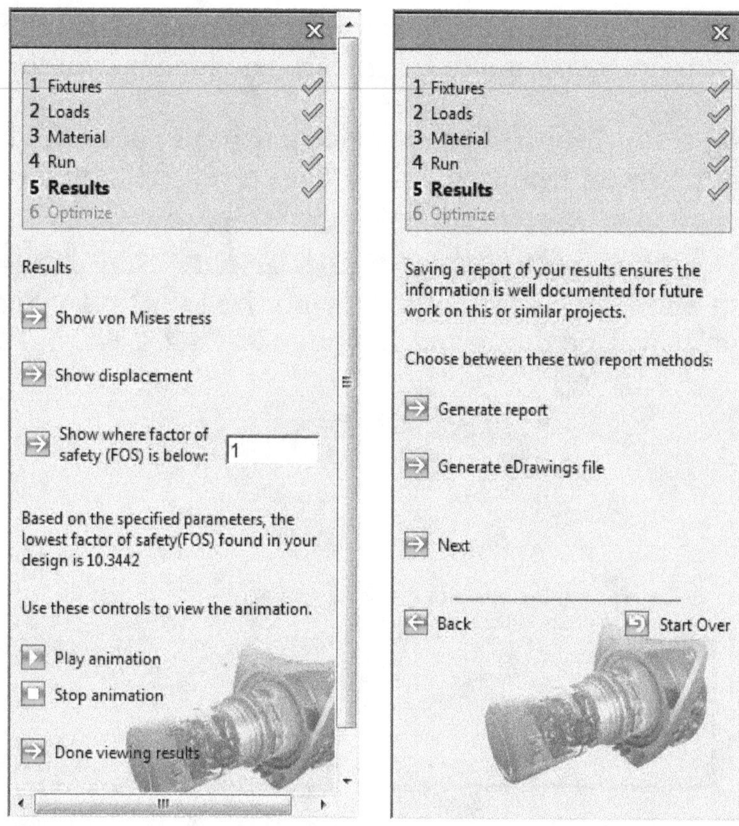

Figure-16. Modified results page of SimulationXpress

Figure-17. Modified Results page

- Select the **Generate report** button to create the report. On selecting this button, the **Report Settings** dialog box will be displayed as shown in Figure-18.

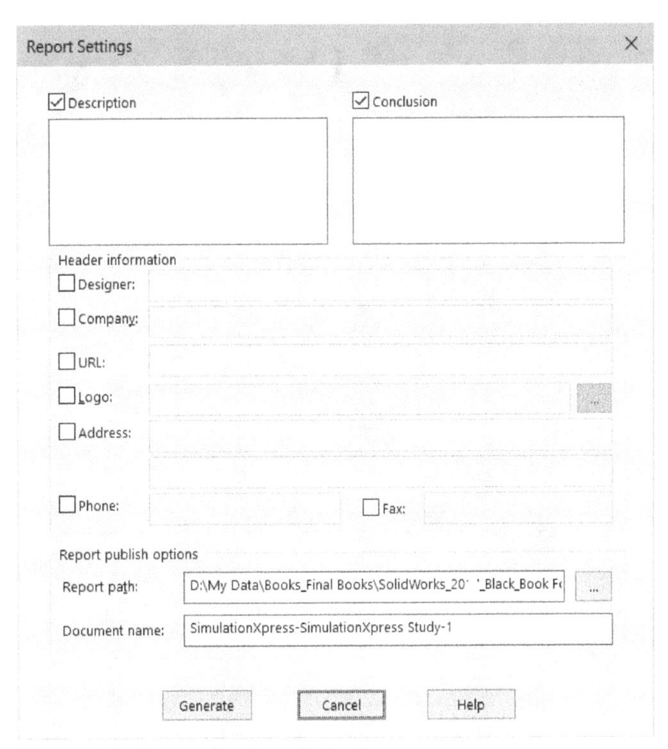

Figure-18. Report Settings dialog box

- Select the check boxes to enter information in the desired field in the dialog box. You can enter the path for saving report in the **Report path** field of the dialog box. After specifying the information, select the **Generate** button to create the report at the specified path. The **Generating Report** dialog box will be displayed as shown in Figure-19. Once the report generation gets completed; a word document will open automatically with the complete report of analysis.

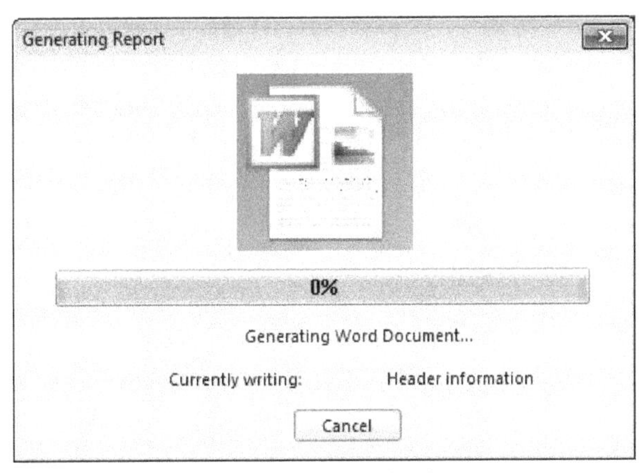

Figure-19. Generating Report dialog box

- Similarly, you can generate an eDrawing of the results report by selecting **Generate eDrawings file** link button. The eDrawings can be opened in the eDrawing software provided with the SolidWorks package; refer to Figure-20.

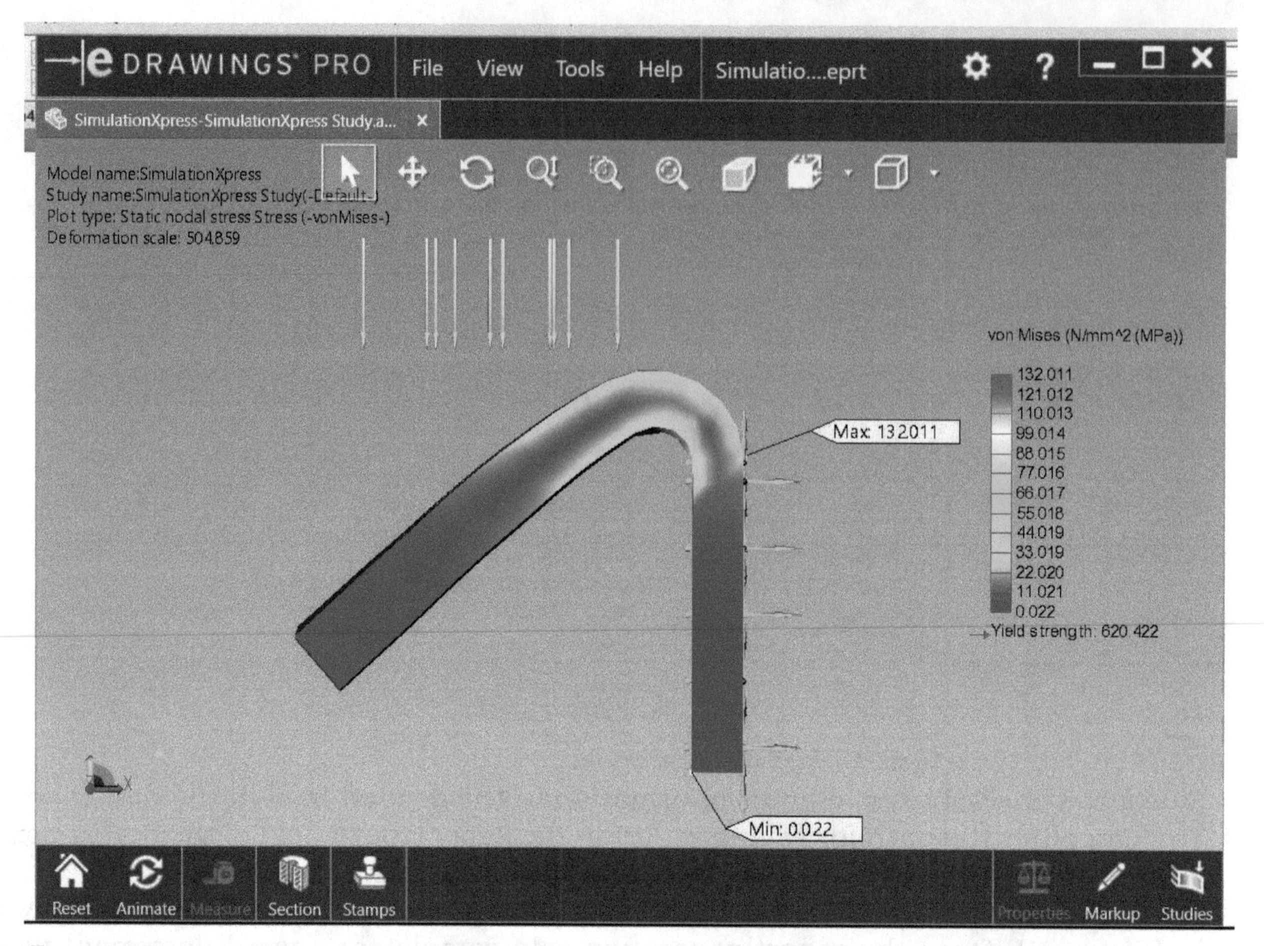

Figure-20. SimulationXpress report in eDrawing

Optimizing

- After checking the results, if you want to optimize the model then select the **Next** link button from the SimulationXpress. The **Optimize** page of **SimulationXpress** Task Pane will be displayed as shown in Figure-21.

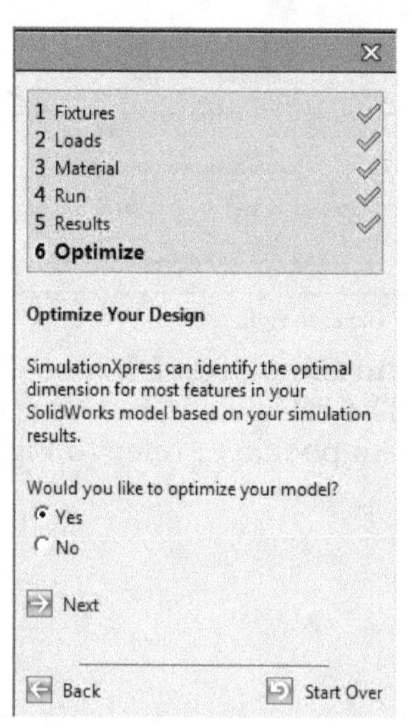

Figure-21. Optimize page of SimulationXpress

- Make sure that the **Yes** radio button is selected in the page and then select the **Next** button from the SimulationXpress. On doing so, the **Parameters** dialog box will be displayed and the driving dimensions will be displayed in the modeling area; refer to Figure-22 and Figure-23.

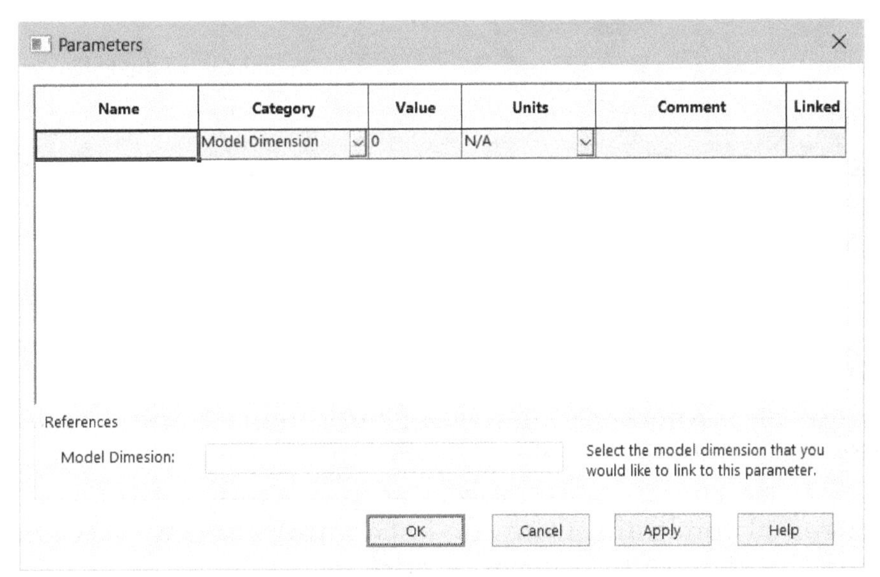

Figure-22. Parameters dialog box

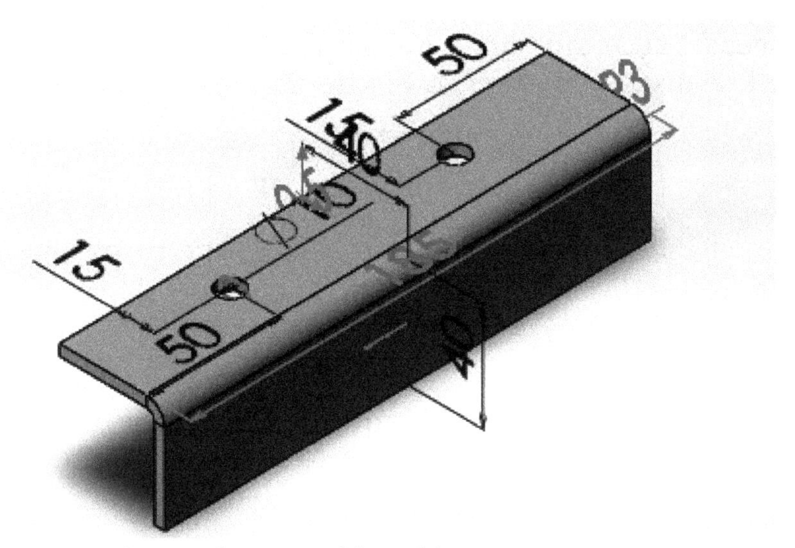

Figure-23. Driving dimensions of the model

- Click on one of the driving dimension from the model that you want to change for optimization (like thickness of part) and then select the **OK** button from the dialog box. The value will be added in the **DesignXpress Study** tab displayed below the modeling area; refer to Figure-24.

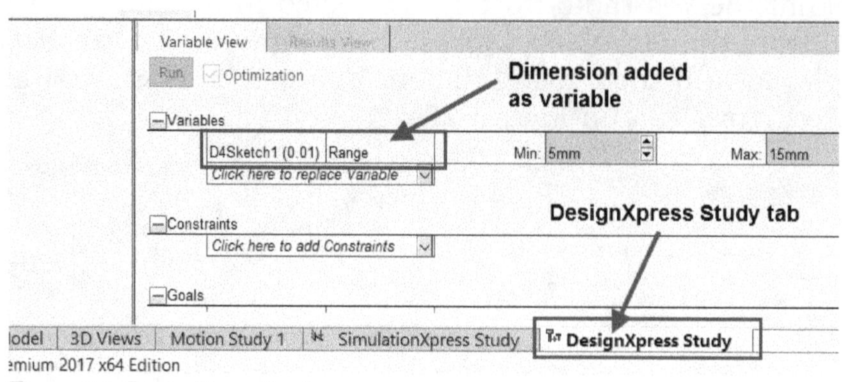

Figure-24. DesignXpress Study

- To replace the variable for optimization, click on the **Click here to replace variables** drop-down under the **Variables** node in the **DesignXpress Study**. On doing so, the **Parameters** dialog box will be displayed again. Click in the blank **Name** field and select another dimension from the modeling area to replace.
- After adding desired variable, select the **OK** button from the dialog box. Now, click on the **Click here to add Constraints** drop-down under the **Constraints** node and select desired constraint from the list. In this case, we have selected **Factor of Safety** as the constraint; refer to Figure-24.
- Set the value for constrain using the spinner.
- After setting all desired parameters; click on the **Run** button displayed at the top in the **DesignXpress Study** tab. On doing so, the **DesignXpress Study in Progress** dialog box will be displayed; refer to Figure-25.

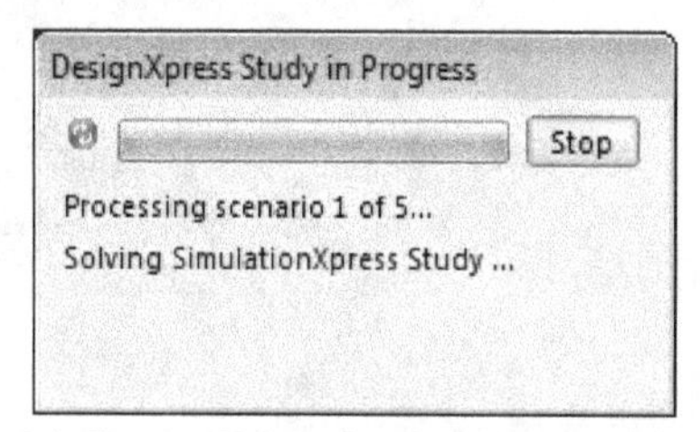

Figure-25. DesignXpress Study in Progress dialog box

- After the study is complete, the optimization results will be displayed in the **Results View** tab of the **DesignXpress Study**; refer to Figure-26.

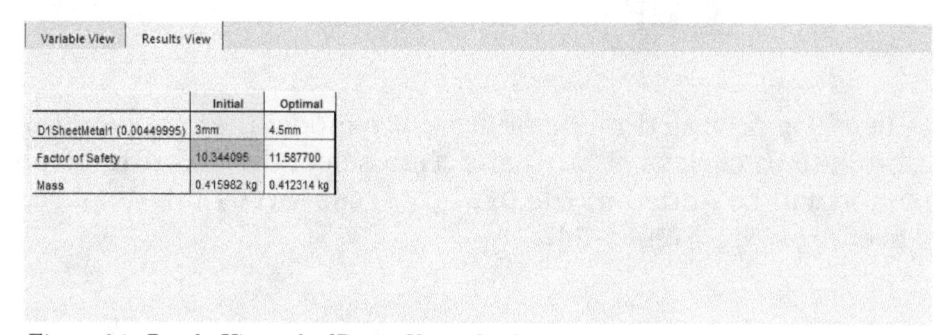

Variable View | Results View

	Initial	Optimal
D1SheetMetal1 (0.00449995)	3mm	4.5mm
Factor of Safety	10.344095	11.587700
Mass	0.415982 kg	0.412314 kg

Figure-26. Results View tab of DesignXpress Study

- In the **Optimize** page of **SimulationXpress**, the options will be modified as shown in Figure-27.

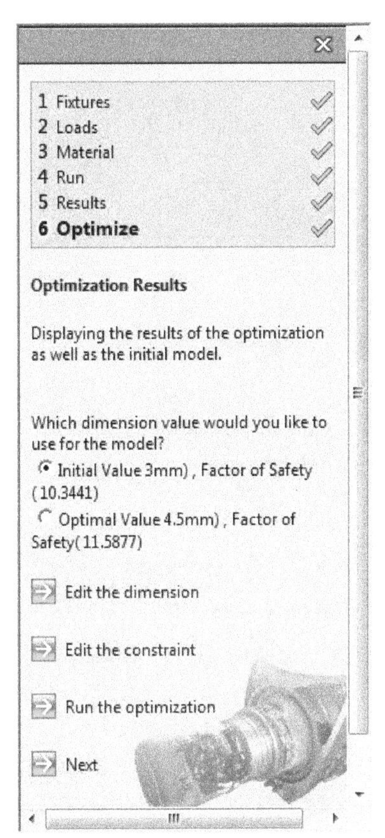

Figure-27. Modified Optimize page of SimulationXpress

- Select the radio button for **Optimal** value and then click on the **Run the optimization** link button again to cross check and then click on the **Next** button to apply the optimization.
- Exit the **SimulationXpress** by clicking on the **Close** button at the top-right corner of it and select **Yes** from the dialog boxes displayed to save results.

FLOXPRESS ANALYSIS

FloXpress Analysis is used to check the flow of a fluid through the designed passage. This passage can be in a solid model or it can be in an assembly. This analysis can be performed by using the **FloXpress Analysis Wizard** button available in the **Evaluate** tab of the **Ribbon**; refer to Figure-28.

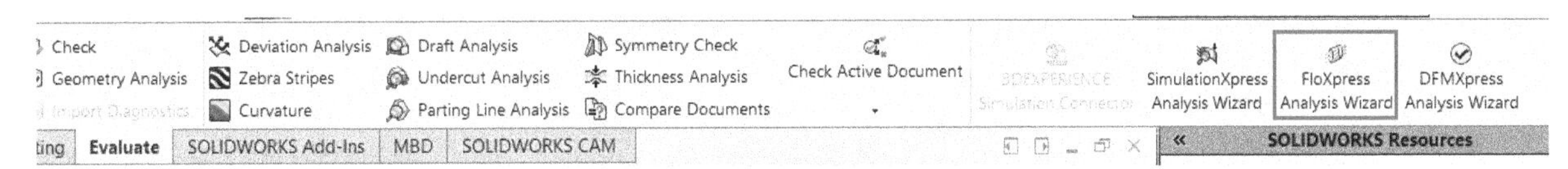

Figure-28. FloXpress Analysis Wizard button

To perform the express flow analysis of a passage, you need to close all the openings in the model with the help of a lid. To do so, perform the following steps.

- Open the model for which you want to perform the flow analysis; refer to Figure-29. Note that using the FloXpress, you can check flow of component having only one inlet and one outlet.

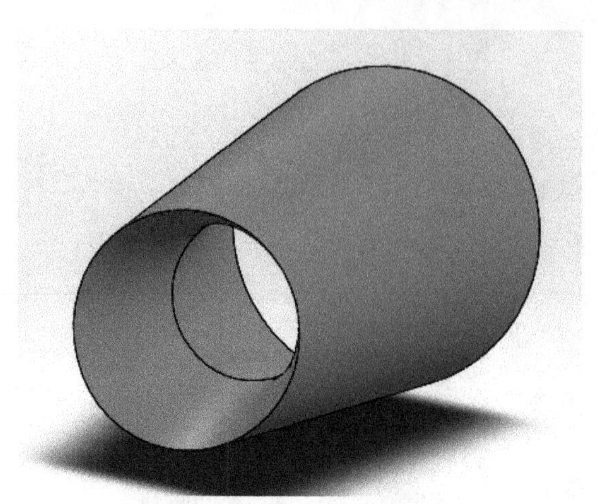

Figure-29. Model for flow analysis

Preparing Model

- Now, we need to close both the ends. To do so, select the outer round edge of one end of the model and then select the **Filled Surface** tool from the **Surfaces CommandManager** in the **Ribbon**; refer to Figure-30. On doing so, the **Filled Surface PropertyManager** will be displayed. Click on the **OK** button from the **PropertyManager**. Similarly, create the filled surface on the other end. The model after creating the filled surfaces will be displayed as shown in Figure-31.

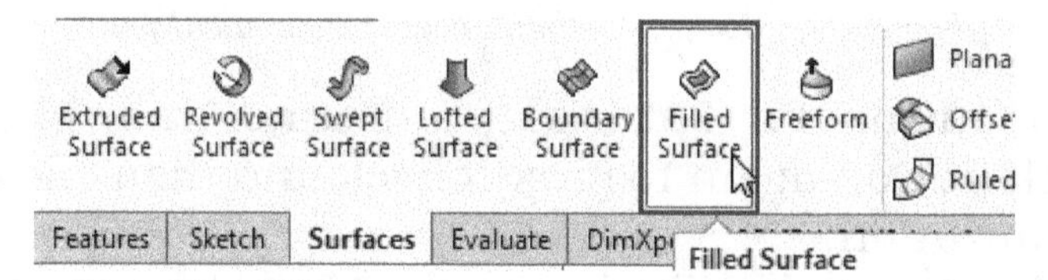

Figure-30. Filled Surface tool

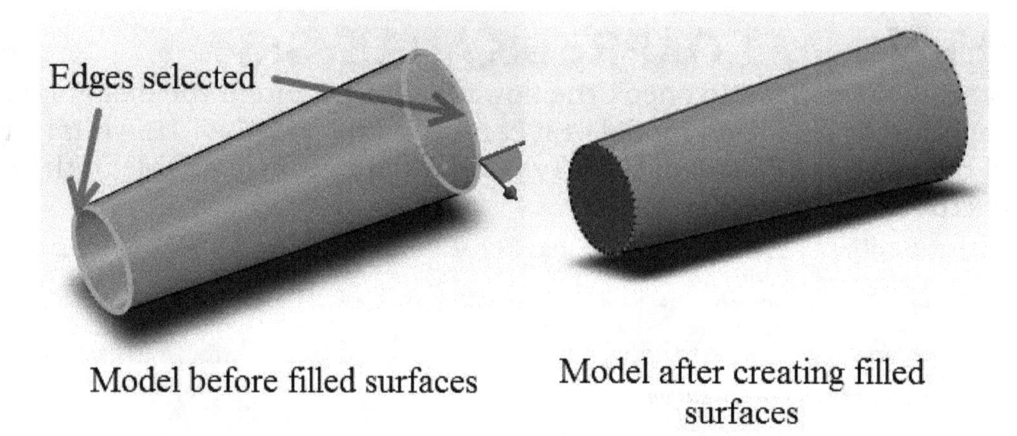

Figure-31. Filled surfaces creation

- Now, we need to thicken these surface to create the lid. To do so, select one of the surfaces and then select the **Thicken** tool from the **Surfaces** tab. On doing so, the **Thicken PropertyManager** will be displayed.
- Select the **OK** button from the **PropertyManager**, the surfaces will be thickened. Similarly, thicken the other surface, the model after thickening will be displayed as shown in Figure-32.

Figure-32. Thickened surfaces to create lids

Starting Flow analysis

- Select the **FloXpress Analysis Wizard** tool from the **Evaluate** tab of the **Ribbon**, the **Welcome PropertyManager** will be displayed in the left of the application window; refer to Figure-33.

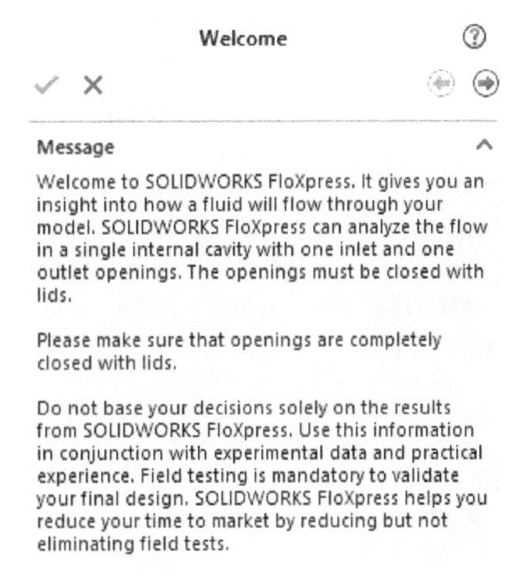

Figure-33. Welcome PropertyManager

- Click on the **Next** button from the **PropertyManager**. On doing so, the **Check Geometry PropertyManager** will be displayed as shown in Figure-34.

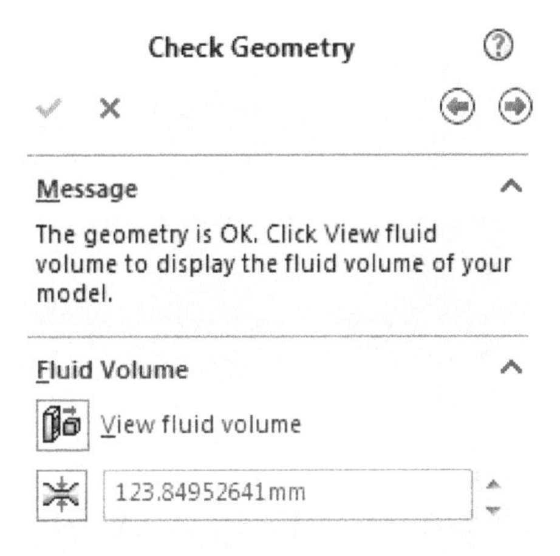

Figure-34. Check Geometry PropertyManager

- Click on the **View fluid volume** to check the fluid volume in designed passage. On doing so, the model will be displayed as shown in Figure-35.

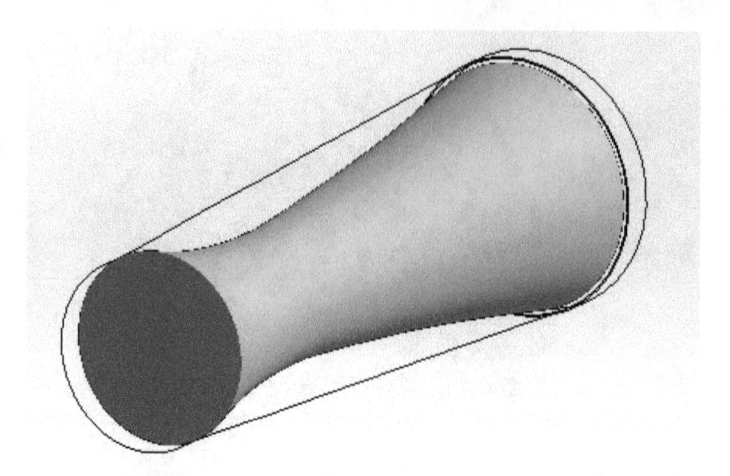

Figure-35. Fluid Volume in model

- You can change the smallest flow passage for the model at this stage. To do so, click on the **Smallest Flow Passage** button and then change the value by using the spinner.
- After specifying desired parameters, click on the **Next** button from the **PropertyManager**. On doing so, the **Fluid PropertyManager** will be displayed and you are prompted to specify the type of fluid for the analysis. You can select either water or air.
- In this case, we select the **Air** radio button. Now, click on the **Next** button from the **PropertyManager.** On doing so, the **Flow Inlet PropertyManager** will be displayed as shown in Figure-36.

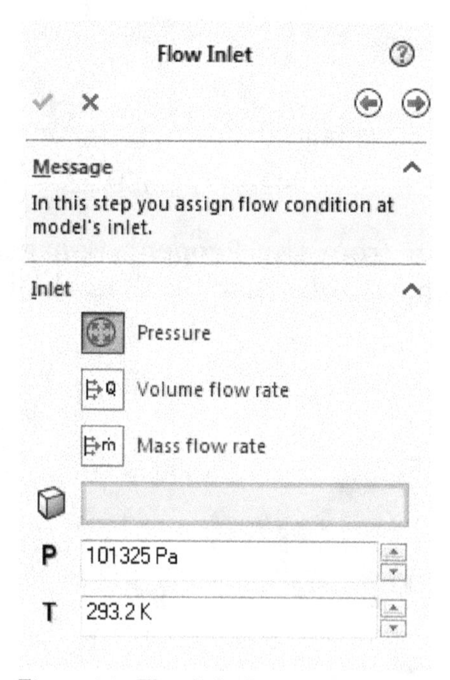

Figure-36. Flow Inlet PropertyManager

- Select desired button and then specify desired values in the edit boxes displayed at the bottom of the **Flow Inlet PropertyManager**.
- Select the inner face of the model as inlet; refer to Figure-37. You might need to right-click on the face and then select the **Select Other** option from the shortcut menu.

Figure-37. Face to be selected as Inlet

- After selecting the face, select the **Next** button from the **PropertyManager**. On doing so, the **Flow Outlet PropertyManager** will be displayed as shown in Figure-38.

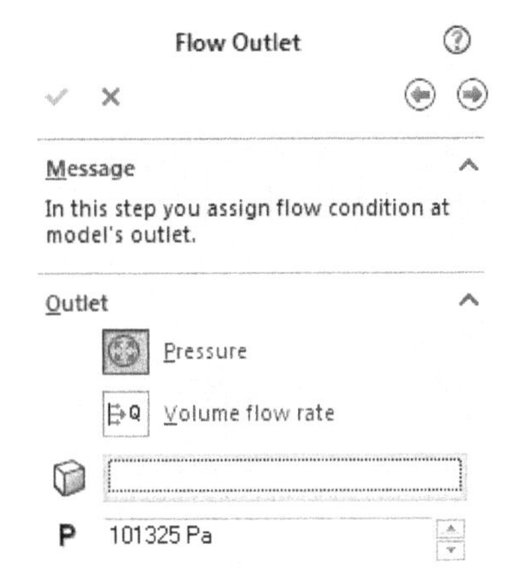

Figure-38. Flow Outlet PropertyManager

- Now, you need to select the outlet for the model. To do so, move the cursor over the flat face of the outlet lid and then right-click; a shortcut menu will be displayed. Now, select the face as shown in Figure-39.

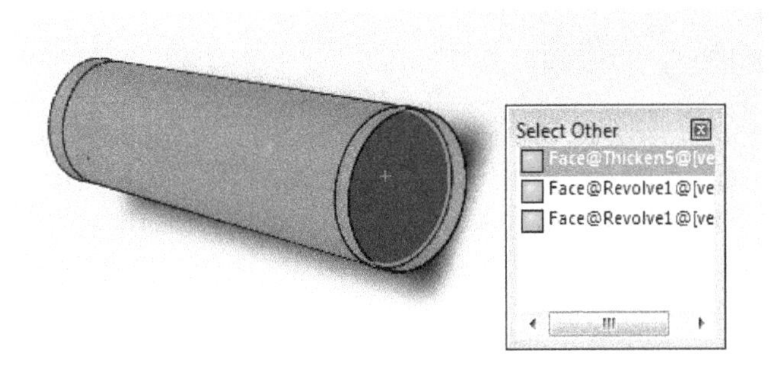

Figure-39. Face to be selected as Outlet

- To check the flow fluid, you need to change the display style of the model. To do so, select the **Wireframe** button from the **Display Style** drop-down; refer to Figure-40.

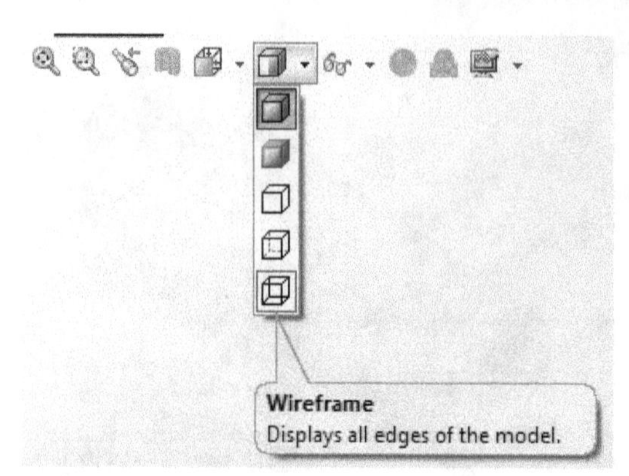

Figure-40. Wireframe button

- After changing the display style, the model will be displayed as shown in Figure-41.

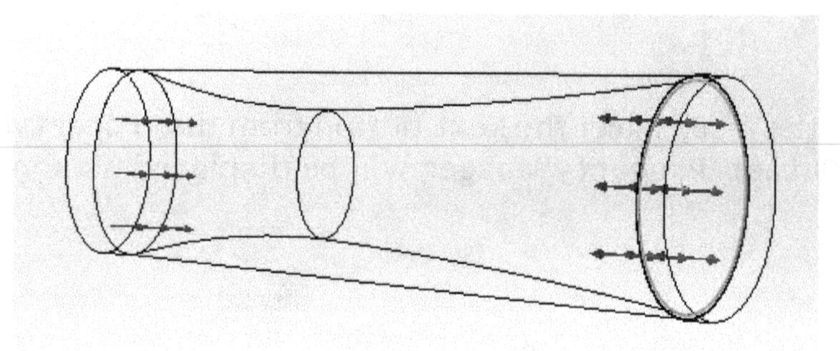

Figure-41. Model in wireframe style

- Click on the **Next** button from the **PropertyManager** to display the **Solve PropertyManager**; refer to Figure-42.

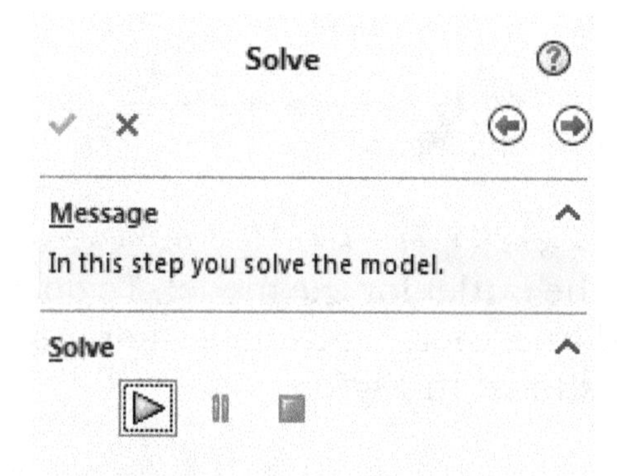

Figure-42. Solve PropertyManager

- Click on the **Solve** button to run the analysis. On doing so, the system will start to solve and after the CFD problem is solved, the solution will be displayed.
- You can generate the report of result by using the **Generate Report** button under the **Report** rollout. On doing so, the word document of report will be generated.

DFMXPRESS ANALYSIS

The DFMXpress analysis is used to check whether the model in the Modeling area is manufacturable or not. Using this analysis, you can check whether the model is manufacturable by Mill/Drill Manufacturing process, Turn with Mill/drill process, Injection Molding process or Sheetmetal process. The procedure is given next.

- To analyze the model for these manufacturing processes, click on the **DFMXpress Analysis Wizard** tool; the **DFMXpress** task pane will be displayed in the right of the application window; refer to Figure-43.

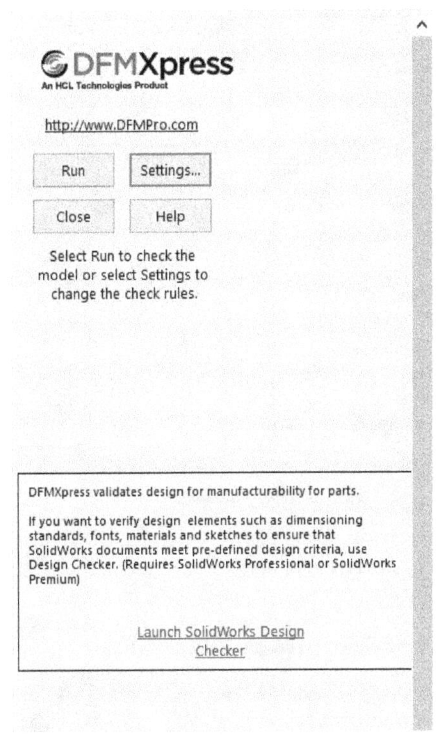

Figure-43. DFMXpress

- Now, click on the **Settings** button to specify the parameters related to the analysis; refer to Figure-44.
- Select the desired process and then set the related parameters in the fields. These parameters are basically the limitations of your machines to manufacture an item.
- After specifying all the parameters, click on the **Run** button to run the analysis. After the analysis gets completed, the results will be displayed at the bottom in the DFMXpress; refer to Figure-45.

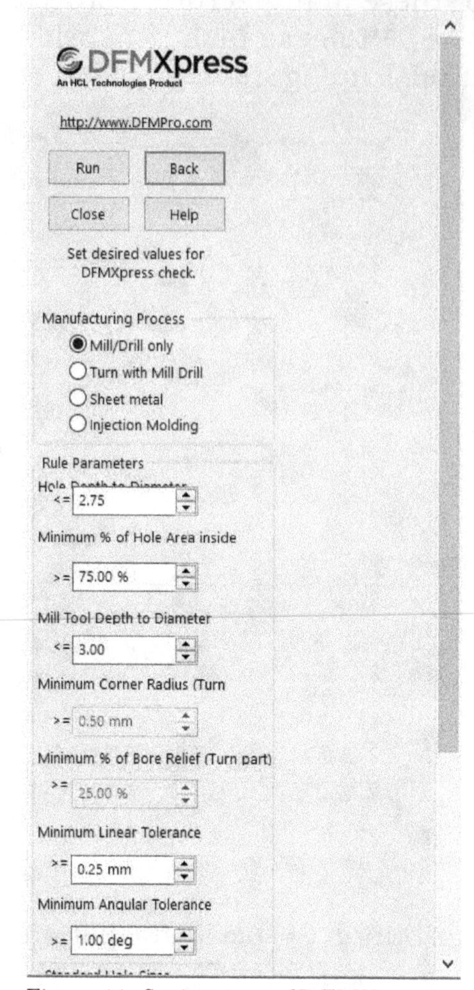

Figure-44. Settings page of DFMXpress

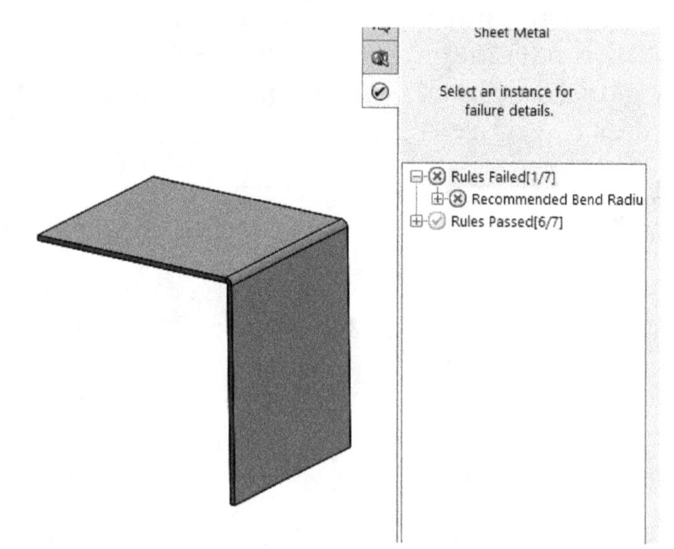

Figure-45. Result of DFMXpress analysis

- Expand the rules which have failed and update your part based on the failed rules.
- Click on the **Close** button from the Task Pane to exit.

COSTING

In real world, it is possible to make same part with two or more processes, like part in Figure-46 can be manufactured by milling, shaping, sheetmetal processes or casting process. But in industry, we are also concerned about the cost of manufacturing and quality of product. In this section, we will learn to estimate the cost of manufacturing the model by various processes.

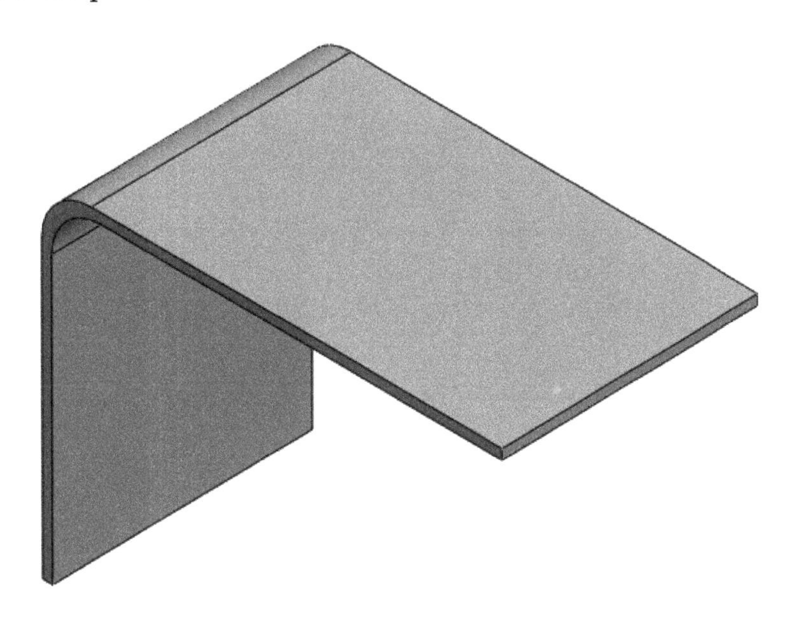

Figure-46. Model for costing

- Click on the **Costing** button from the **Evaluate CommandManager** in the **Ribbon**. The Costing interface will be displayed; refer to Figure-47.

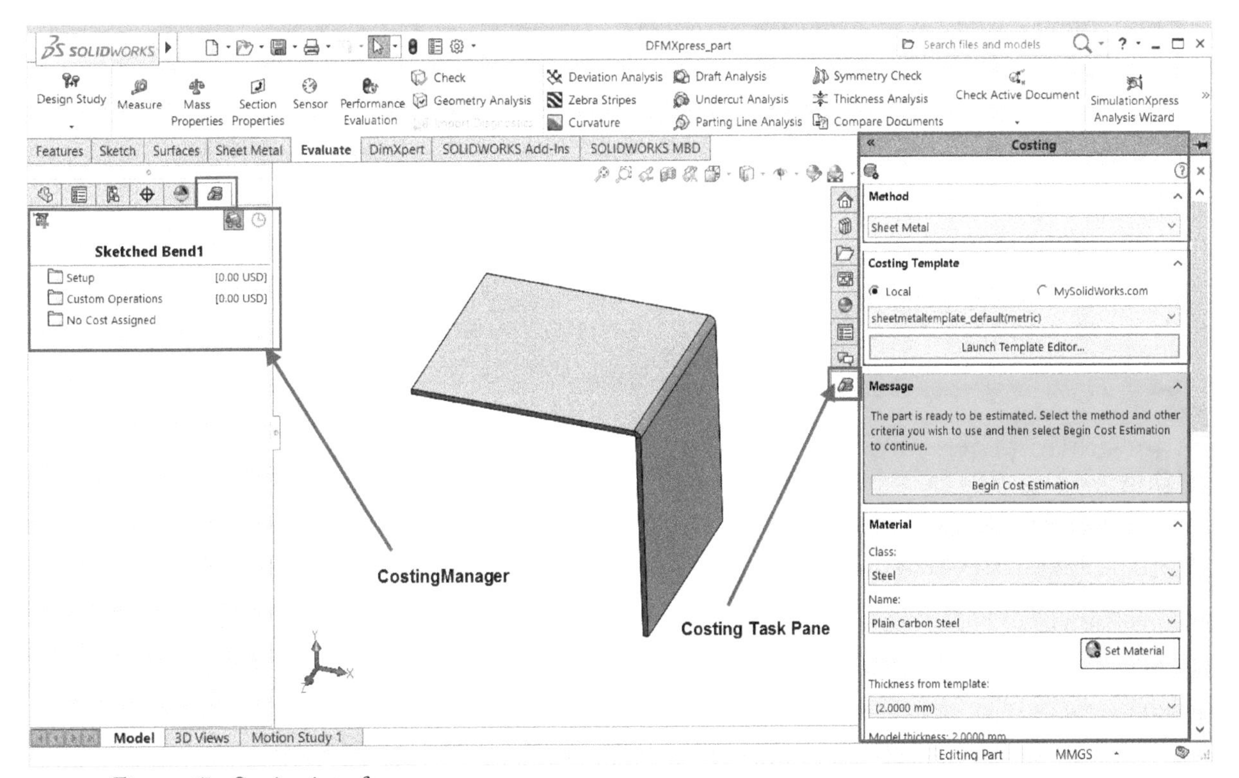

Figure-47. Costing interface

- Select desired process from **Method** drop-down; refer to Figure-48.

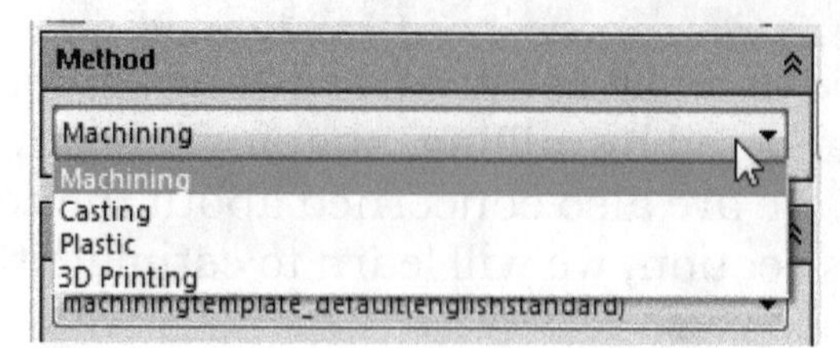

Figure-48. Method drop-down

- Click in the **Template** drop-down of **Costing** task pane to select the desired template. Using the **Launch Template Editor** button, you can launch the **Costing Template Editor** to edit the template; refer to Figure-49. Note that using the Editor, you can change the cost and time of various manufacturing process with respect to the selected material. Like, you can change the feed rate for milling a steel material by clicking on the **Mill** tab in right of the **Costing Template Editor** dialog box. Changing the feed rate will change the cycle time and hence the cost of machining.

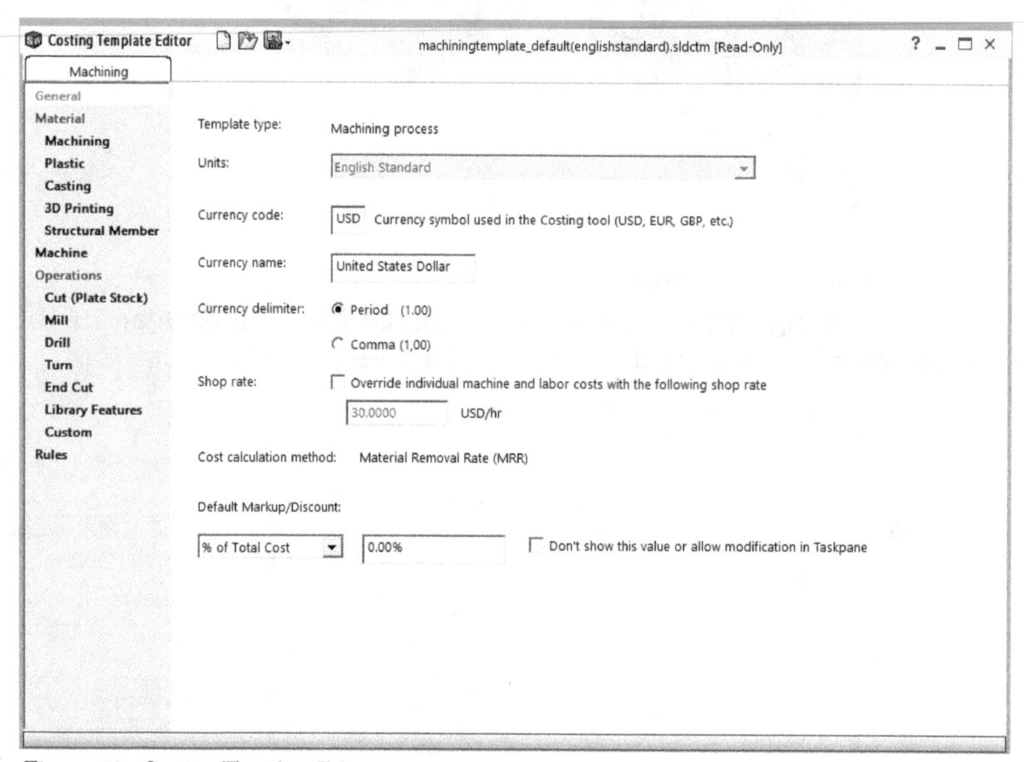

Figure-49. Costing Template Editor

- After editing the template, save the file and exit the editor.
- Click on the **Begin Cost Estimate** button from the **Message** rollout of the **Costing Task Pane**. Cost estimate will be displayed task pane and breakup of cost will be displayed in the **CostManager**.
- It might be possible that you have some processes which are not having price rate. In that case, you need to manually specify the price. Such processes are displayed under **No Cost Assigned** node; refer to Figure-50.

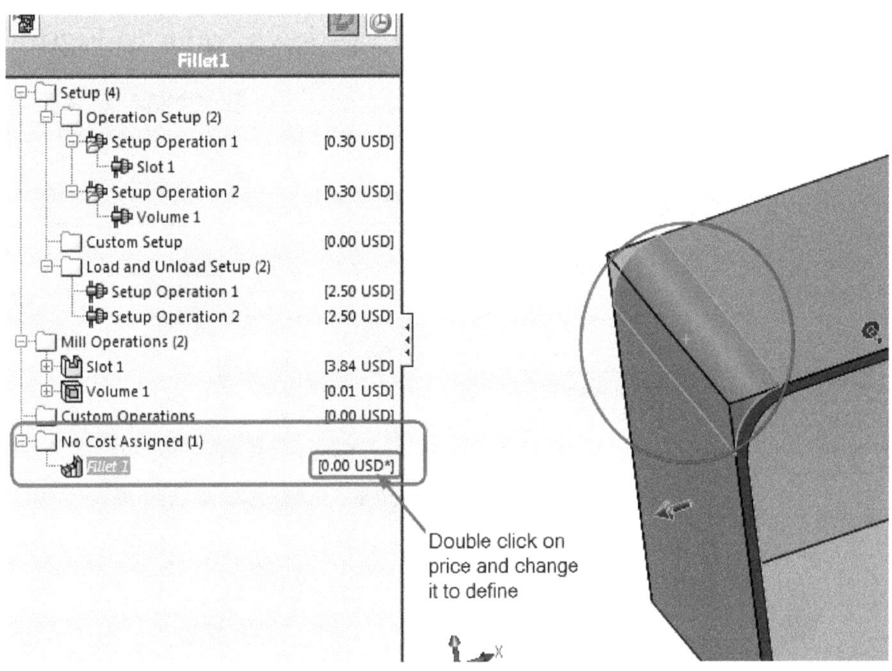

Figure-50. No cost defined

- After specifying all the costs, change the material to the desired one and also update the per kg price of material in the **Material** rollout in task pane; refer to Figure-51.

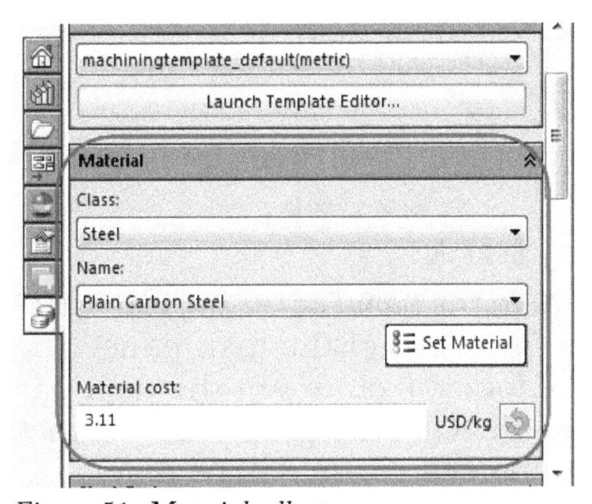

Figure-51. Material rollout

- Check the price of manufacturing in the **Estimated Cost per Part** area of **Costing** task pane and get the optimum process for manufacturing your model.

SUSTAINABILITY

The **Sustainability** tool is used to perform screening-level life cycle assessment (LCA) directly on individual part designs to help you understand the environmental impacts of your design decisions. The procedure to use this tool is given next.

- Click on the **Sustainability** tool from the **Evaluate CommandManager** in the **Ribbon**. The information box about use of sustainability application will be displayed. Click on the **Continue** button. The **Sustainability Task Pane** will be displayed in the right of application window; refer to Figure-52.
- Select desired material from the **Material** rollout of the task pane.

- Scroll down to **Manufacturing** rollout and select the region in which you are manufacturing the part from the **Region** drop-down; refer to Figure-53.

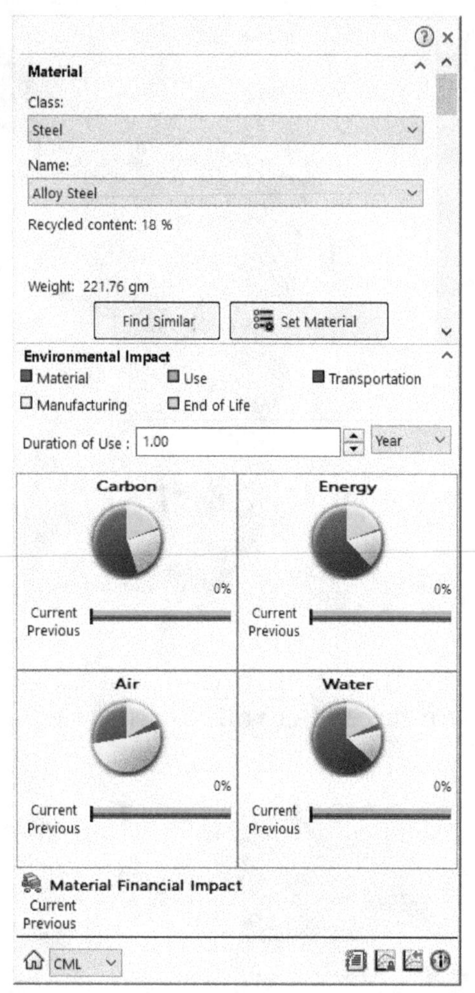

Figure-52. Sustainablility Task Pane

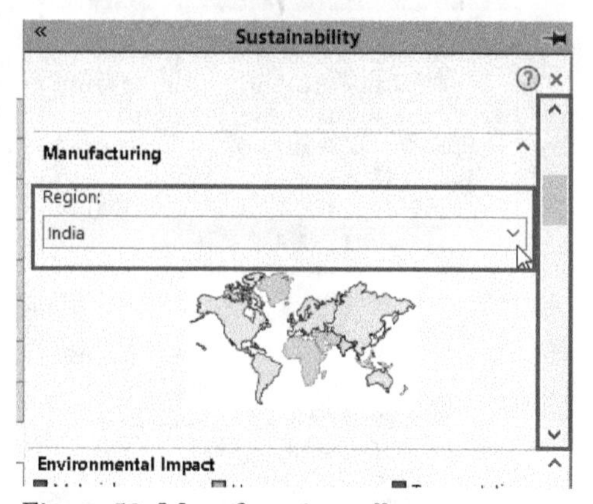

Figure-53. Manufacturing rollout

- Set the other parameters in the **Manufacturing** rollout, **Use** rollout, **Transportation** rollout, and **End of life** rollout of the task pane. The result of analysis will be displayed in the form of four pie charts in the task pane.
- Move the cursor of over the **Current** graph line in the **Carbon** area of the graph. The current value of carbon emission will be displayed; refer to Figure-54.

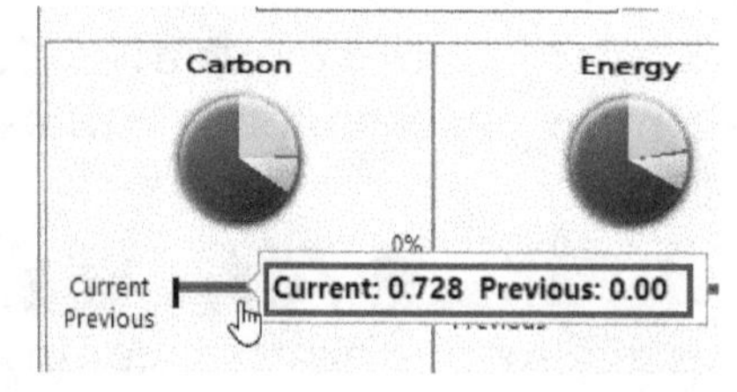

Figure-54. Current value of carbon emission

- Move the cursor over Carbon pie chart to check the percentage of emission done by various stages of part.
- Similarly, you can check the other environmental impacts of part.
- Change the material and other manufacturing aspects to compare the results.
- Click on the **Close** button on the task pane to exit the tool.

DRIVEWORKSXPRESS

The **DriveWorksXpress Wizard** tool is used to perform rule based design automation. The procedure to use this tool is given next.

- Open the model file in SolidWorks which you want to use for DriveWorksXpress application.
- Click on the **DriveWorksXpress Wizard** tool from the **Evaluate CommandManager** in the **Ribbon**. The DriveWorksXpress task pane will be displayed at the right in the application window.
- Click on the **Create/Change Database** button to start a new database for part/ assembly. The **Open** dialog box will be displayed.
- Specify desired name of database to be created in the **File name** edit box in the dialog box.
- Browse to desired location where you want to save the database file and click on the **Open** button. The **Projects** page will be displayed; refer to Figure-55.

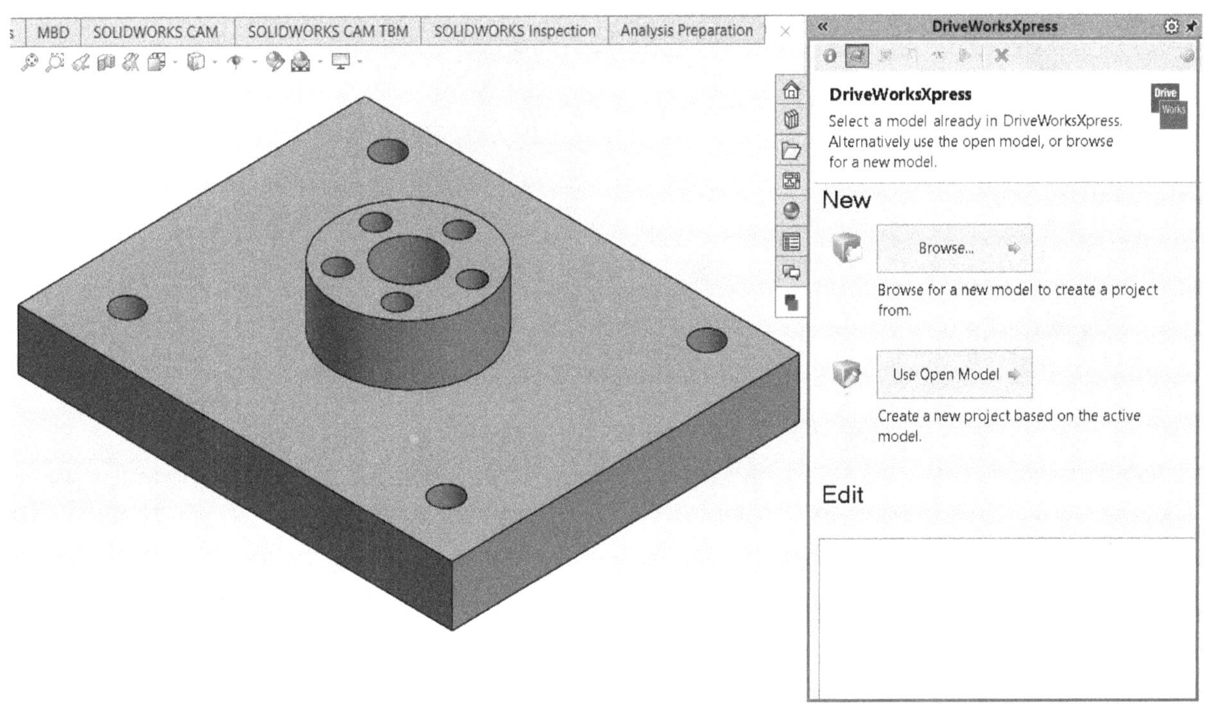

Figure-55. Projects page of DriveWorksXpress task pane

- Click on the **Use Open Model** button to use currently open model. The **Capture** page of task pane will be displayed.
- Click on the **Captured Models** node and select the check boxes for models to be included in **DriveWorks**. Note that this option is useful for assemblies.
- Click on the **Dimensions and Features** node of the page to select dimensions and features to be varied using database; refer to Figure-56.

Figure-56. Dimensions and Features page

- Select the feature from the model like pattern, holes, and so on to define related parameter from the **FeatureManager Design Tree**. Specify desired name for feature in the task pane; refer to Figure-57. Click on the **Add** button to add the feature in database.

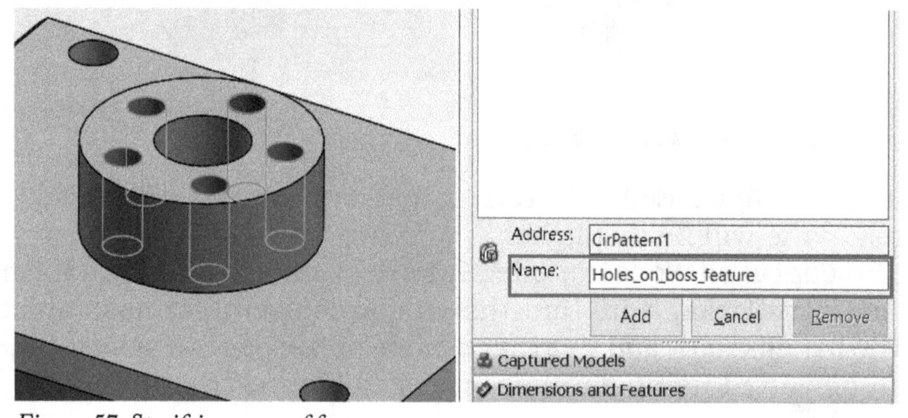

Figure-57. Specifying name of feature

- Double-click on the feature whose dimensions are to be included in the database. One by one select the desired dimension and add it to database; refer to Figure-58.

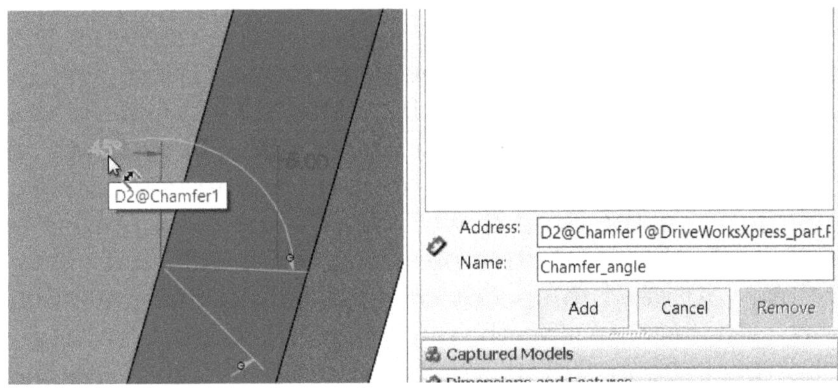

Figure-58. Adding dimension to database

- Click on the **Form** button at the top in the task pane to define how the parameters will be specified in the database while creating the model instances. The page will be displayed as shown in Figure-59.

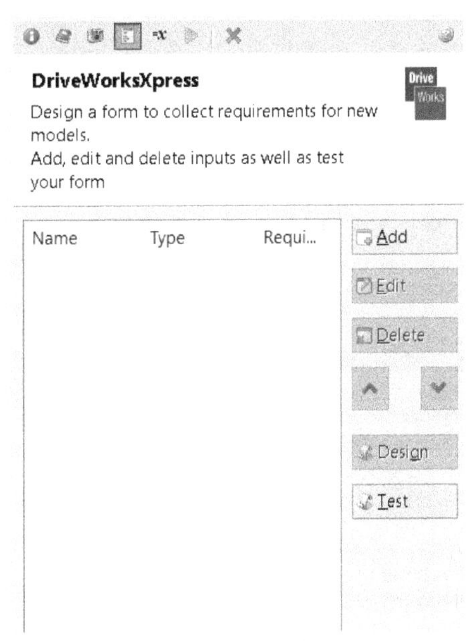

Figure-59. Form page in task pane

- Click on the **Add** button to add an input for earlier selected feature or dimension. The options to create input will be displayed; refer to Figure-60.

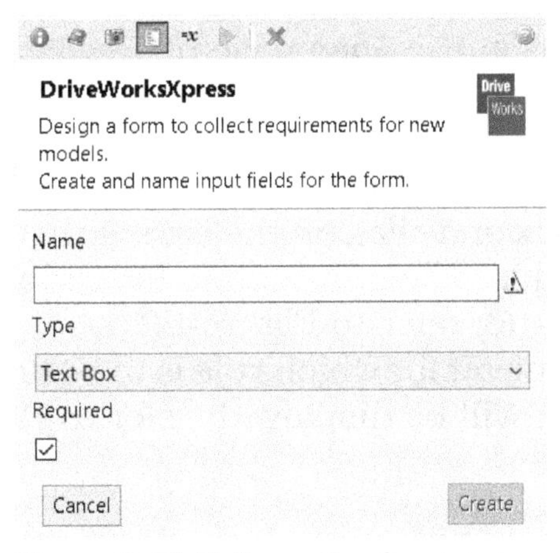

Figure-60. Fields for creating form input

- Specify the desired name for input in **Name** edit box like if there is pattern of holes for which you want to create input then specify the name as Number of holes. Select the desired option from the **Type** drop-down. There are 5 options in this drop-down. Select the **Text Box** option from the drop-down if you want to specify text like for name of file. Select the **Numeric Text Box**, **Drop Down**, and **Spin Button** option to specify numeric values for various dimensions and features. Select the **Check Box** option from the drop-down if you want to suppress or un-suppress the features. Specify the respective parameters for selected option. For **Numeric Text Box** option, you need to specify minimum and maximum values. For **Spin Button** option, you need to specify minimum-maximum values and increment by spinner. For **Drop-Down** option, you need to specify multiple values separated by "|" like 10|12|14|20. The **Text Box** and **Check Box** options do not need any parameters specified on this page. After specifying desired parameter, click on the **Create** button to create input field in form. You need to create as many input fields as you have dimensions and features selected for modifying model; refer to Figure-61.

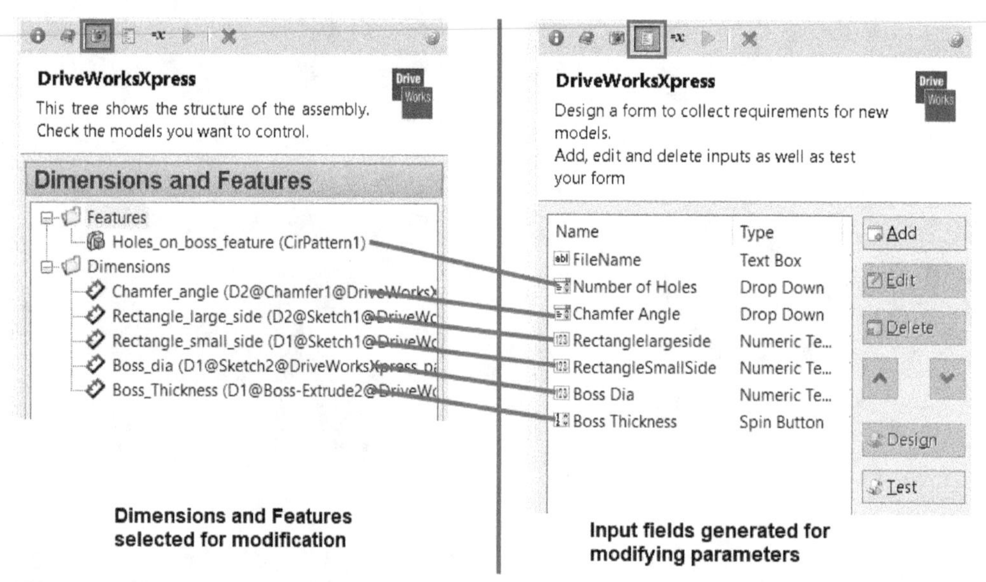

Figure-61. Input fields created for modifying parameters

Note that at any time you can switch between various pages of **DriveWorksXpress Task Pane** by clicking on respective button in the Task Pane.

- Click on the **Test** button in the **Form** page of Task Pane to check how your form is designed.
- Click on the **Rules** button to open **Rules** page of the Task Pane. The **Rules** page will be displayed; refer to Figure-62.
- Select the **Total** check box and click on the **Edit Rules** button to match parameters with respective input fields. The parameters for which rules are to be defined are displayed in the Task Pane; refer to Figure-63.
- Double-click on the parameter for which rule is to be defined. The **DriveWorksXpress Rule Builder** dialog box will be displayed; refer to Figure-64.

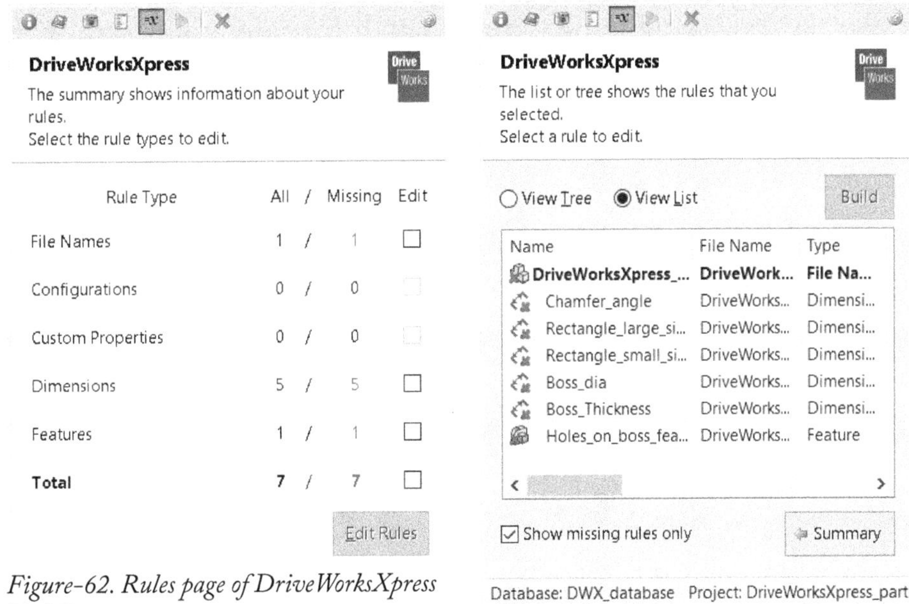

Figure-62. Rules page of DriveWorksXpress Task Pane

Figure-63. Parameters with missing rules

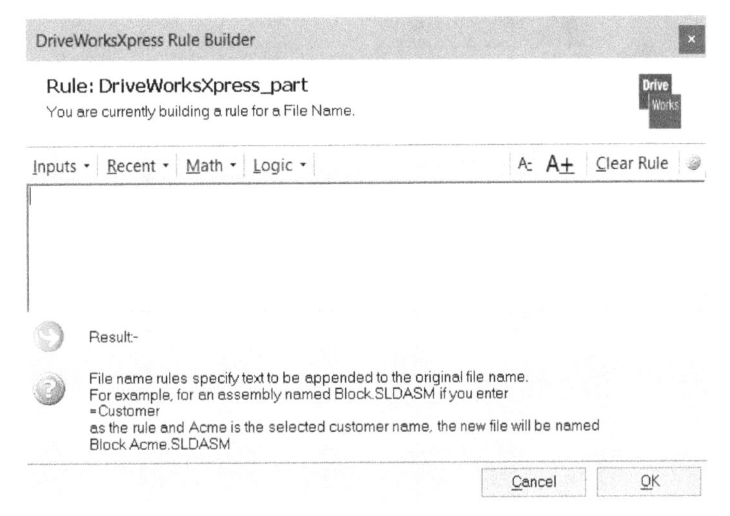

Figure-64. DriveWorksXpress Rule Builder dialog box

- Click on the **Inputs** drop-down and select desired input created for selected parameter. (In our case, we have created **FileName** input field for File Name parameter; refer to Figure-65.)

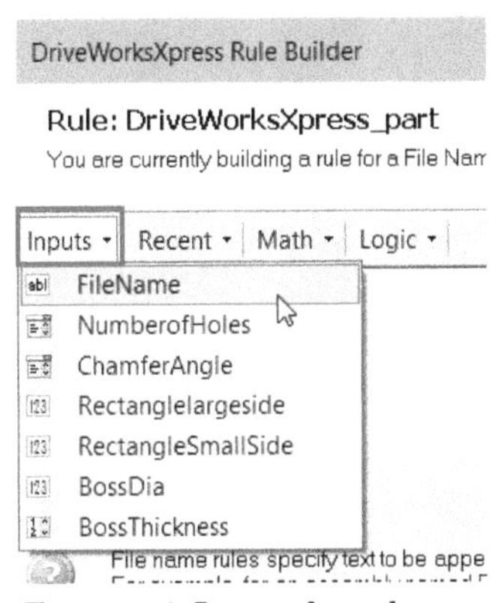

Figure-65. Inputs drop-down

- Similarly, set the input fields for all the parameters and click on the **Run** button from the top in the Task Pane. The form to create instance of model will be displayed; refer to Figure-66.

Note that when you hover the cursor on a field then tooltip will be displayed telling you the range, increment, and other information for current field.

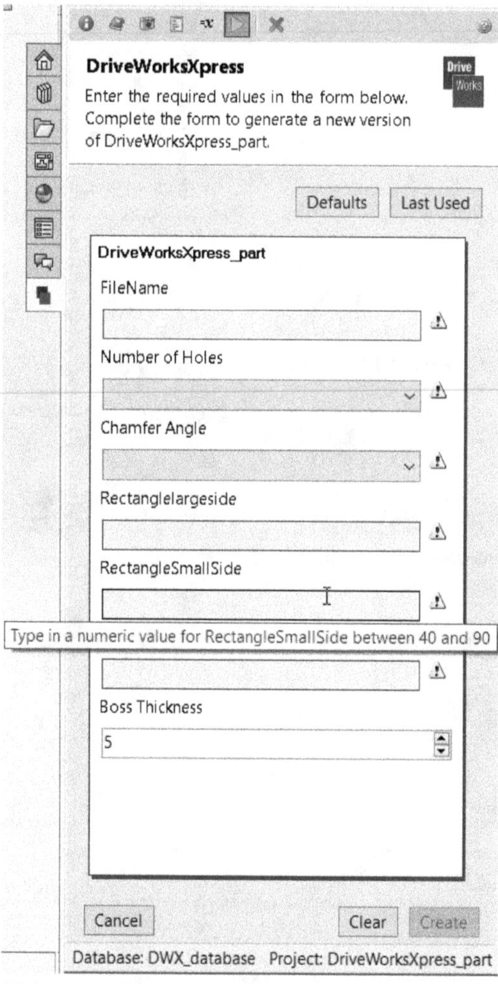

Figure-66. Form generated for drive works

- Specify the desired parameters in the field and click on the **Create** button to create an instance of part. After creating one instance, click on the **Finish** button and create another instance. Repeat the step to create desired number of instances. Note that the instances will be created in the same folder where you have saved the original file.
- Click on the **Close** button from the top in the Task Pane to exit DriveWorksXpress.

PRACTICE 1

Design a wall bracket to bear a load of 100 kg. There should be two counterbore holes for M10 bolts to mount it on wall. Material for wall bracket is Cast Iron. Estimate the cost of manufacturing wall bracket using casting process.

SELF-ASSESSMENT

Q1. Which of the following analysis should be used to test the load bearing capacity of a component?

a. FloXpress Analysis
b. SimulationXpress Analysis
c. StainabilityXpress Analysis
d. DFMXpress Analysis

Q2. Which of the following analysis is used to check the flow of a fluid through the designed passage?

a. FloXpress Analysis
b. SimulationXpress Analysis
c. StainabilityXpress Analysis
d. DFMXpress Analysis

Q3. Which of the following analysis is used to check whether the model in the Modeling area is manufacturable or not?

a. FloXpress Analysis
b. SimulationXpress Analysis
c. StainabilityXpress Analysis
d. DFMXpress Analysis

Q4. FloXpress Analysis can be performed with water as well as air. (T/F)

Q5. Different manufacturing processes hardly affect the production of a component in manufacturing facility. (T/F)

Answer to Self-Assessment:

1. (b) SimulationXpress Analysis, **2.** (a) FloXpress Analysis, **3.** (d) DFMXpress Analysis, **4.** (T), **5.** (F)

FOR STUDENT NOTES

Chapter 12

Mold Tools

Topics Covered

The major topics covered in this chapter are:

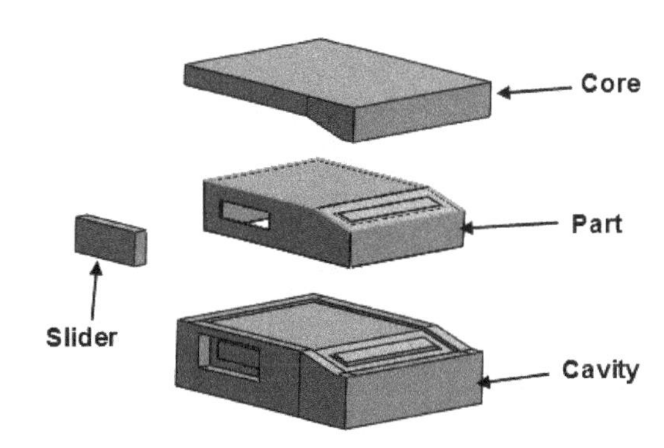

- ***Starting the Mold tools***
- ***Analyzing the model for molding***
- ***Preparing the model for mold***
- ***Starting the Mold project***
- ***Creating the parting line***
- ***Creating the Shutoff surfaces***
- ***Creating the Parting surface***
- ***Creating the Splitting the Core and Cavity from Tooling***

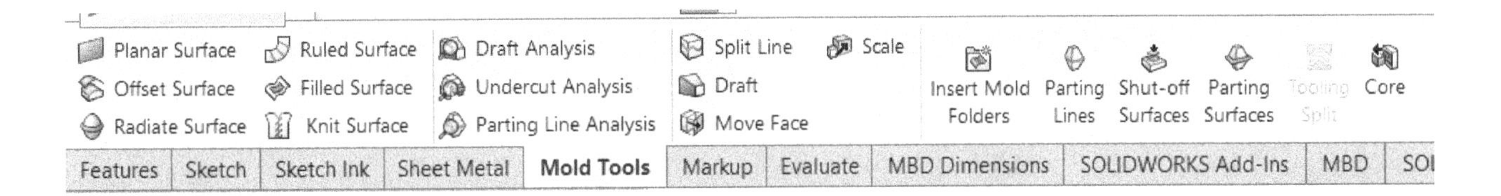

STARTING THE MOLD TOOLS

To start with the mold tools, first you need to open or import a model. To open/ import a model, follow steps given next.

- Go to the **File** menu and click on the **Open** option or click on the **Open** button from the **Menu Bar** or click on the **Open a Document** link in the **SolidWorks Resources Task Pane**; refer to Figure-1. The **Open** dialog box will be displayed as Figure-2.

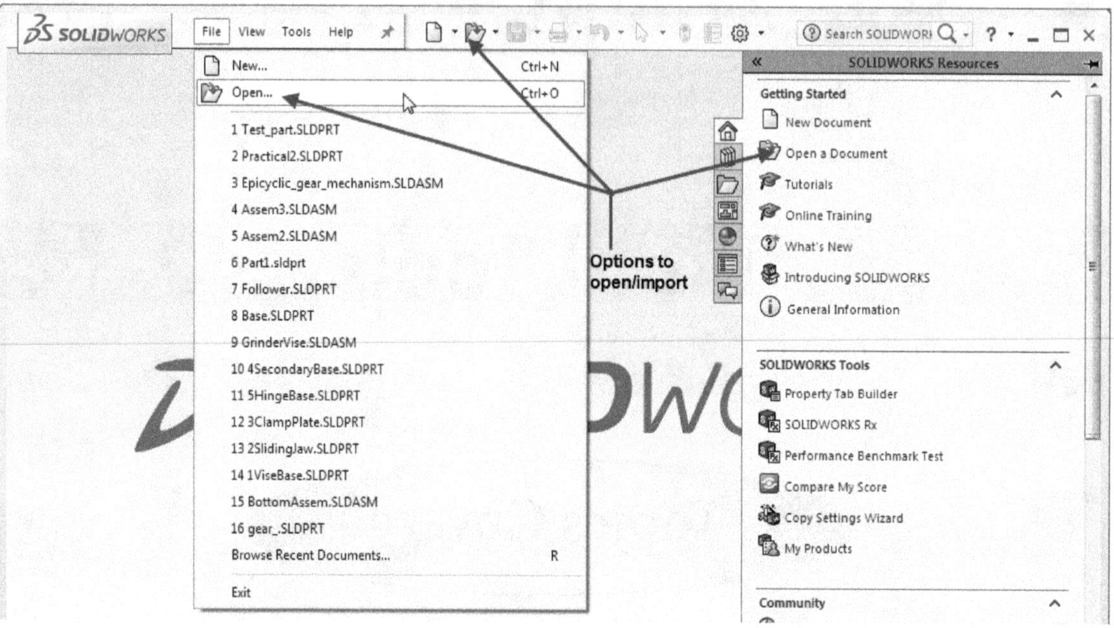

Figure-1. Open or Import methods

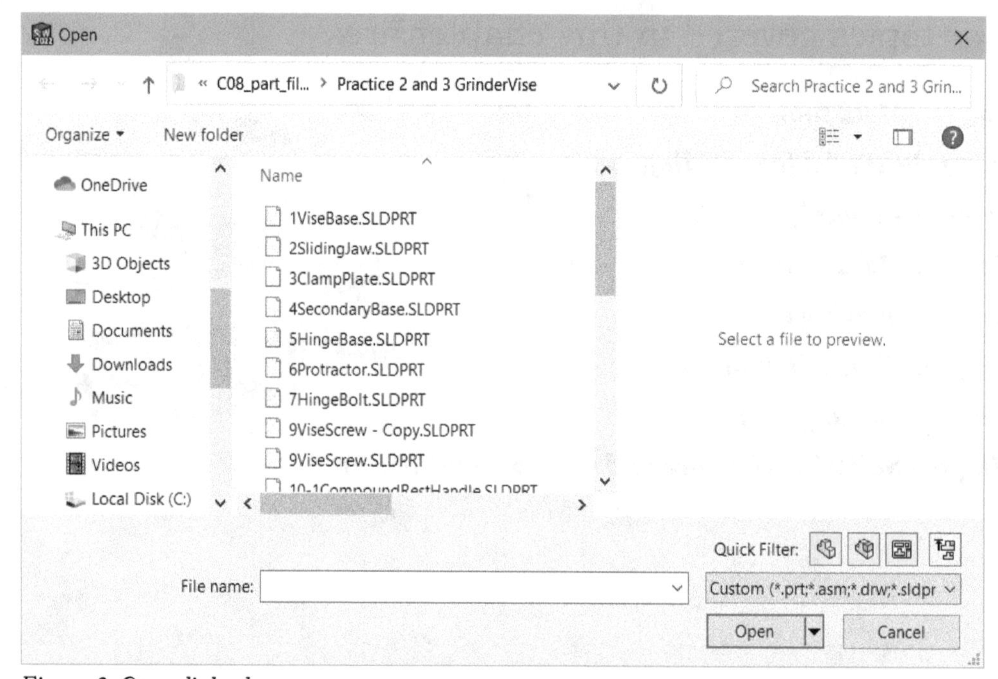

Figure-2. Open dialog box

- Click on the **SolidWorks Files** drop-down displayed at the bottom right in the dialog box. List of file types supported by SolidWorks will be displayed; refer to Figure-3.

- Select the type for your file from the list and browse to the folder in which you have placed the file.
- Double-click on the file, you want to open or import. The file will open in the application; refer to Figure-4.

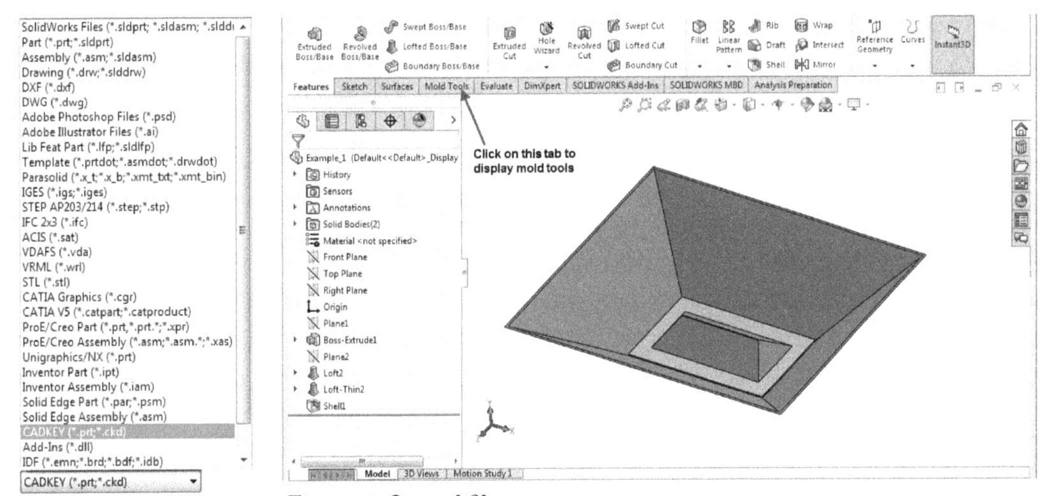

Figure-3. File types supported

Figure-4. Opened file

- Click on **Mold Tools CommandManager** of **Ribbon** to display tools of molding. If the **Mold Tools CommandManager** is not displaying by default, then right-click on any of the tab displaying in the **Ribbon** and move cursor over **Tabs** cascading menu. The list of **CommandManagers** will be displayed; refer to Figure-5.

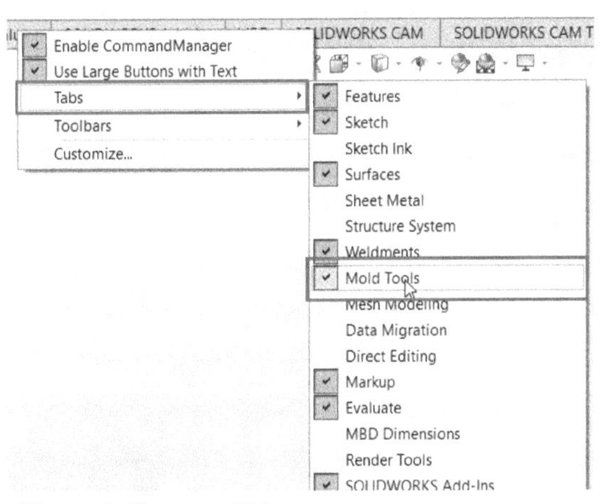

Figure-5. CommandManager menu

- Click on the **Mold Tools** option from this list, the **Mold Tools CommandManager** will get added in the **Ribbon**. On clicking the **Mold Tools CommandManager**, the Mold Tools will be displayed as shown in Figure-6.

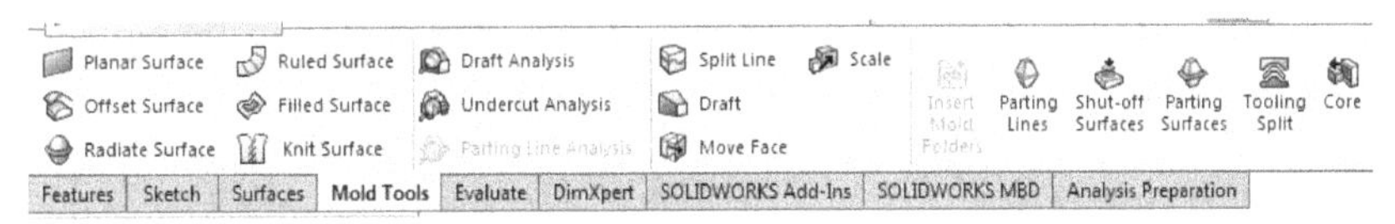

Figure-6. Mold Tools CommandManager

- Most of the tools have already been discussed in the book. Now, you will learn about the tools that are specific to making mold tools.

The first step for molding a part is to analyze whether it can be produced by molding process or not. So, we need to analyze it for the possibility of its mold design.

ANALYZING THE MODEL

There are three tools available in the **Mold Tools** tab to analyze the model for molding: **Draft Analysis**, **Undercut Analysis**, and **Parting Line Analysis**. These tools are discussed next.

Draft Analysis

The **Draft Analysis** tool is used to check the draft angles of various faces in the model. Draft is an important requirement of Molding. Draft is the taper angle given to various faces of the mold part for easy and safe ejection from the mold tooling (core and cavity). The angle value of draft depends on the material and geometry of the mold part. Typically 1° to 3° of draft is given on all faces of the mold part. If you have steps at parting line, then 5° to 7° of draft is required for shutoff. (**Shutoff** is the surface where core and cavity meet each other.) To perform the draft analysis, follow the steps given next.

- Click on the **Draft Analysis** tool in the **CommandManager**, the **Draft Analysis PropertyManager** will be displayed in the left of the screen; refer to Figure-7.
- A box will be highlighted in blue color in the **Analysis Parameters** rollout of the **PropertyManager**. You are asked to select a face with respect to which the angles will be measured and the pull direction will be defined.
- Select a flat face from the model. Note that this face will become the mating face of core and cavity later. Figure-8 shows the face selected as neutral face.

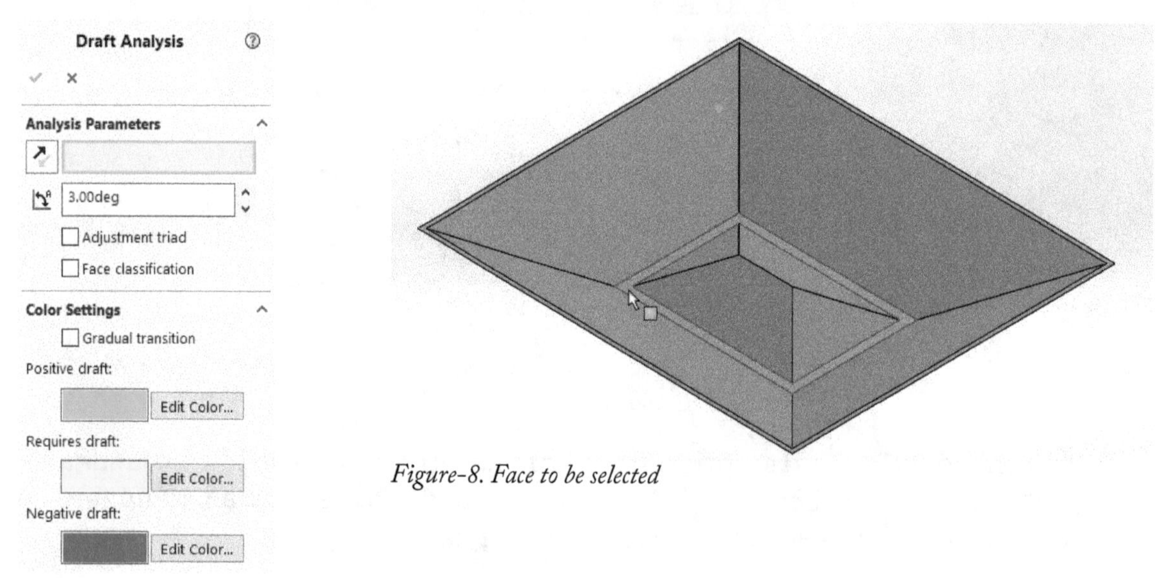

Figure-8. Face to be selected

Figure-7. Draft_Analysis_PropertyManager

- As you select the face, all the faces of the model will be painted with the colors specified in the lower area of the **PropertyManager**. By default, the green colored faces are created on core steel and red colored faces are created on cavity steel. The yellow colored faces fall in the undercut categories. These yellow colored faces should either be split up to make them in core & cavity or they should be created on sliders.

- You can change the colors of the faces as per your requirement by selecting the **Edit Color** buttons displayed in the lower area of the **PropertyManager**.
- Note that when you move cursor over the faces of the model, the draft angle value of the current face will be displayed along with the cursor; refer to Figure-9.

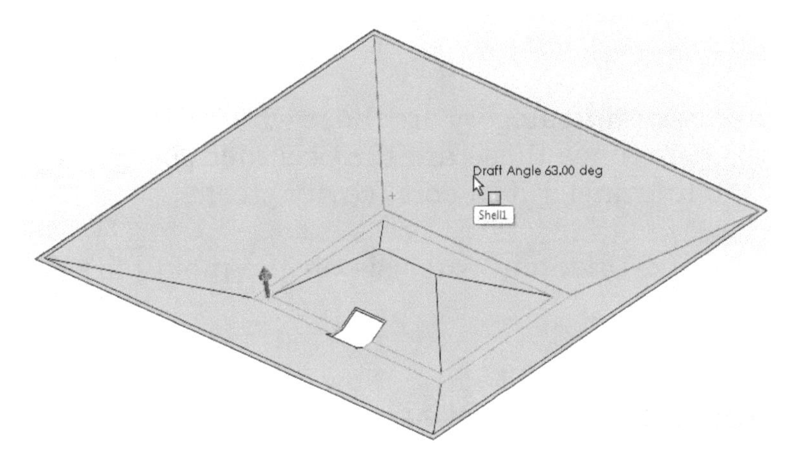

Figure-9. Draft angle over the faces

- If you need to adjust the pull direction, then select the **Adjustment Triad** check box. On doing so, the triad will be displayed around the pull direction arrow; refer to Figure-10. Using this triad, you can change the direction of pull.

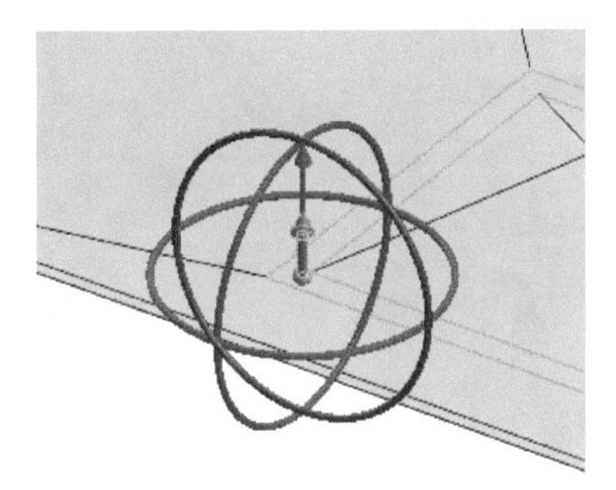

Figure-10. Adjustment triad

- Select the **Face classification** check box to classify faces. Now, check the colors of the model. By default, **Green** color denotes the positive draft which means that the faces are going to be in the **Core**. **Red** color denotes negative draft which means that the faces are going to be in the **Cavity**. **Yellow** color denotes that the faces do not have draft angle applied. **The faces that are painted yellow need to be worked on**. You need to apply desired draft angle to these faces. The method to apply draft will be discussed later in this chapter.
- Click on the **OK** button from the **PropertyManager** to keep colored faces. To exit the colored faces mode, click again on the **Draft Analysis** button from the **Ribbon**.

Undercut Analysis

The **Undercut Analysis** tool is used to check the faces of the mold part that behave as undercuts. SolidWorks classify the undercut faces in five categories: **Direction1 undercut**, **Direction2 undercut**, **Occluded undercut**, **Straddle undercut**, and **No undercut**.

Direction1 undercut: The faces in this category are imprinted on Core tooling.

Direction2 undercut: The faces in this category are imprinted on Cavity tooling.

Occluded undercut: The faces in this category are those faces which can neither be included in core nor in cavity. These faces require sliders. In SolidWorks, sometimes these faces are shifted to No undercut category.

Straddle undercut: The faces in this category are those which you can put in any of the two: the core steel or the cavity steel.

No undercut: The faces in this category are those which are not counted as undercut. But in SolidWorks, sometimes these are the occluded faces. Most of the time, you need to split the face to transfer it in core/cavity steel.

Now, we will perform the Undercut Analysis on the mold part. The steps are given next.

- Exit all the other analyses if still active. Now, click on the **Undercut Analysis** tool. The **Undercut Analysis PropertyManager** will be displayed; refer to Figure-11.
- You can select a pull face or you can select the parting line (if available).
- On selecting the pull face, the mold part will be displayed as shown in Figure-12.

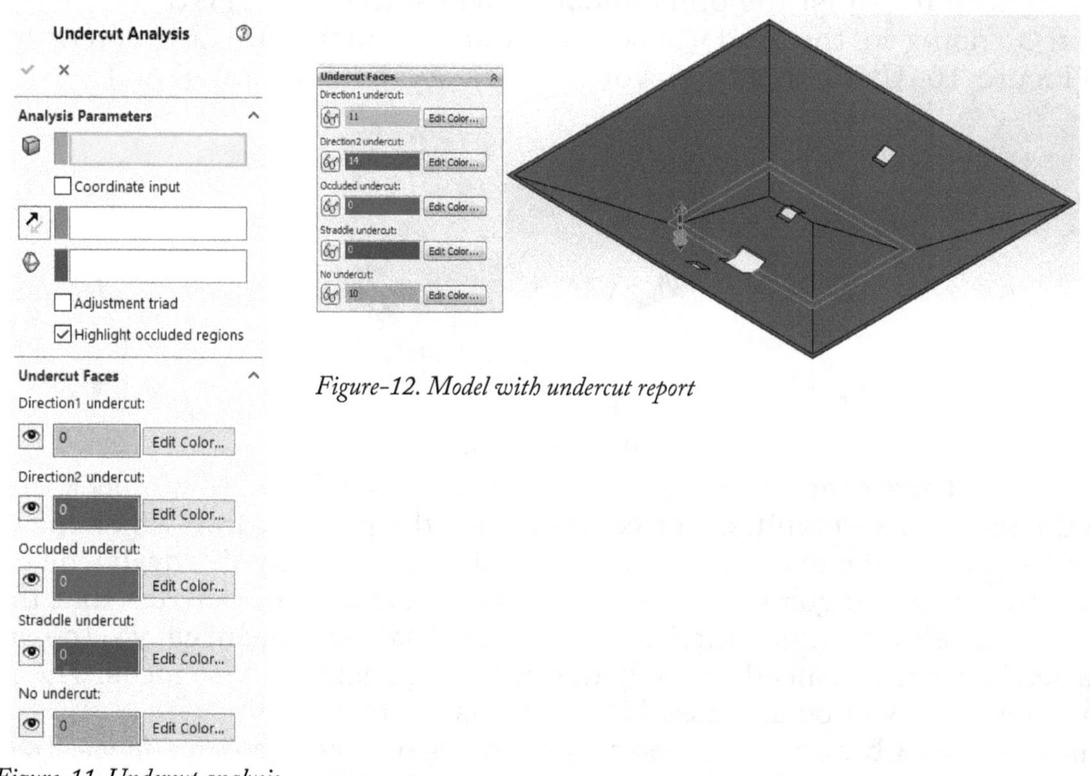

Figure-11. Undercut analysis PropertyManager

Figure-12. Model with undercut report

- You can see from the model and report that there are 10 faces that come in **No undercut** category. If you want to hide all the other faces and want to display only **No undercut** faces, then click on the **Show/Hide** button next to all other categories in the **PropertyManager**; refer to Figure-13.

Figure-13. Mold part after hiding other categories

- Sometimes, we need to split these faces to include them in their respective die steel. You will learn about the splitting later in this chapter.

Parting Line Analysis

The **Parting Line** tool is used to check the possible parting line for the mold part. Using this tool, you can check the parting line for multiple pull directions. The steps to perform Parting Line Analysis are given next.

- Check the part given in Figure-14 and try to find out the parting lines manually (Parting line is the line at which the core and cavity meet.)

Figure-14. Part for parting line check

- Now, we will check what should be the parting lines by which we can divide the mold part into core and cavity. To do so, click on **Parting Line Analysis** tool to start the Analysis. The **Parting Line Analysis PropertyManager** will be displayed as shown in Figure-15.
- Select the faces perpendicular to which the pull direction is to be defined. Refer to Figure-16.

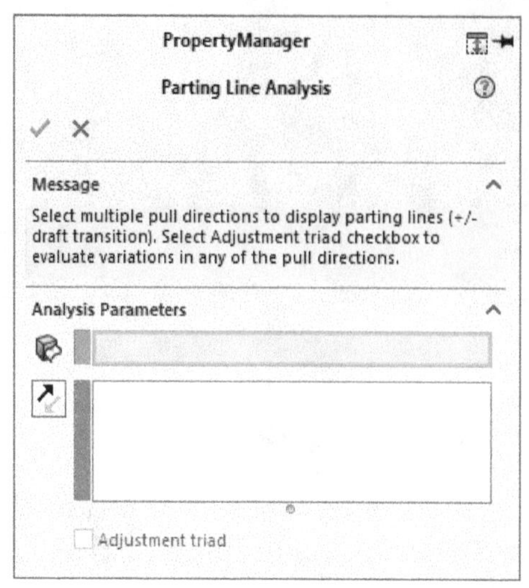

Figure-15. Parting_Line_Analysis_PropertyManager

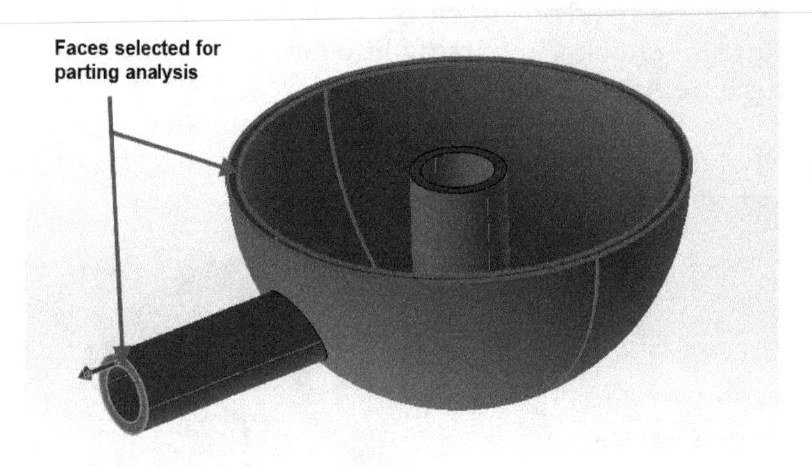

Figure-16. Faces selected for parting line analysis

- Check the dark and light lines. In the above figure, there are three types of lines: dark lines, dashed lines, and dotted lines. Dark lines denote main core and cavity. Dashed lines and dotted lines denote sub-inserts for the mold or the outline where the mold part needs to be divided. (Sub-Inserts are the parts that are assembled in the main Core/Cavity to create a mold part.)

PREPARING MODEL FOR MOLD

After performing the above three analyses, we need to modify the mold part so that molding becomes feasible. There are four options available in **Mold Tools CommandManager** to prepare the model for molding: **Split Line**, **Draft**, **Move Face**, and **Scale**. These tools are discussed next.

Splitting Faces using Split Line tool

In some of the cases, a face of mold part cannot be completely allotted to core or cavity. In those cases, you need to divide that face into two or more parts that can fit for core and cavity. The steps given next explain the use of splitting for mold.

- Perform the **Undercut Analysis** to find out the areas where you need to split the faces to accommodate the face in core/cavity.
- After finding the areas that require splitting, click on the **Split Line** tool from the **Ribbon**. The **Split Line PropertyManager** will be displayed as shown in Figure-17.

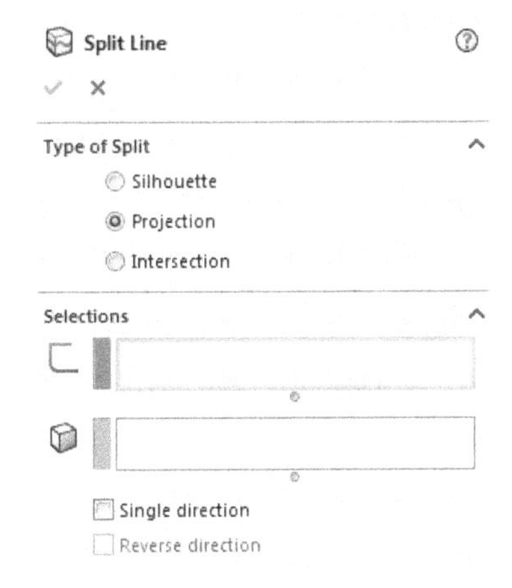

Figure-17. Split Line PropertyManager

- By default, **Projection** radio button is selected and you are supposed to select a sketch for dividing the face. Figure-18 shows a mold part that need to be divided from mid.

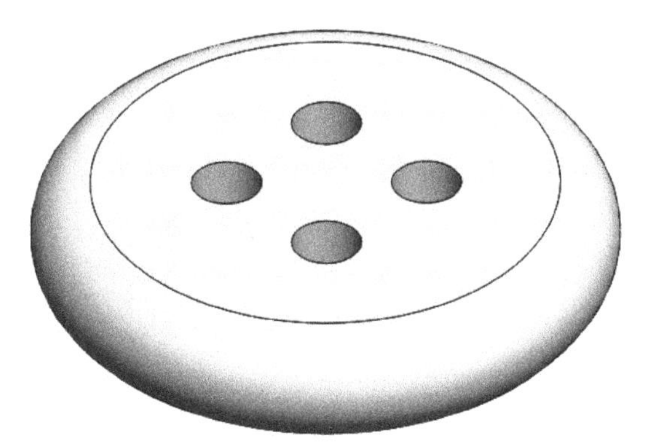

Figure-18. Mold part to be splitted

- A line sketch is drawn at the center of the mold part so that it can divide the part into two pieces; refer to Figure-19. Note that line should be drawn on the face to be divided.

Note that to draw the sketch, you need to cancel the **Split Line** tool, select the **Sketch** tab from the **Ribbon** and after you draw the sketch, you need to select the **Split Line** tool again.

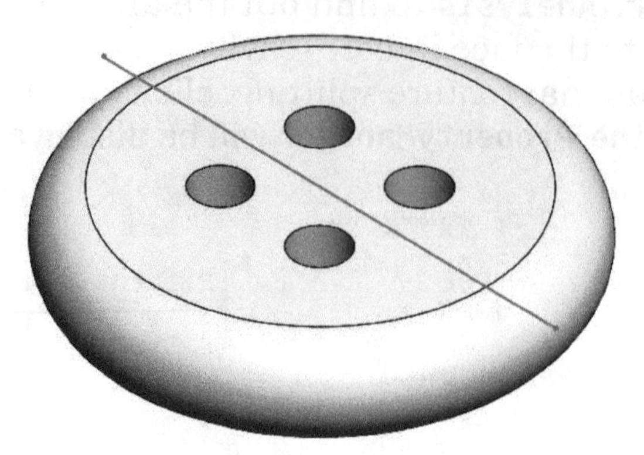

Figure-19. Sketched line drawn at the center

- Select the line sketch, it will be displayed in the first selection box in the **Split Line PropertyManager**.
- Click in the next selection box. You will be asked to select the faces to be split up.
- Select all the faces of the mold part that you want to split by using the projection of sketch; refer to Figure-20.
- Select the **OK** button from the **PropertyManager**, the mold part will be displayed as shown in Figure-21.

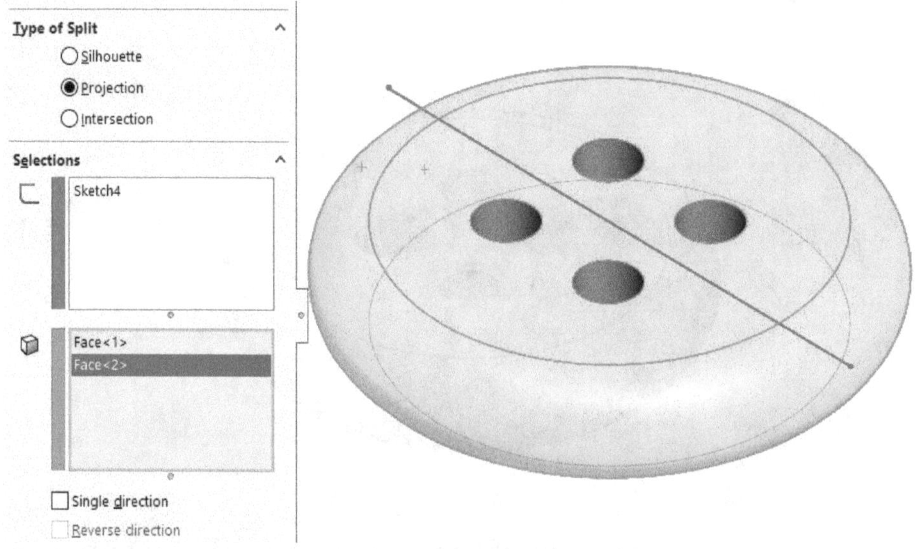

Figure-20. Faces selected

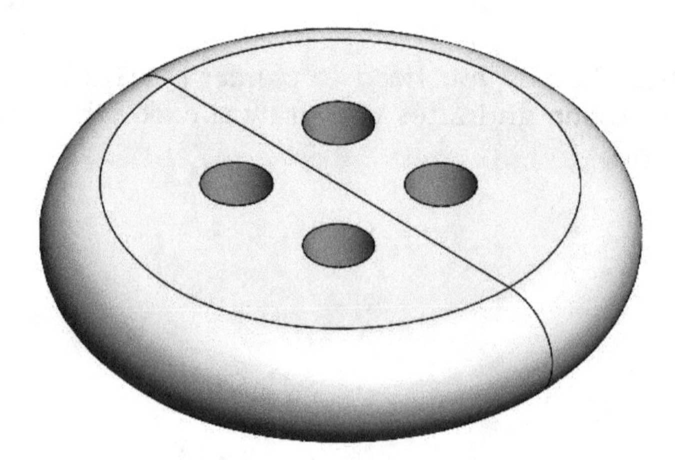

Figure-21. Mold part after splitting

- You can also split this part by using the **Silhouette** radio button. Select this radio button, the first selection box will change with **Direction of Pull** selection box.
- Select a plane or face or edge that defines the direction of pull.
- Click in the next selection box and select all the required faces of the mold part that you want to split; refer to Figure-22.

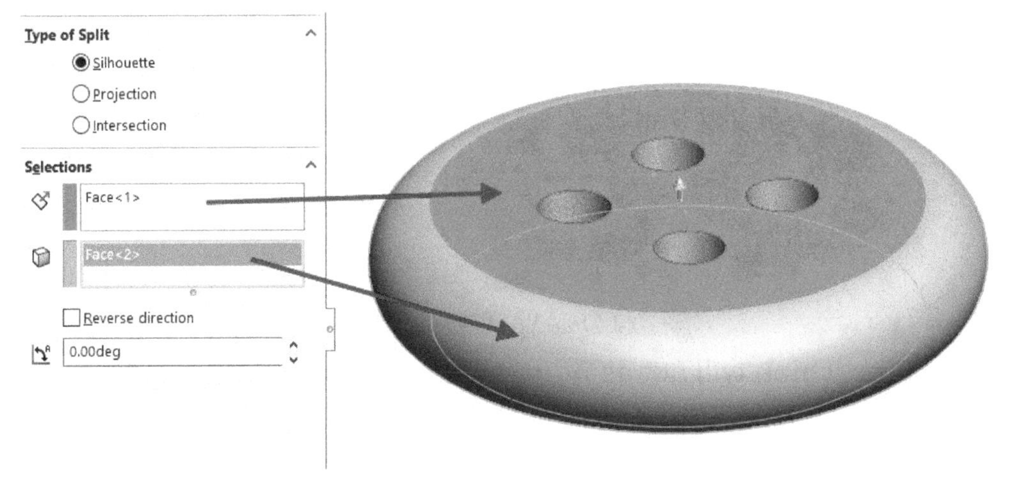

Figure-22. Selection for silhouette split

- Click on the **OK** button, the part will split based on undercut analysis and pull direction; refer to Figure-23.

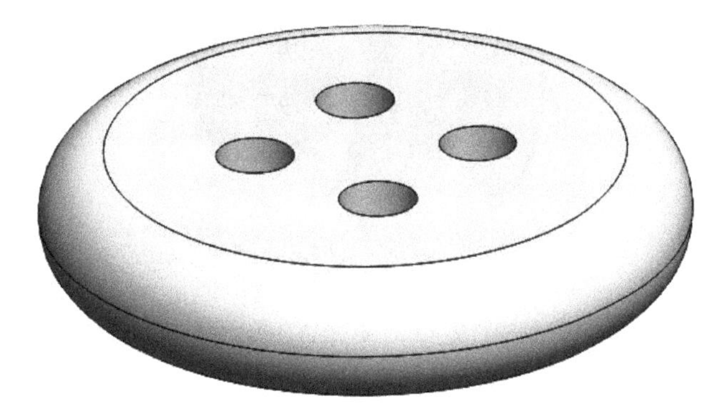

Figure-23. Silhouette split

- In the same way, you can split the mold part by using the **Intersection** radio button.

Note that you can also use 3D sketch to create split lines/curves.

Applying draft using Draft tool

After you perform the draft analysis, some of the faces of the mold part will be displayed in yellow color which means that they require the draft. In such cases, we use the **Draft** tool to apply draft angle at those faces. The following steps explain the procedure to apply draft angle.

- Click on the **Draft** tool to apply draft. The **DraftXpert PropertyManager** will be displayed as shown in Figure-24.

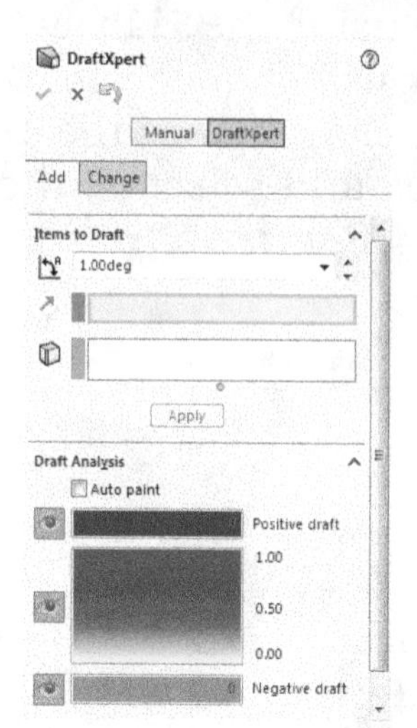

Figure-24. DraftXpert PropertyManager

- Select the plane which will act as neutral plane for the draft angle and then select the faces on which you want to apply draft angle; refer Figure-25.

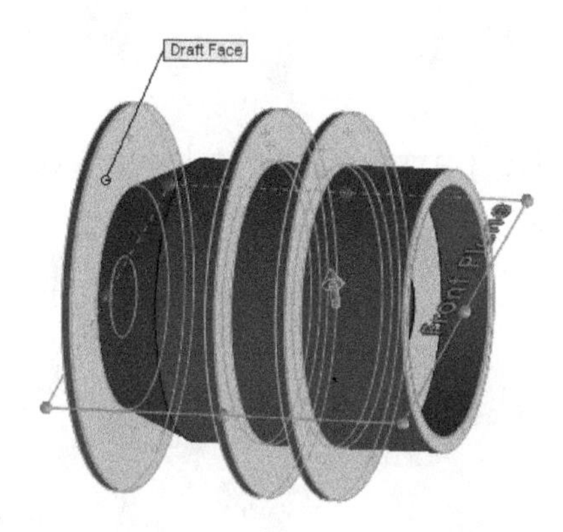

Figure-25. Faces selected for applying draft

- Change the angle value by using the spinner in the **PropertyManager** and then click on the **OK** button to apply the draft.

Increasing/Decreasing thickness of walls using the Move Face tool

Since it is recommended to have a uniform thickness in the mold part, at some places you may require to change the thickness of part. The steps to change the thickness of wall are given next.

- Click on the **Move Face** tool from the **Ribbon**. The **Move Face PropertyManager** will be displayed as shown in Figure-26.

Figure-26. Move Face PropertyManager

- Select the face that you want to move and then specify the distance value in the **ΔX**, **ΔY**, or **ΔZ** spinners as per your requirement.
- If you want to rotate the faces then click on the **Rotate** radio button and then specify the angle values in the respective spinners at the bottom of the **PropertyManager**.

Figure-27 shows a model while increasing its wall thickness.

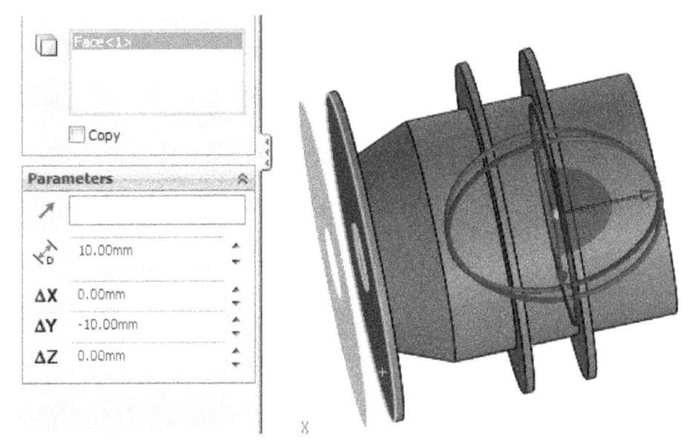

Figure-27. Model while increasing wall thickness

Scaling the model to allow shrinkage in part

While creating mold for a part, we need to increase the size of model by certain percentage so that it do not get undersized when it comes out of mold after cooling. (During the cooling of mold part, its plastic shrinks by a certain amount. This certain amount is called shrinkage allowance). The steps to scale a mold part are given next.

- Click on the **Scale** tool, the **Scale PropertyManager** will be displayed as shown in Figure-28.
- Select desired solid and surface bodies to be scaled from the **Graphics area** or from the **FeatureManager Design Tree**.
- By default, **Centroid** is selected in the **Scale about** drop-down. You can select **Origin** or a **Coordinate System** as reference for scaling by using respective option from the **PropertyManager**.

- After selecting desired option from the drop-down, specify the value of scale in the spinner.

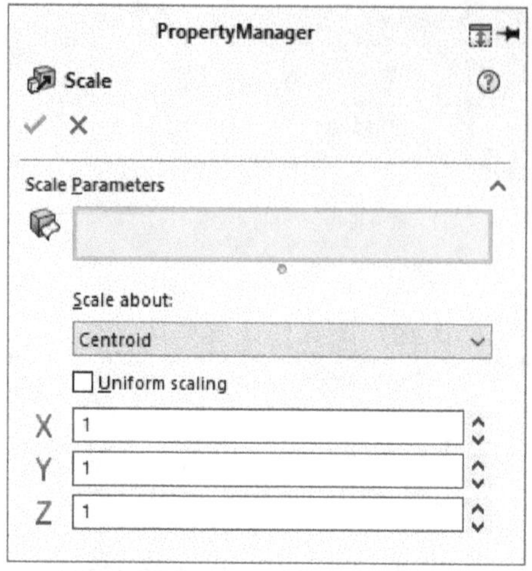

Figure-28. Scale_PropertyManager

- Shrinkage value is given by mold designer or material supplier. You need to add the shrinkage value supplied by you in addition to 1. For example, the shrinkage value for ABS plastic is 0.004 supplied to you. In this case, specify the scale factor as 1.004 in the spinner.

After performing the analyses and doing the required operation, we are ready to start the mold project.

INSERTING MOLD FOLDER

This is the first step when you start a mold project in SolidWorks. To start the project, click on the **Insert Mold Folders** button; the folders that are required for various components of mold will be created in the current project.

PARTING LINE

In this step, we design the parting line by using the standard identification of SolidWorks based on draft analysis. Follow the steps given below to create parting line.

- Click on the **Parting Lines** tool from the **Ribbon**. The **Parting Line PropertyManager** will display; refer to Figure-29.

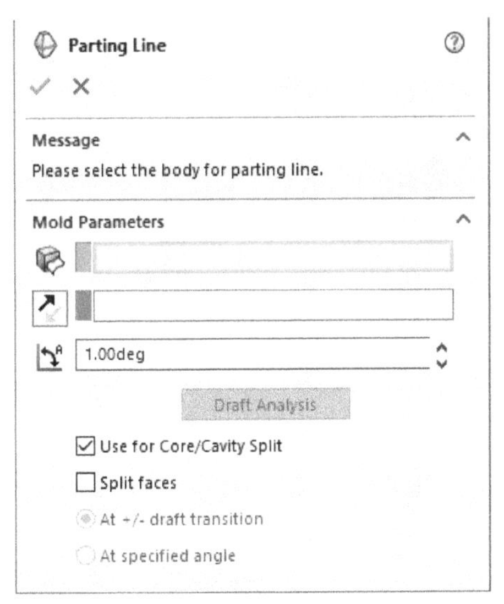

Figure-29. Parting_Line_PropertyManager

- Click on the neutral plane i.e. a flat face with respect to which the draft angles will be measured.
- Set the draft value in the spinner and click on the **Draft Analysis** button. The model will be colored for core and cavity. Also, the parting line will display in violet color; refer to Figure-30.

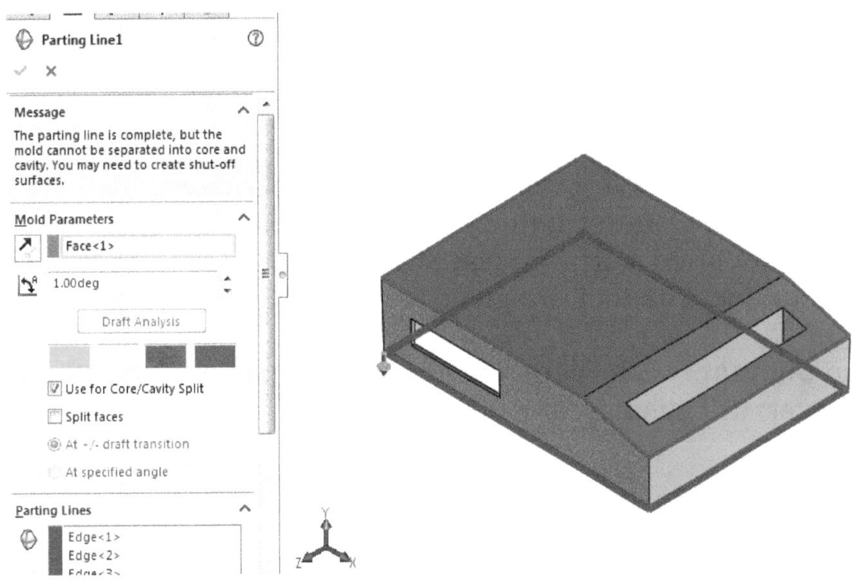

Figure-30. Parting line preview

- Click on the **OK** button from the **Parting Line PropertyManager** to create the parting line.
- If you want to manually create parting line then clear the selection in the **Parting Lines** rollout by using right-click shortcut menu; refer to Figure-31. Now, one by one select the consecutive edges of the model that you want to use as parting line; refer to Figure-32.

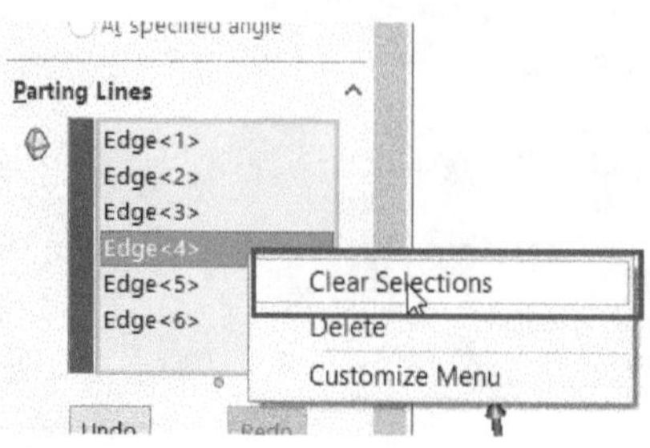

Figure-31. Clear Selections option

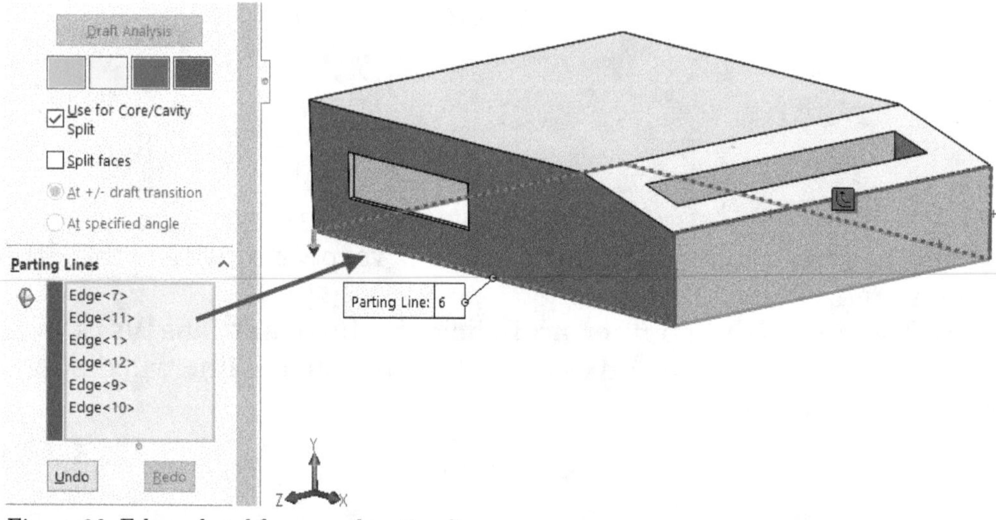

Figure-32. Edges selected for manual parting line

SHUT-OFF SURFACES

Any opening in the model must be closed by a surface so that the core steel and cavity steel meet at defined surface. The **Shut-off Surfaces** tool performs this job for us. The steps to use this tool are given next.

- Click on the **Shut-off Surfaces** tool from the **Ribbon**. The **Shut-off Surface PropertyManager** will display and the open loop will be selected automatically; refer to Figure-33.

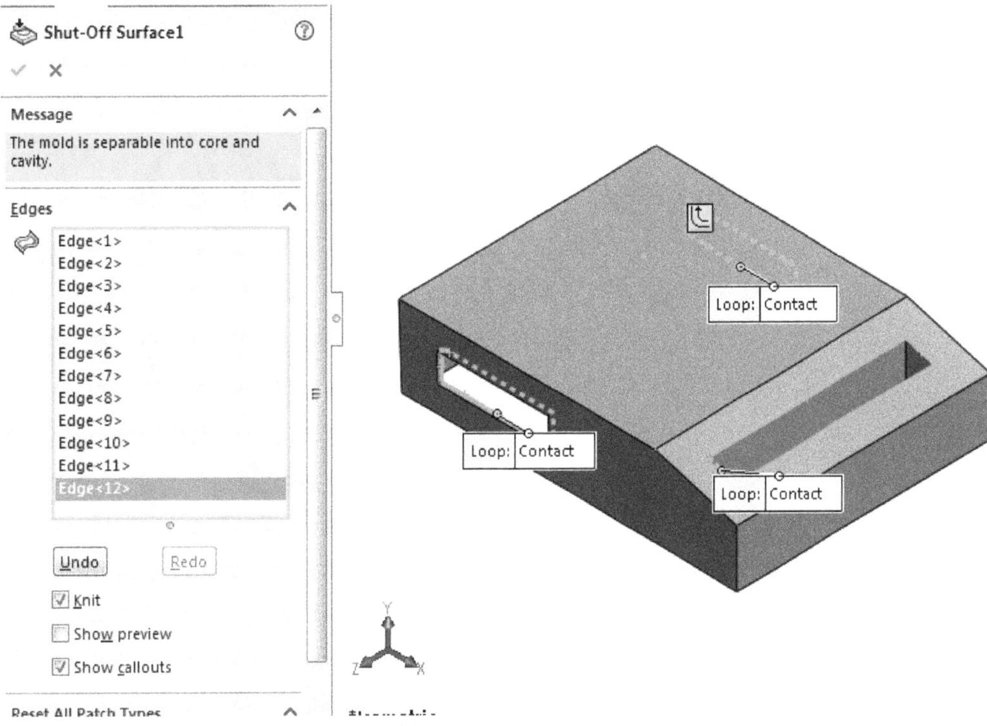

Figure-33. Shut-Off Surface PropertyManager

- Click on the **All Tangent** button from the **Reset All Patch Types** rollout, to create straight patches.
- Click on the **OK** button to create the shut-off surfaces. The surfaces will be created; refer to Figure-34.
- Note that you can also select the edges manually for creating shutoff surfaces in the same way as discussed for parting line.

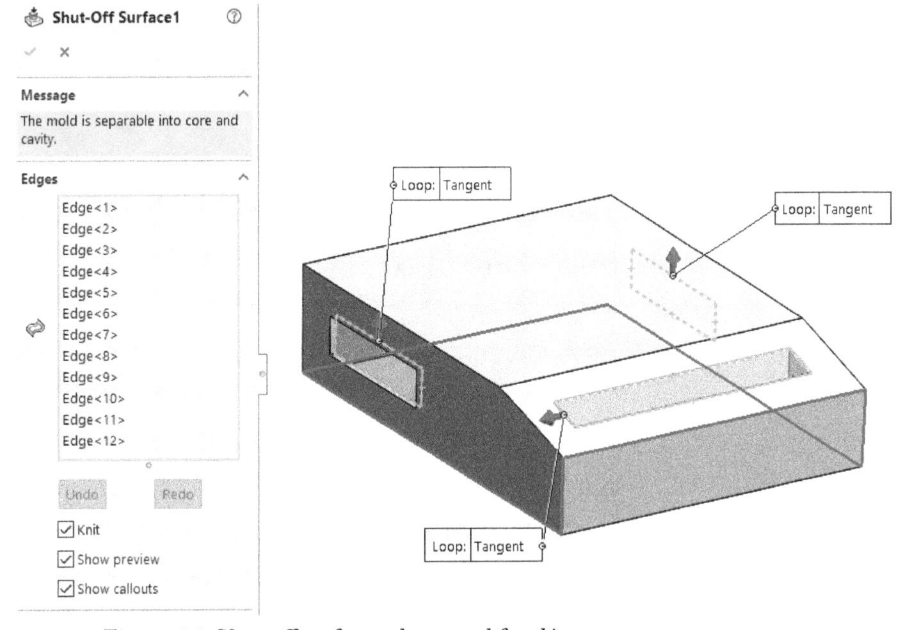

Figure-34. Shut-off surface to be created for this case

PARTING SURFACES

Parting surface is the surface by which the core and cavity is separated. Parting surfaces are created on the basis of parting lines. Note that parting surface is always a continuous surface. The steps to create parting surface are given next.

- Click on the **Parting Surfaces** tool from the **Ribbon**. The **Parting Surface PropertyManager** will display. Also, the preview of parting surface will display; refer to Figure-35.

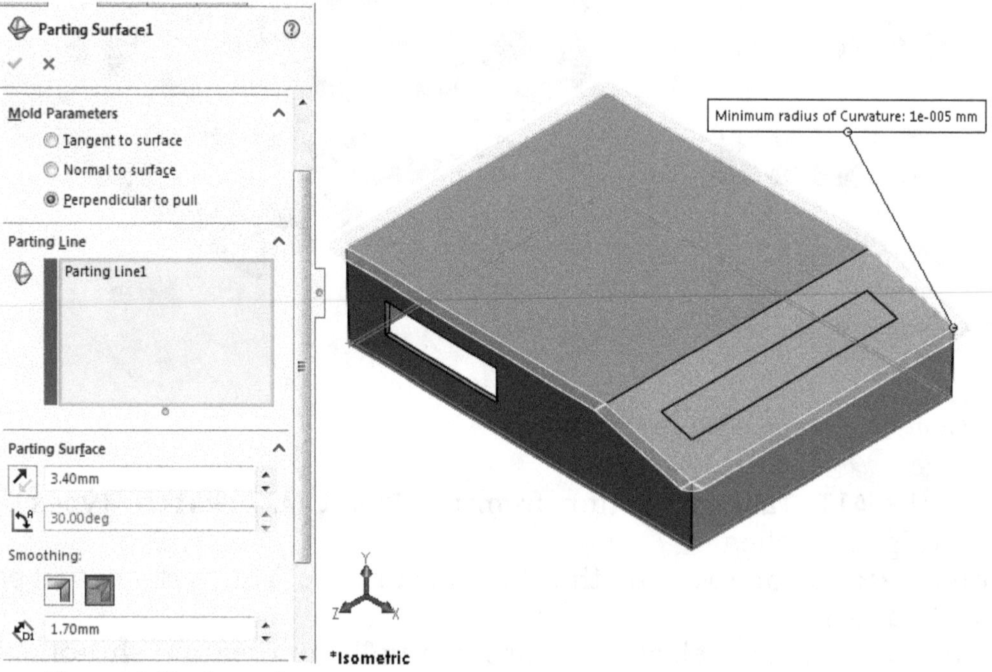

Figure-35. Parting surface

- Increase the length of parting surface by using the first spinner in the **Parting Surface** rollout in the **PropertyManager**.
- You can change the direction of parting surface by using the radio buttons in the **Mold Parameters** rollout.
- If the **Perpendicular to pull** radio button is selected then click on the **Manual Mode** check box to manually change the direction of surface by dragging the key points; refer to Figure-36.
- Click on the **OK** button from the **PropertyManager**.

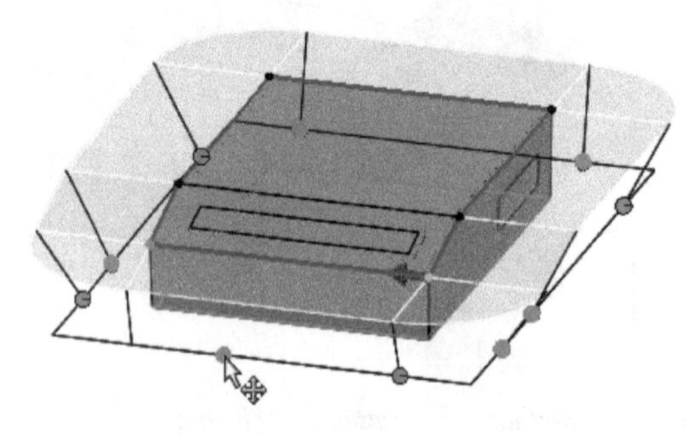

Figure-36. Manual editing of parting surface

TOOLING SPLIT

Now, we want to extract core and cavity from the model by using the parting surface. To split the tooling to generate core and cavity, follow the steps given below.

- Click on the **Tooling Split** tool from the **Ribbon**. You are asked to create sketch for the tooling from which core and cavity will be extracted.
- Select a plane parallel to the parting surface to create tooling; refer to Figure-37.

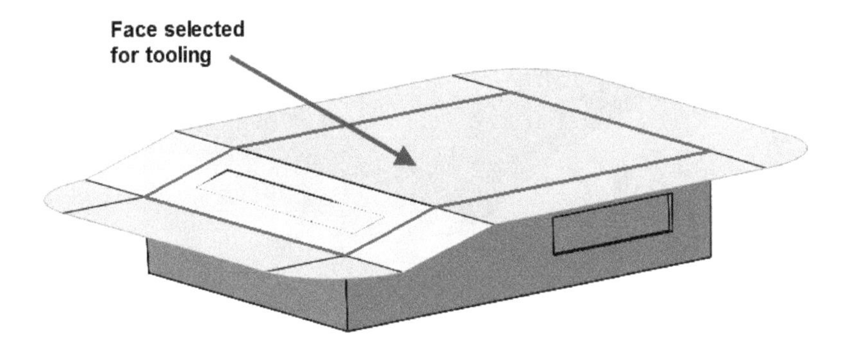

Figure-37. Face selected for tooling

- On selecting the face, the sketching environment will be displayed.
- Create the sketch for the tooling. You can change the display style to **Hidden Lines Visible** for clear view; refer to Figure-38. Note that the sketch of tooling should be big enough to contain the whole size of component.

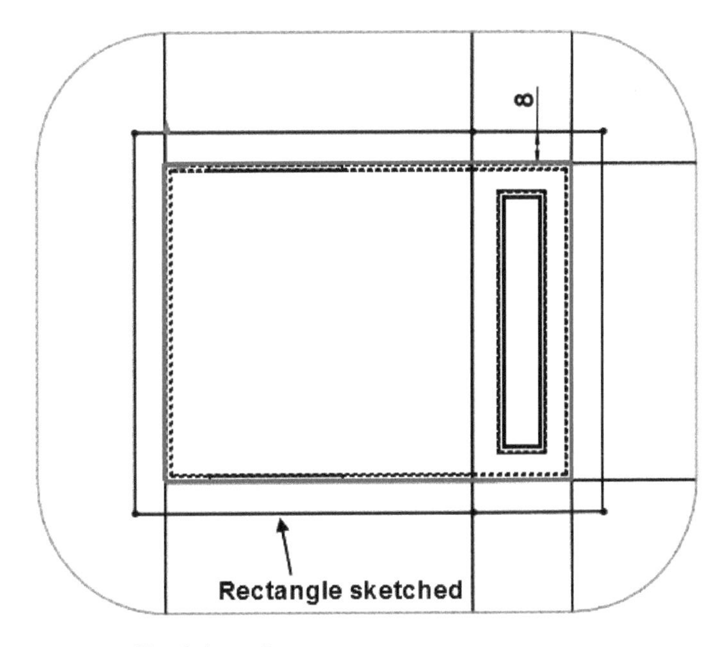

Figure-38. Sketch for tooling

- Exit the sketch environment and change the display style to shaded and orientation to isometric; refer to Figure-39.

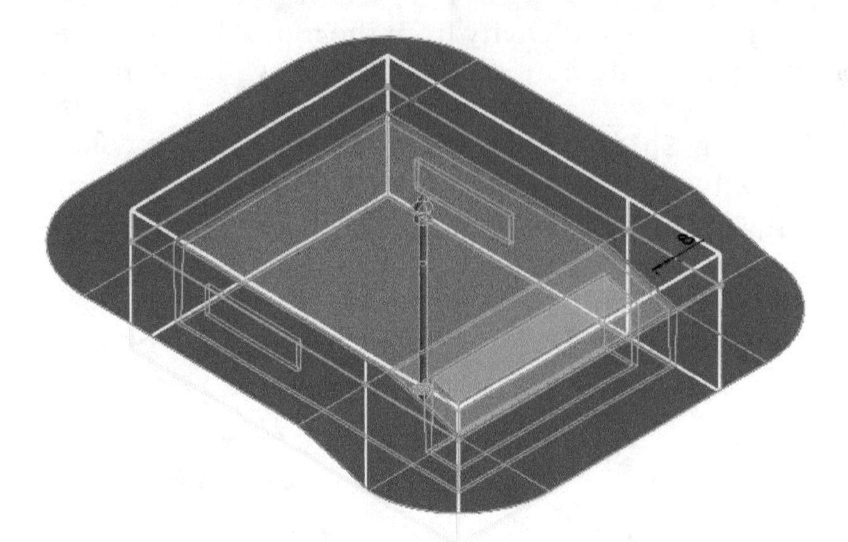

Figure-39. Extrusion tooling split

- Specify desired extrusion height in the spinners.
- Click **OK** from the **PropertyManager** to creating the tooling split.
- The core and cavity are added in the **Solid Bodies** folder of **FeatureManager Design Tree**; refer to Figure-40.

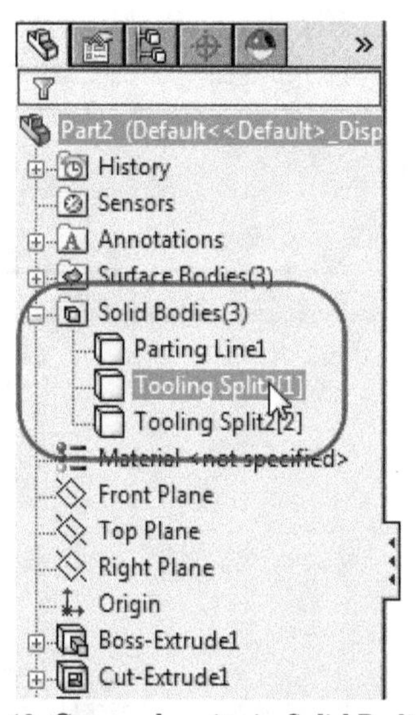

Figure-40. Core and cavity in Solid Bodies folder

Right-click on any of the tooling split and select the **Isolate** option from the menu to check the part separately.

CORE

There are a few portions in the core or cavity that are to be inserted from sides. Such parts are called sliders in engineering. To create sliders, follow the steps given below.

- Isolate the tooling from which you want to extract sliders.
- Click on the **Core** tool from the **Ribbon**.

- Select the side face of the tooling from which you want to create slider. Change the display style to **Hidden Lines Visible** for clear view; refer to Figure-41.

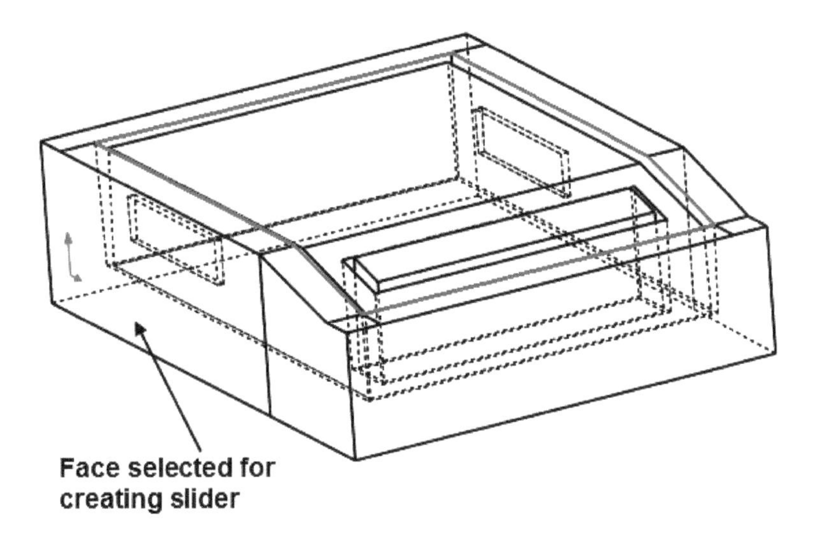

Figure-41. Face selection for slider

- Create sketch for the slider; refer to Figure-42. You can use offset tool for easy creation of sketch.

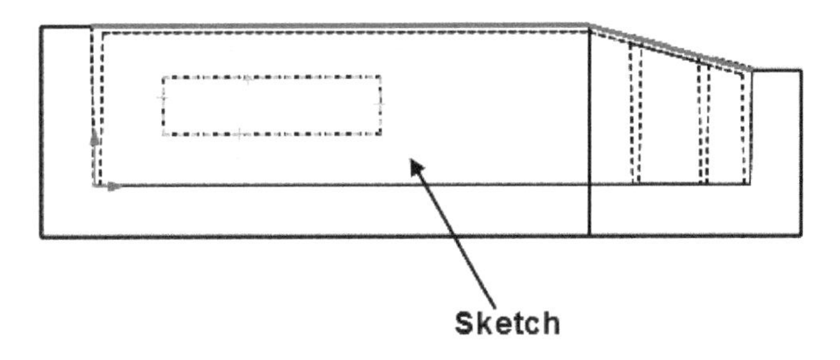

Figure-42. Sketch for slider

- Exit the sketch environment and extrude the sketch till the face you want to include in slider.
- Click on the **OK** button from the **PropertyManager** to create the slider; refer to Figure-43.

To move the individual parts of mold, click on the **Move/Copy** option from the **Features** cascading menu of **Insert** menu; refer to Figure-44. The **Move/Copy Body PropertyManager** will display. Select the part that you want to move. Triad will be displayed providing you the options to move the part. Click on the **Translate/Rotate** button at the bottom in the **PropertyManager** if the triad is not displayed. Move the selected part to desired location and click **OK** from the **PropertyManager**. Similarly, you can move other parts of the tool split. Refer to Figure-45.

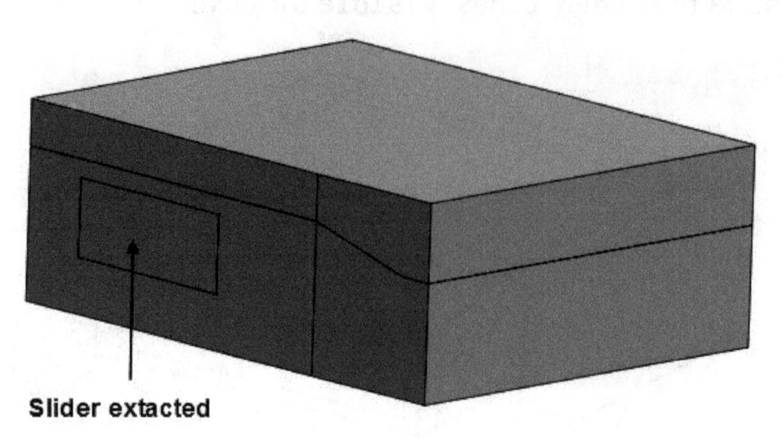

Figure-43. Slider created

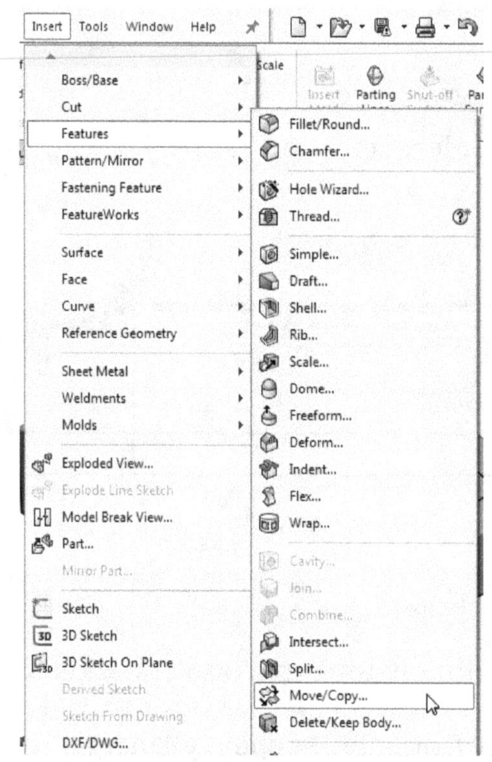

Figure-44. Move or Copy tool

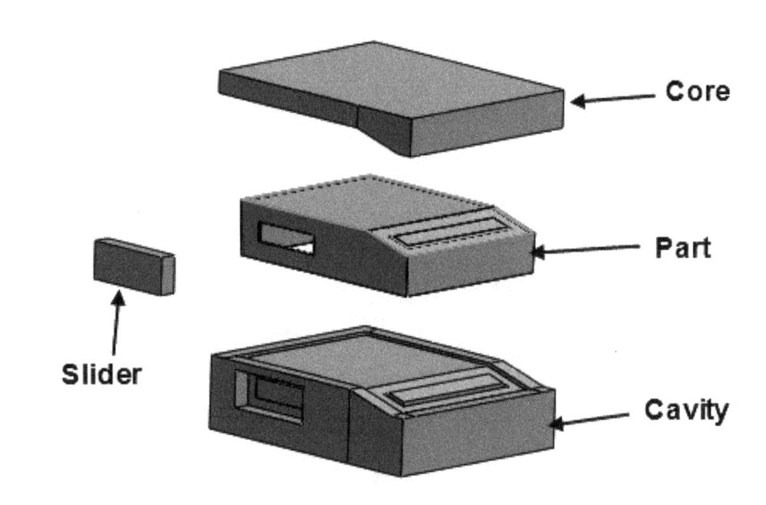

Figure-45. Tool split parts after moving

PRACTICE 1

Make a mold core and cavity of the model shown in Figure-46 out of a 380x440x110 mm^3 block. The part file is available in resources of the book.

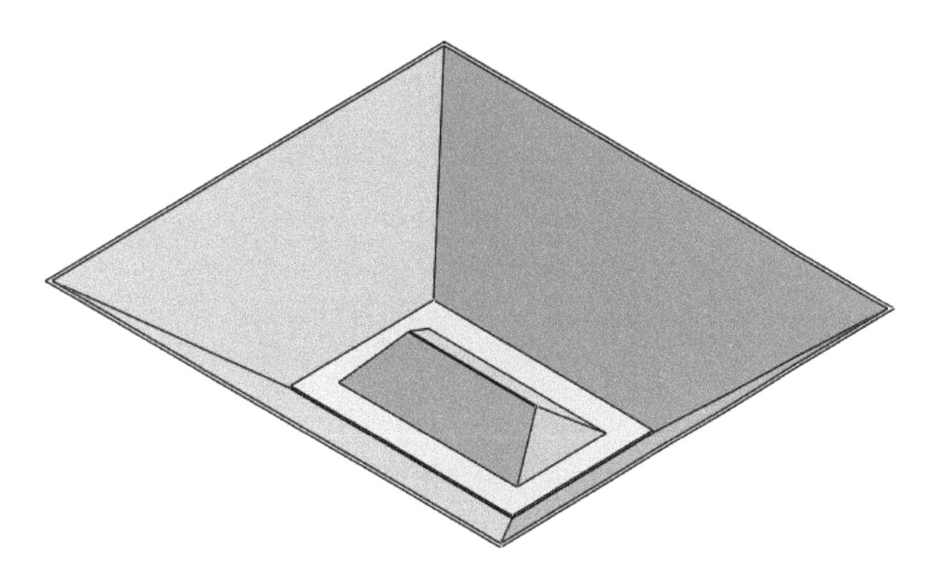

Figure-46. Practice 1 model

PRACTICE 2

Make a mold core and cavity of the model shown in Figure-47. The part file is available in the resources of the book.

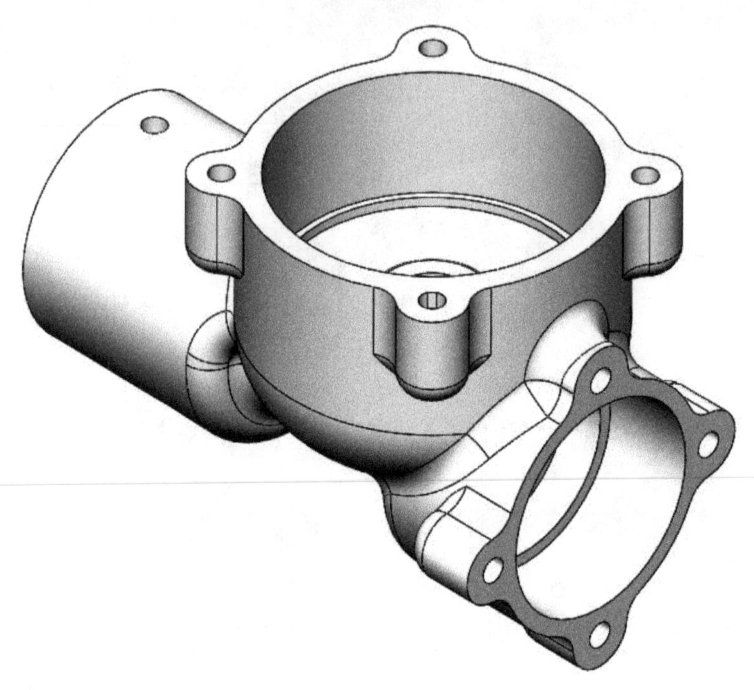

Figure-47. Model for Practice 2 molding

SELF ASSESSMENT

Q1. Which of the following analysis is not performed on the part for checking possibility of mold design?

a. Draft Analysis
b. Stress Analysis
c. Undercut Analysis
d. Parting Line Analysis

Q2. If there are steps at parting line, then draft required for shutoff is

a. 1 degree to 3 degree
b. 3 degree to 5 degree
c. 5 degree to 7 degree
d. 7 degree to 9 degree

Q3. To compensate for shrinkage of molded part, we need to use .... tool.

a. **Draft**
b. **Scale**
c. **Move Face**
d. **Split Line**

Q4. Any opening in the model can be closed by .... tool so that the core steel and cavity steel meet at defined surface.

a. **Shut-off Surfaces**
b. **Parting Surfaces**
c. **Move Face**
d. **Draft**

Q5. .... is the surface by which the core and cavity is separated.

a. Shut-off surface
b. Parting surface
c. Ruled surface
d. Offset surface

FOR STUDENT NOTES

Answer to Self-Assessment:
1. b, **2.** c, **3.** b, **4.** a, **5.** b

Chapter 13

Sheetmetal and Practice

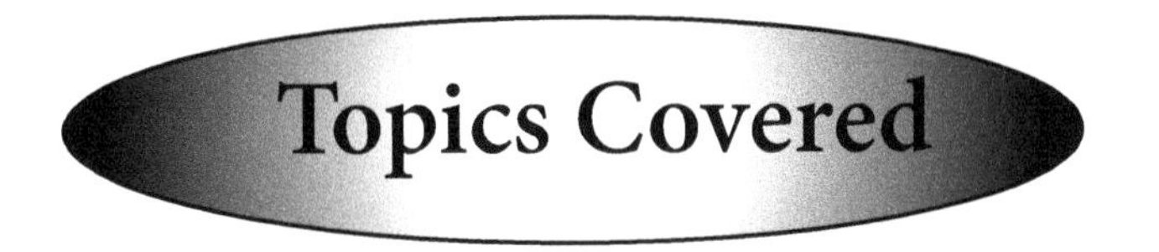

The major topics covered in this chapter are:

- ***Sheet metal Introduction***
- ***Sheet metal Creation Tools***
- ***Sheet metal Design Terms***
- ***Tools for applying cut***
- ***Corners Modification***
- ***Vents, Rip, and Bend insertion***
- ***Flat Pattern***
- ***Solid to Sheet metal***
- ***Practice***

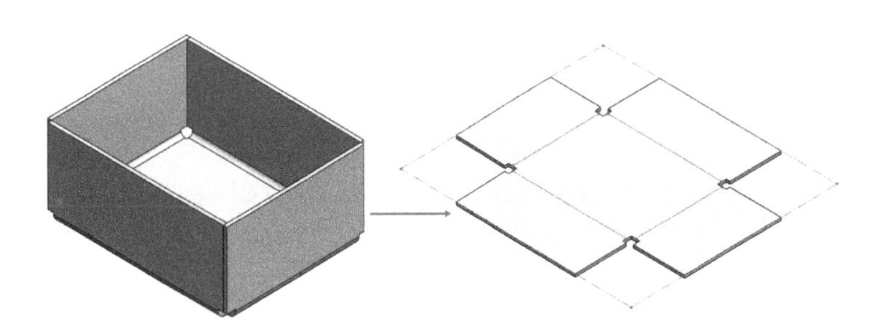

SHEET METAL INTRODUCTION

Sheet metal is used when you need a component of thickness in the range of 0.16 mm to 12.70 mm and do not require conventional cutting machines. The components that can be created by Punch-press and bending machines are designed in Sheet Metal environment. In SolidWorks, there is a separate **CommandManager** to design sheet metal components named **Sheet Metal**; refer to Figure-1.

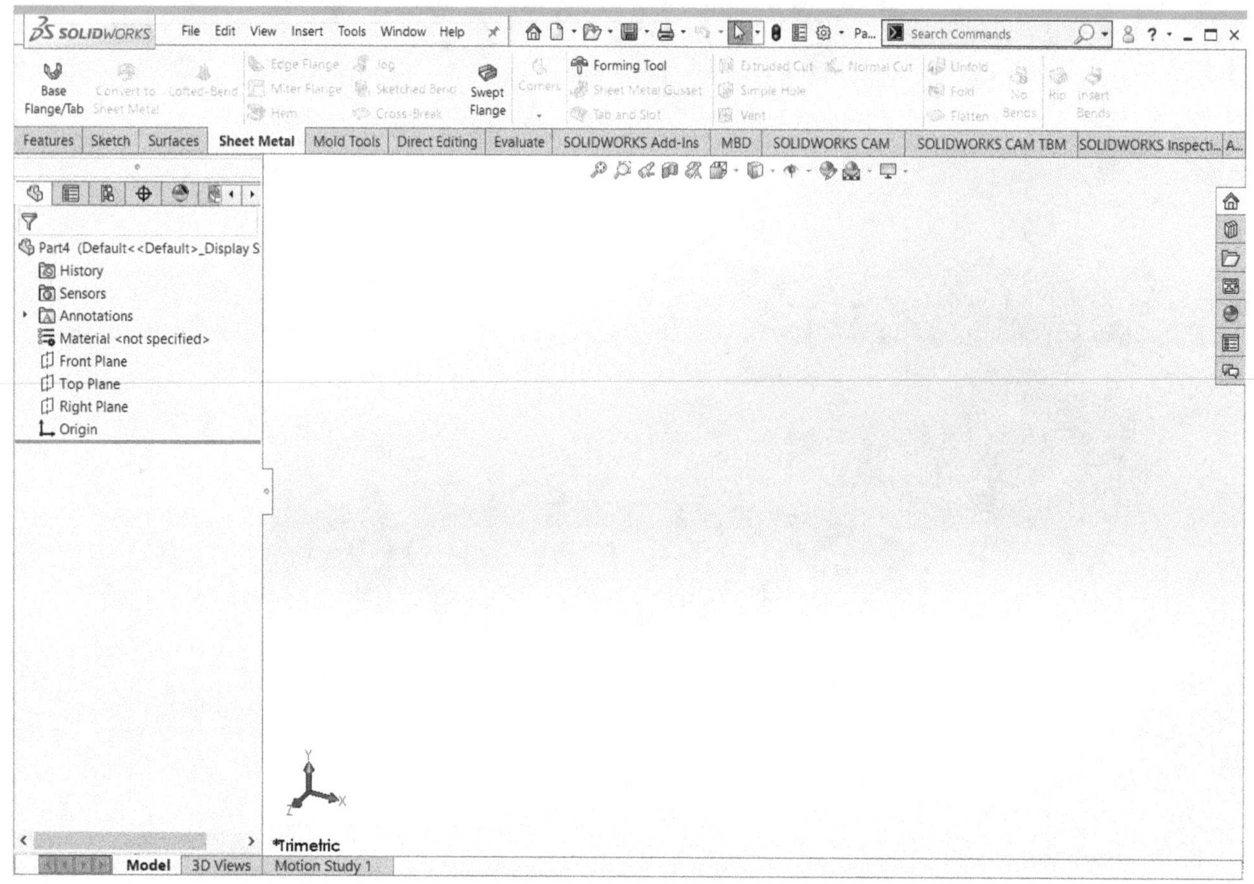

Figure-1. Sheet metal environment

The tools to create sheet metal designs are discussed next.

CREATING BASE FLANGE/TAB

The **Base Flange/Tab** tool is used to create base feature of the sheet metal component. All the other features will be created on this base flange. The steps to create base flange/tab are given next.

- Click on the **Base Flange/Tab** tool. You will be asked to select the sketching plane to draw sketch of the base flange.
- Select a plane and draw the sketch.
- Exit the sketch environment. Preview of the flange will be displayed; refer to Figure-2. The **Base Flange PropertyManager** will display; refer to Figure-3 and Figure-4.

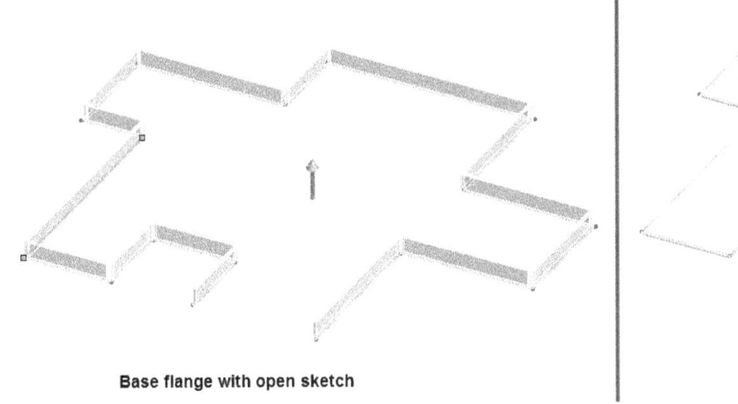

Figure-2. Base flange preview

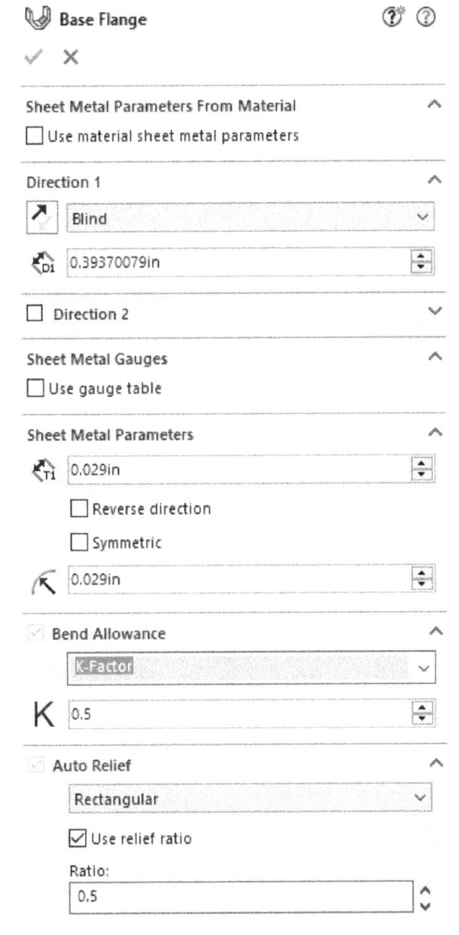

Figure-3. Base_Flange_PropertyManager_with_open_sketch_selected

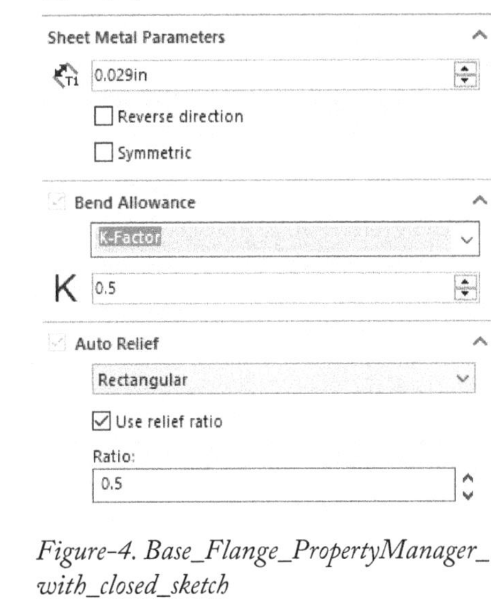

Figure-4. Base_Flange_PropertyManager_with_closed_sketch

Setting Parameters for Base Flange/Tab with Open sketch

- Specify desired length of the flange/tab in direction 1 and direction 2 by using the options in the **Direction 1** and **Direction 2** rollouts in the **PropertyManager**.
- Set desired thickness and bend radius in the edit boxes available in **Sheet Metal Parameters** rollout or you can use the gauge table for defining parameters.
- To use the gauge table, select the **Use gauge table** check box from the **Sheet Metal Gauges** rollout. You are asked to select a sample table.
- Select desired sample table from the drop-down; refer to Figure-5.

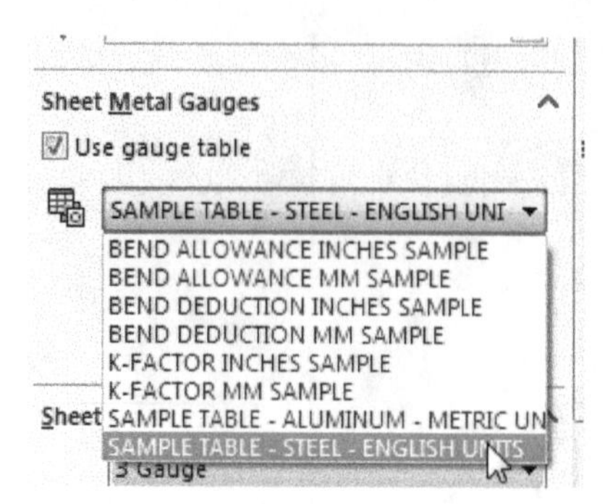

Figure-5. Sample table selected

- Select desired gauge of sheet from the **Gauge** drop-down in the **Sheet Metal Parameters** rollout.
- Select **Symmetric** check box from **Sheet Metal Parameters** rollout to add an equal amount of material to both sides of the sketch.
- Select desired option from the **Bend Allowance** drop-down in **Bend Allowance** rollout. Select the **Bend Table** option from the drop-down to use a bend table for setting allowances. Select the **K-Factor** option from the drop-down to set bend allowance based on specified K-factor value. Select the **Bend Allowance** option to specify desired value of bend allowance. Select the **Bend Deduction** option to specify the value of bend deduction parameter for allowance. Note that bend deduction is the length by which flat pattern is deducted to get final value. Select the **Bend Calculation** option from the drop-down to use tables or samples for bend allowance calculation.
- The option in **Auto Relief** drop-down are used to provide relief cut when there is a possibility of material overlap at intersections or bends.
- Set the parameters as desired and click on the **OK** button from the **PropertyManager** to create the base flange/tab.

CREATING LOFTED-BEND

The **Lofted-Bend** tool is used to create sheet metal component by joining two or more sketch sections. This tool works in the same way as the **Lofted Surfaces** or **Lofted Boss/Base** tool works. Figure-6 shows a lofted bend. Procedure to use this tool is given next.

- Make sure you have two or more open sketches in the graphics area and then click on the **Lofted-Bend** tool from the **Ribbon**. The **Lofted-Bends PropertyManager** will be displayed; refer to Figure-7.

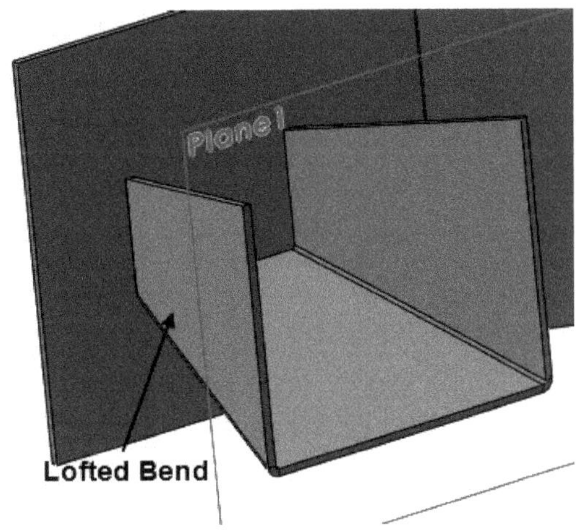

Figure-6. Lofted bend

- Select the **Bent** radio button if you want to create real physical bends, rather than formed geometry and approximated bend lines in a flat pattern. Bent lofted bends form a realistic transition between two profiles to facilitate instructions for press brake manufacturing.
- Select the **Formed** radio button if you want to create a formed transition between two profiles assuming that a forming tool is used to create lofted bend. In this case, you cannot have sharp corners in the profiles; refer to Figure-8.

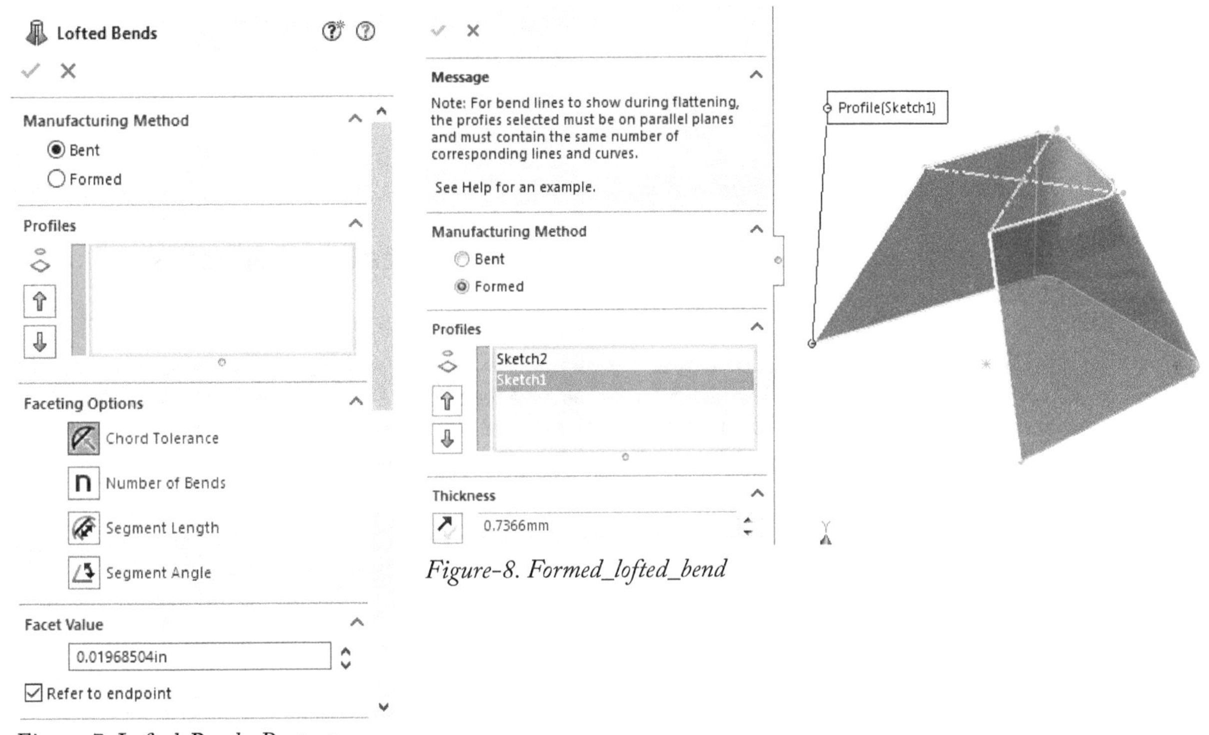

Figure-7. Lofted-Bends_Property-Manager

Figure-8. Formed_lofted_bend

- In most of the cases, you will be using the **Bent** radio button. Select the **Bent** radio button and then select the profiles.
- Select the **Chord Tolerance** button from the **Faceting Options** rollout and specify desired tolerance for cord created during bend transition from point to arc of sections; refer to Figure-9. Similarly, you can use the **Number of Bends**, **Segment Length**, and **Segment Angle** buttons to modify faces.

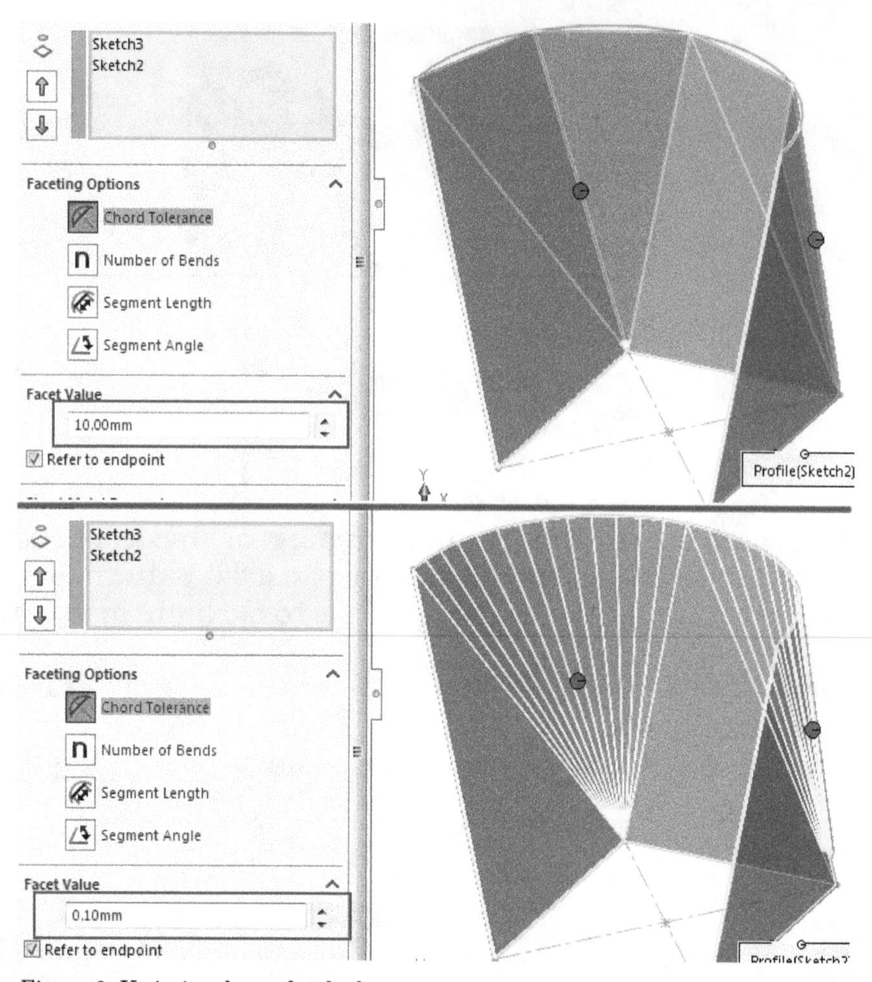

Figure-9. Variation due to chord tolerance

- Specify the sheetmetal parameter in the **Sheet Metal Parameters** rollout as discussed earlier.

SHEET METAL DESIGN TERMS

While going forward in the chapter, you will come across some technical terms are used in sheet metal industry. Here, we will discuss about these technical terms and their effects on production.

Sheet Metal Definition

Metal that has been rolled into a sheet having a thickness between foil and plate. The thickness of metal can vary from fraction of millimeters to 12.5 mm, in general.

Gauge

Gauge is a traditional measurement unit of sheet thickness commonly used in USA, India, and many other parts of world. It is a non-linear unit. Higher the gauge number the thinner is the sheet metal. Like gauge 00 means 12.7 mm thickness and gauge 38 means 0.16 mm thickness. You can find a table on gauge to mm conversion easily in local stores. Use of Gauge to designate sheet metal thickness is discouraged by numerous international standards organizations. For example, ASTM states in specification ASTM A480-10a "The use of gauge number is discouraged as being an archaic term of limited usefulness not having general agreement on meaning."

Bend Allowance

When the sheet metal is put through the process of bending the metal around the bend is deformed and stretched. As this happens you gain a small amount of total length in your part. Likewise, when you are trying to develop a flat pattern, you will have to make a deduction from your desired part size to get the correct flat size. The **Bend Allowance** is defined as the material you will add to the actual leg lengths of the part in order to develop a flat pattern. The leg lengths are the part of the flange which is outside of the bend radius. In our example a part with flange lengths of 2" and 3" with an inside radius of .250" at 90° will have leg lengths of 1.625" and 2.625" respectively, refer to Figure-10. When we calculate the Bend Allowance, we find that it equals .457". In order to develop the flat pattern we add .457" to 1.625" and 2.625" to arrive at 4.707". In SolidWorks, you don't need to specify the value of bend allowance as it is automatically calculated based on K factor.

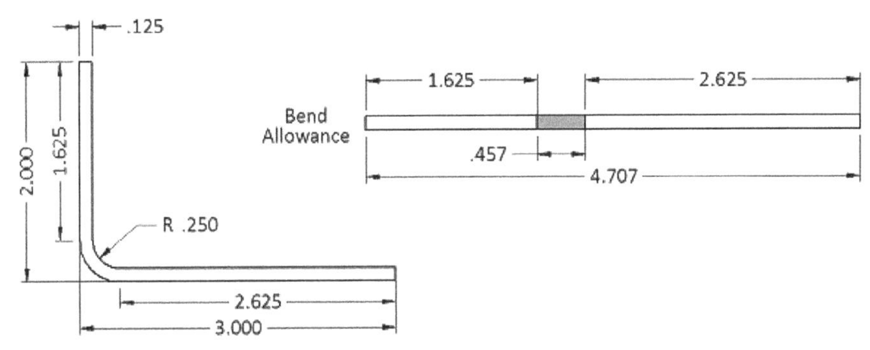

Figure-10. Bend Allowance example

K-Factor

The K-Factor in sheet metal designing is the ratio of the neutral axis to the material thickness. When metal is bent the upper section is going to compress and the lower section is going to stretch. The line where the transition from compression to stretching occurs is called the neutral axis. The location of the neutral axis varies and is based on the material's physical properties and its thickness. The K-Factor is the ratio of the Neutral Axis' Offset (t) and the Material Thickness (MT). Figure-11 shows how the top of the bend is compressed and the bottom is stretched.

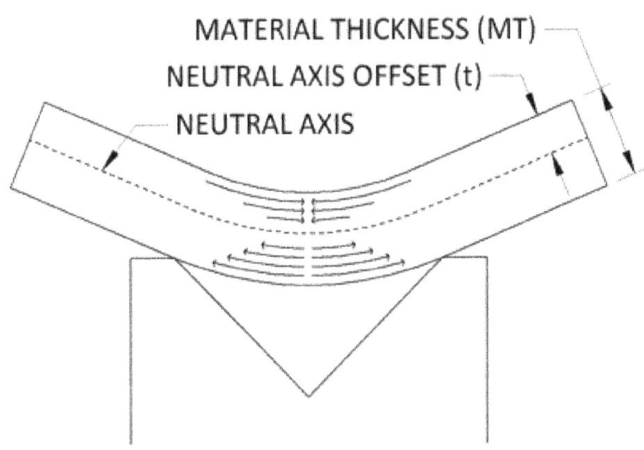

Figure-11. Neutral axis of bent model

$$K = t/MT$$

Generally, K is taken as 0.33 for soft materials and 0.4 for hard materials as a thumb rule. Since, K factor is not just mathematical term, you need experiments to find exact value of K for your material situations. We have given the general steps to find out K-factor by experiment.

Calculating the K-Factor

Since the K-Factor is based on the property of the metal and its thickness, there is no simple way to calculate it ahead of the first bend. Typically the K-Factor is going to be between 0 and .5. In order to find the K-Factor, you will need to bend a sample piece and deduce the Bend Allowance. The Bend Allowance is then plugged into the equation to find the K-Factor.

- Begin by preparing sample blanks which are of equal and known sizes. The blanks should be at least a foot long to ensure an even bend, and a few inches deep to make sure you can sit them against the back stops. For our example let's take a piece that is 14 Gauge, .075", 4" Wide and 12" Long. The length of the piece won't be used in our calculations. Preparing at least 3 samples and taking the average measurements from each will help
- Set up your press brake with desired tooling you'll be using to fabricate this metal thickness and place a 90° bend in the center of the piece. For our example this means a bend at the 2" mark.
- Once you've bent your sample pieces carefully measure the flange lengths of each piece. Record each length and take the average of lengths. The length should be something over half the original length. For our example the average flange length is 2.073"
- Second measure the inside radius formed during the bending. A set of radius gauges will get you fairly close to finding the correct measurement, however to get an exact measurement an optical comparator will give you the most accurate reading. For our example the inside radius is measured at .105"
- Now that you have your measurements, we'll determine the Bend Allowance. To do this first determine your leg length by subtracting the material thickness and inside radius from the flange length. (Note this equation only works for 90° bends because the leg length is from the tangent point.) For our example the leg length will be 2.073 – .105 – .075 = 1.893.
- Subtract twice the leg length from the initial length to determine the Bend Allowance. 4 – 1.893 * 2 = .214.
- Plug the Bend Allowance (BA), the Bend Angle (A), Inside Radius (R) and Material Thickness (T) into the below equation to determine the K-Factor (K).

$$K = \frac{-R + \frac{BA}{\pi A/180}}{T}$$

CREATING EDGE FLANGE

The **Edge Flange** tool is used to create a flange by using selected edge. The steps to do so are given next.

- Click on the **Edge Flange** tool. The **Edge Flange PropertyManager** will be displayed.

- Select an edge using which you want to create the flange wall. The flange end will get attached to cursor; refer to Figure-12.

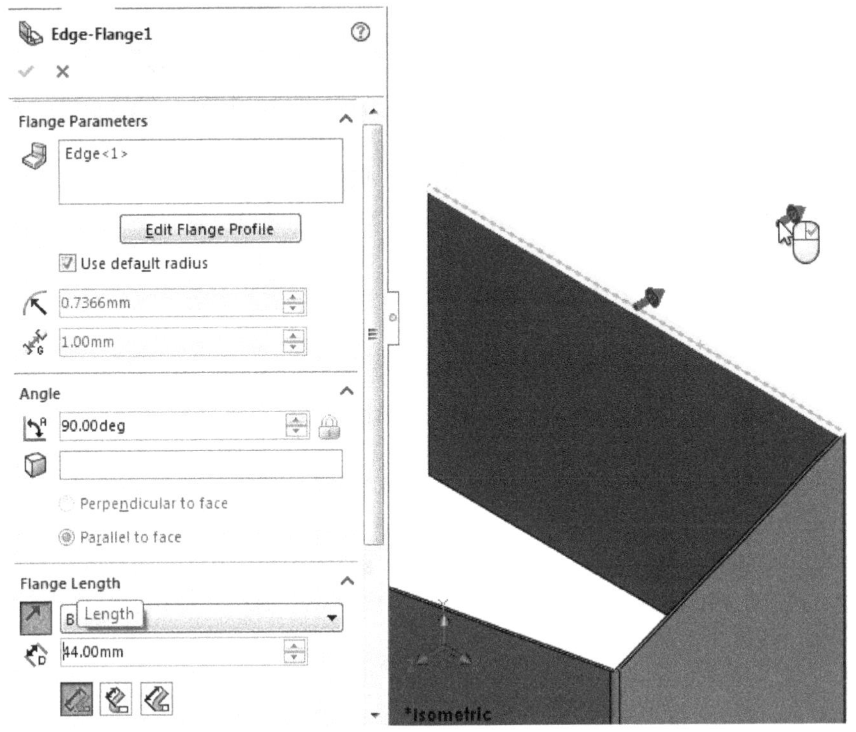

Figure-12. Edge Flange PropertyManager and preview of flange

- Click to specify the end point of the flange.
- You can change the length of the flange by using the spinner in the **Flange Length** rollout.

Some of the important options in the **PropertyManager** are explained next.

Flange Parameters Rollout

The options in the **Flange Parameters** rollout are used to define the edge reference to be used for creating the edge flange, the bending radius, and the profile of the edge flange. These options are discussed next.

Edge

The **Edge selection** box is used to select the edges to create the edge flange.

Edit Flange Profile

The **Edit Flange Profile** button is chosen to edit the profile of the edge flange. By default, the edge flange is created along the entire length of the selected edge. To edit the profile of the edge flange, choose the **Edit Flange Profile** button; the **Profile Sketch** dialog box will be displayed informing you that the sketch is valid. Also, the sketching environment will be invoked in the background. Edit the sketch of the profile of the edge flange using the sketching tools.

You will also notice that while editing the sketch of the edge flange, the **Profile Sketch** dialog box informs you whether the sketch is valid for creating the edge flange or not. If the status of the sketch is shown valid in the **Profile Sketch** dialog box, the preview of the flange will be displayed in the drawing area. You can drag the end points of the line create on selected edge for flange to reduce/increase the width of flange. After editing the profile, choose the **Finish** button from the **Profile Sketch** dialog box; the flange will be created and the **Edge-Flange PropertyManager** will be automatically closed. Note that if you want to modify other parameters of the flange, choose the **Back** button from the **Profile Sketch** dialog box. Figure-13 shows the edge flange created along the entire length of the selected edge. Figure-14 shows the modified profile sketch and resulting edge flange.

Angle Rollout

The **Angle** rollout is used to define the angle of the flange. The default angle of the flange is 90 degrees. You can define any other angle of the flange by using the **Flange Angle** spinner. The angle of the edge flange can be greater than 0-degree and less than 180 degrees. You can also select a face and specify whether the resulting flange will be parallel or normal to it. Figure-15 shows an edge flange created at an angle of 45 degrees. Figure-16 shows an edge flange created at an angle of 135 degrees.

Figure-13. Edge flange created along the entire length of the edge

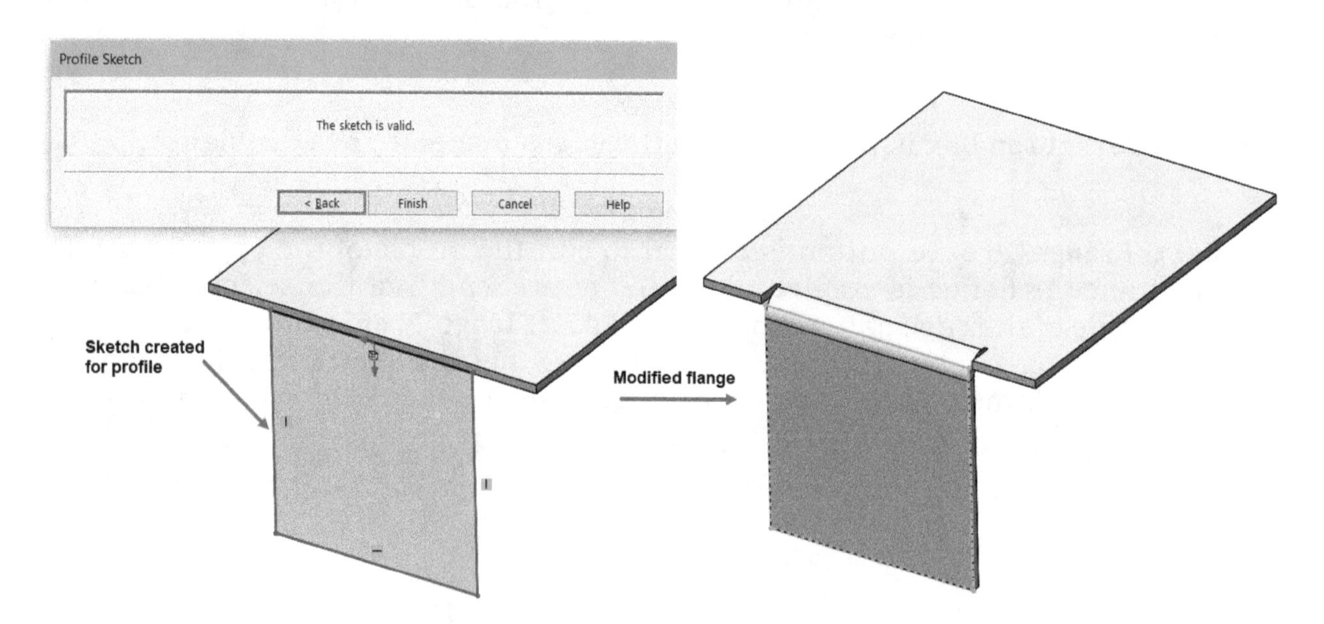

Figure-14. Modified profile sketch and edge flange created

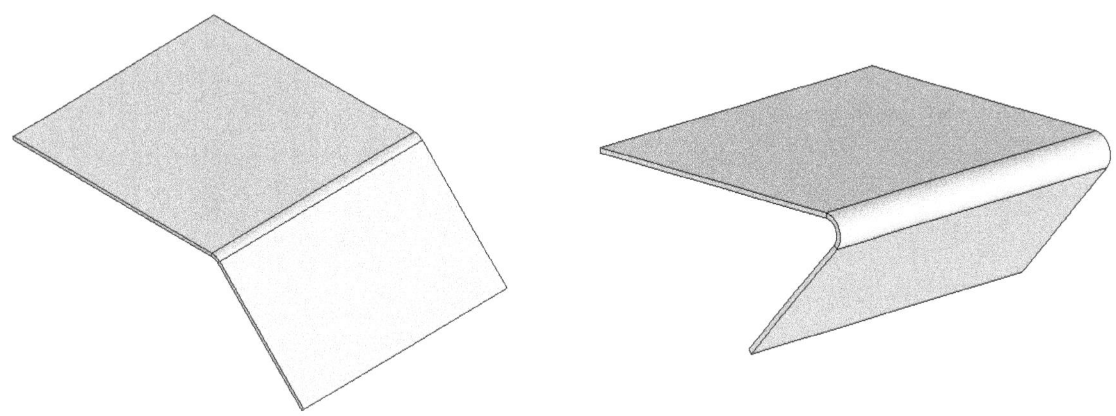

Figure-15. Edge flange created at an angle of 45 degrees *Figure-16. Edge flange created at an angle of 135 degrees*

Flange Length Rollout

The **Flange Length** rollout is used to define the length of the flange. In other words, the options for feature termination are available in this rollout. These options are the same as discussed earlier. The other two options provided in this rollout are discussed next.

Outer Virtual Sharp

The **Outer Virtual Sharp** button is used to define the length of the flange from the outer virtual sharp. The outer virtual sharp is an imaginary vertex created by extending the tangent lines virtually from the outer radius of the bend, as shown in Figure-17.

Inner Virtual Sharp

The **Inner Virtual Sharp** button is chosen by default and is used to define the length of the flange from the inner virtual sharp. The inner virtual sharp is an imaginary vertex created by extending the tangent lines virtually from the inner radius of the bend, as shown in Figure-17.

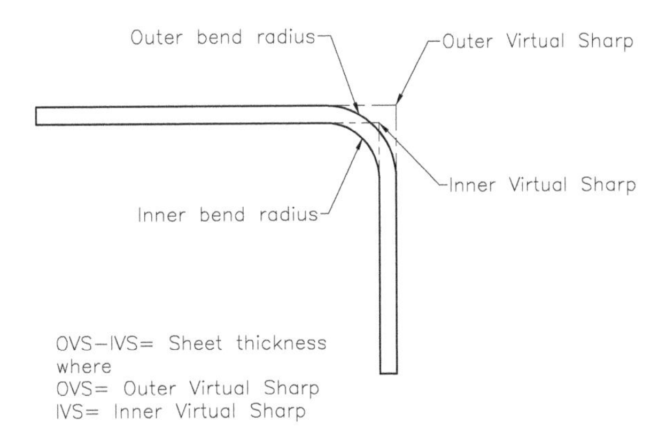

Figure-17. Outer Virtual Sharp and Inner Virtual Sharp

Tangent Bend

The **Tangent Bend** button is used to define the length of the flange from the imaginary line that is created by extending the tangent line from the outer radius of the bend and parallel to the end edge of the flange to be created, refer to Figure-18. This button will be available only for the flange to be created whose bend radius is greater than 90-degree.

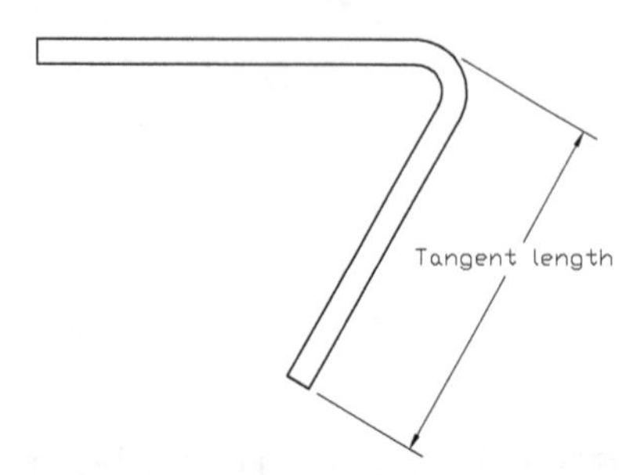

Figure-18. Tangent length of Tangent Bend in base flange

Flange Position Rollout

The **Flange Position** rollout is used to define the position of the flange on an edge. The options in this rollout are discussed next.

Material Inside

The **Material Inside** button is used to create the edge flange in such a way that the material of the flange after the bend lies inside the maximum limit of sheet. Figure-19 shows the edge flange created with the Material Inside button chosen.

Material Outside

The **Material Outside** button is chosen by default and the edge flange is created such that the material of the flange after the bend lies outside the maximum limit of the sheet. Figure-20 shows the edge flange created with the **Material Outside** button chosen.

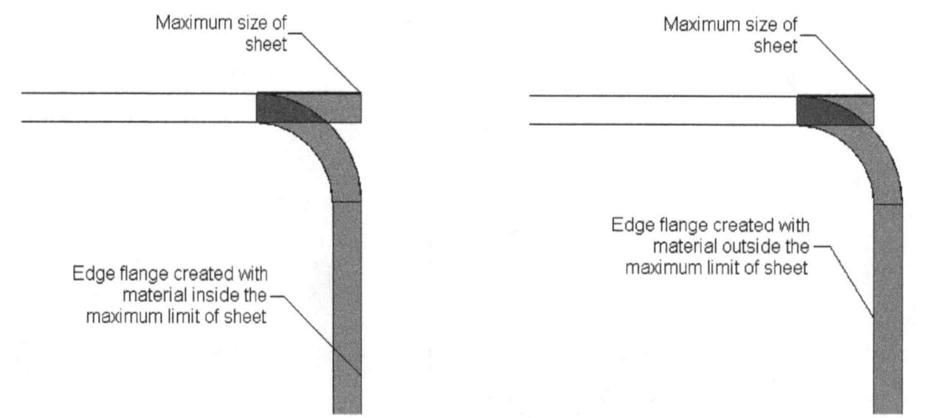

Figure-19. Edge flange created with the Material Inside button chosen

Figure-20. Edge flange created with the Material Outside button chosen

Bend Outside

The **Bend Outside** button is used to create an edge flange such that the bending of the sheet starts from the point that is beyond the maximum limit of the sheet, as shown in Figure-21.

Bend from Virtual Sharp

The **Bend from Virtual Sharp** button is used to create an edge flange with the bending of the sheet starting from the virtual sharp. The position of the flange depends on whether you choose the **Outer Virtual Sharp** button, **Inner Virtual Sharp** button, or the **Tangent Bend** button from the **Flange Length** rollout. Figure-22 shows the edge flange created with the **Inner Virtual Sharp** and **Bend from Virtual Sharp** buttons chosen.

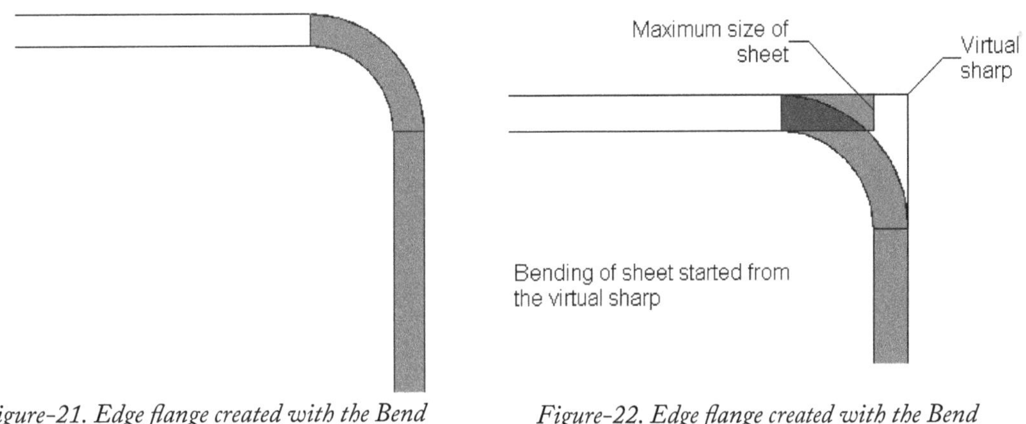

Figure-21. Edge flange created with the Bend Outside button chosen

Figure-22. Edge flange created with the Bend from Virtual Sharp button chosen

Tangent to Bend

The **Tangent to Bend** button is used to create the edge flange in such a way that the material of the flange after bending lies tangent to the maximum limit of the sheet, refer to Figure-23. Note that this option is not valid for the bend angle less than 90 degree.

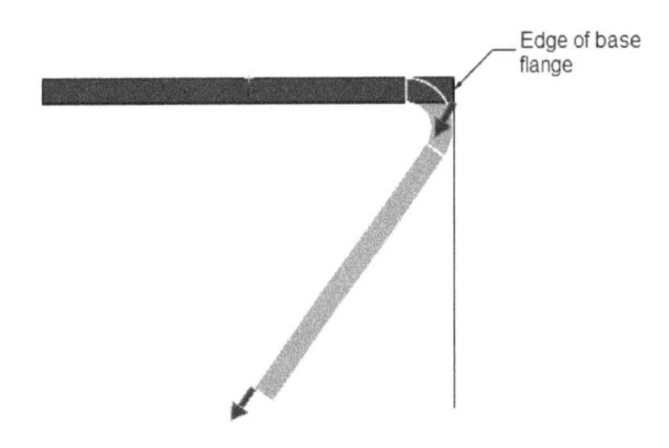

Figure-23. Edge flange created with the Tangent to Bend button chosen

Trim side bends

Select the **Trim side bends** check box to trim extra materials in the bends surrounding the current edge flange. By default, this check box is not selected. Figure-24 shows the edge flange created with the **Trim side bends** check box cleared. Figure-25 shows the edge flange created with the **Trim side bends** check box selected.

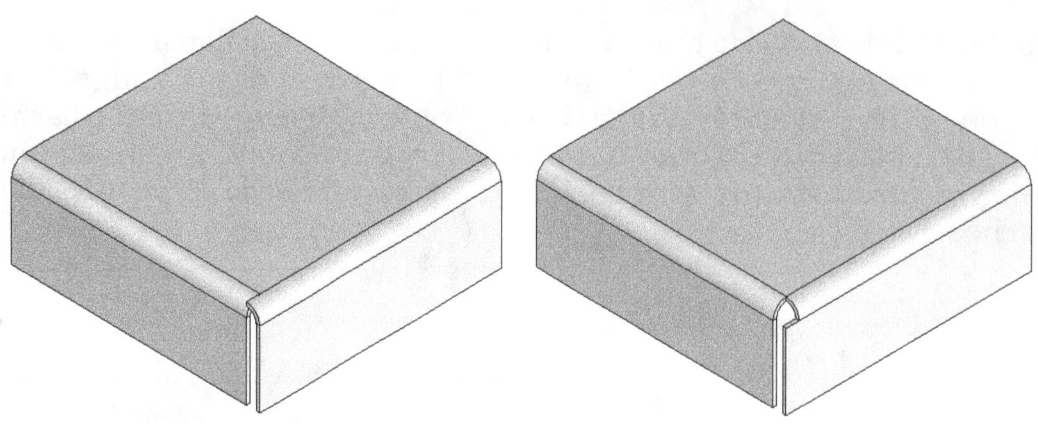

Figure-24. Edge flange created with the Trim side bends check box cleared

Figure-25. Edge flange created with the Trim side bends check box selected

Offset

The **Offset** check box is available only when you create an edge flange using the **Material Inside**, **Material Outside**, **Bend Outside**, or **Tangent to Bend** options. This check box is used to create an edge flange at an offset distance from the selected edge reference. On selecting the **Offset** check box, the **Offset End Condition** drop-down and the **Offset Distance** spinner will be displayed. Specify the offset distance using the spinner. Figure-26 shows the edge flange created with the **Offset** check box cleared. Figure-27 shows the edge flange created with the **Offset** check box selected and the offset distance specified in the **Offset Distance** spinner.

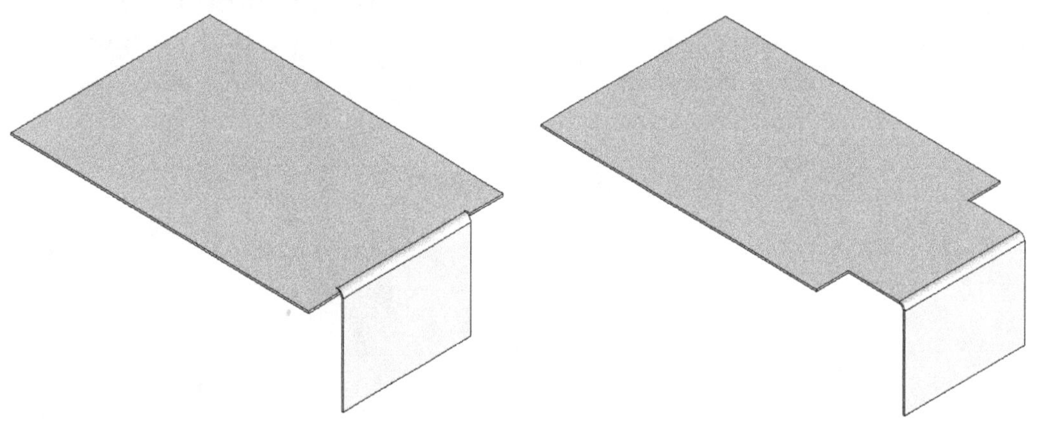

Figure-26. Edge flange created with the Offset check box cleared

Figure-27. Edge flange created with the Offset check box selected

Custom Bend Allowance Rollout

The **Custom Bend Allowance** rollout is used to define the bend allowance other than the default bend allowance that you defined while creating the base flange. To apply the custom bend allowance, expand this rollout by selecting the **Custom Bend Allowance** check box. Then use the options in this rollout to define the bend allowance for the current bend as discussed earlier.

Custom Relief Type Rollout

The **Custom Relief Type** rollout is used to define the type of relief other than the default that was defined while creating the base flange. To apply the custom relief, expand this rollout by selecting the check box in the title bar of the **Custom Relief Type** rollout, as shown in Figure-28.

The types of reliefs that can be defined for a sheet metal component are discussed next.

Obround Relief

The **Obround** option is used to provide the obround relief such that the edges of the relief merging with the sheet are rounded. The **Use relief ratio** check box is selected by default. Therefore, you can modify the value of the relief ratio by setting the value in the **Relief Ratio** spinner. If you clear the **Use relief ratio** check box, the **Relief Width** and **Relief Depth** spinners will be displayed, as shown in Figure-29. You can modify the relief width and relief depth individually by using these two spinners.

Figure-30 shows the edge flange created by providing the obround relief with the default relief ratio. Figure-31 shows the edge flange created by providing obround relief after modifying the relief ratio.

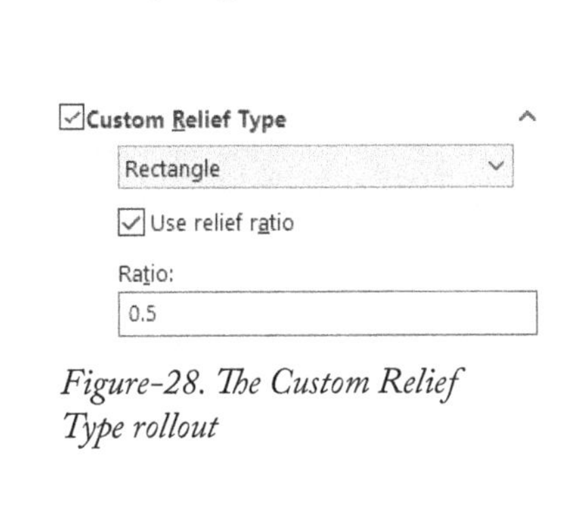

Figure-28. The Custom Relief Type rollout

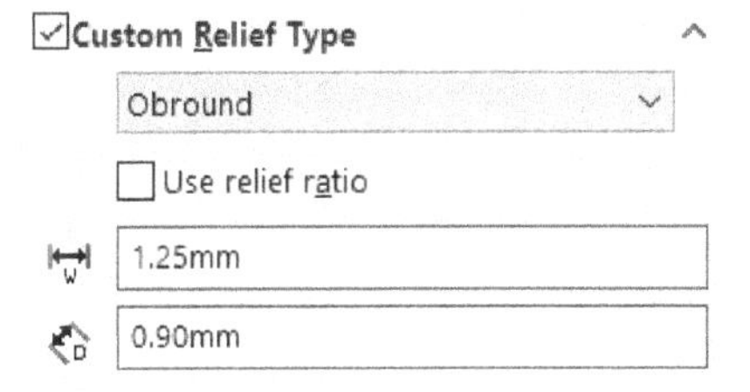

Figure-29. The Relief Width and Relief Depth spinners displayed in the Custom Relief Type rollout

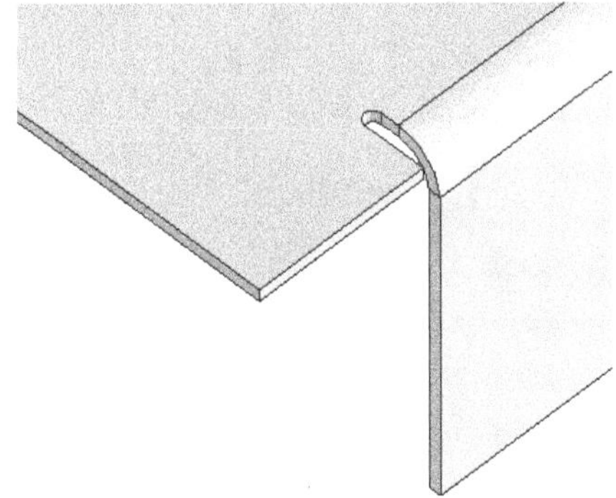

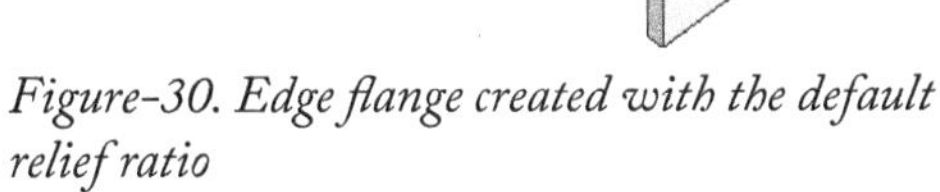

Figure-30. Edge flange created with the default relief ratio

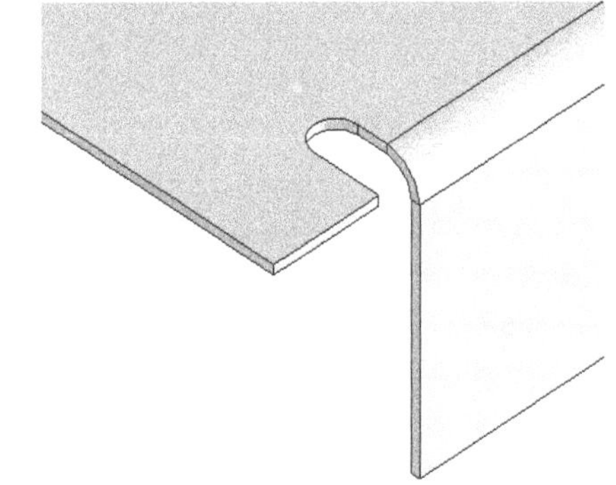

Figure-31. Edge flange created after modifying the relief ratio

Rectangle Relief

The **Rectangle** option is selected by default in this rollout. This option is used to provide the rectangular relief to the sheet metal components. The options for defining the rectangular relief are the same as discussed in the previous paragraph. Figure-32 shows an edge flange created by providing the rectangular relief with the default relief ratio. Figure-33 shows an edge flange created by providing rectangular relief after modifying the relief ratio.

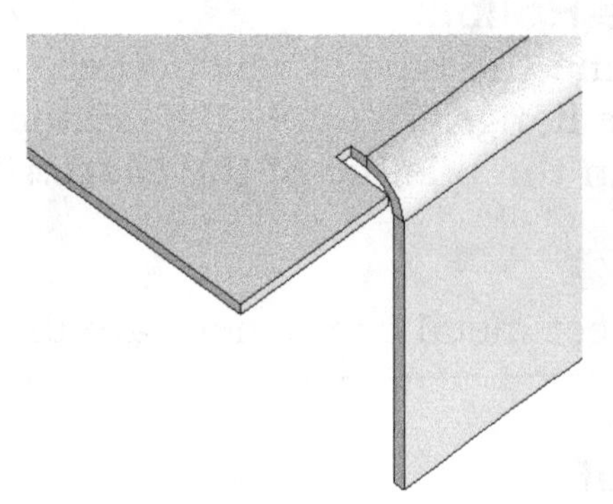

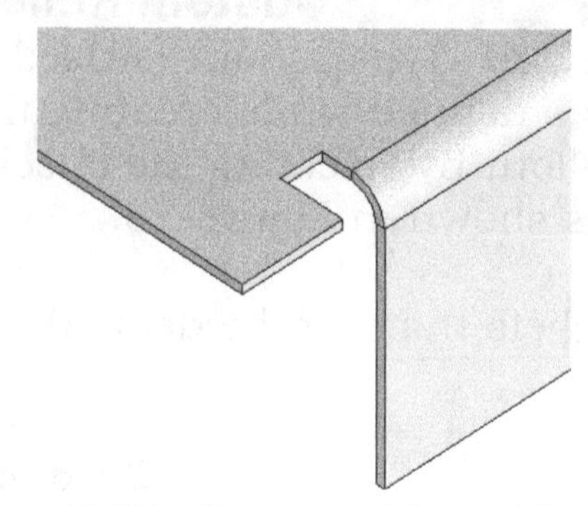

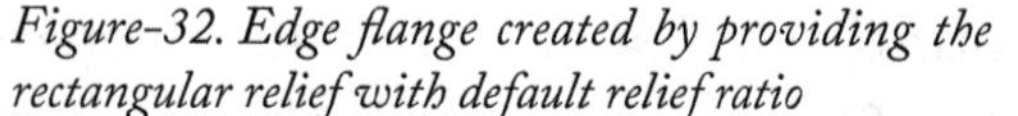

Figure-32. Edge flange created by providing the rectangular relief with default relief ratio

Figure-33. Edge flange created by providing the rectangular relief after modifying the relief ratio

Tear Relief

You can provide the tear relief to an edge flange by using the **Tear** option. The tear relief will tear the sheet in order to accommodate the bending of the sheet. When you select the **Tear** option from the **Relief Type** drop-down list, all the other options are replaced by the **Rip** and **Extend** buttons, as shown in Figure-34.

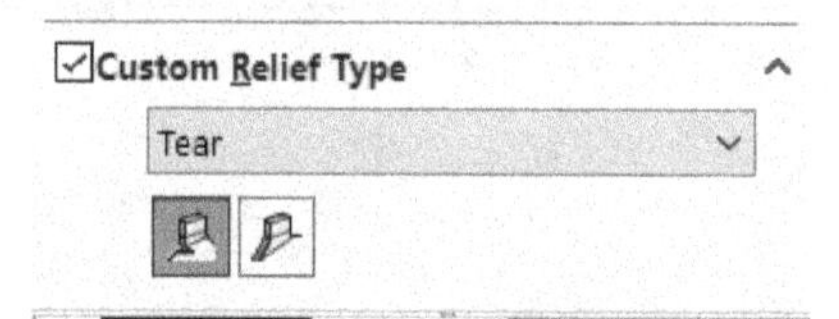

Figure-34. The Custom Relief Type rollout with the Tear option selected from the Relief Type drop-down list

The **Rip** button is chosen by default. This option rips or tears the sheet to accommodate the bending of the sheet, as shown in Figure-35. When the **Extend** button is chosen, the outer faces of the bend will be extended to the outer faces of the sheet on which you create the edge flange, as shown in Figure-36.

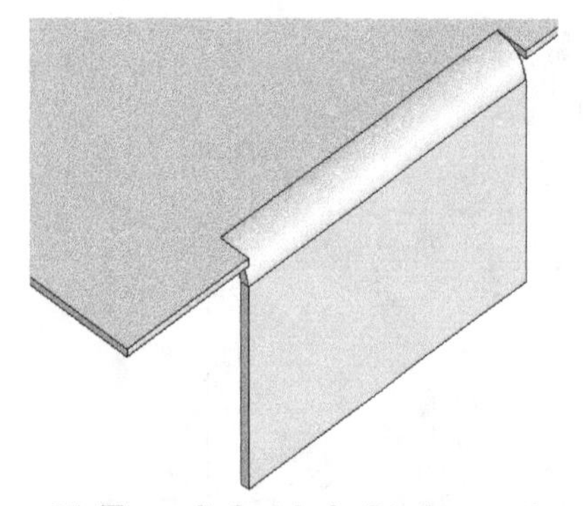

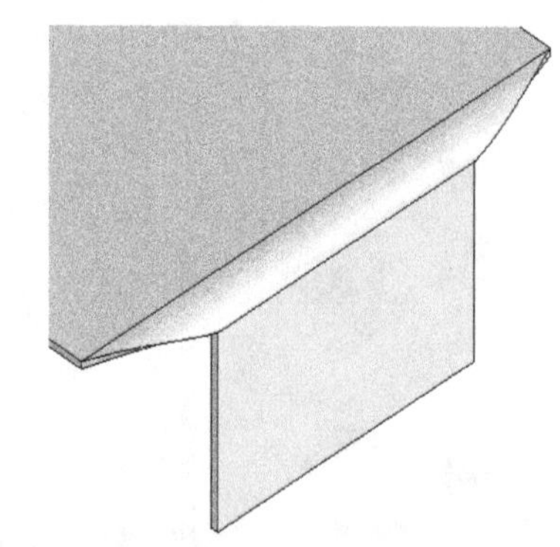

Figure-35. Tear relief with the Rip button chosen

Figure-36. Tear relief with the Extend button chosen

MITER FLANGE

The **Miter Flange** tool is used to create a flange of specified shape. This type of flange is best used in creating tray type shapes. The steps to create miter flange are given next.

- Click on the **Miter Flange** tool from the **Ribbon**. You are asked to select a plane/edge to define sketching plane.

- Select the edge on which you want to create flange. The sketching environment will display.
- Create an open sketch to define shape of flange; refer to Figure-37 and exit the sketching environment. Preview of flange will display; refer to Figure-38.

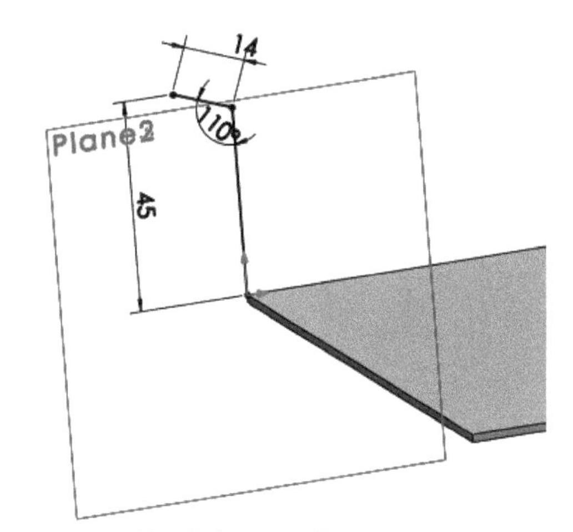

Figure-37. Sketch for miter flange

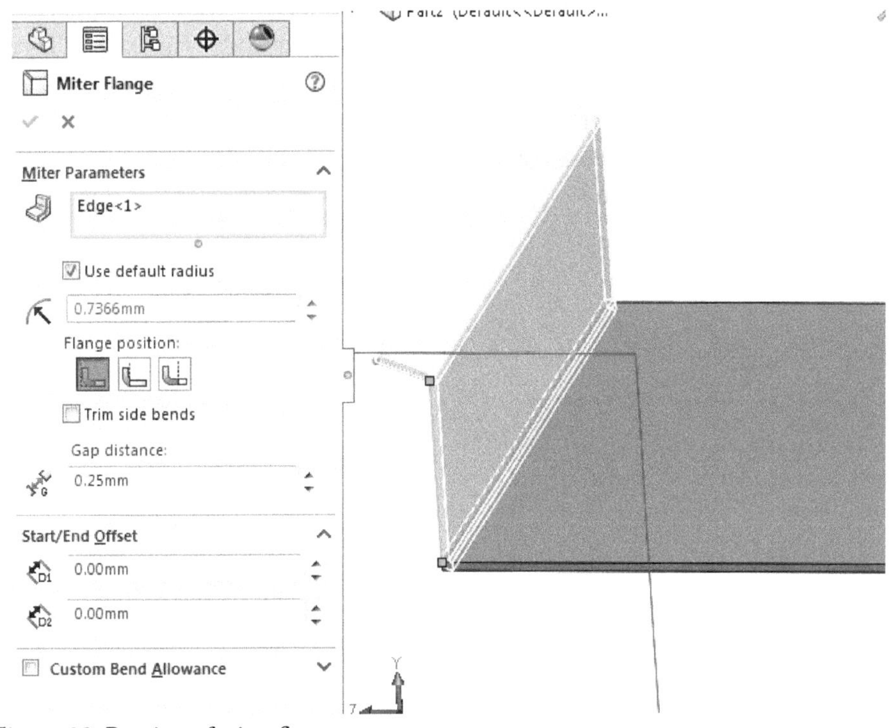

Figure-38. Preview of miter flange

- If you want to create tray then select all the edges of the current loop; refer to Figure-39.
- Click on the **OK** button to create the flange.

Some of the important options in the **PropertyManager** are explained next.

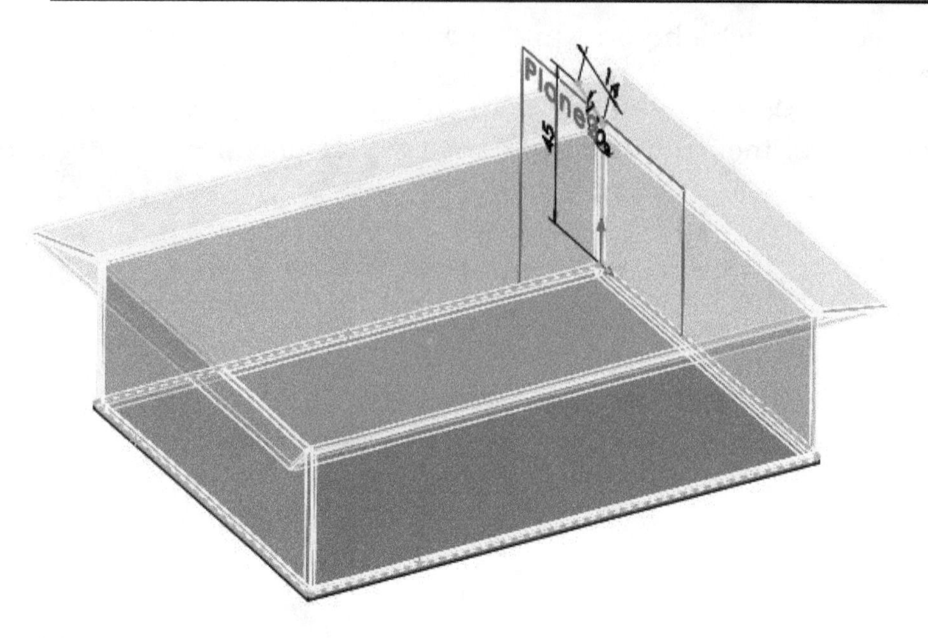

Figure-39. Closed loop of miter flange

Gap distance Area

The spinner in this area is used to define the rip distance between two consecutive flanges. Set the value in the **Rip Gap** spinner to modify the distance value of the rip. Figure-40 shows the miter flange created using the default distance value and Figure-41 shows the miter flange created using the modified rip distance. In case of error enter the higher gap value.

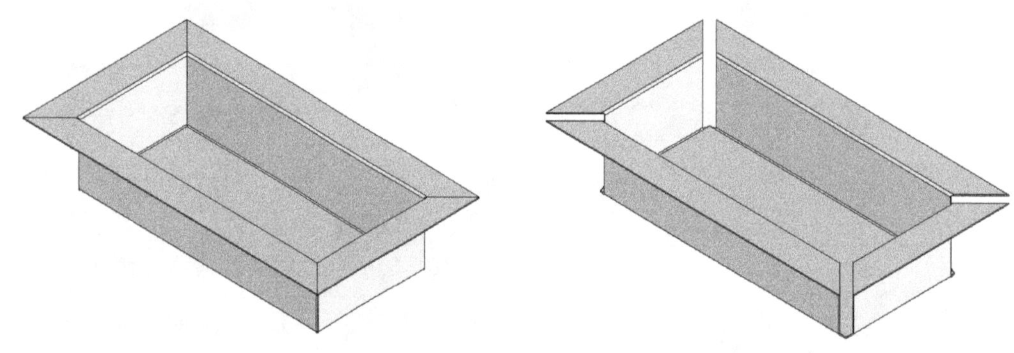

Figure-40. Miter flange with the default rip distance

Figure-41. Miter flange with the modified rip distance

Start/End Offset Rollout

You can specify the start and end offset distances of the miter flange by using the options in the **Start/End Offset** rollout. The **Start Offset Distance** spinner is used to specify the offset distance from the start face of the miter flange. The **End Offset Distance** spinner is used to specify the offset distance from the end face of the miter flange. If the start and end offset distances are applied to the miter flange created on the continuous edges of the base flange, the start offset distance will be applied to the first edge and the end offset distance will be applied to the edge selected at last. Figure-42 shows the miter flange created on a single edge with the start and end offsets. Figure-43 shows the offsets applied to the miter flange created by selecting all the edges of the base flange.

Figure-42. Miter flange created at an offset distance on a single edge

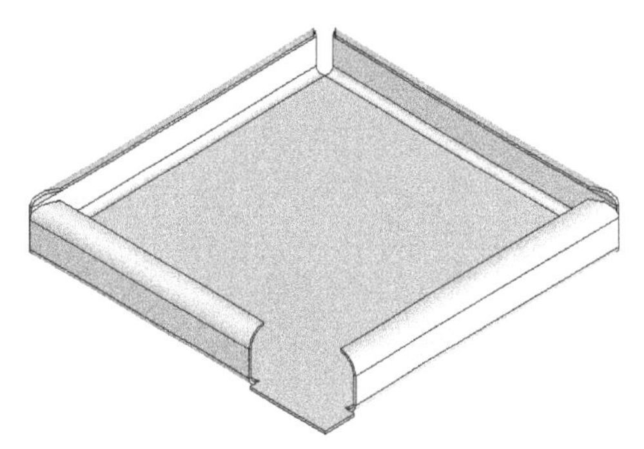

Figure-43. Miter flange created at an offset distance on all edges

HEM

The **Hem** tool is used to create bend at the end edge of the sheet. The steps to create hems are given next.

- Click on the **Hem** tool from the **Ribbon**. The **Hem PropertyManager** will display; refer to Figure-44.

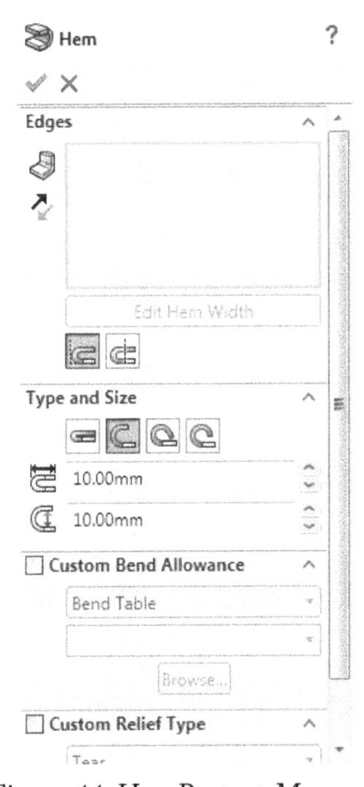

Figure-44. Hem PropertyManager

- Select the edge on which you want to create the hem. Preview of hem will display; refer to Figure-45.
- Select desired shape using the **Closed**, **Open**, **Tear Drop**, or **Rolled** button from the **Type and Size** rollout.
- Specify the size and other parameters and click on the **OK** button from the **PropertyManager** to create the hem.

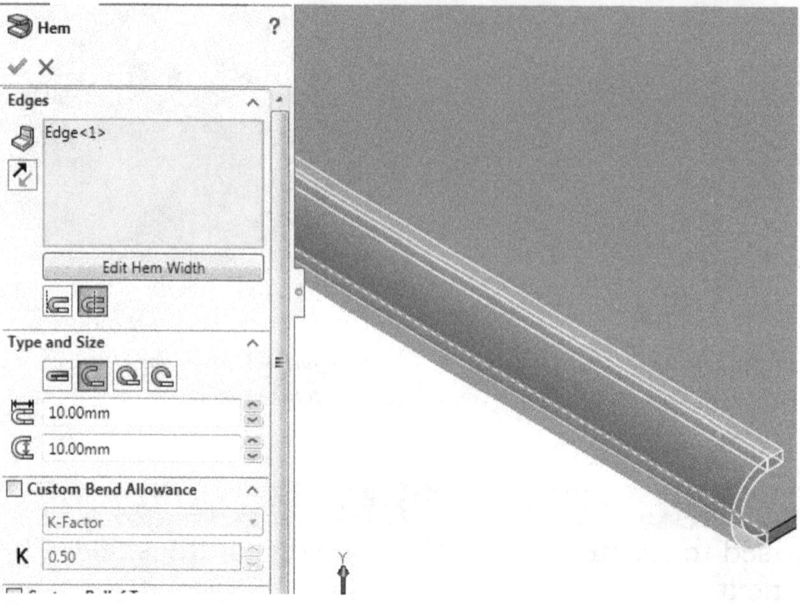

Figure-45. Preview of hem

JOG

The **Jog** tool can be used to create double bend in the sheet. The steps to create jog are given next.

- Click on the **Jog** tool from the **Ribbon**. You are asked to select a plane for creating sketch.
- Sketch the bend line on the sheet metal face and exit the sketching environment. You are asked to select a face that you want to be fixed.
- Select the fixed face. The preview of jog will display; refer to Figure-46.

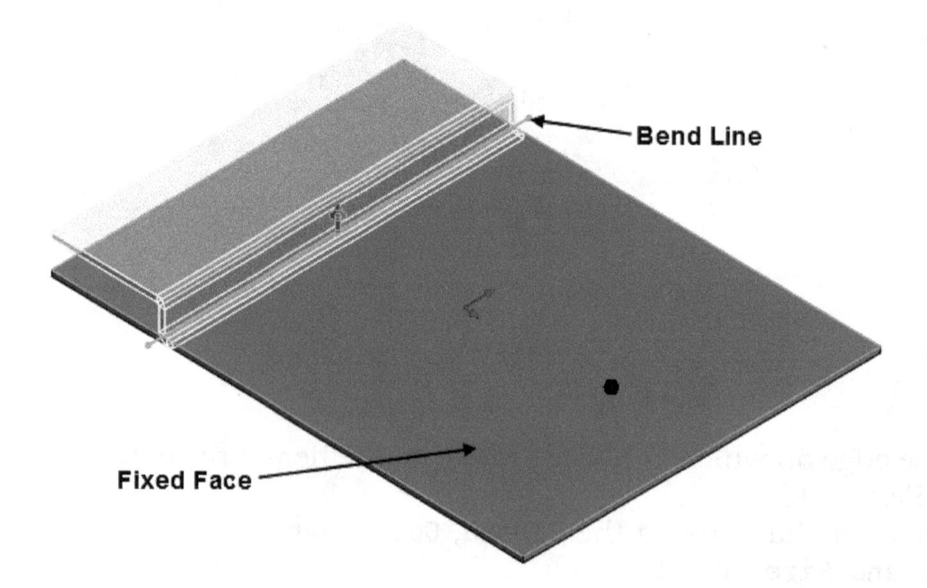

Figure-46. Preview of Jog

- Specify desired parameters in the **PropertyManager** displayed and click on the **OK** button to create the bend.

SKETCHED BEND

The **Sketched Bend** tool is used to bend a sheet metal part by specified angle at selected bend line. The steps to bend a part are given next.

- Click on the **Sketched Bend** tool from the **Ribbon**. You are asked to select a flat face to bend.
- Select the face of sheet metal part. The sketching environment will display.
- Draw the bend line in such a way that it intersect to the edges of the sheet metal part's face.
- Exit the sketch and specify the parameters in the **PropertyManager**.
- Click on the **OK** button from the **PropertyManager**. The bend will be created with specified parameters; refer to Figure-47.

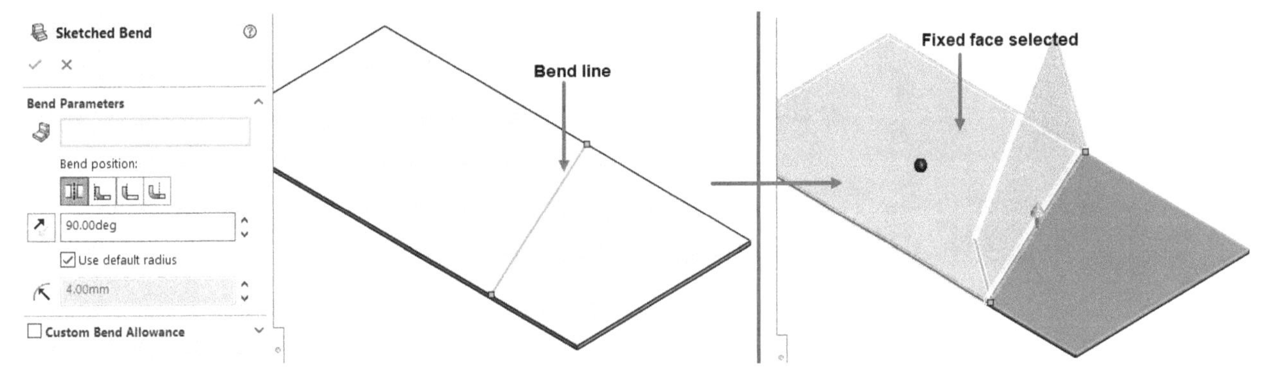

Figure-47. Sketched Bend

CROSS-BREAK

The **Cross-Break** tool is used to create cross-break in the sheet metal part. In HVAC or other duct works, cross-breaks are provided to stiffen the sheet. The procedure to use this tool is given next.

- Click on the **Cross-Break** tool from the **Sheet Metal** tab in the **Ribbon**. The **Cross Break PropertyManager** will be displayed; refer to Figure-48.
- Select the face on which you want to apply the cross-break. Preview of the cross-break will be displayed; refer to Figure-49.

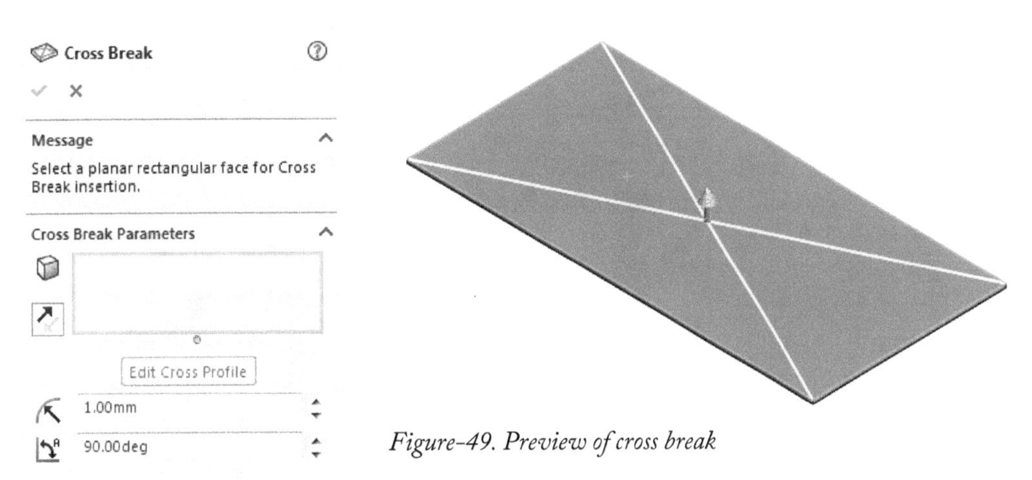

Figure-48. Cross Break Property-Manager

Figure-49. Preview of cross break

- Set desired parameters and click on the **OK** button from the **PropertyManager**.

SWEPT FLANGE

The **Swept Flange** tool is used to create a flange by sweeping sketch along selected path. This tools works in the same way as **Swept Boss/Base** tool does. The procedure to use this tool is given next.

- For using the **Swept Flange** tool, you need a sketch for section and a sketch or edge for path. So, make sure you have both section and path sketches already created.
- Click on the **Swept Flange** tool from the **Sheet Metal CommandManager** in the **Ribbon**. The **Swept Flange PropertyManager** will be displayed; refer to Figure-50.

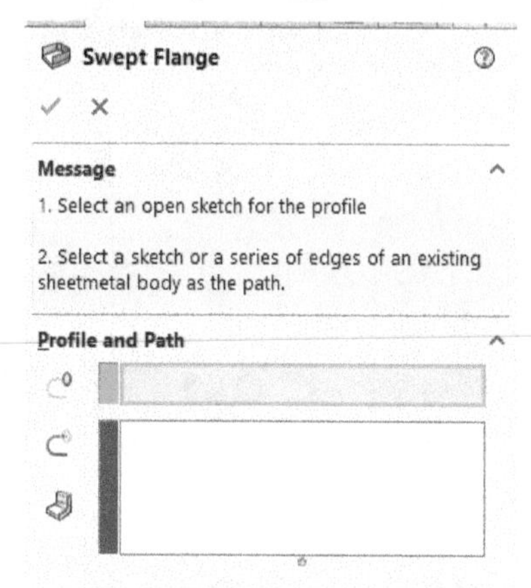

Figure-50. Swept Flange PropertyManager

- Select the sketch created for profile. Note that the sketch for profile should be an open sketch. On selecting the sketch, you will be asked to select an edge(s) or sketch for path.
- Select desired open/closed sketch or edge(s). The preview of swept flange will be displayed; refer to Figure-51. Note that you can select multiple edges if they form continuous chain.

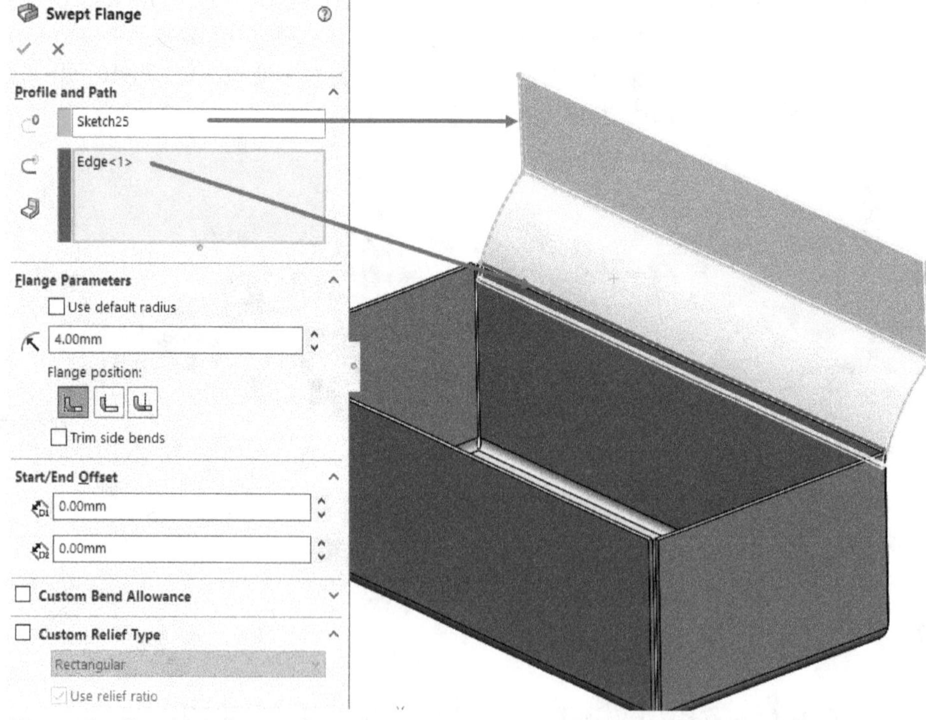

Figure-51. Preview of swept flange

- Set the other parameters as discussed earlier and click on the **OK** button from the **PropertyManager**.

CLOSE CORNER

The **Close Corner** tool is used to change the closing of flange walls at corners. The steps to use this tool are given next.

- Click on the **Closed Corner** tool from **Corners** drop-down in the **Ribbon**. The **Closed Corner PropertyManager** will be displayed.
- Select the face at corner that you want to close. Preview will be displayed; refer to Figure-52.

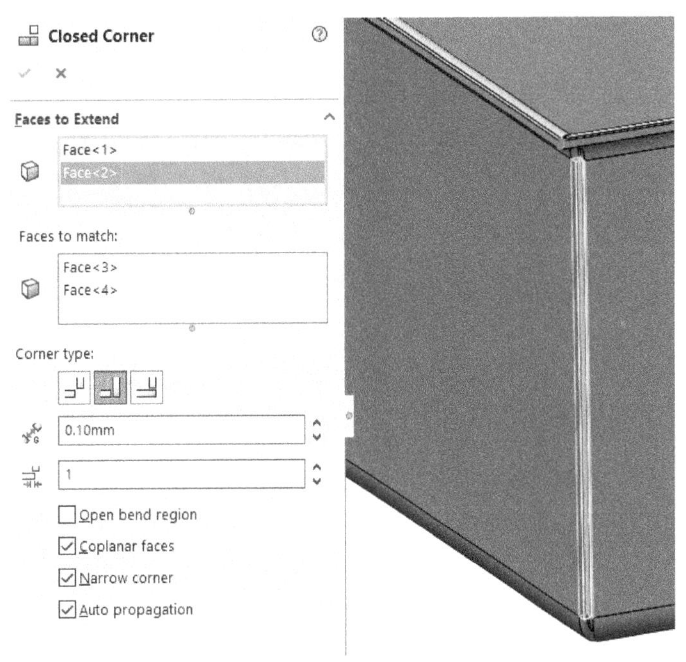

Figure-52. Preview of closed corner

- Set desired options in the **PropertyManager** and click on the **OK** button to create closed corner.

WELDED CORNER

The **Welded Corner** tool is used to weld the gap between two adjacent walls. The procedure to use this tool is similar to **Close Corner** tool. The procedure is discussed next.

- Click on the **Welded Corner** tool from the **Corners** drop-down in the **Ribbon**; refer to Figure-53. The **Welded Corner PropertyManager** will be displayed; refer to Figure-54.

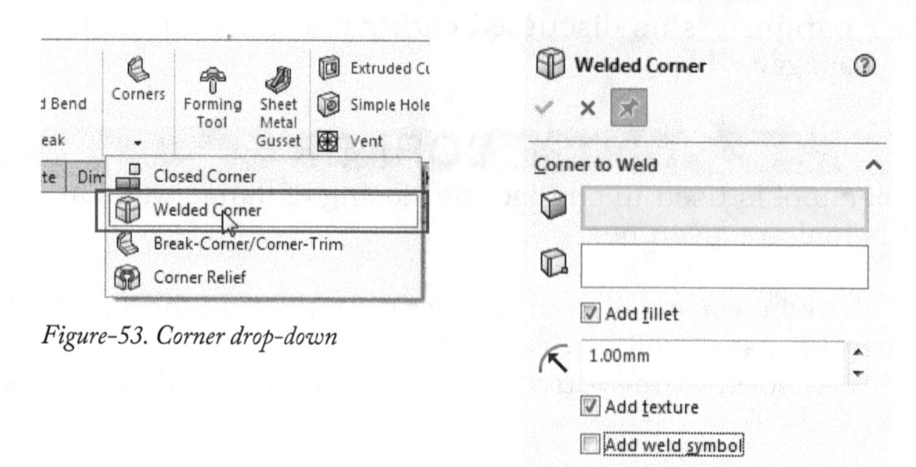

Figure-53. Corner drop-down

Figure-54. Welded Corner PropertyManager

- Select the side face of sheet metal wall to be closed by welding bead. Preview of the welded corner will be displayed; refer to Figure-55.

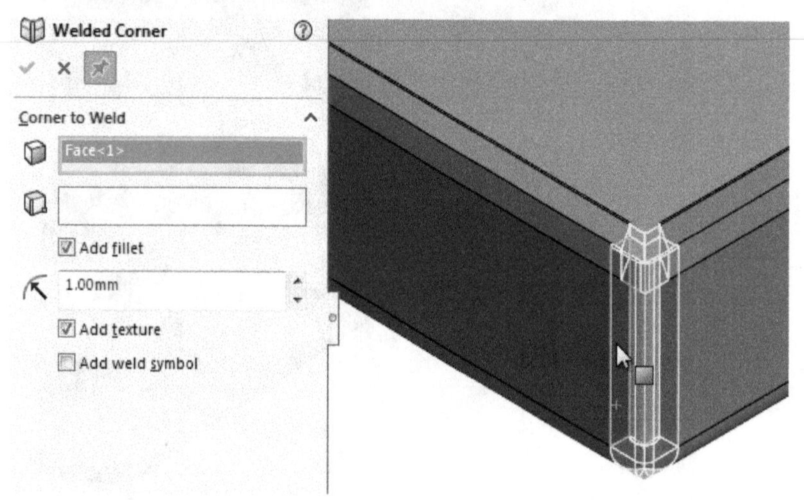

Figure-55. Preview of welded corner

- Set the fillet radius and other parameters in the **PropertyManager** and click on the **OK** button to create the welded corner.

BREAK-CORNER/CORNER-TRIM

The **Break-Corner/Corner-Trim** tool is used to chip-off the sharp corners of the sheet metal component. The procedure to use this tool is given next.

- Click on the **Break-Corner/Corner-Trim** tool from the **Corners** drop-down in the **Ribbon**. The **Break Corner PropertyManager** will be displayed; refer to Figure-56.
- Select the sharp edge of the sheet metal component. Preview of break corner will be displayed; refer to Figure-57.

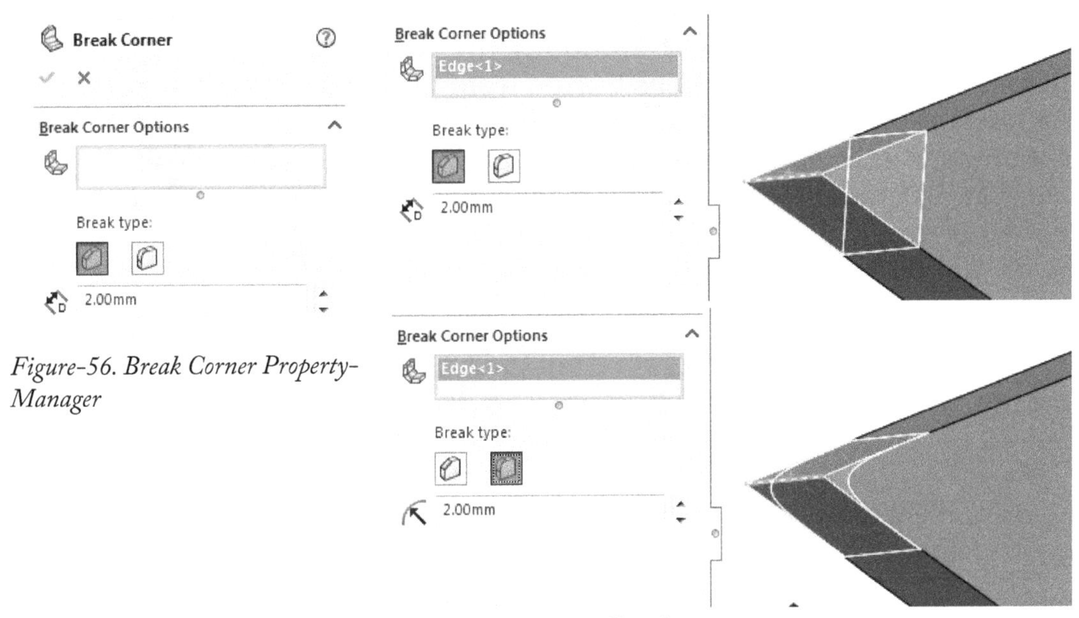

Figure-56. Break Corner PropertyManager

Figure-57. Preview of break corner

- Set desired parameters and click on the **OK** button from the **PropertyManager** to create the corner break.

CORNER RELIEF

The **Corner Relief** tool is used to apply bend relief at the corners of sheet metal parts. The procedure to use this tool is given next.

- Click on the **Corner Relief** tool from the **Corners** drop-down in the **Ribbon**. The **Corner Relief PropertyManager** will be displayed; refer to Figure-58.

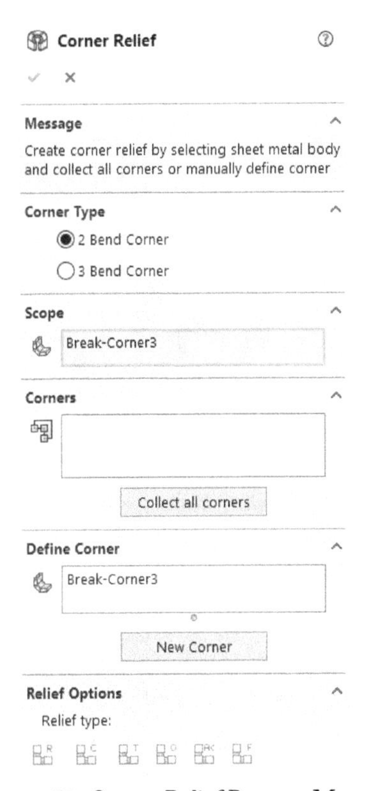

Figure-58. Corner Relief PropertyManager

- Select the **2 Bend Corner** or **3 Bend Corner** radio button from the **Corner Type** rollout of the **PropertyManager** to modify the respective type of corners.
- Click on the **Collect all corners** button from the **PropertyManager**. All the corners in the sheet metal model will be displayed in the selection box in **PropertyManager**.
- Select one or all the corners from the selection box and select desired button from the **Relief Options** rollout in the **PropertyManager**. The respective type of corner relief will be applied to the sheet metal body.
- Click on the **OK** button from the **PropertyManager** to apply selected corner relief.

FORMING TOOL

The **Forming Tool** option is used to used to convert the model into a sheetmetal forming tool for layer use. The procedure to use this tool is given next.

- Click on the **Forming Tool** from the **Sheet Metal CommandManager**. The **Form Tool PropertyManager** will be displayed; refer to Figure-59.

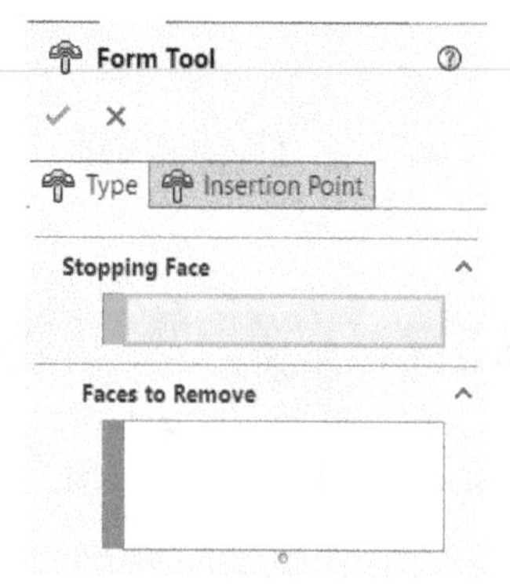

Figure-59. Form Tool PropertyManager

- Select the stopping face and faces to be removed; refer to Figure-60.

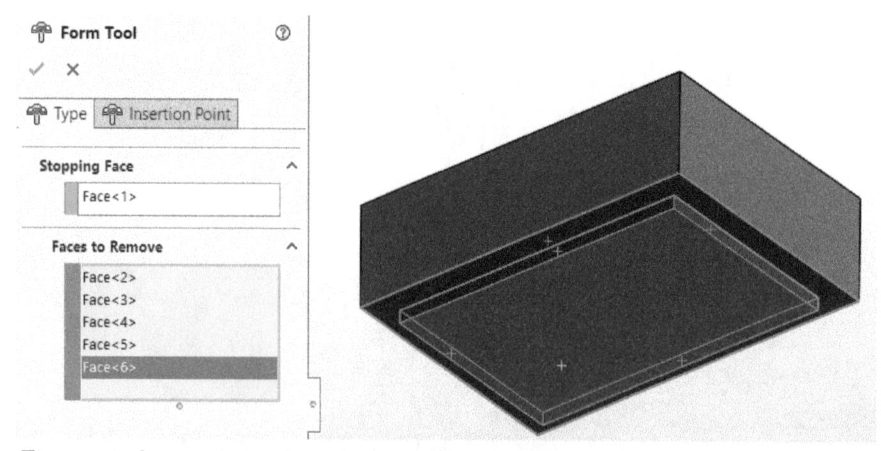

Figure-60. Converting model to forming tool

- Click on the **OK** button from the **PropertyManager** to convert the model into forming tool.

SHEET METAL GUSSET

The **Sheet Metal Gusset** tool is used to add rib to support any sheet metal flange. The steps to create sheet metal gusset are given next.

- Click on the **Sheet Metal Gusset** tool from the **Ribbon**. The **Sheet Metal Gusset PropertyManager** will display; refer to Figure-61.
- Select the two faces (one of base and other of flange). The preview of gusset will display; refer to Figure-62.
- Set desired parameters and click on the **OK** button from the **PropertyManager** to create the gusset.

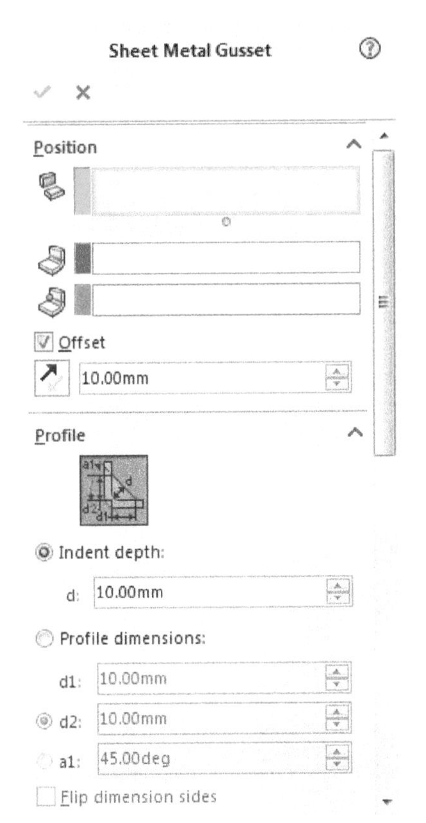

Figure-61. Sheet Metal Gusset PropertyManager

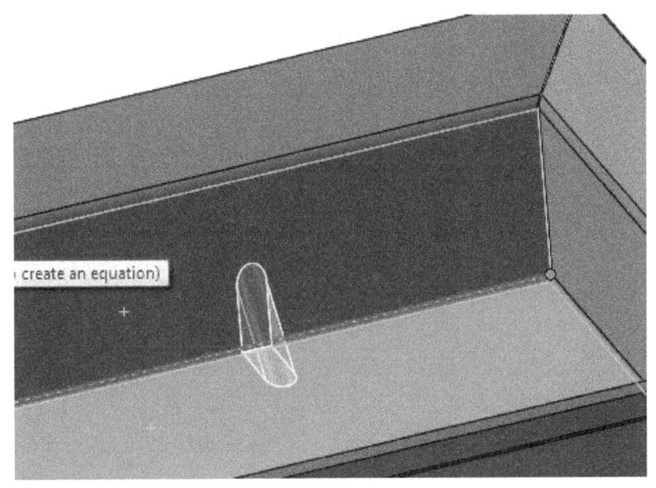

Figure-62. Preview of gusset

TAB AND SLOT

The **Tab and Slot** tool is used to create tab and slot in the sheet metal flanges to connect them in assembly. The procedure to create tab and slot is given next.

- Click on the **Tab and Slot** tool from the **Sheet Metal CommandManager** in the **Ribbon**. The **Tab and Slot PropertyManager** will be displayed; refer to Figure-63.
- Set desired parameters in various rollouts of the **PropertyManager** before selecting references like shape & size of tabs and slots.
- Click on the edge at which you want to create tabs and slots; refer to Figure-64.

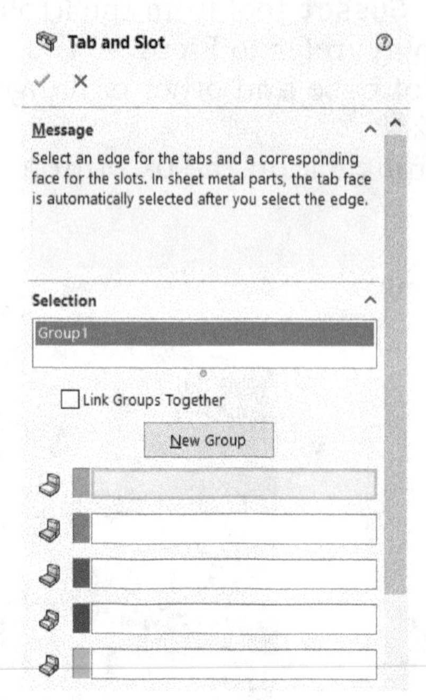

Figure-63. Tab and Slot PropertyManager

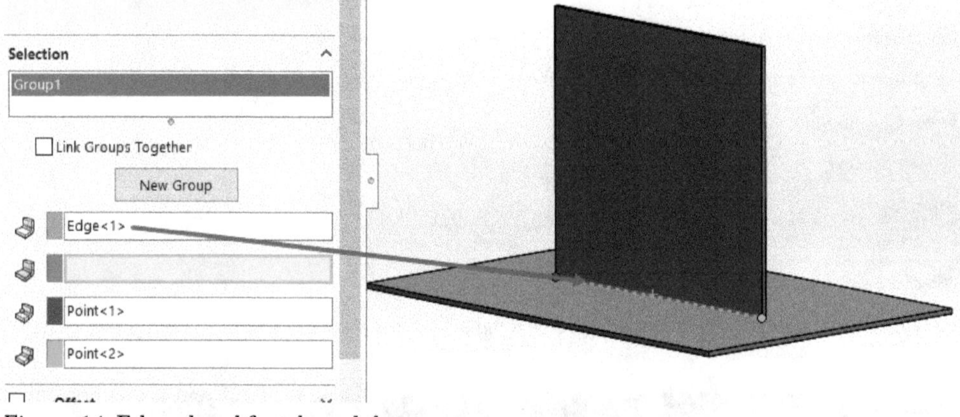

Figure-64. Edge selected for tabs and slots

- Select the face on which you want to create slots. Preview of the tabs and slots will be displayed; refer to Figure-65.

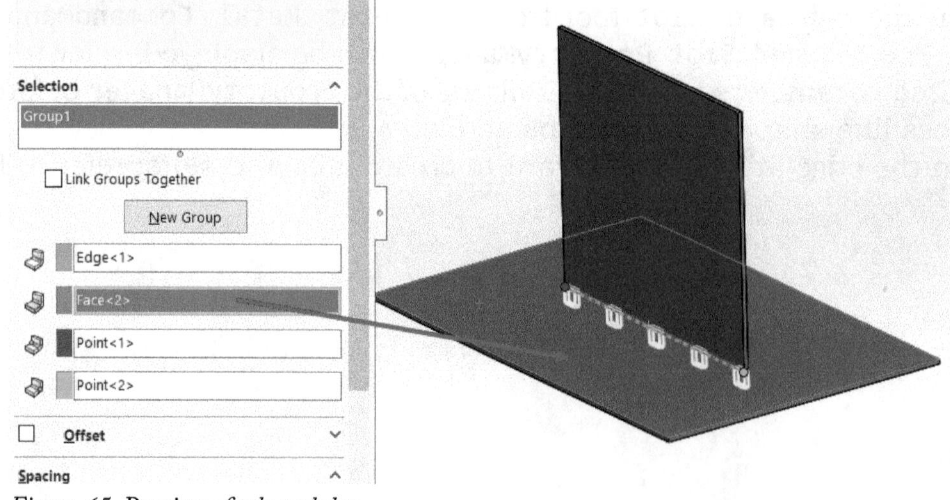

Figure-65. Preview of tabs and slots

- Click on the **OK** button from the **PropertyManager** to create the tabs and slots.

EXTRUDE CUT

The **Extrude Cut** tool is used to create cuts in the sheet metal part by extrusion. The steps to create extruded cut are given next.

- Click on the **Extruded Cut** tool from the **Ribbon**. You are asked to select the face of sheet metal part.
- Click on the face and draw the closed sketch of the cut.
- Exit the sketch and click on the **OK** button to create the cut.

VENT

The **Vent** tool is used to create vents in the sheet metal part for air/material circulation. The procedure to use this tool is given next.

- Create a sketch for vent feature on the sheet metal face which should have geometry for boundary, ribs, spars, and fill-in boundary; refer to Figure-66.

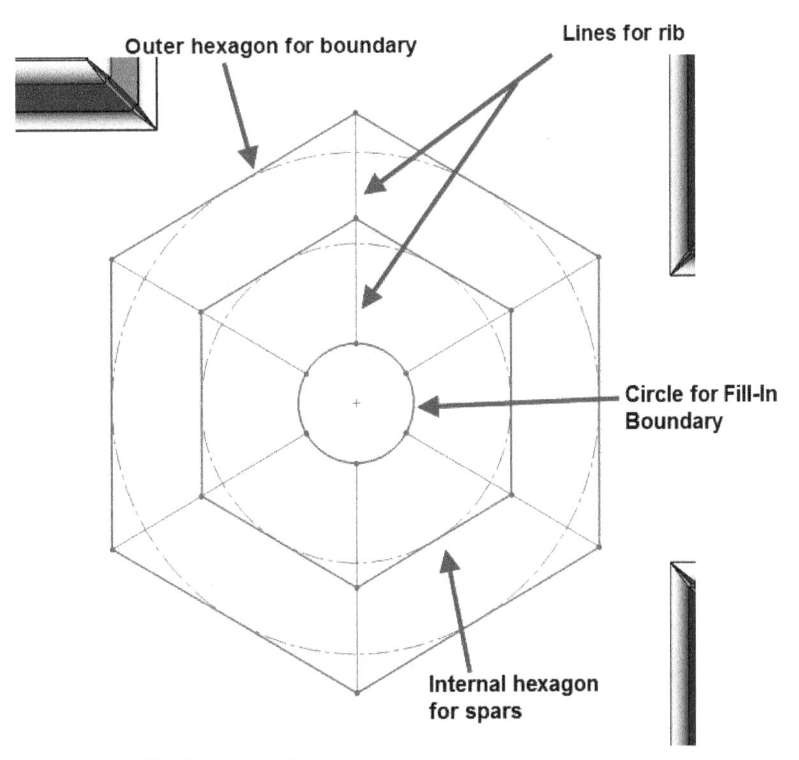

Figure-66. Sketch for vent feature

- Click on the **Vent** tool from the **Sheet Metal CommandManager** in the **Ribbon**. The **Vent PropertyManager** will be displayed; refer to Figure-67. Also, you will be asked to select the entities to be used as boundary of vent.

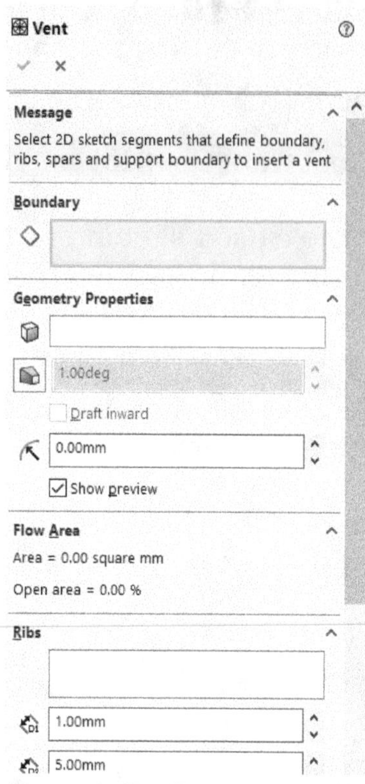

Figure-67. Vent PropertyManager

- Select the closed sketch entity to be used as boundary of vent.
- Set the radius and draft values in the **Geometry Properties** rollout if you wish to.
- Scroll-down in the **PropertyManager** and click in the **Ribs** selection box. Click on the open sketches to create rib of vent; refer to Figure-68.

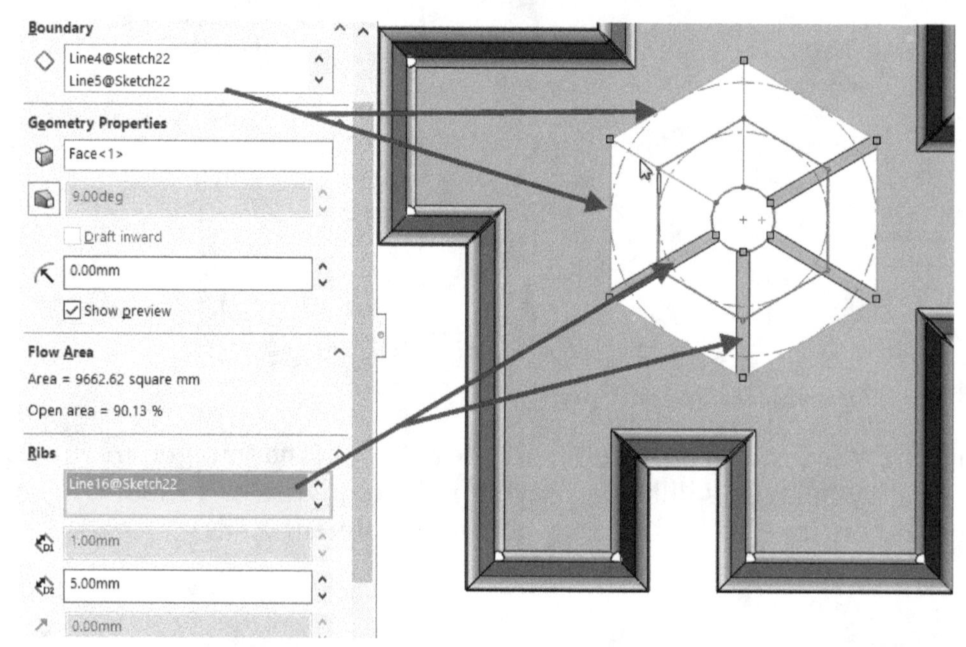

Figure-68. Preview of vent ribs

- Set the width of rib in the edit box of **Ribs** rollout in **PropertyManager**.
- Click in the **Spars** selection box and select the sketch(s)for spars. Preview will be displayed. Set desired width of spars.

- Click in the **Fill-In Boundary** selection box and select the closed region created for fill-in boundary; refer to Figure-69.

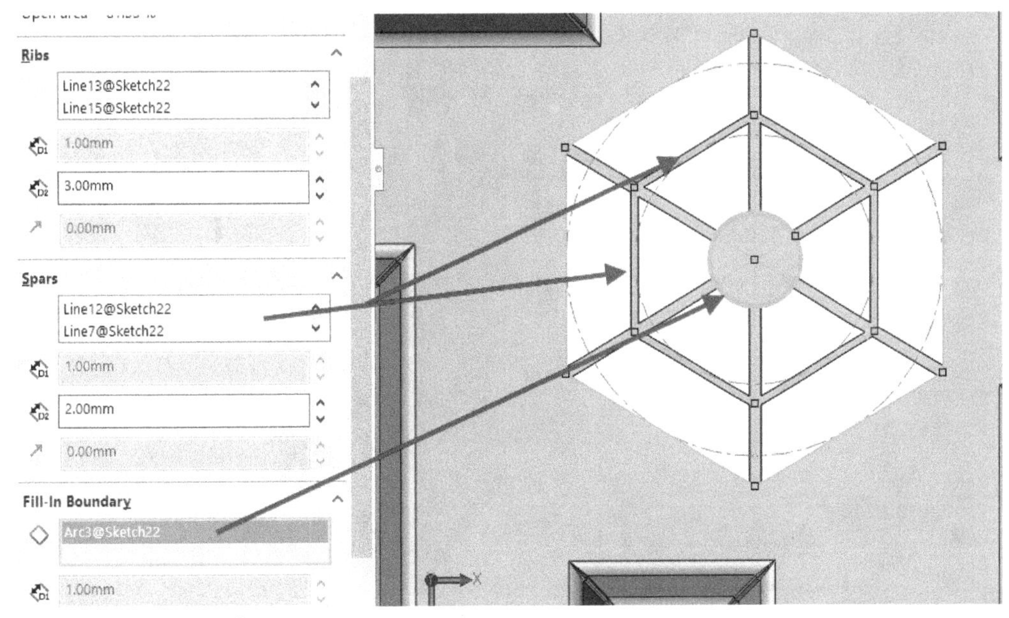

Figure-69. Preview of vent

- Set desired parameters and click on the **OK** button to create the vent.

NORMAL CUT

The **Normal Cut** tool is used to make cut created on bends normal to the faces. The procedure to use this tool is given next.

- Click on the **Normal Cut** tool from the **Sheet Metal CommandManager** in the **Ribbon**. The **Normal Cut PropertyManager** will be displayed; refer to Figure-70.

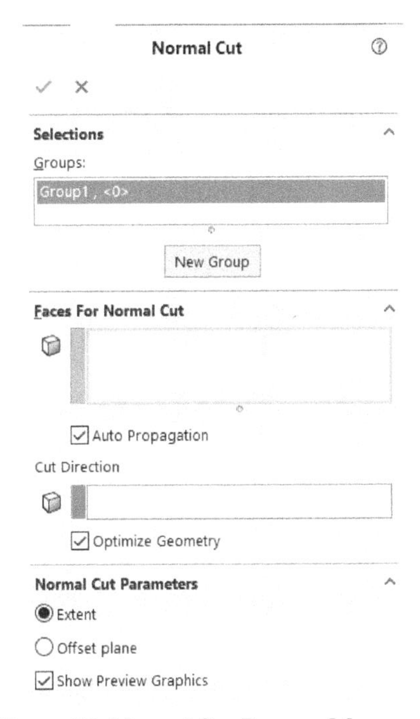

Figure-70. Normal Cut PropertyManager

- Select the face of cut created on the bend. Preview of cut will be displayed; refer to Figure-71.

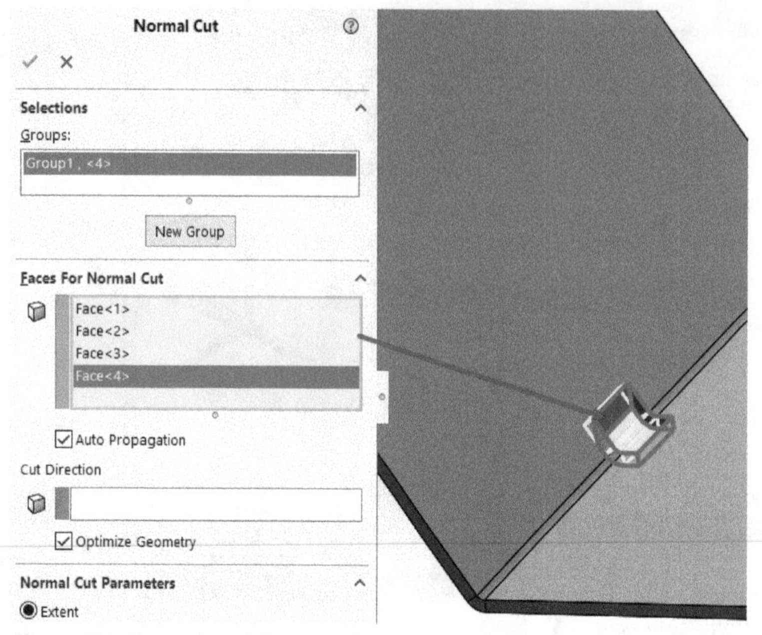

Figure-71. Faces selected for normal cut

- Set the other parameters as desired and click on the **OK** button from the **PropertyManager**.

UNFOLD TOOL AND FOLD TOOL

The **Unfold** tool is used to unbend the selected bends. While the **Fold** tool does the reverse of unfold tool. After unbending, you can perform editing which will be reflected in the folded part. In most of the cases, the **Unfold** and **Fold** tools work together. The procedure to use unfold tool is given next.

Unfold Tool

- Click on the **Unfold** tool from the **Sheet Metal CommandManager** in the **Ribbon**. The **Unfold PropertyManager** will be displayed as shown in Figure-72.

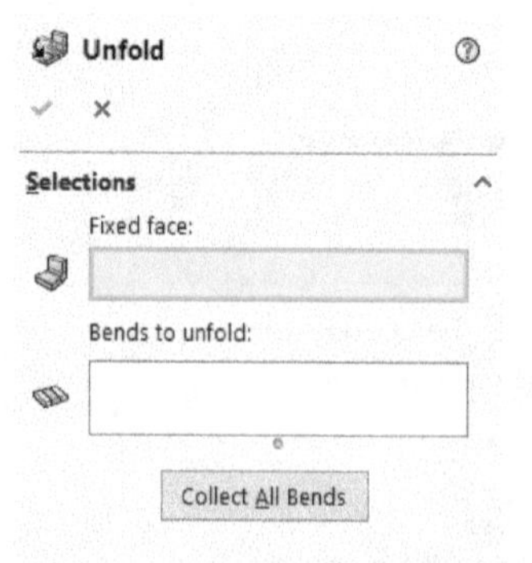

Figure-72. Unfold PropertyManager

- Select the face that you want to be fixed while unfolding. You will be asked to select the bends to be unfolded.
- Select the bends and then click on the **OK** button; refer to Figure-73. The sheet metal part will be unfolded.

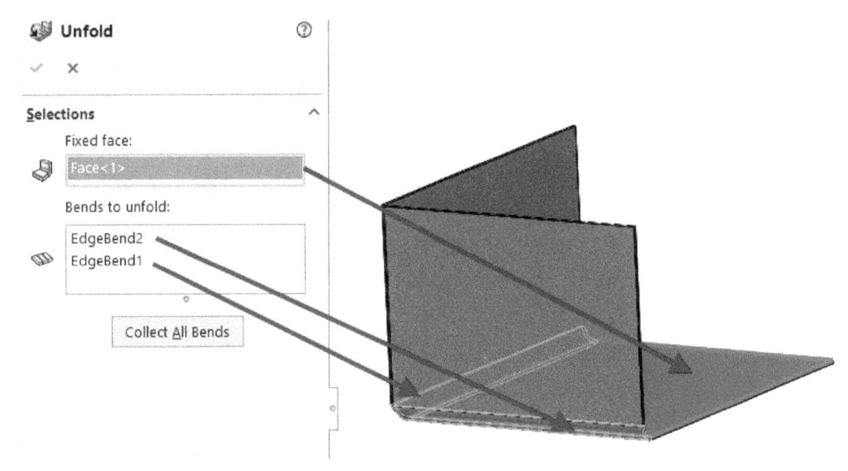

Figure-73. Selection for unfolding

Fold Tool

After unfolding, we have made some cuts in the part at the bends. We will now fold back the unfolded sheet metal as given next.

- Click on the **Fold** tool from the **Sheet Metal CommandManager** in the **Ribbon**. The **Fold PropertyManager** will be displayed; refer to Figure-74.

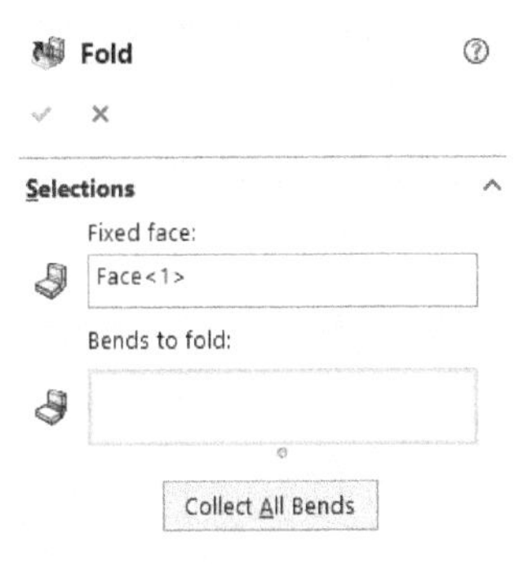

Figure-74. Fold PropertyManager

- Select the fixed face if not selected by default and then select the unfolded bends.
- Click on the **OK** button from the **PropertyManager** to fold the unfolded bends; refer to Figure-75.

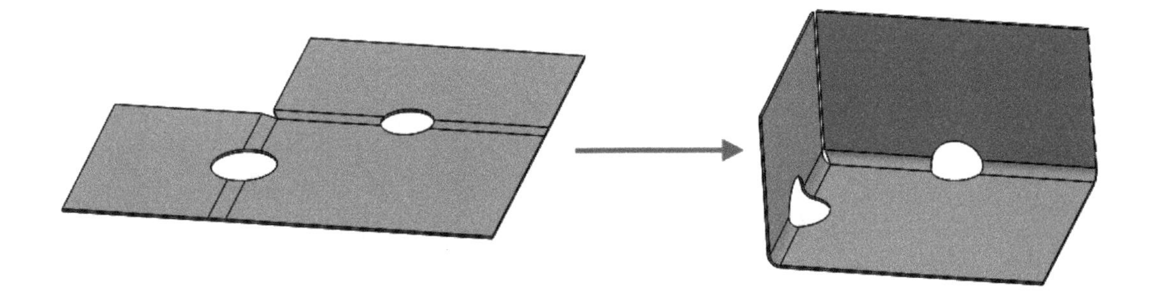

Figure-75. Folding unfolded bends

CONVERT TO SHEET METAL

The **Convert to Sheet Metal** tool is used to convert a solid body into sheet metal component. The procedure to use this tool is given next.

- Make sure you have a solid object in the drawing area and then click on the **Convert to Sheet Metal** tool from the **Sheet Metal CommandManager** in the **Ribbon**. The **Convert to Sheet Metal PropertyManager** will be displayed; refer to Figure-76. Also, you will be asked to select the base of sheet metal part.

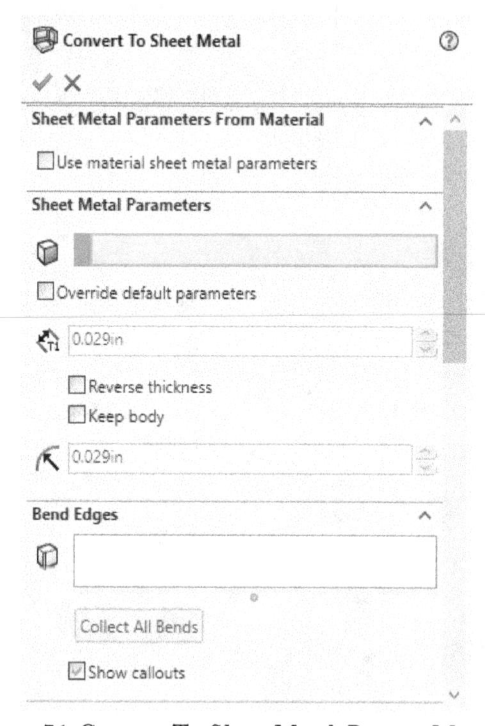

Figure-76. Convert_To_Sheet_Metal_PropertyManager

- Select a flat face of the model. Specify the sheet thickness and default bend radius in the respective spinners in **Sheet Metal Parameters** rollout of the **PropertyManager**; refer to Figure-77.

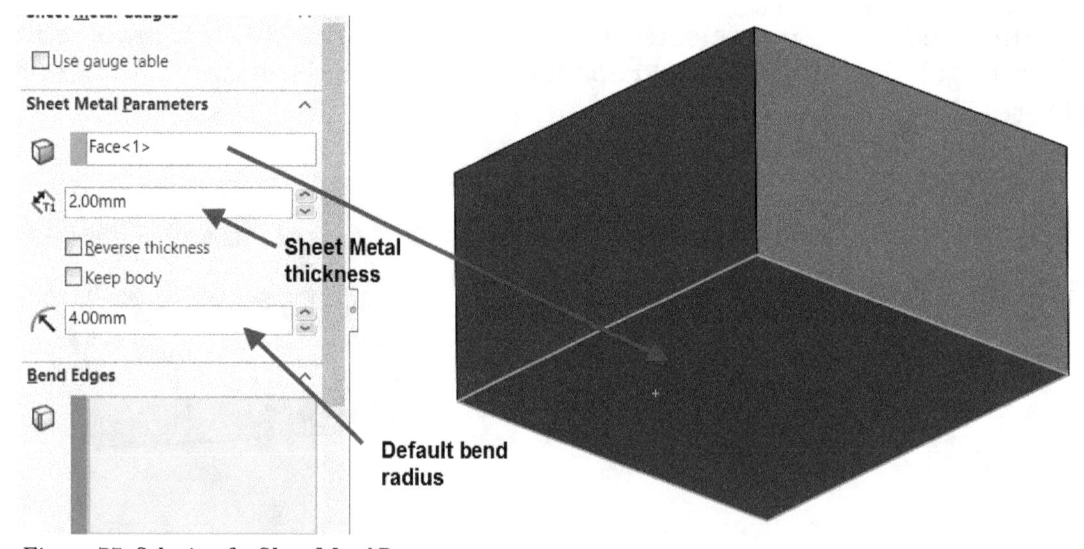

Figure-77. Selection for Sheet Metal Parameters

- Click in the **Bend Edges** selection box and select the edges of the model where bends are to be applied; refer to Figure-78.
- Scroll-down in the **PropertyManager** and specify desired parameters for corners, bend allowance, and auto relief.
- Click on the **OK** button from the **PropertyManager** to convert the solid into Sheet Metal part; refer to Figure-79.

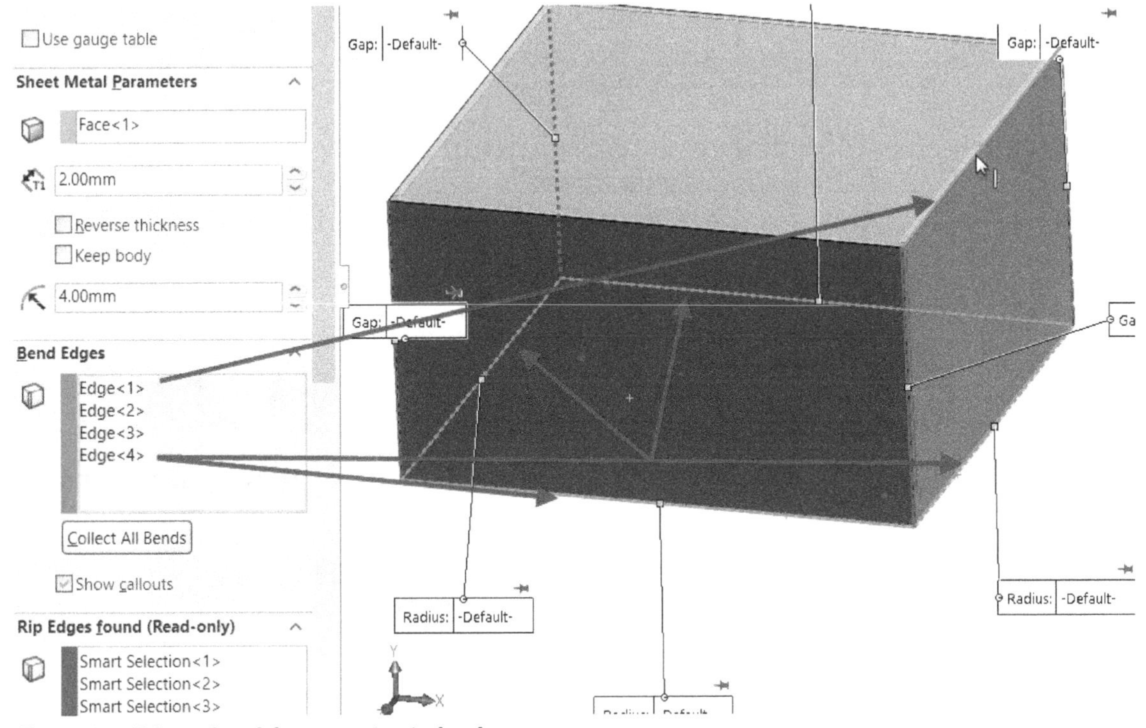

Figure-78. Edges selected for conversion in bends

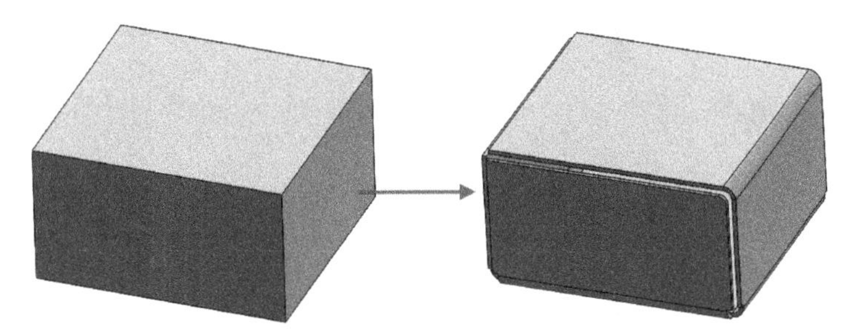
Figure-79. Solid to sheet metal conversion

RIP TOOL

The **Rip** tool is used to remove sharp edges and replace then with specified gap. This tool is mainly used when you need to convert solid parts into sheet metal. The procedure to use this tool is given next.

- Click on the **Rip** tool from the **Sheet Metal CommandManager**. The **Rip PropertyManager** will be displayed; refer to Figure-80.

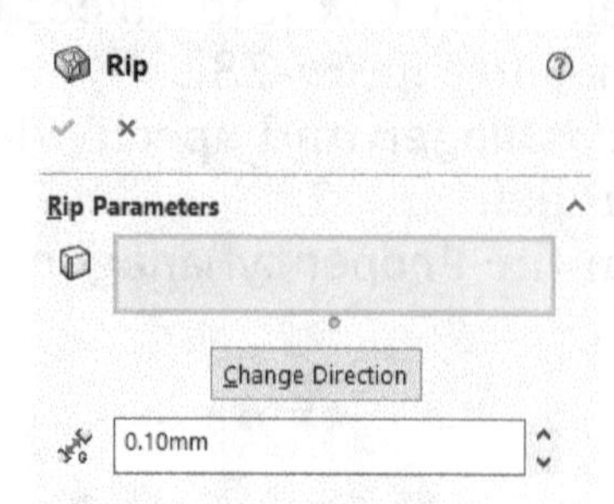

Figure-80. Rip PropertyManager

- Select the edge(s) of the model that you want to be ripped. Set the gap size in the **Rip Gap** edit box of the **PropertyManager**.
- Click on the **OK** button from the **PropertyManager** to create the rip feature; refer to Figure-81.

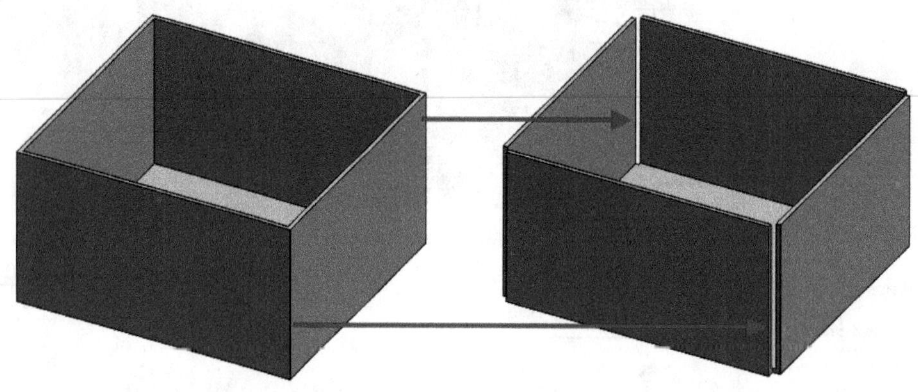

Figure-81. Edges ripped

INSERT BENDS TOOL

The **Insert Bends** tool is used to insert bends in the solid model so that it can be used in sheet metal work. The procedure to use this tool is given next.

- Click on the **Insert Bends** tool from the **Sheet Metal CommandManager**. The **Bends PropertyManager** will be displayed; refer to Figure-82. Also, you will be asked to select the fixed face of the model.
- Select the flat face that you want to be fixed.
- Specify the parameters like bend radius, bend allowance, relief etc. in **PropertyManager**.
- Click in the **Edges to Rip** selection box in the **Rip Parameters** rollout of the **PropertyManager** and select the edges that you want to be ripped.

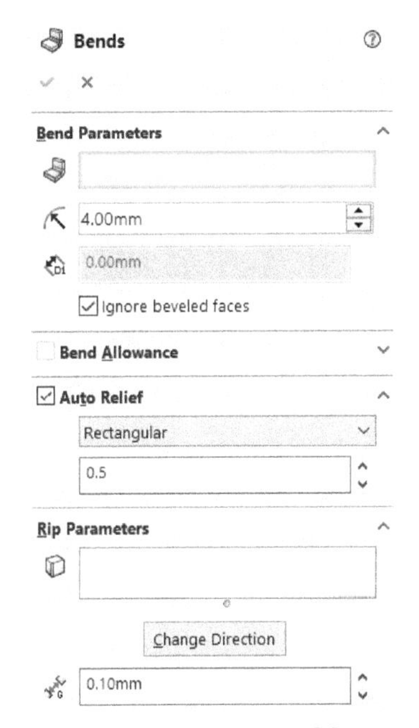

Figure-82. Bends PropertyManager

- Click on the **OK** button from the **PropertyManager**. Bends will be applied automatically to the sharp edges.

FLATTEN

The **Flatten** tool is used to create flat pattern of the sheet metal part. Click on the tool from the **Ribbon** and Flat pattern will be displayed. In some cases, you may be asked to specify the base. In those cases, select the flat face that you want to be fixed. Figure-83 shows a sheet metal part and its flat pattern.

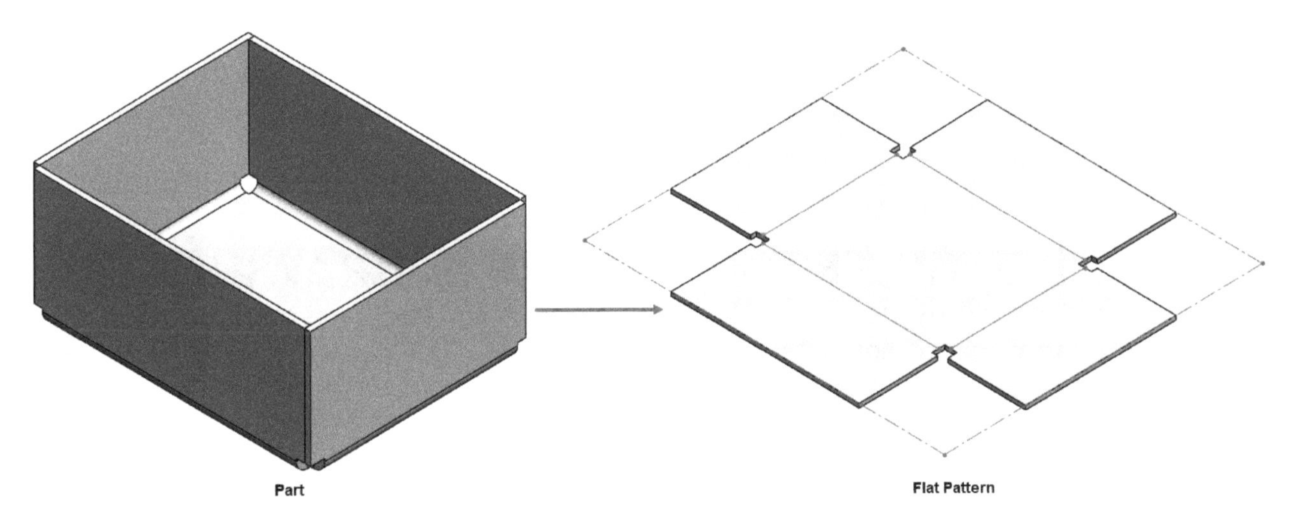

Figure-83. Part and its flat pattern

Flat Pattern Properties

After creating flat pattern, you can modify the properties of flat pattern. To do so, right-click on the **Flat-Pattern1** feature from the **FeatureManager Design Tree** and select the **Edit Feature** option from the shortcut menu; refer to Figure-84. The **Flat Pattern PropertyManager** will be displayed; refer to Figure-85.

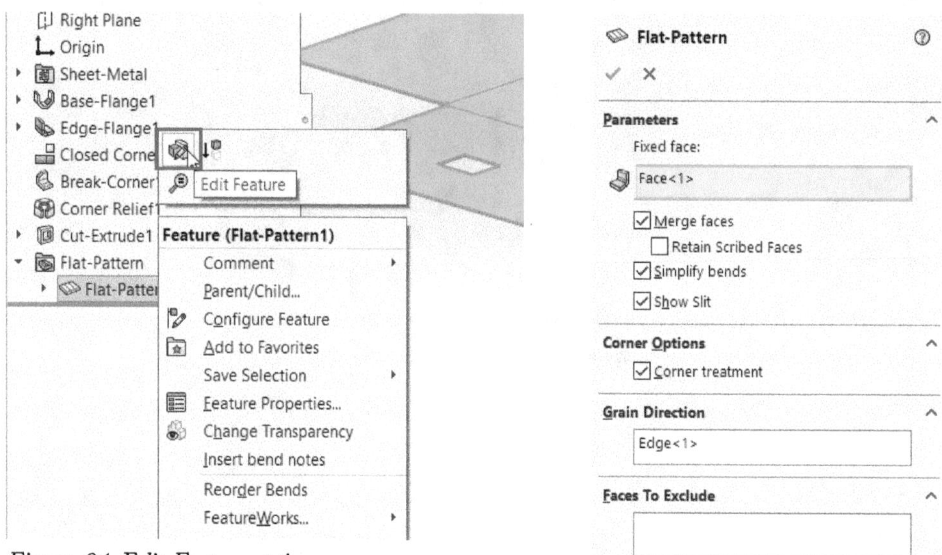

Figure-84. Edit Feature option

Figure-85. Flat Pattern PropertyManager

- If you want to change the fixed face then click in the **Fixed face** selection box of **PropertyManager** and select desired face.
- Select the **Merge faces** check box to merge all the faces which are planar and coincident in flat pattern.
- Select the **Retain Scribed Faces** check box to include text and other scribed features on the flat pattern.
- Select the **Simplify bends** check box to make the irregular bends as straight line bends.
- Select the **Show Slit** check box to display reliefs in flat pattern at corners.
- Select the **Corner treatment** check box to apply smooth edges in flat pattern.
- Click in the **Grain Direction** selection box and select an edge or line to define grain direction of sheet metal for flat pattern.
- Click in the **Faces To Exclude** selection box and select the faces that you want to exclude in flat pattern. For example, if there is a gusset attached to two faces at bend then you can exclude the face from flat pattern.
- After setting desired parameters, click on the **OK** button to create flat pattern.

INSERTING FLAT PATTERN IN DRAWING

- Start the drawing by using the **Make Drawing from Part/Assembly** button from the **New** drop-down of **Menu Bar**; refer to Figure-86. The drawing mode will open.

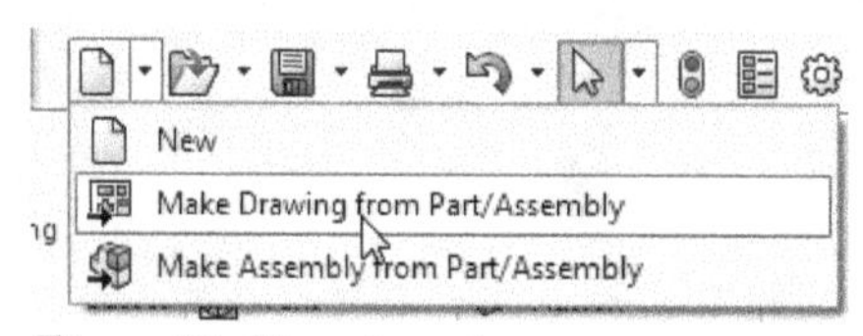

Figure-86. New drop-down

- Select the sheet size from the list and click on the **OK** button from the dialog box displayed. The **View Palette** will be displayed in the right of the drawing; refer to Figure-87.

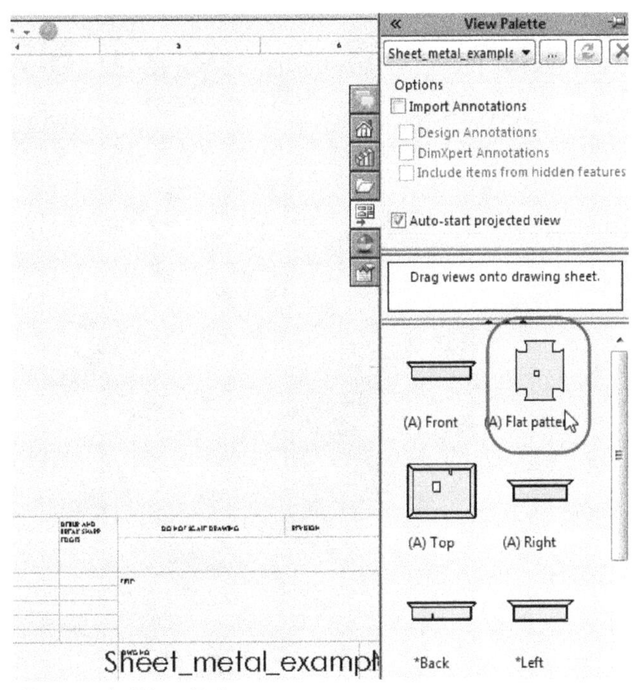

Figure-87. View Palette

- Drag the **Flat pattern** view from the palette and place it at desired location in the drawing; refer to Figure-88.

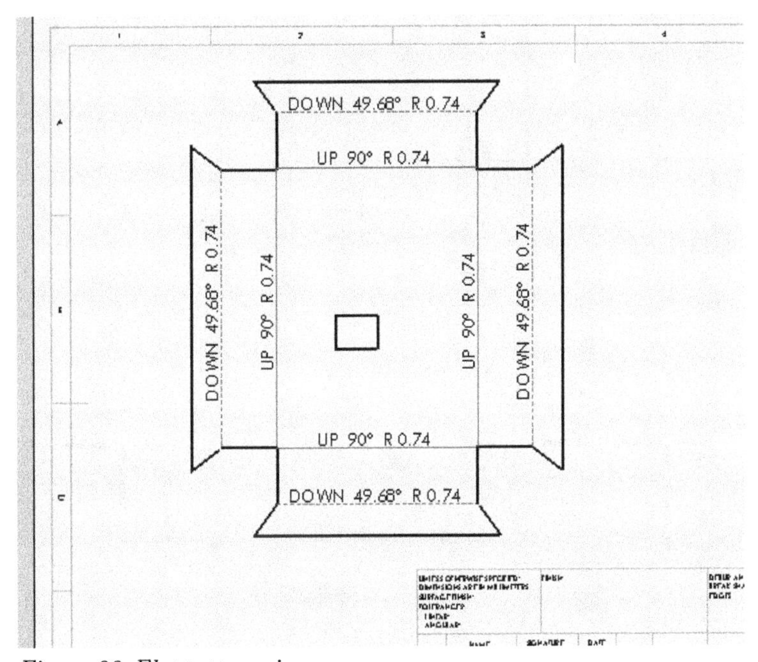

Figure-88. Flat pattern view

PRACTICE 1

Create the sheet metal model as shown in Figure-89. Dimensions are given in Figure-90.

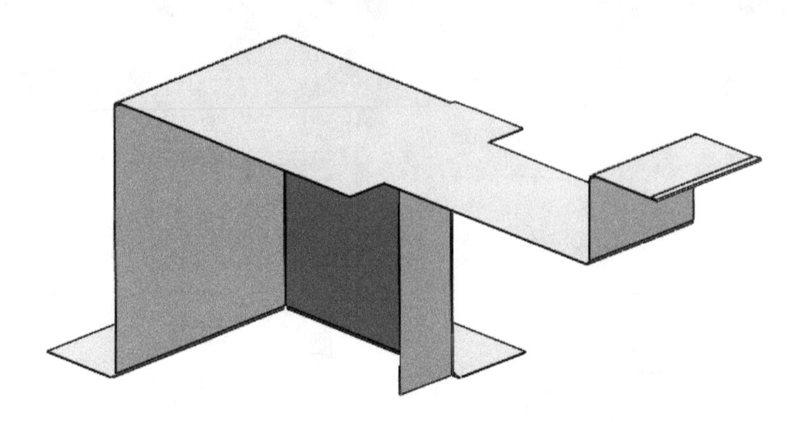

Figure-89. Sheet metal model

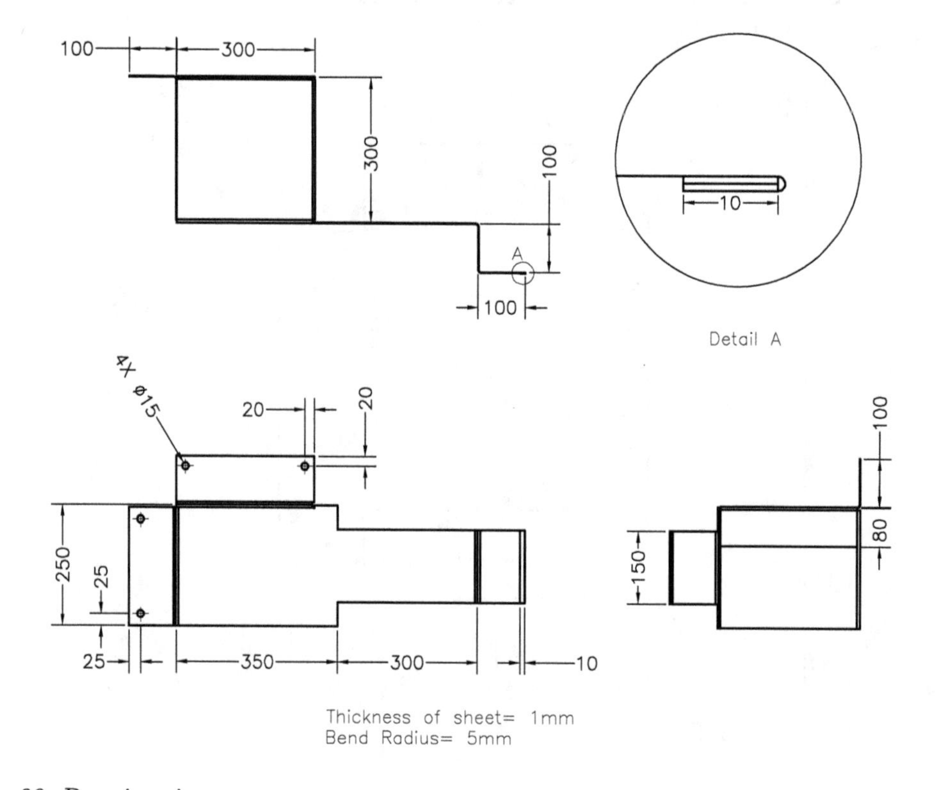

Figure-90. Drawing views

PRACTICE 2

Create the sheet metal model of the bend as shown in Figure-91. Dimensions are given in the figure.

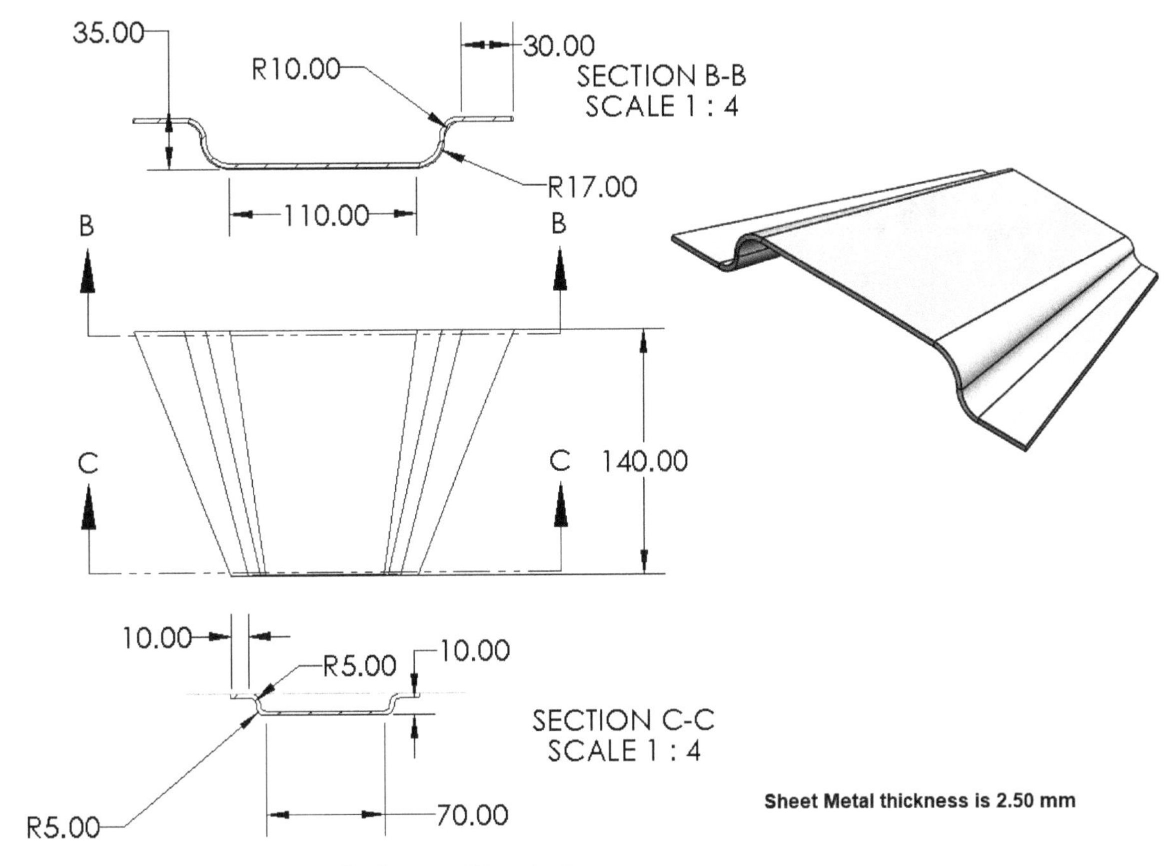

Figure-91. Drawing and model for sheetmetal Practice 2

PRACTICE 3

Create the sheet metal model of Hooper as shown in Figure-92.

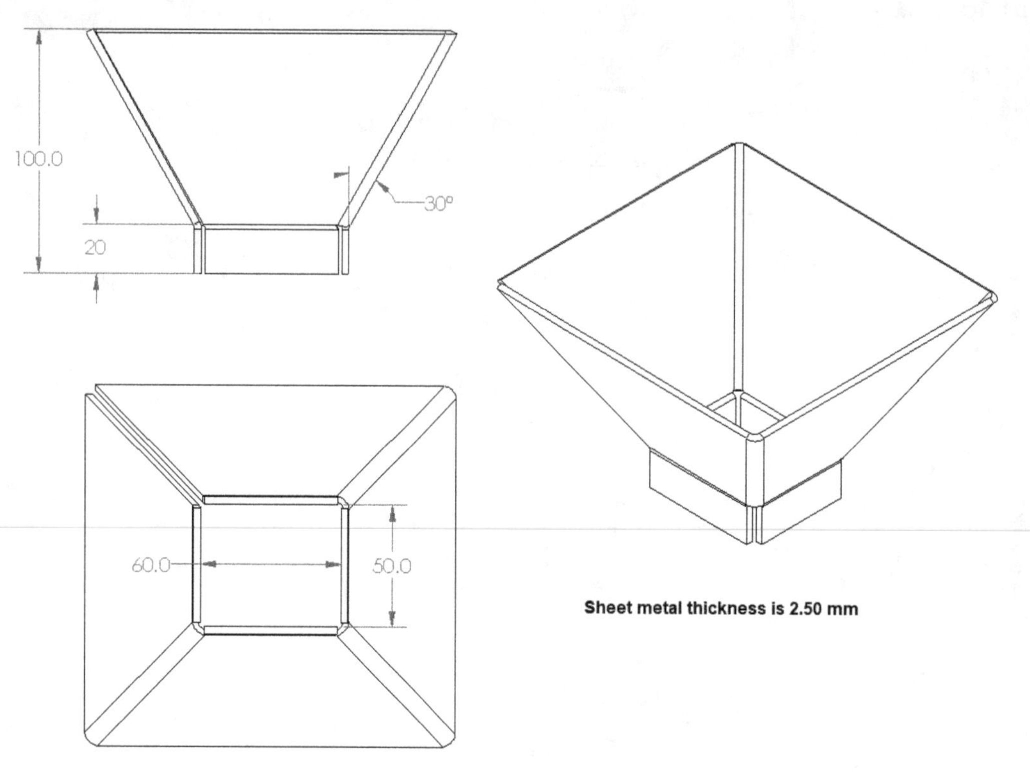

Figure-92. Hooper sheetmetal drawing

PRACTICE 4

Create the sheet metal model of a cover holder shown in Figure-93.

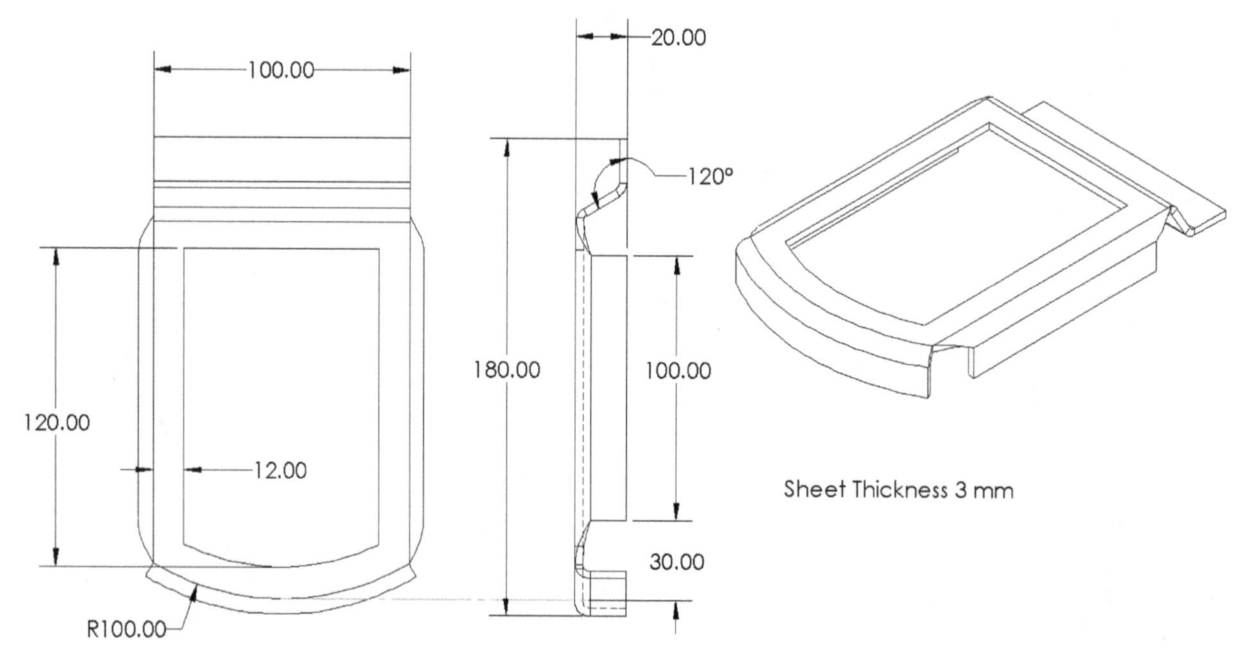

Figure-93. Cover holder sheetmetal Practice 4

To get more drawings for practice, write us at **cadcamcaeworks@gmail.com**

SELF ASSESSMENT

Q1. The...........tool is used to create base feature of the sheet metal component.

Q2. The.............tool is used to create sheet metal component by joining two or more sketch sections.

Q3. The............relief will tear the sheet in order to accommodate the bending of the sheet.

Q4. The..............tool is used to create a flange of desired shape.

Q5. The............tool is used to create bend at the end edge of the sheet.

Q6. The **Jog** tool can be used to create double bend in the sheet. (T/F)

Q7. In HVAC or other duct works, cross-breaks are provided to stiffen the sheet. (T/F)

Q8. The **Close Corner** tool is used to make all the corners of a sheet metal part closer. (T/F)

FOR STUDENT NOTES

Answer to Self-Assessment:

1. Base Flange/Tab, **2.** Lofted-Bend, **3.** Tear, **4.** Miter Flange, **5.** Hem, **6.** T, **7.** T, **8.** F

Chapter 14

Weldments and Markup Tools

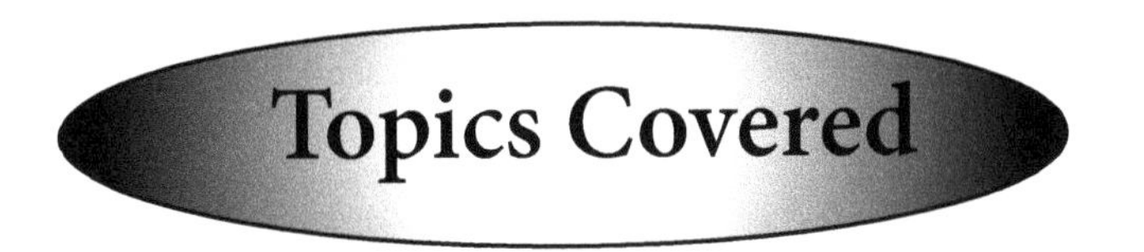

The major topics covered in this chapter are:

- ***Introduction.***
- ***Weldment tool.***
- ***Structural members.***
- ***End Cap.***
- ***Weld Beads.***
- ***Welding Symbols in Drawing.***

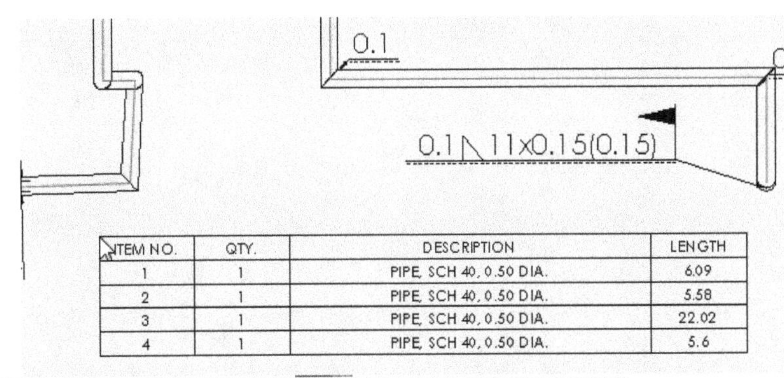

ITEM NO.	QTY.	DESCRIPTION	LENGTH
1	1	PIPE, SCH 40, 0.50 DIA.	6.09
2	1	PIPE, SCH 40, 0.50 DIA.	5.58
3	1	PIPE, SCH 40, 0.50 DIA.	22.02
4	1	PIPE, SCH 40, 0.50 DIA.	5.6

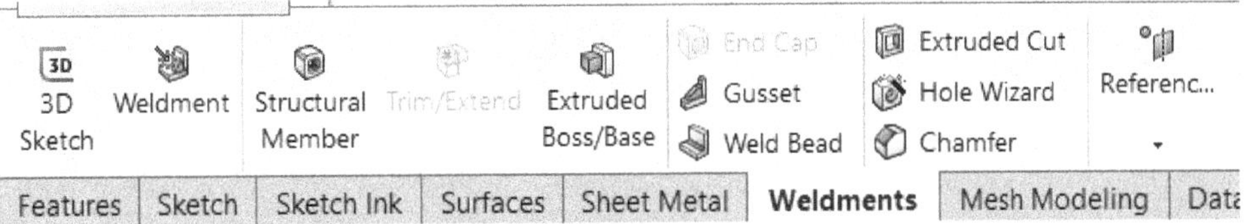

INTRODUCTION

This chapter is dedicated to welding joints, also called weldments. Welding is a method to permanently join parts with the help of welding beads. In SolidWorks, we can display the welding beads in model as well as in drawing. In model, we can display a solid bead of weld around the selected faces/edges. We can also attach the welding bead symbol there. In drawing, we can attach the welding symbol to the affected edges/faces. We can also insert the cut list to tabulate the components used in the model. But, before we start using SolidWorks for weldments, its important to revise some basics of welding.

WELDING SYMBOLS AND REPRESENTATION IN DRAWING

The symbols to represent various type of welds are given next.

Butt/Groove Weld Symbols

Various symbols that come under this category are given next. Refer to Figure-1 and Figure-2.

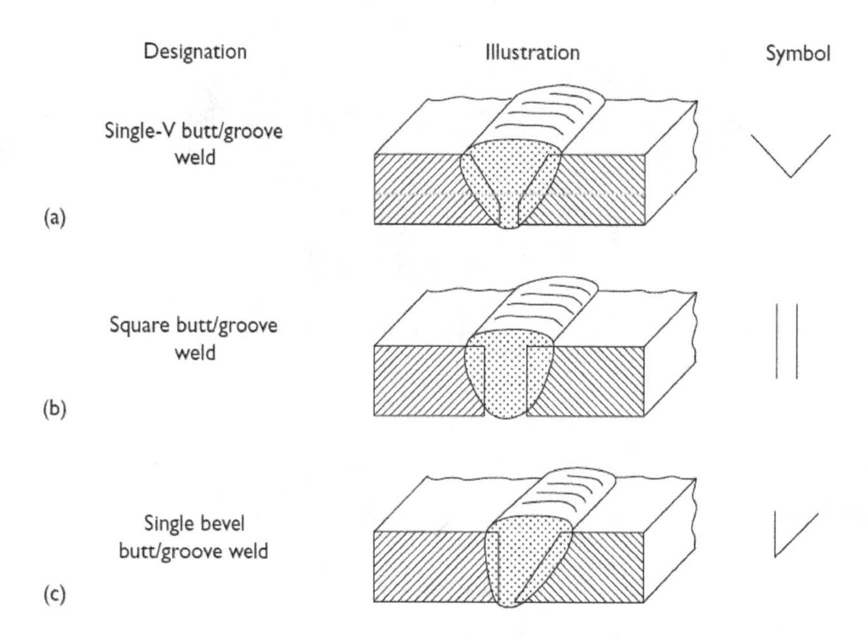

Figure-1. Welding symbols list 1

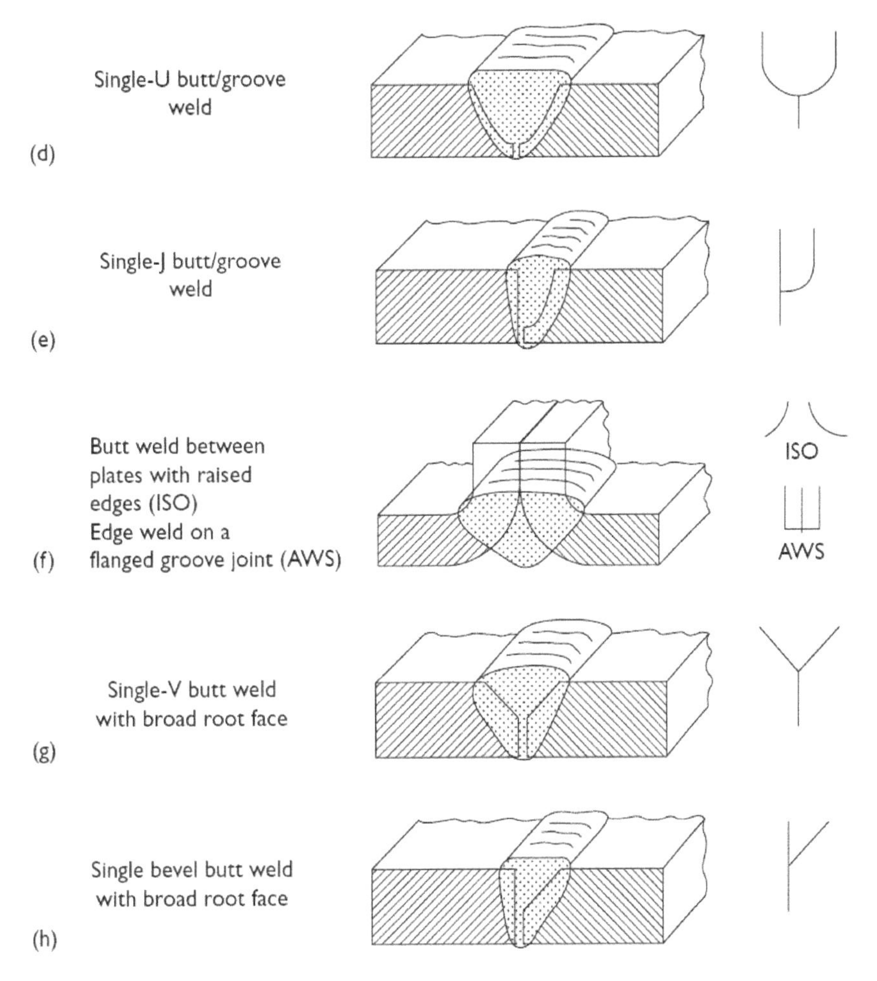

Figure-2. Welding symbols list 2

Fillet and Edge Weld Symbols

Various symbols that come under this category are given next. Refer to Figure-3.

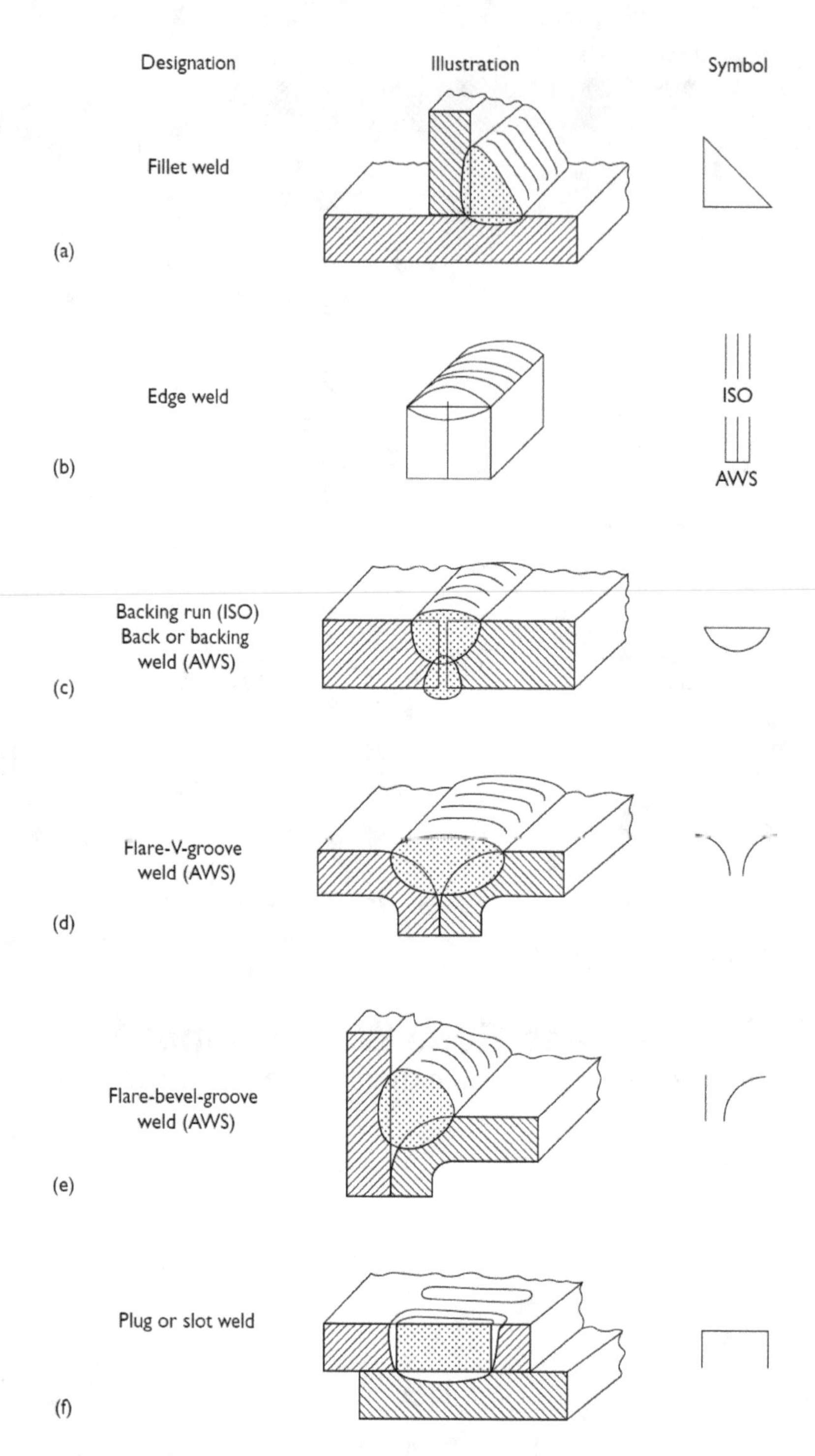

Figure-3. Welding symbols list 3

Miscellaneous Weld Symbols

Various symbols that come under this category are given next. Refer to Figure-4.

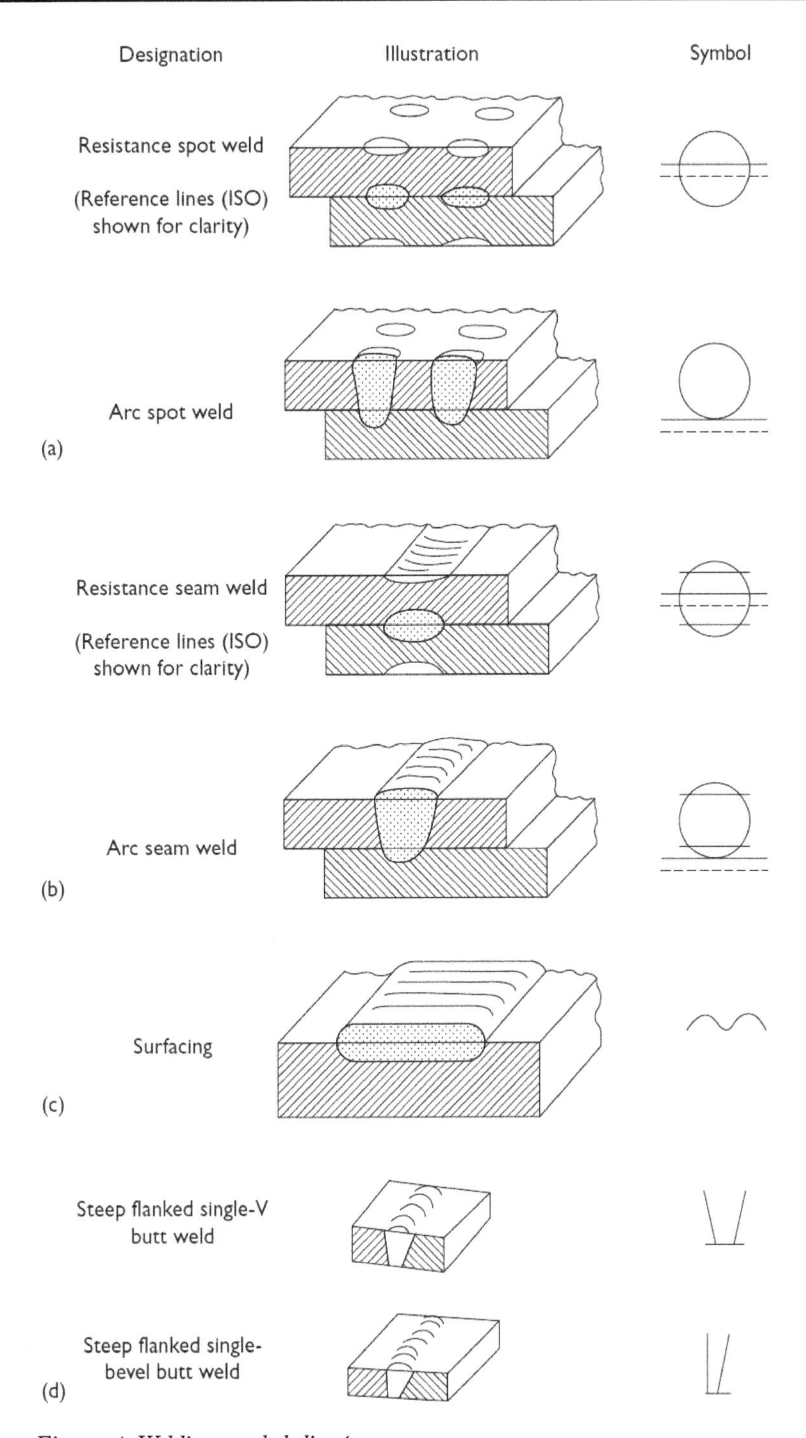

Figure-4. Welding symbols list 4

Now, we know various symbols used in welding drawings but keep a note that placement of welding symbol along the arrow decides the side on which the welding will be done on the object; refer to Figure-5.

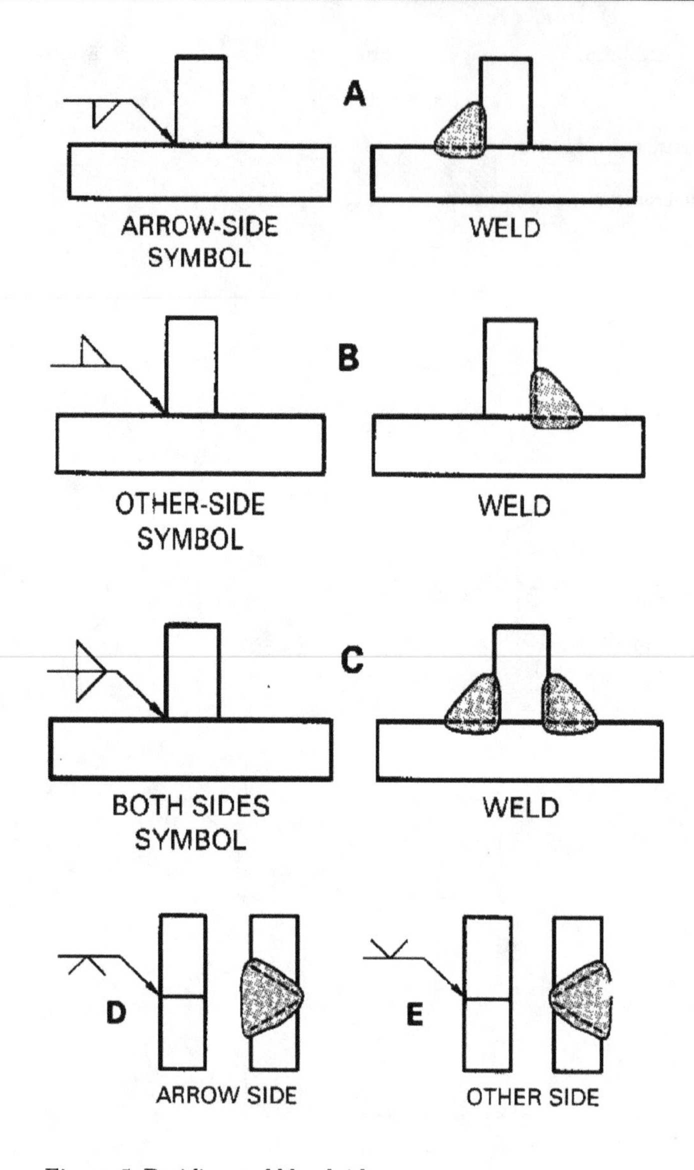

Figure-5. Deciding weld bead side

Dimensioning a weld bead

Alike the other measurements, weld is also measured with respect to various references so that we can control its quality. Figure-6 shows the information required for dimensioning a weld bead.

Till this point, we have learned the basics of weld symbol representations in drawings. So, we are ready to dive into SolidWorks for creating welding representations.

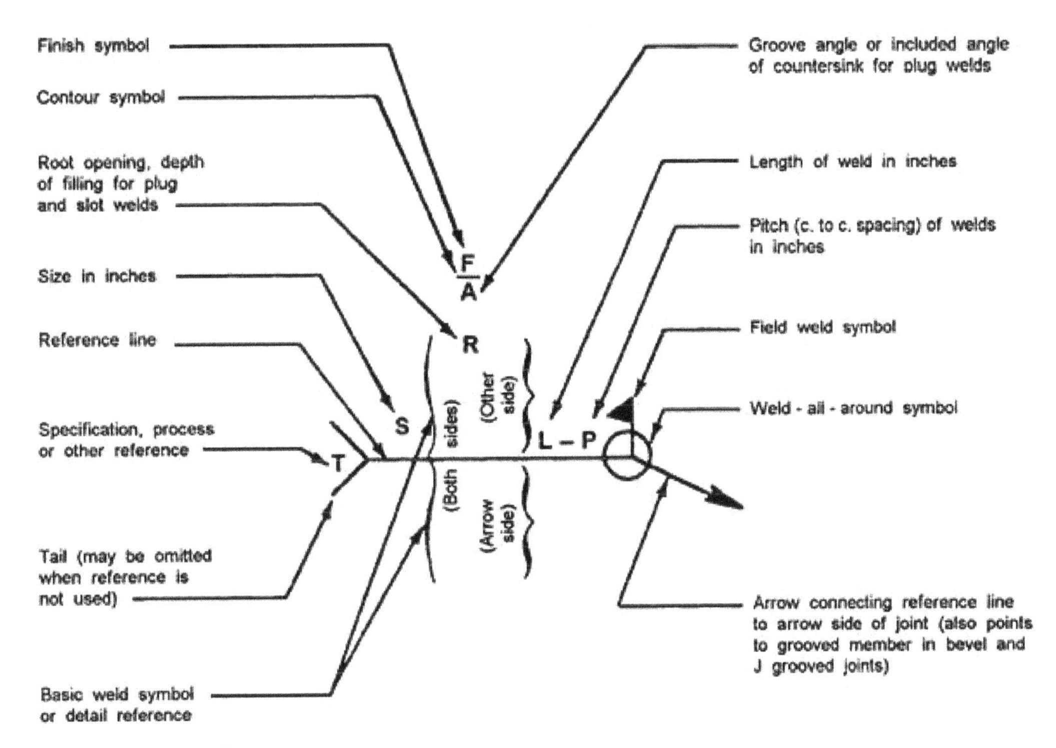

Figure-6. Welding dimension

The tools to add weldments are available in the **Weldments CommandManager** of the **Ribbon**; refer to Figure-7. If this **CommandManager** is not displayed by default, then you can add it by right-clicking on any of the **CommandManager** in the **Ribbon** and then selecting the **Weldments** option from the shortcut menu displayed; refer to Figure-8.

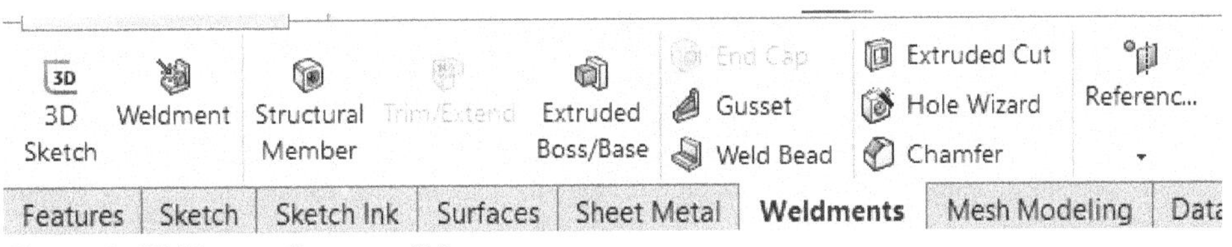

Figure-7. Weldments CommandManager

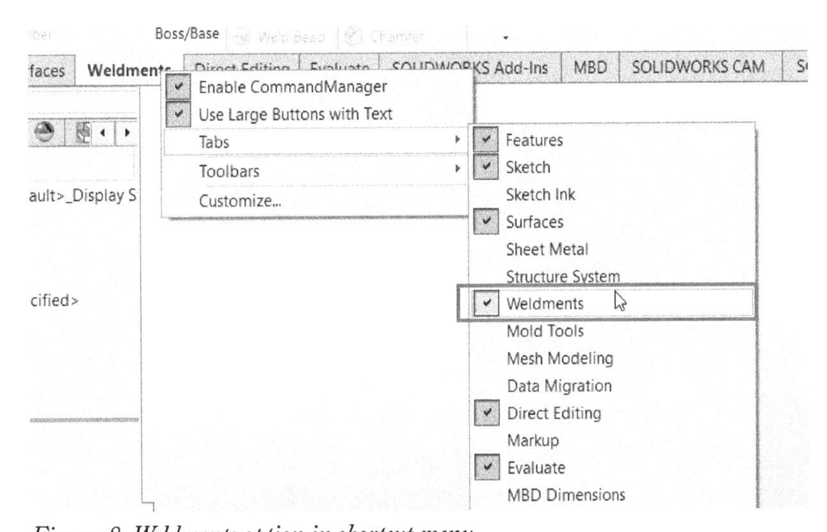

Figure-8. Weldments option in shortcut menu

Most of the tools in this tab have been discussed earlier. The remaining tools are discussed next.

WELDMENT TOOL

The **Weldment** tool is used to start multi-body environment in which you can create parts without merging them into a single body. In this way, we are allowed to join two entities in the modeling environment with the help of weld beads. The **Weldment** tool does not do a function like **Extrude** or **Revolve** tool but it starts **Weldment** environment. Click on this tool to start the environment.

STRUCTURAL MEMBER

The **Structural Member** tool is used to create structural member for welding, like angle, C channel, pipe, and so on. The procedure to use this tool is given next.

- Click on the **Structural Member** tool from the **Ribbon**. The **Structural Member PropertyManager** will be displayed; refer to Figure-9.

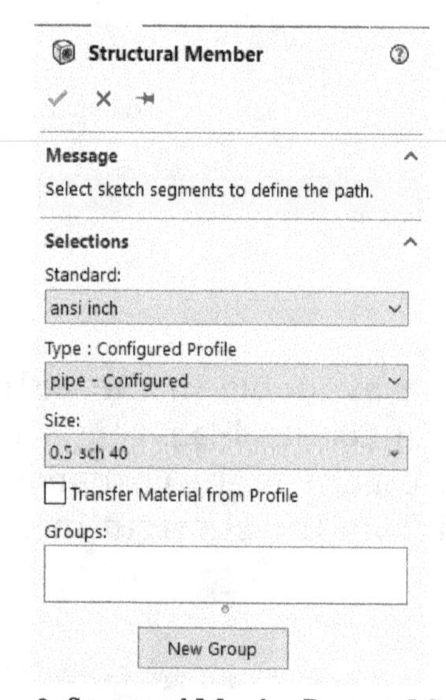

Figure-9. Structural Member PropertyManager

- Select desired standard, type and size of structural member from the **PropertyManager**.
- One by one click on the line member of the sketch. Preview of the structural member will be displayed; refer to Figure-10.
- After selecting the entities, click on the **New Group** button to make new selection set for differently oriented entities.
- Select the **Apply corner treatment** check box to modify intersection of structure at corner. The buttons to modify corner treatment will be displayed. The use of different corner treatment buttons is shown in Figure-11. You can select whether it is simple cut or cope cut using the respective buttons if butt connection is being created.
- Select the **Allow protrusion** check box if you want one side of structure to be extended.
- You can mirror the profile by using the selecting the **Mirror profile** check box. The radio buttons to define mirror axis will be displayed below the check box. Select the **Horizontal axis** or **Vertical axis** radio button to use respective axis for mirroring.

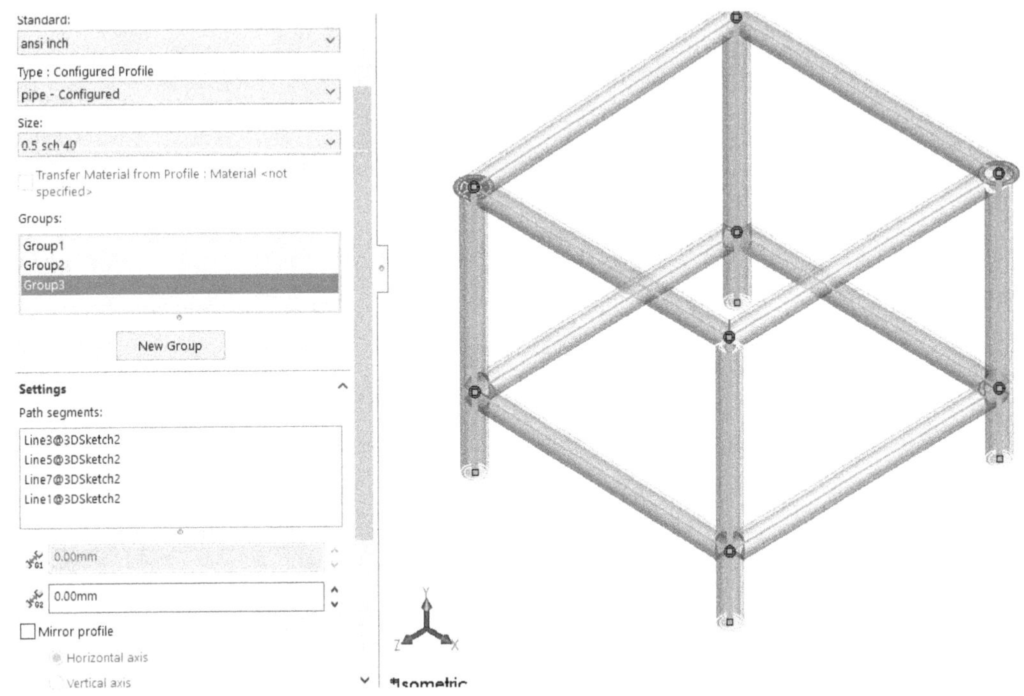

Figure-10. Preview of the structure

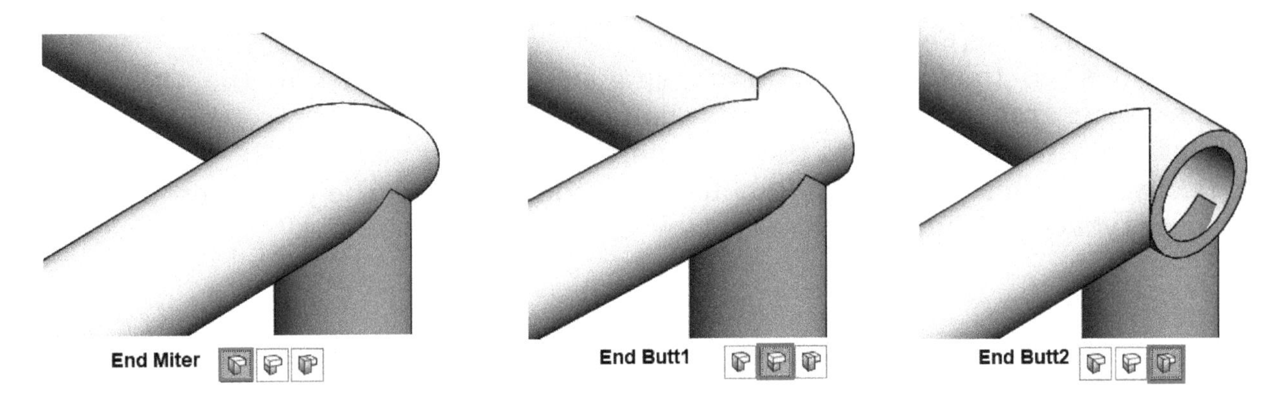

Figure-11. Corner Treatment buttons

- Specify desired angle in the **Rotation Angle** edit box to rotate the structural member.
- Select desired configuration of structural members in the **Configurations** rollout.
- Set desired parameters and click on the **OK** button from the **PropertyManager** to create the structure.

END CAP TOOL

The **End Cap** tool is used to close the ends of an open structural member by using a lid. The procedure to use this tool is given next.

- Click on the **End Cap** tool from the **Ribbon**. The **End Cap PropertyManager** will be displayed; refer to Figure-12.

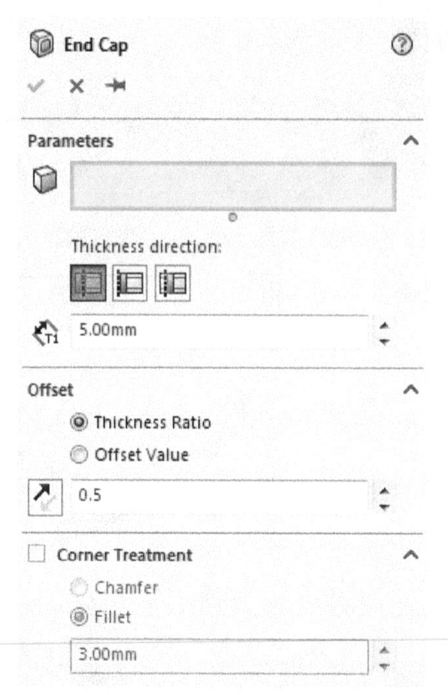

Figure-12. End Cap PropertyManager

- Select the end face of the structural member. Preview of the end cap will be displayed; refer to Figure-13.
- Select desired thickness direction using the three buttons in the **Thickness direction:** area of the **Parameters** rollout.
- Specify desired thickness in the **Thickness** edit box.
- Specify the other required parameters and then click on the **OK** button from the **PropertyManager** to create the end cap.

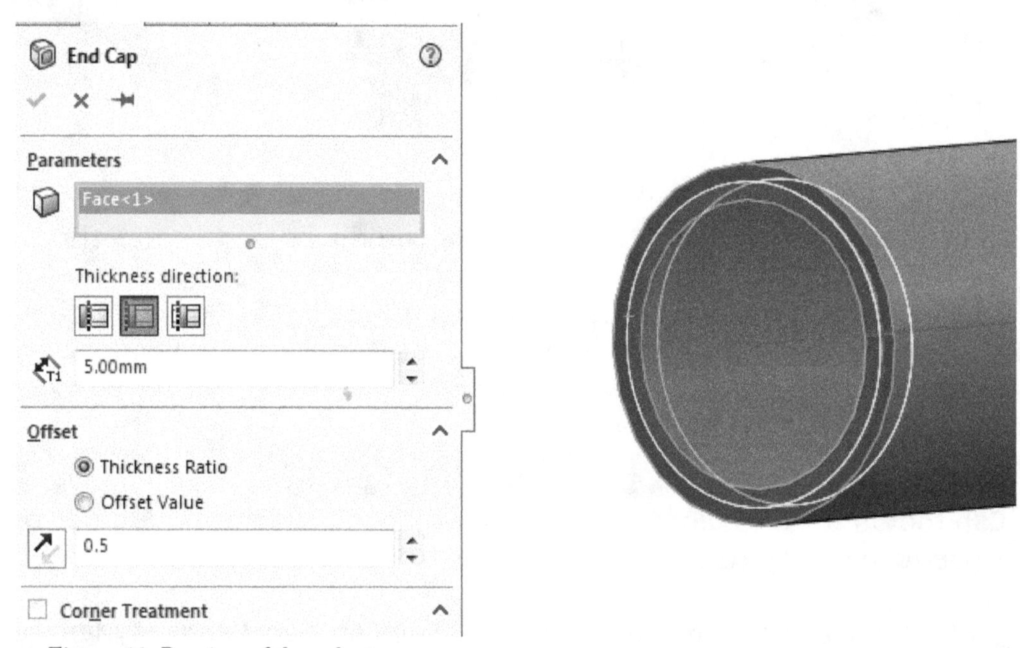

Figure-13. Preview of the end cap

The **Gusset** tool has been discussed earlier.

WELD BEAD

The **Weld Bead** tool is used to apply desired type of welding bead on the selected edges. The procedure to use this tool is given next.

- Click on the **Weld Bead** tool from the **Ribbon**. The **Weld Bead PropertyManager** will be displayed; refer to Figure-14.

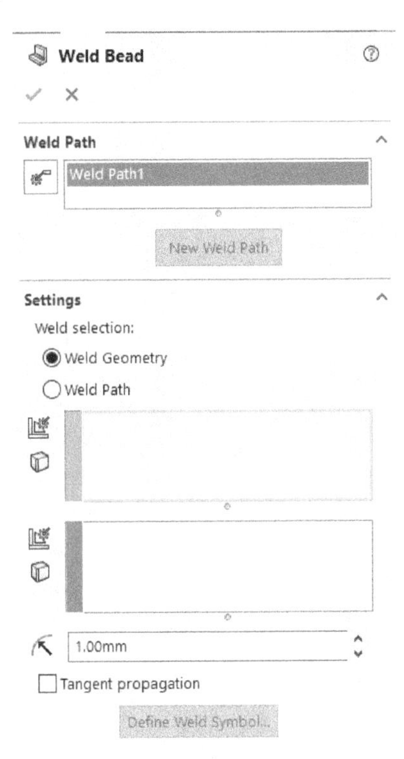

Figure-14. Weld Bead PropertyManager

- If the **Weld Geometry** radio button is selected in the **Weld selection** area of the **Settings** rollout then select the face sets of the two joining parts; refer to Figure-15.

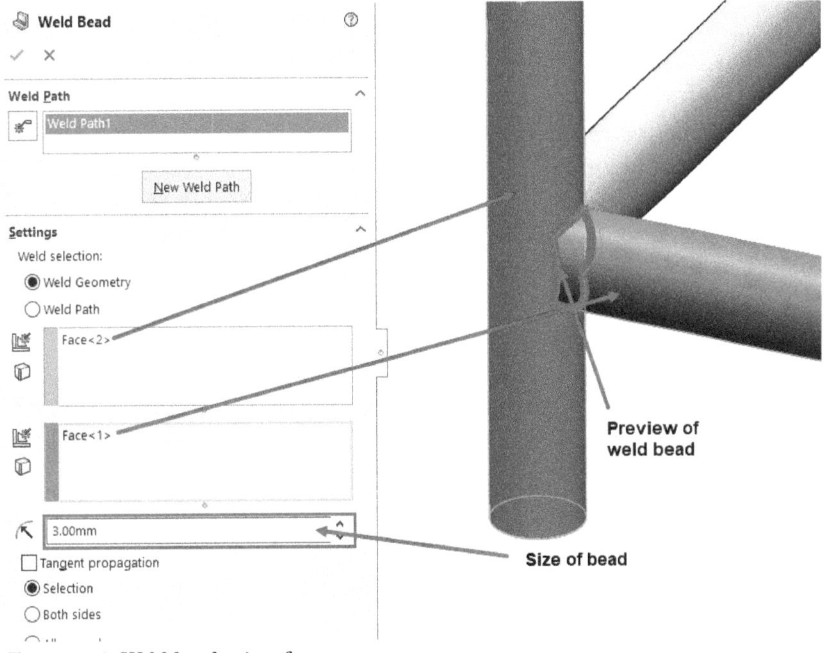

Figure-15. Weld bead using faces

- If you select the **Weld Path** radio button from the **Weld selection** area of the **Settings** rollout then you need to select the edge/edges along which you want to represent welding bead; refer to Figure-16.

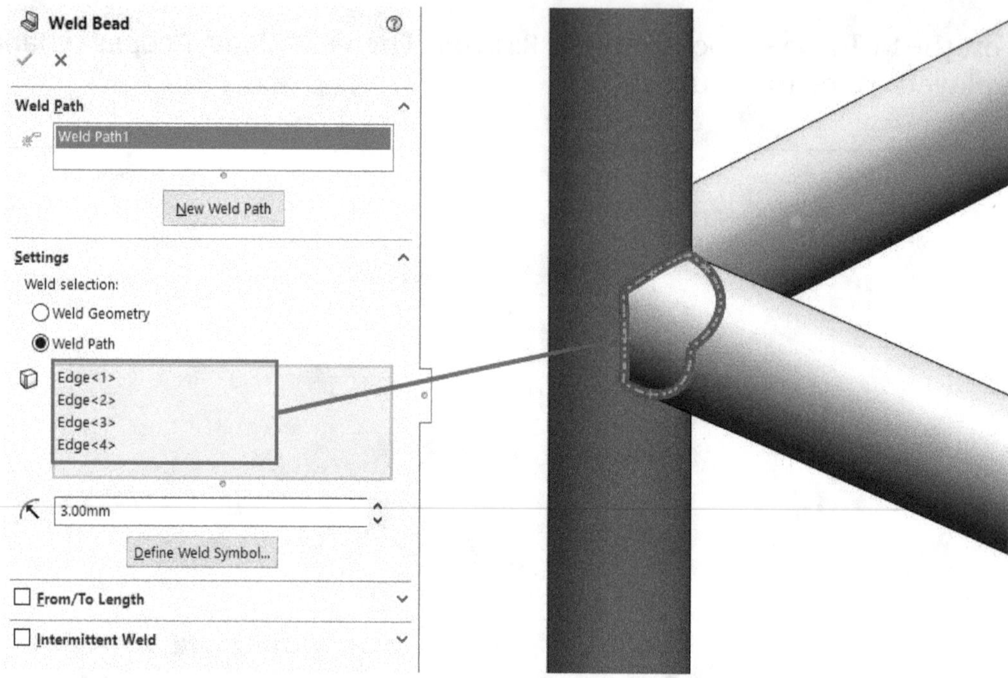

Figure-16. Weld bead using edges

- After creating every closed bead, you need to start a new weld path by clicking on the **New Weld Path** tool from the **Weld Path** rollout in the **PropertyManager**.
- To display the weld symbol on the model, click on the **Define Weld Symbol** button from the **Settings** rollout. The dialog box to create weld bead symbol will be displayed; refer to Figure-17. Note the standard used to create weld symbol depends on the drafting standard selected in the **Document Properties** tab of **Options** dialog box; refer to Figure-18.
- Now, this is the dialog box in which you need to feed all the information regarding the weld bead. You can go back to the starting of this chapter for symbol reference and then specify the dimensions for weld bead.
- Click on the **Weld Symbol** button in the dialog box and select desired welding symbol; refer to Figure-19. Click on the **More Symbols** option from the flyout if desired symbol is not listed.

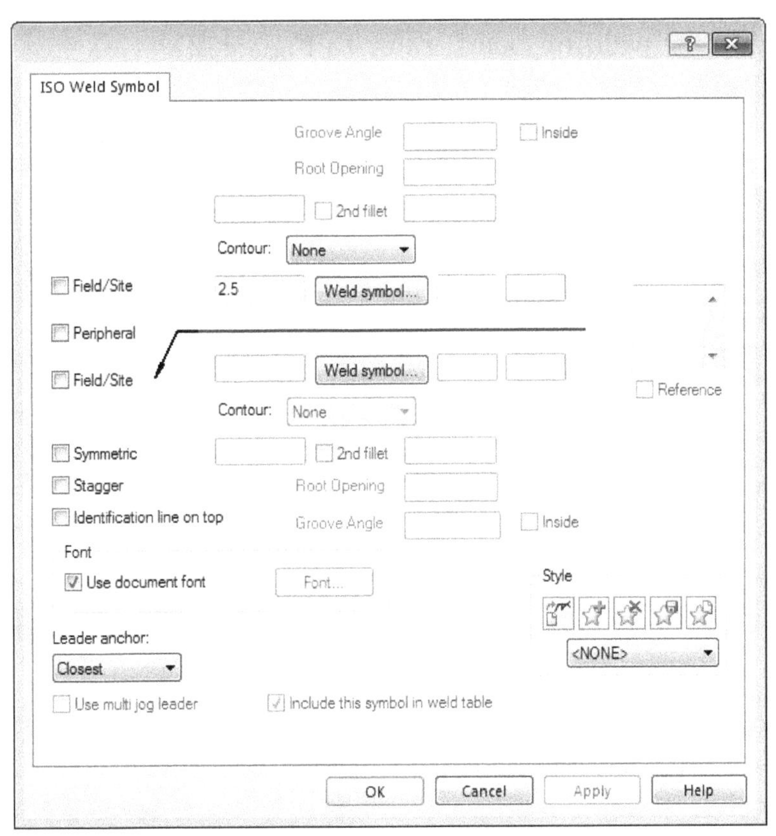

Figure-17. Weld Symbol dialog box

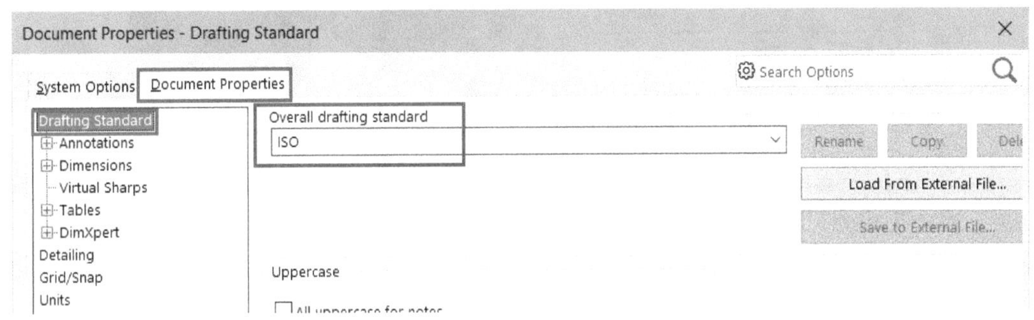

Figure-18. Drafting standard option

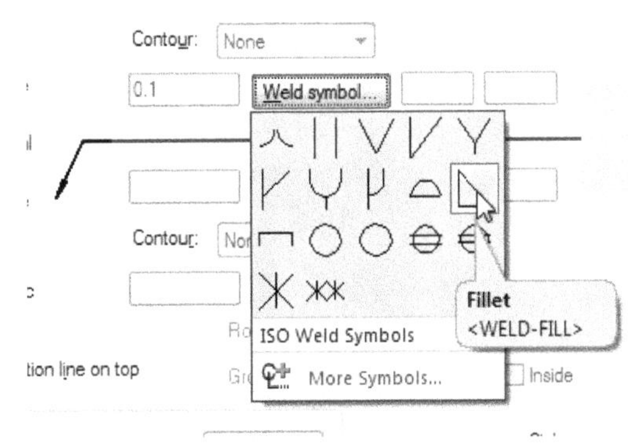

Figure-19. Weld symbols

- After specifying desired dimension, click on the **OK** button from the dialog box.
- To set desired length and start point of weld bead, select the **From/To Length** check box and specify desired parameters for bead; refer to Figure-20.

- Similarly, you can select the **Intermittent Weld** check box and specify the intermediate gap for welding bead; refer to Figure-21.

Figure-20. Setting length of weld bead

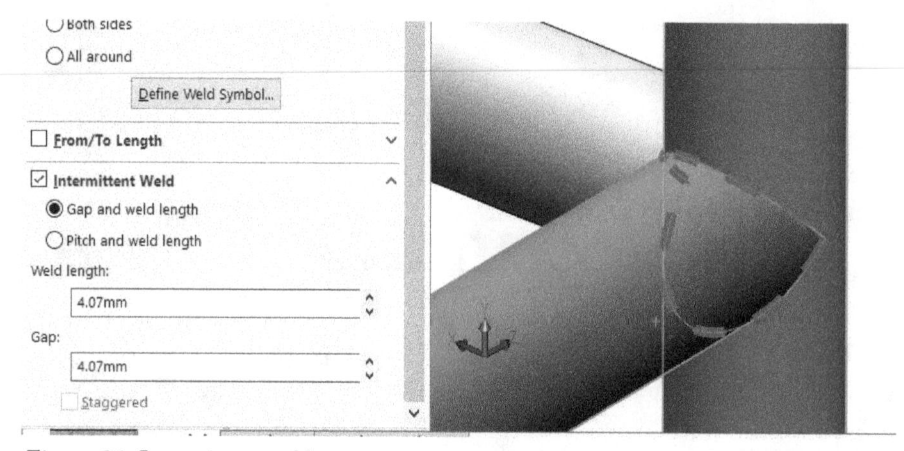

Figure-21. Intermittent weld

- After specifying desired parameters, click on the **OK** button from the **PropertyManager** to create the bead. The bead will be added in the **Weld Folder** in **FeatureManager Design Tree**; refer to Figure-22.

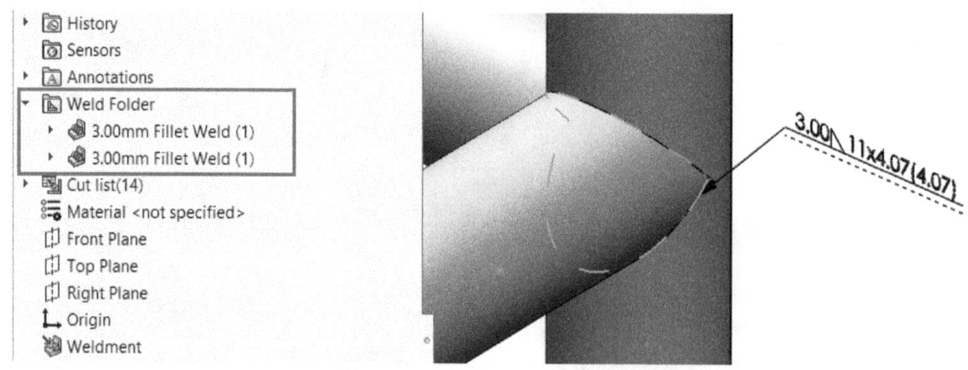

Figure-22. Weld Folder

INSERTING WELDING DATA IN DRAWING

There is no use of assigning welding symbols unless the manufacturer/fabricator does not get them in the drawing. The procedure to insert the welding data in the drawing is given next.

- After creating the welding model, click on the **Make Drawing from Part** option from the **File** menu; refer to Figure-23. The drawing environment will be displayed.

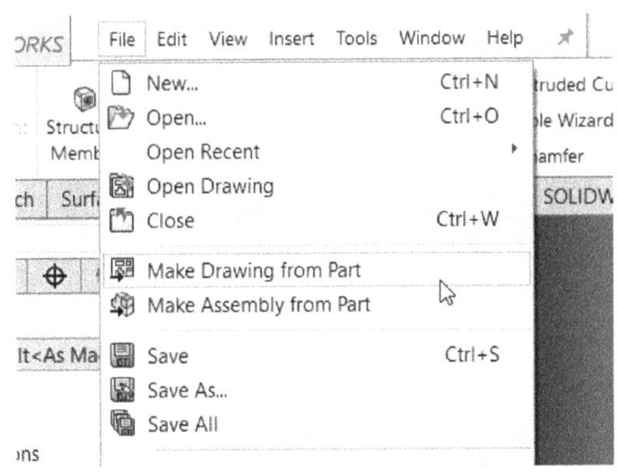

Figure-23. Make Drawing from Part option

- Select desired sheet size and create the basic views; refer to Figure-24.
- Select the view/views in which you want to display the annotations and click on the **Model Items** button from the **Annotations** tab in the **Ribbon**. The **Model Items PropertyManager** will be displayed.
- Select the **Weld Symbols** button from the **Annotations** rollout in the **PropertyManager**; refer to Figure-25.

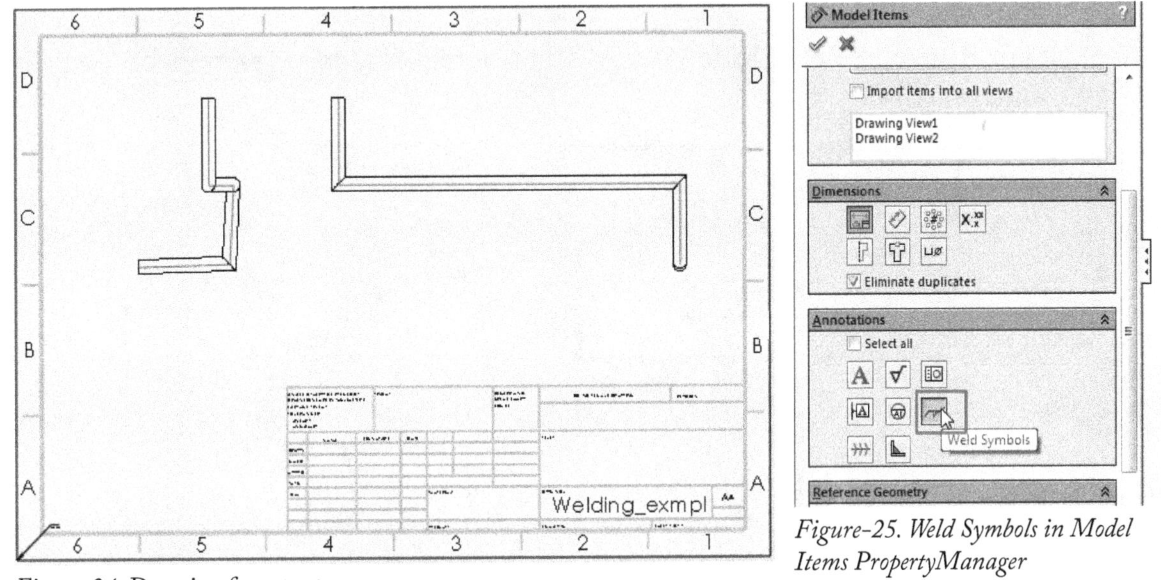

Figure-24. Drawing from part

Figure-25. Weld Symbols in Model Items PropertyManager

- Click on the **OK** button from the **PropertyManager**. The welding symbols will be assigned in the drawing; refer to Figure-26.

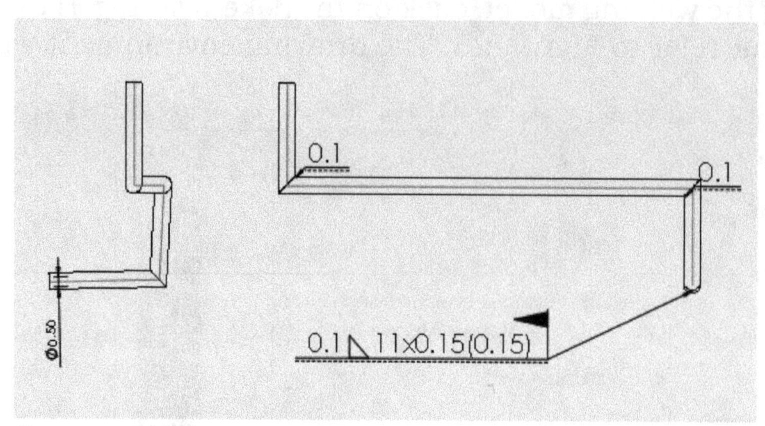

Figure-26. Drawing with welding symbols

Inserting the Cut list

Cut list is a kind of part list similar to bill of material. The cut list is used to identify the structural members that are being joined by using the welding bead. The procedure to insert the cut list is given next.

- Click on the **Weldment Cut List** option from the **Tables** drop-down in the **Annotation CommandManager** of the **Ribbon**; refer to Figure-27. You are asked to select a view for which you want to create the cut list.
- Select desired view. The **Weldment Cut List PropertyManager** will be displayed; refer to Figure-28.

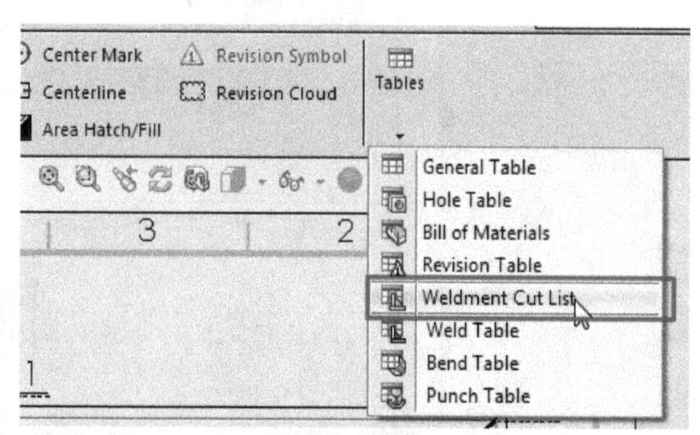

Figure-27. Weldment cut list option

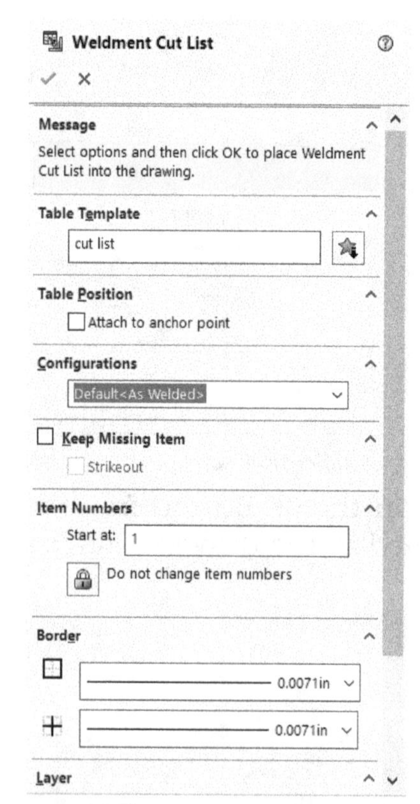

Figure-28. Weldment Cut List PropertyManager

- Specify desired options, if any. Next, click on the **OK** button from the **PropertyManager**. The cut list will get attached to the cursor; refer to Figure-29.

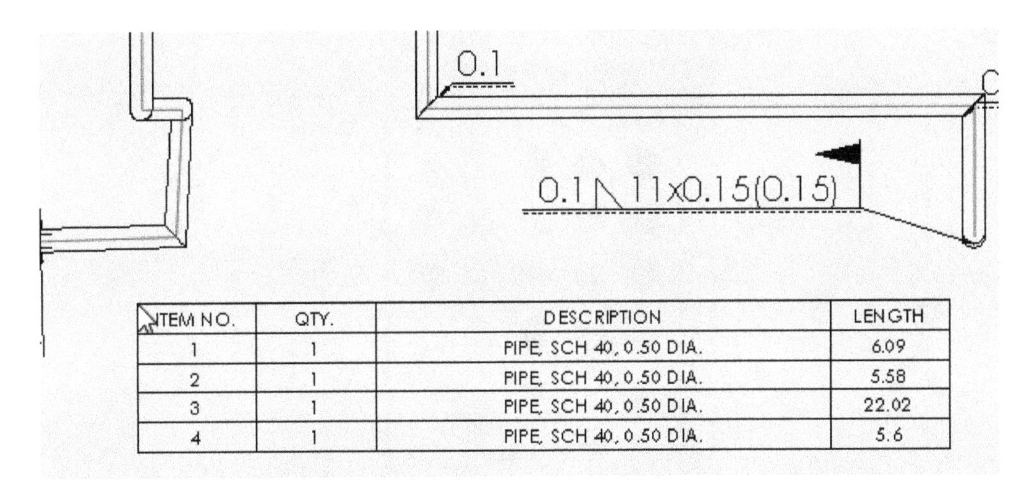

Figure-29. Cut list attached to the cursor

- Click at desired location to place the cut list. Similarly, you can place the Weld table to specify the welding length and parameters.

MARKUP TOOLS

The markup tools are used to mark areas of attention for your team and apply notes. The markup tools are available in the **Markup CommandManager** in the **Ribbon**; refer to Figure-30. If **Markup CommandManager** is not displayed by default then you can display the **CommandManager** by procedure discussed earlier using right-click shortcut menu.

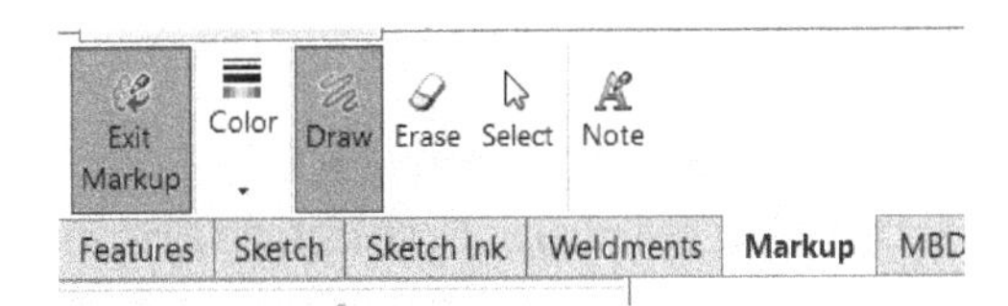

Figure-30. Markup CommandManager

Click on the **Markup** tool from the **Markup CommandManager** in the **Ribbon** to activate the markup tools.

Various tools in this **CommandManager** are discussed next.

Setting Color and Thickness for Markup

- Click on the **Color** button from the **Markup CommandManager**. The options to modify color and thickness for markup will be displayed; refer to Figure-31.

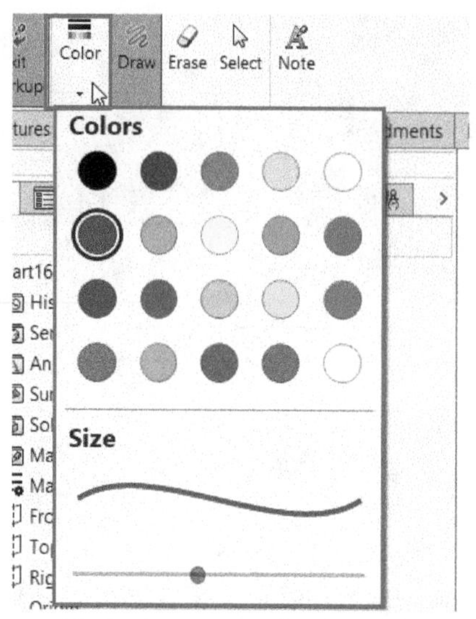

Figure-31. Options to modify color and thickness

- Set desired color from the thumbnails and use the slider to define thickness for the markup.

Drawing Markup

Markup allow you to create freehand sketch for annotation. Select the **Draw** button from the **Markup CommandManager** in the **Ribbon** to activate draw mode. Press LMB, hold the button, and drag to create free hand sketch for markup.

Erasing Markup

Click on the **Erase** button from the **Markup CommandManager** in the **Ribbon** to activate the erase mode for markups. Drag the cursor using LMB over the markups to be erased.

Click on the **Select** button from the **Markup CommandManager** to select markups. After selecting markup, you can change the color and thickness for selected markup.

The **Note** tool in **Markup CommandManager** works in the same way as discussed earlier.

Click on the **Exit Markup** button from **CommandManager** to exit markup mode.

SELF ASSESSMENT

Q1. Draw the symbol of Square butt/groove weld.

Q2. Draw the symbol of Single bevel butt weld with broad root face.

Q3. Draw the symbol of Fillet weld.

Q4. Draw the symbol of Plug weld.

Q5. Draw the symbol of Single-J butt/groove weld.

Q6. Which of the following figure shows correct annotation of welding symbol?

a. 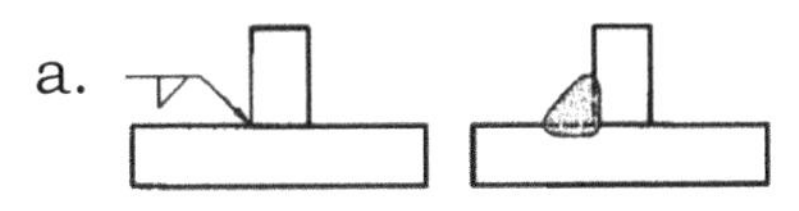

b. 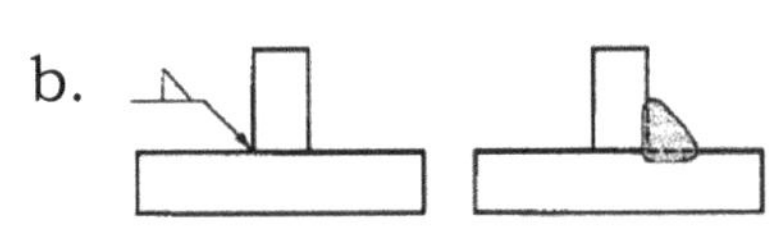

c. Both of the above.
d. None of the above.

Q7. Which of the following is not a type of structural member in the **Structural Member PropertyManager**?

a. pipe
b. round bar
c. c channel
d. s section.

For Student Notes

Answer to Self-Assessment:
6. c, **7.** b,

Chapter 15

3D Printing and Model Based Definition (MBD)

Topics Covered

The major topics covered in this chapter are:

- ***3D Printing and Processes***
- ***Print 3D Tool***
- ***MBD (Model Based Definition)***
- ***DimXpert Tools***
- ***3D PDF generation***
- ***3D PMI Compare***

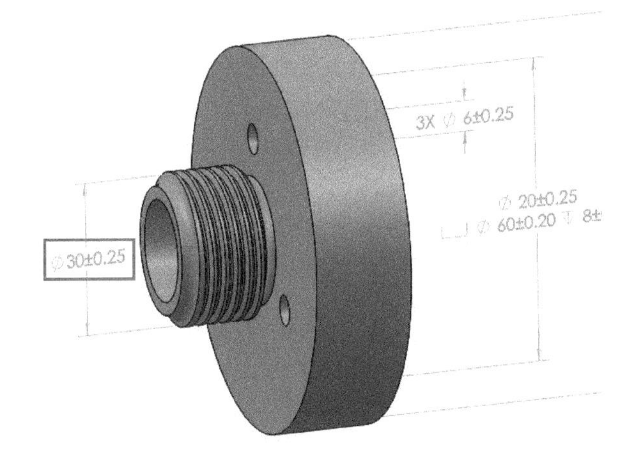

3D PRINTING

3D Printing also known as Additive Manufacturing is not a new concept as it was developed in 1981 but since then 3D Printing technology is continuously evolving. In early stages, the 3D printers were able to create only prototypes of objects using the polymers. But now a days, 3D printers are able to produce final products using metals, plastics and biological materials. 3D printers are being used for making artificial organs, architectural art pieces, complex design objects etc. Although, 3D Printing technique was created for manufacturing industry but now, it has found more applications in medical field.

In SolidWorks, there is a very simple and robust mechanism for 3D printing. The procedure of 3D Printing itself is not difficult but it is important to prepare your part well for 3D printing. We will first discuss the part preparation for 3D Printing and then we will use SolidWorks tools for performing 3D print.

PART PREPARATION FOR 3D PRINTING

Part preparation is very important step for 3D Printing. If your part is not stable in semi molten state then it is less suitable for 3D printing. Stability of model is directly dependent on the material you are using for 3D Printing. We will learn more about part preparation but before that it is important to understand different type of processes available in 3D Printing.

3D Printing Processes

Not all 3D printers use the same technology. There are several ways to print and all those available are additive, differing mainly in the way layers are build to create the final object.

Some methods use melting or softening material to produce the layers. Selective laser sintering (SLS) and fused deposition modeling (FDM) are the most common technologies using this way of 3D printing. Another method is when we talk about curing a photo-reactive resin with a UV laser or another similar power source one layer at a time. The most common technology using this method is called stereolithography (SLA).

In 2010, the American Society for Testing and Materials (ASTM) group "ASTM F42 – Additive Manufacturing", developed a set of standards that classify the Additive Manufacturing processes into 7 categories according to Standard Terminology for Additive Manufacturing Technologies. These seven processes are:

1. Vat Photopolymerisation
2. Material Jetting
3. Binder Jetting
4. Material Extrusion
5. Powder Bed Fusion
6. Sheet Lamination
7. Directed Energy Deposition

Brief introduction to these processes is given next.

Vat Photopolymerisation

A 3D printer based on the Vat Photopolymerisation method has a container filled with photopolymer resin which is then hardened with a UV light source; refer to Figure-1.

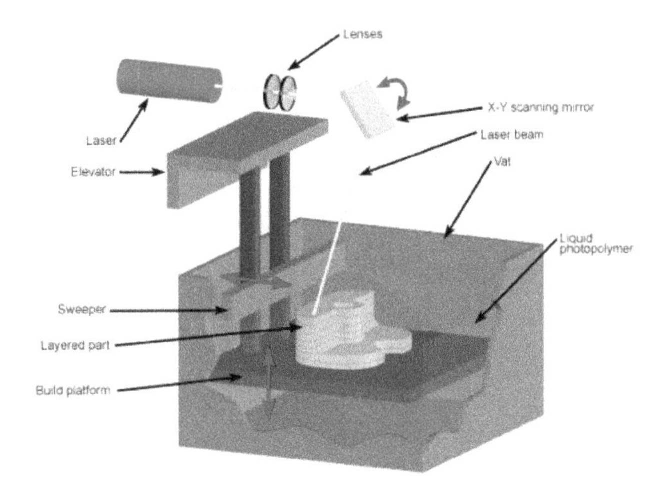

Figure-1. 3D Printing via vat-photopolymerisation

The most commonly used technology in this processes is Stereolithography (SLA). This technology employs a vat of liquid ultraviolet curable photopolymer resin and an ultraviolet laser to build the object's layers one at a time. For each layer, the laser beam traces a cross-section of the part pattern on the surface of the liquid resin. Exposure to the ultraviolet laser light cures and solidifies the pattern traced on the resin and joins it to the layer below.

After the pattern has been traced, the SLA's elevator platform descends by a distance equal to the thickness of a single layer, typically 0.05 mm to 0.15 mm (0.002" to 0.006"). Then, a resin-filled blade sweeps across the cross section of the part, re-coating it with fresh material. On this new liquid surface, the subsequent layer pattern is traced, joining the previous layer. The complete three dimensional object is formed by this project. Stereolithography requires the use of supporting structures which serve to attach the part to the elevator platform and to hold the object because it floats in the basin filled with liquid resin. These are removed manually after the object is finished.

Material Jetting

In this process, material is applied in droplets through a small diameter nozzle, similar to the way a common inkjet paper printer works, but it is applied layer-by-layer to a build platform making a 3D object and then hardened by UV light; refer to Figure-2.

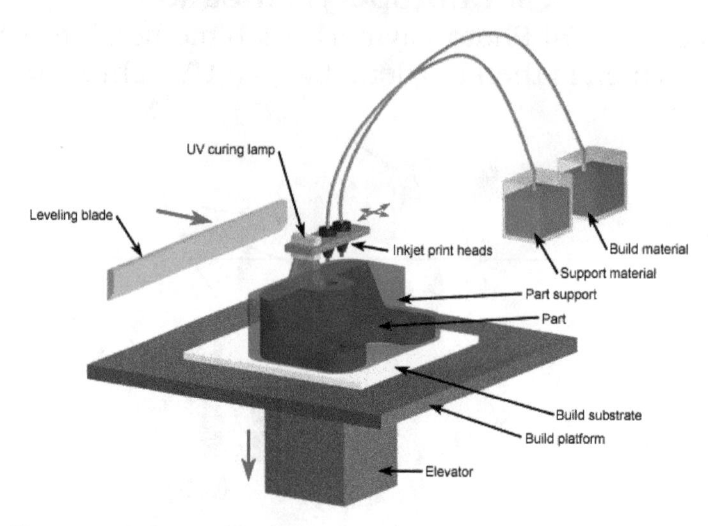

Figure-2. 3D Printing via Material-Jetting

Binder Jetting

With binder jetting two materials are used: powder base material and a liquid binder. In the build chamber, powder is spread in equal layers and binder is applied through jet nozzles that "glue" the powder particles in the shape of a programmed 3D object; refer to Figure-3. The finished object is "glued together" by binder remains in the container with the powder base material. After the print is finished, the remaining powder is cleaned off and used for 3D printing the next object.

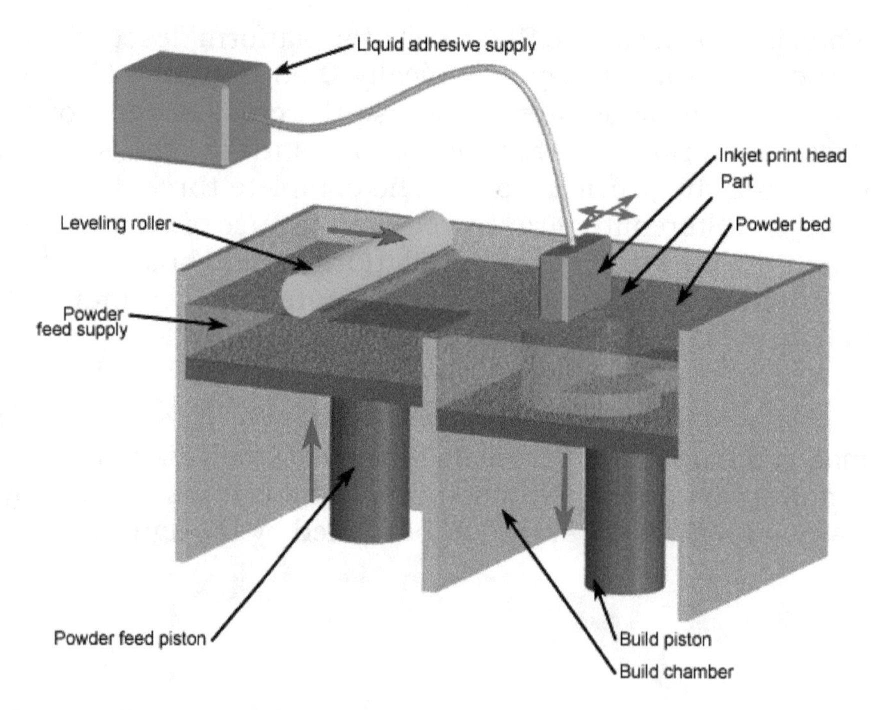

Figure-3. 3D Printing via binder-jetting

Material Extrusion

The most commonly used technology in this process is Fused deposition modeling (FDM). The FDM technology works using a plastic filament or metal wire which is unwound from a coil and supplying material to an extrusion nozzle which can turn the flow on and off. The nozzle is heated to melt the material and can be moved in both horizontal and vertical directions by a numerically controlled mechanism, directly controlled by a computer-aided manufacturing (CAM) software package; refer to Figure-4. The object is produced by extruding melted material to form layers as the material hardens immediately after extrusion from the nozzle. This technology is most widely used with two plastic filament material types: ABS (Acrylonitrile Butadiene Styrene) and PLA (Polylactic acid) but many other materials are available ranging in properties from wood filed, conductive, flexible etc.

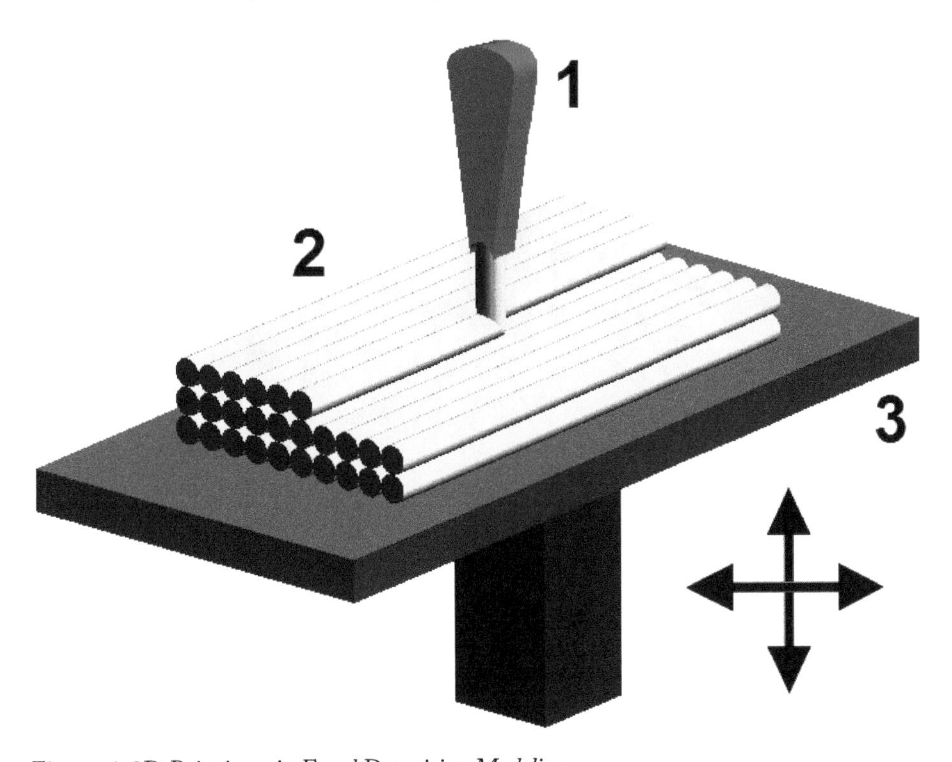

Figure-4. 3D Printing via Fused Deposition Modeling

In the above figure:
1 – nozzle ejecting molten material (plastic),
2 – deposited material (modelled part),
3 – controlled movable table.

Powder Bed Fusion

The most commonly used technology in this processes is Selective laser sintering (SLS). This technology uses a high power laser to fuse small particles of plastic, metal, ceramic or glass powders into a mass that has desired three dimensional shape. The laser selectively fuses the powdered material by scanning the cross-sections (or layers) generated by the 3D modeling program on the surface of a powder bed; refer to Figure-5. After each cross-section is scanned, the powder bed is lowered by one layer thickness. Then a new layer of material is applied on top and the process is repeated until the object is completed.

All untouched powder remains as it is and becomes a support structure for the object. Therefore there is no need for any support structure which is an advantage over SLS and SLA. All unused powder can be used for the next print.

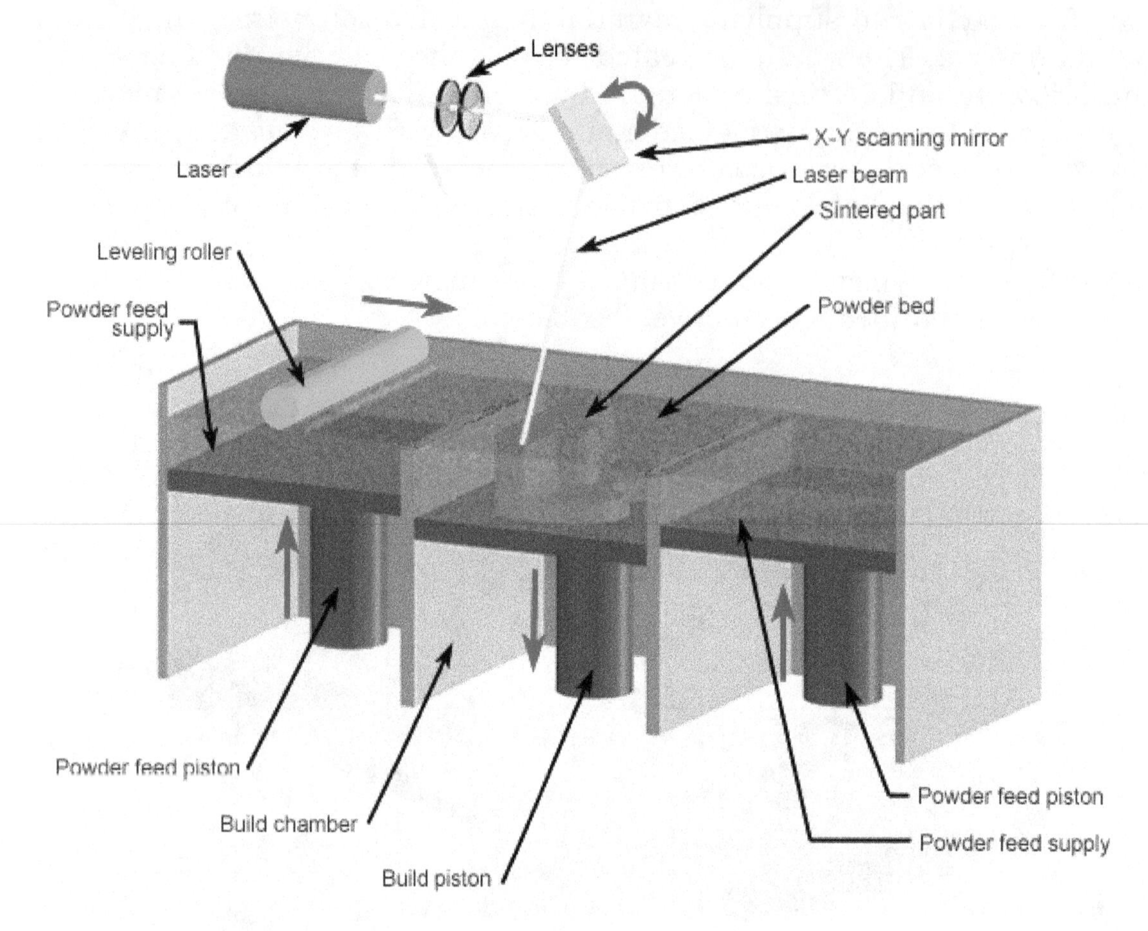

Figure-5. 3D Printing via Selective Laser Sintering

Sheet Lamination

Sheet lamination involves material in sheets which is bound together with external force. Sheets can be metal, paper or a form of polymer. Metal sheets are welded together by ultrasonic welding in layers and then CNC milled into a proper shape; refer to Figure-6. Paper sheets can be used also, but they are glued by adhesive glue and cut in shape by precise blades.

Directed Energy Deposition

This process is mostly used in the high-tech metal industry and in rapid manufacturing applications. The 3D printing apparatus is usually attached to a multi-axis robotic arm and consists of a nozzle that deposits metal powder or wire on a surface and an energy source (laser, electron beam or plasma arc) that melts it, forming a solid object; refer to Figure-7.

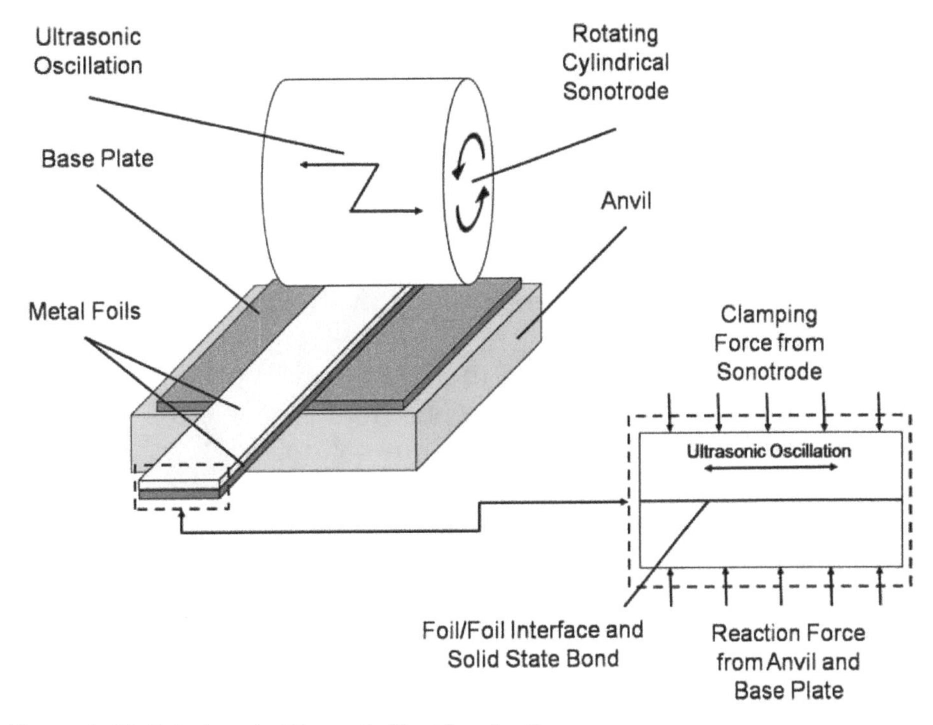

Figure-6. 3D Printing via Ultrasonic Sheet Lamination

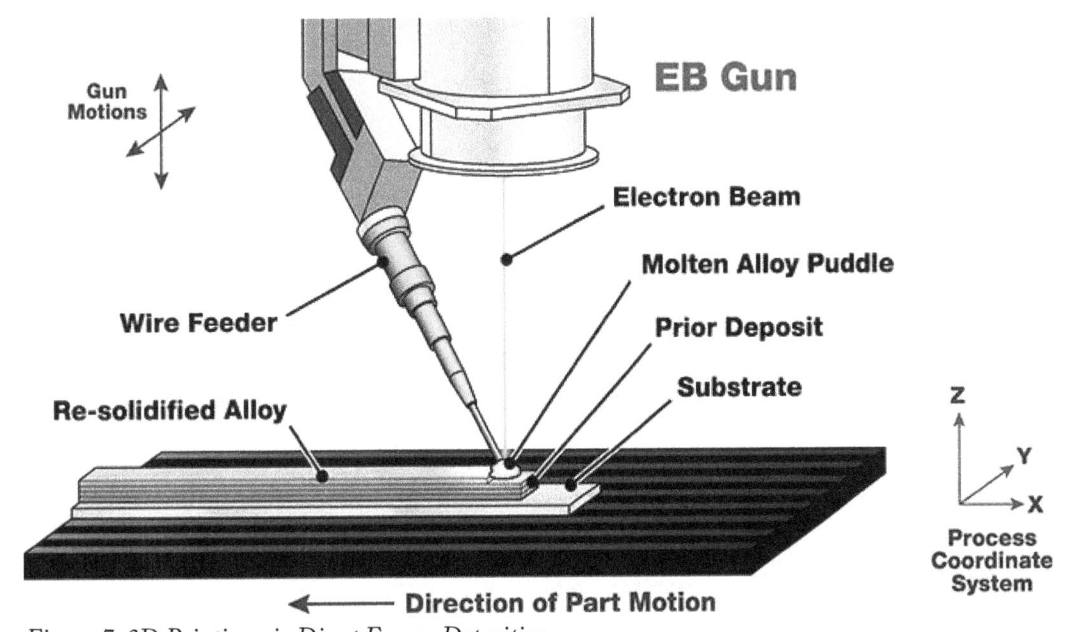

Figure-7. 3D Printing via Direct Energy Deposition

Now, you know about different 3D Printing techniques so it is clear from different processes that the main work of 3D printer is to solidify material at your will, through different techniques. Now, we will learn about the points to be taken care of while preparing model for 3D printing.

Part Preparation for 3D Printing

Various important points to remember while preparing part for 3D printing are given next.

- You should avoid holes in the areas where model is not supported. Holes can cause material to flow out in various 3D printing processes.

- Use the Solid modeling tools. It does not mean that you cannot use surface modeling tools but you should avoid surfaces in your model. Thicken your surfaces after performing modeling operations.
- Make sure that you have not left any unwanted piece inside the model enclosure after performing boolean operations.
- Shell your model after creating it. You can 3D print solid model but if a hollow box can do your work then why to waste material on solid cube. Less material means cost efficient.
- If you want to write text on your model then check the specification of your printer for possible font range.
- The color and textures you apply on model in SolidWorks will not be exported for 3D printing so do not waste your time on them.
- 3D printing is closer to mesh modeling than solid modeling. So, check your mesh model by exporting solid model in .stl file.
- If your file has quite a bit of text and multiple emboss/engrave features, try exporting the file as a vector. Vector files are more appropriate for extremely complex files. Try .iges or .step.
- Orientation is particularly important when it comes to 3D printing in order to determine the interior and exterior of an object. Orient your part the same way as your want it in 3D Printing.
- Don't make a multi-body model for 3D printing. Your printer may die thinking what to do when two separate bodies overlap each other!!
- SolidWorks do not has capability to edit mesh model so you should prepare your model carefully to get desired mesh model.

We know the basic guidelines for preparing 3D model so it is time to export or print model on 3D Printer.

PRINT3D

The **Print3D** tool is used to print the model in drawing area on the 3D Printer. The procedure to use this tool is given next.

- After preparing model, click on the **Print3D** tool from the **File** menu. The **Print3D PropertyManager** will be displayed; refer to Figure-8. Also, you will be asked to select a flat face of the model to place it on machine bed. Note that you should have Windows 8.1 or higher installed in system to use **Print3D PropertyManager**.
- Select desired face of the model to place it on the machine bed. The model will be displayed in a transparent box. This transparent box is the limit of your machine to 3D print; refer to Figure-9.
- Select desired printer from the drop-down in the **Printer** rollout of the **PropertyManager**.

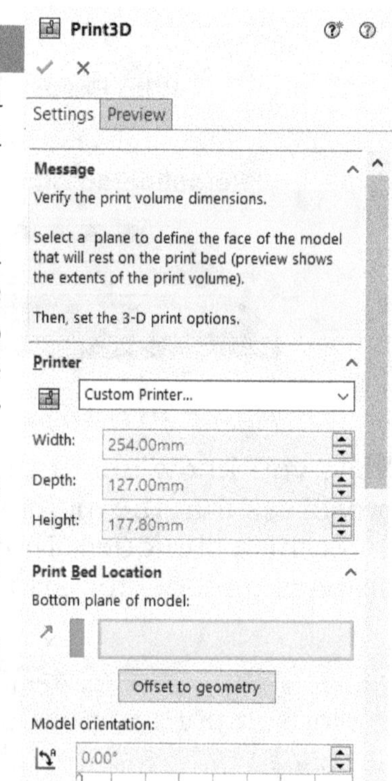

Figure-8. Print3D PropertyManager

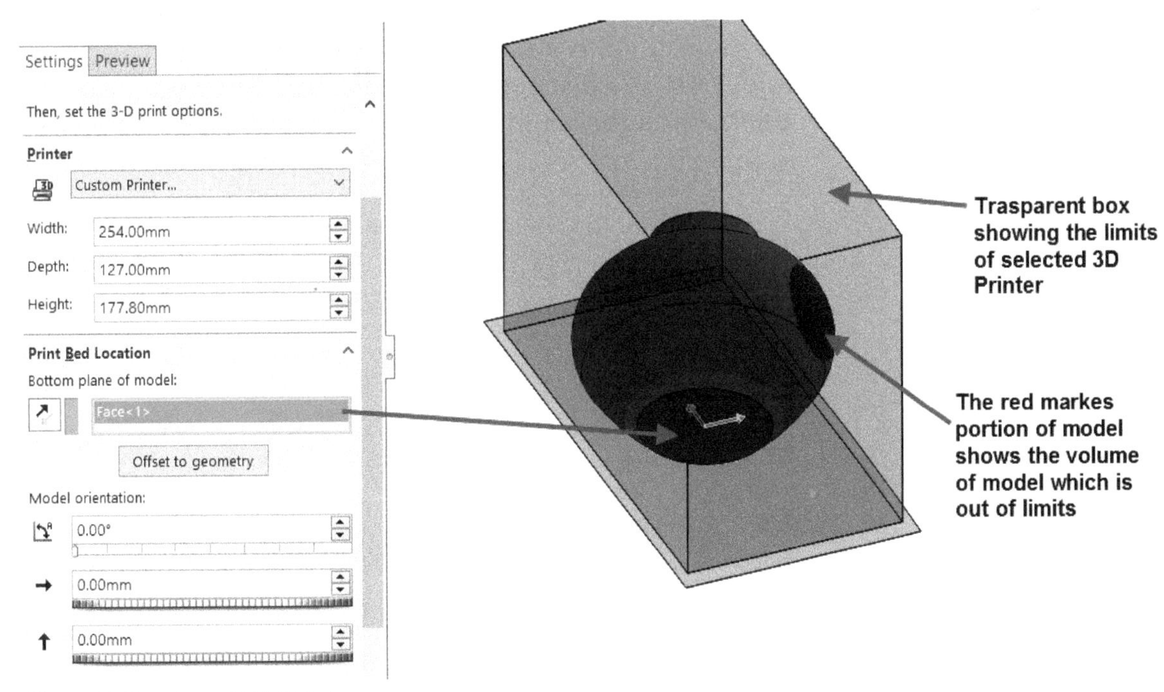

Figure-9. Preview of 3D Printer

- Set the orientation and location of part on the printer bed using the options in the **Print Bed Location** rollout of the **PropertyManager**.
- Scale your model to fit properly in printer volume by using the **Scale** edit box or **Scale to Fit** button in the **Scale** rollout.
- Set the quality of printing in the **Job quality** drop-down of **Options** rollout. Select desired percentage of material to be filled inside the object using the **Infill percentage** drop-down. If 0 is selected in this drop-down then the part will be created hollow.
- Select the **Include supports** check box to automatically add support in the part during 3D Printing. Support is applied where material can fall down during 3D Printing.
- Select the **Include raft** check box to include raft in the 3D Printed model. This raft of disposable material can be removed after job of printing is complete from the base plate.
- Click on the **Printer Properties** button to set properties of 3D Printer.
- To save the current model in 3D Printing format file, expand the **Save To File** rollout at the bottom of **PropertyManager**. The rollout will be displayed as shown in Figure-10.

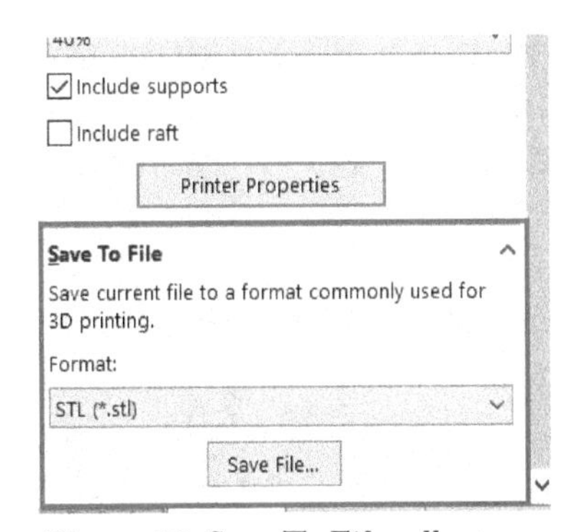

Figure-10. Save To File rollout

- Select desired format from the drop-down and click on the **Save File** button. The **Save As** dialog box will be displayed. Save the file at desired location.
- To run analysis on the object to be printed, click on the **Preview** tab in the **PropertyManager**. The **PropertyManager** will be displayed as shown in Figure-11.

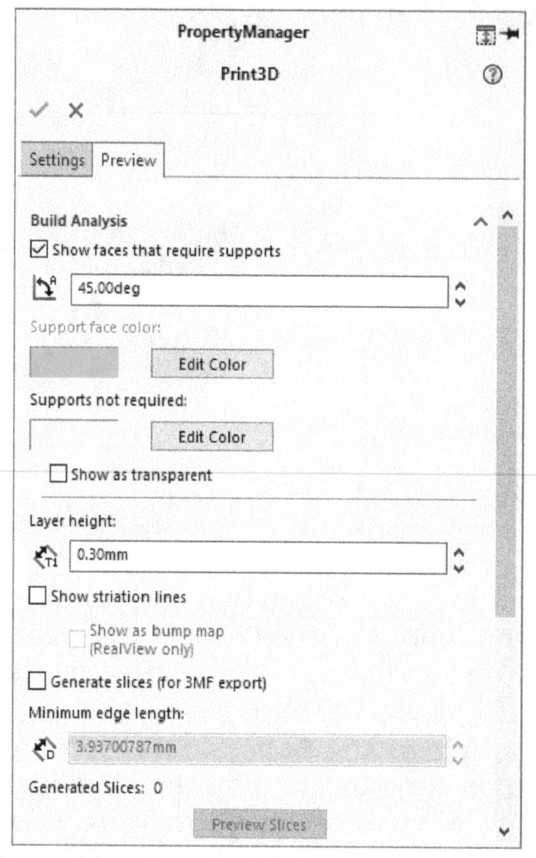

Figure-11. Preview_tab_of_Print3D_PropertyManager

- Set desired parameters in the **PropertyManager** and then click on the **Calculate** button. By default, the areas which need support are displayed in green color. You can change the color by using **Edit Color** buttons. Select the **Show as transparent** check box in the **Build Analysis** rollout of **PropertyManager** to check the inside of part.
- Click on the **OK** button from the **PropertyManager** to start printing.

MODEL BASED DEFINITION (MBD)

MBD is an integrated drawing-less manufacturing solution for SOLIDWORKS. It helps to define, organize, and publish 3D Product Manufacturing Information (PMI) including 3D model data in industry standard file formats, such as eDrawings, 3D PDF and STP242. Unlike traditional 2D drawings, **MBD** guides the manufacturing process directly in 3D, which helps streamline production, cut cycle time, reduce errors, and support industry standards such as Military-Standard-31000A, ASME Y14.41, ISO 16792, and DIN ISO 16792. The purpose of using **MBD** is to reduce time of creating 2D drawings and hence reducing engineering cost. Using MBD, you can display the annotations in the model itself while designing. If you are going to create 2D drawings after creating model then using MBD will not be much beneficial to you. The tools for MBD are available in the **MBD CommandManager** of the **Ribbon**; refer to Figure-12.

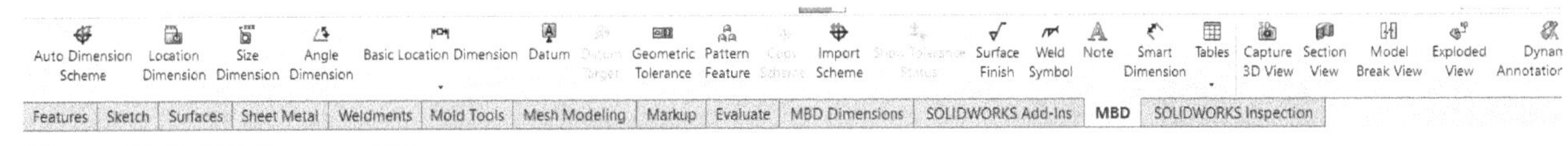

Figure-12. MBD CommandManager

Various tools in this **CommandManager** are discussed next.

Auto Dimension Scheme

The **Auto Dimension Scheme** tool is used to display all the dimensions you have earlier applied to the model during sketching and other operations. The procedure to use this tool is given next. (Note: we have used the model of **Chapter 09** of this book as example for this tool.)

- Make sure you have a model in the drawing area with dimensions applied to it in sketching and other operations. Click on the **Auto Dimension Scheme** tool from the **MBD CommandManager** in the **Ribbon**. The **Auto Dimension Scheme PropertyManager** will be displayed as shown in Figure-13.
- From the **Part type** area of **Settings** rollout, select the **Prismatic** radio button if you want the dimensions and geometric tolerances to be displayed in the form of callouts; refer to Figure-14. Select the **Turned** radio button if you want the dimensions and geometric tolerances to displayed by standard dimensioning scheme; refer to Figure-14.
- From the **Tolerance type** area of the **Settings** rollout, select the **Plus and Minus** radio button to display the dimensions with plus-minus tolerances; refer to Figure-14. Select the **Geometric** radio button to display the tolerances in geometric tolerance form; refer to Figure-15.

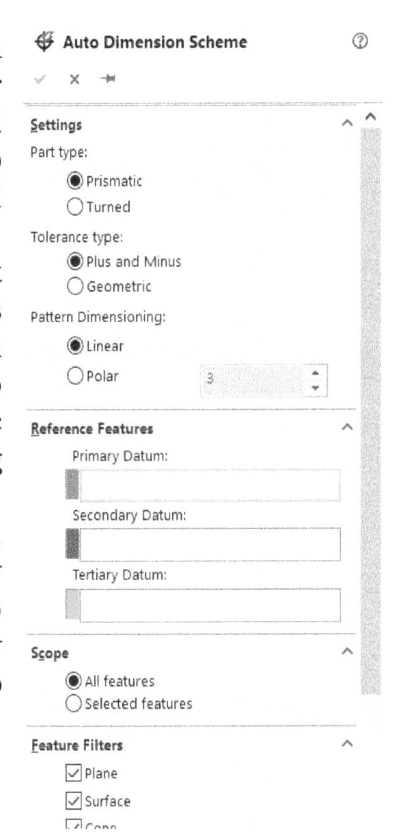

Figure-13. Auto Dimension Scheme PropertyManager

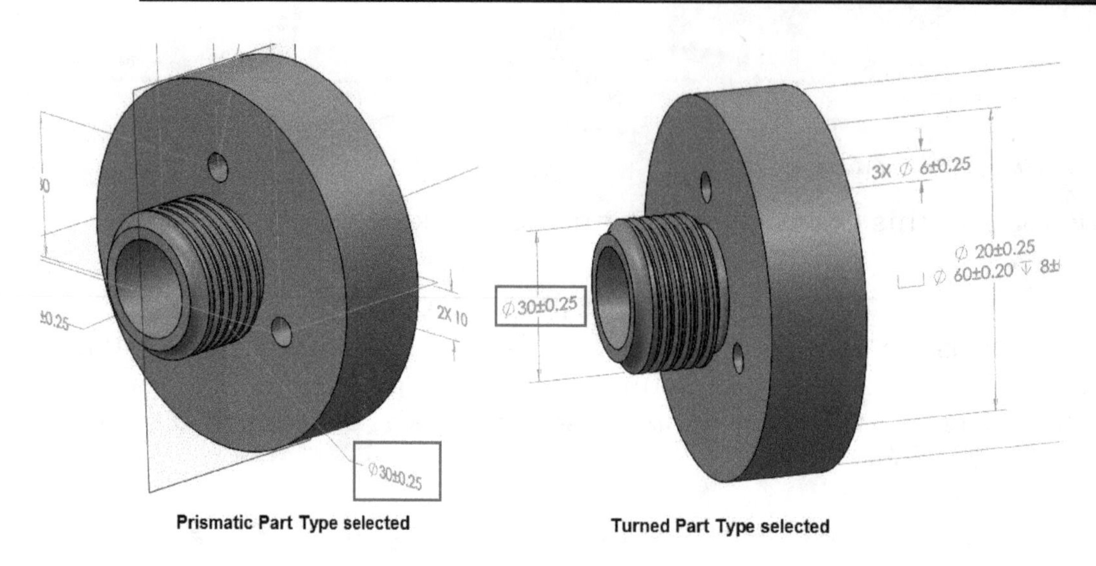

Figure-14. Dimensioning by different part types

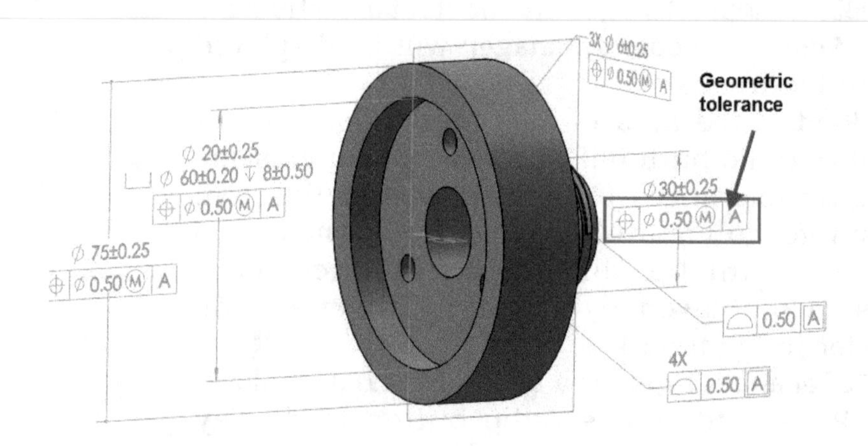

Figure-15. Dimensioning by geometric tolerance with Turned part type

- Select the **Linear** radio button from **Datum Selection** area to display the linear pattern in dimensioning. In this way, polar patterns will be dimensioned as linear instances. Select the **Polar** radio button to display polar pattern in dimensioning; refer to Figure-16. After selecting the **Polar** radio button, set the minimum number of instances to be taken as pattern in the spinner next to radio button.

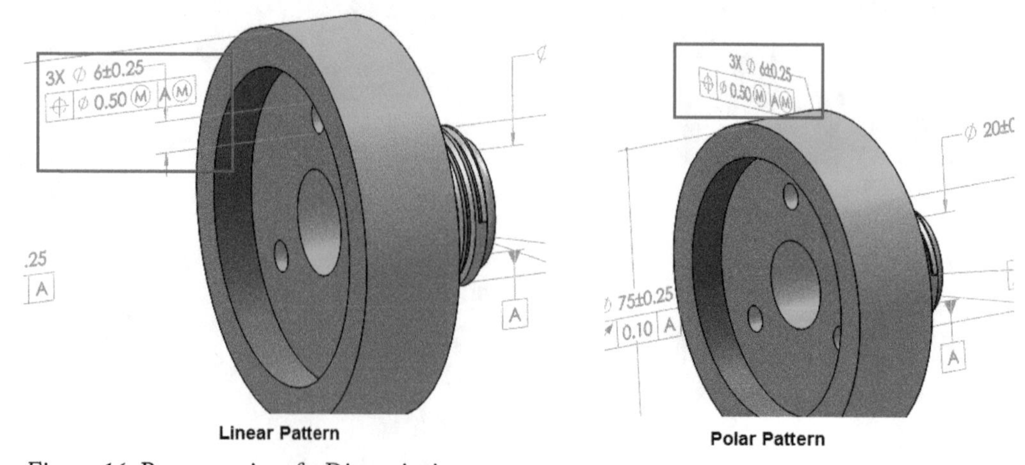

Figure-16. Pattern options for Dimensioning

- Click in the **Primary Datum** selection box of **Datum Selection** rollout and select the reference face/plane/axis to be the primary datum for geometric tolerance. Similarly, select the Secondary and Tertiary datums; refer to Figure-17.

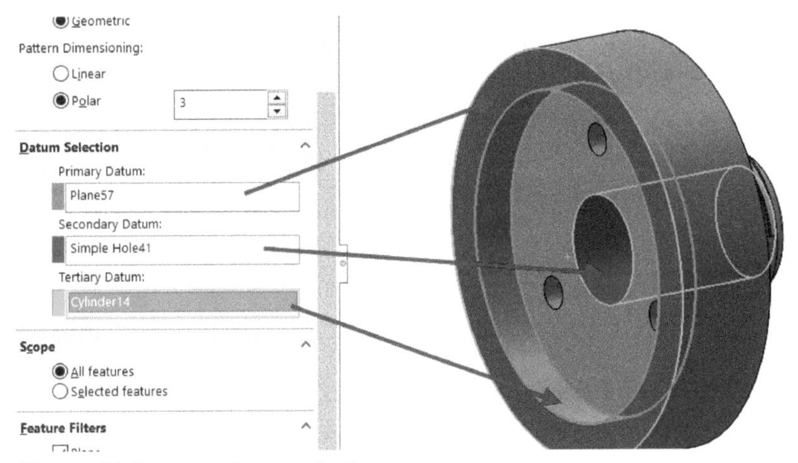

Figure-17. Datum reference selection

- Note that on selecting reference, a small toolbar is displayed along the cursor. Select the appropriate button from toolbar while selecting datum reference.
- Set the scope for dimensioning from **Scope** rollout and select check boxes to include features for dimensioning from the **Feature Filters** rollout.
- Click on the **OK** button from the **PropertyManager** to apply dimensions. Note that the dimension history will be displayed in the **DimXpertManager** at the right in the application window.

Location Dimension

The **Location Dimension** tool is used to create location dimension for various features in the model. The procedure to use this tool is given next.

- Click on the **Location Dimension** tool from the **MBD CommandManager**. You will be asked to select the face(s) to apply dimension.
- Select the faces/axes to be dimensioned. The dimension will get attached to cursor; refer to Figure-18.

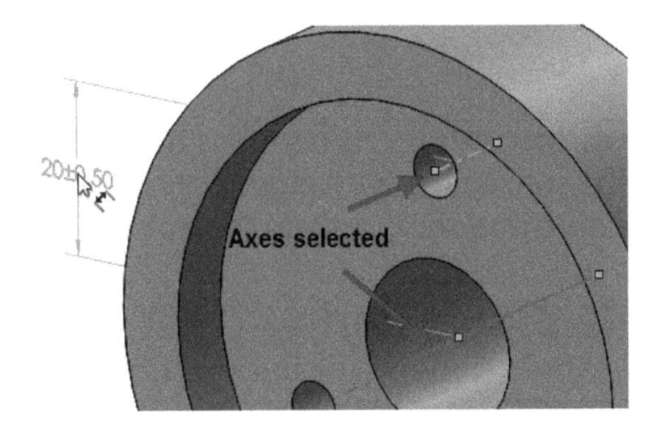

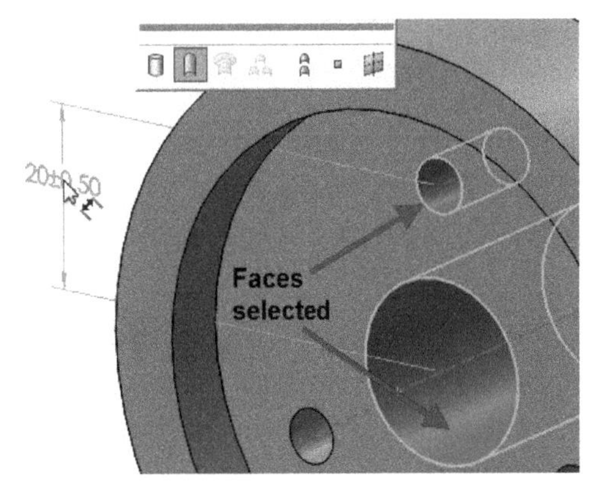

Figure-18. Location Dimensioning

- Click at desired location to place the dimension. The dimension will be created and **DimXpert PropertyManager** will be displayed; refer to Figure-19.
- Set desired parameters and click on the **OK** button. The options in this **PropertyManager** are same as discussed for **Dimension PropertyManager** in previous chapters.

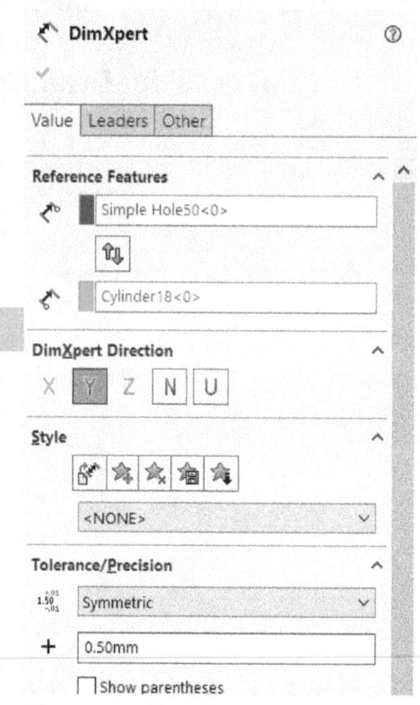

Figure-19. DimXpert Property-Manager

Size Dimension

The **Size Dimension** tool is used to apply size dimensions to various features in the model. The procedure to use Size Dimension tool is given next.

- Click on the **Size Dimension** tool from the **MBD CommandManager**. You will be asked to select the faces/ feature to be dimensioned.
- Select the feature to be dimensioned. The dimension will get attached to cursor and a toolbar will be displayed; refer to Figure-20.

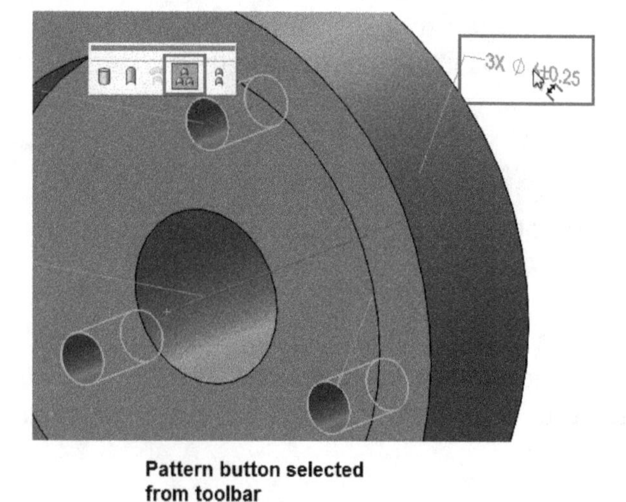

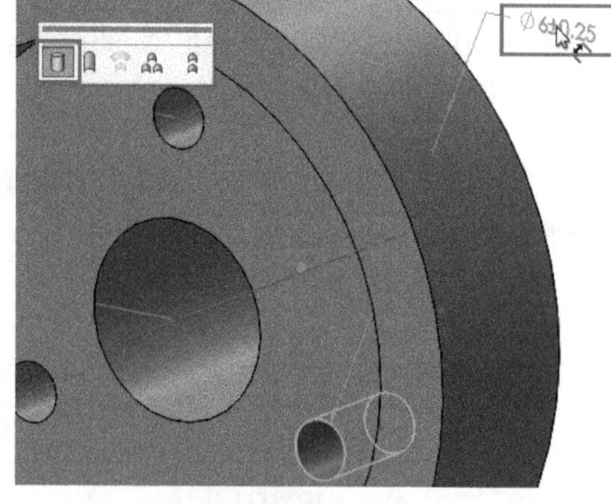

Figure-20. Creating size dimension

- Select desired button from the toolbar to modify the dimension accordingly.
- Click in the drawing area to place the dimension. The **DimXpert PropertyManager** will be displayed. Set desired options and then click on the **OK** button.

Applying Angle Dimension

The **Angle Dimension** tool is used to apply angle dimension between two faces. The procedure to use this tool is given next.

- Click on the **Angle Dimension** tool from the **MBD CommandManager** in the **Ribbon**. You will be asked to select faces for applying dimension.
- Select two faces between which angle dimension will be applied. The angle dimension will get attached to cursor; refer to Figure-21.

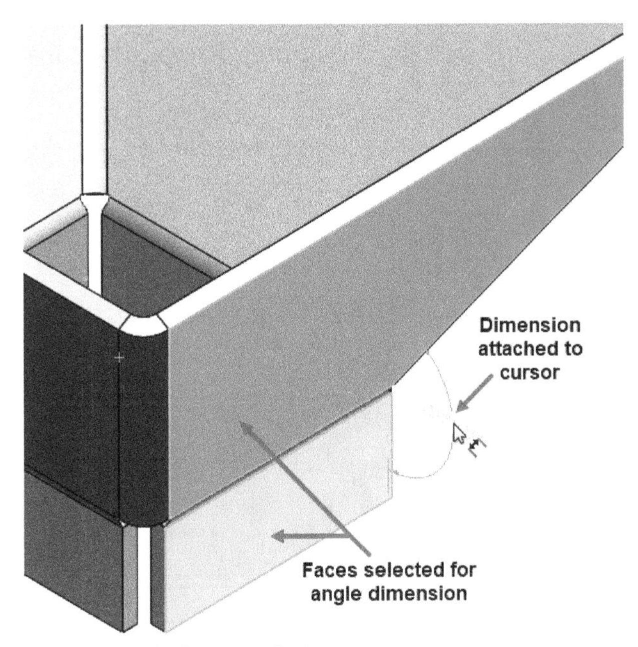

Figure-21. Applying angle dimension

- Click at desired location to place the dimension. The **DimXpert PropertyManager** will be displayed.
- Set the parameters as discussed earlier in **PropertyManager** and click on the **OK** button.

Basic Location Dimension/Basic Size Dimension

The **Basic Location Dimension** and **Basic Size Dimension** tools work in the same way as **Location Dimension** and **Size Dimension** tools respectively. Basic dimensions by nature are theoretically exact values; however, the feature(s) of a part they define as ideal or exact do need to have tolerances to permit acceptable levels of imperfection during manufacturing.

When a feature is defined with basic dimensions, the tolerance for that feature must be expressed through a geometric tolerance. Most often, the geometric tolerance is indicated directly to the feature or feature of size on the face of the drawing; however, some companies include a general geometrical tolerance (such as a position tolerance or profile of a surface tolerance) in the drawing's general notes. This can be an effective tool when the note is carefully written.

Drawings based on ISO standards frequently use a class of general geometrical tolerances standardized in ISO 2768-2 :1989. The three classes are identified through the use of the upper case letters H, K, or L after the ISO 2768 indication on the drawing. You would need a copy of the standard to interpret the amount of tolerance available.

Datum

The **Datum** tool is used to apply datum reference symbol to selected face/axis. The procedure to use this tool is given next.

- Click on the **Datum** tool from the **MBD CommandManager**. The Datum symbol will get attached to cursor and the **Datum Feature PropertyManager** will be displayed; refer to Figure-22.

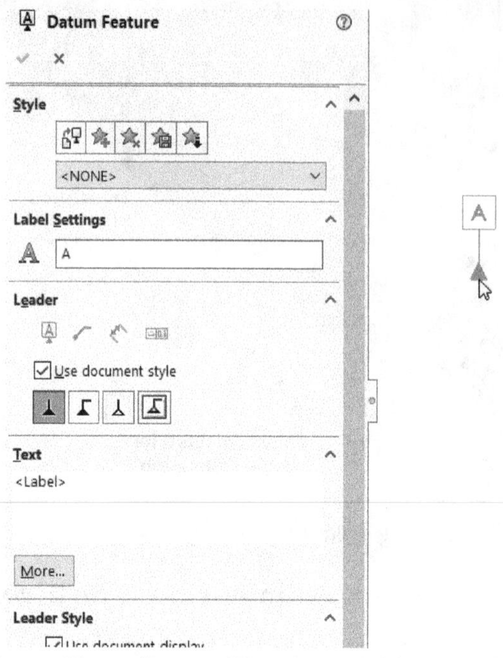

Figure-22. Datum Feature PropertyManager with Datum symbol attached to cursor

- The options in the **PropertyManager** are same as discussed in Chapter 10. Click on the face/axis of model and click at desired location to place the symbol.
- Press **ESC** to exit the tool.

Geometric Tolerance

The **Geometric Tolerance** tool is used to apply geometric dimensioning and tolerance to the selected feature. The tool works in the same way as discussed in Chapter 10 of this book.

Pattern Feature

The **Pattern Feature** tool is used to form a pattern group of same features at different locations. For example, you have created 5 holes on a plate at different locations without using pattern tools. Out of which two holes are in one hole group and three holes are in another hole group; refer to Figure-23. Now if you generate size dimension for holes then there will be two dimensions as shown in Figure-24 (a). If you combine all the holes in one pattern feature then the size dimension will be generated as shown in Figure-24 (b).

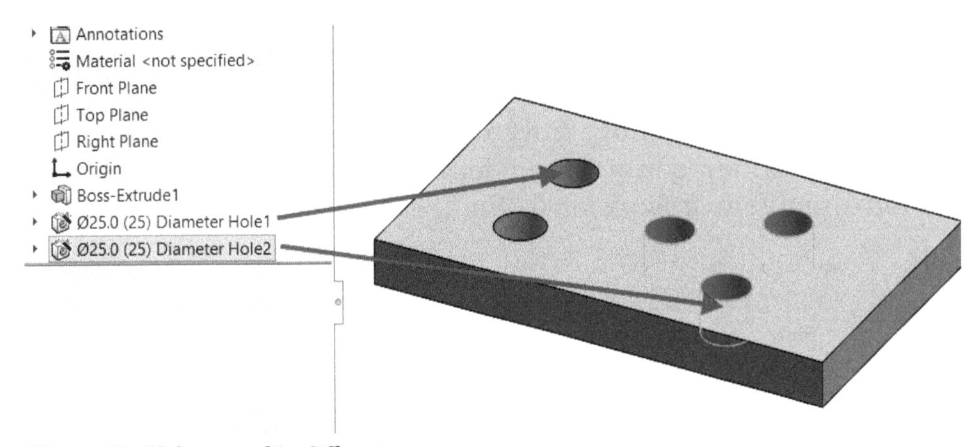

Figure-23. Holes created in different groups

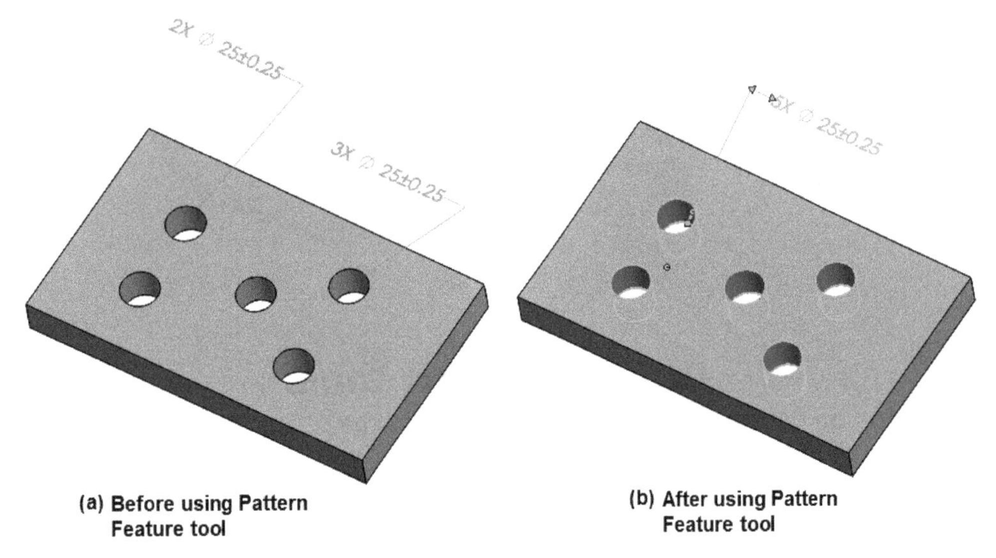

Figure-24. Use of Pattern Feature tool

The procedure to use this tool is given next.

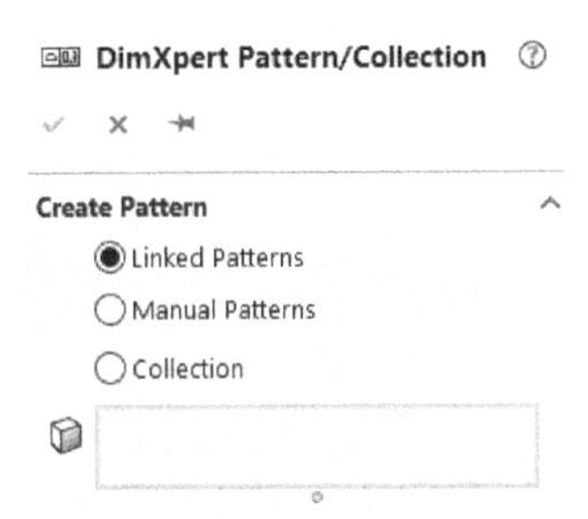

Figure-25. DimXpert Pattern Collection PropertyManager

- Click on the **Pattern Feature** tool from the **CommandManager**. The **DimXpert Pattern/Collection PropertyManager** will be displayed; refer to Figure-25.
- Select the **Linked Patterns** radio button to select one by one the already created patterns to combine them. Select the **Manual Patterns** radio button to select each of the feature individually. On selecting the **Manual Patterns** radio button, the **Find all on same face** check box is displayed. Select this check box if you want to select all the features of same time on the current face. Select the **Collection** radio button to make a collection of features selected. Note that collection is just grouping of features, it does not reflect in dimensioning.
- After selecting desired radio button, select the features from the model and click on the **OK** button.

Show Tolerance Status

The **Show Tolerance Status** tool is used to highlight the features/faces to which tolerances have been applied. Click on this button to toggle tolerance status. The **Green** colored features/faces show fully constrained features, the **Yellow** colored features/faces show under constrained features, and **Red** colored features/faces show over constrained features; refer to Figure-26.

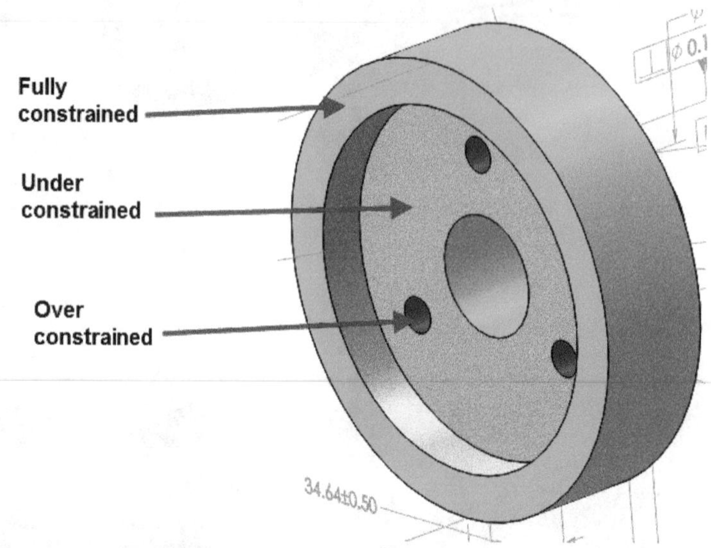

Figure-26. Tolerance status

Other DimXpert Tools

The other tools in the **CommandManager** viz. **Datum Target**, **Surface Finish**, **Weld Symbol**, **Balloon**, **Stacked Balloons**, and **Note** work in the same way as they do in Drawing environment. These tools have been discussed in **Chapter 10** of the book. The **Smart Dimension** tool in **MBD CommandManager** is used to invoke other dimensioning tools of **DimXpert**.

Inserting Tables

All the tools in **Tables** drop-down of the **CommandManager** work in the same way. Here, we will discuss the method to insert Title Block Table, you can use the same method for inserting other tables. The procedure to insert Title Block Table is given next.

- Click on the **Title Block Table** tool from the **Tables** drop-down in the **MBD CommandManager**. The **Title Block Table PropertyManager** will be displayed; refer to Figure-27.
- Click on the **Open template for Title Block Table** button (highlighted by cursor in Figure-27) from the **PropertyManager**. The Open dialog box will be displayed.
- Select desired template of title block from the dialog box and click on the **Open** button. The title block template will be added in the **Table Template** edit box of the **PropertyManager**.
- Set the thickness of table borders and internal lines using the drop-downs in the **Border** rollout of the **PropertyManager**.
- Click on the **OK** button from the **PropertyManager**. If your part is not oriented to any standard view viz. **Front**, **Top** or **Right** then the **Select Annotation View** dialog box will be displayed; refer to Figure-28.

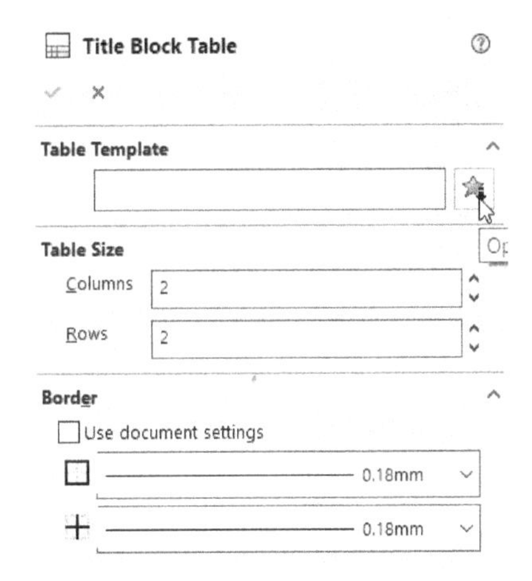

Figure-27. Title Block Table PropertyManager

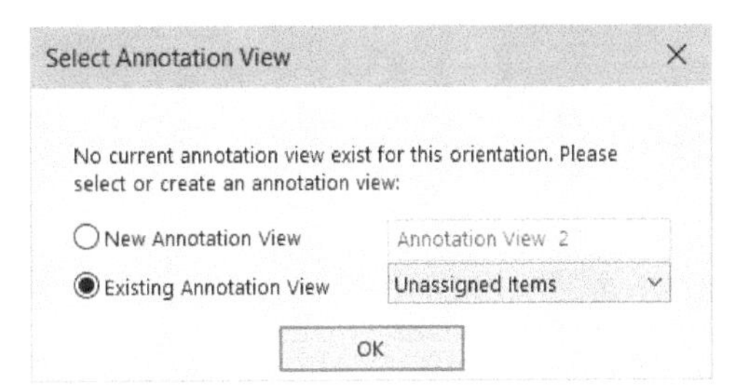

Figure-28. Select Annotation View dialog box

- Click on the drop-down next to **Existing Annotation View** radio button and select desired plane in which you want to place the table.
- Click on the **OK** button from the dialog box. The table will get attached to cursor.
- Click at desired location to place the table. Double-click in the fields of table and enter desired text; refer to Figure-29.

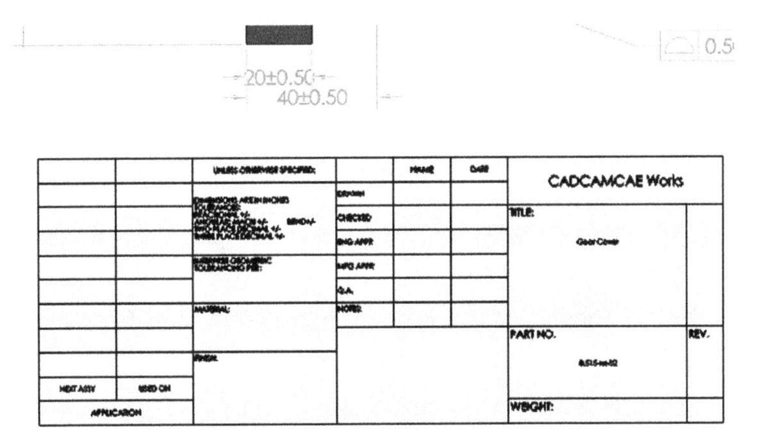

Figure-29. Title block inserted in front plane

Section View

The **Section View** tool as the name suggests, is used to section the part/assembly. Section view is sometimes very important to see the internal functioning of an assembly like in the engines. The procedure to use **Section View** tool is discussed next.

- Make sure you have a part/assembly in the drawing area. Click on the **Section View** tool from the **MBD CommandManager**. The **Section View PropertyManager** will be displayed with preview of section of an assembly; refer to Figure-30.

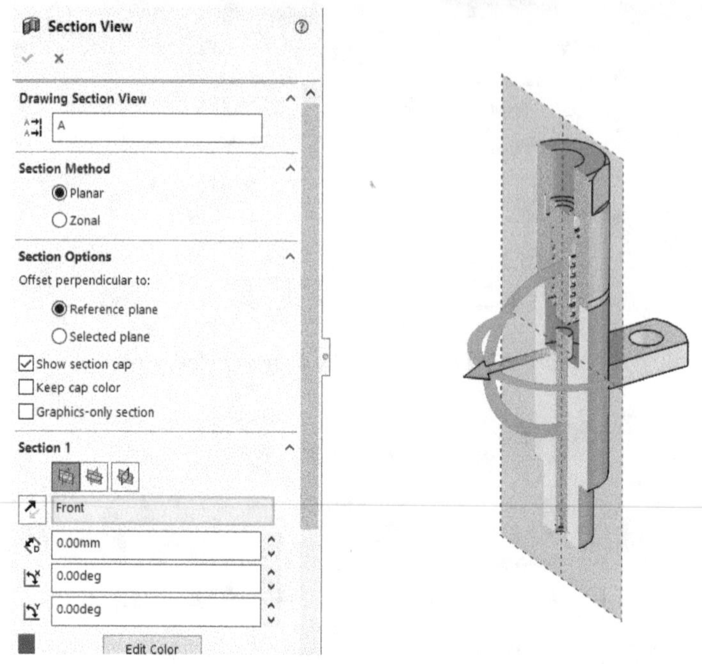

Figure-30. Section View PropertyManager with preview of sectioned assembly

- Specify the naming letter for section view in the **Drawing Section View** edit box in the **PropertyManager**.
- Select the **Show section cap** to make sure that the section view is colored and do not look like surface model. Select the **Keep cap color** check box if you want to display color on the cap applied to section view.
- Select desired place from the **Section 1** rollout to change the plane of section. You can select three buttons in the rollout or you can select any face from the model to set the section plane.
- Similarly, set the other options in the rollout. If you want to create section of model from two sides then select the **Section 2** check box. The **Section 2** rollout will expand. Set desired parameters in the rollout as discussed earlier. In the same way, you can use **Section 3** check box to select third plane/face for sectioning.
- Click on the **OK** button to create the section view; refer to Figure-31. Click on the **Section View** button from the **CommandManager** again to exit the section view.

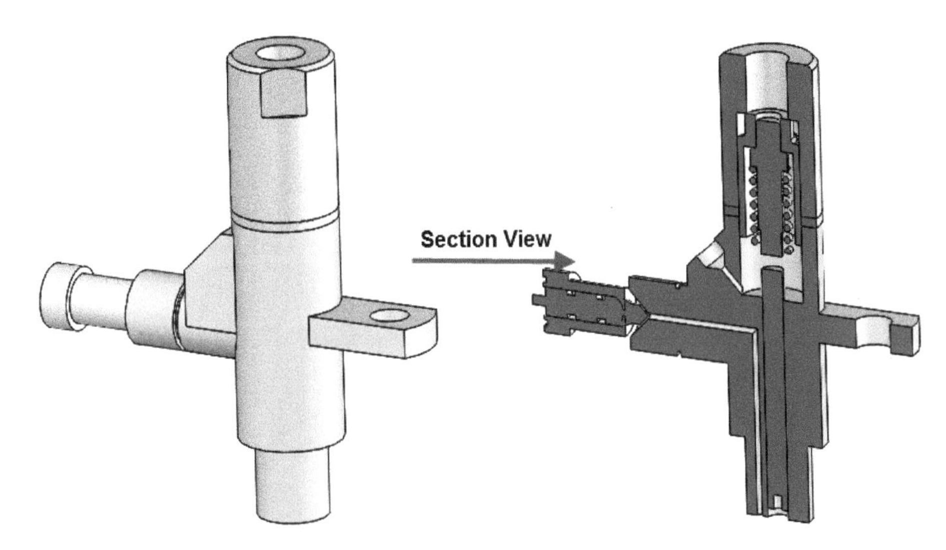

Figure-31. Section view created

Model Break View

The **Model Break View** tool is used to display the model broken by specified plane-gap. The procedure to use this tool is given next.

- Click on the **Model Break View** tool from the **CommandManager**. The **Model Break View PropertyManager** will be displayed; refer to Figure-32.

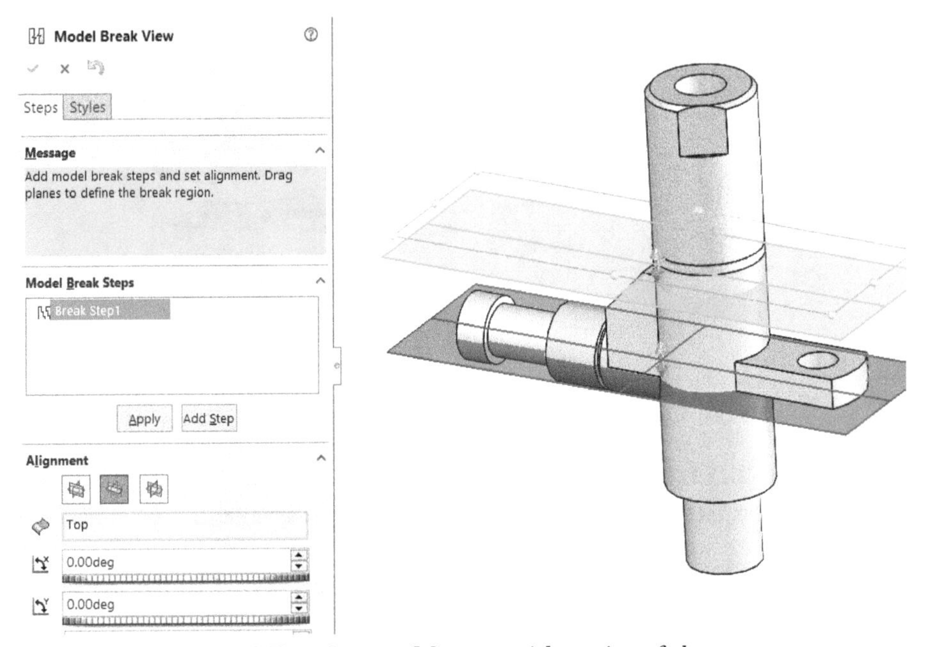

Figure-32. Model Break View PropertyManager with preview of planes

- Select desired plane from the **Alignment** rollout of the **PropertyManager** and enter the parameters like angle along X or Y of the plane, or distance between the two planes displayed in preview. Note that the area between the two planes will be cut.
- Select the **Preview** check box to see the preview while editing values.
- Click on the **Apply** button to apply current break step. If you want to create multiple breaks then click on the **Add Step** button from the **PropertyManager**; refer to Figure-33.

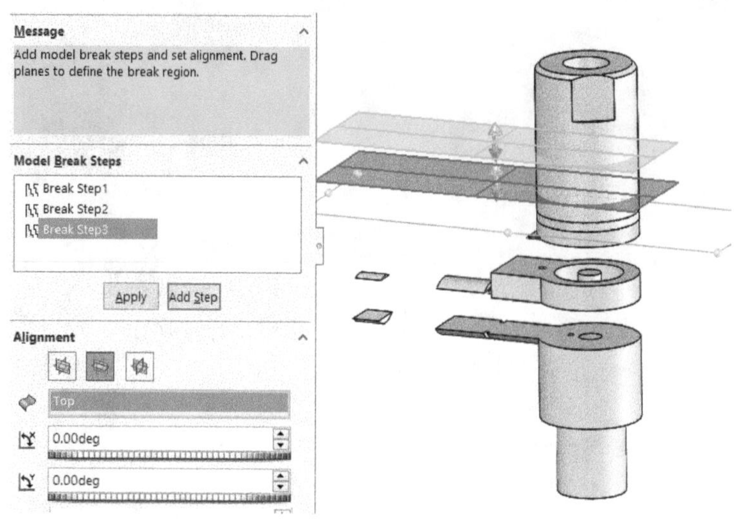

Figure-33. Multiple break created in model

- Click on the **Styles** tab in the **PropertyManager** and select desired style button to check the break view style; refer to Figure-34.

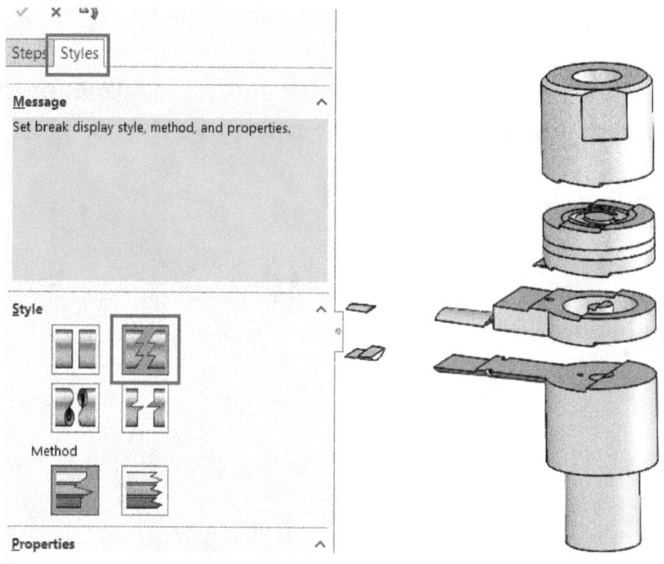

Figure-34. Styles tab in PropertyManager

- Click on the **OK** button from the **PropertyManager** to create the model break view. The new view will be added in the **ConfigurationManager**.

The **Exploded View** tool works in the same way as discussed in **Chapter 7** of this book.

Capture 3D View

While going through previous chapters, you have created 2D views of model like front view, right view, top view, and so on. In SolidWorks, you can create 3D views to represent various aspects of model in 3D. For example, you can save a model view with sections done on the assembly or you can create exploded view of model in 3D. These type of 3D views can be created by **Capture 3D View** tool. The procedure to use this tool is given next.

- Orient the part as you want it to be in the 3D view. Click on the **Capture 3D View** tool from the **MBD CommandManager**. The **Capture 3D View PropertyManager** will be displayed; refer to Figure-35.

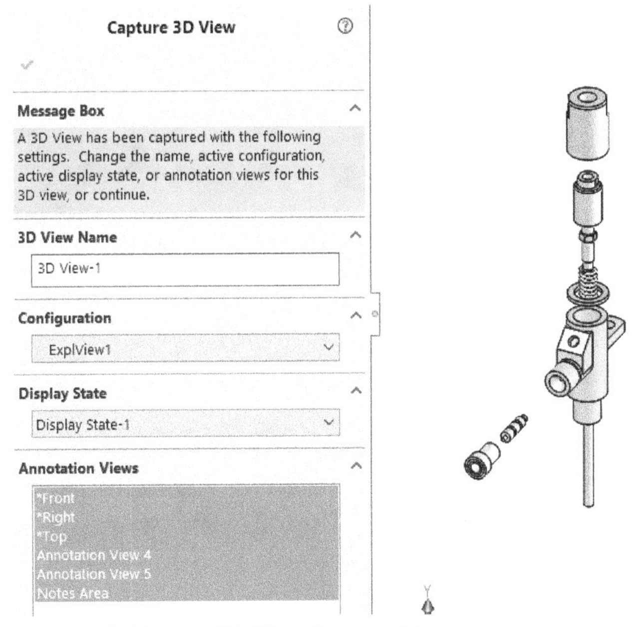

Figure-35. Capture 3D View PropertyManager

- Specify the name of view in the **3D View Name** edit box of the **PropertyManager**.
- Select the views from **Annotation Views** rollout to include their annotations in the current 3D view.
- Click on the **OK** button. A new 3D view will be added to the **3D Views** tab at the bottom in the application window; refer to Figure-36.

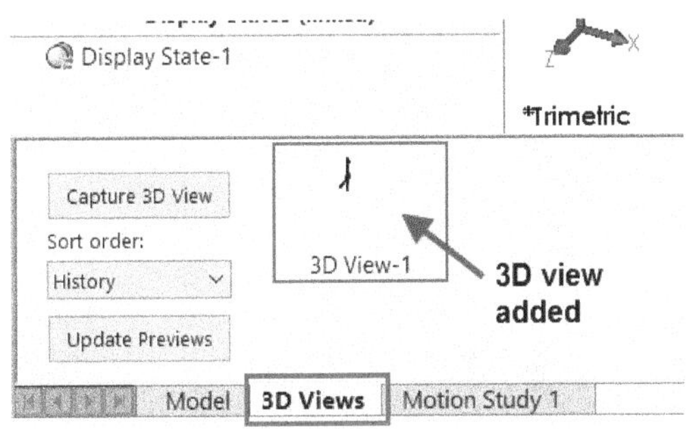

Figure-36. 3D view added

- Change the orientation or make a section view and then click on the **Capture 3D View** tool again to create another 3D view.

Dynamic Annotation Views

The **Dynamic Annotation Views** tool is used to make the annotations dynamic with respect to selected view. For example, if you have applied annotations to the model in front and right view, and you activate the **Dynamic Annotation Views** tool then you will not see the annotation of right view when model is in front view. Click on the **Dynamic Annotation Views** tool from the **MBD CommandManager** to toggle and rotate the model in various views to check annotations relevant to current view only; refer to Figure-37.

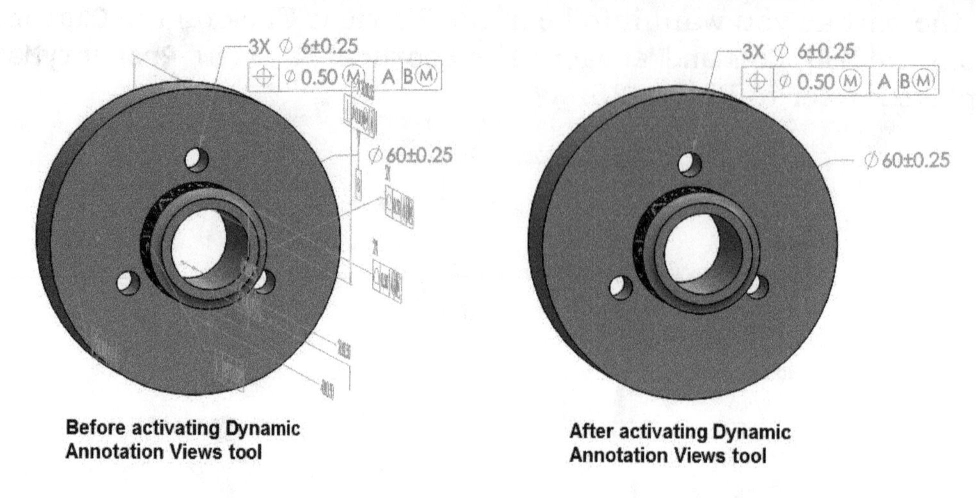

Figure-37. Result of selecting Dynamic Annotation Views tool

Publish to 3D PDF

The **Publish to 3D PDF** tool is used to create 3D PDF file from the model which can be distributed to clients who are not having CAD software in their system. Note that you need to capture a 3D view in **3D Views** tab of SolidWorks to create 3D PDF. The procedure to use this tool is given next.

- Click on the **Publish to 3D PDF** tool from the **MBD CommandManager**. The **Template Selection** dialog box will be displayed; refer to Figure-38.

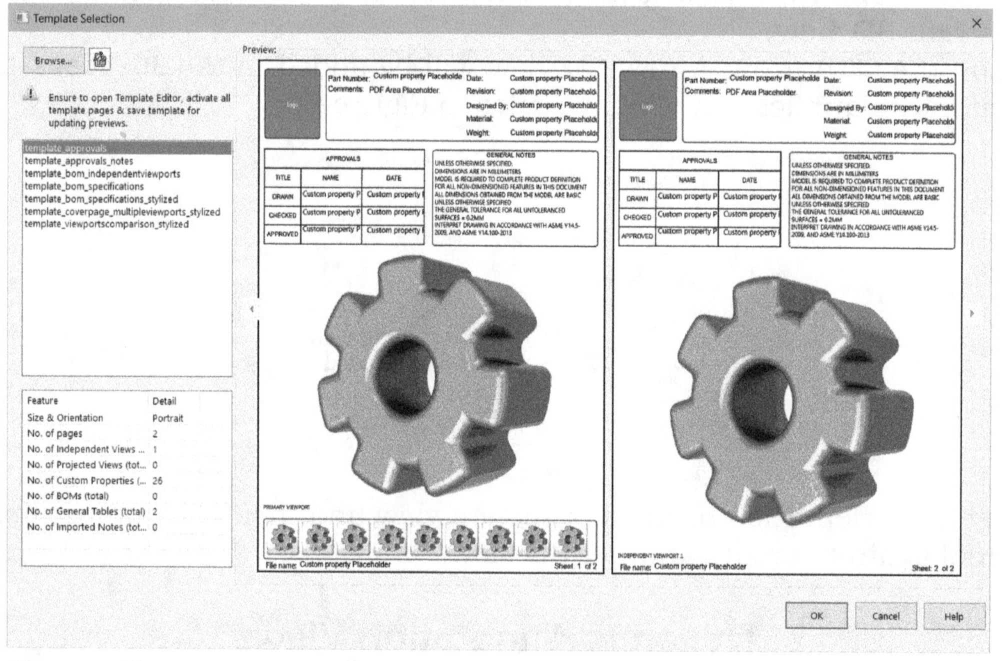

Figure-38. Template Selection dialog box

- Select desired template from the left area of the dialog box and click on the **OK** button. The PDF will be created; refer to Figure-39 and **Publish to 3D PDF PropertyManager** will be displayed; refer to Figure-40.

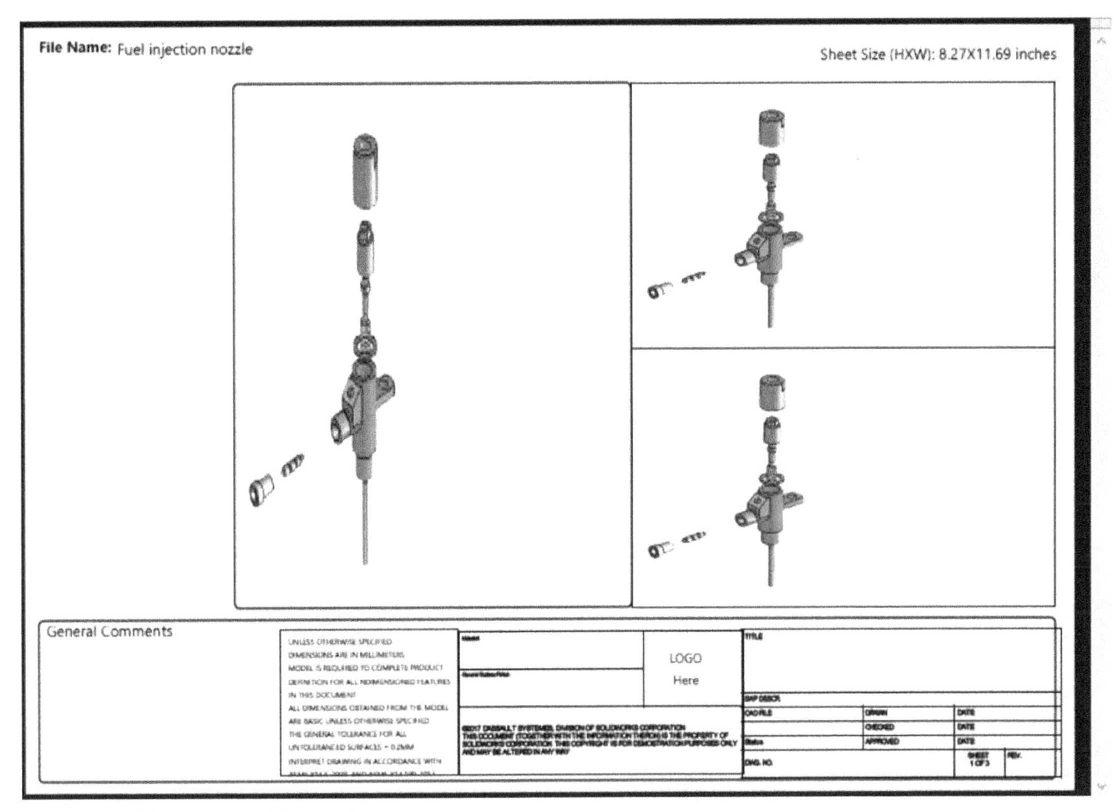

Figure-39. Temporary pdf created

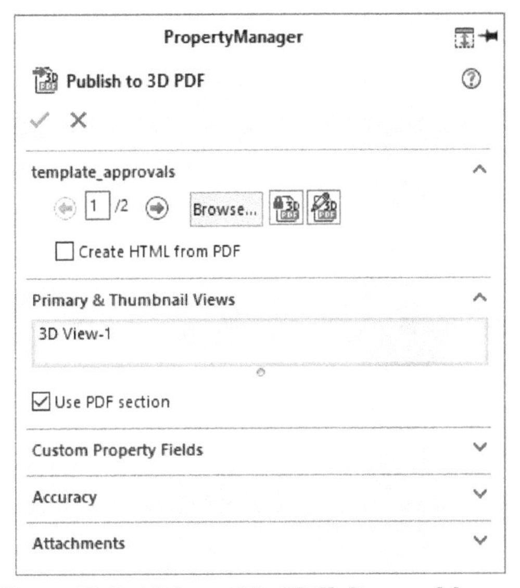

Figure-40. Publish_to_3D_PDF_PropertyManager

- Expand the **Custom Property Fields** rollout from the **PropertyManager** and set desired parameters like Revision date, designed by, and so on.
- Expand the **Attachments** rollout and click on the **Attach Files** button. The **Open** dialog box will be displayed.
- Select desired file and click on the **Open** button to attach it to 3D PDF.
- Set the other properties as desired and click on the **OK** button from the **PropertyManager**. The **Save As** dialog box will be displayed; refer to Figure-41.

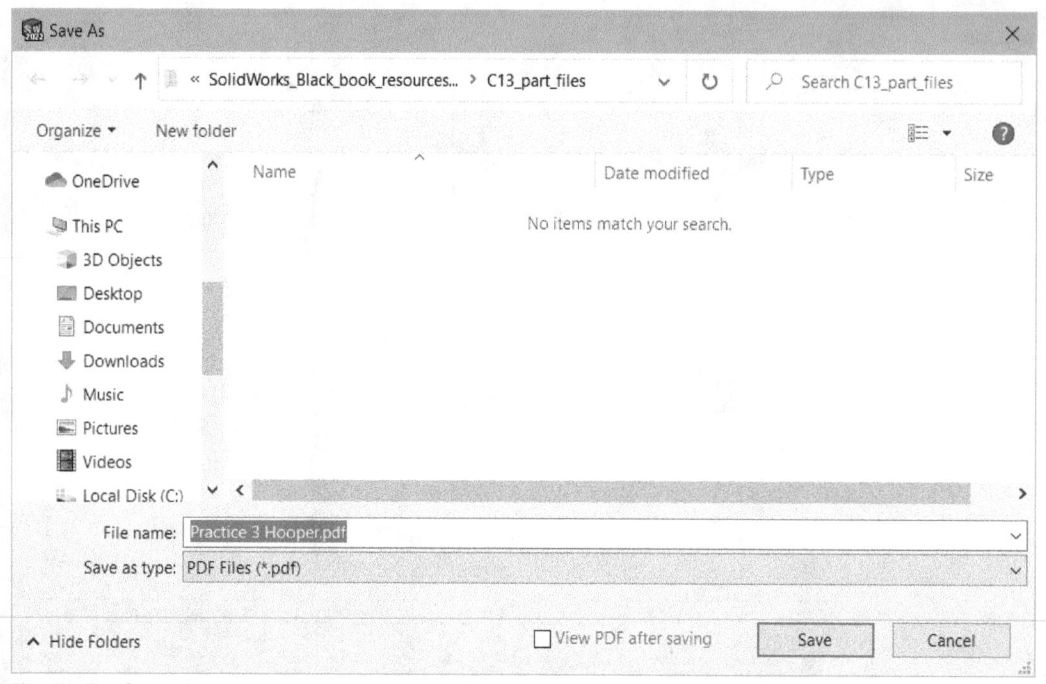

Figure-41. Save As dialog box

- Select the **View PDF after saving** check box to open the pdf after creation.
- Specify desired name of file, move to desired directory, and click on the **Save** button to save the file. The 3D PDF file will be created; refer to Figure-42.

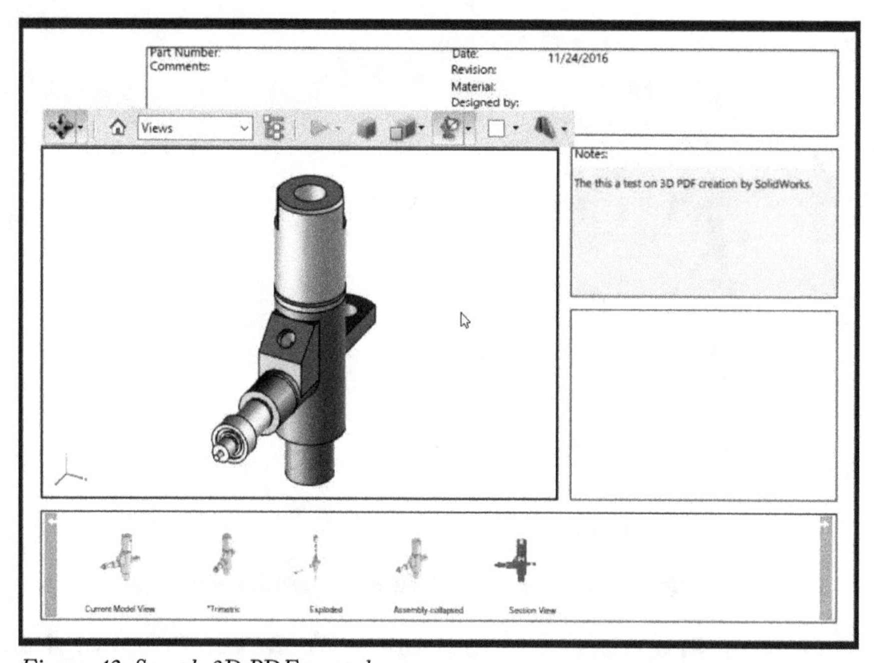

Figure-42. Sample 3D PDF created

- You can perform various actions in the 3D PDF like zoom, pan, rotate, and spin the model. You can also measure various parameters in the PDF itself using the tools in PDF reader.

3D PDF Template Editor

The **3D PDF Template Editor** tool is used to edit/create the templates using which the 3D PDFs are created. The procedure to use **3D PDF Template Editor** tool is given next.

- Click on the **3D PDF Template Editor** tool from the **CommandManager**. The **Template Editor** window will be displayed; refer to Figure-43.

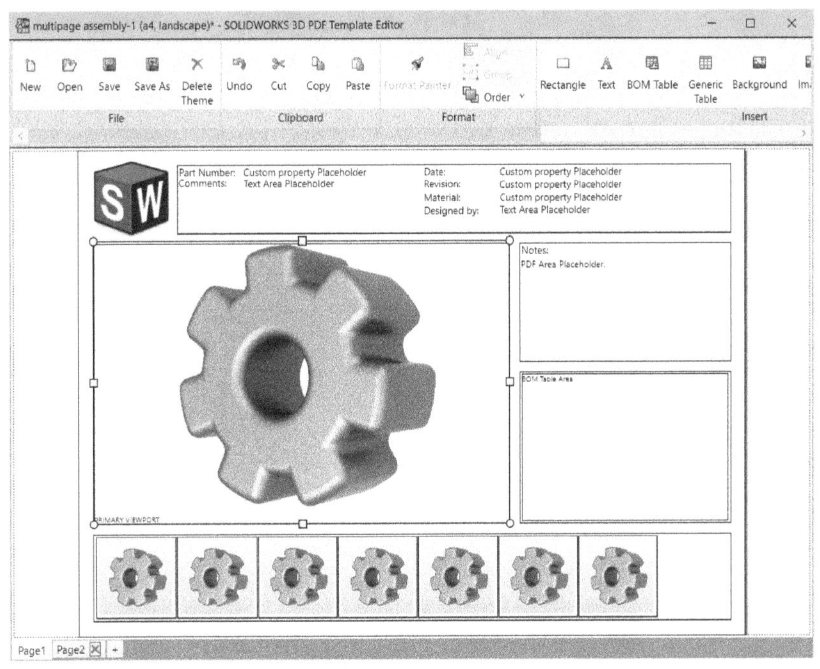

Figure-43. Template Editor window

- Click at the item you want to change. The respective editing box will be displayed; refer to Figure-44.

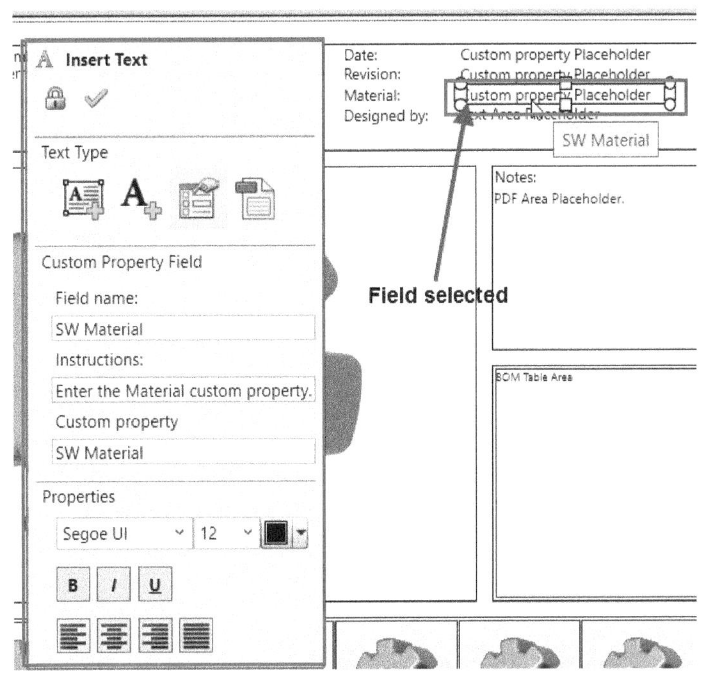

Figure-44. Editing box for fields in 3D PDF template

- Set desired parameters and click on the **Save/Save As** button to save the template.

Now, you can use the newly created template to generate 3D PDFs. Note that you need to start SolidWorks with Administrator rights to save the template file in system directories.

In the same way, you can publish eDrawings file and STEP 242 file.

3D PMI Compare

The **3D PMI Compare** tool is used to compare the 3D product and manufacturing information (PMI) of two documents. It compares DimXpert annotations, reference dimensions, and other annotations between two part documents. The procedure to use this tool is given next.

- Click on the **3D PMI Compare** tool from the **CommandManager**. The **3D PMI Compare** task pane will be displayed on the right in the application window; refer to Figure-45.
- Click on the browse button next to **Reference document** drop-down in the task pane and open the model you want to compare.
- In the same way, select the document for **Modified document** drop-down. The **Run Comparison** button will become active.
- Click on the **Run Comparison** button. The PMI compare list will be displayed in the task pane; refer to Figure-46.

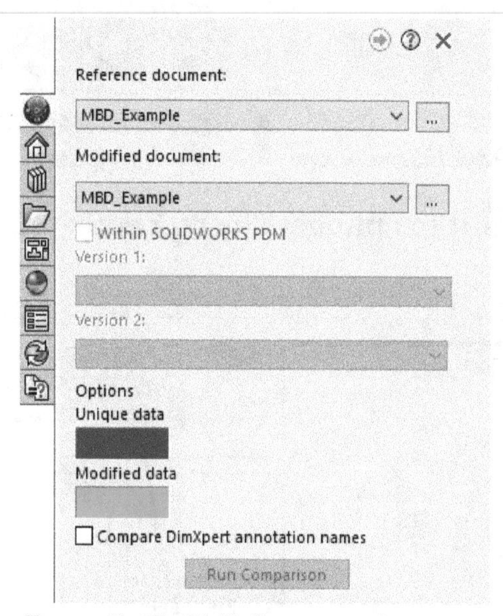

Figure-45. 3D_PMI_Compare_task_pane

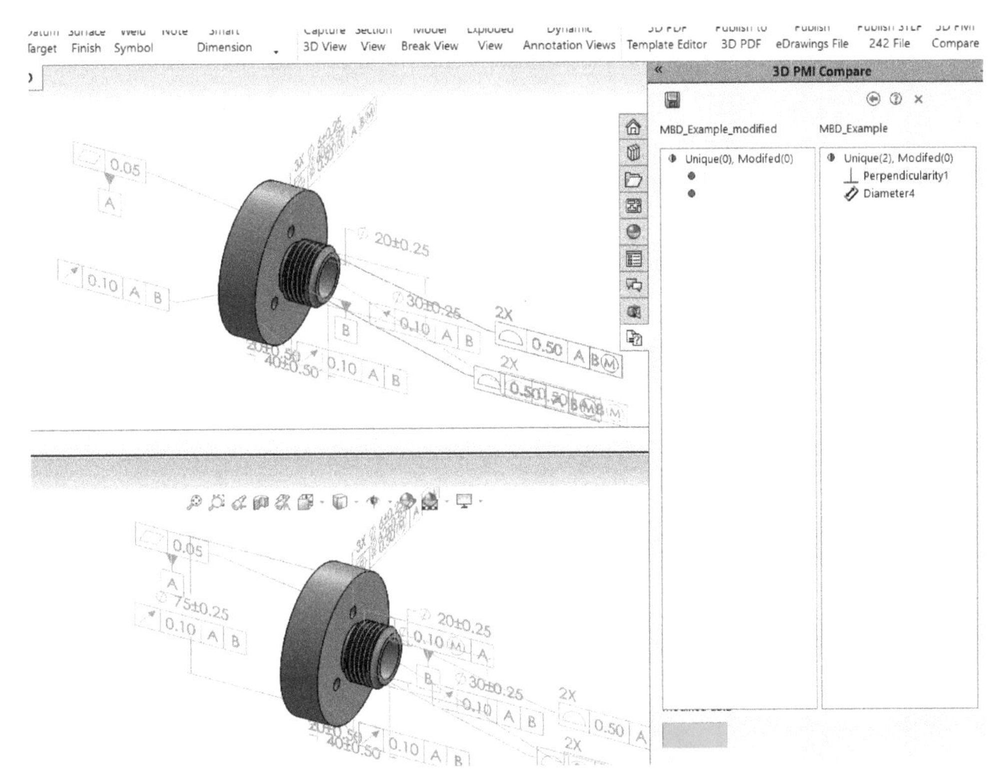

Figure-46. PMI Compare list

- Save the report by using **Save Report** button in the task pane, if required.
- Click on the **Close** button in the task pane to exit.

SOLIDWORKS INSPECTION

The SolidWorks Inspection is an add-in used for creating inspection sheet and documentation used by production and quality department in a manufacturing firm. The tools to create inspection document are available in **SOLIDWORKS Inspection CommandManager**; refer to Figure-47.

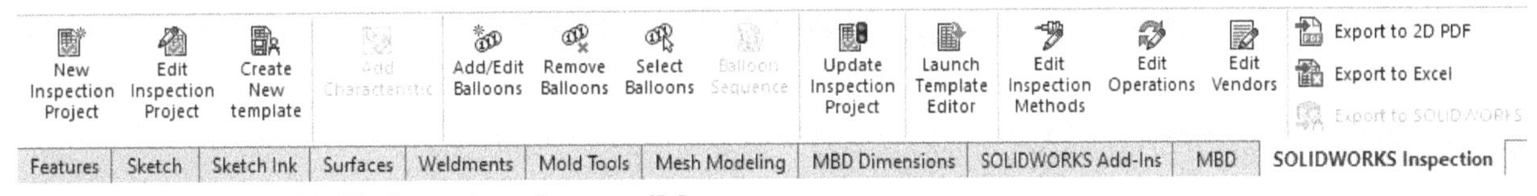

Figure-47. SOLIDWORKS_Inspection_CommandManager

Various tools of this **CommandManager** are discussed next.

STARTING NEW INSPECTION PROJECT

The **New Inspection Project** tool is used to start a new project for inspection. Note that you will need a model dimensioned by MBD or in Drawing for creating inspection sheet The procedure to use this tool is given next.

- Click on the **New Inspection Project** tool from the **SOLIDWORKS Inspection CommandManager** in the **Ribbon**. The **Project Template Selection PropertyManager** will be displayed; refer to Figure-48.

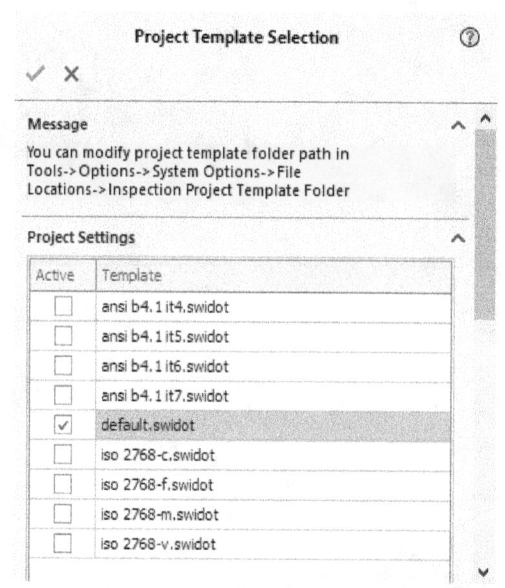

Figure-48. Project_Template_Selection_PropertyManager

- Select desired template and click on the **OK** button from the **PropertyManager**. The **Create Inspection Project PropertyManager** will be displayed; refer to Figure-49.

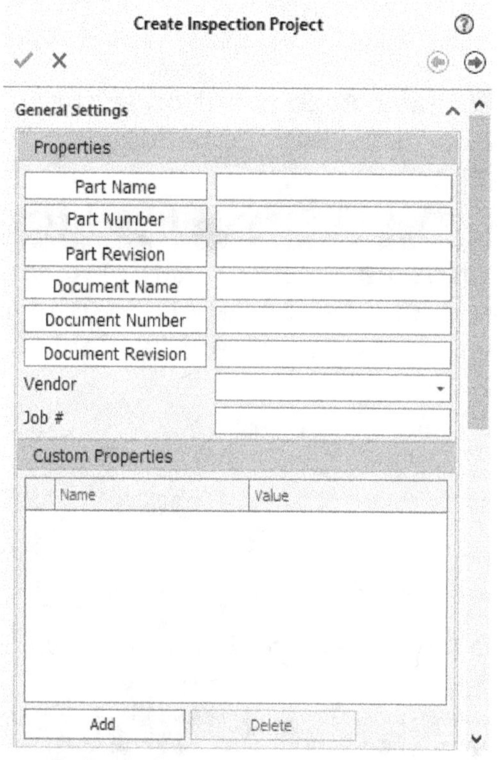

Figure-49. Create_Inspection_Project_PropertyManager

- Specify desired name of part, part number, revision number, document name, vender, and so on in the **PropertyManager**.
- Click on the **Next** button in **PropertyManager**. The options to extract information for inspection sheet will be displayed in the **PropertyManager**; refer to Figure-50.

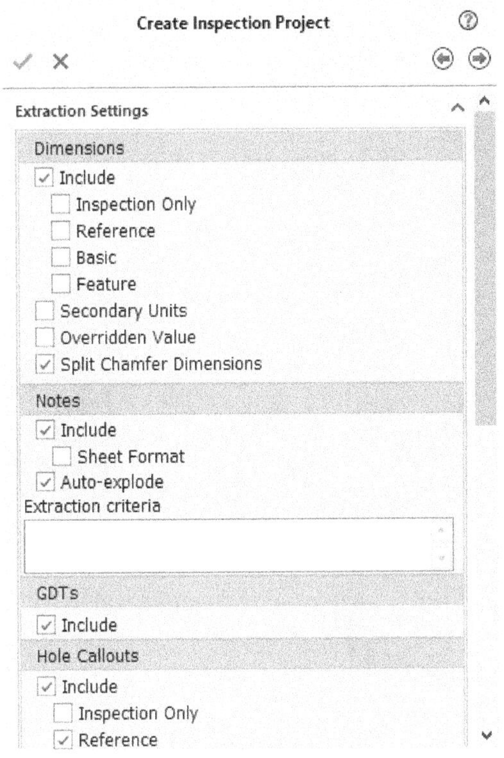

Figure-50. Extraction_Settings_for_inspection_project

- Select check boxes for parameters to be included in the inspection sheet and then click on the **Next** button. The **Tolerance Settings** page of **PropertyManager** will be displayed; refer to Figure-51.

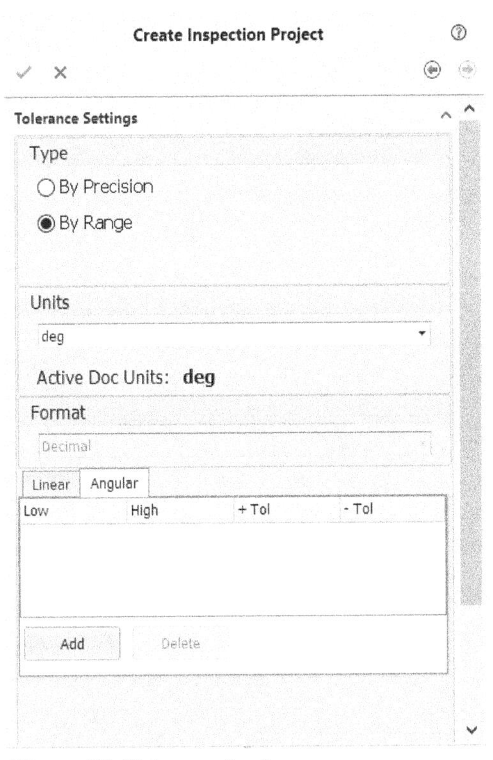

Figure-51. Tolerance Settings page

- Set desired tolerance for linear and angular dimensions. After setting desired parameters, click on the **OK** button from the **PropertyManager**. The inspection table will be generated based on dimensions applied in MBD and drawing views; refer to Figure-52. Note that the balloons will also be generated for inspection dimensions.

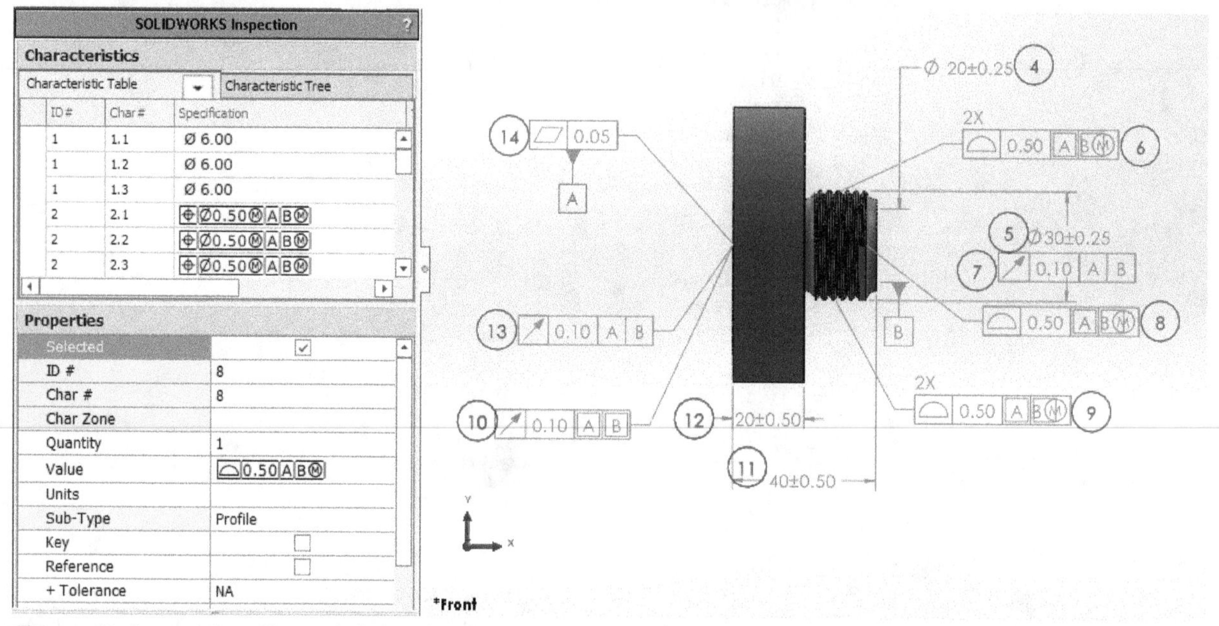

Figure-52. Inspection_table_with_balloons

You can edit any of the inspection dimension by selecting it. The parameters will be displayed in the **Properties** section of **SOLIDWORKS Inspection**; refer to Figure-53.

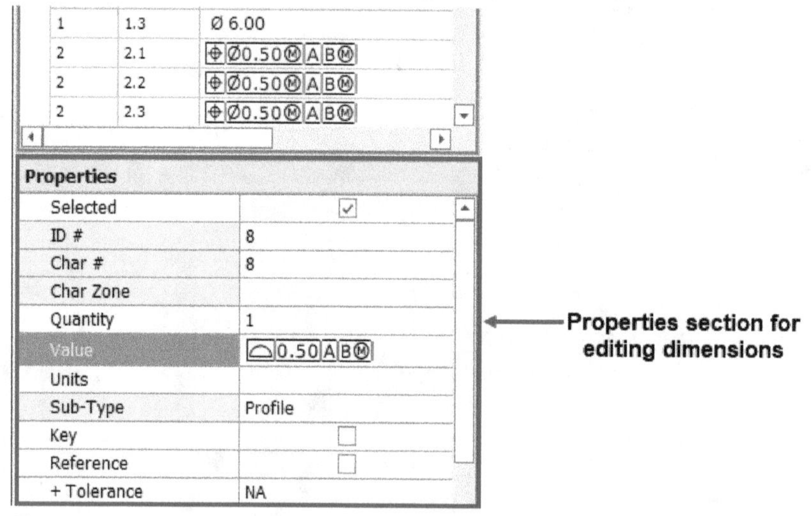

Figure-53. Properties_section

The **Edit Inspection Project** tool in **CommandManager** is used to modify the parameters of current inspection project. On selecting this tool, the **Create Inspection Project PropertyManager** will be displayed which has been discussed earlier.

The **Create New Template** tool is used to create a new template for inspection project based on current project.

ADDING/EDITING INSPECTION BALLOONS

The **Add/Edit Balloons** tool is used to add balloons or edit the earlier created balloons. Note that using this tool, you can modify all the balloons earlier created. The procedure to use this tool is given next.

- Click on the **Add/Edit Balloons** tool from the **SOLIDWORKS Inspection CommandManager** in the **Ribbon**. The **Ballooning Settings PropertyManager** will be displayed; refer to Figure-54.

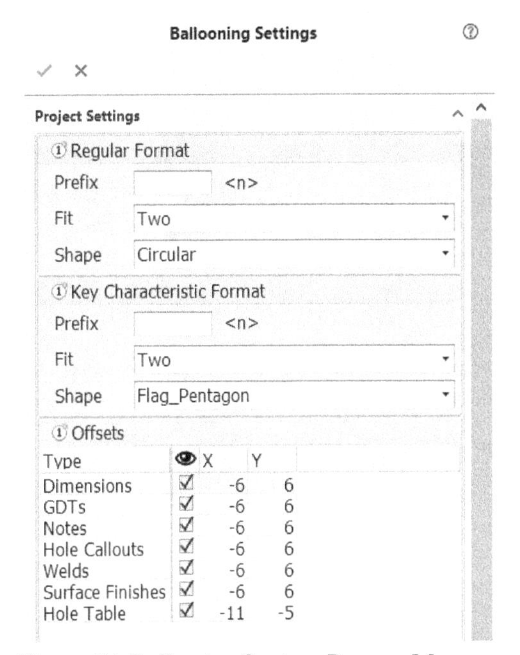

Figure-54. Ballooning Settings PropertyManager

- Specify desired prefix value in the **Prefix** edit box.
- In the **Fit** drop-down, select desired option to define size of balloon based on number of digits.
- Select desired option from the **Shape** drop-down to define the shape of balloon.
- Similarly, you can set the parameters for key characteristics.
- Select check boxes for dimensions and parameters to be used for ballooning from the table.
- After setting desired parameters, click on the **OK** button from the **PropertyManager**.

REMOVING INSPECTION BALLOONS

The **Remove Balloons** tool is used to remove all the balloons applied for inspection dimensions. Click on the **Remove Balloons** tool and all the balloons will be removed.

SELECTING INSPECTION BALLOONS

The **Select Balloons** tool is used to select all the balloons applied on inspection dimensions.

SEQUENCING INSPECTION BALLOONS

The **Balloon Sequence** tool is used to rearrange all inspection balloons of the currently active drawing in numerical sequence.

UPDATING INSPECTION PROJECT

The **Update Inspection Project** tool from the **SOLIDWORKS Inspection PropertyManager** is used to update the inspection dimensions based on changes in the model.

LAUNCHING TEMPLATE EDITOR

The **Launch Template Editor** tool is used to edit template for exporting inspection sheet. The procedure to use this tool is given next.

- Click on the **Launch Template Editor** tool from the **Ribbon**. The default excel file editor will be displayed with **Select an inspection sheet template** dialog box; refer to Figure-55.

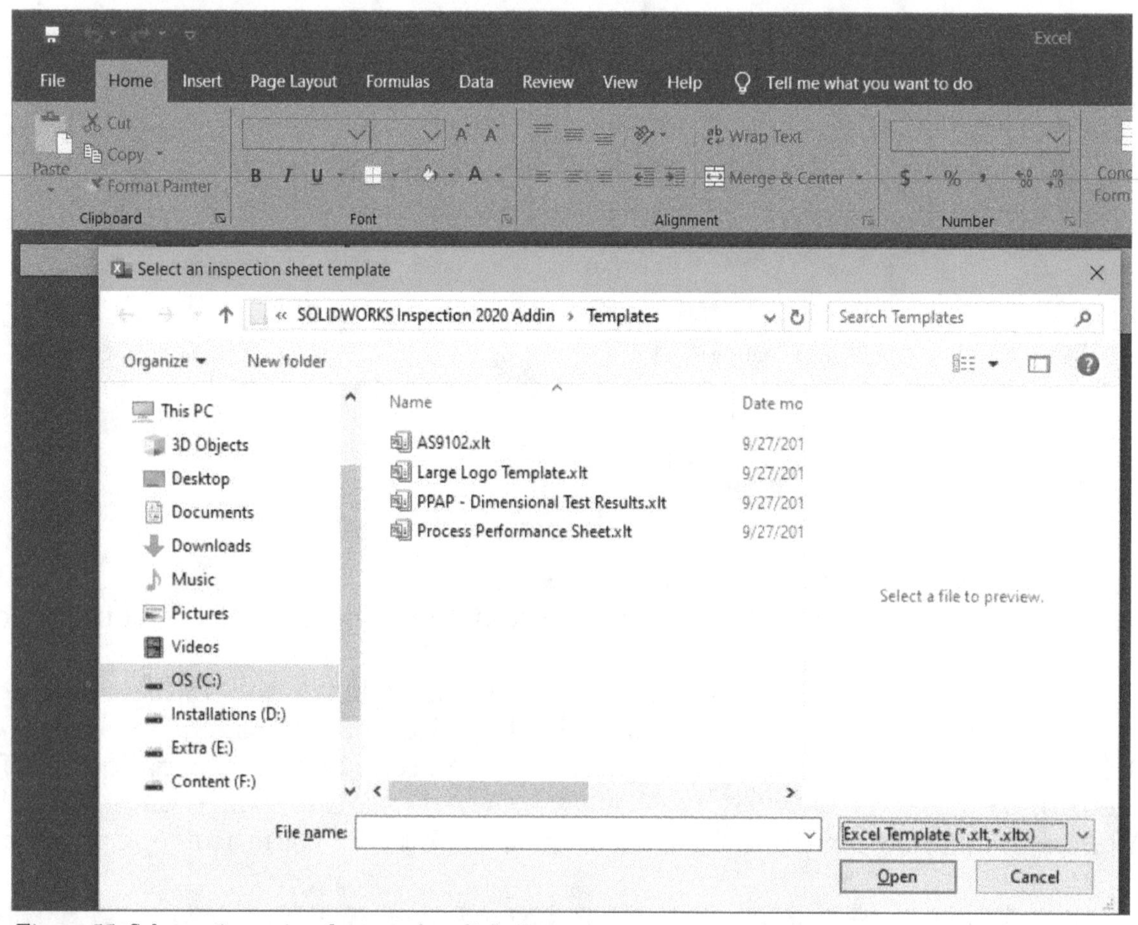

Figure-55. Select an inspection sheet template dialog box

- Select desired template sheet from the dialog box and click on the **Open** button. The file will open in excel file editor and **SOLIDWORKS Inspection - Template Editor** dialog box will be displayed; refer to Figure-56.
- Select desired radio button from the dialog box and select the parameter that you want to insert in the excel file.
- Click in the field where you want to place the new parameter and click on the **Insert** button from the **SOLIDWORKS Inspection - Template Editor** dialog box.
- After adding desired parameters and modifying them, click on the **Finished** button. The **Save your inspection template** dialog box will be displayed; refer to Figure-57.

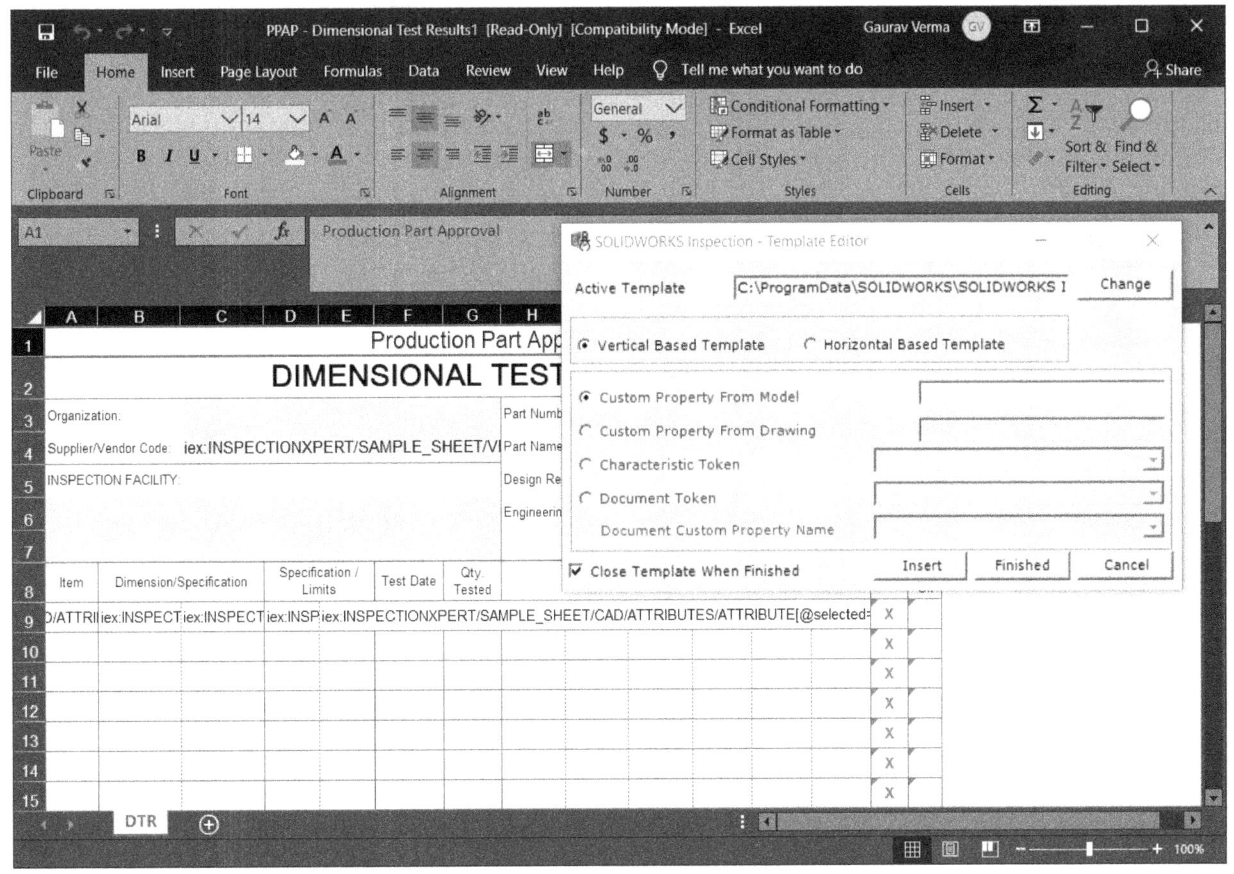

Figure-56. Template opened in Excel editor

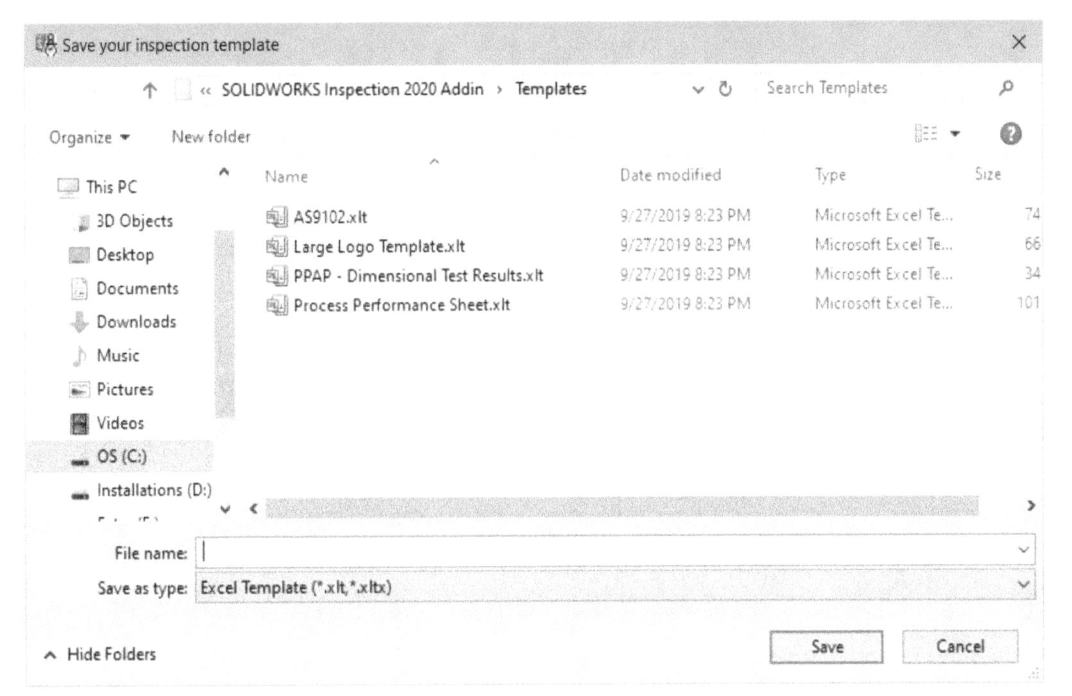

Figure-57. Save your inspection template dialog box

- Specify desired name for new template in the **File name** edit box and click on the **Save** button. The new template will be created at specified location.

DEFINING INSPECTION METHODS

The **Edit Inspection Methods** tool is used to modify inspection method for selected dimension. While working on the machine shop, it is important to note down measuring method along with respective dimension in the inspection sheet. The procedure specify inspection method is given next.

- Select the dimension for which you want to define inspection method in **Characteristic Tree** and click on the **Edit Inspection Methods** tool from the **SOLIDWORKS Inspection CommandManager** in the **Ribbon**. The **Inspection Methods PropertyManager** will be displayed; refer to Figure-58.

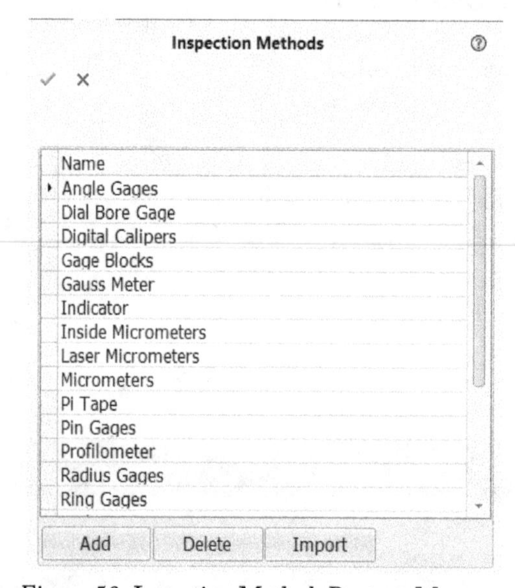

Figure-58. Inspection Methods PropertyManager

- Select desired measurement instrument/method.
- If you want to add a new method then click on the **Add** button and specify desired name for the method.
- You can edit any of the method by double-clicking on it in the table.
- If you want to delete any method then select it and click on the **Delete** button. Similarly, you can use the **Import** button to import a list of methods in CSV format.
- After setting desired parameter and selecting method, click on the **OK** button from the **PropertyManager**.

You need to specify inspection method for each of the dimension by using this procedure.

EDITING OPERATIONS FOR INSPECTION DIMENSIONS

Using the **Edit Operations** tool, you can specify the operation to which the selected dimension belongs to. The procedure to use this tool is given next.

- Select the dimension for which you want to specify the operation and click on the **Edit Operations** tool from the **Ribbon**. The **Edit Operations PropertyManager** will be displayed; refer to Figure-59.

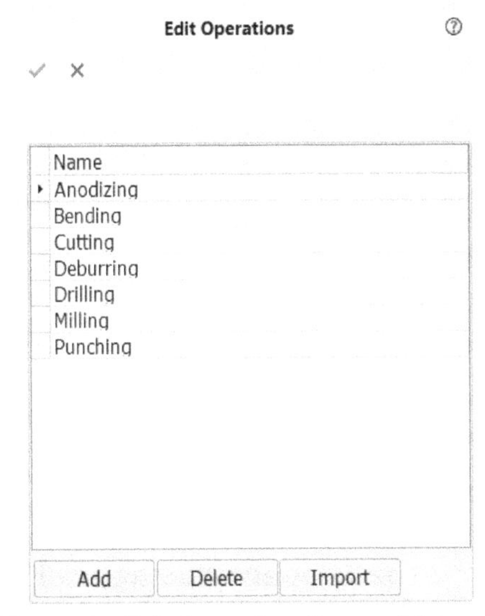

Figure-59. Edit Operations PropertyManager

- Select desired operation for list and click on the **OK** button. You can also set operation for the selected dimension from the **Operation** field of **Properties** section.
- Similarly, you can set operation for other dimensions.

Similarly, you can use the **Edit Vendors** tool from the **Ribbon** to modify vendor for current model.

Specify the other parameters in the **Properties** section to fill all the details of selected dimension. Note that if you have added more parameters in the template of current inspection project then they will also be appear here.

EXPORTING INSPECTION SHEET

There are 5 tools available in **SOLIDWORKS Inspection CommandManager** to export inspection sheet viz. **Export to 2D PDF**, **Export to Excel**, **Export to 3D PDF**, **Export to eDrawing**, and **Export to SOLIDWORKS Inspection Project**. Here, we will discuss the procedure of using **Export to Excel** tool since it is the most used option. You can apply the same procedure to other tools for exporting.

- After creating the inspection sheet and specifying all the dimension properties, click on the **Export to Excel** tool from the **Ribbon**. The **Exporting PropertyManager** will be displayed; refer to Figure-60.
- Select the check boxes for document templates to be generated from the **Multisheet** column and select a check box for active sheet from the **Active** column in the **PropertyManager**.

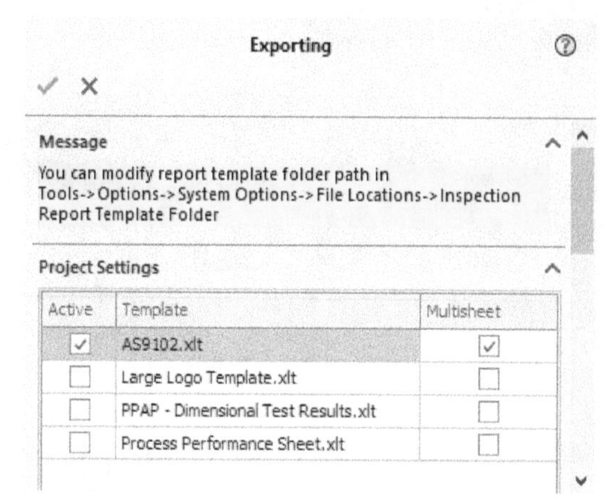

Figure-60. Exporting_PropertyManager

- Click on the **OK** button from the **PropertyManager**. The **Save As** dialog box will be displayed asking you to specify the location for file.
- Specify desired location and name of file. Click on the **Save** button. The file will be displayed in default Excel file editor; refer to Figure-61.

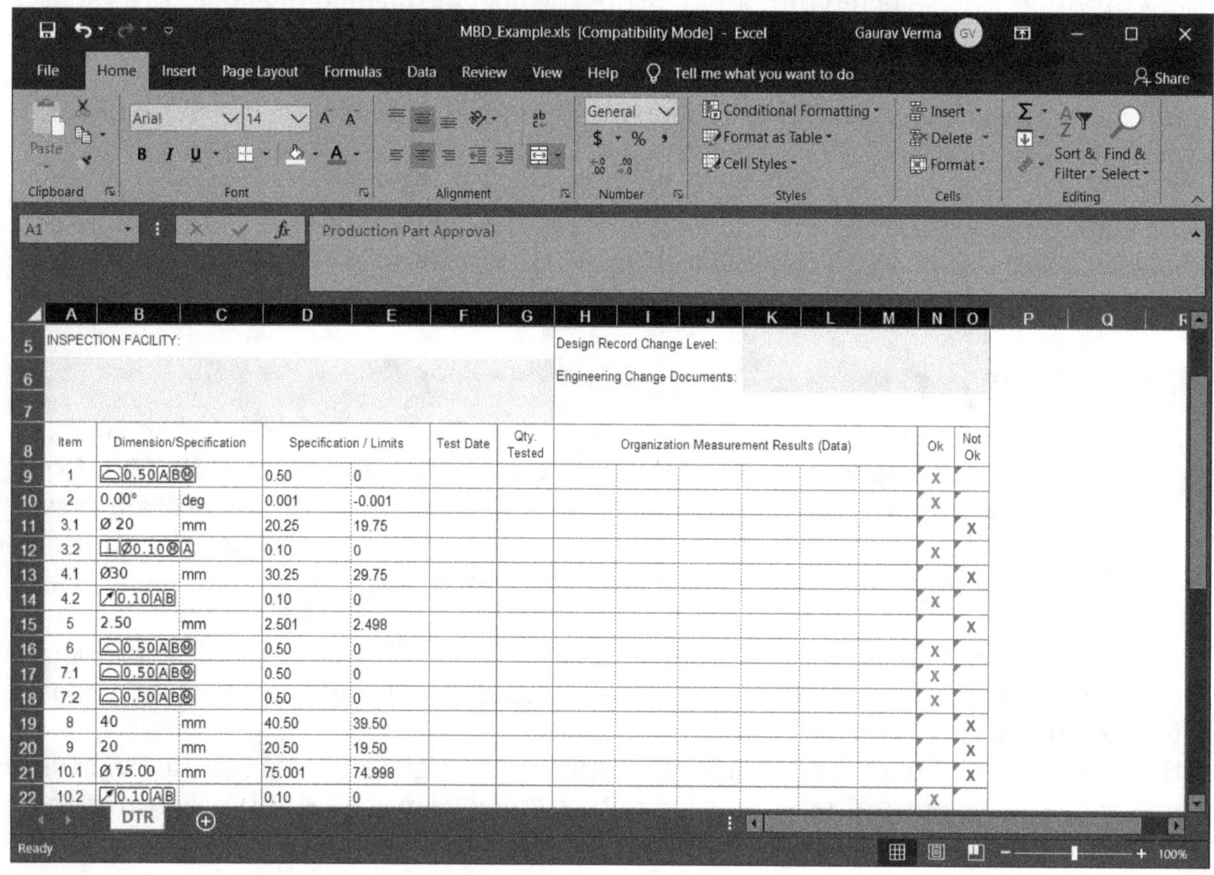

Figure-61. Inspection sheet exported to excel

If you have access to Net-Inspect quality management system then you can export the inspection data to Net-Inspect server by using the **Net-Inspect** tool.

SELF ASSESSMENT

Q1. Name the three most common technologies used in 3D Modeling.

Q2. Write the procedure of inserting **Title Block Table**.

Q3. Which of the following Additive Manufacturing Processes use Stereolithography technology for 3D Printing?

a) Vat Photopolymerisation
b) Material Extrusion
c) Power Bed Fusion
d) All of the above

Q4. Which of the following value should be selected from **Infill Percentage** drop-down in the **Print 3D PropertyManager**?

a) 8
b) 5
c) 2
d) 0

Q5. The **Show Tolerance Status** tool is used to highlight the .... to which tolerance have been applied.

Q6. The .... is an Add-in used for creating inspection sheet and documentation used by production and quality department in a manufacturing firm.

Q7. You cannot use surface modeling tools for preparing part for 3D Printing. (T/F)

Q8. In **Auto Dimension Scheme PropertyManager**, the **Prismatic** radio button should be selected if you want the dimensions and geometric tolerances to be displayed in the form of callouts. (T/F)

Answer to Self-Assessment:
1. Selective Laser Sintering (SLS), Fused Deposition Modeling (FDM), Stereolithography (SLA) **3.** (a) **4.** (d) **5.** Features/Faces **6.** SolidWorks Inspection **7.** False **8.** True

FOR STUDENT NOTES

Chapter 16

Introduction to SolidWorks CAM

Topics Covered

The major topics covered in this chapter are:

- ***Defining Machine***
- ***Setting Stocks and Parts***
- ***Extracting Machinable Features***
- ***Performing Operations***
- ***Generating CNC Program***

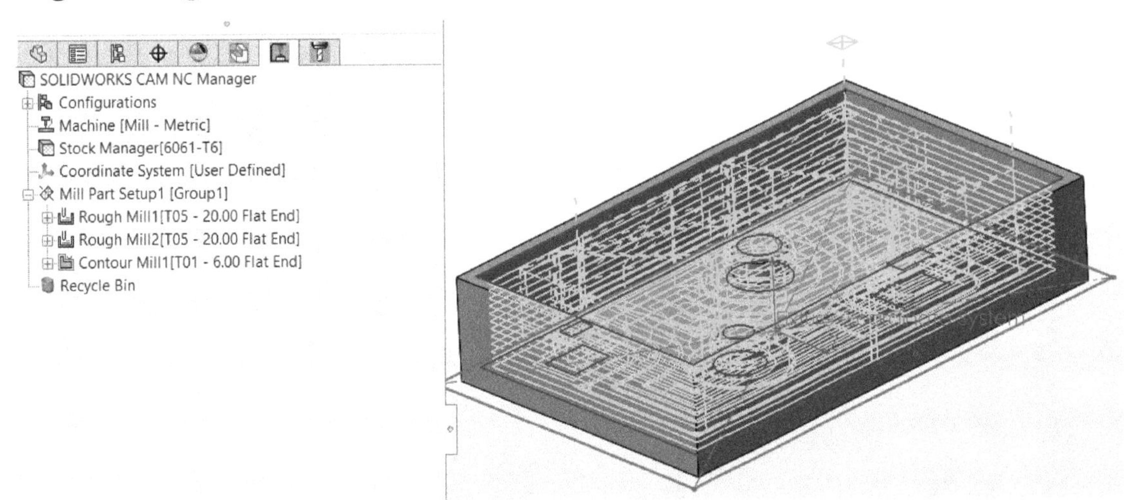

INTRODUCTION

SolidWorks CAM was first introduced in SolidWorks 2018. SolidWorks CAM is based on CAMWorks software. This software add-in is available as Add-In for SolidWorks so you can activate or deactivate it as discussed earlier. Being an Add-In it provides a well integrated mode of operation within SolidWorks. The software recognizes geometric tolerance and other dimensional features of the model created in SolidWorks. By default the **SolidWorks CAM** Add-In is active in SolidWorks and the tools related to CAM are available in **SOLIDWORKS CAM CommandManager**; refer to Figure-1.

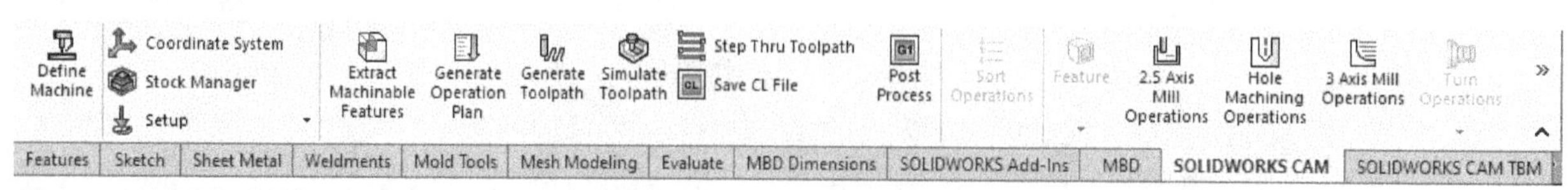

Figure-1. SOLIDWORKS CAM CommandManager

Note that various trees are available for SolidWorks CAM in the left area of the application window; refer to Figure-2.

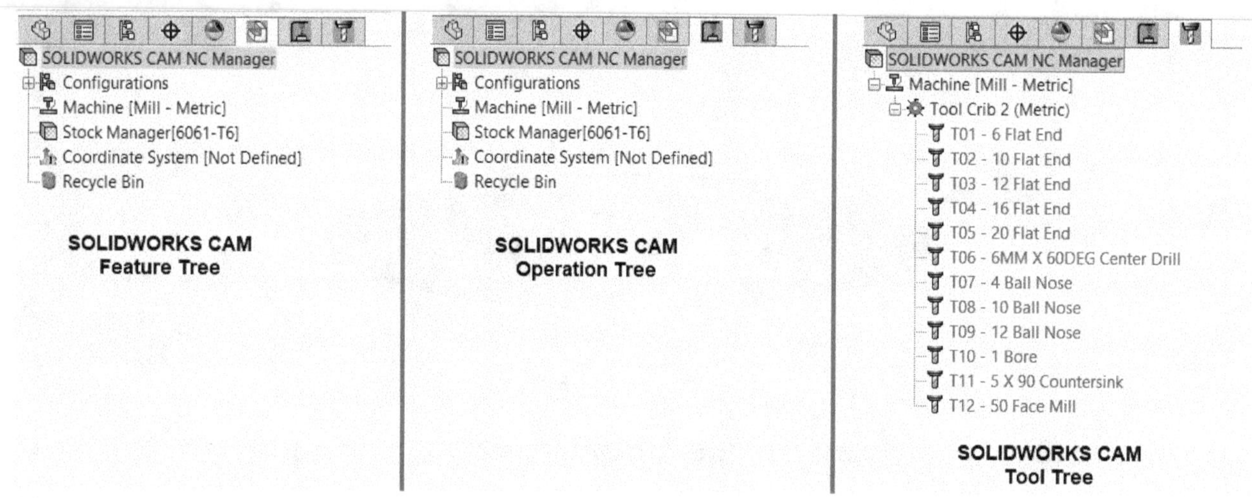

Figure-2. SOLIDWORKS CAM Tree

Various tools and procedures of SolidWorks CAM are discussed next.

DEFINING MACHINE

The **Define Machine** tool is used to define the type of machine and other related parameters for creating machine setup as per the job. The procedure to define machine is given next.

- After opening the part file, right-click on the **Machine [Mill-Metric]** option from the **SOLIDWORKS CAM Feature Tree** and select **Edit Definition** option; refer to Figure-3. The **Machine** dialog box will be displayed; refer to Figure-4.

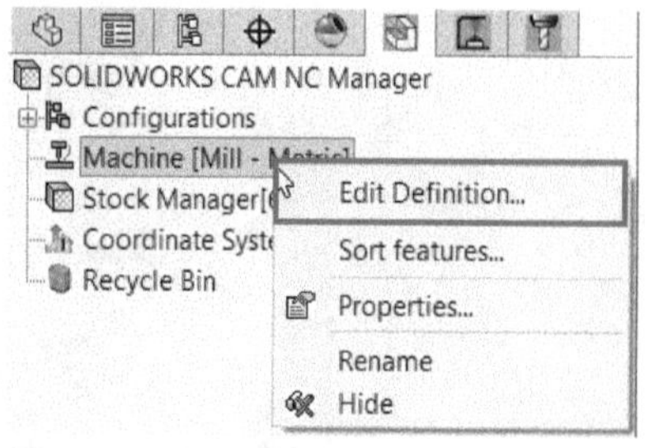

Figure-3. Edit Definition option

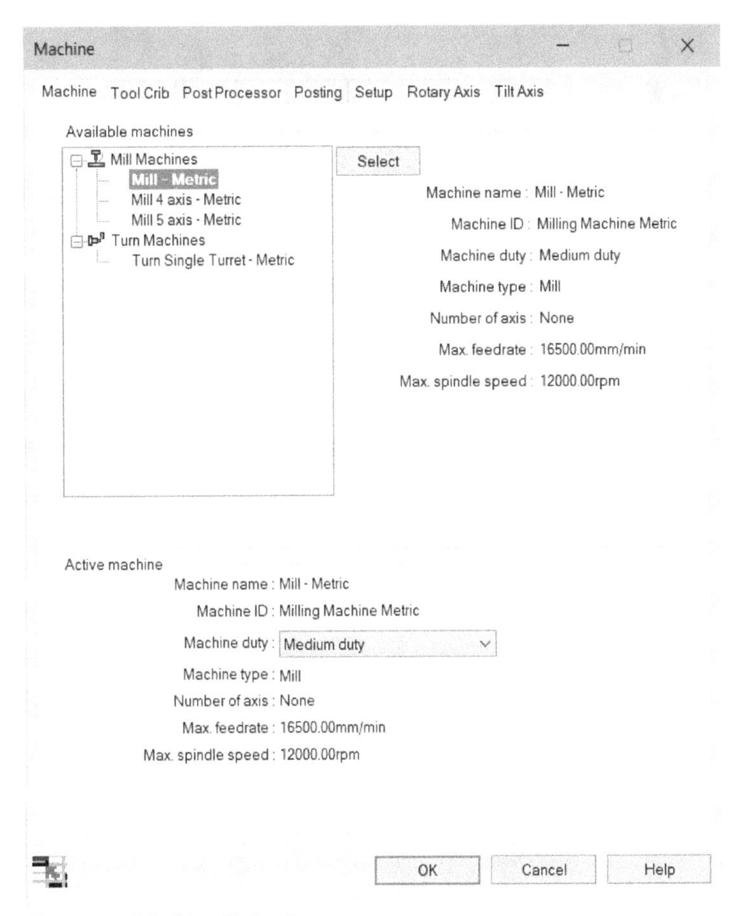

Figure-4. Machine dialog box

- Select desired machine from the list of machine. Note that the machines will be active only based on your part. Set the parameters related to machine.

Setting Tool Crib

- Click on the **Tool Crib** tab from the dialog box. The dialog box will be displayed as shown in Figure-5.
- There are 12 tools by default in the tool crib. You can add or remove the tools based on your machine's tool cassette.

Adding Tools

- To add a tool, click on the **Add Tool** button from the dialog box. The **Tool Select Filter** dialog box will be displayed; refer to Figure-6.

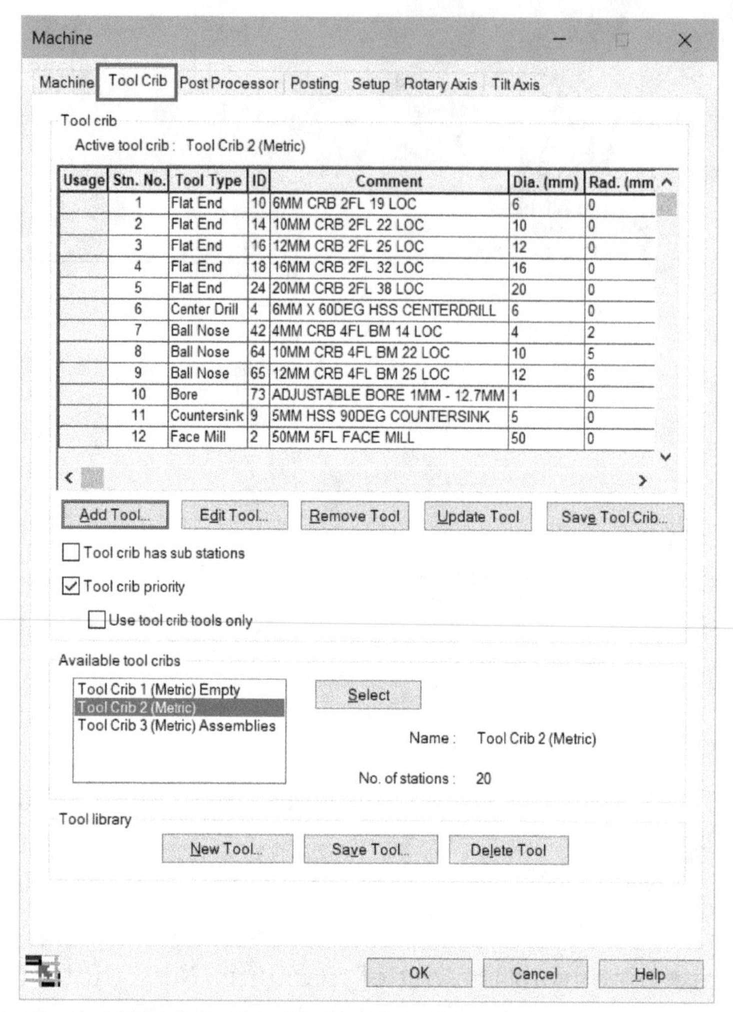

Figure-5. Tool Crib tab in Machine dialog box

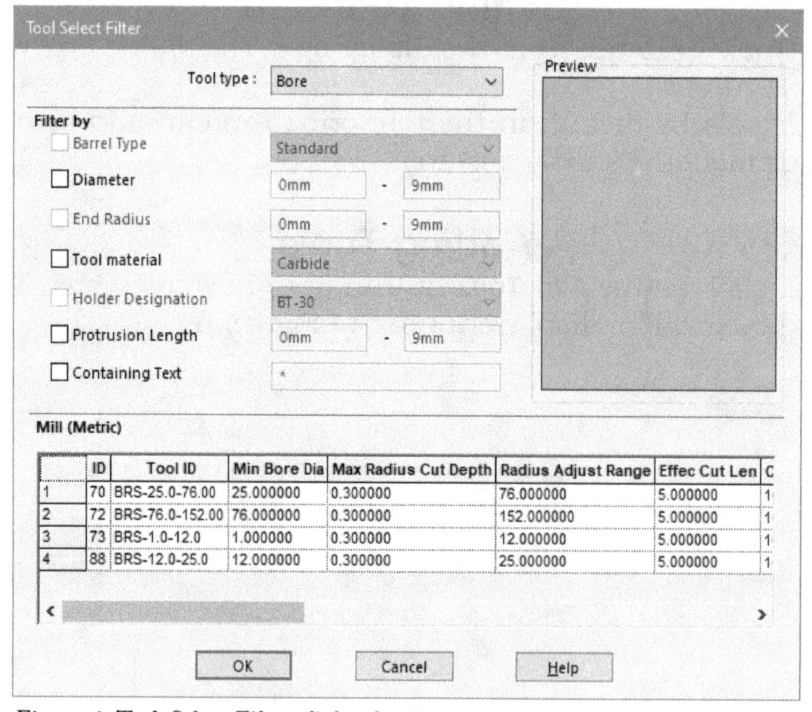

Figure-6. Tool_Select_Filter_dialog_box

- Click in the **Tool type** drop-down and select desired type of tool to be added in the crib. Like if you want to use Ball Nose cutting tool then select the Ball Nose option from the drop-down.
- Set desired filters for the tool and select desired tool to be added.
- Click on the **OK** button from the dialog box. The tool will be added in the list and the preview of tool will be displayed; refer to Figure-7.

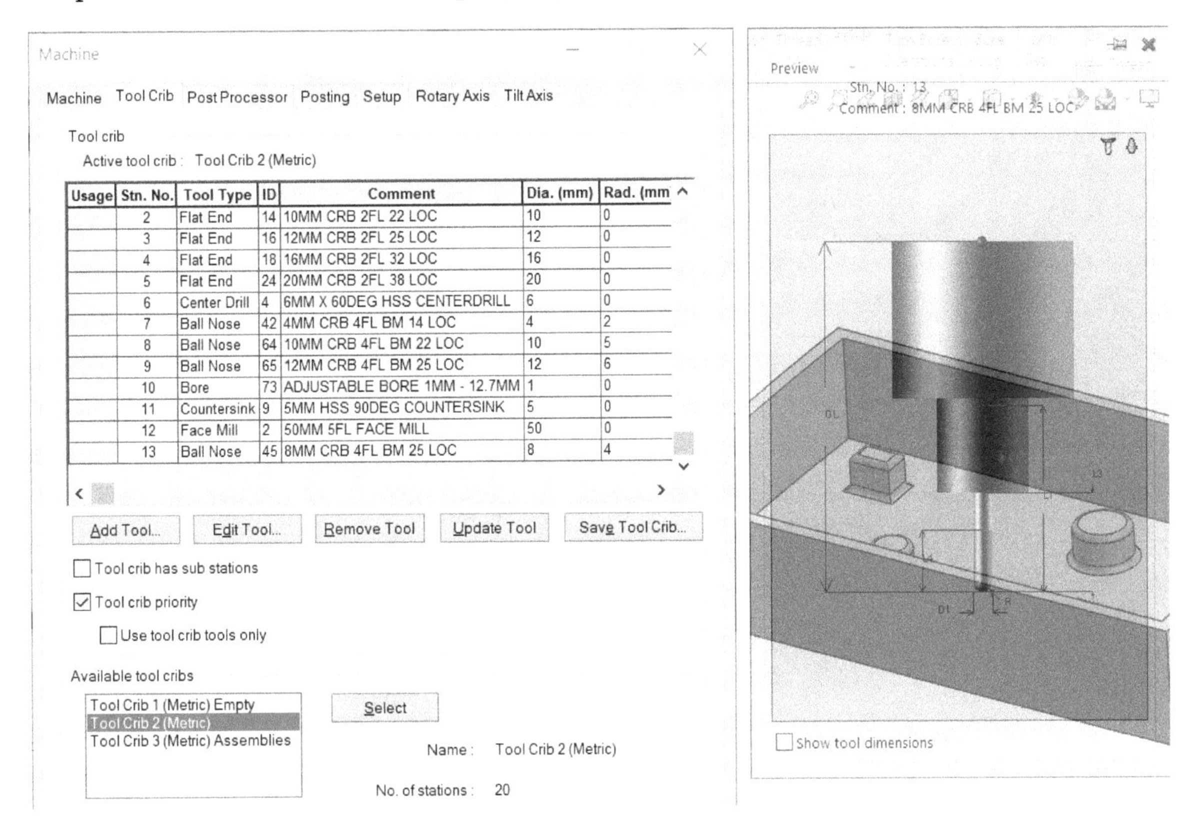

Figure-7. Preview of the tools

Editing Tool

- Select the tool you want to edit from the list and click on the **Edit Tool** button. The **Edit Tool Parameters** dialog box will be displayed; refer to Figure-8.
- Set desired parameters in the dialog box. Note that you can set the tool holder and tool station as well for your tool.
- After setting the tool, click on the **OK** button.

Removing Tool from the Crib

- Select the tool that you want to remove from crib and click on the **Remove Tool** button from the dialog box. The tool will be removed.

Updating tool in Database

Based on the parameters defined by you in the dialog box, you can update the definitions of the tools in the central database of SolidWorks. To do so, after editing tools; click on the **Update Tool** button from the dialog box.

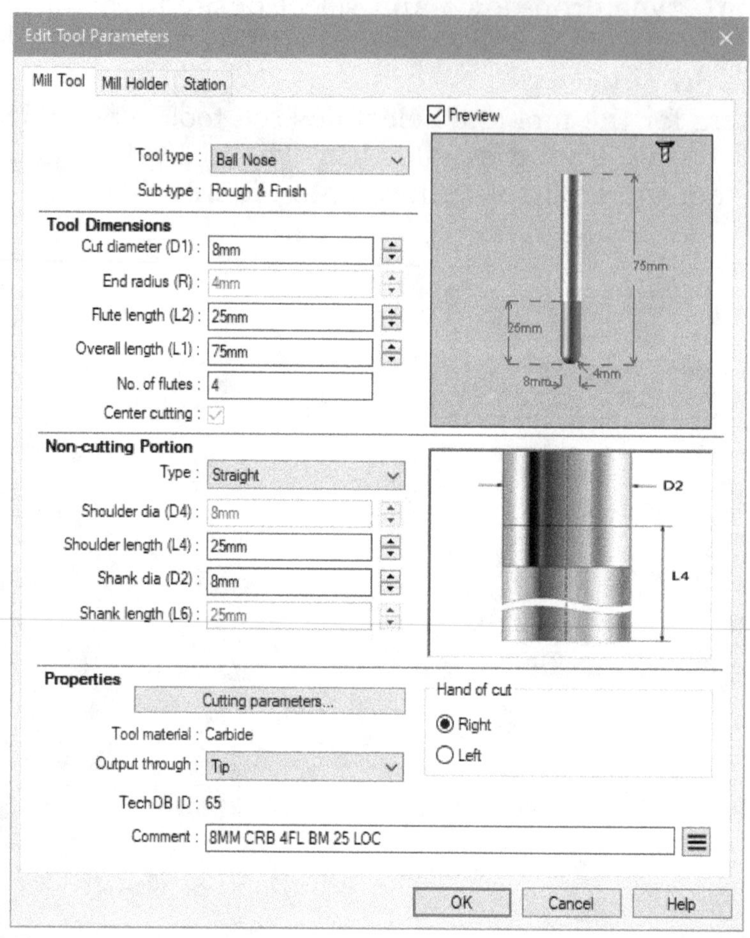

Figure-8. Edit_Tool_Parameters_dialog_box

Saving Tool Crib

After creating desired tool crib, you can save it for later use. The procedure is given next.

- Click on the **Save Tool Crib** button from the dialog box. The **Save to Database** dialog box will be displayed; refer to Figure-9.
- Set desired name and parameters and then click on the **Save** button.

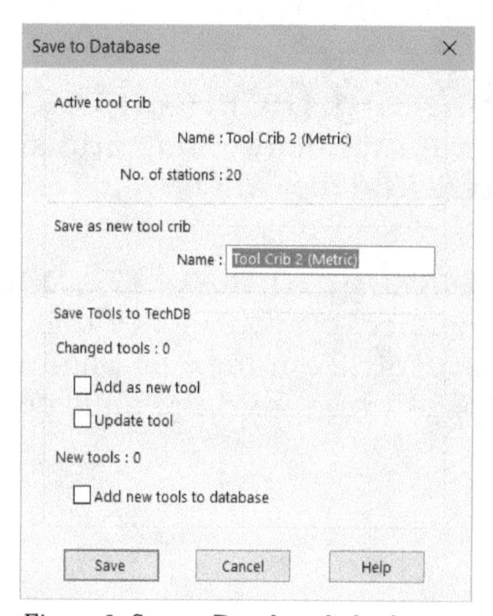

Figure-9. Save to Database dialog box

Creating and Saving New Tool in Library

- Click on the **New Tool** button from the **Tool library** area of the dialog box. The **New Tool** dialog box will be displayed similar to **Edit Tool Parameters** dialog box; refer to Figure-10.

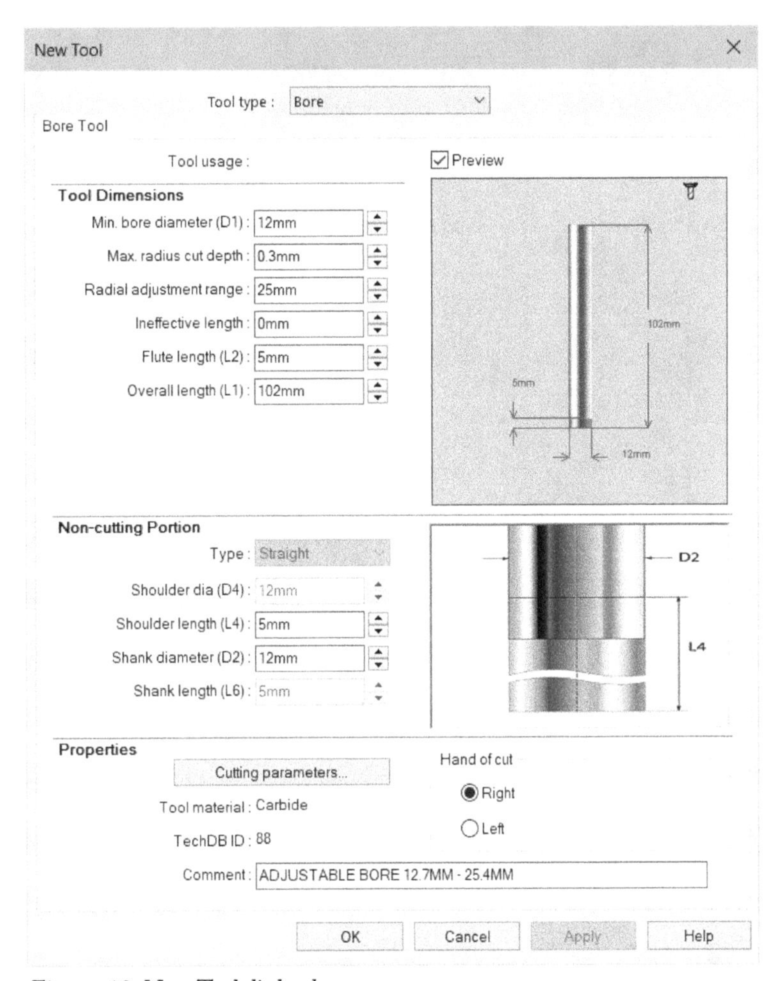

Figure-10. New Tool dialog box

- Set the parameters as discussed earlier and click on the **OK** button to create the tool. The tool will be created and added to the library as well as tool crib.
- If a tool is not added in the library, then select it from the crib table and click on the **Save Tool** button from the **Tool library** area.

Setting Postprocessor Parameters

The post processor is used to translate the codes generated by CAM program to machine readable codes based on machine controller. The options for post processor are available in the **Post Processor** tab of the **Machine** dialog box; refer to Figure-11.

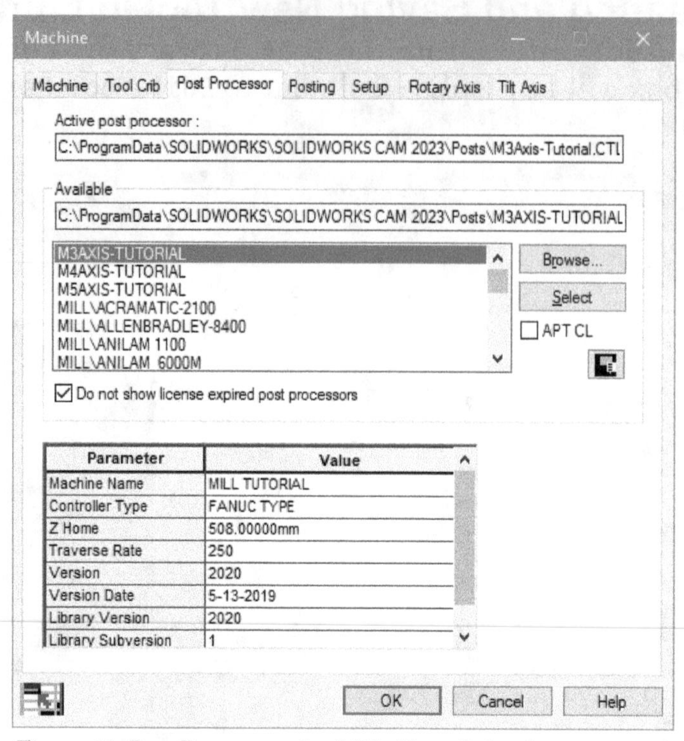

Figure-11. Post_Processor_tab_of_Machine_dialog_box

- Select desired post processor from the list and click on the **Select** button.
- If you have an updated post processor then click on the **Browse** button. The **Open** dialog box will be displayed.
- Select desired post processor and click on the **Open** button.

Posting Options

The options related to posting coolant offset etc. are available in the **Posting** tab; refer to Figure-12.

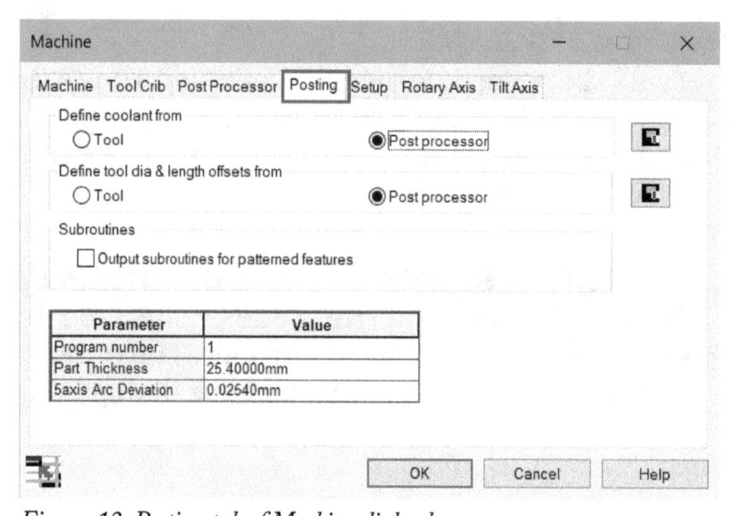

Figure-12. Posting tab of Machine dialog box

- Set desired parameters and click on the **OK** button to apply the parameters.

You can set the other parameters of the dialog box as discussed earlier.

SETTING COORDINATE SYSTEM

The coordinate system is an important part of CNC machines programming. All the coordinates for machine programming are referenced to the coordinate system selected. The procedure to set coordinate system is given next.

- Right-click on the **Coordinate System** option from the **SOLIDWORKS CAM Feature Tree** and select the **Edit Definition** option from the shortcut menu; refer to Figure-13. The **Fixture Coordinate System PropertyManager** will be displayed; refer to Figure-14. Also, you will be asked to select a point to define the coordinate system.

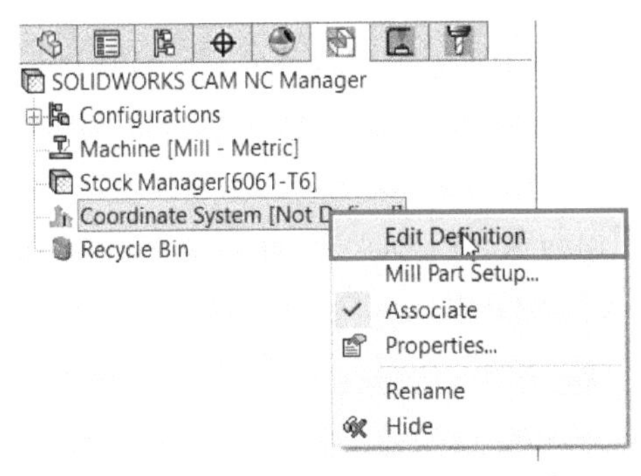

Figure-13. Edit Definition for Coordinate System

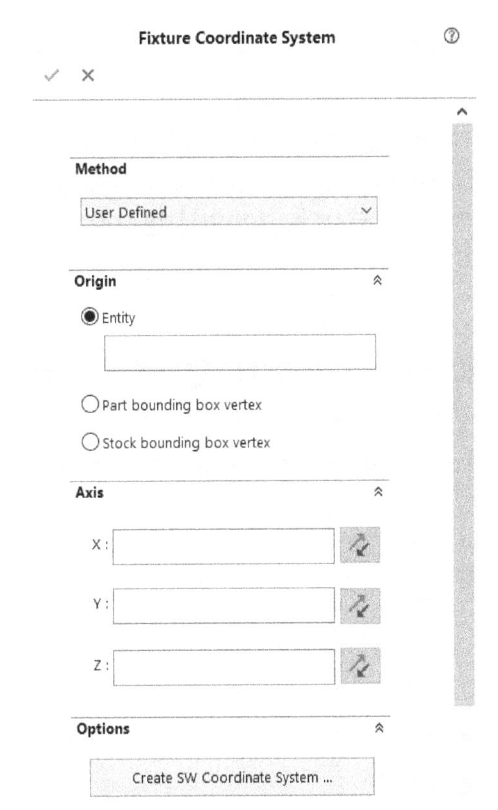

Figure-14. Fixture Coordinate System PropertyManager

- Select a point on the model that you want to be coordinate system. To align the coordinate system at desired orientation, click in the respective direction selection box of the **Axis** rollout and select desired reference.
- If you want to use any vertex of the part bounding box then select the **Part bounding box vertex** radio button. Vertices will be displayed; refer to Figure-15.

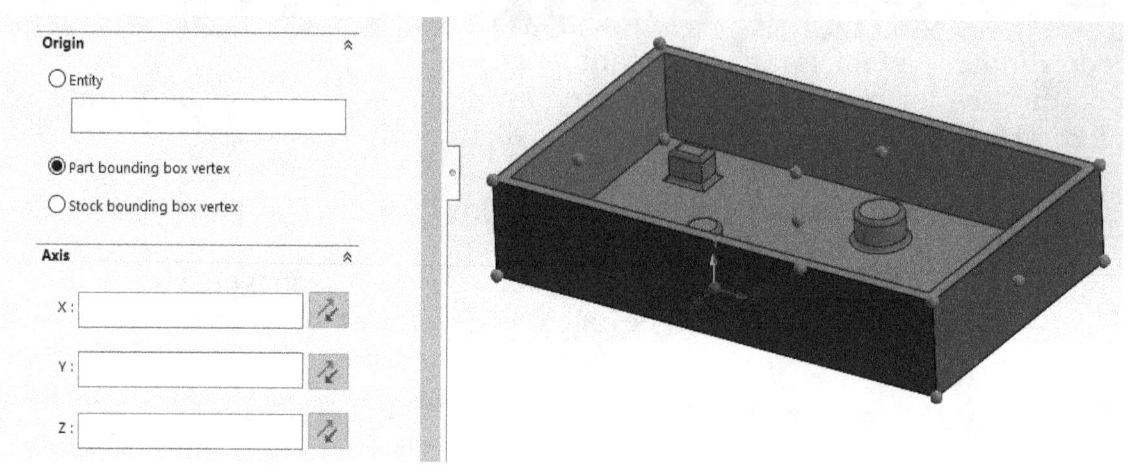

Figure-15. Bounding box vertices

- Select desired vertex to place coordinate system.
- If you want to use a vertex of bounding box of stock as coordinate system reference then select **Stock bounding box vertex** radio button.
- After selecting the point and setting the direction references, click on the **OK** button from the **PropertyManager** to create the coordinate system.

DEFINING OR EDITING STOCK

Stock is the pile of material to be removed after machining to produce the part. The procedure to define/edit the stock is given next.

- Click on the **Stock Manager** tool from the **SOLIDWORKS CAM CommandManager** in the **Ribbon**. The **Stock Manager PropertyManager** will be displayed; refer to Figure-16.

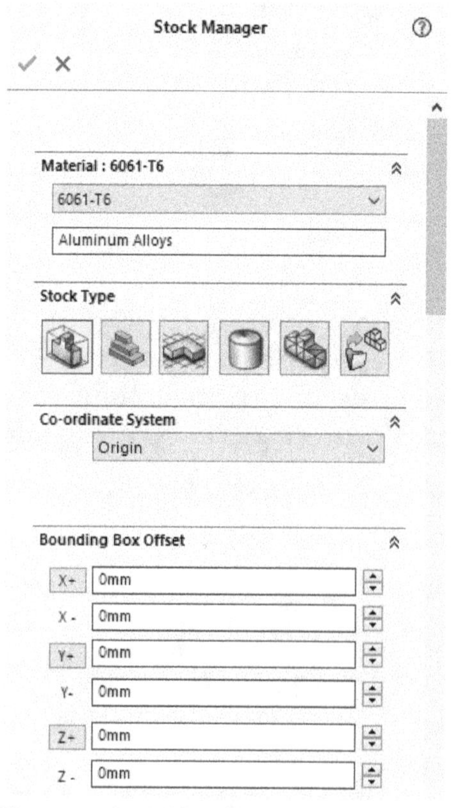

Figure-16. Stock Manager PropertyManager

- Select desired material from the drop-down.
- There are various options to set the stock type in the **Stock Type** rollout; **Bounding Box**, **Pre-defined Bounding Box**, **Extruded Sketch**, **Cylindrical**, **STL File**, and **Part File**.
- Select the **Bounding Box** button if you want to create a stock bounding the whole workpiece.
- Select the **Pre-defined Bounding Box** button if you want to create bounding box stock with specified dimensions and offset values.
- Select the **Extruded Sketch** button if you want to create stock by extruding the selected sketch; refer to Figure-17.
- Select the **Cylindrical** button if you want to define a cylindrical stock for mill parts and assemblies by specifying parameters such as the center axis, origin, diameter, length of the cylinder, and offsets.
- Select the **STL File** button and click on the **Browse** button in the **STL File** rollout. The **Open** dialog box will be displayed; refer to Figure-18. Select desired file and click on the **Open** button.

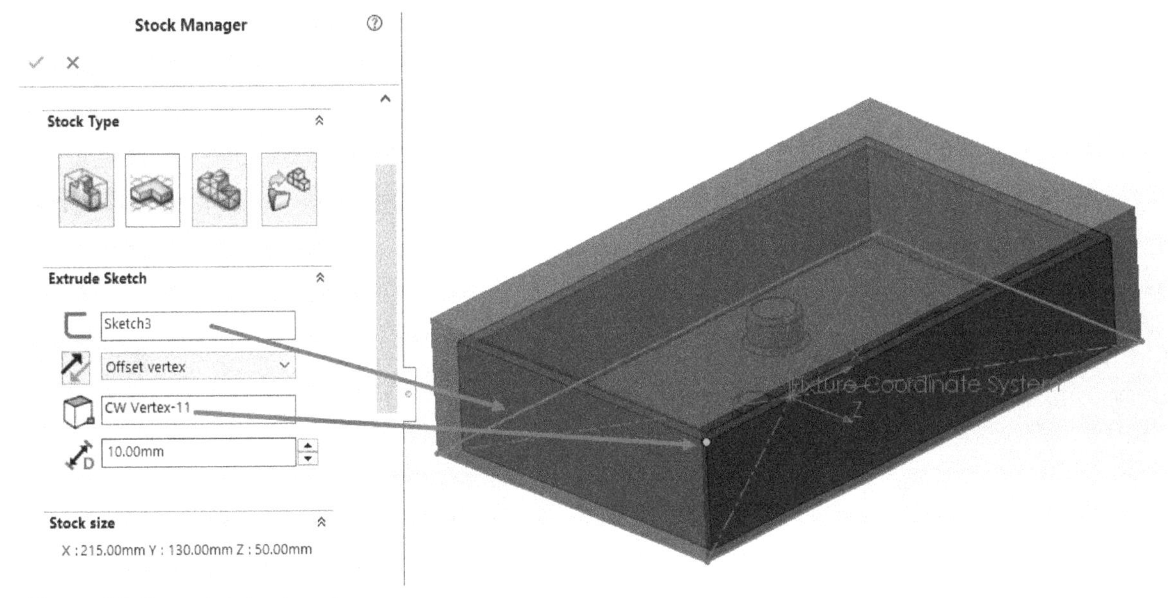

Figure-17. Extruded sketch stock

- Click on the **Part File** button from the rollout if you want to use a SolidWorks file as stock. After selecting the button, click on the **Browse** button. The **Open** dialog box will be displayed. Select desired file and click on the **Open** button from the dialog box. The part will be placed at the origin.
- After creating the stock and specifying desired parameters, click on the **OK** button. The stock will be created.

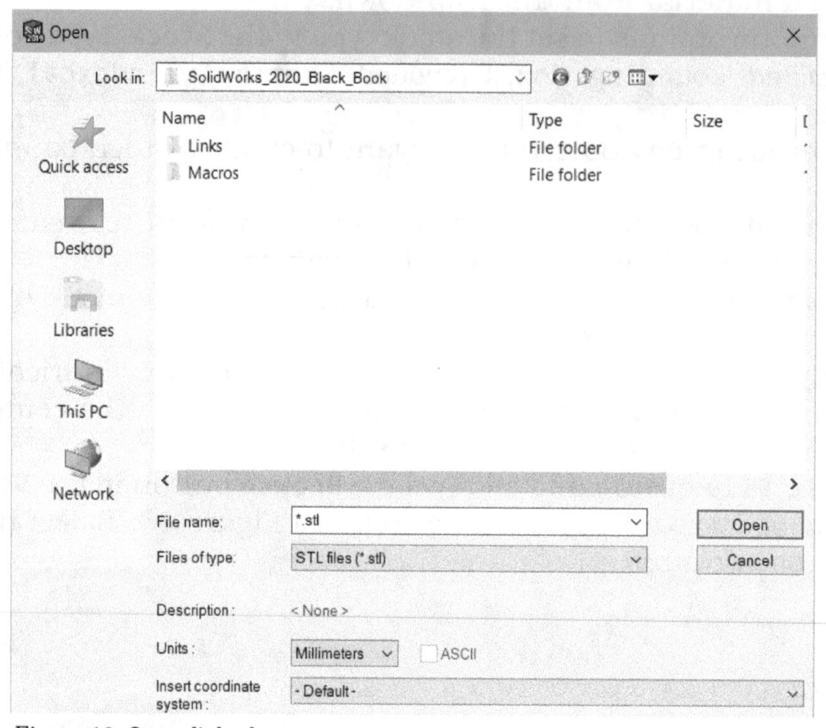

Figure-18. Open dialog box

SETTING MILLING OPERATION PARAMETERS

The milling operations require a few parameters to be set before performing cutting operations. The procedure to set the parameters are given next.

- Click on the **Mill Setup** tool from the **Setup** drop-down in the **SOLIDWORKS CAM CommandManager**. The **Mill Setup PropertyManager** will be displayed; refer to Figure-19.

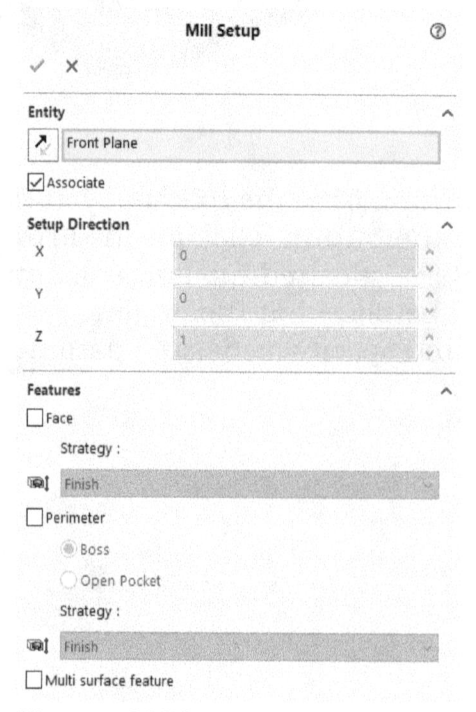

Figure-19. Mill Setup PropertyManager

- Click in the selection box of **Entity** rollout and select the face/plane perpendicular to which the tool will be aligned while cutting. In simple words, select desired face/plane to define cutting direction.
- Select the **Associate** check box if you want the setup to be modified with the model if any changes are made to it.
- Select desired feature from the **Features** rollout. Like, select the **Face** check box if you want SolidWorks to automatically identify the face features for machining. Select the **Perimeter** check box if you boss/open pocket features to be identified. Select the **Multi surface feature** check box if you want to identify multi surface features.
- Click on the **OK** button from the **PropertyManager** to apply the settings.

EXTRACTING MACHINABLE FEATURES

The **Extract Machinable Features** tool is used to extract all the machinable features based on the cutting strategies in the database. The procedure to use this tool is given next.

- Right-click on **SOLIDWORKS CAM NC Manager** option from the **SOLIDWORKS CAM Feature Tree** and select the **Extract Machinable Features** option from the shortcut menu; refer to Figure-20. The identified features will be displayed; refer to Figure-21.

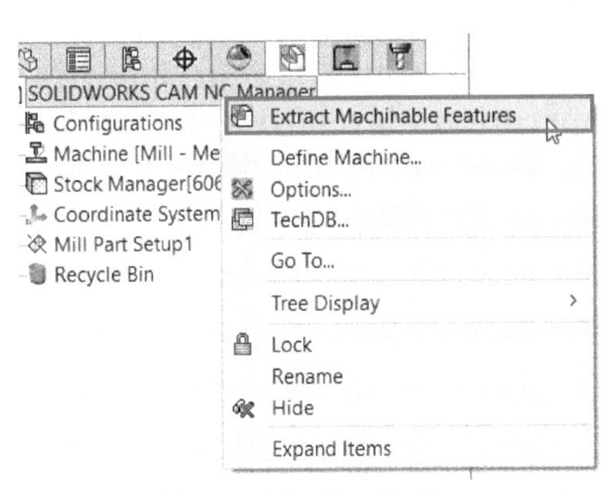

Figure-20. Extract Machinable Features option

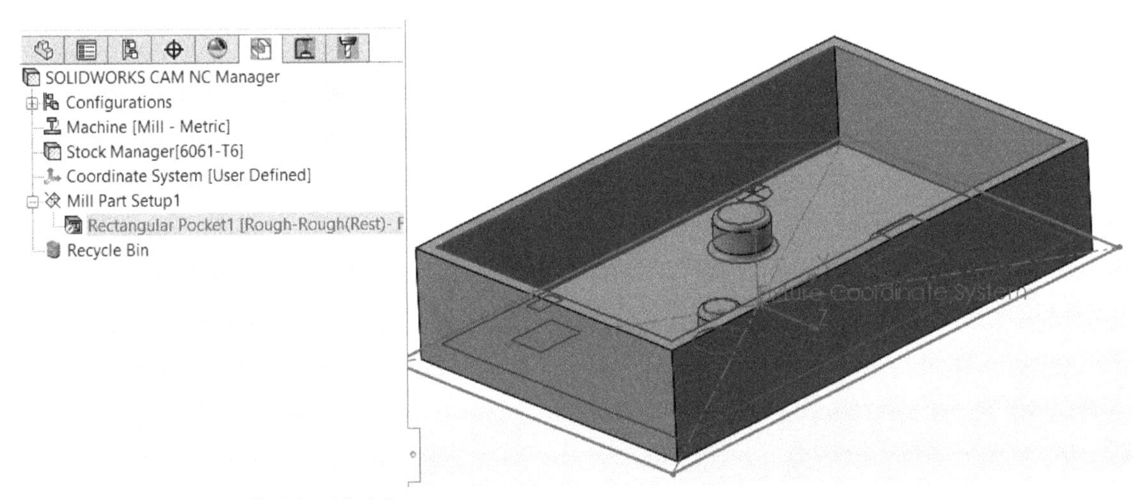

Figure-21. Automatically identified feature

GENERATING OPERATION PLAN

The **Generate Operation Plan** option is used to create cutting strategies based on identified features. To generate cutting strategies, right-click on desired machining feature and click on the **Generate Operation Plan** option from the shortcut menu. The material cutting strategies will be created; refer to Figure-22.

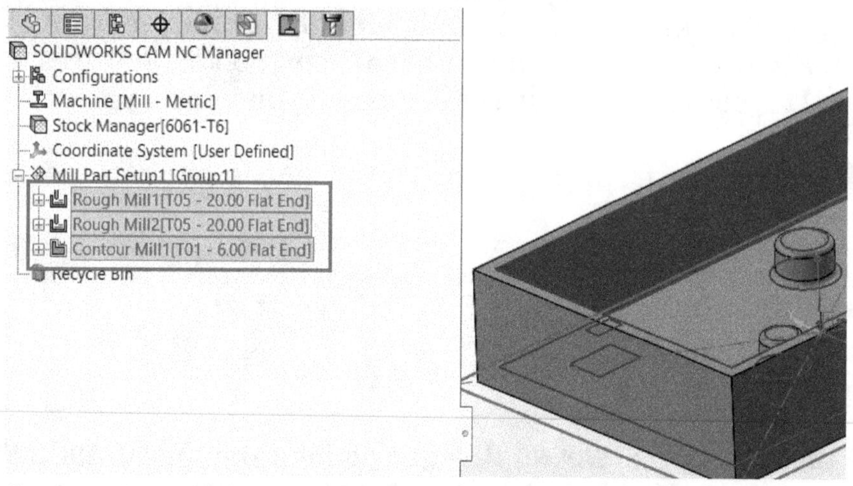

Figure-22. Material cutting strategies created

GENERATING TOOL PATHS

The **Generate Toolpath** option is used to create tool paths based on earlier generated operation plan. Select the operation plans from the **SOLIDWORKS CAM Operation Tree** and right-click on it. A shortcut menu will be displayed; refer to Figure-23. Select the **Generate Toolpath** option from the shortcut menu. The tool paths will be generated; refer to Figure-24.

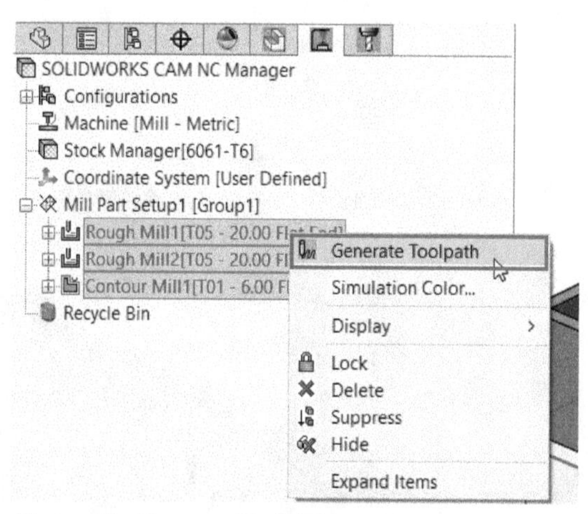

Figure-23. Generate Toolpath option

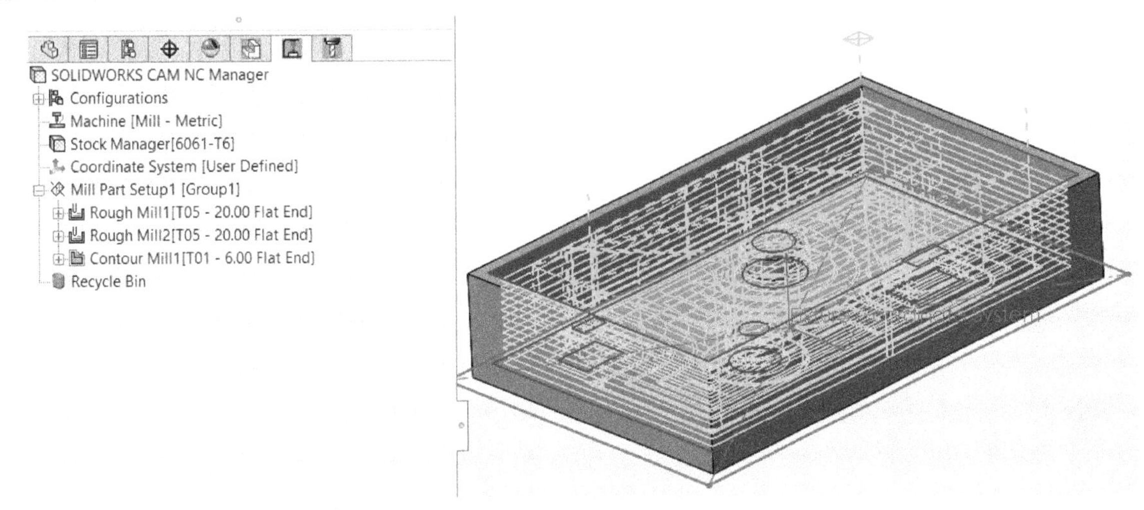

Figure-24. Toolpaths generated

SIMULATING TOOL PATHS

To simulate the toolpaths, select desired toolpath from the **SOLIDWORKS CAM Operation Tree** and right-click on it. A shortcut menu will be displayed. Select the **Simulate Toolpath** option from the shortcut menu; refer to Figure-25. The **Simulate Toolpath PropertyManager** will be displayed; refer to Figure-26.

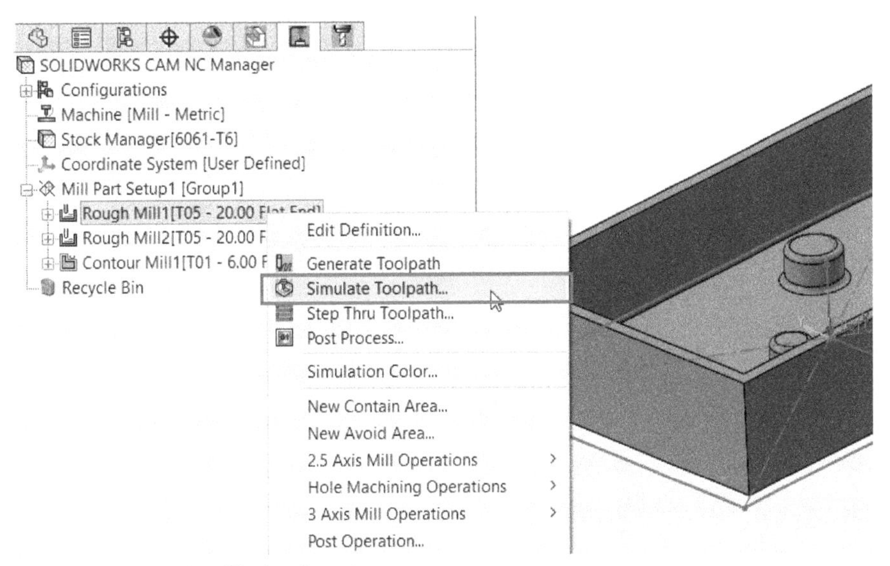

Figure-25. Simulate Toolpath option

- Click on the **Play** button to simulate the toolpath. Set desired options in the **PropertyManager** and click on the **OK** button to exit the tool.

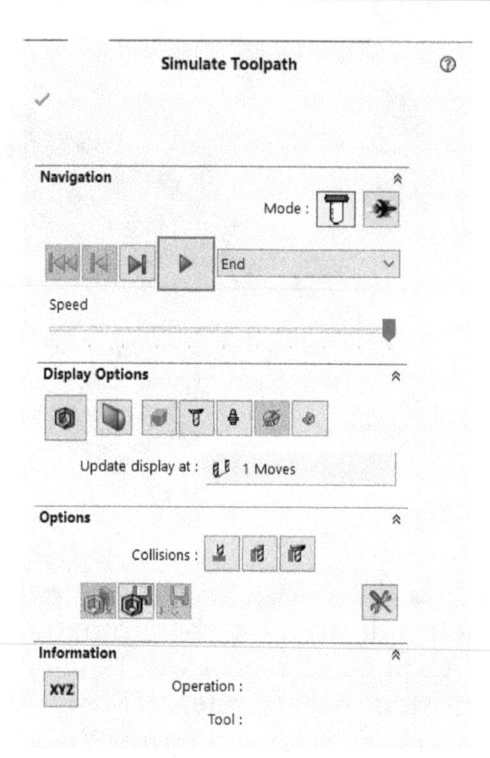

Figure-26. Simulate Toolpath PropertyManager

VISUALIZING TOOLPATHS

The **Step Thru Toolpath** option is used to check toolpaths at different instances of time. Select desired toolpath from the **SOLIDWORKS CAM Operation Tree** and right-click on the toolpath. A shortcut menu will be displayed; refer to Figure-25. Select the **Step Thru Toolpath** option from the shortcut menu. The **Step Through Toolpath PropertyManager** will be displayed; refer to Figure-27.

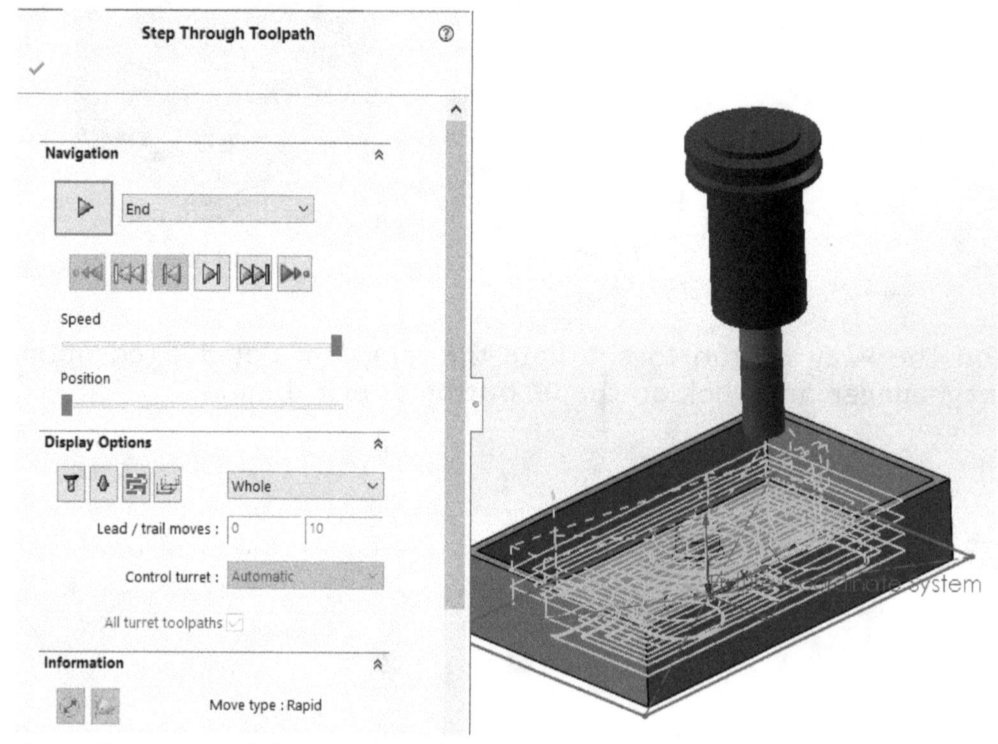

Figure-27. Step Through Toolpath PropertyManager

Click on the **Play** button from the **PropertyManager** to check each of the toolpath simulation.

SAVING TOOLPATHS

The **Save CL File** tool is used to save the toolpaths in CL file format. The procedure is given next.

- Click on the **Save CL File** tool from the **SOLIDWORKS CAM** cascading menu of **Tools** menu. The **Save As** dialog box will be displayed.
- Specify desired name and location. Click on the **Save** button to save the file.

POST PROCESSING TOOLPATHS

Post processing is the process of generating G codes from the toolpaths. The procedure to do so is given next.

- Click on the **Post Process** tool from the **SOLIDWORKS CAM** cascading menu of the **Tools** menu. The **Post Output File** dialog box will be displayed; refer to Figure-28.

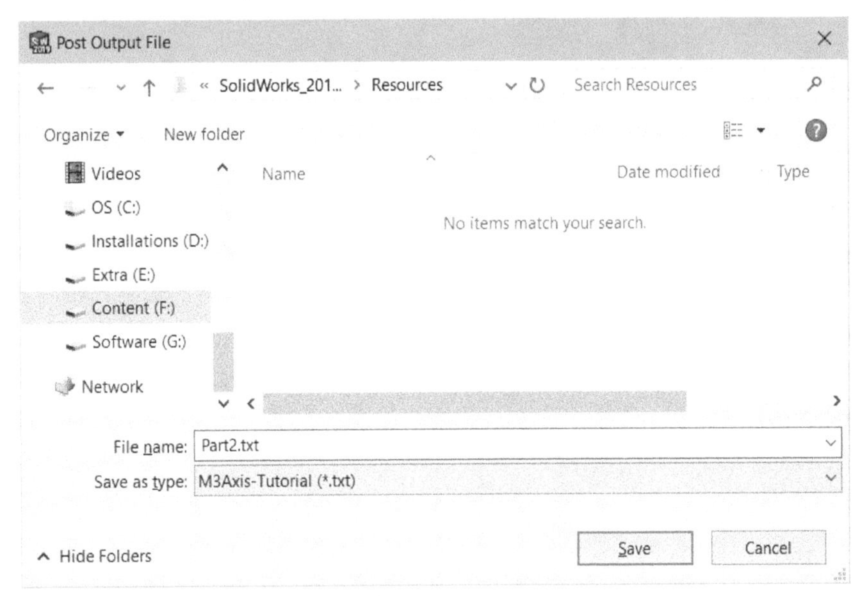

Figure-28. Post Output File dialog box

- Specify desired name and click on the **Save** button. The **Post Process PropertyManager** will be displayed; refer to Figure-29.
- Select the **Open G-Code file in** check box if you want to edit the G-codes after creating them.
- Click on the **Play** button from the **PropertyManager** to start creating G-codes. The codes will be displayed in the **NC Codes** box.
- Click on the **OK** button from the **PropertyManager**. The **SOLIDWORKS CAM NC Editor** will be displayed if you have selected the **Open G-Code file in** check box; refer to Figure-30.
- Modify the file as required and save it. Close the application

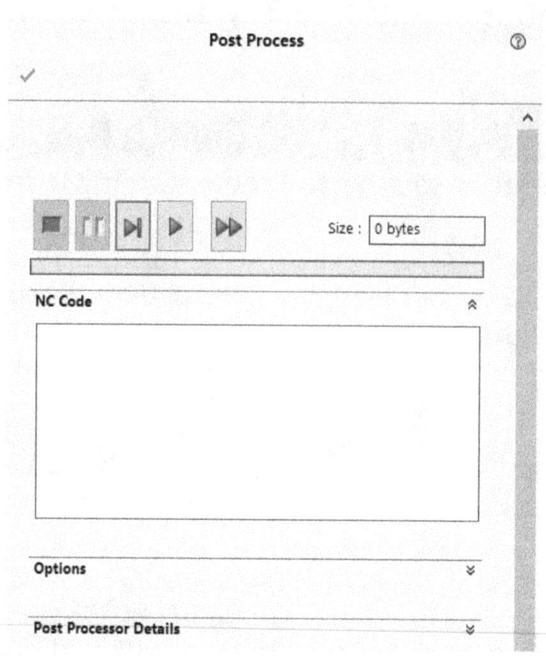

Figure-29. Post Process PropertyManager

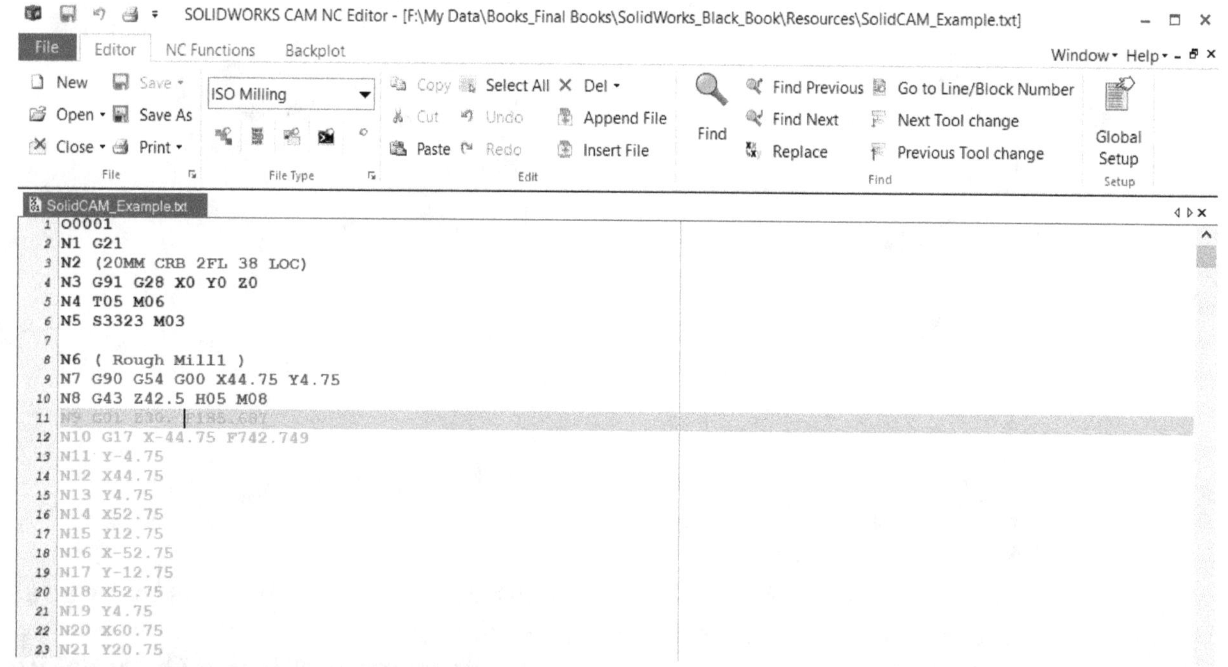

Figure-30. SOLIDWORKS CAM NC Editor

MANUALLY CREATING MILL OPERATIONS

In previous sections, you have learned to automatically create milling operations. Now, we will discuss the procedure of creating different milling operations manually. We will work with Rough Mill Operation and you can apply the same procedure to other tools.

- Click on the **2.5 Axis Mill Operations** tool from the **SOLIDWORKS CAM CommandManager** in the **Ribbon**. The **New Operation : Rough Mill PropertyManager** will be displayed; refer to Figure-31.

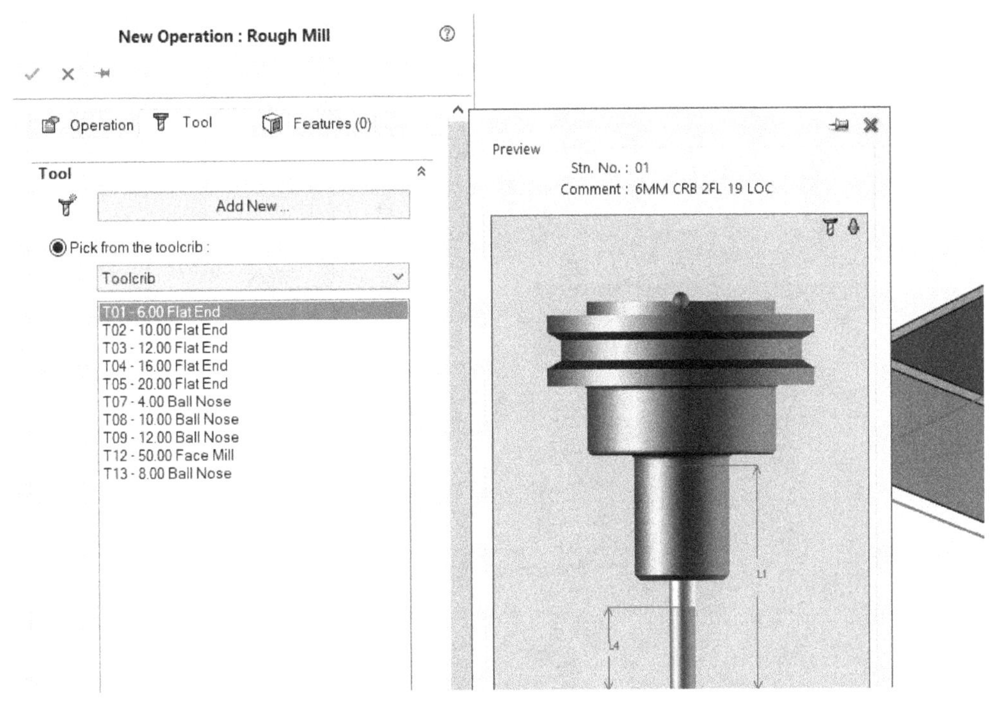

Figure-31. New Operation : Rough Mill PropertyManager

- Select desired operation type from the drop-down at the top in the **Operation** tab of the **PropertyManager**. Select the **Rough Mill** option to perform rough milling operation for large stock. Select the **Contour Mill** option to remove material around the edges of boss/pocket features. Select the **Face Mill** option to remove material from the top face of a part. Select the **Thread Mill** option to machine threads on boss or hole features. We are using the **Rough Mill** option in our case; refer to Figure-32.

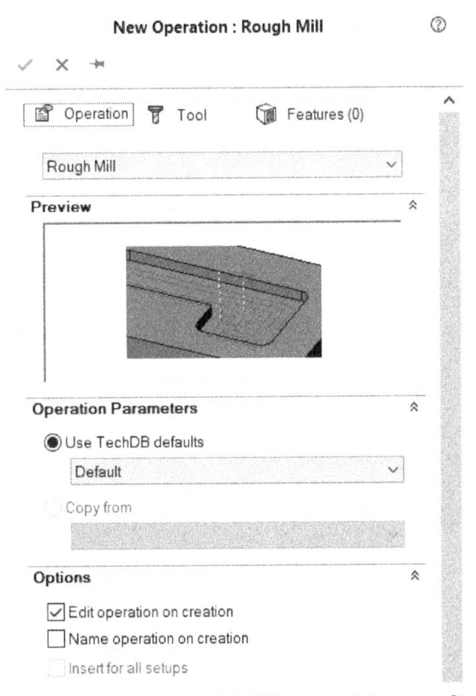

Figure-32. New Operation Rough Mill PropertyManager Operation tab

- Click on the **Tool** tab in the **PropertyManager**. The options will be displayed as shown in Figure-31.

- Select desired tool from the list.
- Click on the **Features** tab in the **PropertyManager**. The options will be displayed as shown in Figure-33.

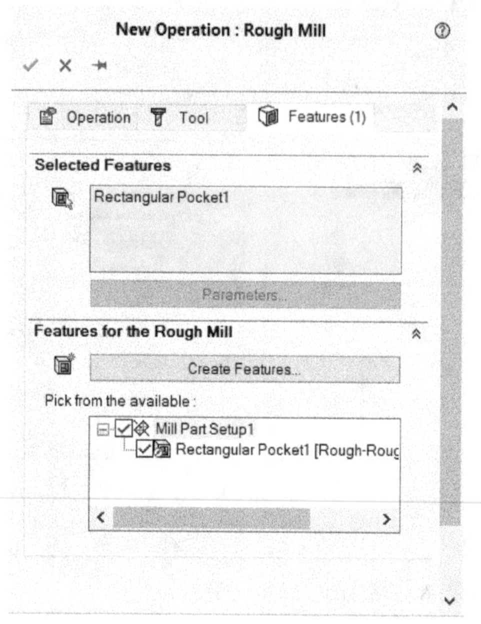

Figure-33. Features tab in New Operation PropertyManager

- Select desired feature from the **Pick from the available** box in the **Features for the Rough Mill** rollout if you have earlier applied **Extract Machinable Features** tool. If you have not used the tool earlier then you can create features now also.
- Click on the **Create Features** button from the **PropertyManager** and then click on the **2.5 Axis Feature** option. The **Setup for 2.5 Axis Feature PropertyManager** will be displayed; refer to Figure-34. Select the reference plane to define cutting direction and click on the New Feature button. The **2.5 Axis Feature: Select Entities PropertyManager** will be displayed; refer to Figure-35.

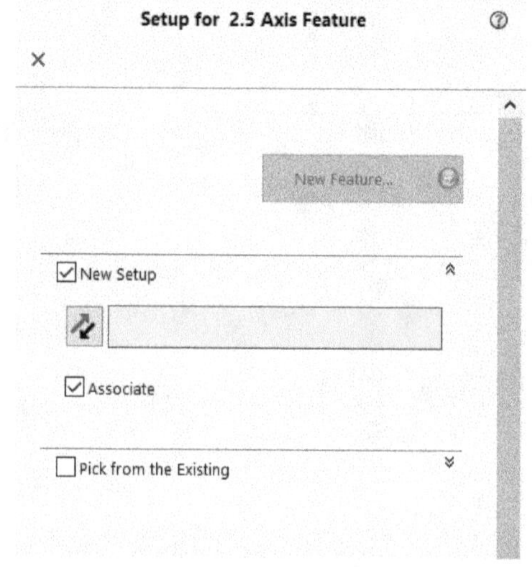

Figure-34. Setup for 2.5 Axis Feature PropertyManager

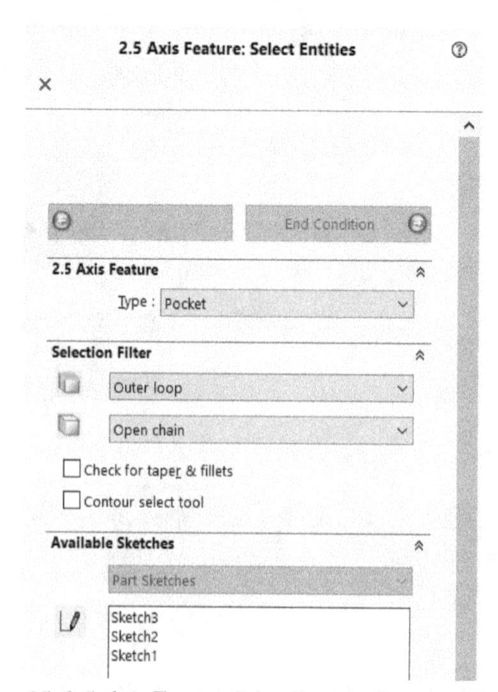

Figure-35. 2.5 Axis Feature Select Entities PropertyManager

- Select desired feature type from the **Type** drop-down of **2.5 Axis Feature** rollout. Set the selection filters as required from the **Selection Filter** rollout.
- Click on the face(s) to be used to create the feature; refer to Figure-36.

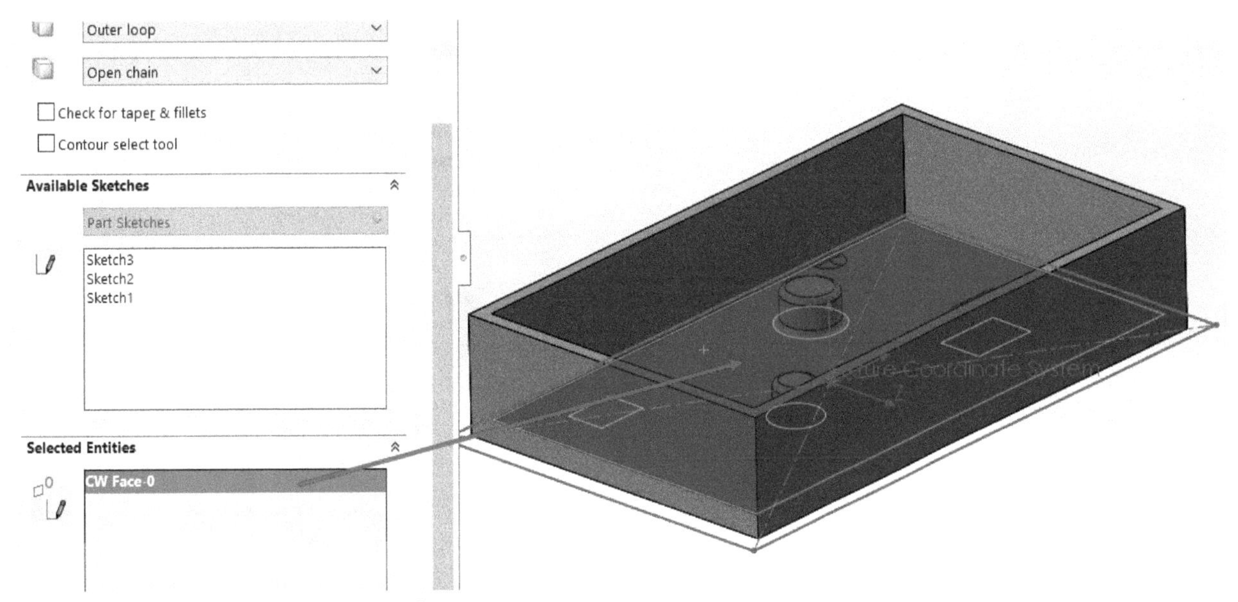

Figure-36. Face selected for creating mill features

- Click on the **End Condition** button at the top in the **PropertyManager**. The options will be displayed as shown in Figure-37.

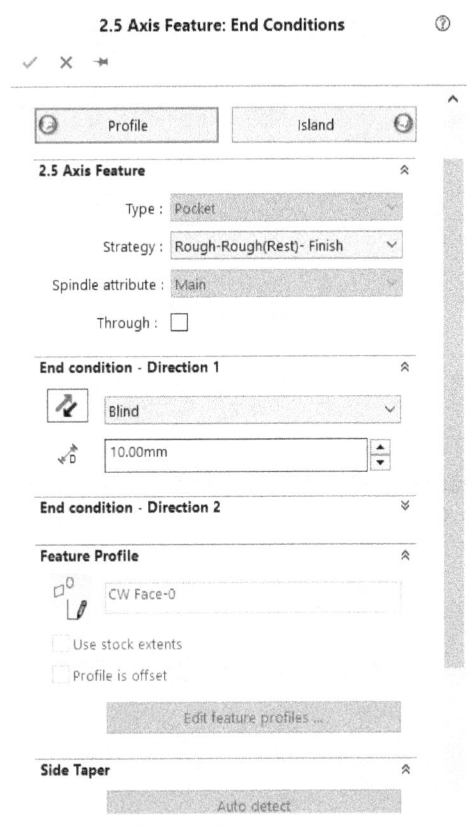

Figure-37. End condition options

- Set desired parameters and click on the **OK** button. The feature will be created.

- After creating the feature and selecting it, click on the **OK** button from the **New Operation PropertyManager**. The cutting operation will be created and the **Operation Parameters** dialog box will be displayed; refer to Figure-38.
- Set the parameters as needed and click on the **OK** button to create the operation.
- You can now generate the toolpath as discussed earlier.

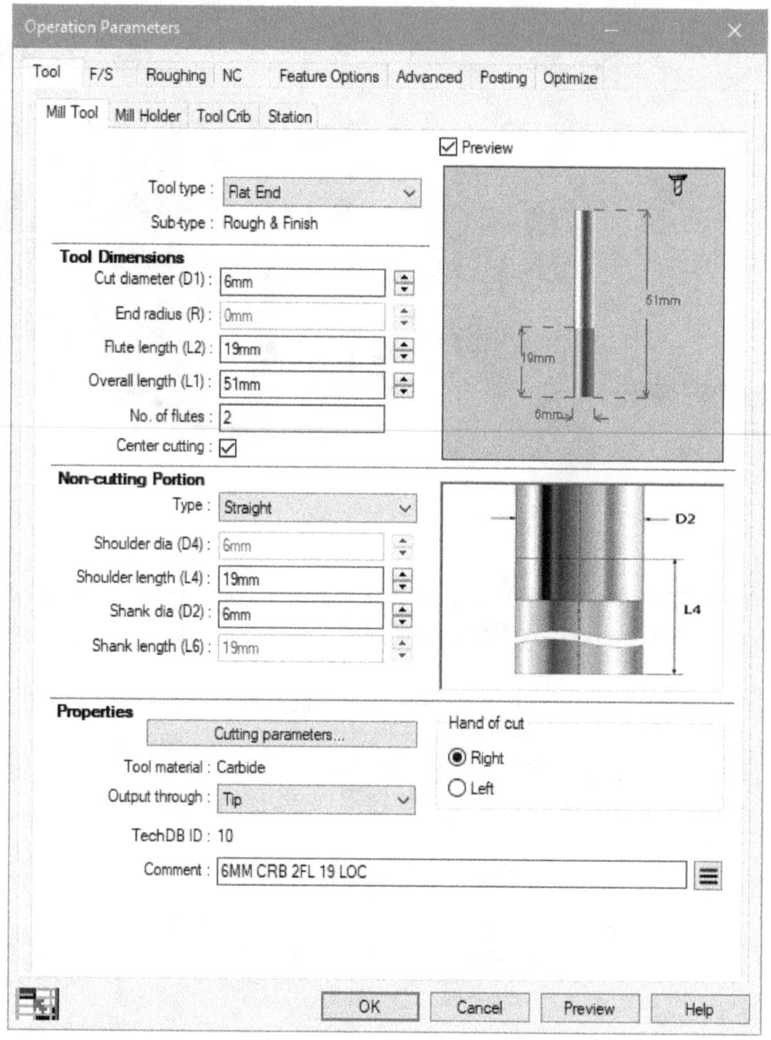

Figure-38. Operation_Parameters_dialog_box

Till this point, we have learned about the basics of SolidWorks. Since the book is not dedicated to SolidWorks CAM so we have given just introduction here. You can refer to our book on SolidWorks CAM viz. SolidWorks CAM 2023 Black Book for detail.

SELF ASSESSMENT

Q1. Write the use and procedure of setting post processor parameters.

Q2. From which of the following tabs in the **Machine** dialog box, you can add the tool for machine set up?

a) Machine
b) Post processor
c) Setup
d) None of the above

Q3. Which of the following options is not the type of stock used for machining?

a) Bounding box
b) Tool Crib
c) Cylindrical
d) STL file

Q4. .... is the pile of material to be removed after machining to produce the part.

Q5. To generate cutting strategies, right click on desired machining feature and click on the .... option from the shortcut menu.

Q6. **Post Processing** is the process of generating T codes from the toolpaths. (T/F)

Q7. In the **New Operation : Rough Mill Property Manager**, select the **Face Mill** option to remove material from the top face of the part. (T/F)

Answer to Self-Assessment:
2. (d) **3.** (b) **4.** Stock **5.** Generate Operation Plan **6.** False **7.** True

FOR STUDENT NOTES

Chapter 17

;

Rendering

Topics Covered

The major topics covered in this chapter are:

- ***Modifying Appearance***
- ***Editing Scene***
- ***Modifying Decal***
- ***Selecting Display State***
- ***Setting Options for Rending***
- ***Rendering Region***
- ***Final Rendering***

INTRODUCTION

Rendering is an important aspect of modern CAD software. Rendering is used to represent CAD models as real world objects with real texture. The tools to generate rendering are available in the **Render Tools CommandManager** of **Ribbon**; refer to Figure-1. The tools on the right-side of separator in **CommandManager** are active only after activating the **PhotoView 360** add-in. The procedure to activate an Add-in has been discussed in Chapter 1 of this book.

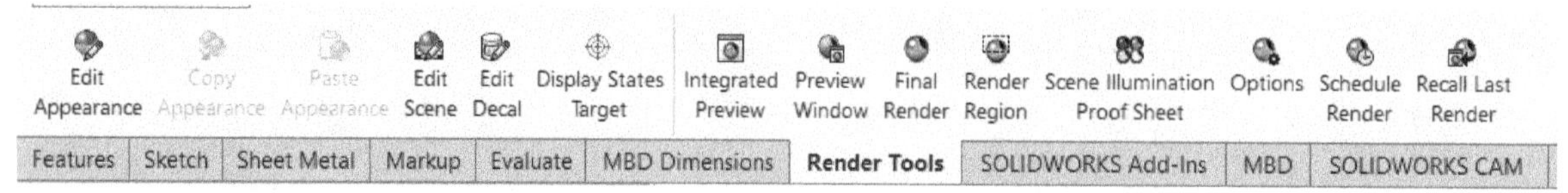

Figure-1. Render Tools CommandManager

Various tools in this **CommandManager** are discussed next.

EDITING APPEARANCE

The **Edit Appearance** tool is used to apply color and texture to the model. The procedure to edit appearance of model is given next.

- Click on the **Edit Appearance** tool from the **Render Tools CommandManager** in the **Ribbon**. The **Color PropertyManager** will be displayed along with **Appearance** task pane; refer to Figure-2.

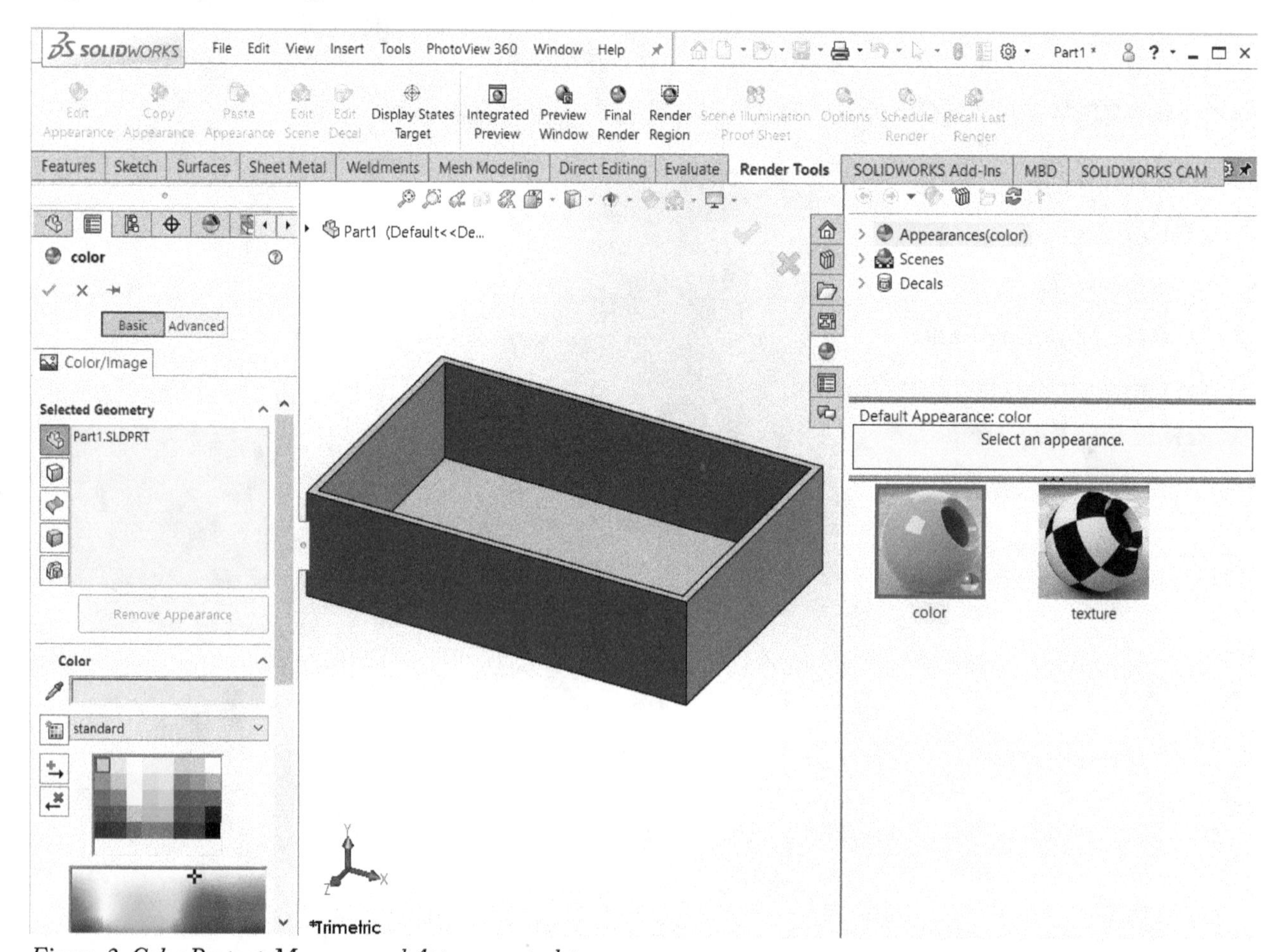

Figure-2. Color PropertyManager and Appearance task pane

- Click on the **Select Bodies** button from the **Selected Geometry** rollout in the **PropertyManager** and select the objects to which you want to apply the color/ texture.
- Select desired color from the **Color** rollout.
- Click on the **texture** option from the **Appearances, Scenes, and Decals** task pane to apply texture to the model; refer to Figure-3.

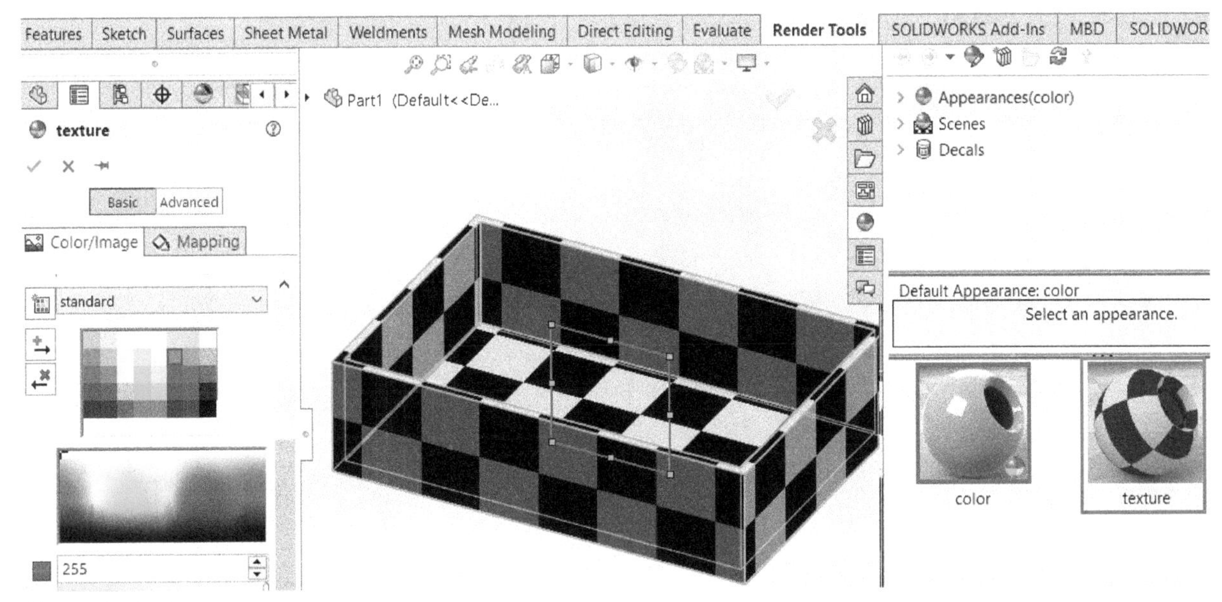

Figure-3. Applying texture

- Click on the **Browse** button from the **Image** rollout to select image for texture. The **Open** dialog box will be displayed.
- Select desired texture image file and click on the **OK** button. The texture will be applied.
- If you want to apply a texture from SolidWorks Library then select desired category from the top section of task pane and double-click on desired texture; refer to Figure-4.

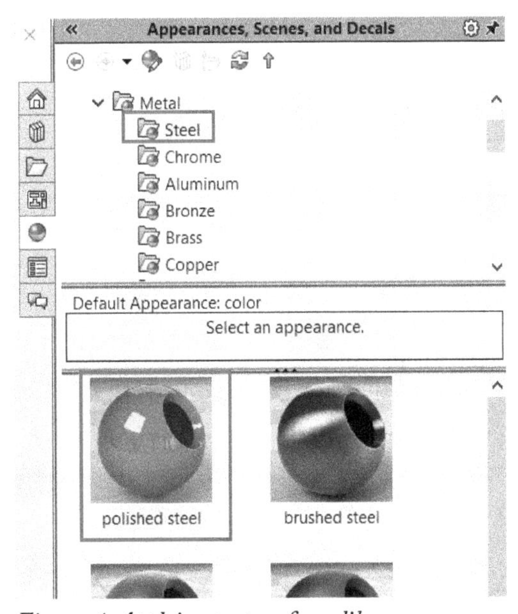

Figure-4. Applying texture from library

- Set desired parameters in the **PropertyManager** and click on the **OK** button.

COPYING AND PASTING APPEARANCES

After selecting the model on which you have applied appearance, click on the **Copy Appearance** tool from the **Render Tools CommandManager**. Appearance of selected model will be copied.

After copying, select the model/body on which you want to apply copied appearance and select the **Paste Appearance** tool from the **CommandManager**.

EDITING SCENE

The **Edit Scene** tool is used to modify scenario in which rendering is to be created. The procedure to edit scene is given next.

- Click on the **Edit Scene** tool from the **Render Tools CommandManager** in the **Ribbon**. The **Edit Scene PropertyManager** will be displayed; refer to Figure-5.

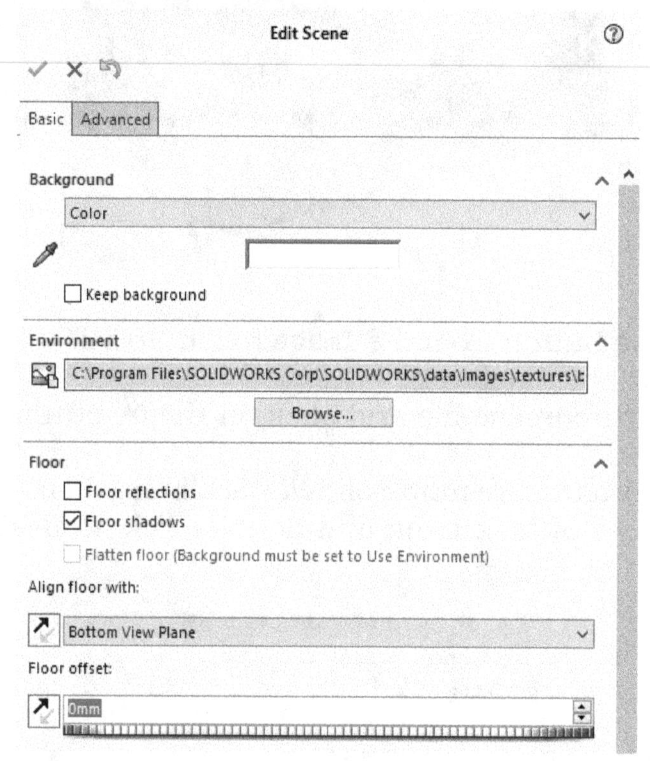

Figure-5. Edit Scene PropertyManager

- Select desired category from the **Scenes** node in the **Appearances, Scenes, and Decals** task pane on the right in the application window and double-click on desired scene.
- Set desired parameters in the **PropertyManager** and click on the **OK** button.

EDITING DECAL

The **Edit Decal** tool is used to apply and modify decals applied on faces of selected model. The procedure to use this tool is given next.

- Click on the **Edit Decals** tool from the **Render Tools CommandManager** in the **Ribbon**. The **Decals PropertyManager** will be displayed and decals will be displayed in the task pane; refer to Figure-6.

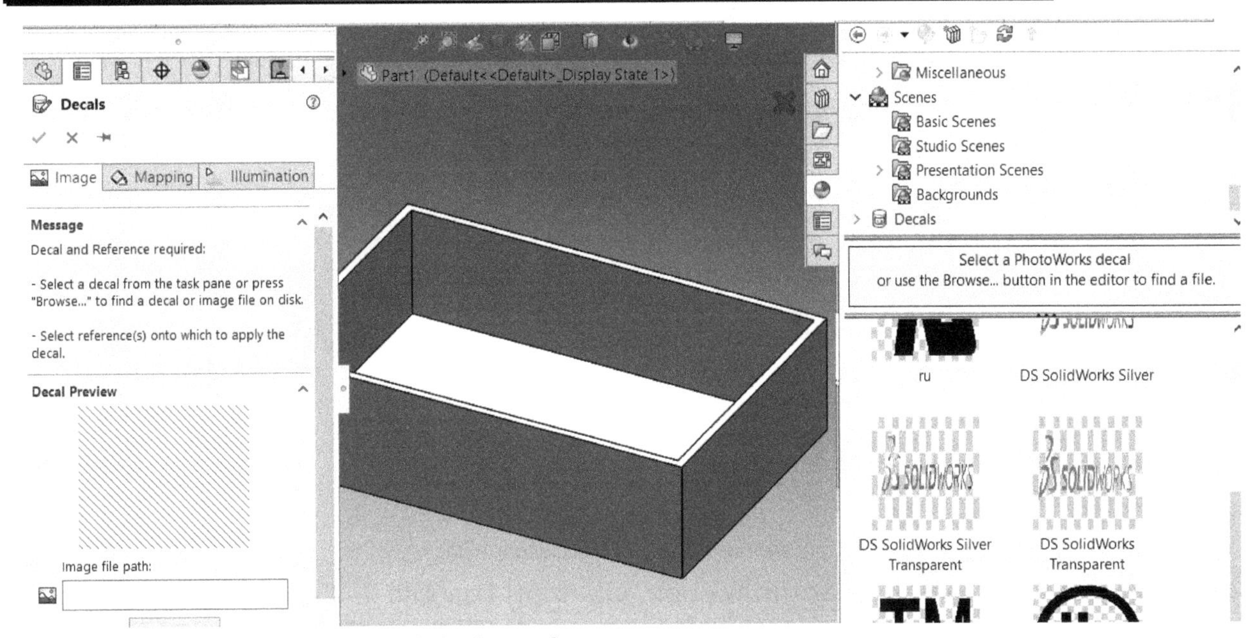

Figure-6. Decals PropertyManager with decals in task pane

- Select desired face on which you want to apply decal and then double-click on desired decal. You can drag and rotate the decal after placing it.
- Set desired parameters in various tabs of **PropertyManager** and click on the **OK** button.

SELECTING DISPLAY STATES TARGET

The **Display States Target** tool is used to select desired display state from display states. The procedure to use this tool is given next.

- Click on the **Display States Target** tool from the **Render Tools CommandManager**. The **Display State Target** dialog box will be displayed; refer to Figure-7.

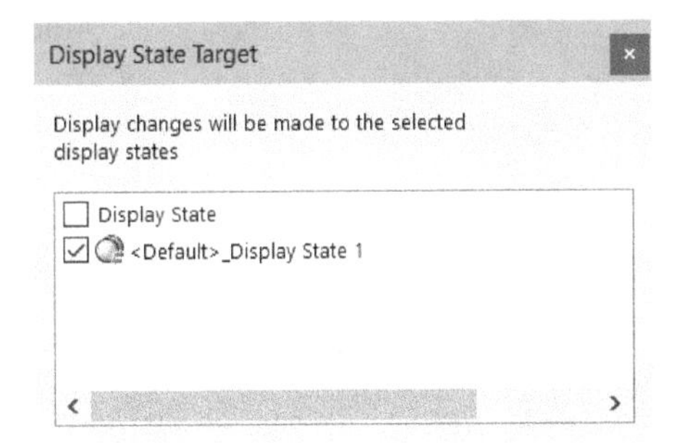

Figure-7. Display State Target dialog box

- Select desired display state from the dialog box and close the dialog box.

INTEGRATED PREVIEW

The **Integrated Preview** tool is used to display the rendering preview. The procedure to use this tool is given next.

- Click on the **Integrated Preview** tool from the **Render Tools CommandManager** in the **Ribbon**. The **Use Perspective Views in Renderings** dialog box will be displayed; refer to Figure-8.

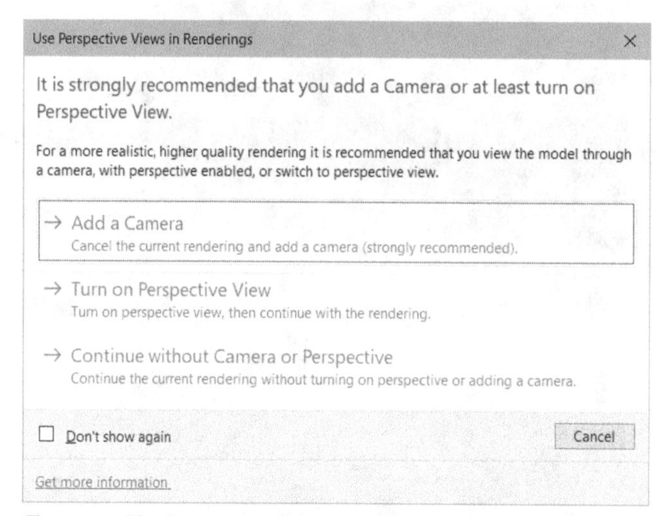

Figure-8. Use Perspective Views in Renderings dialog box

There are three options in this dialog box. Click on the **Add a Camera** option to add a camera from where rendering image will be created. Click on the **Turn on Perspective View** option from the dialog box to create rendering in perspective view. Click on the **Continue without Camera or Perspective** option to create rendering of model at its current state. These options are discussed later.

- After setting desired parameters if you click on the **Integrated Preview** button from the **Ribbon** then the render preview of model will be displayed; refer to Figure-9.

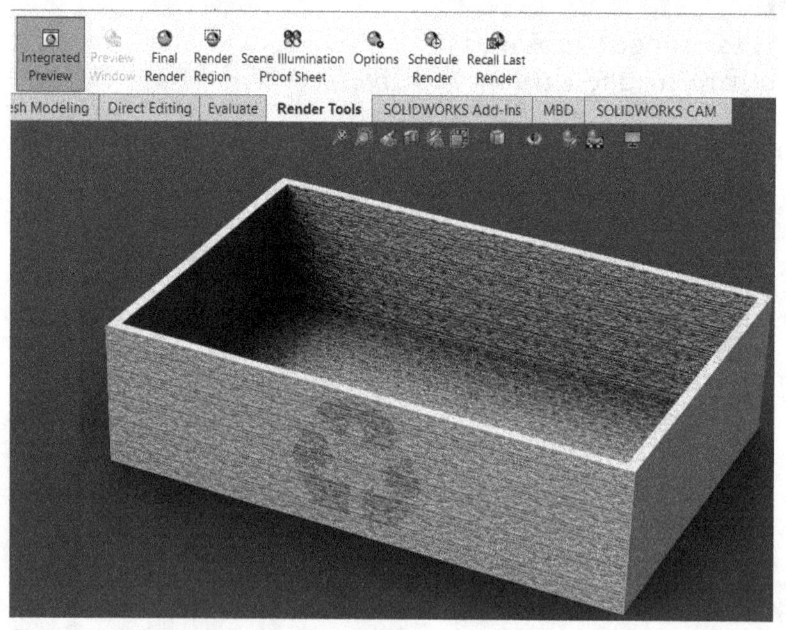

Figure-9. Integrated preview of rendering

Adding a Camera

The **Add Camera** option is used to add a camera for creating rendering image. The options to add and modify camera are available in the **Scene, Lights, and Cameras DisplayManager**; refer to Figure-10. The procedure to add a camera is given next.

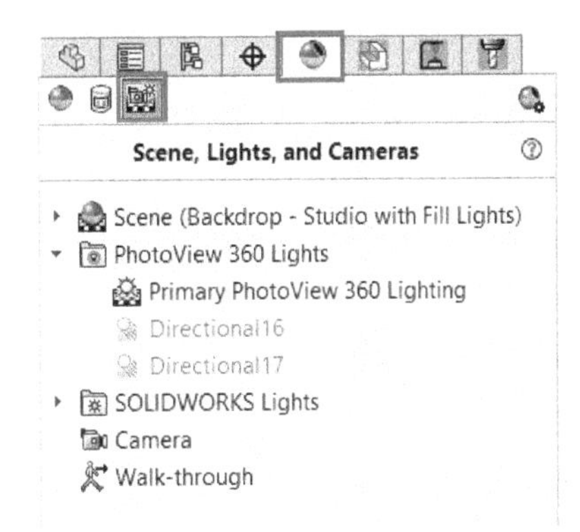

Figure-10. Scene, Lights, and Cameras DisplayManager

- Click on the **Add a Camera** option from the **Use Perspective Views in Renderings** dialog box or right-click on the **Camera** option from the **Scene, Lights, and Cameras DisplayManager** and select the **Add Camera** option from the shortcut menu. The **Camera PropertyManager** will be displayed; refer to Figure-11.

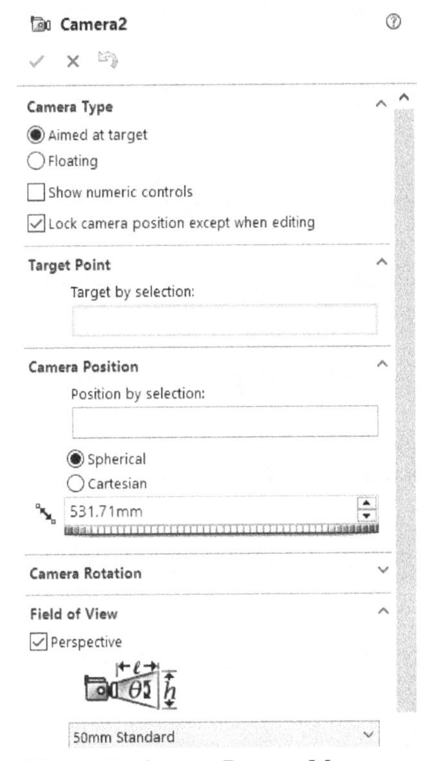

Figure-11. Camera PropertyManager

- Select the **Aimed at target** radio button if you want to make camera targeted on desired location. Select the **Floating** radio button if you want to make camera float in random direction while dragging it. We will use **Aimed at target** option in our case.
- After selecting the **Aimed at target** radio button, click at desired location where you want to focus the camera.
- Click in the **Position by selection** box in the **Camera Position** rollout and select desired location to place camera or use the slider to move the camera. You can also drag the camera by using arrows in triad displayed on camera.
- Set the other parameters as desired and click on the **OK** button.

Activating Perspective View

- Click on the **Perspective** option from the **View Settings** drop-down in **Heads-Up View Toolbar**; refer to Figure-12. The model will be displayed in perspective view.

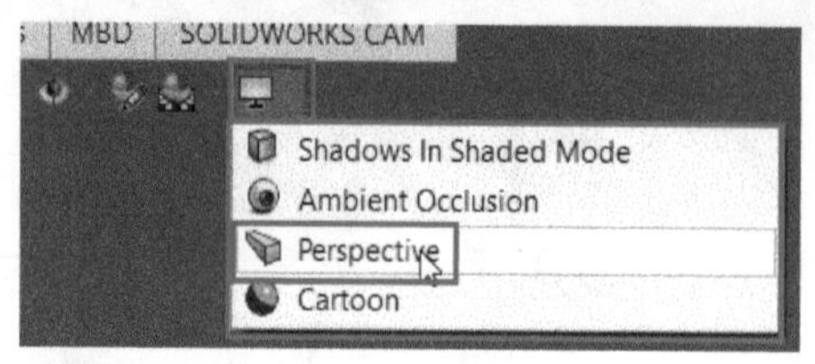

Figure-12. Perspective option

RENDERING OPTIONS

The **Options** tool is used to edit rendering options of PhotoView 360. The procedure to use this tool is given next.

- Click on the **Options** tool from the **Render Tools CommandManager** in the **Ribbon**. The **PhotoView 360 Options PropertyManager** will be displayed; refer to Figure-13.

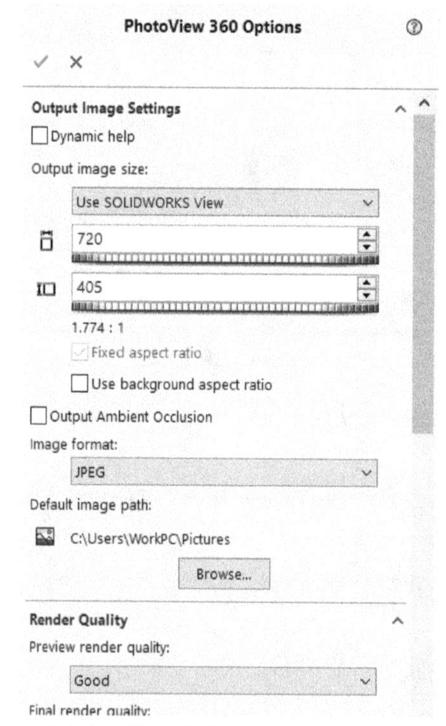

Figure-13. PhotoView 360 Options PropertyManager

- Select desired option from the **Output image size** drop-down in the **PropertyManager** to define size of rendered image.
- Select the **Fixed aspect ratio** check box to keep the fixed aspect ratio between horizontal and vertical image size.
- Select the **Use background aspect ratio** check box to set the aspect ratio same as background image.
- Set desired image format from the **Image format** drop-down and click on the **Browse** button from the **Output Image Settings** rollout in the **PropertyManager**. The **Select Folder** dialog box will be displayed.
- Select desired location where you want to save the image files and click on the **OK** button.

- Select desired option from the drop-downs in the **Render Quality** rollout to set quality level of rendering.
- Set desired value of Gamma to lighten or darken the render image.
- Select the **Bloom** check box to define glow around brighter objects in rendering and set the bloom set point & bloom extent. Bloom set point is the threshold point of brightness after which blooming of image will kick in. The bloom extent is the distance from source object up to which glow will expand.
- If you want to perform a contour or cartoon rendering then select the **Contour/ Cartoon Rendering** check box. Select desired option from the drop-down in the rollout and define the parameters like thickness of contour lines, render type and so on.
- Select the **Direct Caustics** check box to include photons generated by refraction, reflection, and bounce of light from light source. Define the quality and number of photons in respective edit boxes.
- Select the **Network Rendering** check box to use other computers connected in network for rendering. Specify the load percentage for client in the edit box. Note that a file will be generated in network directory which can be used by client computers for network rendering.
- Set the other parameters as desired and click on the **OK** button from the **PropertyManager**.

PREVIEW WINDOW

The **Preview Window** tool is used to check preview of rendering in a window; refer to Figure-14.

Figure-14. Rendering preview window

RENDER REGION

The **Render Region** tool is used to create a region to be used for rendering. The procedure to create render region is given next.

- Click on the **Render Region** tool from the **Render Tools CommandManager** in the **Ribbon**. The render region box will be displayed in the drawing area; refer to Figure-15.

Figure-15. Render region window

- Drag the key points of region box to resize it.
- Click on the Final Render, Integrated Preview, or Preview Window tool to check the rendering of region.

FINAL RENDER

The **Final Render** tool is used to create final rendering image file based on specified parameters. The procedure to use this tool is given next.

- Click on the **Final Render** tool from the **Render Tools CommandManager** in the **Ribbon**. The **Final Render** window will be displayed; refer to Figure-16.
- Click on the **Save Image** button from the window to save the current rendered image. The **Save Image** dialog box will be displayed.
- Set desired name & location of file and click on the **Save** button.
- Click on the **Close** button to close the window.

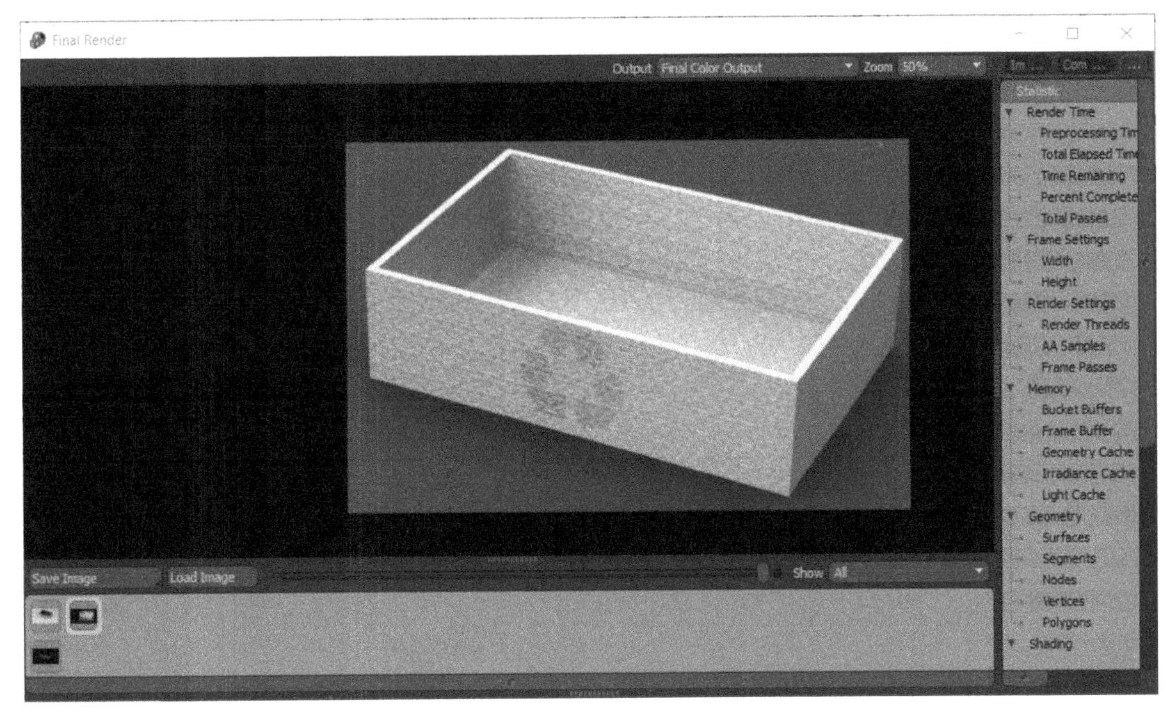

Figure-16. Final Render dialog box

SELF ASSESSMENT

Q1. What do you mean by **Bloom setpoint** while editing Rendering options?

Q2. Which of the following tool is used to apply color and texture to the model?

a) Edit appearance
b) Edit scene
c) Edit decal
d) None of the above

Q3. The **Integrated Preview** tool is used to display the .... preview.

Q4. The **Options** tool is used to edit rendering options of ....

Q5. If you want to perform a contour or cartoon rendering, then select the **Network Rendering** check box in the **PhotoView 360 Options Property Manager**. (T/F)

Q6. The **Final Render** tool is used to create final rendering image file based on specified parameters.. (T/F)

FOR STUDENT NOTES

Answer to Self-Assessment:

2. (d) **3.** Rendering **4.** PhotoView 360 **5.** False **6.** True

Index

Symbols

2.5 Axis Feature option 16-20
3D PDF Template Editor tool 15-26
3D PMI Compare tool 15-28
3D Printing 15-2
3D Printing Processes 15-2
3D Sketch button 4-2
3D Sketching 4-2
3D Texture tool 6-15
3 Point Arc 2-17
3 Point Arc Slot 2-15
3 Point Center Rectangle tool 2-13
3 Point Corner Rectangle tool 2-12

A

Add a fixture button 11-4
Add a force or Add a pressure link button 11-6
Add dimensions 2-33
Add/Edit Balloons tool 15-33
Add or Update a Style button 3-3
Add/Remove button 6-27
Add Tool button 16-3
Advanced Angle 7-17
Advanced Distance 7-16
Air radio button 11-18
Aligned Text 3-13
AlongZ relation 2-54, 2-55
Annotation CommandManager 1-11
Arc 2-16
Area Hatch/Fill tool 2-48
Assembly button 1-5, 7-2
Assembly CommandManager 1-10
Assembly Constraints (Mates) 7-5
Assembly Mode CommandManagers 1-10
Assembly with Magnetic Mates 8-18
Asset Publisher tool 8-15
Auto Dimension Scheme tool 15-11
Auto Shape button 3-15
Auto Sketch Entities button 3-15
Auto-space components check box 7-23
Auto Spline button 3-16
Auxiliary View button 10-9
Axis 4-20
Axis of Revolution box 4-11
Axis PropertyManager 4-20

B

Base Flange/Tab tool 13-2
Basic Location Dimension 15-15
Basic option 3-4
Basic Settings of SolidWorks 1-25
Basic Size Dimension tool 15-15
Bearing Calculator tool 8-14
Begin Assembly PropertyManager 7-2
Belt/Chain tool 8-5
Bend Allowance 13-7
Bilateral option 3-5
Bill of Material 10-39
Bill of Materials 7-24
Bill of Materials button 7-24
Bill of Materials PropertyManager 10-39
Blocks cascading menu 6-24
Bodies to Mirror rollout 5-13
Body Compare tool 6-34
Boundary Boss/Base tool 4-30
Boundary PropertyManager 4-30
Boundary Surface tool 9-5
Break tool 10-19
Broken Leader 3-13
Browse button 7-3
Butt/Groove Weld Symbols 14-2

C

Calculate () button 7-29
Cam 7-18
Cams tool 8-11
Cap Ends check box 4-8
Capture 3D View tool 15-22
Centerline 2-9
Centerline Parameters rollout 4-29
Center of Face button 4-24
Center of Mass 4-24
Centerpoint Arc 2-16
Centerpoint Arc Slot 2-15
Centerpoint Straight Slot 2-14
Center Rectangle tool 2-12
Chamfer Dimension tool 10-24
Chamfer tool 5-7
Check Geometry PropertyManager 11-17
Choose Material link button 11-7
Circle 2-15
Circle PropertyManager 2-15
Circular Cam 8-12

Circular Pattern tool 5-11, 5-13
Circular Sketch Pattern tool 2-36
Closed Corner tool 13-23
Closing a Document 1-12
Coincident 7-6
Coincident relation 2-56
Collinear relation 2-52
CommandManagers 1-6
Concentric 7-9
Concentric relation 2-55
ConfigurationManager 8-16
ConfigurationManager tab 8-16
Configurations 8-16
Conic 2-25
Convert Entities tool 4-3
Convert to Mesh Body tool 6-14
Convert to Sheet Metal tool 13-34
Coordinate System PropertyManager 4-22
Coordinate System tool 4-22
Copy Entities tool 2-39
Coradial relation 2-53
Core tool 12-20
Corner Rectangle tool 2-11
Costing button 11-23
Create Inspection Project PropertyManager 15-30
Create New Template tool 15-32
Create Stud on a Surface option 4-46
Custom sheet size radio button 10-4
Cut with Surface tool 9-19

D

Data Migration CommandManager 1-8
Datum Feature button 10-32
Datum Target button 10-33
Datum Target PropertyManager 10-33
Datum tool 15-15
Decimate Mesh Body tool 6-31
Define Machine tool 16-2
Deform tool 6-4
Delete a Style button 3-3
Delete Face tool 9-14
Delete/Keep Body tool 6-13
Design Table 5-32
DesignXpress Study 11-13
Detail View PropertyManager 10-16
DFMXpress Analysis 11-2, 11-20
DFMXpress Analysis Wizard tool 11-21
Dimensional Constraints (Dimensions) 2-49
Dimension Equations 5-31
Dimensioning a line 2-49
Dimensioning a weld bead 14-6
Dimensioning elliptical arcs/ellipse 2-50
Dimensioning inclined line 2-50
Dimension PropertyManager 3-2, 10-26
Dimension Text rollout 3-9
dimetric or trimetric orientation 10-7
DimXpert CommandManager 1-7
Direct Editing CommandManager 1-8
Display Style 2-7
Distance 7-10
Document Properties tab 1-26
Dome tool 6-2
Draft Analysis tool 12-4
Draft PropertyManager 5-23
Draft tool 5-22, 5-24, 12-11
DraftXpert PropertyManager 5-22, 5-24, 12-11
Draw button 14-18
Drawing button 1-5
Drawing Mode CommandManagers 1-11
Drawing Sheet Selection 10-2
Drawing View PropertyManager 10-6
DriveWorksXpress 11-2
DriveWorksXpress Wizard tool 11-27
Dual Dimension Rollout 3-10
Dynamic Annotation Views tool 15-23
Dynamic Mirror tool 2-45

E

Edge-Flange PropertyManager 13-10
Edge Flange tool 13-8
Edge selection box 13-9
Edit Block option 6-27
Edit Flange Profile button 13-9
Edit Inspection Methods tool 15-36
Edit Inspection Project tool 15-32
Edit Operations tool 15-36
Edit tool 6-30
Edit Vendors tool 15-37
eDrawing 11-10
Ellipse 2-22
End Cap tool 14-9
Equal relation 2-55
Equation Driven Curve 2-21
Equations tool 5-28
Erase button 14-18

Evaluate CommandManager 1-8
Evaluate tab 11-2
Explode Block tool 6-28
Exploded View 7-22
Exploded View of Assembly 10-39
Explode Line Sketch button 7-23
Export to Excel tool 15-37
Extend Entities 2-32
Extend Surface tool 9-16
Extract Machinable Features tool 16-13
Extruded Boss/Base tool 4-5
Extruded Cut 4-33
Extruded Cut tool 13-29
Extruded Surface 9-2
Extrude PropertyManager 4-33

F

Face fillet radio button 5-5
Factor of Safety 11-14
Favorite rollout 4-37
Features CommandManager 1-6
Filled Surface FeatureManager 11-16
Filled Surface tool 9-6, 11-16
Fillet and Edge Weld Symbols 14-3
Fillet PropertyManager 5-2
Fillet tool 5-2
Fillet Type rollout 5-3
Fill Surface PropertyManager 9-6
first angle projection 5-37
Fit option 3-7
Fit (tolerance only) option 3-8
Fit with tolerance 3-8
Fix relation 2-56
Fixture Coordinate System PropertyManager 16-9
Fixture FeatureManager 11-4
Fixture Setting 11-4
Flange Length rollout 13-11
Flange Parameters rollout 13-9
Flat pattern view 13-39
Flex tool 6-8
Flow Inlet PropertyManager 11-18
Flow Outlet PropertyManager 11-19
FloXpress Analysis 11-2, 11-15
FloXpress Analysis Wizard button 11-15
Fold Tool 13-33
Follow Path option 4-14
Freeform tool 6-3, 9-7
Front Plane 2-2
Full round fillet radio button 5-6
Fully Defined Sketch 2-57
Fully Define Sketch button 2-57

G

Gauge 13-6
Gear 7-20
Generate Operation Plan option 16-14
Generate report link button 11-10
Generate Toolpath option 16-14
Geometric Constraints 2-52
Geometric Tolerance symbol 10-37
Global Variables 5-30
Grooves tool 8-10

H

Heads-up View Toolbar 2-5
Hem tool 13-19
Hide/Show Items 2-7
Hinge 7-19
Hole Callout 10-38
Hole Fit 3-7
Hole Position PropertyManager 8-2
Hole Series tool 8-2
Hole Wizard tool 4-37
Horizontal relation 2-52
Horizontal Text 3-13

I

Imported tool 6-17
Indent tool 6-7
Insert Bends tool 13-36
Inside button 3-12
Inspection Dimension button 3-9
Intermittent Weld check box 14-14
Intersection relation 2-55
Intersect PropertyManager 5-27
Intersect tool 5-27

J

Jog Line tool 2-44
Jog tool 13-20

K

K-Factor 13-7
Knit Surface tool 9-18

L

Launch Template Editor tool 15-34

Layer Rollout 10-30
Layout CommandManager 1-10
Leader Style rollout 3-12
Leader Thickness drop-down 10-30
Limit 3-5
Limiting reference type drop-down 4-7
Linear Cam 8-13
Linear/Linear Coupler 7-15
Linear Motor (Actuator) button 7-29
Linear Pattern tool 5-9
Linear Sketch Pattern 2-35
Line tool 2-7
Lip/Groove tool 6-22
Load Style button 3-4
Location Dimension tool 15-13
Lock 7-9
Lofted-Bend tool 13-4
Lofted Boss/Base tool 4-27
Lofted Surface tool 9-4
Loft PropertyManager 4-27

M

Macro 6-28
Magnetic mates 8-15
Magnetic Mates 8-15
Make base construction 2-34
Make Drawing from Part/Assembly button 13-38
Make tool 6-24
Markup CommandManager 14-17
Markup tool 14-17
Mass Properties tool 5-35
Mate Controller tool 7-25
Mate Reference tool 4-25
Material dialog box 5-35
Material Setting 11-7
Mate tool 7-5
maximum tolerance 3-6
Merge Points relation 2-56
Merge Result check box 4-30
Mesh Modeling CommandManager 6-31
Midpoint relation 2-56
Mill Setup tool 16-12
minimum tolerance 3-6
Mirror Entities 2-34
Mirror PropertyManager 2-34
Mirror tool 5-12
Miter Flange tool 13-16
Model Break View tool 15-21
Model Items PropertyManager 10-25
Model Items tool 10-25
Mold Folder 12-14
Mold Tools 12-2
Mold Tools CommandManager 1-7
Mold Tools option 12-3
MotionManager 7-29
Motion Study 7-27
Motor button 7-28
Motor PropertyManager 7-28
Mounting Boss tool 6-17
Move/Copy tool 6-12
Move Entities tool 2-38
Move Face PropertyManager 12-12
Move Face tool 12-12

N

Net-Inspect tool 15-38
New button 1-4
New Inspection Project tool 15-29
New SolidWorks Document dialog box 1-4
Note PropertyManager 10-26
Note tool 10-26
nuts and bolts 8-35

O

Obround Relief 13-15
Offset Distance 2-33
Offset Entities 2-33
Offset Surface tool 9-9
Offset Text button 3-9
Open G-Code file in check box 16-17
Opening a Document 1-11
Optimize page of SimulationXpress 11-12
Optimizing 11-12
Ordinate dimensioning 2-50
Ordinate Dimension tool 2-50
Orientation rollout 2-8
Orientation/twist type drop-down 4-14
O-Ring Grooves 8-10
orizontal/Vertical Dimensioning between Points 3-13
Outside button 3-11
Over Defined message 2-59
Override Units Rollout 3-13

P

Parabola 2-24
Parallel 7-7

Parallelogram tool 2-13
Parallel relation 2-54
ParallelYZ relation 2-54
ParallelZX relation 2-54
Part button 1-5
Partial Ellipse 2-23
Parting Line 12-14
Parting Line Analysis tool 12-7
Parting Surfaces tool 12-18
Part Mode CommandManagers 1-6
Part Reviewer 11-2
Path Mate 7-13
Pattern Feature tool 15-16
Pen button 3-15
Perimeter Circle 2-16
Perpendicular 7-8
Perpendicular relation 2-53
Pierce relation 2-56
Planar Surface tool 9-8
Plane 4-17
Plane PropertyManager 4-17
Playing Motion Study 7-29
Point 2-29
Point PropertyManager 4-23
Point radio button 6-4
Point tool 4-23
Polygon 2-17
Post Processor tab 16-7
Post Process tool 16-17
Power Modify button 3-17
Previous View 2-6
Print3D tool 15-8
Profile Center button 7-11
Projected View tool 10-9
Projection radio button 12-9
projection type (First angle or Third angle) 10-3
Publish to 3D PDF tool 15-24

R

Rack Pinion 7-21
Record tool 6-29
Rectangle 2-10
Rectangle Relief 13-15
Reference Geometry 4-17
Relations 2-49
Relations/Snaps 1-26
Remove Balloons tool 15-33
Replace Entity tool 2-46
Replace Face tool 9-14
Results 11-10
Results View tab 11-14
Revolved Boss/Base tool 4-10
Revolved Cut tool 4-34
Revolved Surface tool 9-3
Right Plane 2-2
Rip button 13-16
Rip tool 13-35
Rotate Entities tool 2-39
Ruled Surface tool 9-10
Run Simulation link button 11-9
Run tool 6-29

S

Save a Style button 3-4
Save CL File tool 16-17
Save tool 6-26
Save Tool Crib button 16-6
Scale Entities tool 2-40
Scale PropertyManager 12-13
Scale rollout 10-8
Scale tool 12-13
Screw 7-21
Section View 2-6
Section View button 10-10
Section View PropertyManager 10-10
Section View tool 15-19
Segment Imported Mesh Body tool 6-32
Segment tool 2-42
Select Balloons tool 15-33
Shaft Fit 3-7
Sheet Format/Size dialog box 10-2
Sheet Metal CommandManager 1-7
Sheet Metal environment 13-2
Sheet Metal Gusset tool 13-27
sheet size 10-3
Sheet tab 10-3
Shell PropertyManager 5-25
Shell tool 5-25
Show button 2-3
Show custom sizing check box 4-37
Show in exploded state check box 10-39
Show Tolerance Status tool 15-18
Shut-off Surfaces tool 12-16
Silhouette radio button 12-11
Simulate Toolpath option 16-15
SimulationXpress Analysis 11-2
SimulationXpress Analysis Wizard 11-2
SimulationXpress Options dialog box 11-3

Size Dimension tool 15-14
Sketch Chamfer 2-27
Sketch CommandManager 1-6, 2-4
Sketched Bend tool 13-21
Sketch Fillet 2-26
Sketch Ink CommandManager 3-14
Sketch option 1-26
Sketch Picture tool 2-47
Sketch Text PropertyManager 2-27
SketchXpert PropertyManager 2-59
slider 12-20
Slot 2-13, 7-18
Smallest Flow Passage button 11-18
Smart button 3-12
Smart Dimension 2-49
Smart Fasteners tool 8-6
Snap Hook Groove tool 6-19
Snap Hook tool 6-18
Solid Leader 3-13
SolidWorks 2016 Installation Manager 1-2
SOLIDWORKS CAM CommandManager 16-2
SolidWorks Download folder 1-2
SOLIDWORKS Inspection CommandManager 15-29
SOLIDWORKS Inspection - Template Editor dialog box 15-34
SOLIDWORKS MBD 15-10
Source/Destination rollout 10-26
Special Surfacing Tools 9-8
Spline 2-18
Split Entities tool 2-44
Split Line tool 12-9
Split tool 6-11
StainabilityXpress 11-2
Standard 3 View button 10-5, 10-17
standard library 8-35
Starting SolidWorks 2016 1-2
Step Thru Toolpath option 16-16
STL File button 16-11
Stock Manager tool 16-10
Straight Slot 2-13
Stretch Entities tool 2-41
Structural Member tool 14-8
Structural Steel Calculator tool 8-9
Stud Wizard tool 4-44
Style drop-down 3-12
Styled Spline 2-20
Style rollout 3-2
Surface Finish tool 10-30
Surface Finish/Weld Symbol/Hole Callout 10-38
Surfaces CommandManager 1-8
Sustainability tool 11-25
Swept Boss/Base tool 4-12
Swept Surface tool 9-4
Symmetric 7-12
Symmetric relation 2-56
symmetric tolerance 3-5
System of units drop-down 11-4
System Options dialog box 1-25
System Options tab 1-26

T

Tangent 7-8
Tangent Arc 2-17
Tangent Bend button 13-12
Tangent relation 2-55
Tear Relief 13-16
Text Fonts Rollout 3-13
Text tool 2-27
Thicken FeatureManager 11-16
Thicken tool 9-18
Thin Feature check box 4-8
third angle projection 5-37
Title Block 10-40
Title Block Table tool 15-18
toggle between three standard planes 4-3
Tolerance Precision drop-down 3-8
Tolerance/Precision rollout 3-4
Tolerance Type drop-down 3-4
toolbox 8-35
Toolbox 8-8
Tool Crib tab 16-3
Tooling Split tool 12-19
Top Plane 2-2
Translate/Rotate button 6-12
Trim Entities 2-30
Trim Surface tool 9-16
Twist Along Path option 4-15

U

Undercut Analysis tool 12-6
Unfold Tool 13-32
Unit Precision drop-down 3-8
Universal Joint 7-22
Untrim Surface tool 9-17
Update all annotations linked to this Style radio button 3-3
Update Inspection Project tool 15-34

V

Variable Radius Parameters rollout 5-4
Vent tool 6-21, 13-29
Vertex radio button 5-7
Vertical relation 2-52
View fluid volume 11-17
View Layout CommandManager 1-11
View Layout tab 10-5
View Orientation 2-6

W

Weld Bead tool 8-4, 14-11
Weldment Cut List option 14-16
Weldments 14-7
Weldments CommandManager 1-8
Weldment tool 14-8
Weld Path radio button 14-12
Weld Symbols button 14-15
Width 7-13
Witness/Leader Display Rollout 3-11
WorkFlow in SolidWorks 1-30
Wrap PropertyManager 5-26
Wrap tool 5-26

Z

Zoom to Area 2-6
Zoom to Fit 2-6

Ethics of an Engineer

- Engineers shall hold paramount the safety, health, and welfare of the public and shall strive to comply with the principles of sustainable development in the performance of their professional duties.
- Engineers shall perform services only in areas of their competence.
- Engineers shall issue public statements only in an objective and truthful manner.
- Engineers shall act in professional manners for each employer or client as faithful agents or trustees, and shall avoid conflicts of interest.
- Engineers shall build their professional reputation on the merit of their services and shall not compete unfairly with others.
- Engineers shall act in such a manner as to uphold and enhance the honor, integrity, and dignity of the engineering profession and shall act with zero-tolerance for bribery, fraud, and corruption.
- Engineers shall continue their professional development throughout their careers, and shall provide opportunities for the professional development of those engineers under their supervision.

Printed in the USA
CPSIA information can be obtained
at www.ICGtesting.com
LVHW081818161023
761249LV00007B/197